Switching Power Supply Design

Third Edition

Abraham I. Pressman

Keith Billings

Taylor Morey

New York Chicago San Francisco
Lisbon London Madrid Mexico City
Milan New Delhi San Juan
Seoul Singapore Sydney Toronto

The McGraw-Hill Companies

Library of Congress Cataloging-in-Publication Data

McGraw-Hill books are available at special quantity discounts to use as premiums and sales promotions, or for use in corporate training programs. To contact a special sales representative, please visit the Contact Us page at www.mhprofessional.com.

Switching Power Supply Design, Third Edition

1234567890 DOC DOC 019

ISBN 978-0-07-148272-1
MHID 0-07-148272-5

Sponsoring Editor
Wendy Rinaldi

Acquisitions Coordinator
Joya Anthony

Production Supervisor
George Anderson

Art Director, Cover
Jeff Weeks

Editorial Supervisor
Janet Walden

Proofreader
Paul Tyler

Composition
International Typesetting and Composition

Cover Designer
Jeff Weeks

Project Editor
LeeAnn Pickrell

Indexer
Ted Laux

Illustration
International Typesetting and Composition

In fond memory of Abraham Pressman, master of the art, 1915–2001.
Immortalized by his timeless writings and his legacy—a gift
of knowledge for future generations.

To Anne Pressman, for her help and encouragement
on the third edition.

To my wife Diana for feeding the brute and allowing him
to neglect her, yet again!

About the Authors

Abraham I. Pressman was a nationally known power supply consultant and lecturer. His background ranged from an Army radar officer to over four decades as an analog-digital design engineer in industry. He held key design roles in a number of significant "firsts" in electronics over more than a half century: the first particle accelerator to achieve an energy over one billion volts, the first high-speed printer in the computer industry, the first spacecraft to take pictures of the moon's surface, and two of the earliest textbooks on computer logic circuit design using transistors and switching power supply design, respectively.

Mr. Pressman was the author of the first two editions of *Switching Power Supply Design.*

Keith Billings is a Chartered Electronic Engineer and author of the *Switchmode Power Supply Handbook*, published by McGraw-Hill. Keith spent his early years as an apprentice mechanical instrument maker (at a wage of four pounds a week) and followed this with a period of regular service in the Royal Air Force, servicing navigational instruments including automatic pilots and electronic compass equipment. Keith went into government service in the then Ministry of War and specialized in the design of special test equipment for military applications, including the UK3 satellite. During this period, he became qualified to degree standard by an arduous eight-year stint of evening classes (in those days, the only avenue open to the lower middle-class in England). For the last 44 years, Keith has specialized in switchmode power supply design and manufacturing. At the age of 75, he still remains active in the industry and owns the consulting company DKB Power, Inc., in Guelph, Canada. Keith presents the late Abe Pressman's four-day course on power supply design (now converted to a Power Point presentation) and also a one-day course of his own on magnetics, which is the design of transformers and inductors. He is now a recognized expert in this field. It is a sobering thought to realize he now earns more in one day than he did in a whole year as an apprentice.

Keith was an avid yachtsman for many years, but he now flies gliders as a hobby, having built a high-performance sailplane in 1993. Keith "touched the face of god," achieving an altitude of 22,000 feet in wave lift at Minden, Nevada, in 1994.

Taylor Morey, currently a professor of electronics at Conestoga College in Kitchener, Ontario, Canada, is co-author of an electronics devices textbook and has taught courses at Wilfred Laurier University in Waterloo. He collaborates with Keith Billings as an independent power supply engineer and consultant and previously worked in switchmode power supply development at Varian Canada in Georgetown and Hammond Manufacturing and GFC Power in Guelph, where he first met Keith in 1988. During a five-year sojourn to Mexico, he became fluent in Spanish and taught electronics engineering courses at the Universidad Católica de La Paz and English as a second language at CIBNOR biological research institution of La Paz, where he also worked as an editor of graduate biology students' articles for publication in refereed scientific journals. Earlier in his career, he worked for IBM Canada on mainframe computers and at Global TV's studios in Toronto.

Contents

Acknowledgments

Worthy of special mention is my engineering colleague and friend of many years, Taylor Morey. He spent many more hours than I did carefully checking the text, grammar, figures, diagrams, tables, equations, and formulae in this new edition. I know he made many thousands of adjustments, but should any errors remain they are entirely my responsibility.

I am also indebted to Anne Pressman for permission to work on this edition and to Wendy Rinaldi and LeeAnn Pickrell and the publishing staff of McGraw-Hill for adding the professional touch.

Many people contribute to a work like this, not the least of these being the many authors of the published works mentioned in the bibliography and references. Some who go unnamed also deserve our thanks. "We see further because we stand on the shoulders of giants."

—Keith Billings

Preface

Not many technical books continue to be in high demand well beyond the natural life of their author. It speaks well to the excellent work done by Abraham Pressman that his book on switching power supply design, first published in 1977, still enjoys brisk sales some eight years after his demise at the age of 86. He leaves us a valuable legacy, well proven by the test of time.

Abraham had been active in the electronics industry for nearly six decades. For 15 years, up to the age of 83, Abraham had presented a training course on switching design. I was privileged to know Abraham and collaborate with him on various projects in his later years. Abe would tell his students that my book was the *second* best book on switching power supplies (not true, but rare and valuable praise indeed from the old master).

When I started designing switching power supplies in the 1960s, very little information on the subject was available. It was a new technology, and the few companies and engineers specializing in this area were not about to tell the rest of world what they were doing. When I found Abraham's book, a veil of secrecy was drawn away, shedding light on this new technology. With the insight provided by Abe, I moved forward with great strides.

When, in 2000, Abe found he was no longer able to continue with his training course, I was proud that he asked me to take over his course notes with a view to continuing his presentation. I found the volume of information to be daunting, however, and too much for me to present in four days, although he had done so for many years. Furthermore, I felt that the notes and overhead slides had deteriorated too much to be easily readable.

I simplified the presentation and converted it to PowerPoint on my laptop, and I first presented the modified, three-day course in Boston in November 2001. There were only two students (most companies had cut back their training budget), but this poor turnout was more than compensated for by the attendance of Abraham and his wife Anne. Abe was very frail by then, and I was so pleased that he lived to see his legacy living on, albeit in a very different form. I think he was a bit bemused by the dynamic multimedia presentation, as I leisurely controlled it from my laptop. I never found out what he really thought about it, but Anne waved a finger and said, "Abe would stand at the blackboard with a pointer to do that!"

When McGraw-Hill asked me to co-author the third edition of Abe's book, I was pleased to agree, as I believe he would have wanted me to do that. In the eight years since the publication of the second edition, there have been many advances in the technology and vast improvements in the performance of essential components. This has altered many of the limitations that Abe mentions, so this was a good time to make adjustments and add some new work.

As I reviewed the second edition, a comment made by an English gardener standing outside his cottage in a country village unchanged for hundreds of years, came to mind. In response to a new arrival, a young yuppie who wanted to modernize things, he said, "Look around you lad, there's not much wrong wi'it, is there?" This comment could well be applied to Abe's previous edition.

For this reason, I decided not to change Abe's well-proven treatise, except where technology has overtaken his previous work. His pragmatic approach, dealing with each topology as an independent entity, may not be in the modern idiom as taught by today's experts, but for the *ab initio* engineer trying to understand the bewildering array of possible topologies, as well as for the more experienced engineer, it is a well-proven and effective method. The state-space averaging models, canonical models, the bilateral inversion techniques, or duality principles so valuable to modern experts in this field were not for Abraham. His book provides a solid underpinning of the fundamentals, explaining not only how but also why we do things. There is time enough later to learn the more modern concepts from some of the excellent specialist books now available (see the bibliography).

Abe's original manuscript was handwritten and painstakingly typed out by his wife Anne over several years. For this third edition, McGraw-Hill converted the manuscript to digital files for ease of editing. This made it easier for Taylor Morey and me to make minor and mainly cosmetic changes to the text and many corrections to equations, calculations, and diagrams, some corrupted by the conversion process. We also made adjustments where we felt such changes would help the flow, making it easier for the reader to follow the presentation. These changes are transparent to the reader, and they do not change Abraham's original intentions.

Where new technology and recent improvements in components have changed some of the limitations mentioned in the second edition, you will find my adjusting notes under the heading **After Pressman**. Where I felt additional explanations were justified, I have inserted a **Tip** or **Note**.

I have also added new sections to Chapter 7 and Chapter 9, where I felt that recent improvements in design methods would be helpful to the reader and also where improvements in IGBT technology made these devices a useful addition to the more limited range of devices previously favored by Abraham. In this way, the original structure of the second edition remains unchanged, and because the index and cross references still apply, the reader will find favorite sections in the same places. Unfortunately, the page numbers did change, as there was no way to avoid this.

Even if you already have a copy of the second edition of Pressman's book, I am sure that with the improvements and additional sections, you will find the third edition a worthwhile addition to your reference library. You will also find my book, *Switchmode Power Supply Handbook,* Second Edition (McGraw-Hill, 1999), a good companion, providing additional information with a somewhat different approach to the subject.

PART 1

Topologies

CHAPTER 1

Basic Topologies

1.1 Introduction to Linear Regulators and Switching Regulators of the Buck Boost and Inverting Types

In this book, we describe many well-known *topologies* (elemental building blocks) that are commonly used to implement linear and switching power supply designs. Each topology has both common and unique properties, and the experienced designer will choose the topology best suited for the intended application. However, for those engineers just starting in this area, the choice may appear rather daunting. It is worth spending some time to develop a basic understanding of the properties, because the correct initial choice will avoid wasting time on a topology that may not be the best for the application.

We will see that some topologies are best used for AC/DC offline converters at lower output powers (say, < 200 W), whereas others will be better at higher output powers. Again some will be a better choice for higher AC input voltages (say, $\geq$ 220 VAC), whereas others will be better at lower AC input voltages. In a similar way, some will have advantages for higher DC output voltages (say, > 200 V), yet others are preferred at lower DC voltages. For applications where several output voltages are required, some topologies will have a lower parts count or may offer a trade-off in parts counts versus reliability, while input or output ripple and noise requirements will also be an important factor. Further, some topologies have inherent limitations that require additional or more complex circuitry, whereas the performance of others can become difficult to analyze in some situations.

So we should now see how helpful it can be in our initial design choice to have at least a working knowledge of the merits and limitations of all the basic topologies. A poor initial choice can result in performance limitation and perhaps in extended design time and cost. Hence it is well worth the time and effort to get to know the basic performance parameters of the various topologies.

In this first chapter, we describe some of the earliest and most fundamental building blocks that form the basis of all linear and switching power systems. These include the following regulators:

- Linear regulator
- Buck regulator
- Boost regulator
- Inverting regulator (also known as *flyback* or *buck-boost*)

We describe the basic operation of each type, show and explain the various waveforms, and describe the merits and limitations of each topology. The peak transistor currents and voltage stresses are shown for various output power and input voltage conditions. We look at the dependence of input current on output power and input voltage. We examine efficiency, DC and AC switching losses, and some typical applications.

1.2 Linear Regulator—the Dissipative Regulator

1.2.1 Basic Operation

To demonstrate the main advantage of the more complex switching regulators, the discussion starts with an examination of the basic properties of what preceded them—the linear or *series-pass* regulator.

Figure 1.1*a* shows the basic topology of the linear regulator. It consists of a transistor *Q*1 (operating in the linear, or non-switching mode) to form an electrically variable resistance between the DC source (V_{dc}) developed by the 60-Hz isolation transformer, rectifiers, and storage capacitor C_f, and the output terminal at V_o that is connected to the external load (not shown).

In Figure 1.1*a*, an error amplifier senses the DC output voltage V_o via a sampling resistor network *R*1, *R*2 and compares it with a reference voltage V_{ref}. The error amplifier output drives the base of the series-pass power transistor *Q*1 via a drive circuit. The phasing is such that if the DC output voltage V_o tends to increase (say, as a result of either an increase in input voltage or a decrease in output load current), the drive to the base of the series-pass transistor is reduced. This increases the resistance of the series-pass element *Q*1 and hence controls the output voltage so that the sampled output continues to track the reference voltage. This negative-feedback loop works in the reverse direction for any decreases in output voltage, such that the error amplifier increases the drive to *Q*1 decreasing the collector-to-emitter resistance, thus maintaining the value of V_o constant.

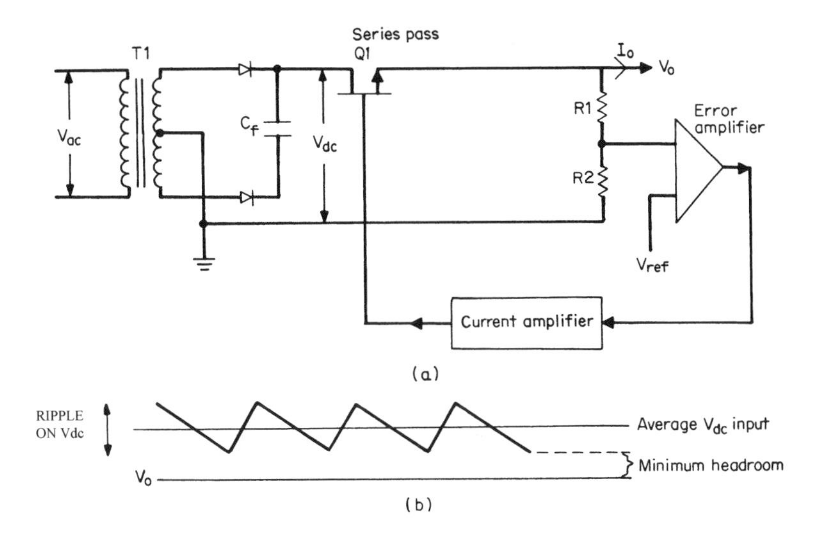

FIGURE 1.1 (*a*) The linear regulator. The waveform shows the ripple normally present on the unregulated DC input (V_{dc}). Transistor *Q*1, between the DC source at C_f and the output load at V_o, acts as an electrically variable resistance. The negative-feedback loop via the error amplifier alters the effective resistance of *Q*1 and will keep V_o constant, providing the input voltage sufficiently exceeds the output voltage. (*b*) Figure 1.1*b* shows the minimum input-output voltage differential (or *headroom*) required in a linear regulator. With a typical NPN series-pass transistor, a minimum input-output voltage differential (headroom) of at least 2.5 V is required between V_o and the bottom of the C_f input ripple waveform at minimum V_{ac} input.

In general, any change in input voltage—due to, for example, AC input line voltage change, ripple, steady-state changes in the input or output, and any dynamic changes resulting from rapid load changes over its designed tolerance band—is absorbed across the series-pass element. This maintains the output voltage constant to an extent determined by the gain in the open-loop feedback amplifier.

Switching regulators have transformers and fast switching actions that can cause considerable RFI noise. However, in the linear regulator the feedback loop is entirely DC-coupled. There are no switching actions within the loop. As a result, all DC voltage levels are predictable and calculable. This lower RFI noise can be a major advantage in some applications, and for this reason, linear regulators still have a place in modern power supply applications even though the efficiency is quite low. Also since the power losses are mainly due to the DC current and the voltage across *Q*1, the loss and the overall efficiency are easily calculated.

1.2.2 Some Limitations of the Linear Regulator

This simple, DC-coupled series-pass linear regulator was the basis for a multi-billion-dollar power supply industry until the early 1960s. However, in simple terms, it has the following limitations:

- The linear regulator is constrained to produce only a lower regulated voltage from a higher non-regulated input.
- The output always has one terminal that is common with the input. This can be a problem, complicating the design when DC isolation is required between input and output or between multiple outputs.
- The raw DC input voltage (V_{dc} in Figure 1.1*a*) is usually derived from the rectified secondary of a 60-Hz transformer whose weight and volume was often a serious system constraint.
- As shown next, the regulation efficiency is very low, resulting in a considerable power loss needing large heat sinks in relatively large and heavy power units.

1.2.3 Power Dissipation in the Series-Pass Transistor

A major limitation of a linear regulator is the inevitable and large dissipation in the series-pass element. It is clear that all the load current must pass through the pass transistor $Q1$, and its dissipation will be $(V_{dc} - V_o)(I_o)$. The minimum differential $(V_{dc} - V_o)$, the headroom, is typically 2.5 V for NPN pass transistors. Assume for now that the filter capacitor is large enough to yield insignificant ripple. Typically the raw DC input comes from the rectified secondary of a 60-Hz transformer. In this case the secondary turns can always be chosen so that the rectified secondary voltage is near $V_o + 2.5$ V when the input AC is at its low tolerance limit. At this point the dissipation in $Q1$ will be quite low.

However, when the input AC voltage is at its high tolerance limit, the voltage across $Q1$ will be much greater, and its dissipation will be larger, reducing the power supply efficiency. Due to the minimum 2.5-volt headroom requirement, this effect is much more pronounced at lower output voltages.

This effect is dramatically demonstrated in the following examples. We will assume an AC input voltage range of ±15%. Consider three examples as follows:

- Output of 5 V at 10 A
- Output of 15 V at 10 A
- Output of 30 V at 10 A

Assume for now that a large secondary filter capacitor is used such that ripple voltage to the regulator is negligible. The rectified secondary voltage range (V_{dc}) will be identical to the AC input voltage range of ±15%. The transformer secondary voltages will be chosen to yield (V_o + 2.5 V) when the AC input is at its low tolerance limit of −15%. Hence, the maximum DC input is 35% higher when the AC input is at its maximum tolerance limit of +15%. This yields the following:

V_o	I_o, A	$V_{dc(min)}$, V	$V_{dc(max)}$, V	Headroom, max, V	$P_{in(max)}$, W	$P_{out(max)}$, W	Dissipation $Q1_{max}$	Efficiency, % $P_o/P_{in(max)}$
5.0	10	7.5	10.1	5.1	101	50	51	50
15.0	10	17.5	23.7	8.7	237	150	87	63
30.0	10	32.5	44.0	14	440	300	140	68

It is clear from this example that at lower DC output voltages the efficiency will be very low. In fact, as shown next, when realistic input line ripple voltages are included, the efficiency for a 5-volt output with a line voltage range of ±15% will be only 32 to 35%.

1.2.4 Linear Regulator Efficiency vs. Output Voltage

We will consider in general the range of efficiency expected for a range of output voltages from 5 V to 100 V with line inputs ranging from ±5 to ±15% when a realistic ripple value is included.

Assume the minimum headroom is to be 2.5 V, and this must be guaranteed at the bottom of the input ripple waveform at the lower limit of the input AC voltages range, as shown in Figure 1.1*b*. Regulator efficiency can be calculated as follows for various assumed input AC tolerances and output voltages.

Let the input voltage range be $\pm T\%$ about its nominal. The transformer secondary turns will be selected so that the voltage at the bottom of the ripple waveform will be 2.5 V above the desired output voltage when the AC input is at its lower limit.

Let the peak-to-peak ripple voltage be V_r volts. When the input AC is at its low tolerance limit, the average or DC voltage at the input to the pass transistor will be

$$V_{dc} = (V_o + 2.5 + V_r/2) \text{ volts}$$

When the AC input is at its high tolerance limit, the DC voltage at the input to the series-pass element is

$$V_{dc(max)} = \frac{1 + 0.01T}{1 - 0.01T}(V_o + 2.5 + V_r/2)$$

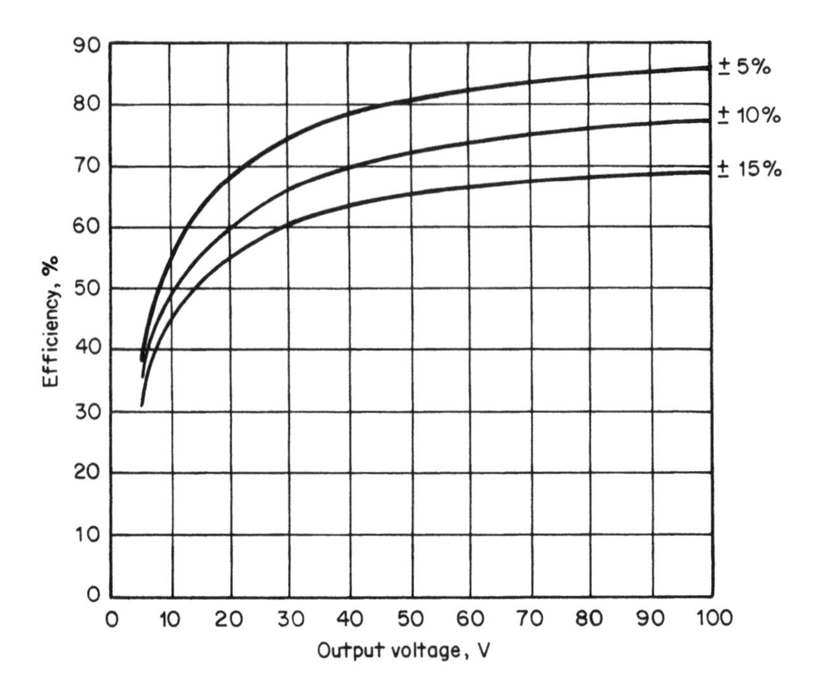

FIGURE 1.2 Linear regulator efficiency versus output voltage. Efficiency shown for maximum V_{ac} input, assuming a 2.5-V headroom is maintained at the bottom of the ripple waveform at minimum V_{ac} input. Eight volts peak-to-peak ripple is assumed at the top of the filter capacitor. (From Eq. 1.2)

The maximum achievable worst-case efficiency (which occurs at maximum input voltage and hence maximum input power) is

$$\text{Efficiency}_{\text{max}} = \frac{P_o}{P_{\text{in(max)}}} = \frac{V_o I_o}{V_{\text{dc(max)}} I_o} = \frac{V_o}{V_{\text{dc(max)}}} \tag{1.1}$$

$$= \frac{1 - 0.01T}{1 + 0.01T} \left(\frac{V_o}{V_o + 2.5 + V_r/2} \right) \tag{1.2}$$

This is plotted in Figure 1.2 for an assumed peak-to-peak (p/p) ripple voltage of 8 V. It will be shown that in a 60-Hz full-wave rectifier, the p/p ripple voltage is 8 V if the filter capacitor is chosen to be of the order of 1000 microfarads (μF) per ampere of DC load current, an industry standard value.

It can be seen in Figure 1.2 that even for 10-V outputs, the efficiency is less than 50% for a typical AC line range of ±10%. In general it is the poor efficiency, the weight, the size, and the cost of the 60-Hz input transformer that was the driving force behind the development of switching power supplies.

However, the linear regulator with its lower electrical noise still has applications and may not have excessive power loss. For example, if a reasonably pre-regulated input is available (frequently the case in some of the switching configurations to be shown later), a liner regulator is a reasonable choice where lower noise is required. Complete integrated-circuit linear regulators are available up to 3-A output in single plastic packages and up to 5 A in metal-case integrated-circuit packages. However, the dissipation across the internal series-pass transistor can still become a problem at the higher currents. We now show some methods of reducing the dissipation.

1.2.5 Linear Regulators with PNP Series-Pass Transistors for Reduced Dissipation

Linear regulators using PNP transistors as the series-pass element can operate with a minimum headroom down to less than 0.5 V. Hence they can achieve better efficiency. Typical arrangements are shown in Figure 1.3.

With an NPN series-pass element configured as shown in Figure 1.3*a*, the base current (I_b) must come from some point at a potential higher than $V_o + V_{be}$, typically $V_o + 1$ volts. If the base drive comes through a resistor as shown, the input end of that resistor must come from a voltage even higher than $V_o + 1$. The typical choice is to supply the base current from the raw DC input as shown.

A conflict now exists because the raw DC input at the bottom of the ripple waveform at the low end of the input range cannot be permitted to come too close to the required minimum base input voltage (say, $V_o + 1$). Further, the base resistor R_b would need to have a very low value to provide sufficient base current at the maximum output current. Under these conditions, at the high end of the input range (when $V_{dc} - V_o$ is much greater), R_b would deliver an excessive drive current; a significant amount would have to be diverted away into the current amplifier, adding to its dissipation. Hence a compromise is required. This is why the minimum header voltage is selected to be typically 2.5 V in this arrangement. It maintains a more constant current through R_b over the range of input voltage.

However, with a PNP series-pass transistor (as in Figure 1.3*b*), this problem does not exist. The drive current is derived from the common negative line via the current amplifier. The minimum header voltage is defined only by the knee of the I_c versus V_{ce} characteristic of the pass transistor. This may be less than 0.5 V, providing higher efficiency particularly for low-voltage, high-current applications.

Although integrated-circuit linear regulators with PNP pass transistors are now available, they are intrinsically more expensive because the fabrication is more difficult.

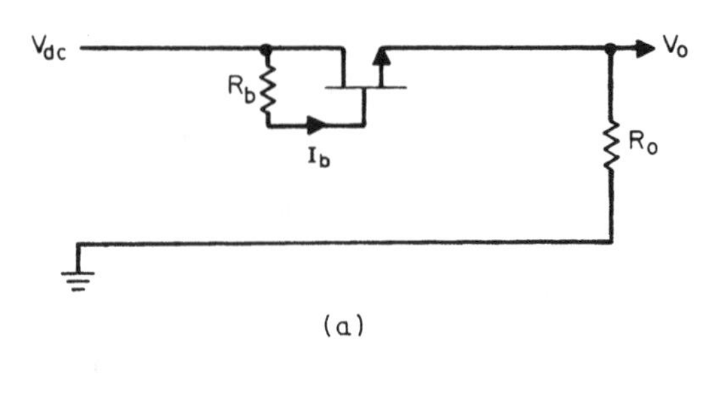

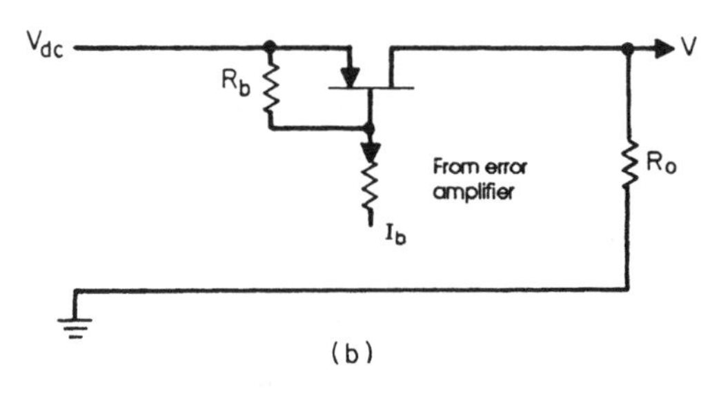

FIGURE 1.3 (*a*) A linear regulator with an NPN series-pass transistor. In this example, the base drive is taken from V_{dc} via a resistor R_b. A typical minimum voltage of 1.5 V is required across R_b to supply the base current, which when added to the base-emitter drop makes a minimum header voltage of 2.5 V. (*b*) Linear regulator with a PNP series-pass transistor. In this case the base drive (I_b) is derived from the negative common line via the drive circuit. The header voltage is no longer restricted to a minimum of 2.5 V, and much lower values are possible.

Similar results can be obtained with NPN transistors by fitting the transistor in the negative return line. This requires the positive line to be the common line. (Normally this would not be a problem in single output supply.)

This completes our overview of linear regulators and serves to demonstrate some of the reasons for moving to the more complicated switching methods for modern, low-weight, small, and efficient power systems.

1.3 Switching Regulator Topologies

1.3.1 The Buck Switching Regulator

The high dissipation across the series-pass transistor in a linear regulator and the large 60-Hz transformer required for line operation made linear regulators unattractive for modern electronic applications.

Further, the high power loss in the series device requires a large heat sink and large storage capacitors and makes the linear power supply disproportionately large.

As electronics advanced, integrated circuits made the electronic systems smaller. Typically, linear regulators could achieve output power densities of 0.2 to 0.3 W/in^3, and this was not good enough for the ever smaller modern electronic systems. Further, linear power supplies could not provide the extended hold-up time required for the controlled shutdown of digital storage systems.

Although the technology was previously well known, switching regulators started being widely used as alternatives to linear regulators only in the early 1960s when suitable semiconductors with reasonable performance and cost became available. Typically these new switching supplies used a transistor switch to generate a square-waveform from a non-regulated DC input voltage. This square wave, with adjustable duty cycle, was applied to a low pass output power filter so as to provide a regulated DC output.

Usually the filter would be an inductor (or more correctly a choke, since it had to support some DC) and an output capacitor. By varying the duty cycle, the average DC voltage developed across the output capacitor could be controlled. The low pass filter ensured that the DC output voltage would be the average value of the rectangular voltage pulses (of adjustable duty cycle) as applied to the input of the low pass filter. A typical topology and waveforms are shown later in Figure 1.4.

With appropriately chosen low pass inductor/capacitor (*LC*) filters, the square-wave modulation could be effectively minimized, and near-ripple-free DC output voltages, equal to the average value of the duty-cycle-modulated raw DC input, could be provided. By sensing the DC output voltage and controlling the switch duty cycle in a negative-feedback loop, the DC output could be regulated against input line voltage changes and output load changes.

Modern very high frequency switching supplies are currently achieving up to 20 W/in^3 compared with 0.3 W/in^3 for the older linear power supplies. Further, they are capable of generating a multiplicity of isolated output voltages from a single input. They do not require a 50/60-Hz isolation power transformer, and they have efficiencies from 70% up to 95%. Some DC/DC converter designers are claiming load power densities of up to 50 W/in^3 for the actual switching elements.

1.3.1.1 Basic Elements and Waveforms of a Typical Buck Regulator

After Pressman *In the interest of simplicity, Mr. Pressman describes fixed-frequency operation for the following switching regulator examples. In such regulators the* on *period of the power device* (T_{on}) *is adjusted to maintain*

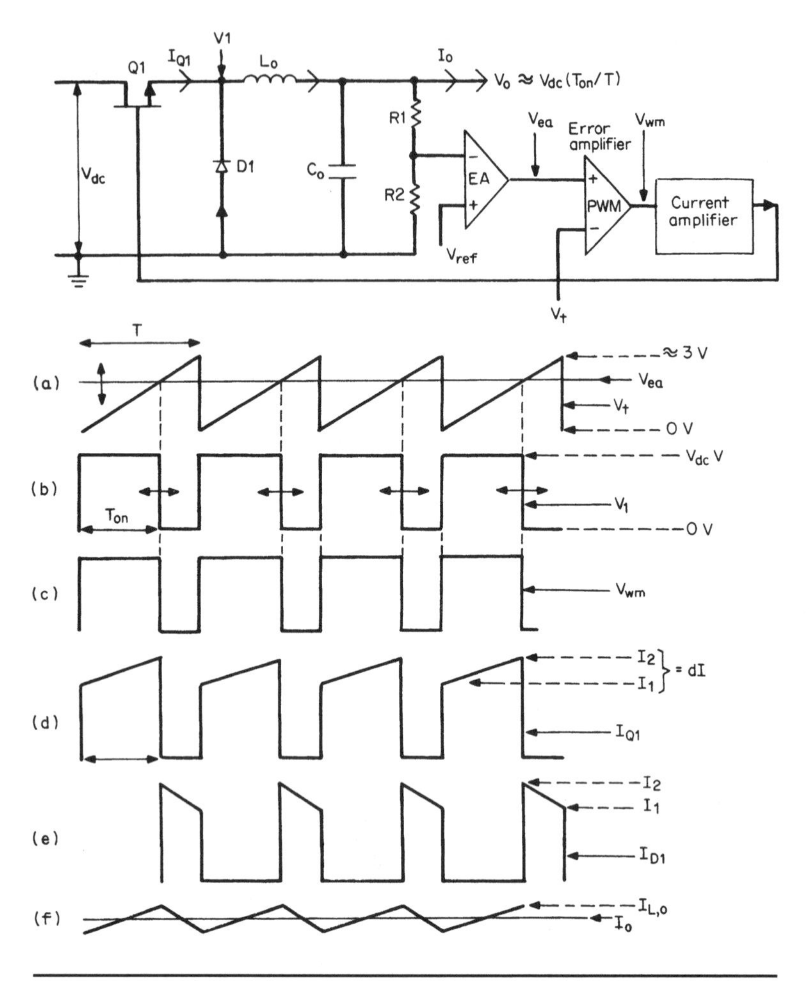

FIGURE 1.4 Buck switching regulator and typical waveforms.

regulation, while the total cycle period (T) is fixed, and the frequency is thus fixed at 1/T.

The ratio T_{on}/T *is normally referred to as the* duty ratio *or* duty cycle (D) *in many modern treatises. In other books on the subject, you may find this shown as* $T_{on}/(T_{on} + T_{off})$, *where* T_{off} *is the* off *period of the power device so that* $T_{on} + T_{off} = T$. *Operators* D *and* M *are also used in various combinations but essentially refer to the same quantity.*

Bear in mind that other modes of operation can be and are used. For example, the on *period can be fixed and the frequency changed, or a combination of both may be employed.*

The terms dI, di, dV, dv, dT *and* dt *are used somewhat loosely in this book and normally refer to the changes* ΔI, ΔV, *and* Δt, *where, for example, in the limit,* $\Delta I/\Delta t$ *goes to the derivative* di/dt, *giving the rate of change of current with time or the slope of the waveform. Since in most cases the waveform slopes are linear the result is the same so this becomes a moot point. ~K.B.*

1.3.1.2 Buck Regulator Basic Operation

The basic elements of the buck regulator are shown in Figure 1.4. Transistor $Q1$ is switched hard "on" and hard "off" in series with the DC input V_{dc} to produce a rectangular voltage at point $V1$. For fixed-frequency duty-cycle control, $Q1$ conducts for a time T_{on} (a small part of the total switching period T). When $Q1$ is "on," the voltage at $V1$ is V_{dc}, assuming for the moment the "on" voltage drop across $Q1$ is zero.

A current builds up in the series inductor L_o flowing toward the output. When $Q1$ turns "off," the voltage at $V1$ is driven rapidly toward ground by the current flowing in inductor L_o and will go negative until it is caught and clamped at about −0.8 V by diode $D1$ (the so-called free-wheeling diode).

Assume for the moment that the "on" drop of diode $D1$ is zero. The square voltage shown in Figure 1.4*b* would be rectangular, ranging between V_{dc} and ground, (0 V) with a "high" period of T_{on}. The average value of this rectangular waveform is $V_{dc}T_{on}/T$. The low pass L_oC_o filter in series between $V1$ and the output V extracts the DC component and yields a clean, near-ripple-free DC voltage at the output with a magnitude V_o of $V_{dc}T_{on}/T$.

To control the voltage, V_o is sensed by sampling resistors $R1$ and $R2$ and compared with a reference voltage V_{ref} in the error amplifier (EA). The amplified DC error voltage V_{ea} is fed to a pulse-width-modulator (PWM). In this example the PWM is essentially a voltage comparator with a sawtooth waveform as the other input (see Figure 1.4*a*). This sawtooth waveform has a period T and amplitude typically in the order of 3 V. The high-gain PWM voltage comparator generates a rectangular output waveform (V_{wm}, see Figure 1.4*c*) that goes high at the start of the sawtooth ramp, and goes low the instant the ramp voltage crosses the DC voltage level from the error-amplifier output. The PWM output pulse width (T_{on}) is thus controlled by the EA amplifier output voltage.

The PWM output pulse is fed to a driver circuit and used to control the "on" time of transistor switch $Q1$ inside the negative-feedback loop. The phasing is such that if V_{dc} goes slightly higher, the EA DC level goes closer to the bottom of the ramp, the ramp crosses the EA output level earlier, and the $Q1$ "on" time decreases, maintaining the output voltage constant. Similarly, if V_{dc} is reduced, the "on" time of $Q1$ increases to maintain V_o constant. In general, for all changes, the "on" time of $Q1$ is controlled so as to make the sampled DC output voltage $V_oR_2/(R_1 + R_2)$ closely track the reference voltage V_{ref}.

1.3.2 Typical Waveforms in the Buck Regulator

In general, the major advantage of the switching regulator technique over its linear counterpart is the elimination of the power loss intrinsic in the linear regulator pass element.

In the switching regulator the pass element is either fully "on" (with very little power loss) or fully "off" (with negligible power loss). The buck regulator is a good example of this—it has low internal losses and hence high power conversion efficiency.

However, to fully appreciate the subtleties of its operation, it is necessary to understand the waveforms and the magnitude and timing of the currents and voltages throughout the circuit. To this end we will look in more detail at a full cycle of events starting when $Q1$ turns fully "on." For convenience we will assume ideal components and steady-state conditions, with the amplitude of the input voltage V_{dc} constant, exceeding the output voltage V_o, which is also constant.

When $Q1$ turns fully "on," the supply voltage V_{dc} will appear across the diode $D1$ at point $V1$. Since the output voltage V_o is less than V_{dc}, the inductor L_o will have a voltage impressed across it of $(V_{dc} - V_o)$. With a constant voltage across the inductor, its current rises linearly at a rate given by $di/dt = (V_{dc} - V_o)/L_o$. (This is shown in Figure 1.4*d* as a ramp that sits on top of the step current waveform.)

When $Q1$ turns "off," the voltage at point $V1$ is driven toward zero because it is not possible to change the previously established inductor current instantaneously. Hence the voltage polarity across L_o immediately reverses, trying to maintain the previous current. (This polarity reversal is often referred to as the *flyback* or *inductive kickback* effect of the inductor.) Without diode $D1$, $V1$ would have gone very far negative, but with $D1$ fitted as shown, as the $V1$ voltage passes through zero, $D1$ conducts and clamps the left side of L_o at one diode drop below ground. The voltage across the inductor has now reversed, and the current in the inductor and $D1$ will ramp down, returning to its original starting value, during the "off" period of $Q1$.

More precisely, when $Q1$ turns "off," the current I_2 (which had been flowing in $Q1$, L_o and the output capacitor C_o and the load just prior to turning "off") is diverted and now flows through diode $D1$, L_o and the output capacitor and load, as shown in Figure 1.4*e*. The voltage polarity across L_o has reversed with a magnitude of $(V_o + 1)$. The current in L_o now ramps down linearly at a rate defined by the equation $di/dt = (V_o + 1)/L_o$. This is the downward ramp that sits on a step in Figure 1.4*e*. Under steady-state conditions, at the end of the $Q1$ "off" time, the current in L_o will have fallen to I_1 and is still flowing through $D1$, L_o and the output capacitor and load.

Note *Notice the input current is discontinuous with a pulse-like characteristic, whereas the output current remains nearly continuous with some relatively small ripple component depending on the value of* L_o *and* C_o. ~ *K.B.*

Now when $Q1$ turns "on" again, it initially supplies current into the cathode of $D1$, displacing its previous forward current. While the current in $Q1$ rises toward the previous value of I_1, the forward $D1$ current will be displaced, and V_1 rises to near V_{dc}, back-biasing $D1$. Because $Q1$ is switched "on" hard, this recovery process is very rapid, typically less than 1 μs.

Notice that the current in L_o is the sum of the $Q1$ current when it is "on" (see Figure 1.4*d*) plus the $D1$ current when $Q1$ is "off." This is shown in Figure 1.4f as $IL_{,o}$. It has a DC component and a triangular waveform ripple component $(I_2 - I_1)$ centered on the mean DC output current I_o. Thus the value of the current at the center of the ramp in Figure 1.4*d* and 1.4*e* is simply the DC mean output current I_o. As the load resistance and hence load current is changed, the center of the ramp (the mean value) in either Figure 1.4*d* or 1.4*e* moves, but the slopes of the ramps remain constant, because during the $Q1$ "on" time, the ramp rate in L_o remains the same at $(V_{dc} - V_o)/L_o$, and during the $Q1$ "off" time, it remains the same at $(V_o + 1)/L$ as the load current changes, because the input and output voltages remain constant.

Because the p-p ripple current remains constant regardless of the mean output current, it will be seen shortly that when the DC current I_o is reduced to the point where the lower value of the ripple current in Figure 1.4*d* and 1.4*e* just reaches zero (the critical load current), there will be a drastic change in performance. (This will be discussed in more detail later.)

1.3.3 Buck Regulator Efficiency

To get a general feel for the intrinsic power loss in the buck regulator compared with a linear regulator, we will start by assuming ideal components for transistor $Q1$ and diode $D1$ in both topologies. Using the currents shown in Figure 1.4*d* and 1.4*e*, the typical conduction losses in $Q1$ and free-wheeling diode $D1$ can be calculated and the efficiency obtained. Notice that when $Q1$ is "off," it operates at a maximum voltage of V_{dc} but at zero current. When $Q1$ is "on," current flows, but the voltage across $Q1$ is zero. At the same time, $D1$ is reverse-biased at a voltage of V_{dc} but has zero current. (Clearly, if $Q1$ and $D1$ were ideal components, the currents would flow through $Q1$ and $D1$ with zero voltage drop, and the loss would be zero.)

Hence unlike the linear regulator, which has an intrinsic loss even with ideal components, the intrinsic loss in a switching regulator with ideal components is zero, and the efficiency is 100%. Thus in the buck regulator, the real efficiency depends on the actual performance of the components. Since improvements are continually being made in semiconductors, we will see ever higher efficiencies.

To consider more realistic components, the losses in the buck circuit are the conduction losses in $Q1$ and $D1$ and the resistive winding loss in the choke. The conduction losses, being related to the mean DC currents, are relatively easy to calculate. To this we must add the AC switching losses in $Q1$ and $D1$, and the AC induced core loss in the inductor, so the switching loss is more difficult to establish.

The switching loss in $Q1$ during the turn "on" and turn "off" transitions is a result of the momentary overlap of current and voltage during the switching transitions. Diode $D1$ also has switching loss associated with the reverse recovery action of the diode, where again there is a condition of voltage and current stress during the transitions. The ripple waveform in the inductor L_o results in hysteretic and eddy current loss in the core material. We will now calculate some typical losses.

1.3.3.1 Calculating Conduction Loss and Conduction-Related Efficiency

By neglecting second-order effects and AC switching losses, the conduction loss can be quite easily calculated. It can be seen from Figure 1.4*d* and 1.4*e* that the average currents in $Q1$ and $D1$ during their conduction times of T_{on} and T_{off} are the values at the center of the ramps or I_o, the mean DC output current. These currents flow at a forward voltage of about 1 V over a wide range of currents. Thus conduction losses will be approximately

$$P_{\text{dc}} = L(Q1) + L(D1) = 1I_o\frac{T_{\text{on}}}{T} + 1I_o\frac{T_{\text{off}}}{T} = 1I_o$$

Therefore, by neglecting AC switching losses, the conduction-related efficiency would be

$$\text{Conduction Efficiency} = \frac{P_o}{P_o + \text{losses}} = \frac{V_o I_o}{V_o I_o + 1I_o} = \frac{V_o}{V_o + 1} \tag{1.3}$$

1.3.4 Buck Regulator Efficiency Including AC Switching Losses

After Pressman *The switching loss is much more difficult to establish, because it depends on many variables relating to the performance of the semiconductors and to the methods of driving the switching devices. Other variables, related to the actual power circuit designs, include the action of any* *snubbers, load line shaping, and energy recovery arrangements. It depends on what the designer may choose to use in a particular design. (See Chapter 11.)*

Unless all these things are considered, any calculations are at best only a very rough approximation and can be far from the real values found in the actual design, particularly at high frequencies with the very fast switching devices now available.

After Pressman *I leave Mr. Pressman's original calculations, shown next, untouched except for some minor editing, because they serve to illustrate the root cause of the switching loss. However, I would recommend that the reader consider using more practical methods to establish the real loss. Many semiconductor manufacturers now provide switching loss equations for their switching devices when recommended drive conditions are used, particularly the modern fast* IGBTs *(Insulated Gate Bipolar Transistors). Some fast digital oscilloscopes claim that they will actually measure switching loss, providing the real-time device current and voltage is accurately provided to the oscilloscope. (Doing this can also be problematical at very high frequency.)*

The method I prefer, which is unquestionably accurate, is to measure the temperature rise of the device in question in a working model. The model must include all the intended snubbers and load line shaping circuits, etc. Replacing the AC current in the device with a DC current to obtain the same temperature rise will provide a direct indication of power loss by simple DC power measurements. This method also allows easy optimization of the drive and load line shaping, which can be dynamically adjusted during operation for minimum temperature rise and hence minimum switching loss. ~ K.B.

Mr. Pressman continues as follows:

Alternating-current switching loss (or voltage/current overlap loss) calculation depends on the shape and timing of the rising and falling voltage and current waveforms. An idealized linear example—*which is unlikely to exist in practice*—is shown in Figure 1.5*a* and serves to illustrate the principle.

Figure 1.5*a* shows the best-case scenario. At the turn "on" of the switching device, the voltage and current start changing simultaneously and reach their final values simultaneously. The current waveform goes from 0 to I_o, and voltage across $Q1$ goes from a maximum of V_{dc} down to zero. The average power during this switching transition is $P(T_{on}) = \int_0^{T_{on}} IV\,dt = I_o V_{dc}/6$, and the power averaged over one complete period is $(I_o V_{dc}/6)(T_{on}/T)$.

Assuming the same scenario of simultaneous starting and ending points for the current fall and voltage rise waveforms at the turn "off" transition, the voltage/current overlap dissipation at this transition is given by $P(T_{off}) = \int_0^{T_{off}} IV\,dt = I_o V_{dc}/6$ and this power averaged over one complete cycle is $(I_o V_{dc}/6)(T_{off}/T)$.

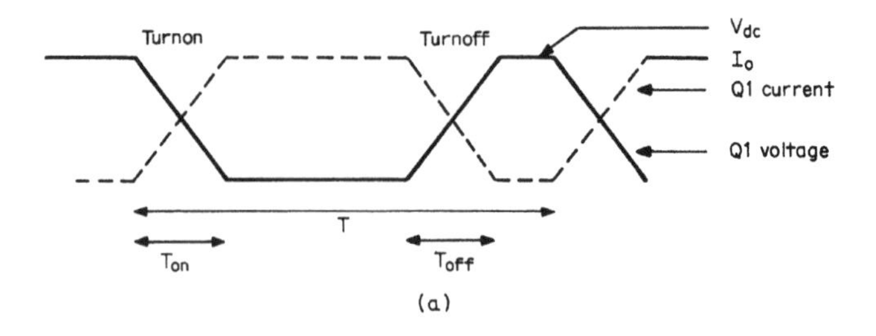

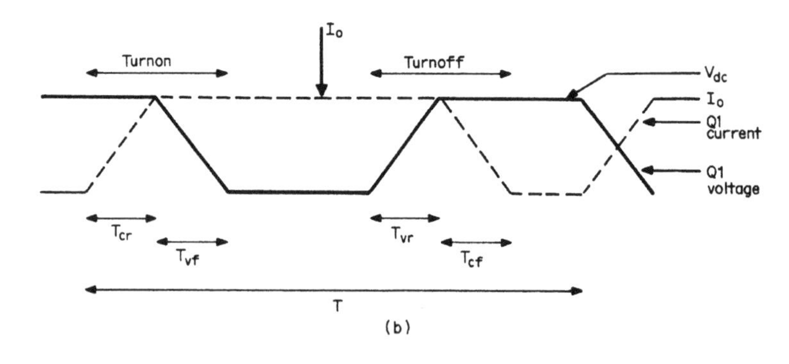

FIGURE 1.5 Idealized transistor switching waveforms. (*a*) Waveforms show the voltage and current transitions starting and ending simultaneously. (*b*) Waveforms show the worst-case scenario, where at turn "on" voltage remains constant at $V_{dc(max)}$ until current reaches its maximum. At turn "off," the current remains constant at I_o until $Q1$ voltage reaches its maximum of V_{dc}.

Assuming $T_{on} = T_{off} = T_s$, the total switching losses (the sum of turn "off" and turn "on" losses) are $P_{ac} = (V_{dc}\, I_o T_s)/3T$, and efficiency is calculated as shown next in Eq. 1.4.

$$\begin{aligned}\text{Efficiency} &= \frac{P_o}{P_o + \text{DC losses} + \text{AC losses}} \\ &= \frac{V_o I_o}{V_o I_o + 1 I_o + V_{dc} I_o T_s/3T} \\ &= \frac{V_o}{V_o + 1 + V_{dc} T_s/3T}\end{aligned} \tag{1.4}$$

It would make an interesting comparison to calculate the efficiency of the buck regulator and compare it with that of a linear regulator. Assume the buck regulator provides 5 V from a 48-V DC input at 50-kHz switching frequency (T = 20 μs).

If there were no AC switching losses and a switching transition period T_s of 0.3 μs were assumed, Eq. 1.3 would give a conduction loss efficiency of

$$\text{Efficiency} = \frac{5}{5+1} = 83.3\%$$

If switching losses for the best-case scenario as shown in Figure 1.5*a* were assumed, for $T_s = 0.3$ μs and $T = 20$ μs, Eq. 1.4 would give a switching-related efficiency of

$$\begin{aligned}\text{Efficiency} &= \frac{5}{5+1+48 \times 0.3/3 \times 20} \\ &= \frac{5}{5+1+0.24} = \frac{5}{5+1.24} \\ &= 80.1\%\end{aligned}$$

If a worst-case scenario were assumed (which is closer to reality), as shown in Figure 1.5*b*, efficiencies would lower. In Figure 1.5*b* it is assumed that at turn "on" the voltage across the transistor remains at its maximum value (V_{dc}) until the on-turning current reaches its maximum value of I_o. Then the voltage starts falling. To a close approximation, the current rise time T_{cr} will equal voltage fall time. Then the turn "on" switching losses will be

$$P(T_{on}) = \frac{V_{de} I_o}{2}\frac{T_{cr}}{T} + \frac{I_o V_{dc}}{2}\frac{T_{vf}}{T}$$

also for $T_{cr} = T_{vf} = T_s$, $P(T_{on}) = V_{dc} I_o (T_s/T)$.

At turn "off" (as seen in Figure 1.5*b*), we may assume that current hangs on at this maximum value I_o until the voltage has risen to its maximum value of V_{dc} in a time T_{vr}. Then current starts falling and reaches zero in a time T_{cf}. The total turn "off" dissipation will be

$$P(T_{off}) = \frac{I_o V_{dc}}{2}\frac{T_{vr}}{T} + \frac{V_{dc} I_o}{2}\frac{T_{cf}}{T}$$

With $T_{vr} = T_{cf} = T_s$, $P(T_{off}) = V_{dc} I_o (T_s/T)$. The total AC losses (the sum of the turn "on" plus the turn "off" losses) will be

$$P_{ac} = 2 V_{dc} I_o \frac{T_s}{T} \tag{1.5}$$

and the total losses (the sum of DC plus AC losses) will be

$$P_t = P_{dc} + P_{ac} = 1 I_o + 2 V_{dc} I_o \frac{T_s}{T} \tag{1.6}$$

and the efficiency will be

$$\begin{aligned}\text{Efficiency} &= \frac{P_o}{P_o + P_t} = \frac{V_o I_o}{V_o I_o + 1I_o + 2V_{\text{dc}} I_o T_s / T} \\ &= \frac{V_o}{V_o + 1 + 2V_{\text{dc}} T_s / T}\end{aligned} \tag{1.7}$$

Hence in the worst-case scenario, for the same buck regulator with $T_s = 0.3\ \mu\text{s}$, the efficiency from Eq. 1.7 will be

$$\begin{aligned}\text{Efficiency} &= \frac{5}{5 + 1 + 2 \times 48 \times 0.3/20} = \frac{5}{5 + 1 + 1.44} \\ &= \frac{5}{5 + 1 + 2.44} \\ &= 67.2\%\end{aligned}$$

Comparing this with a linear regulator doing the same job (bringing 48 V down to 5 V), its efficiency (from Eq. 1.1) would be $V_o / V_{\text{dc(max)}}$, or 5/48; this is only 10.4% and is clearly unacceptable.

1.3.5 Selecting the Optimum Switching Frequency

We have seen that the output voltage of the buck regulator is given by the equation $V_o = V_{\text{dc}} T_{\text{on}} / T$. We must now decide on a value for this period and hence the operating frequency.

The initial reaction may be to minimize the size of the filter components L_o, C_o by using as high a frequency as possible. However, using higher frequencies does not necessarily minimize the overall size of the regulator when all factors are considered.

We can see this better by examining the expression for the AC losses shown in Eq. 1.5, $P_{\text{ac}} = 2V_{\text{dc}} I_o \frac{T_s}{T}$. We see that the AC losses are inversely proportional to the switching period T. Further, this equation only shows the losses in the switching transistor; it neglects losses in the free-wheeling diode $D1$ due to its finite *reverse recovery time* (the time required for the diode to cease conducting reverse current, measured from the instant it has been subjected to a reverse bias voltage). The free-wheeling diode can dissipate significant power and should be of the ultrafast soft recovery type with minimum recovered charge. The reverse recovery time will typically be 35 ns or less.

In simple terms, the more switching transitions there are in a particular period, the more switching loss there will be. As a result there is a trade-off—decreasing the switching period T (increasing the switching frequency) may well decrease the size of the filter elements, but it will also add to the total losses and may require a larger heat sink.

In general, although the overall volume of the buck regulator will be lower at a higher frequency, the increase in the switching loss and the more stringent high-frequency layout and component-selection requirements make the final choice a compromise among all the opposing elements.

Note *The picture is constantly changing as better, lower cost, and faster transistors and diodes are developed. My choice at the present stage of the technology is to design below 100 kHz, as this is less demanding on component selection, layout, and transformer/inductor designs. As a result it is probably lower cost. Generally speaking, higher frequencies absorb more development time and require more experience. However, efficient commercial designs are on the market operating well into the MHz range. The final choice is up to the designer, and I hesitate to recommend a limit because technology is constantly changing toward higher frequency operation. ~ K.B.*

1.3.6 Design Examples

1.3.6.1 Buck Regulator Output Filter Inductor (Choke) Design

Note *The output inductor and capacitor may be considered a low pass filter, and it is normally treated in this way for transfer function and loop compensation calculations.*

However, at this stage, the reader may prefer to look upon the inductor as a device that tends to maintain the current reasonably constant during the switching action. (That is, it stores energy when the power device is "on" and transfers this energy to the output when the power device is "off.")

I prefer the term choke *for the power inductor, because in this application it must support an element of DC current as well as the applied AC voltage stress. It will be shown later (Chapter 7) that the design of pure inductors (with zero DC current component) is quite different from the design of chokes, with their relatively large DC current component.*

In the following section Mr. Pressman outlines the parameters that control the design and selection of this critical part. ~ K.B.

The current waveform of the output inductor (choke) is shown in Figure 1.4f, and its characteristic "dual ramp" shape is defined in Section 1.3.2. Notice that the current amplitude at the center of the ramp is the mean value equal to the DC output current I_o.

We have seen that as the DC output load current decreases, the slope of the ramp remains constant (because the voltage across L_o remains constant). But as the mean load current decreases, the ripple current waveform moves down toward zero.

At a load current of half the peak-to-peak magnitude of the ramp, $I_o = (I_2 - I_1)/2dI$, the lower point of the ramp just touches zero.

At this point, the current in the inductor is zero and its stored energy is zero. (The inductor is said to have "run dry.") If the load current is further reduced, there will be a period when the inductor current remains at zero for a longer period and the buck regulator enters into the "discontinuous current" operating mode. This is an important transition because a drastic change occurs in the current and voltage waveforms and in the closed loop transfer function.

This transition to the discontinuous mode can be seen in the real-time oscilloscope picture of Figure 1.6*a*. This shows the power switch current waveforms for a buck regulator operating at 25 kHz with an input voltage of 20 V and an output of 5 V as the load current is reduced from a nominal current of 5 A down to about 0.2 A.

The top two waveforms have the characteristic ramp-on-a-step waveshape with the step size reducing as the load current is reduced. The current amplitude at the center of the ramp indicates the effective DC output current.

In the third waveform, where $I_o = 0.95$ A, the step has gone and the front end of the ramp starts at zero current. This is the *critical load current* indicating the start of the discontinuous current mode (or run-dry mode) for the inductor. Notice that in the first three waveforms, the $Q1$ "on" time is constant, but decreases drastically as the current is further reduced, moving deeper into the discontinuous mode.

In this example, the control loop has been able to maintain the output voltage constant at 5 V throughout the full range of load currents, even after the inductor has gone discontinuous. Hence it would be easy to assume that there is no problem in permitting the inductor to go discontinuous. In fact there are changes in the transfer function (discussed next) that the control loop must be able to accommodate. Further, the transition can become a major problem in the boost-type topologies discussed later.

For the buck regulator, however, the discontinuous mode is not considered a major problem. For load currents above the onset of the discontinuous made, the DC output voltage is given by $V_o = V_1 T_{\text{on}}/T$. Notice the load current is not a parameter in this equation, so the voltage remains constant with load current changes without the need to change the duty ratio. (The effective output resistance of the buck regulator is very low in this region.) In practice the "on" time changes slightly as the current changes, because the forward drop across $Q1$ and the inductor resistance change slightly with current, requiring a small change in T_{on}.

If the load is further reduced so as to enter discontinuous mode, the transfer function changes drastically and the previous equation for output voltage ($V_o = V_1 T_{\text{on}}/T$) no longer applies. This can be seen in the bottom two waveforms of Figure 1.6*a*. Notice the "on" time of $Q1$ has decreased and has become a function of the DC output current.

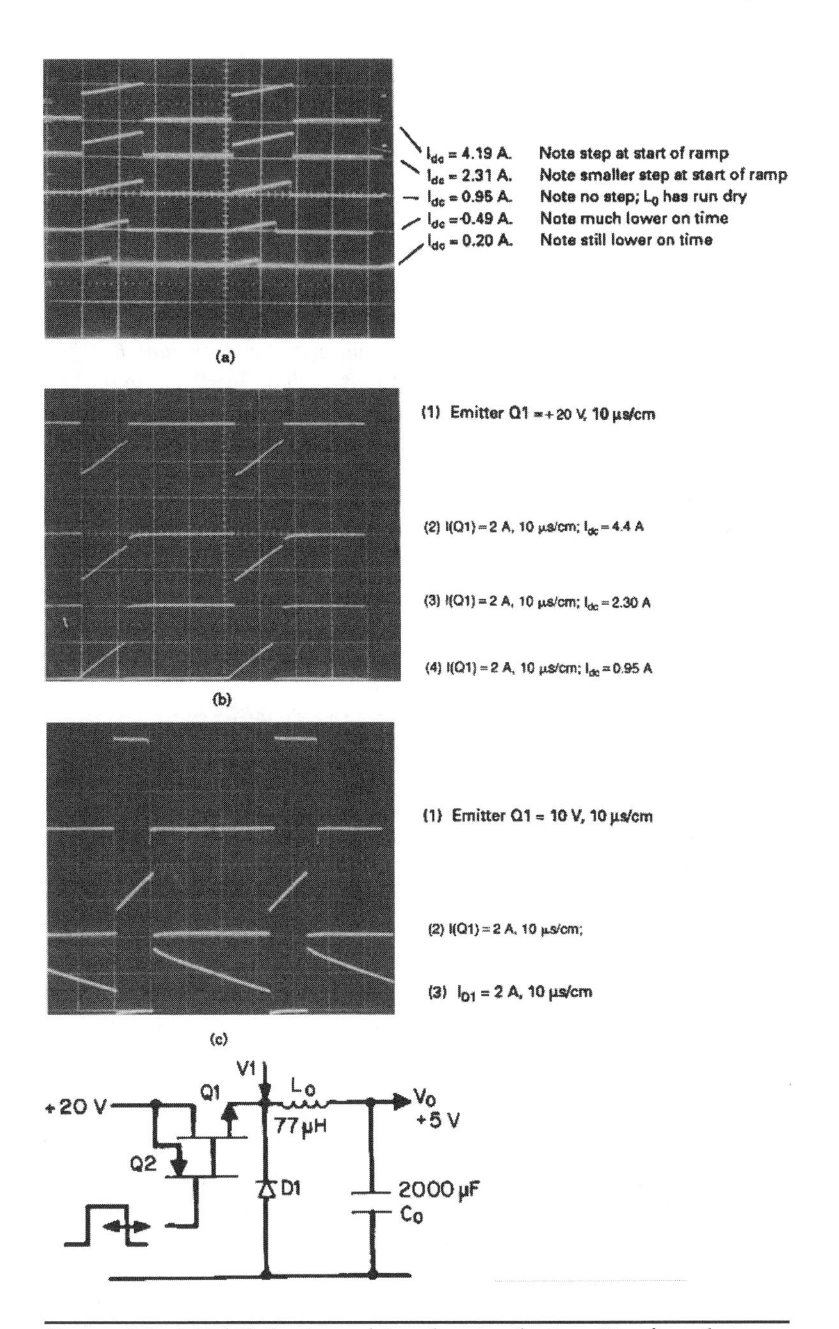

FIGURE 1.6 A 25-kHz buck regulator, showing the transition from the continuous mode to the discontinuous mode at the critical load current, with the inductor L_0 running dry. Note, in Figure 1.6*a*, line three above, that the "on" time remains constant only so long as the inductor is in the continuous mode.

TIP *The ratio* T_{on}/T *is normally referred to as the duty ratio D. The voltage formula for continuous operation is simply* $V_o = V_1.D$. *However, for discontinuous operation, the duty ratio becomes a function of the load current, and the situation is much more complicated. In the discontinuous mode, the output voltage* V_o *is given by the formula*

$$V_o = \frac{V_1.2D}{D + (D^2 + (8L/RT))^{1/2}}$$

Since the control loop will maintain the output voltage constant, the effective value of the load resistance R will be inversely proportional to the load current. Hence by holding V_o, V_1, *L, and T constant, to maintain the voltage constant, requires that the remaining variable (the duty ratio D) must change with load current.*

At the critical transition current, the transfer function will change from continuous mode in which the duty ratio remained constant with load change (zero output impedance) to the discontinuous mode in which the duty ratio must change with reducing load current (a finite output impedance). Hence in the discontinuous mode, the control loop must work much harder, and the transient performance will be degraded. ~ *K.B.*

Dynamically, at load currents above the onset of the discontinuous mode, the output L/C filter automatically accommodated output current changers by changing the amplitude of the step part of the ramp-on-step waveforms shown in the *Q*1 and *D*1 waveforms of Figures 1.4*d* and 1.4*e*. To the first order, it could do this without changing the *Q*1 "on" time.

The DC output current is the time average of the *Q*1 and *D*1 ramp current. Notice that in Figure 1.6*a*, line three and line four, that at lower currents where the inductor has gone discontinuous and the step part of the latter waveforms has gone to zero, the only way the current can decrease further is to decrease the *Q*1 "on" time. The negative-feedback loop automatically adjusts the duty ratio to achieve this.

The dramatic change in the waveforms can be seen very clearly between Figure 1.7*a* (for the critical current condition) and Figure 1.7*b* (for the discontinuous condition). Figure 1.7*b*(2) shows the *D*1 current going to zero just before *Q*1 turns "on" (the inductor has dried out and gone discontinuous). With zero current in L_o, the output voltage will seek to appear at the emitter of *Q*1. However, the sudden transition results in a decaying voltage "ring," at a frequency determined by L_o and the distributed capacitance looking into the *D*1 cathode and *Q*1 emitter junction at point *V*1. This is shown in Figure 1.7*b*(1).

TIP *Although the voltage ring is not damaging, in the interest of RFI reduction, it should be suppressed by a small R/C snubber across D1.* ~ *K.B.*

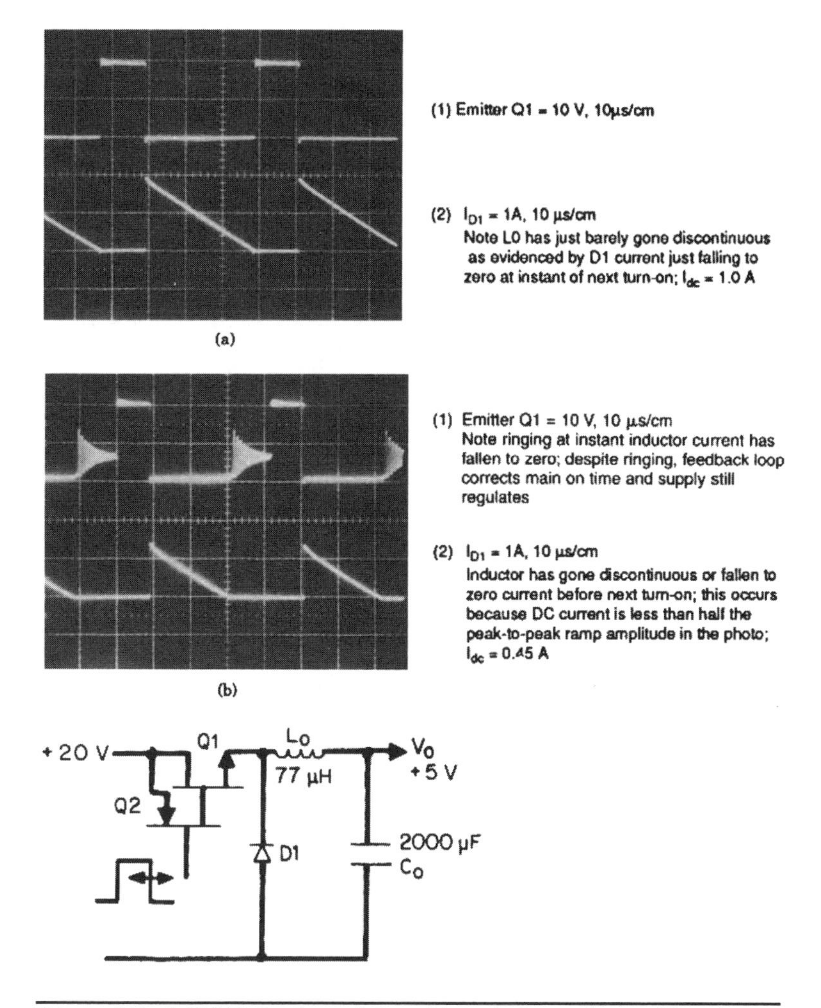

FIGURE 1.7 A 25-kHz buck regulator with typical waveforms. *Q*1 emitter voltage waveforms and *D*1 current waveforms for continuous conduction at the critical current (*a*) and in the discontinuous mode (*b*).

1.3.6.2 Designing the Inductor to Maintain Continuous Mode Operation

Although we have shown that operating in the discontinuous mode is not necessarily a major problem in the buck regulator, it can become a problem in some applications, particularly in boost-type topologies. The designer has the option to design the inductor so that it remains in the continuous mode for the full range of expected (but limited) load currents, as described next.

In this example the inductor will be chosen so that the current remains continuous if the DC output current stays above a specified minimum value. (Typically this is chosen to be around 10% of the rated load current, or 0.1 I_{on}, where "I_{on}" is defined as the nominal output current.)

The inductor current ramp is $dI = (I_2 - I_1)$, as shown in Figure 1.4*d*. Since the onset of the discontinuous mode occurs at a DC current of half this amplitude, then

$$I_o(\text{min}) = 0.1 I_{on} = (I_2 - I_1)/2 \quad \text{or} \quad (I_2 - I_1) = dI = 0.2 I_{on}$$

Also

$$dI = V_L T_{on}/L = (V_1 - V_o) T_{on}/L$$

where V_1 is voltage at the input of $Q1$ and is very close to V_{dc}, then

$$L = \frac{(V_{dc} - V_o) T_{on}}{dI} = \frac{(V_{dc} - V_o) T_{on}}{0.2 I_{on}}$$

where $T_{on} = V_o T / V_{dc}$ and V_{dcn} and I_{on} are nominal values, then

$$L = \frac{5(V_{dcn} - V_o) V_o T}{V_{dcn} I_{on}} \tag{1.8}$$

Thus, if L is selected from Eq. 1.8, then

$$dI = (I_2 - I_1) = 0.2 I_{on}$$

where I_{on} is the center of the inductor current ramp at nominal DC output current.

Since the inductor current will swing ±10% around its center value I_{on}, the inductor must be designed so that it does not significantly saturate at a current of at least 1.1 I_{on}.

Chapter 7, Section 7.6 provides information for the optimum design of inductors and chokes.

1.3.6.3 Inductor (Choke) Design

In the preceding example, continuous mode operation is required, so the current must not reach zero for the full range of load currents. Thus the inductor must support a DC current component and should be designed as a choke.

Well-designed chokes have a low, but relatively constant, inductance under AC voltage stress and DC bias conditions. Typically chokes use either gapped ferrite cores or composite cores of various powdered ferromagnetic alloys, including powdered iron or Permalloy, a magnetic alloy of nickel and iron. Powdered cores have a distributed air-gap because they are made from a suspension of powdered ferromagnetic particles, embedded in a nonmagnetic carrier to provide a uniformly distributed air-gap. The inductor value calculated by Eq. 1.8 must be designed so that it does not saturate at the specified peak current (110% of I_{on}). The design of such chokes is described in more detail in Chapter 7, Section 7.6.

The maximum range of current in the buck regulator will be determined by the choke design, the ratings of the power components, and the DC and AC losses given by Eq. 1.6. To remain in continuous conduction, the minimum current must not go below 10% of the rated I_{on}. Below this the load regulation will degrade slightly.

This wide (90%) industry standard dynamic load range results in a relatively large choke, which may not be acceptable. However, the designer has considerable flexibility of choice with some trade-offs. If a smaller choke is chosen (say, half the value given by Eq. 1.8), it will go discontinuous at one-fifth rather than one-tenth of the nominal DC output current. This will degrade the load regulation slightly, commencing at the higher minimum current. But since it has less inductance, the buck regulator will respond more quickly to dynamic load changes.

1.3.7 Output Capacitor

The output capacitor (C_o) shown in Figure 1.4 is chosen to satisfy several requirements. C_o will not be an ideal capacitor, as shown in Figure 1.8. It will have a parasitic resistance R_o and inductance L_o in series with its ideal pure capacitance C_o as shown. These are referred to as the *equivalent series resistance* (ESR) and *equivalent series inductance* (ESL). In general, if we consider the bulk ripple current amplitude in the series choke L_f, we would expect the majority of this ripple current to flow into the output capacitor C_o. Hence the output voltage ripple will be determined by the value of the output filter capacitor, C_o, its equivalent series resistance (ESR), R_o, and its equivalent series inductance (ESL), L_o.

For low-frequency ripple currents, L_o can be neglected and the output ripple is mainly determined by R_o and C_o.

Note *The actual transition frequency depends on the design of the capacitor, and manufacturers are constantly improving. Typically it will be above 500 kHz. ~ K.B*

So below about 500 kHz, L_o can normally be neglected. Typically C_o is a relatively large electrolytic, so that at the switching frequency,

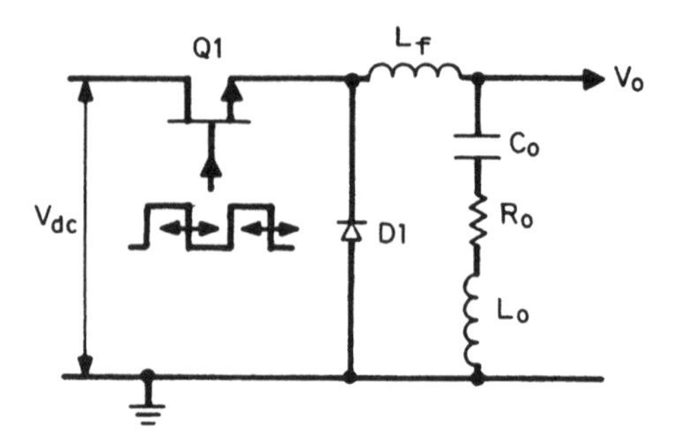

FIGURE 1.8 Output capacitor C_o showing parasitic components.

the ripple voltage component contributed by C_o is small compared with that contributed by R_o. Thus at the mid-frequencies, to the first order, the output ripple is closely given by the AC ripple current in L_f times R_o.

More precisely, there are two ripple components due to each of R_o and C_o. They are not in phase because that generated by R_o is proportional to $I_2 - I_1$ (the peak-to-peak inductor ramp current of Figure 1.4f) and that due to C_o is proportional to the integral of that current. However, for a worst-case comparison we can assume that they are in phase.

To obtain these ripple voltage components and to permit capacitor selection, it is necessary to know the values of the ESR R_o, which are seldom given by capacitor manufacturers. An examination of a number of manufacturers' catalogs shows that for the older types (aluminum electrolytic) for a large range of voltage ratings and capacitance values, R_oC_o tends to be constant. It ranges from 50 to $80 \times 10^{-6}\Omega F$.

After Pressman *Modern low-ESR electrolytic capacitors are now designed for this application, and the ESR values are provided by the manufacturers. If the low-ESR types are chosen, then clearly the lower ESR values should be used in the following calculations. ~ K.B.*

It is instructive to calculate the capacitive and resistive ripple components for a typical buck regulator.

Design Example:

Assume a design for a 25-kHz buck regulator with a step down from 20 V to 5 V with a load current $I_{on} = 5$ A. Let's require the ripple voltage to be below 50 millivolts with continuous conduction down to 10% load.

Assuming the minimum load is to be 10%, then $I_{o(\min)} = 0.1I_{\text{on}} =$ 0.5 A. We will calculate L from Eq. 1.8:

$$L = \frac{5(V_{\text{dcn}} - V_o)V_o T}{V_{\text{dcn}} I_{\text{on}}} = \frac{5(20-5)5 \times 40 \times 10^{-6}}{20 \times 5} = 150\,\mu\text{H}$$

Now dI (the peak-to-peak ramp amplitude) is $(I2 - I1) = 0.2I_{\text{on}} = 1$ A.

If we assume the majority of the output ripple voltage will be produced by the capacitor ESR (R_o), we can simply select a capacitor value such that the ESR will satisfy the ripple voltage as follows:

With a resistive ripple component of $V_{\text{rr}} = 0.05$ V peak-to-peak, then the required ESR $R_o = V_{\text{rr}}/dI = 0.05/(I2 - I1)$ and $R_o = 0.05\,\Omega$.

Using the preceding typical ESR/capacitance relationship ($R_o C_o =$ 50×10^{-6}):

$$C_o = 50 \times 10^{-6}/0.05 = 1000\,\mu\text{F}$$

Note *Clearly, for modern low ESR capacitors, we would use the published ESR values. ~ K.B.*

We will now calculate the ripple voltage contribution from the capacitance, ($C_o = 1000\,\mu$F).

Calculating the capacitive ripple voltage V_{cr} from Figure 1.4*d*, it is seen that the ripple current is positive from the center of the "off" time to the center of the "on" time or for one-half of a period, or 20 μs in this example. The average value of this triangle of current is $(I_2 - I_1)/4 = 0.25$ A. This current produces a ripple voltage across the pure capacitance part C_o of

$$V_{\text{cr}} = \frac{It}{C_o} = \frac{0.25 \times 20 \times 10^{-6}}{1000 \times 10^{-6}} = 0.005 \text{ V}$$

The ripple current below the I_o line in Figure 1.4f yields another 0.005-V ripple for a total peak-to-peak capacitive ripple voltage of 0.01 V (only 10 millivolts compared with the resistive component of 50 millivolts). Thus, in this particular case, the ripple due to the capacitance is relatively small compared with that due to the ESR resistor R_o and to the first order may be ignored.

In the preceding example, the filter capacitor was chosen to yield the desired peak-to-peak ripple voltage by choosing a capacitor with a suitable ESR R_o from

$$R_o = \frac{V_{\text{or}}}{I_2 - I_1} = \frac{V_{\text{or}}}{0.2I_{\text{on}}} \tag{1.9}$$

Using the typical relationship that the R_oC_o product will be near 65×10^{-6}:

$$C_o = \frac{65 \times 10^{-6}}{R_o} = (65 \times 10^{-6})\frac{0.2I_{\text{on}}}{V_{\text{or}}} \tag{1.10}$$

The justification for this approach is demonstrated more generally in the paper by K.V. Kantak.[1] He shows that if R_oC_o is larger than half the transistor "on" time and half the transistor "off" time—which is the more usual case—the output ripple is determined by the ESR resistor as shown above.

1.3.8 Obtaining Isolated Semi-Regulated Outputs from a Buck Regulator

Very often, low-power ancillary outputs are required for various control functions. This can be done with few additional components as shown in Figure 1.9. The regulation in the additional outputs is typically of the order of 2 to 3%.

It can be seen in Figure 1.4 that the return end of the regulated output voltage is common with the return end of the raw DC input. In Figure 1.9, a second winding with $N2$ turns is added to the output filter choke. Its output is peak-rectified with diode $D2$ and capacitor $C2$. The start of the $N1$, $N2$ windings is shown by the dots. When $Q1$ turns "off," the finish of $N1$ goes negative and is caught at one diode drop below ground by free-wheeling diode $D1$. Since the main output V_o is regulated against line and load changes, the reverse voltage across

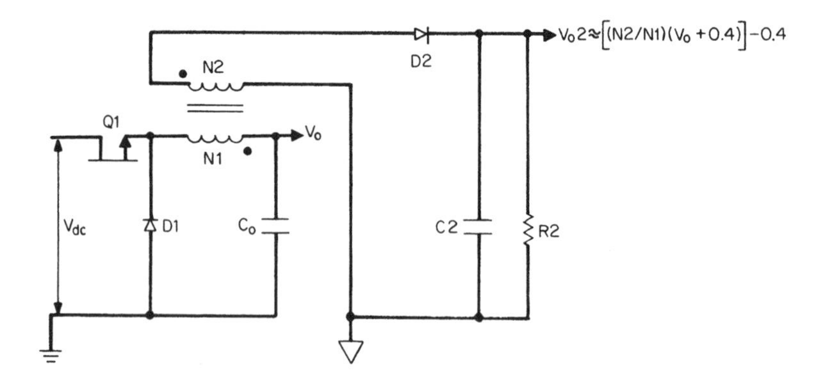

FIGURE 1.9 Showing how a second isolated output can be derived from a buck regulator by using the output choke as a transformer. The second output is DC-isolated from input ground and is regulated to within about 2 to 3%, as its primary is powered from the regulated V_o output and the fixed clamped voltage at the cathode of $D1$ when $Q1$ turns "off."

$N1$ is constant as long as the free-wheeling diode $D1$ continues to conduct. Using a low-forward-drop Schottky diode for $D1$, its forward drop remains constant at about. 0.4 V over a large range of DC output current.

Thus when $Q1$ turns "off," the voltage across N_2 is relatively constant at N_2/N_1 $(V_o + 0.4)$ volts with its dot end positive. This is peak rectified by $D2$ and $C2$ to yield $V_o2 = N_2/N_1(V_o + 0.4) - 0.4$ if $D2$ is also a Schottky diode. This output is independent of the supply voltage V_{dc} as $D2$ is reverse biased when $Q1$ turns "on." Capacitor C_2 should be selected to be large enough that the ancillary voltage does not decay too much during the maximum $Q1$ "on" time. Since $N2$ and $N1$ are isolated from each another, the ancillary output can be isolated or referenced to any other part of the circuit.

TIP *This can be a useful technique, but use it with care; notice the ancillary power is effectively stolen from the main output during the reverse recovery of the choke. Hence the main output power needs to be much larger than the total ancillary power to maintain* D1 *in conduction. A minimum load will be required on the main output if the ancillary outputs are to be maintained. Notice that using the ancillary outputs to power essential parts of the control circuit can have problems, as the system may not start.* ~ *K.B.*

1.4 The Boost Switching Regulator Topology

1.4.1 Basic Operation

The buck regulator topology shown in Figure 1.4 has the limitation that it can only produce a lower voltage from a higher voltage. For this reason it is often referred to as a *step-down* regulator.

The boost regulator (Figure 1.10) shows how a slightly different topology can produce a higher regulated output voltage from a lower unregulated input voltage. Called a *boost regulator* or a *ringing choke,* it works as follows.

An inductor $L1$ is placed in series with V_{dc} and a switching transistor $Q1$ to common. The bottom end of $L1$ feeds current to $Q1$ when $Q1$ is "on" or the output capacitor C_o and load resistor through rectifying diode $D1$ when $Q1$ is "off."

Assuming steady-state conditions, with the output voltage and current established, when $Q1$ turns "on" (for a period T_{on}), $D1$ will be reverse biased and does not conduct. Current ramps up linearly in $L1$ to a peak value $I_p = V_{dc}T_{on}/L1$.

During the $Q1$ "on" time, the output current is supplied entirely from C_o, which is chosen to be large enough to supply the load current for the time T_{on} with the specified minimum droop.

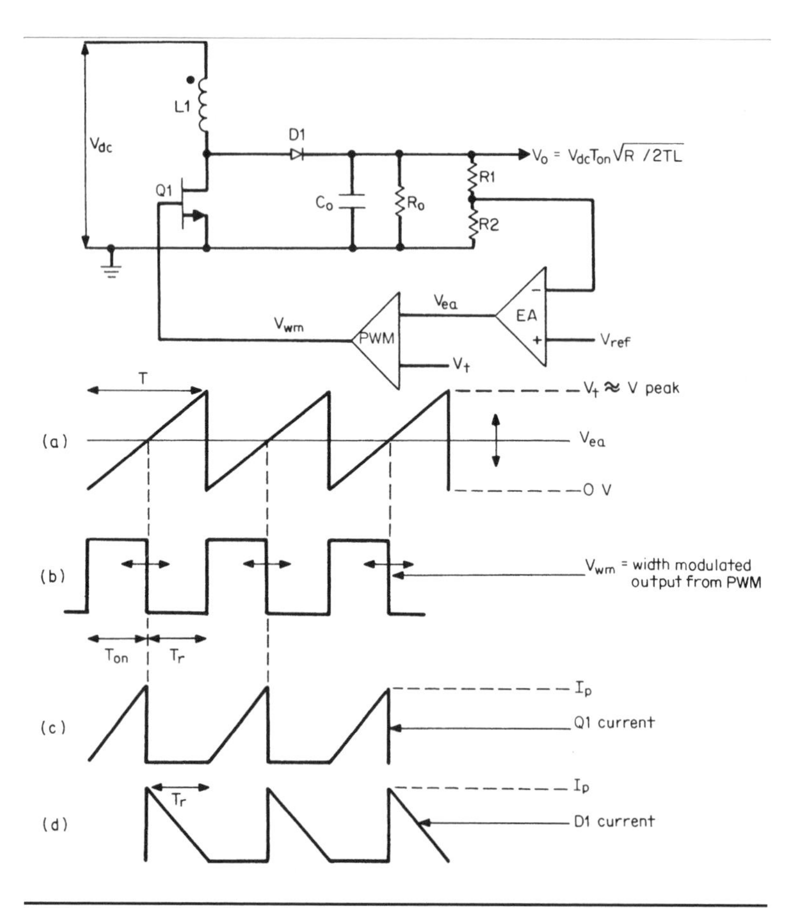

FIGURE 1.10 Boost regulator and critical waveforms. Energy stored in $L1$ during the $Q1$ "on" time is delivered to the output via $D1$ at a higher output voltage when $Q1$ turns "off" and the polarity across $L1$ reverses.

When $Q1$ turns "off," since the current in an inductor cannot change instantaneously, the voltage across $L1$ reverses in an attempt to maintain the current constant. Now the lower end of $L1$ goes positive with respect to the input voltage. With the output voltage V_o higher than the input V_{dc}, $L1$ delivers its stored energy to C_o via $D1$. Hence C_o is *boosted* to a higher voltage than V_{dc}. This energy replenishes the charge drained away from C_o when $D1$ was not conducting. At the same time current is also supplied to the load from V_{dc} via $L1$ and $D1$ during this action.

In simple terms, the output voltage is regulated by controlling the $Q1$ "on" time in a negative-feedback loop. If the load current increases, or the input voltage decreases, the "on" time of $Q1$ is automatically increased to deliver more energy to the load, or the converse. Hence, in normal operation the "on" period of $Q1$ is adjusted to maintain the output voltage constant.

1.4.2 The Discontinuous Mode Action in the Boost Regulator

TIP *The boost regulator has two quite different modes of operation depending on the conduction state of the inductor. If the inductor current reaches zero at the end of a cycle, it is said to operate in a* discontinuous *mode. If there is some current remaining in the inductor at the end of a cycle, it is said to be in a* continuous *mode of operation.*

When speaking about switching regulators, the output filter capacitor is not normally included in the analysis of the converter. The output current of a switching regulator is, therefore, not the DC output current to the load, but rather the combined current that flows in the output capacitor and the load in parallel.

Notice that unlike the buck regulator, the boost regulator has a continuous input current (with some ripple current) but a discontinuous output current for all modes of operation. Hence the terms continuous *and* discontinuous mode *refer to what is going on in the inductor.*

There is a dramatic difference in the transfer function between the two modes of operation that significantly changes the transient performance and intrinsic stability. This is explained more fully in Chapter 12. ~ K.B.

We will consider in more detail the action for discontinuous mode operation, in which the energy in the inductor is completely transferred to the output during the "off" period of $Q1$, and we will establish some power and control equations.

We have seen that when $Q1$ turns "on," the current ramps up linearly in $L1$ to a peak value $I_p = V_{dc}T_{on}/L1$. Thus energy is stored in $L1$, and at the end of the "on" period, this stored energy will be

$$E = 0.5 L_1 I_p^2 \tag{1.11}$$

where E is in joules, L is in henries, and I_p is in amperes.

If the current through $D1$ (and hence $L1$) has fallen to zero before the next $Q1$ turn "on" action, all the energy stored in $L1$ (Eq. 1.11) during the previous $Q1$ "on" period will have been delivered to the output load, and the circuit is said to be operating in the discontinuous mode.

The energy E in joules delivered to the load per cycle, divided by the period T in seconds, is the output power in watts. Thus if all the energy of Eq. 1.11 is delivered to the load once per period T, the power to the load from $L1$ alone (assuming for the moment 100% efficiency) would be

$$P_L = \frac{\frac{1}{2}L(I_p)^2}{T} \tag{1.12}$$

However, during the "off" time of $Q1$ (T_r in Figure 1.10*d*), the current in $L1$ is ramping down toward zero, and the same current is also flowing from the supply V_{dc} via $L1$ and $D1$ and is contributing to the load power P_{dc}. This is equal to the average current during T_r multiplied by its duty cycle and V_{dc} as follows:

$$P_{dc} = V_{dc}\frac{I_p}{2}\frac{T_r}{T} \tag{1.13}$$

The total power delivered to the load is then the sum of the two parts as follows:

$$P_t = P_L + P_{dc} = \frac{\frac{1}{2}L_1(I_p)^2}{T} + V_{dc}\frac{I_p}{2}\frac{T_r}{T} \tag{1.14}$$

But $I_p = V_{dc}T_{on}/L_1$. Substituting for I_p, in 1.14 we get

$$P_t = \frac{(\frac{1}{2}L_1)(V_{dc}T_{on}/L_1)^2}{T} + V_{dc}\frac{V_{dc}T_{on}}{2L_1}\frac{T_r}{T}$$

$$= \frac{V_{dc}^2 T_{on}}{2TL_1}(T_{on} + T_r) \tag{1.15}$$

To ensure that the current in $L1$ has ramped down to zero before the next $Q1$ turn "on" action, we set $(T_{on} + T_r)$ to kT, where k is a fraction less than 1. (That is, the period T is made greater than the inductor conduction period.) Then

$$P_t = \left(V_{dc}^2 T_{on}/2\,TL_l\right)(kT)$$

But for an output voltage V_o and output load resistor R_o,

$$P_t = \frac{V_{dc}^2 T_{on}}{2TL_1}(kT) = \frac{V_o^2}{R_o}$$

or

$$V_o = V_{dc}\sqrt{\frac{kR_oT_{on}}{2L1}} \tag{1.16}$$

Thus the negative-feedback loop keeps the output constant against input voltage changes and output load R_o changes in accordance with Eq. 1.16. As V_{dc} and R_o (the load current) go down or up, the loop will increase or decrease T_{on} so as to keep V_o constant.

1.4.3 The Continuous Mode Action in the Boost Regulator

As mentioned in the previous section, if the $D1$ current (the inductor current) falls to zero before the next turn "on" action, the circuit is said to operate in the discontinuous mode (see Figure 1.10*d*).

However, if the current in $D1$ and $L1$ has not fallen to zero at the end of the "on" period, the inductor current will not be zero at the next $Q1$ turn "on" action. Hence the current in $Q1$ will have a front-end step as shown in Figure 1.11. The current in the inductor cannot change instantaneously. Currents in $Q1$ and $D1$ will have the characteristic ramp-on-a-step waveshape as shown in Figure 1.11.

The circuit is now said to be operating in the continuous mode because the inductor current does not reach zero during a cycle of operation.

Assuming the feedback loop maintains the output voltage constant, as R_o or V_{dc} decreases, the feedback loop increases the $Q1$ "on" period T_{on} to maintain the output voltage constant. As the load current increases, R_o or V_{dc} continues to decrease, a point is reached such that T_{on} is so large that the decaying current through $L1$ and $D1$ will not have fallen to zero before the next turn "on" action, and the action moves into the continuous mode as shown in Figures 1.10 and 1.11.

Now an error-amplifier circuit, which had successfully stabilized the loop while it was operating in the discontinuous mode, may not

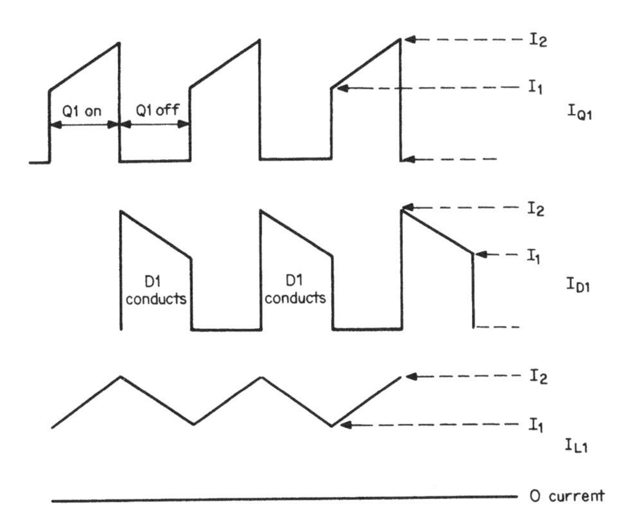

FIGURE 1.11 Typical current waveforms in $Q1$, $D1$, and $L1$ for a boost regulator operating in continuous mode. Note that inductor $L1$ has not had enough time to transfer all its energy to the load before the next $Q1$ turn "on" action.

be able to keep the loop stable in the continuous mode and may oscillate. In traditional feedback-loop analysis, the continuous-mode boost regulator has a *right-half-plane-zero in the transfer function.*[2] The only way to stabilize a loop with a right-half-plane-zero is to drastically reduce the error-amplifier bandwidth.

TIP *In simple terms, in the discontinuous mode, there is a short period when there is zero current in the inductor and zero current in* D1. *That is, there is a small time-gap between the energy transfer period (when* Q1 *is "off" and* D1 *is conducting) and the energy storage period (when* Q1 *is "on" and* D1 *is not conducting). This time margin (*dead time*) is critical to the way the power system behaves and does not exist in the continuous mode.*

It is very important to fully understand the difference between the two modes of operation, because in any switching topology that has a boost-type behavior, the effect will be evident. To better understand this, we will consider a transient load increase in a continuous mode boost topology and follow the sequence of events as the circuit responds to the load change.

Consider a continuous-mode buck system, running in steady-state conditions, with a stabilized output voltage and a load current that maintains the inductor in continuous conduction. We now apply a sudden increase in load current. The output voltage will tend to fall, and the control loop will increase the "on" period of Q1 *to initiate an increase in current in* L1. *However, it takes several cycles before the current in* L1 *will increase very much (depending on the value of the inductor, the input voltage, and the actual increase in the* Q1 *"on" time).*

It is important to notice that the immediate effect of increasing the "on" period is to decrease the "off" period (because the total period is fixed). Since D1 *only conducts during the "off" period of* Q1 *(and this period is immediately reduced), the mean output current will initially decrease, rather than increase as was required. Hence we have a situation where we tried to increase the output current, but the immediate effect was to reduce the output current. This will correct itself slowly as the current in the inductor increases over a few cycles.*

From a control theory perspective, for a short time this effect introduces an additional 180° of phase shift into the closed loop control system during the transient period when the L1 *current is increasing. In terms of control theory this translates to a zero in the right half-plane of the transfer function; it is the cause of the right-half-plane-zero in the small signal transfer function.*

Notice that the effect is related to the dynamic behavior of the power components and cannot be changed by the control circuit. In fact, a perfect high-gain fast-response control circuit would result in the "on" period going to the full pulse width on the first pulse, and there would be zero output current for a short period. Hence, the right-half-plane-zero cannot be eliminated by the loop compensation network. The only option is to slow down the rate of change of pulse width to allow the output to keep up without too much droop.

(In control theory parlance, the control loop must be rolled off at a frequency well below the right-half-plane-zero crossover frequency.)

In the discontinuous mode the performance is quite different. The small time-gap margin allows the "on" period to increase without the need to reduce the "off" period (within the limits of the margin), so the problem is not present, providing the margin is large enough to accommodate the change in pulse width.

Be aware that in the continuous conduction mode, the right-half-plane-zero effect will be found in any switching converter (or combination of converters and transformers) that has a boost-type action in any part of the circuit. The flyback converter is a typical example of this. The mathematics of this effect will be found in Chapter 12 and reference 2. ~ K.B.

1.4.4 Designing to Ensure Discontinuous Operation in the Boost Regulator

For the preceding reasons, the designer may prefer to ensure that the boost regulator remains fully within the discontinuous mode for the full range of operating conditions.

In Figure 1.10*d* we see that the decaying *D*1 current just comes down to zero at the start of the next turn "on" action. This is the threshold between discontinuous and continuous mode operation.

This threshold is seen from Eq. 1.16 to occur at certain combinations of V_{dc}, T_{on}, R_o, *L*1, and *T* that result in the *L*1, *D*1 current just falling to zero prior to the next turn "on" action of *Q*1. It can be seen from Figure 1.10*a* that any further decrease in V_{dc} or R_o (increase in load current) will force the circuit into the continuous mode such that oscillation can occur unless the error amplifier has been rolled off at a very low frequency.

To avoid this problem, we will see from Eq. 1.16 that T_{on} must be selected so that when it is a maximum (which is when V_{dc} and R_o are at their minimum specified values) and the current in *D*1 has fallen back to zero, there is a usable working dead-time margin (T_{dt}) before *Q*1 turns "on" again.

At the same time, we must ensure that by the time the current in *D*1 returns to zero, the *L*1 core will have been restored to its previous starting place on its hysteresis loop, shown as *B*1 in Figure 1.12. If the core is not fully restored to *B*1, then after many such cycles, the starting point will drift up the hysteresis loop and saturate the core. Since the impedance of a saturated core drops to its winding resistance only (because it cannot sustain voltage), the voltage at the transistor collector will suddenly move up to the supply voltage, and with negligible resistance in the path, the transistor will be destroyed.

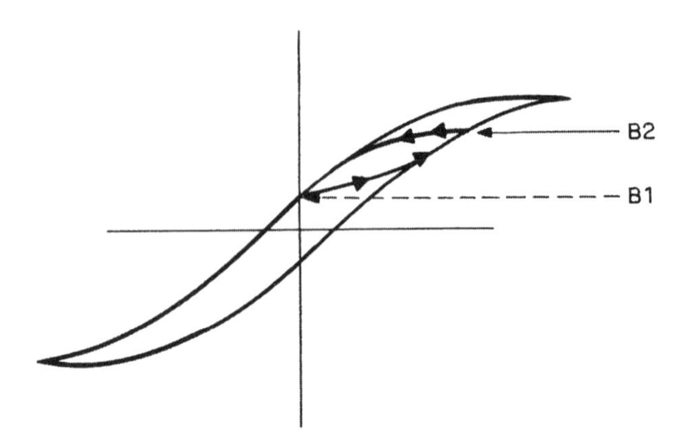

FIGURE 1.12 The working B/H loop. A choke core must not be allowed to walk up or down its hysteresis loop. If it is driven from, say, $B1$ to $B2$ by a given forward volt-second product, it must be subjected to an equal volt-second product in the opposite direction to restore it to $B1$ before the next "on" period.

In this example, to ensure that the circuit remains in the discontinuous mode, a dead-time T_{dt} of 20% of a full period will be provided. Hence we must ensure that the sum of the maximum "on" time of $Q1$ plus the core reset time plus the dead time will equal a full period, as shown in Figure 1.13. This will ensure that the stored current in $L1$ will have fallen to zero well before the next $Q1$ turn "on" action.

Hereafter, a line appearing *below* a term will indicate the *minimum* permitted or specified or required value of that term, and a line appearing *over* a term will indicate the *maximum* value of that term.

Then $\overline{T_{\mathrm{on}}} + T_r + T_{\mathrm{dt}} = T$, $\overline{T_{\mathrm{on}}} + T_r + 0.2T = T$,

or

$$\overline{T_{\mathrm{on}}} + T_r = 0.8T \tag{1.17}$$

From Eq. 1.16, the maximum "on" time $\overline{T_{\mathrm{on}}}$ occurs at minimum $\underline{V_{\mathrm{dc}}}$ and minimum $\underline{R_o}$. Then for the "on" or set volt-second product to equal the "off" or reset volt-second product at minimum $\underline{R_o}$:

$$\underline{V_{\mathrm{dc}}}\,\overline{T_{\mathrm{on}}} = (V_o - \underline{V_{\mathrm{dc}}})T_r \tag{1.18}$$

Now Eqs. 1.17 and 1.18 have only two unknowns, $\overline{T_{\mathrm{on}}}$ and T_r, and thus both are determined. $\overline{T_{\mathrm{on}}}$ is then

$$\overline{T_{\mathrm{on}}} = \frac{0.8T(V_o - \underline{V_{\mathrm{dc}}})}{V_o} \tag{1.19}$$

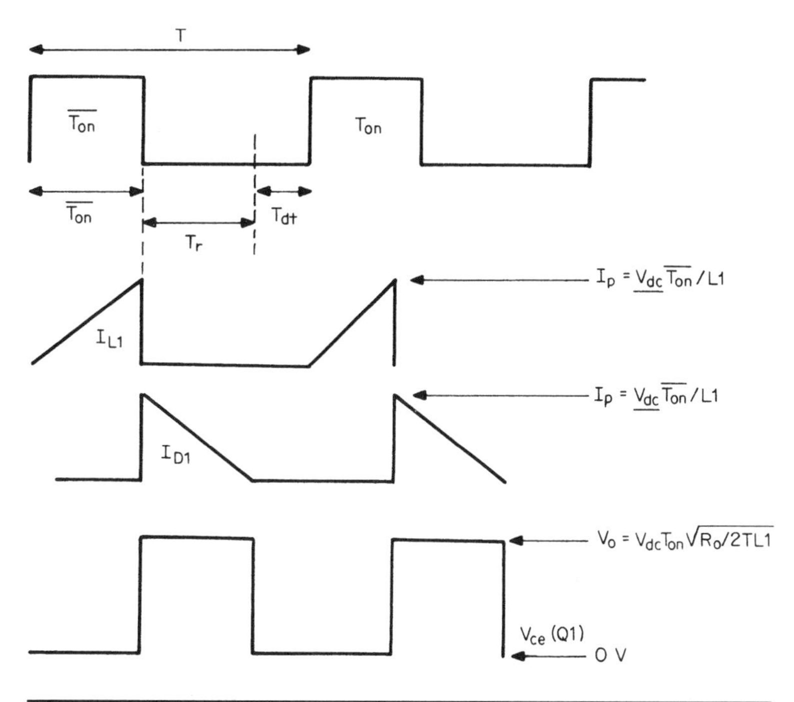

FIGURE 1.13 Boost regulator waveforms in the discontinuous mode with 20% dead-time margin. For discontinuous-mode operation, the current in $D1$ (see Figure 1.10) must have decayed to zero before the next turn "on" action. To ensure this, the inductor $L1$ is chosen such that $T_{on(max)} + T_r = 0.8T$, leaving a dead time T_{dt} of $0.2T$.

Now in Eq. 1.16, with $\underline{V_{dc}}$ and $\underline{R_o}$ (maximum load current) specified, $\overline{T_{on}}$ is calculated from Eq. 1.19 and $k[= (T_{on} + T_r)/T)] = 0.8$ from Eq. 1.17.

Inductor $L1$ is fixed so the circuit is guaranteed not to enter the continuous mode. However, if the output load current is increased beyond its specified maximum value (R_o decreased below its specified minimum) or V_{dc} is decreased below its specified minimum, the feedback loop will attempt to increase T_{on} to keep V_o constant. This will eat into the dead time, T_{dt}, and move the circuit closer to continuous mode. To avoid this, we must limit the maximum "on" time or a maximum peak current must be provided.

TIP *A good method that accounts for all variables is to inhibit the turn "on" of Q1 until the inductor current reaches zero. For fixed-frequency operation this limits the load current. Alternatively it can be set up to provide variable-frequency operation, which is often preferred.* $\sim$ *K.B.*

With $L1$ determined earlier from Eq. 1.16, V_{dc} specified, and $\overline{T_{on}}$ calculated from Eq. 1.19, the peak current in $Q1$ can be calculated from Eq. 1.14, and a transistor selected to have adequate gain at I_p.

The boost regulator is frequently used at low power levels in non-isolated applications due to the very low parts count. A typical application would be on a printed-circuit board where it is desired to step up a 5-V computer logic level supply to, say, 12 or 15 V for operational amplifiers.

Frequently at higher power levels in battery-supplied power supplies, as the battery discharges, its output voltage drops significantly. Many systems whose prime power is a nominal 12- or 28-V battery will present problems when the battery voltage falls to about 9 or 22 V. Boost regulators are frequently used in such applications to boost the voltages back up to the 12- and 28-V level. Power requirements in such applications can be in the range 50 to 200 watts.

1.4.5 The Link Between the Boost Regulator and the Flyback Converter

The boost regulator has been treated in great detail because boost action appears in many converter combinations. For example, by replacing the inductor $L1$ with a transformer (more correctly a choke with an additional secondary winding), a very similar, valuable, and widely used topology, the *flyback converter,* is realized.

Like the boost, the flyback stores energy in its magnetics during the "on" period of the power device and transfers the energy to the output load during the "off" period.

Because the secondary windings can be isolated from the input, the outputs are not constrained to share a common return line. Also by using multiple secondaries, a multiple output power supply is possible. The outputs may be higher or lower voltage than the input, and may be common or isolated as required.

The problems of discontinuous or continuous operation and the design relationships and procedures for the flyback are similar to those of the boost regulator and will be discussed in more detail in Chapter 4.

1.5 The Polarity Inverting Boost Regulator

1.5.1 Basic Operation

Figure 1.14 shows a different arrangement of the boost regulator that provides polarity inversion. It uses the same basic principle as the previous boost regulator in that energy is stored in the inductor during the "on" period of $Q1$, which is then transferred to the output load and C_o in the "off" period of $Q1$.

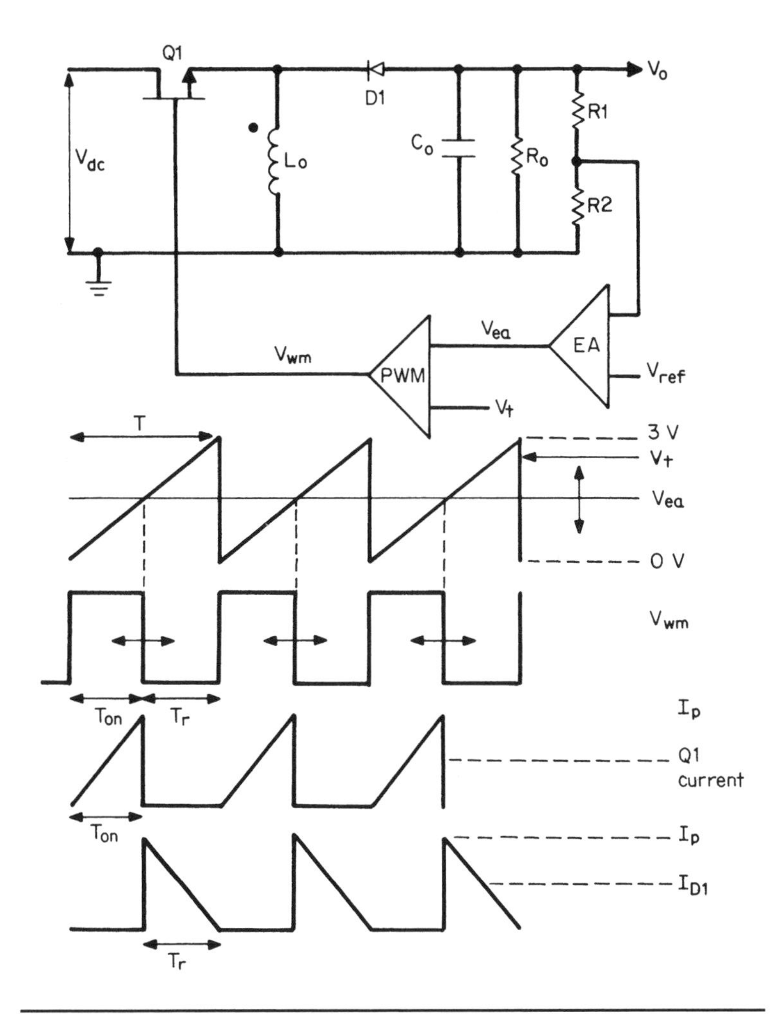

FIGURE 1.14 The polarity-inverting boost regulator and typical waveforms.

Comparing Figures 1.14 and 1.10, it will be seen that the transistor and inductor have changed places. In the reverse polarity inverter, the transistor is above the inductor rather than below it as it was in the boost circuit. Also the rectifying diode has been reversed.

When *Q*1 turns "on," diode *D*1 is reverse biased because its cathode is at V_{dc} (assuming to a close approximation that the voltage drop across *Q*1 is zero). Also, assuming steady-state conditions, such that C_o has charged down to some negative voltage, then *D*1 remains reverse biased throughout the *Q*1 "on" period. A fixed-voltage V_{dc} will be impressed across the inductor L_o, and the current in it ramps up linearly at a rate $di/dt = V_{dc}/L_o$.

After an "on" period T_{on}, the current in L_o will have reached $I_p = V_{dc}T_{on}/L_o$, and the energy stored in L_o (in joules) is $E = .5L_o I_p^2$. When $Q1$ turns "off," the voltage polarity across L_o reverses in an attempt to maintain its current constant. Thus at the instant of turn "off," the same inductor current I_p (which was flowing through $Q1$ before it turned "off") now continues to flow down through L_o to common, pulling the current through $D1$ from C_o. This current charges the top end of C_o to a negative voltage.

After a number of cycles, when the required output voltage is developed, the error amplifier adjusts the $Q1$ "on" period T_{on} so that the sampled output voltage $V_o R2/(R1 + R2)$ is equal to the reference voltage V_{ref}. Further, if all the energy stored in L_o is delivered to the load before the next $Q1$ turn "on" action (that is, I_{D1} has fallen to zero), then the circuit operates in the discontinuous mode, and the power delivered to the load will be

$$P_t = \frac{\frac{1}{2} L_o I_p^2}{T} \tag{1.20}$$

It should be noted that unlike the case of the boost regulator, when $Q1$ turns "off," the inductor current does not flow from the supply source (see Eq. 1.13). Hence the only power to the load is that given by Eq. 1.20. Thus assuming 100% efficiency, the output power would be

$$P_o = \frac{V_o^2}{R_o} = \frac{\frac{1}{2} L_o I_p^2}{T} \tag{1.21}$$

and for $I_p = V_{dc}T_{on}/L_o$,

$$V_o = V_{dc}T_{on}\sqrt{\frac{R_o}{2TL_o}} \tag{1.22}$$

1.5.2 Design Relations in the Polarity Inverting Boost Regulator

As in the previous boost circuit, it is desirable to keep the circuit operating in the discontinuous mode by ensuring that the current stored in L_o during the $Q1$ maximum "on" period has decayed to zero at the end of the "off" period T_r. To ensure this action, we will provide a dead time T_{dt} margin of $0.2T$ before the next $Q1$ turn "on" action. Thus if $\overline{T_{on}} + T_r + T_{dt} = T$, then for $T_{dt} = 0.2T$ we obtain

$$\overline{T_{on}} + T_r = 0.8T \tag{1.23}$$

In addition, as in the boost regulator, the "on" volt-second product must equal the reset volt-second product to prevent the core from saturating. Since (as can be seen from Eq. 1.22) the maximum $\overline{T_{on}}$ occurs for minimum V_{dc} and minimum R_o (maximum current), it follows that

$$\underline{V_{dc}}T_{on} = V_o T_r \tag{1.24}$$

Thus both Eqs. 1.23 and 1.24 have two unknowns: $\overline{T_{on}}$ and T_r. This fixes $\overline{T_{on}}$ at

$$\overline{T_{on}} = \frac{0.8 V_o T}{\underline{V_{dc}} + V_o} \tag{1.25}$$

Now, with $\overline{T_{on}}$ calculated from Eq. 1.25 and $\underline{V_{dc}}$, $\underline{R_o}$, V_o, and T specified, Eq. 1.22 defines L_o such that $I_p = \underline{V_{dc}}\overline{T_{on}}/L_o$, and transistor $Q1$ is selected to have adequate gain at I_p.

References

1. K. V. Kantak, "Output Voltage Ripple in Switching Power Converters," in *Power Electronics Conference Proceedings*, Boxborough, MA, pp. 35–44, April 1987.
2. K. Billings, *Switchmode Power Supply Handbook*, New York: McGraw-Hill, 1999, Chap. 9.

CHAPTER 2

Push-Pull and Forward Converter Topologies

2.1 Introduction

In the three switching regulator topologies discussed in the previous chapter, the output returns were all common with the input returns, and multiple outputs were not possible (except for the special case discussed in Section 1.3.8).

In this chapter we look at some of the most widely used fully isolated switching regulator topologies. These topologies—the push-pull, single-ended forward converter, and the double-ended and interleaved forward converters—are similar, so we consider them a single family. All these topologies deliver their power to the loads via a high-frequency transformer; hence outputs may be DC-isolated from the input, and multiple outputs are possible.

2.2 The Push-Pull Topology

2.2.1 Basic Operation (With Master/Slave Outputs)

A push-pull topology is shown in Figure 2.1. It consists of a transformer $T1$ with multiple secondaries. Each secondary delivers a pair of 180° out-of-phase square-wave power pulses whose amplitude is fixed by the input voltage and the number of primary and secondary turns.

The pulse widths for all secondaries are identical, as determined by the control circuit and the negative-feedback loop around the *master* output. The control circuit is similar to the buck and boost regulators shown previously in Figures 1.4 and 1.10, except that two equal

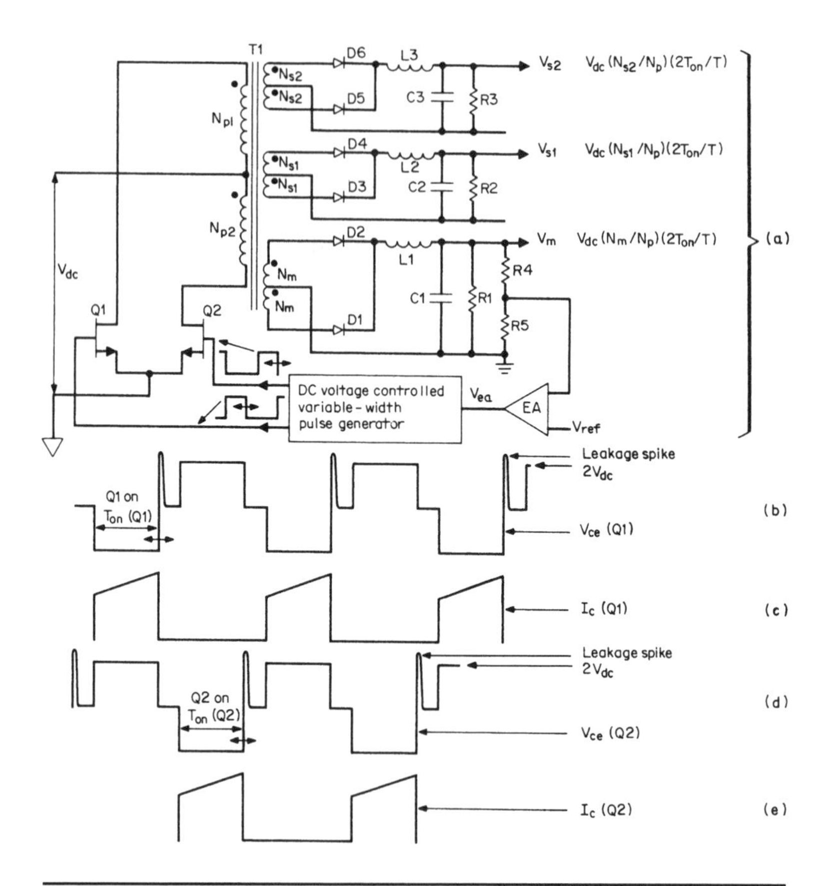

FIGURE 2.1 Push-pull width-modulated converter. Transistors $Q1$ and $Q2$ receive 180 out-of-phase, pulse-width modulated drive signals. The master output is V_{sm}, and there are two slaves, V_{s1} and V_{s2}. The feedback loop is closed around V_{sm}, and the pulse width T_{on} is controlled to regulate the master output against line and load changes. It will be seen that the slaves are regulated against line changes, but only partially against load changes.

adjustable pulse-width, 180°-out-of-phase pulses drive the bases of $Q1$, $Q2$. The additional secondaries N_{s1}, N_{s2} are referred to as *slaves.*

Transistor base drives at turn "on" are sufficient to bring the switched end of each half primary down to $V_{ce(sat)}$, typically about 1 V, over the full specified current range. Hence as each transistor turns "on," it applies a square-voltage pulse to its half primary of magnitude $V_{dc} - 1$.

On the secondary side of the transformer, there will be flat-topped square waves of amplitude $(V_{dc} - 1)(N_s/N_p) - V_d$ with a duration T_o, where V_d is an output rectifier forward drop, taken as 1 V for a conventional fast-recovery diode, and 0.5 V for a Schottky diode. The output pulses at the rectifier cathodes have a duty cycle of $2T_{\text{on}}/T$ because there are two pulses per period.

Thus the waveforms at the inputs to the *LC* filters shown in Figure 2.1 are very much like that at the input to the buck regulator *LC* filter of Figure 1.4, which has a flat-topped amplitude and adjustable width. The *LC* filters of Figure 2.1 serve the same purpose as that of Figure 1.4. They provide a DC output that is the average of the square wave voltage at the input of the filter. The analysis of the inductor and capacitor functions proceeds exactly as for the buck regulator, and the method of calculating their magnitudes is exactly the same as follows.

The DC or average voltage at the V_m output in Figure 2.2 (assuming *D*1, *D*2 are 0.5-V forward-drop Schottky diodes) will be

$$V_m = \left[(V_{\text{dc}} - 1)\left(\frac{N_m}{N_p}\right) - 0.5\right]\frac{2T_{\text{on}}}{T} \qquad (2.1)$$

The waveforms at the V_m output rectifiers are shown in Figure 2.2. If the negative-feedback loop is closed around V_m as shown in Figure 2.1, T_{on} and V_m will be regulated against DC input voltage and load

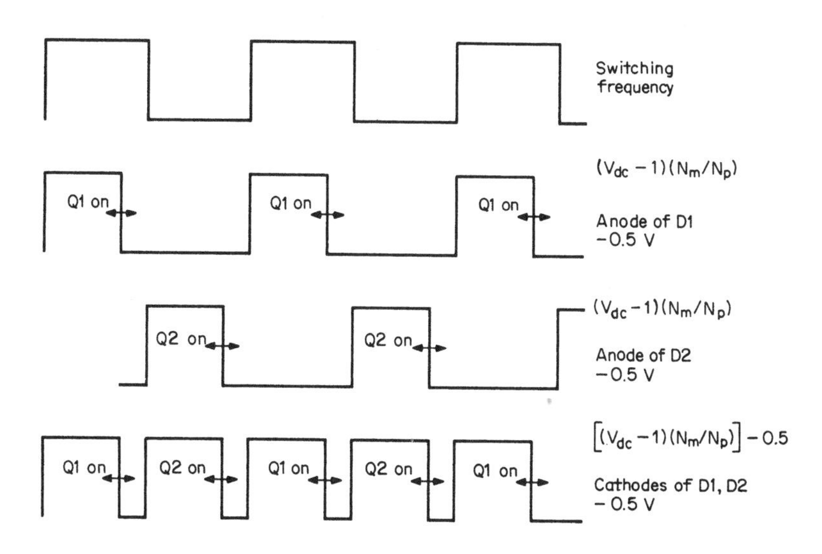

FIGURE 2.2 Voltage waveforms (N_m) at the master secondary winding. The output *LC* averaging filter yields a DC output voltage.

$$V_m = [(V_{\text{dc}} - 1)(N_m/N_p) - 0.5](2T_{\text{on}}/T)$$

As V_{dc} varies, the negative-feedback loop corrects T_{on} in the direction to keep V_m constant.

current changes. Although load current does not appear in Eq. 2.1, a current change will cause V_m to change, that change will be sensed by the error amplifier, and T_{on} will be altered to correct it. Providing the current in $L1$ (see Figure 2.1) does not go discontinuous, changes in T_{on} will be small, and the absolute value of T_{on} will be given by Eq. 2.1 for any turns ratio N_m/N_p, input voltage V_{dc}, and period T.

For the slave secondaries, the voltages at the cathodes of the rectifying diodes are fixed by the number of secondary turns, and the T_{on} duration of the square waves is the same as defined by the master feedback loop. Thus the slave output voltages with normal diodes will be

$$V_{s1} = \left[(V_{dc} - 1)\frac{N_{s1}}{N_p} - 1\right]\frac{2T_{on}}{T} \tag{2.2}$$

$$V_{s2} = \left[(V_{dc} - 1)\frac{N_{s2}}{N_p} - 1\right]\frac{2T_{on}}{T} \tag{2.3}$$

2.2.2 Slave Line-Load Regulation

It can be seen from Eqs. 2.1, 2.2, and 2.3 that the slaves are regulated against V_{dc} input changes by the negative-feedback loop that keeps V_m constant, in accordance with Eq. 2.1. The same equation,

$$V_m = (V_{dc} - 1)T_{on}$$

also appears in Eqs. 2.2 and 2.3, and thus V_{s1}, V_{s2} are also kept constant as V_{dc} changes.

Notice that if load current in the master (V_m) changes, the drops across its rectifying diodes and winding resistance will change slightly. Thus the negative-feedback loop will correct for V_m load change effects and alter T_{on} to keep V_m constant.

For the slave outputs, T_{on} will now change without corresponding changes in V_{dc}, and from Eqs. 2.2 and 2.3, it can be seen that changes in V_{s1}, V_{s2} will result. Such changes in the slave output voltages due to changes in the master output current are referred to as *cross regulation*.

Slave output voltages will also change as a result of changes in their own output currents. In a similar way slave current changes will cause voltage drop changes in their rectifying diodes and winding resistances, lowering the peak voltages slightly. These changes are not corrected by the main feedback loop, which senses only V_m.

However, providing the currents in the slave output inductors $L2$, $L3$, and especially in the master inductor $L1$ do not go discontinuous, slave output voltages can be depended on to vary within only $\pm$ 5 to $\pm$ 8%.

TIP *Much better cross regulation can be obtained by using coupled output inductors (where all outputs share a common inductor core).*[1] *~ K.B.*

2.2.3 Slave Output Voltage Tolerance

Although changes in slave output voltages are relatively small, the absolute values of output voltage are not accurately adjustable. As seen in Eqs. 2.2 and 2.3, they are fixed by T_{on} and their corresponding secondary turns N_{s1}, N_{s2}. But T_{on} is nearly constant, defined by the feedback loop to keep the master voltage constant. Further, since the turns can be changed only by integral numbers, the absolute value of slave output voltage is not finely settable. The change in secondary voltage for a single turn change in N_s is given by $V_m \cdot T_{on}/N_p$.

In most cases, the absolute values of slave output voltage are not too important. Slaves usually drive operational amplifiers or motors, and most often these can tolerate DC voltages within about 2 V of a desired value. If the absolute magnitude is important, the output voltage is usually designed to be higher than required and brought down to a desired exact value with a linear or buck regulator. Because a slave output is semi-regulated, a linear regulator is reasonably efficient.

2.2.4 Master Output Inductor Minimum Current Limitations

The selection of the output inductor for a buck regulator was discussed in Section 1.3.6. It was mentioned that at the average current in which the step at the front of the inductor current waveform has fallen to zero (see Figures 1.6*a* and 1.6*b*), the inductor is said to run dry or to go discontinuous. Below this average current, the feedback loop maintains the buck regulator's output voltage constant by reducing the "on" period; this results in reduction of slave output voltages.

In Figure 1.6*a*, however, it can be seen that at currents above going discontinuous, the "on" time is very nearly constant over large output current changes. Below run-dry, the "on" time changes drastically. In the buck regulator this does not pose a major problem because only one output is involved and the feedback loop keeps this output voltage constant. But in the push-pull width-modulated converter with a master and some slaves, the slave output voltages are directly proportional to the master "on" time, as shown by Eqs. 2.2 and 2.3.

Hence, when slaves are involved it is important that the average master output inductor current not be permitted to go discontinuous above its specified minimum. If the master minimum output current is specified at one-tenth its nominal value for example, a minimum output inductor value must be selected from Eq. 1.8. The slave output voltages will vary within about 5% above the master inductor discontinuous current. Below this critical current, the feedback loop will keep the master output voltage constant by decreasing T_{on} significantly, followed by the slave output voltages.

Further, the slave outputs must not be permitted to go discontinuous above their own specified minimum currents. Slave output inductors should also be selected from Eq. 1.8. Clearly, larger minimum currents imply smaller inductors.

TIP *This problem is also eliminated by using coupled output inductors.*[1]
~ *K.B.*

The push-pull converter is one of the oldest topologies and is still popular. It can provide multiple outputs whose returns are DC-isolated from input ground and from one another. Output voltages can be higher or lower than the input voltage. The master is regulated against line and load variations. The slaves are equally well regulated against line changes and can be within about 5% for load changes as long as output inductors are not permitted to go discontinuous.

2.2.5 Flux Imbalance in the Push-Pull Topology (Staircase Saturation Effects)

The designer needs to be aware of a rather subtle failure mode in push-pull converters, known as *staircase saturation,* caused by a possible flux imbalance in the transformer core.

This effect can best be understood by examination of a typical hysteresis loop of a ferrite core material used in the power transformer as shown in Figure 2.3.

In normal operation, core flux excursions are between levels such as B_1 and B_2 gauss in Figure 2.3. It is important to stay on the linear part of the hysteresis loop below about $\pm$ 2000 G. At frequencies up to 25 kHz or so, core losses are low and these maximum excursions are permissible. As discussed in Section 2.2.9.4, however, core losses go up rapidly with frequency, and above 100 kHz conservative design limits peak flux density to 1200 or even 800 G.

It can be seen in Figure 2.1 that when $Q1$ is "on," the no-dot end of N_{p1} is positive with respect to the dot end, and the core moves up the hysteresis loop—say, from B_1 toward B_2. The actual amount it moves up is proportional to the product of the voltage across N_{p1} and $Q1$ "on" time (from Faraday's law; see Eq. 1.18). When $Q1$ turns "off" and $Q2$ turns "on," the dot end of N_{p2} is positive with respect to the no-dot end, and the core moves back down from B_2 toward B_1. The actual amount it moves down is proportional to the voltage across N_{p2} and the $Q2$ "on" time.

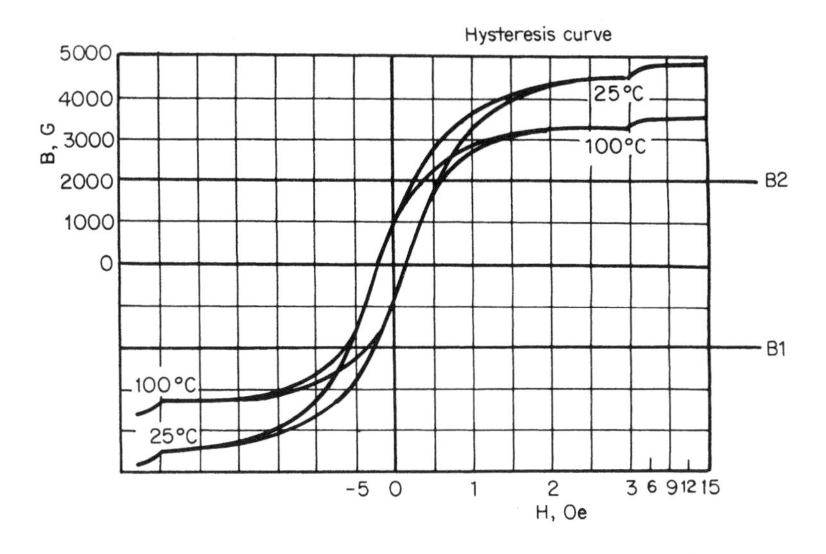

FIGURE 2.3 Hysteresis loop of a typical ferrite core material (Ferroxcube 3C8). Flux excursions are generally limited to ± 2000 G up to about 30 kHz by requirement to stay on the linear part of the loop. At higher frequencies of 100 to 300 kHz, peak flux excursions must be reduced to about ± 1200 or ± 800 G because of core losses. Material 3C8 is a ferrite from Ferroxcube Corporation. Other materials from this or other manufacturers are very similar, differing mainly in core losses and Curie temperature.

Further, if the volt-second product across N_{p1} while $Q1$ is "on" is equal to the volt-second product across N_{p2} while $Q2$ is "on", after one complete period the core will have moved up from B_1 to B_2 and returned exactly to B_1. But if those volt-second products differ by only a few percent and the core has not returned to its exact starting point each cycle, after a number of periods the core will "walk" or "staircase" up or down the hysteresis loop into saturation. In saturation, of course, the core cannot sustain voltage, and the next time a transistor turns "on," it will be destroyed by high current and high voltage.

A number of factors can cause the "on" volt-second product to be different from the "off" or reset volt-second product. The $Q1$ and $Q2$ collector voltages and "on" times may not be exactly equal even if their base drive "on" times are equal. If $Q1$, $Q2$ are bipolar transistors, they have "storage" times that effectively keep the collector "on" after base drive is removed. Storage times can range from 0.3 to 6 μs and have large production spreads. They are also temperature-dependent, increasing significantly as temperature increases. Even if $Q1$ and $Q2$ have equal storage times, they may become unequal if located on a heat sink such that they operate at different temperatures.

Hence if one transistor has a volt-second product only slightly larger than the other, it will start the core progressively drifting off-center toward saturation with each cycle. This will cause one transistor to draw slightly more current than the other as the core moves onto the curved part of the hysteresis loop (see Figure 2.3). As a result, the core magnetizing current on that half-period starts to become a significant part of the load current. The transistor that draws more current will now run slightly warmer, increasing its storage time. With a longer storage time in that transistor, the volt-second product it applies to the core in its "on" half period increases, the current in that half period increases, and storage time in that transistor increases still further. Thus a runaway condition arises that quickly drives the core into saturation and destroys the transistor.

The "on" volt-second products of $Q1$ and $Q2$ also can differ because of their initially unequal "on" or $V_{ce(sat)}$ voltages, which have a significant production spread. As described earlier, with bipolar transistors, any initial difference in "on" voltage is magnified because the "on" voltage of bipolars decreases as temperature increases.

If $Q1$, $Q2$ are MOSFETs (Metal-Oxide-Semiconductor Field-Effect Transistors), the flux-imbalance problem is much less serious. To start with, MOSFETs have no storage time, and with equal input "on" (gate) times, output (drain) times are equal and, importantly, the "on" voltage of a MOSFET transistor increases as temperature increases. Thus the runaway condition described earlier is reversed, providing some compensation. If there were any initial volt-second inequality, one FET current would be greater as the core started moving up the curved part of the hysteresis loop. The FET with the larger current would run warmer, and its "on" voltage would increase and rob voltage from its half primary. This would decrease the volt-second product in that half-period and bring the transistor current back down, providing some compensation.

2.2.6 Indications of Flux Imbalance

The earlier description might imply that any slight imbalance in volt-second product between half cycles causes certain failure, but this is not necessarily so. A push-pull converter can continue to operate reliably with a small amount of flux imbalance without immediately saturating its core and destroying its transistors. Many low power, low voltage push-pull converter designs run quite reliably in spite of the apparent problems.

Notice that with a small volt-second imbalance, if there were not an inherent corrective mechanism, core saturation and transistor failure would always occur after a few switching cycles. Thus, if there were an initial volt-second imbalance of say 0.01% (which would be practically impossible to achieve), it would take only 10,000 cycles until the core would move from a low starting point of B_1 (see Figure 2.3) to a saturating point of B_2, and the transistors would probably be destroyed before that.

One corrective mechanism, that may permit the converter to survive, is the primary winding resistance. If there is an initial volt-second imbalance, the transistor taking more current produces a larger voltage drop across its half primary winding resistance. That voltage drop robs volt-seconds from the winding and tends to restore the volt-second balance.

Thus the converter can remain in an unbalanced state without immediately going into runaway and completely saturating the core. An indication of where the core is working on the hysteresis loop can be obtained by placing a current probe in the transformer center tap as shown in Figure 2.4*d*.

The waveform indicating volt-second balance is shown in Figure 2.4*a*, where alternate current peaks are equal. Primary load current pulses have the characteristic shape of a ramp on a step just as for the buck regulator in Figure 1.4*d*. They have this shape because all the secondaries have output *LC* filters that generate such waveshapes as described in Section 1.3.2.

The primary load current is the sum of all the secondary currents reflected into the primary by their respective turn ratios. However, the total primary current is the sum of these secondary currents plus the primary magnetizing current. The magnetizing current is the current drawn by the *magnetizing inductance,* which is the inductance seen looking into the primary with all secondaries open-circuited. This inductance is always present and effectively is in parallel with the primary winding. This current is added to the secondary currents reflected into the primary as in Figure 2.4*e*.

The waveshape of the total primary current is then the sum of the ramp-on-a-step reflected load currents and the magnetizing current. But providing the core is working in the linear area of the B/H loop, the magnetizing current will be a linear ramp starting from zero current each cycle.

When a transistor turns "on," it applies a step of voltage of approximately $V_{dc} - 1$ across the magnetizing inductance L_{pm}. Magnetizing current then ramps up linearly at a rate

$$dI/dt = (V_{dc} - 1)/L_{pm} \tag{2.4}$$

and for the transistor "on" time of T_{on} it reaches a peak of

$$I_{pm} = \frac{(V_{dc} - 1)(T_{on})}{L_{pm}} \tag{2.5}$$

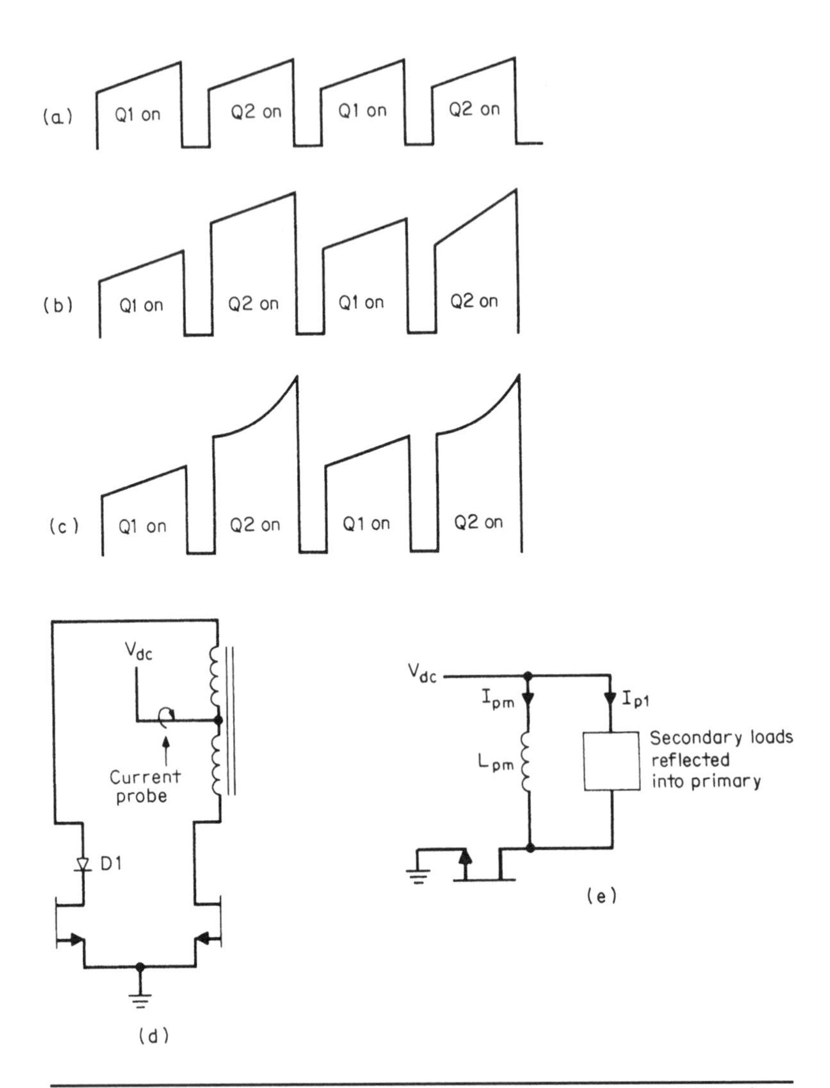

FIGURE 2.4 Current waveforms in the transformer center tap. (*a*) Waveform shows equal volt-second product on the two halves of transformer primary. (*b*) Unequal volt-second product on the two halves of transformer primary. Core is not yet on curved part of hysteresis loop. (*c*) Unequal volt-second product. Upward concavity indicates dangerous situation. Core is far up on curved part of hysteresis loop. (*d*) Adding a diode in series with one side of primary to test how serious a volt-second inequality exists. (*e*) Total primary current is the sum of the ramp-on-a-step reflected secondary load currents plus the linear ramp of magnetizing current.

The magnetizing current I_{pm} is kept small compared with the sum of the load currents reflected into the primary by ensuring that L_{pm} in Eq. 2.5 is large. By design, the peak magnetizing current should be no greater than 10% of the primary load current.

When added to the ramp-on-a-step load current, the ramp of magnetizing current is small, and it simply increases the slope of the latter slightly. Also, if the volt-seconds are equal on alternate half cycles, the peak currents will also be equal on each half cycle as in Figure 2.4*a*, because operation is centered around the origin of the hysteresis loop of Figure 2.3.

However, if the volt-second products on alternate half cycles are unequal, core operation is not centered on the origin of the hysteresis loop. Since the horizontal scale (H oersteds) is proportional to magnetizing current, this shows up as a DC current bias as in Figure 2.4*b*, making alternate current pulses unequal in amplitude.

As long as the DC bias does not drive the core up the hysteresis loop appreciably, the slope of the ramp still remains linear (Fig. 2.4*b*) and operation is still reasonably safe. Primary wiring resistance may keep the core from moving further up into saturation.

But if there is a large inequality in volt-seconds on alternate half cycles, the core is biased closer toward saturation and enters the curved part of the hysteresis loop. Now the magnetizing inductance, which is proportional to the slope of the hysteresis loop, decreases and magnetizing current increases significantly. This shows up as an upward concavity in the current slope in Figure 2.4*c*.

This is a dangerous and imminent failure situation. Now even a small temperature increase can bring on the runaway scenario described earlier. The core will be driven hard into saturation and destroy the power transistor. A push-pull converter design should certainly not be considered safe if current pulses in the primary center tap show any upward concavity in their ramps. Even linear ramps as in Figure 2.4*b* with anything greater than 20% inequality in peak currents are unsafe and should not be accepted.

Note *A more damaging effect can occur if there is a sudden transient load change, because the extra current can take the core immediately into saturation.* ~ *K.B.*

2.2.7 Testing for Flux Imbalance

A simple test to determine how close to a dangerous flux-imbalance situation a push-pull converter may be operating is shown in Figure 2.4*d*. Here a silicon diode with about 1 V forward drop is placed in series with one half of the transformer primary. Now in the "on" state, that half with the diode in series has 1 V less voltage across it than the other half, and there is an artificially produced volt-second unbalance. The center tap waveform will then look like either Figure 2.4*b* or 2.4*c*. The current ramp corresponding to the side that does not have the diode will have the larger volt-second product and the larger peak current. By switching the diode to the other side, the larger peak current will be seen to switch to the opposite transformer half primary.

Now the closeness of the circuit to the upward concave situation of Figure 2.4*c* can be determined. If one series diode can make a current ramp go concave, the circuit is too close to imminent failure. Placing two series diodes on one side will give an indication of how much margin there is.

It should be noted that primary magnetizing current contributes no power to the secondaries. It will not appear in the secondaries. It simply swings the magnetic core across the hysteresis loop.

In Figure 2.3, the magnetizing force H in oersteds (Oe) is related to the current by the fundamental magnetic relation

$$H = \frac{0.4\pi N_p I_m}{l_m} \tag{2.6}$$

where N_p is the number of primary turns
I_m is the magnetizing current in amperes
l_m is the magnetic path length in cm

2.2.8 Coping with Flux Imbalance

Flux imbalance can become a major problem at high voltages and high powers. There are a number of ways to circumvent the problem, but most involve increased cost or component count. Some schemes to combat flux imbalance are described in the following subsections.

2.2.8.1 Gapping the Core

Flux imbalance becomes serious when the core moves out onto the curved part of the hysteresis loop (see Figure 2.3) and magnetizing current starts increasing exponentially as in Figure 2.4*c*. This effect can be reduced by moving the curved part of the hysteresis loop to a higher current by tilting the hysteresis loop. The core can then tolerate a larger DC current bias or volt-second product inequality.

An air gap introduced into the magnetic path of the core has the effect shown in Figure 2.5. It tilts the slope of the hysteresis loop. An air gap of 2 to 4 mils (thousandths of an inch) brings the curved portion of the loop much further away from the origin so that the core can accept a reasonably large offset in H (current imbalance). This can help at higher power levels. It has the disadvantage of reducing the inductance so that the critical current must be larger to prevent discontinuous-mode operation.

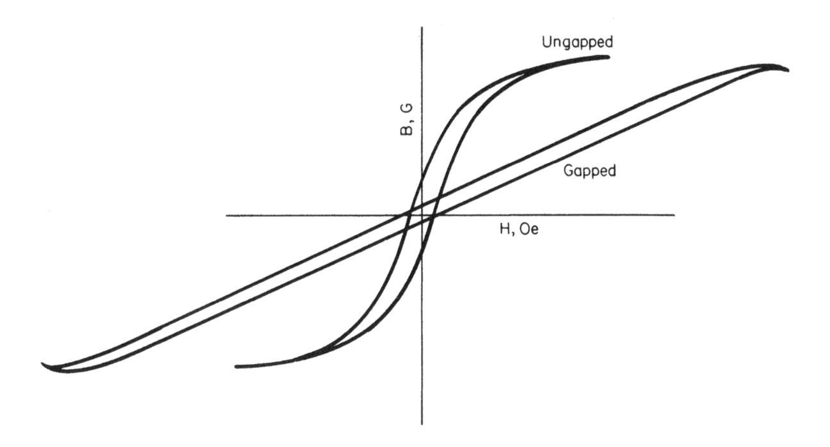

FIGURE 2.5 How a gap in the core reduces the slope of the hysteresis loop.

The air gap for a prototype EE or cup core is easily effected with plastic shims in the center and outer legs. Since the flux passes through the center leg and returns through the outer legs, the total gap is twice the shim thickness. In a production transformer, it is not very much more expensive to have the center leg ground down to twice the shim thickness. This will achieve pretty much the same effect as shims in the center and outer legs, but is preferable as the gap will not change with changes in the thickness of the plastic and results in less magnetic radiation and hence reduced RFI interference.

2.2.8.2 Adding Primary Resistance

It was pointed out in Section 2.2.6 that primary wiring resistance keeps the core from being driven rapidly into saturation if there is a volt-second inequality. If there is such an inequality, the half primary with the larger volt-second product draws a larger peak current. That larger current causes a larger voltage drop across the wiring resistance and robs volt-seconds from that half primary, restoring the current balance.

This effect can be augmented by adding additional resistance in series with both primary halves. The added resistors can be located in either the collectors or emitters of the power transistors. The value is best determined empirically by observing the current pulses in the transformer center tap. The required resistors are usually under 0.25 Ω. They will, of course, increase power loss and reduce efficiency.

2.2.8.3 Matching Power Transistors

Since volt-second inequality arises mainly from an inequality in storage time or voltage in the power transistors, if those parameters are matched, it adds confidence that together with the earlier two "fixes" there will be no problem with flux imbalance.

This is not a good solution and would be an expensive fix as it is quite expensive to match transistors in two parameters. To do such matching requires a specialized test setup that would not be available if field replacements become necessary.

It also must be ascertained that if the matching is done at certain load currents and temperature, the matching still holds when these vary. Further, a storage time match is difficult to make credible, as it depends strongly on forward and reverse base input currents in the bipolar transistors. Generally any matching is done by matching V_{ce} and V_{be} (the "on" collector-to-emitter and base-to-emitter voltages) at the maximum operating current. Hence matching is not a viable solution for high-volume commercial supplies.

2.2.8.4 Using MOSFET Power Transistors

Since most of the volt-second inequality arises from storage time inequality between the two bipolar power transistors, the problem largely disappears if MOSFETs are used, because they have no storage time.

There is an added advantage, as the "on" voltage of a MOSFET transistor increases with temperature. Thus if one half primary tends to take a large current, its transistor runs somewhat warmer and its "on" voltage increases and steals voltage from the winding. This reduces the volt-second product on that side and tends to restore balance. This, of course, is qualitatively in the right direction, which is helpful but cannot be depended on to solve the flux-imbalance problem reliably at all power levels and with a worst-case combination.

However, with power MOSFETs at power levels under 100 W and low input voltages (as in most DC/DC converter applications), push-pull converters can be and are built with a high degree of confidence.

2.2.8.5 Using Current-Mode Topology

By far the best solution to the flux-imbalance problem is to use current-mode control. This completely and reliably solves the flux-imbalance problem; also it has significant additional advantages of its own.

In conventional push-pull, there is always a residual concern that despite all the fixes, a flux-imbalance problem will arise in some worst-case situation and a transistor will be destroyed. Current-mode topology solves this problem by monitoring the current in each of the push-pull transistors on a pulse-by-pulse basis. The control circuit then forces alternate current pulses to have equal amplitude, maintaining the working point very near the center of the B/H loop. Details of current-mode topology will be discussed in Chapter 5.

2.2.9 Power Transformer Design Relationships

Note *The design of wound components is a specialized subject and is covered in more detail in Chapter 7. The correct design of transformers, inductors, and chokes is essential for optimum performance of the equipment. The engineer who takes the time to become fully competent in this area will get much better results, so the reader is urged to study Chapter 7 before proceeding with any real designs.*

After Pressman *In the following section Mr. Pressman shows an iterative method for selecting the core size and winding parameters. It serves as a good example of the rather lengthy process required if this method is used. The reader will do well to study this process, which shows the interaction between the various parameters. However, in practice, optimum designs normally start by defining the maximum permitted temperature rise (typically 30°C), and one of the nomogram-assisted methods or computer programs would be used to provide a much faster solution with a defined result, avoiding the tedious iterative procedure. ~ K.B.*

2.2.9.1 Core Selection

The design of a transformer starts with the initial selection of a core to satisfy the desired total output power. The available output power from a particular core depends on the operating frequency, the operating flux density swing (B_1 and B_2 in Figure 2.3), the core's area A_e, the bobbin winding window area A_b, and the current density in each winding.

Decisions on each of these parameters are interrelated, and choices are made to minimize the transformer size and its temperature rise. In the magnetics section of Chapter 7 an equation is derived showing a recommended output power for a given core as a function of the parameters mentioned earlier.

The equation can be used in a set of iterative calculations, first making a tentative selection of a specific core, peak flux density, and operating frequency, and calculating the available output power. Then if the available power is insufficient, a larger-sized core is selected and the calculations repeated until a core with the required output power is found.

This is a long and cumbersome procedure; instead the equation is turned into a set of charts that permit a core and operating frequency to be selected at a glance for any desired output power. Such equations and charts will be found for most of the commonly used topologies in Chapter 7.

We will assume that these charts will be used to select a specific core so that the area A_e is known. The rest of the transformer design involves calculation of the number of turns on the primary and secondaries, selection of wire sizes, calculation of core and copper losses, and finally the calculation of transformer temperature rise.

The optimum arrangement of the various layers of wire on the core bobbin is important in improving coupling between the windings and in reducing copper losses due to "skin" and "proximity" effects. Winding arrangements, skin, and proximity effects will be discussed in Chapter 7.

For this example, the design will proceed using the core chosen from the selection charts described earlier, providing a known value of the core area A_e.

2.2.9.2 Maximum Power Transistor On-Time Selection

Equation 2.1 has shown that the converter keeps the output voltage V_m constant by increasing T_{on} as V_{dc} decreases. Thus the maximum "on" time T_{on} occurs at the *minimum* specified DC input voltage $\underline{V_{dc}}$. But in this type of converter the maximum "on" time must not exceed half the switching period T. If it were to do so, the reset volt-second product would be less than the set volt-second product (see Section 2.2.5), and after a very few cycles, the core would drift into saturation and destroy a power transistor.

Moreover, because of the inevitable storage time in bipolar transistors, the base drive "on" time cannot be as large as a full half period, as the storage time would cause an overlap with the opposite transistor. This would result in immediate failure, because the two power transistors would effectively short out the winding. Each transistor would take large currents at the full supply voltage and would rapidly be destroyed.

Thus, to ensure that the core will always be reset within one period and eliminate any possibility of simultaneous conduction, whenever the DC input voltage is at its minimum $\underline{V_{dc}}$ and the feedback loop is trying to increase T_{on} to maintain V_m constant, the maximum "on" time will be constrained by some kind of a clamp so as to never be more than 80% of a half period. Then in Eq. 2.1, for the specified $\underline{V_{dc}}$, T and for $\overline{T_{on}} = 0.8T/2$, the ratio N_m/N_p will be fixed to yield the desired output V_m.

TIP *Modern drive and control ICs provide adjustable (so-called) "dead time" to prevent power device overlap. In some designs, dynamic methods are provided such that the state of conduction of the power devices is monitored and the drive signal is delayed until the previous active power device has turned fully "off," before the next is allowed to turn "on." This allows the full* *range of duty cycle to be utilized while completely eliminating any possibility of overlap. ~ K.B.*

2.2.9.3 Primary Turns Selection

The number of primary turns is determined by Faraday's law (see Eq. 1.17). From it N_p is fixed by the minimum voltage across the primary $(\underline{V_{dc}} - 1)$ and the maximum "on" time, which, as earlier, is to be no more than $0.8T/2$. Then

$$N_p = \frac{(\underline{V_{dc}} - 1)(0.8T/2) \times 10^8}{A_e d B} \tag{2.7}$$

Since A_e in Eq. 2.7 is fixed by the selected core, $\underline{V_{dc}}$ and T are specified and the number of primary turns is fixed as soon as dB (the desired flux change in $0.8T/2$) is decided on. This decision is made as follows.

TIP *The reader may prefer to use a dimensionally modified version of Faraday's law that provides turns directly as follows:*

$$N = \frac{VT_{on}}{A_e \Delta B}$$

Where N = *turns*
 V = *voltage across the winding* (V_{dc})
 T_{on} = *maximum "on" period, microseconds*
 ΔB = *flux density swing, teslas (1 tesla = 10,000 gauss)*
 A_e = *effective core area, mm²*

For all magnetic calculations, I prefer to work in the preceding modified SI units, as these yield immediate solutions, avoiding the unwieldy exponents, thus reducing errors. ~ K.B.

2.2.9.4 Maximum Flux Change (Flux Density Swing) Selection

From Eq. 2.7, it is seen that the number of primary turns is inversely proportional to dB, the flux swing. It would seem desirable to maximize dB so as to minimize N_p, since fewer turns would mean that a larger wire size could be used, resulting in higher permissible currents and more output from a given core. Also, fewer turns would result in a less expensive transformer and lower stray parasitic capacities.

From the hysteresis loop of Figure 2.3, however, it is seen that in ferrite cores, the loop enters the curved portion above $\pm$ 2000 G. It is desirable to stay below this point, where the magnetizing current starts increasing rapidly. So initially a good choice would appear to be $\pm$ 2000 G (0.2 tesla). But we must also consider core losses.

Ferrite core losses increase at about the 2.7^{th} power of the peak flux density and at about the 1.6^{th} power of the operating frequency.

Hence, up to about 50 kHz, core losses do not prohibit operation to $\pm$ 2000 G, and it may appear desirable to operate at that flux level.

However, to prevent core saturation under transient conditions, it is better to provide a wider margin. We will see shortly that it is preferable to restrict operation to $\pm$ 1600 G even at frequencies where core losses are not prohibitive. Faraday's law solved for the flux change dB is

$$dB = \frac{(V_{dc} - 1)(T_{on}) \times 10^8}{N_p A_e} \tag{2.8}$$

Equation 2.8 says that if N_p is chosen for a given dB—say, from −2000 to +2000 G, or a dB of 4000 G, then as long as the product of $(V_{dc} - 1)(T_{on})$ is constant, dB will be constant at 4000 G. Further, if the feedback loop is working and keeping the output voltage V_m constant, Eq. 2.1 says that $(V_{dc} - 1)(T_{on})$ is constant and dB will truly remain constant. So providing the feedback loop always ensures that whenever V_{dc} is a minimum, that T_{on} is at a maximum, then T_{on} and V_{dc} can never be simultaneously maximum.

However, in some transient or fault conditions, if T_{on} has been at maximum for a single, or possibly even a few cycles, and V_{dc} had a transient step to 50% above its normal value, the feedback loop may fail to reduce the "on" time rapidly enough (as normally required by Eq. 2.1), and there may exist a short period when V_{dc} and T_{on} would be maximum at the same time. In this event, Equation 2.8 shows that dB would be 1.5(4000) or 6000 G.

Then if the core had started from the −2000-G point, at the end of that "on" time the core would have been driven 6000 G above that, or to +4000 G. The hysteresis loop (see Figure 2.3) shows that at temperatures somewhat above 25°C, it would be deep in saturation and could not support the applied voltage. The transistor would be subject to high current and high voltage and would rapidly fail.

It will be seen in the feedback analysis section of Chapter 12 that the error amplifier has a delay in its response time, because its bandwidth is limited to stabilize the feedback loop. Hence, it is always possible for both the input voltage and "on" time to be maximum for a transient period due to the inevitable delay in the response of the error amplifier, although the error amplifier will eventually correct the "on" time so as to keep the product $(V_{dc} - 1)(T_{on})$ constant in accordance with Eq. 2.1. If the core is subjected to maximum input voltage and maximum "on" time as a result of error-amplifier delay, even for a single cycle, it may saturate the core and destroy a transistor.

However if N_p in Eq. 2.8 is chosen to yield dB of 3200 G at $\underline{V_{dc}}$ and $\overline{T_{on}}$, the design is safer and can tolerate a 50% transient step in input voltage. With $dB = 3200$ G, if the error amplifier is too slow to correct the "on" time, the transformer dB will be 1.5(3200) or 4800 G; and if the core started from its normal minimum flux of −1600 G, it will be driven up to only −1600 + 4800 or +3200 G. The hysteresis loop of Figure 2.3 shows that the core can tolerate that even at 100°C.

Thus the number of primary turns is selected from Eq. 2.7 for $dB =$ 3200 G even at lower frequencies where a large flux may not cause excessive core losses. Above 50 kHz, the core losses increase rapidly and force a lower flux density selection. At 100 to 200 kHz, the peak flux density may be limited to 1200 or even 800 G to achieve an acceptably low core temperature rise.

2.2.9.5 Secondary Turns Selection

The turns for the main and slave outputs are calculated from Eqs. 2.1, 2.2, and 2.3 in accordance with the specified, or calculated, voltage requirements. We see that the input voltage $\underline{V_{dc}}$ and T have been specified. The maximum "on" time $\overline{T_{on}}$ has been arbitrarily set at $0.8T/2$, and N_p has been calculated from Faraday's law (see Eq. 2.7) for the known A_e for the selected core. Flux swing dB has been set at 3200 G for frequencies under 50 kHz and to minimize core losses. Lower values will be used at higher frequencies as discussed earlier.

2.2.10 Primary, Secondary Peak and rms Currents

In this example, wire sizes will be selected on the basis of a conservative operating current density. Current density is given in terms of rms current in amps per circular mil* of wire cross-sectional area.

Hence, before we can start selecting wire sizes for any winding, we require a knowledge of the rms currents in each winding.

2.2.10.1 Primary Peak Current Calculation

Current drawn from the DC input source V_{dc} may be monitored in the transformer center tap and has the waveform shown in Figures 2.1*b* and 2.1*d*. The pulses have the characteristic ramp-on-a-step waveshape because the secondaries all have output LC filters as discussed in Section 1.3.2. The primary current is simply the sum of all the secondary ramp-on-a-step currents reflected into the primary by their turns ratios, plus the magnetizing current.

As discussed in Section 2.2.9.2, at minimum $\underline{V_{dc}}$ input voltage, the transistor "on" times will be 80% of a half period. Further, since there is one pulse for each half period, the duty cycle of the pulses in Figure 2.1

---

*A circular mil is the area of a circle 1 mil in diameter. Thus, area in square inches = $(\pi/4)10^{-6}$ (area in circular mils).

is 0.8 at $\underline{V_{dc}}$. To simplify calculation, the pulses in the figure are assumed to have an equivalent flat-topped waveshape whose amplitude I_{pft} is the value of the current at the center of the ramp.

Then the input power at V_{dc} is that voltage times the average current, which is $0.8I_{pft}$, and assuming 80% efficiency (which is usually achievable up to 200 kHz), $P_o = 0.8P_{in}$ or

$$P_{in} = 1.25P_o = \underline{V_{dc}}0.8I_{pft}$$

Then

$$I_{pft} = 1.56\frac{P_o}{\underline{V_{dc}}} \tag{2.9}$$

This is a useful relation, as it gives the equivalent flat-topped primary current pulse amplitude in terms of what is known—the output power and the specified minimum DC input voltage. It allows selection of a primary wire size from the calculated primary rms current. It also allows a transistor with an adequate current rating to be selected.

2.2.10.2 Primary rms Current Calculation and Wire Size Selection

Each half primary carries only one of the I_{pft} pulses per period, and hence its duty cycle is $(0.8T/2)/T$ or 0.4. It is well known that the rms value of a flat-topped pulse of amplitude I_{pft} at a duty cycle D is

$$I_{rms} = I_{pft}\sqrt{D} = I_{pft}\sqrt{0.4}$$

or

$$I_{rms} = 0.632I_{pft} \tag{2.10}$$

and from Eq. 2.9

$$I_{rms} = 0.632\frac{1.56P_o}{\underline{V_{dc}}} = \frac{0.986P_o}{\underline{V_{dc}}} \tag{2.11}$$

This gives the rms current in each half primary in terms of the known parameters: output power and the specified minimum DC input voltage.

A conservative practice in transformer design is to operate the windings at a current density of 500 circular mils per rms ampere. There is nothing absolute about this; current densities of 300 circular mils per rms ampere are frequently used for windings with only a few turns. As a general rule, however, densities greater than 300 circular mils per rms ampere should be avoided, as that will cause excessive copper losses and temperature rise.

Thus at 500 circular mils per rms ampere, the required number of circular mils for the half primaries is

$$\text{Circular mils} = 500\frac{0.986\,P_o}{V_{dc}} = 493\frac{P_o}{V_{dc}} \tag{2.12}$$

Notice that this is also in terms of known values—output power and specified minimum DC input voltage. Proper wire size can then be chosen from wire tables at the circular mils given by Eq. 2.12.

2.2.10.3 Secondary Peak, rms Current, and Wire Size Calculation

Currents in each half secondary are shown in Figure 2.6. Note the ledge at the end of the transistor "on" time. This ledge of current exists because there is no free-wheeling diode *D*1 at the input to the filter inductor as in the buck regulator of Figure 1.4. In the buck, the free-wheeling diode was essential as a return path for inductor current when the transistor turned off. When the transistor turned off, the polarity across the output inductor reversed, and its input end would have gone disastrously negative if it had not been caught by the free-wheeling diode at about 1 V below ground. Inductor current then continued to flow through the free-wheeling diode *D*1 of Figure 1.4*e*. This problem does not exist in the rectifier circuit shown in Figure 2.6.

In the push-pull output rectifier stage, the function of the free-wheeling diode is performed by the output rectifier diodes *D*1 and *D*2. When either transistor turns "off," the input end of the inductor tries to go negative. As soon as it goes about one diode drop below ground, both rectifiers conduct, each drawing roughly half the total current the inductor had been drawing just prior to turn "off" (see Figures 2.6*d* and 2.6*e*). Since the impedance of each half secondary is small, there is negligible drop across them, and the rectifier diode cathodes are caught at about 1 V below ground.

Thus if half-secondary rms currents are to be calculated exactly, the ledge currents during the 20% dead time should be taken into account. However, in this example it can be seen that they are only about half the peak inductor current and have a duty cycle of $(0.4T/2)/T$ or 0.2. With such small amplitudes and duty cycle they can be ignored in this example. Each half secondary can then be considered to have the characteristic ramp-on-a-step waveform, which at minimum DC input comes out to a duty cycle of $(0.8T/2)/T$ or 0.4. The magnitude of the current at the center of the ramp is the DC output current I_{dc}, as can be seen from Figure 2.6*f*.

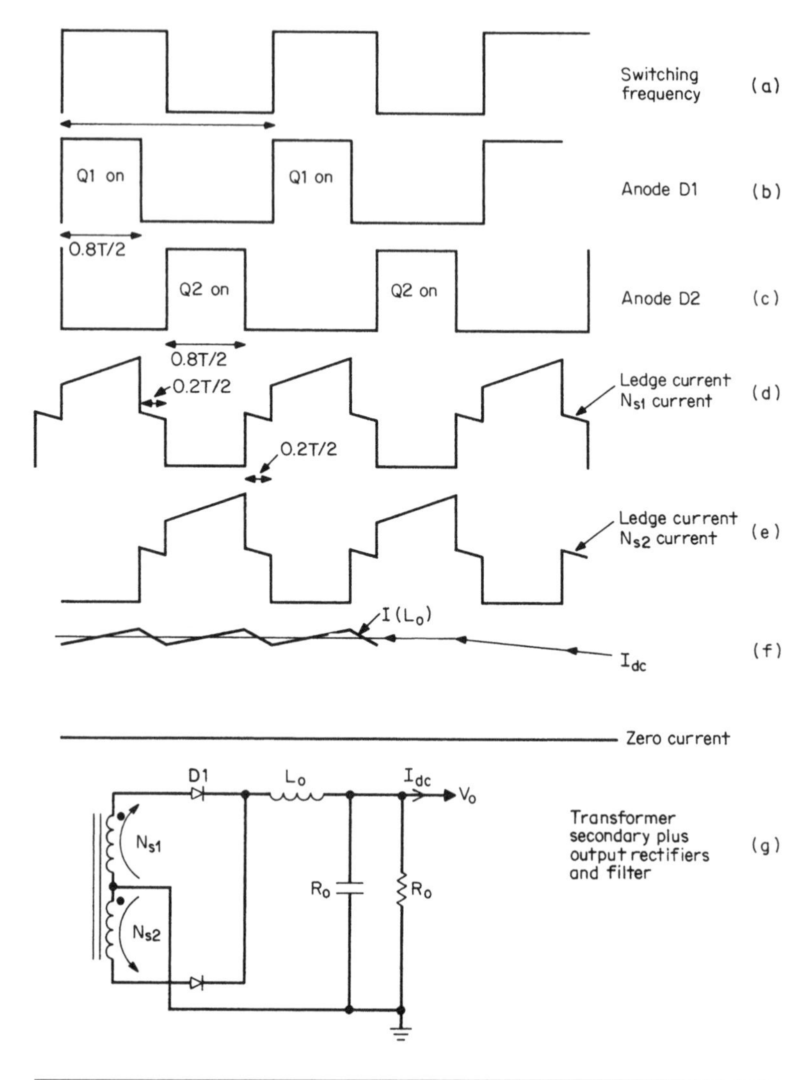

FIGURE 2.6 Output rectifiers $D1$ and $D2$ serve as free-wheeling diodes in a push-pull rectifier circuit. Each secondary winding carries half the normal free-wheeling "ledge" during the 20% dead time. This should be considered in estimating secondary copper losses.

2.2.10.4 Primary rms Current, and Wire Size Calculation

To simplify the primary current rms calculations, the ramp-on-a-step pulses will be approximated by "equivalent flat-topped" pulses I_{aft}, whose amplitude is that at the center of the ramp or the DC output current I_{dc} with a duty cycle of 0.4.

Thus rms current in each half secondary is

$$I_{s(\mathrm{rms})} = I_{\mathrm{dc}}\sqrt{D} = I_{\mathrm{dc}}\sqrt{0.4} = 0.632 I_{\mathrm{dc}} \tag{2.13}$$

At 500 circular mils per rms ampere, the required number of circular mils for each half secondary is

$$\begin{aligned}\text{Secondary circular mil requirement} &= 500(0.632) I_{\mathrm{dc}} \\ &= 3.16 I_{\mathrm{dc}}\end{aligned} \tag{2.14}$$

2.2.11 Transistor Voltage Stress and Leakage Inductance Spikes

It can be seen from the polarities of the transformer primary windings in Figure 2.1 that when either transistor is "on," the opposite transistor's collector is subject to at least twice the DC supply voltage, since both half primaries have an equal number of turns and are in series, with the center tap connected to the supply.

However, the maximum stress is somewhat more than twice the input voltage. An additional contribution comes from the so-called leakage inductance spikes shown in Figures 2.1*a* and 2.1*c*. These come about because there is effectively a small inductance (leakage inductance L_l) in series with each half primary as shown in Figure 2.7*a*.

At the instant of turn "off," current in the transistor falls rapidly at a rate dI/dT, causing a positive-going spike of amplitude $e = L_l\, dI/dT$ at the bottom end of the leakage inductance. Conservative design practice assumes the leakage inductance spike may increase the stress voltage by as much as 30%, more than twice the maximum DC input voltage. Hence the transistors should be chosen so that they can tolerate with some safety margin a maximum voltage stress V_{p} of

$$V_{\mathrm{p}} = 1.3(2V_{\mathrm{dc}}) \tag{2.15}$$

The magnitude of the leakage inductance is not easily calculable. It can be minimized by use of a transformer core with a long center leg and by sandwiching the secondary windings (especially the higher current ones) in between halves of the primary. A good transformer should have leakage inductance of no more than 4% of its magnetizing inductance.

TIP *The leakage inductance of any winding can be easily measured by short-circuiting all other windings and measuring the residual inductance on the required winding. ~ K.B.*

Leakage inductance spikes can be minimized by addition of a snubber circuit (a capacitor, resistor, and diode combination) connected to the transistor collector as shown in Figure 2.7*a*. Such

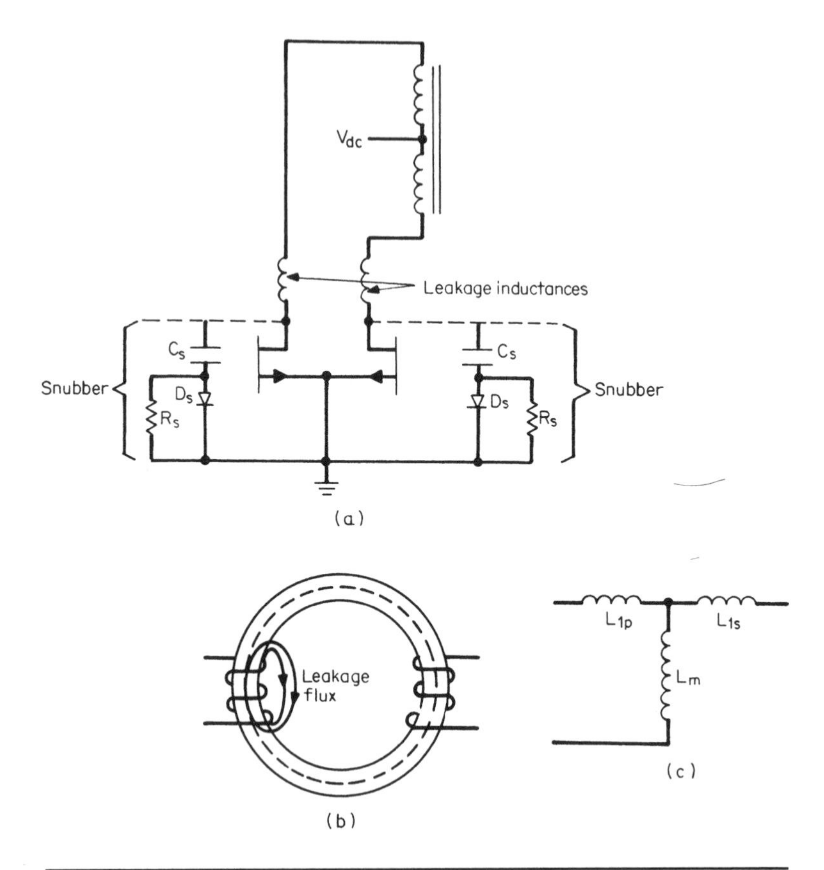

FIGURE 2.7 (*a*) How leakage inductances cause spikes on the collectors of the power devices. (*b*) How leakage inductance stems from the fact that some of the magnetic flux lines return through a local air path rather than linking the secondary through the core. (*c*) The low-frequency equivalent circuit of a transformer showing magnetizing inductance L_m and primary and secondary leakage inductances L_{1p}, L_{1s}.

configurations also serve the important function of reducing AC switching losses by load line shaping (phase shifting the overlap of falling transistor current and rising voltage at the collector). Detailed design of snubbers and some associated penalties they incur are discussed in Chapter 11.

Leakage inductance arises from the fact that some of the primary's magnetic flux lines do not return through the core and couple with the secondary windings. Instead, they return around the primary winding through a local air path as seen in Figure 2.7*b*.

The equivalent circuit of a core with its magnetizing L_m (see Section 2.2.6) and primary L_{1p} leakage inductances is shown in Figure 2.7*c*.

Secondary leakage inductance arises from the fact that some of the secondary current's magnetic flux lines also do not couple with the primary but instead link the secondary windings via a local air path. But in most cases, there are fewer turns on the secondary than on the primary, and L_{1s} can be neglected.

The transformer equivalent circuit shown in Figure 2.7*c* is a valuable tool in the understanding of many unexpected circuit effects and can be used up to about 300 to 500 kHz, where shunt parasitic capacitors across and between windings must also be taken into account.

2.2.12 Power Transistor Losses

2.2.12.1 AC Switching or Current-Voltage "Overlap" Losses

Leakage inductance in the power transformer allows a very rapid collector voltage fall time because for a short time when a transistor turns on, the leakage inductance has a very high impedance. Since the current cannot change instantaneously through an inductor, the collector current rises slowly during the turn "on" edge. Thus there is very little overlap of falling voltage and rising current at turn "on" and negligible switching loss.

At turn "off," however, the inductance tends to maintain the previous current constant. Hence there is significant overlap and a worst-case scenario may be assumed, such as that shown for the buck regulator of Figure 1.5*b*. The exact situation is shown in Figure 2.8,

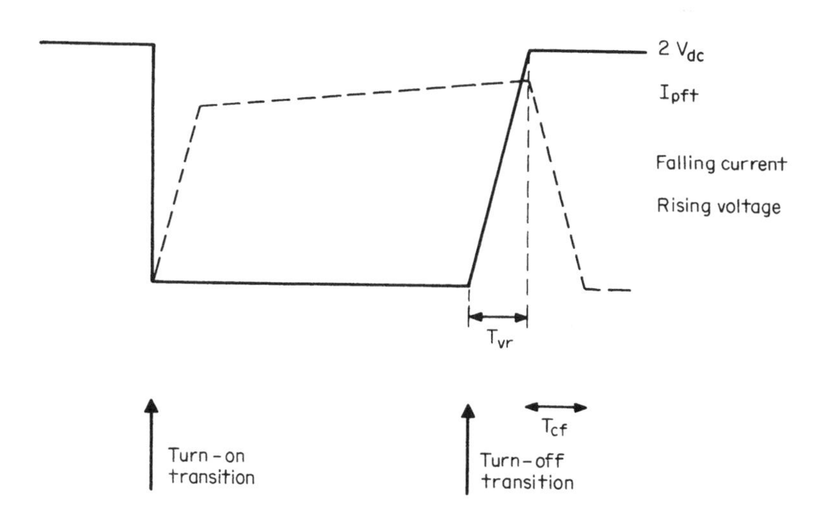

FIGURE 2.8 Switching loss due to current/voltage overlap.

where it is assumed that the current hangs on at its equivalent flat-topped peak value I_{pft} (see Section 2.2.10.1) for the time it takes the voltage to rise from near zero to its maximum value of $2\overline{V_{dc}}$. The voltage then remains at $2\overline{V_{dc}}$ during the time, T_{cf}, it takes the current to fall from I_{pft} to zero. Assuming $T_{vr} = T_{cf} = T_s$ and a switching period T, the total switching dissipation per transistor per period $P_{t(ac)}$ is

$$\begin{aligned} P_{t(ac)} &= I_{pft}\frac{2\overline{V_{dc}}}{2}\frac{T_s}{T} + 2\overline{V_{dc}}\frac{I_{pft}}{2}\frac{T_s}{T} \\ &= 2(I_{pft})(\overline{V_{dc}})\frac{T_s}{T} \end{aligned}$$

and from Eq. 2.9, $I_{pft} = 1.56(P_o/\underline{V_{dc}})$:

$$P_{t(ac)} = 3.12\frac{P_o}{\underline{V_{dc}}}\overline{V_{dc}}\frac{T_s}{T} \tag{2.16}$$

Notice there are negligible switching losses at turn "on" because transformer leakage inductance causes a very fast voltage fall and a slow current rise. This results in very little turn "on" loss. However, worst-case scenario is shown at turn "off." The current remains constant at its peak I_{pft} until voltages rises to $2\overline{V_{dc}}$. The voltage remains at $2\overline{V_{dc}}$ for the duration of the current fall time T_{cf}, producing a large turn "off" loss.

2.2.12.2 Transistor Conduction Losses

The conduction losses are simply the transistor "on" voltage multiplied by the "on" current for each device averaged over a cycle, or

$$P_{dc} = I_{pft}V_{on}\frac{0.8T/2}{T} = 0.4I_{pft}V_{on}$$

It will be seen in Chapter 8 that a technique called *Baker clamping* can be used to reduce transistor storage times for bipolar base drives. This forces the collector "on" potential V_{ce} to be about 1 V over a large range of current. Then for I_{pft} from Eq. 2.9 we obtain

$$P_{dc} = 0.4\frac{1.56P_o}{\underline{V_{dc}}} = \frac{0.624P_o}{\underline{V_{dc}}} \tag{2.17}$$

and total losses per transistor are

$$\begin{aligned} P_{total} &= P_{t(ac)} + P_{dc} \\ &= 3.12\frac{P_o}{\underline{V_{dc}}}\overline{V_{dc}}\frac{T_s}{T} + \frac{0.624P_o}{\underline{V_{dc}}} \end{aligned} \tag{2.18}$$

2.2.12.3 Typical Losses: 150-W, 50-kHz Push-Pull Converter

It will be instructive to calculate the dissipation per transistor in a 150-W push-pull converter at 50 kHz operating from a 48-volt power source.

The standard telephone industry power sources provide a nominal voltage of 48 V, with a minimum ($\underline{V_{dc}}$) of 38 V and maximum ($\overline{V_{dc}}$) of 60 V. It will be assumed that at 50 kHz, bipolar transistors will be used, and a reasonable value of the switching time (T_s as defined earlier) of 0.3 μs.

The DC conduction losses from Eq. 2.17 are

$$P_{dc} = \frac{0.624 \times 150}{38} = 2.46 \text{ W}$$

but the AC switching losses from Eq. 2.16 are much larger at

$$P_{t(ac)} = 3.12 \times \frac{150}{38} \times 60 \times \frac{0.3}{20} = 11.8 \text{ W}$$

Thus the AC overlap or switching losses are about 4.5 times greater than the DC conduction losses. If MOSFET transistors are considered with switching times T_s of about 0.05 μs, it can be seen that switching losses would be negligible in this example.

2.2.13 Output Power and Input Voltage Limitations in the Push-Pull Topology

Aside from the flux-imbalance problem in the push-pull topology, which does not exist in the current-mode controlled version, limitations include the useful power working area as defined in Eq. 2.9, and input voltage in Eq. 2.15.

Equation 2.9 gives the peak current required of the transistor for a desired output power, and Eq. 2.15 gives the maximum voltage stress on the transistor in terms of the maximum DC input voltage. These requirements limit the power rating of the push-pull topology to around 500 W when using bipolar transistors. Above that, it is difficult to find transistors that can meet the peak current and voltage stress while being fast enough with adequate gain.

The technology is constantly improving, and without doubt a faster MOSFET with adequately high voltage and current ratings and sufficiently low "on" voltages would extend this power range.

As an example, we will consider a 400-W push-pull converter operating from telephone industry prime voltage source that is 48 V (nominal), 38 V (minimum), and 60 V (maximum).

Equation 2.9 gives the peak current requirement as $I_{pft} = 1.56P_o/V_{dc} = 1.56(400)/38 = 16.4$ A, and Eq. 2.15 gives the maximum "off" voltage stress as $V_p = 2.6V_{dc} = 2.6 \times 60 = 156$ V. To provide a margin of safety, a transistor with at least a 200-V rating would be selected.

A possible candidate would be the MJ13330 bipolar transistor. It has a 20-A peak current rating, a V_{ceo} rating of 200 V, and V_{cer} rating of 400 V (the voltage it can sustain when it has a negative bias of −1 to −5 V at turn "off"). It can thus meet the peak voltage and current stresses.

At 16 amps, it has a maximum "on" saturation voltage of about 3 V, a minimum gain of about 5, and a storage time of 1.3 to 4 μs. However, with these limitations, it would have high DC and AC switching losses, have difficulty with flux imbalance (unless the current-mode version of push-pull were used), and would have difficulty operating above 40 kHz because of the long storage times.

A potential MOSFET for such an application is the MTH30N20. This is a 30-A, 200-V device that at 16 A would have only 1.3 V "on" state voltage drop and hence half the DC conduction losses of the preceding bipolar transistor. With its fast switching times it would have quite low switching losses, but this and similar devices can be quite expensive.

For offline converters, the push-pull topology is not very attractive due to the large voltage stress of $2.6V_{dc}$ (see Eq. 2.15). For example, with a 120-V AC line input and ± 10% tolerance, the peak rectified DC voltage is $1.41 \times 1.1 \times 120 = 186$ V. Hence during turn "off" at the top of the leakage spike, Eq. 2.15 gives a peak stress of $2.6 \times 186 = 484$ V.

We must also allow for transients in the supply above the maximum steady-state values. Transients are seldom specified for commercial power supplies, but conservative design practice assumes stress at least 15% above the maximum steady-state value, increasing the maximum stress to 1.15×484 or 557 V.

Input voltage transients in special cases can be even greater, for example, the specifications on military aircraft given by Military Standard 704. Here the nominal voltage is 113 V AC but with a 10-ms transient to 180 V AC, the peak "off" stress from Eq. 1.42 would be $180 \times 1.41 \times 2.6$ or 660 V. Although there are many fast bipolar transistors that can safely sustain voltages as high as 850 V with reverse input bias, clearly it is not good practice to use a topology that subjects the transistors to high voltage transients.

Some topologies subject the transistors to only the normal maximum DC input voltage stress with no leakage spike. These are a better choice for high voltage and "offline" applications, not only because of the lesser voltage stress, but also because the smaller voltage excursion at turn "off" produces less EMI (electromagnetic interference).

2.2.14 Output Filter Design Relations

2.2.14.1 Output Inductor Design

It was pointed out in Section 2.2.4 that in both master and slave outputs, the output inductors should not be permitted to go discontinuous. Remember, the discontinuous-mode situation commences at the critical current where the inductor current ramp of Figure 1.6*b* has dropped to zero. This occurs when the DC current has dropped to half the ramp amplitude *dI* (see Section 1.3.6). Then

$$dI = 2\underline{I_{dc}} = V_L \frac{T_{on}}{L_o} = (V_1 - V_o)\frac{T_{on}}{L_o} \tag{2.19}$$

Figure 2.9 shows the output rectifier circuit for calculation of L_o, C_o. When V_{dc} is at its minimum, N_s will be chosen so that as $V1$ is at its

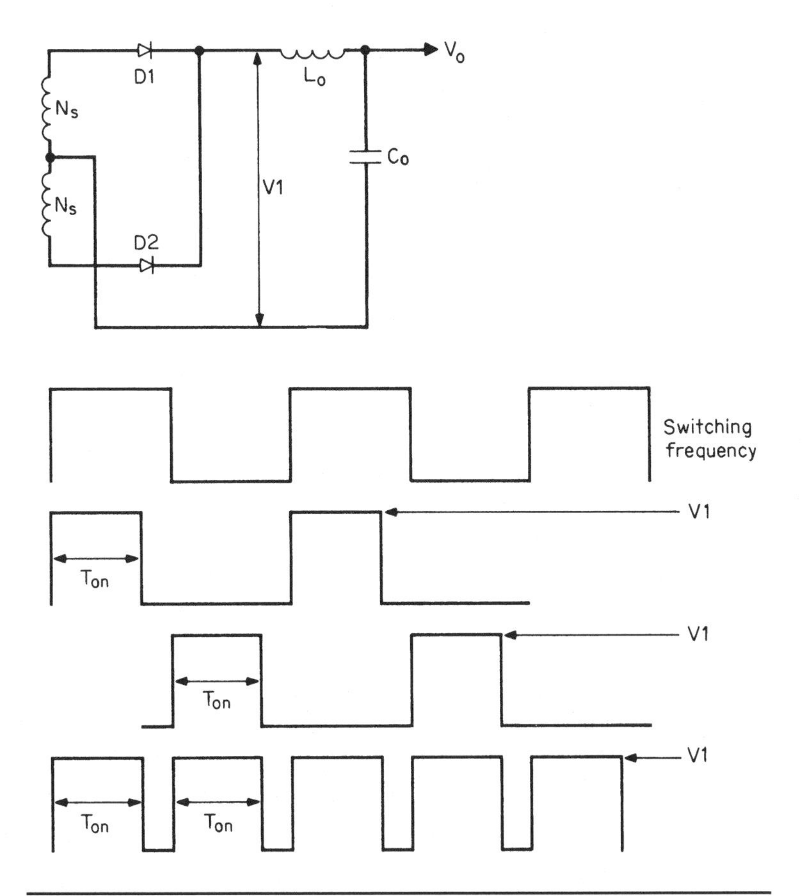

FIGURE 2.9 Output rectifier circuit and waveforms.

minimum, T_{on} will not have to be greater than $0.8T/2$ to yield the specified value of V_o.

But $V_o = V_1(2T_{on}/T)$. Then

$$T_{on} = \frac{V_o T}{2V_1}$$

But N_s will be chosen so that T_{on} will be $0.8T/2$ when V_{dc} and consequently, V_1, are at their minimum so that

$$\overline{T_{on}} = \frac{0.8T}{2} = \frac{V_o T}{2\underline{V_1}} \quad \text{or} \quad \underline{V_1} = 1.25V_o$$

and

$$dI = \frac{(1.25V_o - V_o)(0.8T/2)}{L_o} = 2\underline{I_{dc}} \quad \text{and} \quad L_o = \frac{0.05V_o T}{\underline{I_{dc}}}$$

Then if the minimum current $\underline{I_{dc}}$ is specified as one-tenth the nominal current I_{on} (the usual case),

$$L_o = \frac{0.5V_o T}{I_{on}} \tag{2.20}$$

where L_o is in henries
V_o is in volts
T is in seconds
I_{dc} is minimum output current in amperes
I_{on} is nominal output current in amperes

2.2.14.2 Output Capacitor Design

The output capacitor C_o is selected to meet the maximum output ripple voltage specification. In Section 1.3.7 it was shown that the output ripple is determined almost completely by the magnitude of the ESR (equivalent series resistance, R_o) in the filter capacitor and not by the magnitude of the capacitor itself. The peak-to-peak ripple voltage V_r is very closely equal to

$$V_r = R_o dI \tag{2.21}$$

where dI is the selected peak-to-peak inductor ramp amplitude.

However, it was pointed out that (for aluminum electrolytic capacitors) the product $R_o C_o$ has been observed to be relatively constant over a large range of capacitor magnitudes and voltage ratings.

For aluminum electrolytics, the product $R_o C_o$ ranges between 50 and 80×10^{-6}. Then C_o is selected as

$$\begin{aligned} C_o &= \frac{80 \times 10^{-6}}{R_o} = \frac{80 \times 10^{-6}}{V_r/dI} \\ &= \frac{(80 \times 10^{-6})(dI)}{V_r} \end{aligned} \tag{2.22}$$

where C_o is in farads for dI in amperes (see Eq. 2.19) and V_r is in volts.

2.3 Forward Converter Topology

2.3.1 Basic Operation

A typical triple output forward converter topology is shown in Figure 2.10. This topology is often chosen for output powers under 200 W with DC supply voltages in the range of 60 to 200 V. Below 60 V, the primary input current becomes uncomfortably large at the higher power levels. Above about 250 V, the maximum voltage stress on the transistors becomes uncomfortably large.

Further, it will be shown that above output powers of 200 W or so, the primary input current becomes too large even at the higher supply voltages. We will see this from the following mathematical analysis.

The topology is similar to the push-pull circuit of Figure 2.1, but does not suffer from the latter's major shortcoming of flux imbalance, since it has one rather than two transistors. Compared with the push-pull, at lower power it is more economical in cost and size.

In Figure 2.10 we see a master output V_{om} and two slaves, V_{s1} and V_{s2}. A negative-feedback loop is closed around the master, and controls the $Q1$ "on" time so as to keep V_{om} constant against line and load changes. With an "on" time fixed by the master feedback loop, the slave outputs V_{s1} and V_{s2} are fully regulated against input voltage changes but only partly (about 5 to 8%) against load changes in themselves or in the master. The circuit works as follows.

If we compare the forward converter with the push-pull of Figure 2.1, we see that one of the transistors has been replaced by the diode $D1$. When $Q1$ is turned "on," the start of the primary winding N_p (the dot end) and the start of all secondaries go positive. Current flows into the dot end of N_p. At the same time, all rectifier diodes $D2$ to $D4$ are forward-biased, and current flows out of the starts of all secondaries into the LC filters and the loads. Note that power flows into the loads when the power transistor $Q1$ is turned "on," hence the term *forward converter.* Both the push-pull and buck regulators deliver power to

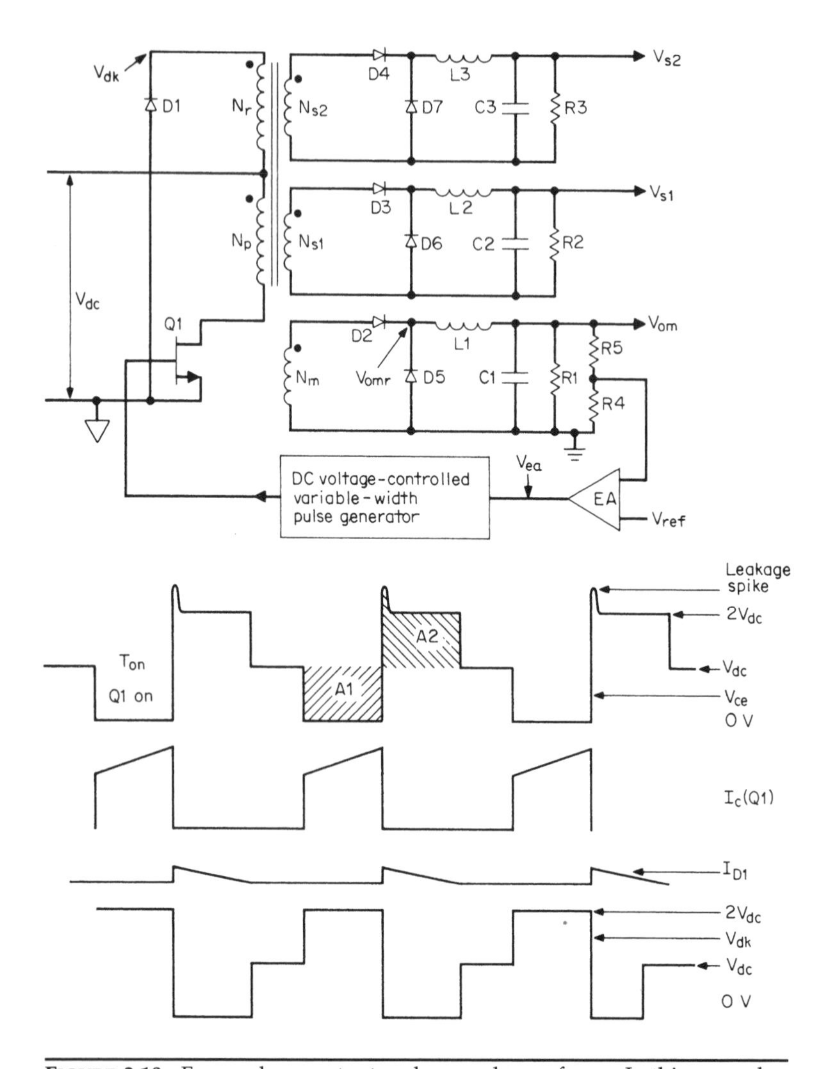

FIGURE 2.10 Forward converter topology and waveforms. In this example the feedback loop is closed around the chosen master output V_{om}, which is regulated against line and load changes. The two semiregulated slaves (V_{s1} and V_{s2}) will be regulated against line changes only.

the loads when the power transistors are "on," so both are forward converters.

In contrast, the boost regulator, the polarity inverter (see Figures 1.10 and 1.14), and the flyback type (which will be discussed in a later chapter) store energy in an inductor or transformer primary when the power transistor is "on" and deliver it to the load when the transistor turns "off." Such energy storage topologies can operate in either the discontinuous or continuous mode. These topologies are fundamentally different from the forward converters and were discussed in Sections 1.4.2 and 1.4.3. They will be taken up again in Chapter 4, which covers the flyback topology.

Consider Figure 2.10: if transistor $Q1$ has an "on" time of T_{on}, the voltage at the master rectifier cathode $D5$ is at a high level for a period of T_{on}. Assuming a 1-V "on" voltage for $Q1$ and a rectifier forward drop of V_{D2}, the high-level voltage V_{omr} is

$$V_{omr} = \left[(V_{dc} - 1)\frac{N_m}{N_p}\right] - V_{D2} \tag{2.23}$$

The circuitry after the rectifier diode cathodes is exactly like that of the buck regulator of Figure 1.4. Diodes $D5$ to $D7$ act like the freewheeling diode $D1$ of that figure. When $Q1$ turns "off," the current established in the magnetizing inductance of $T1$ while $Q1$ was "on" (recall the equivalent circuit of a transformer as in Figure 2.7*c*) reverses the polarity of the voltage across N_p. Now all the starts (dot ends) of primary and secondary windings go negative. Without the "catch" action of diode $D1$, the dot end of N_r would go very far negative; since N_p and N_r usually have equal turns, the no-dot end of N_p would go sufficiently positive to avalanche $Q1$ and destroy it.

However, with the catch action of diode $D1$, the dot end of N_r will be clamped at one diode drop below ground. If there were no leakage inductance in $T1$ (recall again the equivalent circuit of a transformer as in Figure 2.7*c*), the voltage across N_p would equal that across N_r. Assuming that the 1-V forward drop across $D1$ can be neglected, the voltage across N_r and N_p is V_{dc}, and the voltage at the no-dot end of N_p and at the $Q1$ collector is then $2V_{dc}$.

We have seen previously that within one cycle, if a core has moved in one direction on its hysteresis loop, it must be restored to exactly its original position on the loop before it can be allowed to move in the same direction again in the next cycle. Otherwise, after many cycles, the core will "staircase" into saturation. If this is allowed to happen, the core will not be able to support the applied voltage, and the transistor will be destroyed.

Figure 2.10 shows that when $Q1$ is "on" for a time T_{on}, N_p is subjected to volt-second product $V_{dc}T_{on}$ with its dot-end positive, that volt-second product is the area $A1$ in Figure 2.10. By Faraday's law (see Eq. 1.17), that volt-second product causes—say, a positive—flux change $dB = (V_{dc}T_{on}/N_pA_e)10^{-8}$ gauss.

When $Q1$ turns "off," and the magnetizing inductance has reversed the polarity across N_p and kept its no-dot end at $2V_{dc}$ long enough for the volt-second area product $A2$ in Figure 2.10 to equal area $A1$, the core has been restored to its original position on the hysteresis loop, and the next cycle can safely start. We can see that the "reset volt-seconds" has equaled the "set volt-seconds."

When $Q1$ turns "off," the dot ends of all secondaries go negative with respect to their no-dot ends. Current in all output inductors $L1$ to $L3$ will try to decrease. Since current in inductors cannot change instantaneously, the polarity across all inductors reverses in an attempt to maintain the current's constant. The input ends of the inductors try to go far negative, but are caught at one diode drop below output ground by free-wheeling diodes $D5$ to $D7$ (see Figure 2.10), and rectifier diodes $D2$ to $D4$ are reverse-biased. Inductor current now continues to flow in the same direction through the output end, returning through the load, partly through the filter capacitor, up through the free-wheeling diode and back into the inductor.

Voltage at the cathode of the main diode rectifier $D2$ is then as shown in Figure 2.11b. It is high at a level of $[(V_{dc}-1)(N_m/N_p)]-V_{D2}$ for time T_{on}, and for a time $T-T_{on}$ it is one free-wheeling diode ($D5$) drop below ground. The LC filter averages this waveform, and assuming that the forward drop across $D5$ equals that across $D2(=V_d)$, the DC output voltage at V_{om} is

$$V_{om} = \left[\left((V_{dc}-1)\frac{N_m}{N_p}\right) - V_d\right]\frac{T_{on}}{T} \tag{2.24}$$

2.3.2 Design Relations: Output/Input Voltage, "On" Time, Turns Ratios

The negative-feedback loop senses a fraction of V_{om}, compares it with the reference voltage V_{ref}, and varies T_{on} so as to keep V_{om} constant for any changes in V_{dc} or load current.

From Eq. 2.24 it can be seen that as V_{dc} changes, the feedback loop keeps the output constant by keeping the product $V_{dc}T_{on}$ constant. Thus maximum $T_{on}(\overline{T_{on}})$ will occur at minimum specified $V_{dc}(\underline{V_{dc}})$, and Eq. 2.24 can be rewritten for minimum DC input voltage as

$$V_{om} = \left[\left((\underline{V_{dc}}-1)\frac{N_m}{N_p}\right) - V_d\right]\frac{\overline{T_{on}}}{T} \tag{2.25}$$

In relation in Eq. 2.25, a number of design decisions must be made in the proper sequence. First, the minimum DC input voltage V_{dc} is specified. Then the maximum permitted "on" time T_{on}, which occurs at $\underline{V_{dc}}$ (minimum V_{dc}), will be set at 80% of a half period.

This margin is included to ensure (see Figure 2.10) that the area $A2$ can equal $A1$. If the "on" time were permitted to go to a full half period, $A2$ would just barely equal $A1$ at the start of the next full cycle. Then any small increase in "on" time due to storage time changes with

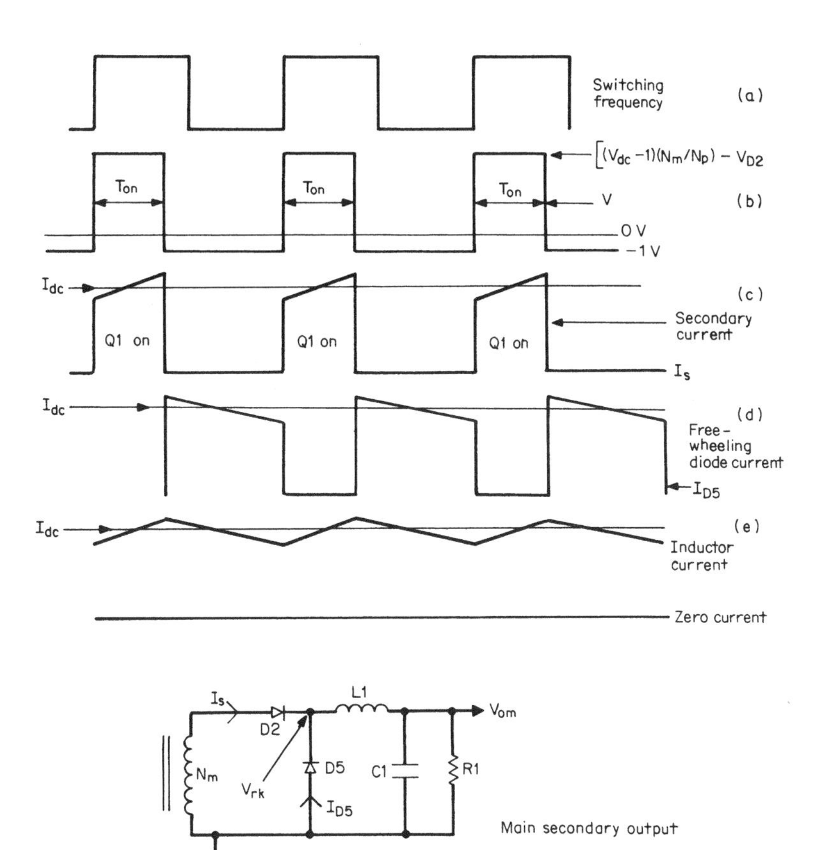

FIGURE 2.11 Critical secondary currents in forward converter. Each secondary has the characteristic ramp-on-a-step waveshape because of the fixed voltage across the output inductor during T_{on} and its constant inductance. Inductor current is the sum of the secondary plus the free-wheeling diode current. It ramps up and down about the DC output current. Primary current is the sum of all the ramp-on-a-step secondary currents, reflected by their turns ratios into the primary. Primary current is therefore also a ramp-on-a-step waveform.

temperature or production spreads would not permit $A2$ to equal $A1$. The core would not be completely reset to its starting point on the hysteresis loop; it would drift up into saturation after a few cycles and destroy the transistor.

Next the number of primary turns N_p is established from Faraday's law (see Eq. 1.17) for V_{dc}, and a certain specified flux change dB in the time T_{on}. Limits on that flux change are similar to those described for the push-pull topology in Section 1.5.9 and will also be discussed later.

Thus, in Eq. 2.25, $\underline{V_{dc}}$, $\overline{T_{on}}$, T, and V_d are specified, and N_p is calculated from Faraday's law. This fixes the number of main secondary turns N_m needed to achieve the required main output voltage V_{om}.

2.3.3 Slave Output Voltages

The slave output filters $L2$, $C2$ and $L3$, $C3$ average the width-modulated rectangular waveforms at their respective rectifier cathodes. The waveform upper levels are $[(V_{dc} - 1)(N_{s1}/N_p)] - V_{d3}$ and $[(V_{dc} - 1)(N_{s2}/N_p)] - V_{d4}$, respectively. The low level voltages are one diode drop below ground. They are at the high level for the same maximum $\overline{T_{on}}$ as is the main secondary, when the input DC input voltage is at the specified minimum $\underline{V_{dc}}$. Again assuming that the forward rectifier and free-wheeling diode drops equal V_d, the slave output voltages at low line $\underline{V_{dc}}$ are

$$V_{s1} = \left[\left((\underline{V_{dc}} - 1)\frac{N_{s1}}{N_p}\right) - V_d\right]\frac{\overline{T_{on}}}{T} \tag{2.26}$$

$$V_{s2} = \left[\left((\underline{V_{dc}} - 1)\frac{N_{s2}}{N_p}\right) - V_d\right]\frac{\overline{T_{on}}}{T} \tag{2.27}$$

By regulating V_{om}, the feedback loop keeps $V_{dc}T_{on}$ constant, but that same product appears in Eqs. 2.26 and 2.27, and hence the slave outputs remain constant as V_{dc} varies.

It can be seen from Eq. 2.24 and Figure 2.14 that the negative-feedback loop keeps the main output constant for either line or load changes by appropriately controlling T_{on} period, so that the sampled output is equal to the reference voltage V_{ref}. This is not so obvious for load changes, since load current does not appear directly in Eq. 2.24, but it does appear indirectly. Load changes will change the "on" voltage of $Q1$ (assumed as 1 V heretofore) and the forward drop in the rectifier diode. Although these changes are small, they will cause small changes in the output voltage that will be sensed and corrected by the error amplifier by making a small change in T_{on}.

Moreover, as can be seen in Eqs. 2.26 and 2.27, any change in T_{on} without a corresponding change in V_{dc} will cause the slave output voltages to change. The slave output voltages also change with changes in their own load currents. As those currents change, the rectifier forward drops also change, causing a change in the peak voltage at the input to the LC averaging filter. So slave output voltages will change the peak voltages to the averaging filters, with no corresponding change in T_{on}. Such changes in the slave output voltages as a result of load changes in the master and slave can be limited to within 5 to 8%.

As discussed in Section 2.2.4, neither master nor slave output inductors can be permitted to go discontinuous at their minimum load currents. This is ensured by choosing appropriately large output inductors, as will be described next.

The number of slave secondary turns N_{s1}, N_{s2} are calculated from Eqs. 2.26 and 2.27, as all parameters there are either specified, or calculated from specified values. The parameters $\underline{V_{dc}}$, T, and V_d are all specified, and $\overline{T_{on}}$ is set at $0.8T/2$ as discussed earlier; N_p is calculated from Faraday's law (see Eq. 1.17) as described earlier.

2.3.4 Secondary Load, Free-Wheeling Diode, and Inductor Currents

Knowledge about the amplitudes and waveshapes of the various output currents is needed to select secondary and output inductor wire sizes and current ratings of the rectifiers and free-wheeling diodes.

As described for the buck regulator in Section 1.3.2, secondary current during the $Q1$ "on" time has the shape of an upward-sloping ramp sitting on a step (see Figure 2.11*c*) because of the constant voltage across the inductor during this time, with its input end positive with respect to the output end.

When $Q1$ turns "off," the input end of the inductor is negative with respect to the output end and inductor current ramps downward. The free-wheeling diode, at the instant of turn "off," picks up exactly the inductor current that had been flowing just prior to turn "off." That diode current then ramps downward (Figure 2.11*d*), as it is in series with the inductor. Inductor current is the sum of the secondary current when $Q1$ is "on" plus the free-wheeling diode current when $Q1$ is "off," and is shown in Figure 2.11*e*. Current at the center of the ramp in any of Figure 2.11*c*, 2.11*d*, or 2.11*e* is equal to the DC output current.

2.3.5 Relations Between Primary Current, Output Power, and Input Voltage

Assume an efficiency of 80% of the total output power from all secondaries to the DC power at the input voltage node. Then $P_o = 0.8P_{in}$ or $P_{in} = 1.25P_o$. Now calculate P_{in} at minimum DC input voltage $\underline{V_{dc}}$,which is $\underline{V_{dc}}$ times the average primary current at minimum DC input.

All secondary currents have the waveshape of a ramp sitting on a step because all secondaries have output inductors. These ramp-on-a-step waveforms have a width of $0.8T/2$ at minimum DC input voltage. All these secondary currents are reflected into the primary by their turns ratios, and hence the primary current pulse is a single ramp-on-a-step waveform of width $0.8T/2$. There is only one such pulse per period (see Figure 2.10) as this is a single-transistor circuit. The duty cycle of this primary pulse is then $(0.8T/2)/T$ or 0.4.

Like the push-pull topology, this ramp-on-a-step can be approximated by an equivalent flat-topped pulse I_{pft} of the same width and whose amplitude is that at the center of the ramp. The average value of this current is then $0.4I_{\text{pft}}$. Then

$$P_{\text{in}} = 1.25P_o = \underline{V_{\text{dc}}}(0.4I_{\text{pft}}) \qquad \text{or} \qquad I_{\text{pft}} = \frac{3.13P_o}{\underline{V_{\text{dc}}}} \tag{2.28}$$

This is a valuable relation. It gives the equivalent peak flat-topped primary current pulse amplitude in terms of what is known at the outset—the minimum DC input voltage and the total output power. This permits an immediate selection of a transistor with adequate current rating and gain if it is a bipolar transistor, or with sufficiently low "on" resistance if it is a MOSFET type.

For a forward converter, Eq. 2.28 shows I_{pft} has twice the amplitude of that required in a push-pull topology (see Eq. 2.9) at the same output power and minimum DC input voltage.

This is obvious, because the push-pull has two pulses of current or power per period as compared with a single pulse in the forward converter. From Eq. 2.25, if the number of secondary turns in the forward converter is chosen large enough, then the maximum "on" time at minimum DC input voltage will not need to be greater than 80% of a half period. Then, as seen in Figure 2.10, the area $A2$ can always equal $A1$ before the start of the next period. The core is then always reset to the same point on its hysteresis loop within one cycle and can never walk up into saturation.

The penalty paid for this guarantee that flux walking cannot occur in the forward converter is that the primary peak current is twice that for a push-pull at the same output power. Despite all the precautions described in Section 2.2.8, however, there is never complete certainty in the push-pull that flux imbalance will not occur under unusual dynamic load or line conditions.

2.3.6 Maximum Off-Voltage Stress in Power Transistor

In the forward converter, with the number of turns on the reset winding N_r equal to that on the power winding N_p, maximum off-voltage stress on the power transistor is twice the maximum DC input voltage plus a leakage inductance spike. These spikes and their origin and minimization have been discussed in Section 2.2.11. Conservative design, even with all precautions to minimize leakage spikes, should assume they may be 30% above twice the maximum DC input voltage. Maximum off-voltage stress is then the same as in the push-pull and is

$$V_{ms} = 1.3(2\overline{V_{dc}}) \tag{2.29}$$

2.3.7 Practical Input Voltage/Output Power Limits

It was stated at the outset in Section 2.3.1 that the practical maximum output power limit for a forward converter whose maximum DC input voltage is under 60 V is 150 to 200 W. This is so because the peak primary current as calculated from Eq. 2.28 becomes excessive, as there is only a single pulse per period as compared with two in the push-pull topology.

Thus consider a 200-W forward converter for the telephone industry in which the specified minimum and maximum input voltages are 38 and 60 V, respectively. Peak primary current from Eq. 2.28 is $I_{pft} = 3.13P_o/\underline{V_{dc}} = 3.13(200)/38 = 16.5$ A, and from Eq. 2.29, maximum off-voltage stress is $\overline{V_{ms}} = 2.6\overline{V_{dc}} = 2.6 \times 60 = 156$ V.

To provide a safety margin, a device with at least a 200-V rating would be used to provide protection against input voltage transients that could drive the DC input above the maximum steady-state value of 60 V.

Transistors with 200-V, 16-A ratings are available, but they all have drawbacks as discussed in Section 2.2.13. Bipolar transistors are slow, and MOSFETs are easily fast enough but expensive. For such a 200-W application, a push-pull version guaranteed to be free from flux imbalance would be preferable; with two pulses of current per period, peak current would be only 8 A. With the resulting lower peak current noise spikes on the ground buses, the radio-frequency interference (RFI) would be considerably lower—a very important consideration for a telephone industry power supply. Such a flux imbalance–free topology is *current mode,* which is discussed later.

The forward converter topology, like the push-pull (discussed in Section 2.2.13), has the same difficulty in coping with maximum voltage stress in an offline converter where the nominal AC input voltage is 120 ± 10%. At high line, the rectified DC input is $1.1 \times 120 \times 1.41 = 186$ V minus 2 V for the rectifier diode drops or 184 V. From Eq. 2.29, the maximum voltage stress on the transistor in the "off" state is $\overline{V_{ms}} = 2.6 \times 184 = 478$ V.

At minimum AC input voltage, the rectified DC output is $\underline{V_{dc}} = (0.9 \times 120 \times 1.41) - 2 = 150$ V, and from Eq. 2.28, the peak primary current is $I_{pft} = 3.13 \times 22/150 = 4.17$ A.

Thus, for a 200-W offline forward converter the problem is more the 478-V maximum voltage stress than the 4.17-A peak primary current stress. As was seen in Section 2.2.13, when a 15% input transient is taken into account, the peak off-voltage stress is 550 V. With a bipolar transistor operating under V_{cev} conditions (reverse input bias of −1 to −5 V at the instant of turn "off"), a voltage stress of even 550 V is not a serious restriction. Many devices have 650- to 850-V V_{cev} ratings and high gain, low "on" drop, and high speed at 4.17 A. But, as discussed in Section 2.2.13, there are preferable topologies, discussed next, that subject the off transistor to only V_{dc} and not twice V_{dc}.

2.3.8 Forward Converter With Unequal Power and Reset Winding Turns

Heretofore it has been assumed that the numbers of turns on the power winding N_p and the reset winding N_r are equal. Some advantages result if N_r is made less or greater than N_p.

The number of primary power turns N_p is always chosen by Faraday's law and will be discussed in Section 2.3.10.2. If N_r is chosen less than N_p, the peak current required for a given output power is less than that calculated from Eq. 2.28, but the maximum $Q1$ off-voltage stress is greater than that calculated from Eq. 2.29. If N_r is chosen larger than N_p, the maximum $Q1$ off-voltage stress is less than that calculated from Eq. 2.29, but the peak primary current for a given output power is greater than that calculated from Eq. 2.28. This can be seen from Figure 2.12 as follows. When $Q1$ turns "off," polarities across N_p and N_r reverse; the dot end of N_r goes negative and is caught at ground by catch diode $D1$. Transformer $T1$ is now an autotransformer. There is a voltage V_{dc} across N_r and hence a voltage $N_p/N_r(V_{dc})$ across N_p. The core is set by the volt-second product by $V_{dc}T_{on}$ during the "on" time and must be reset to its original place on the hysteresis loop by an equal volt-second product. That reset volt-second product is $N_p/N_r(V_{dc})T_r$.

When N_r equals N_p, the reset voltage equals the set voltage, and the reset time is equal to the set time (area $A1$ = area $A2$) as seen in Figure 2.12*b*. For $N_r = N_p$, the maximum $Q1$ "on" time that occurs at minimum DC input voltage is chosen as $0.8T/2$ to ensure that the core is reset before the start of the next period; $T_{on} + T_r$ is then $0.8T$.

Now if N_r is less than N_p, the resetting voltage is larger than V_{dc} and consequently T_r can be smaller (area $A3$ = area $A4$) as shown in Figure 2.12*c*. With a shorter T_r, T_{on} can be longer than $0.8T/2$, and $T_{on} + T_r$ can still be $0.8T$ so that the core is reset before the start of the next period. With a longer T_{on}, the peak current is smaller for the same average current and the same average output power. Thus in Figure 2.12*c*, a

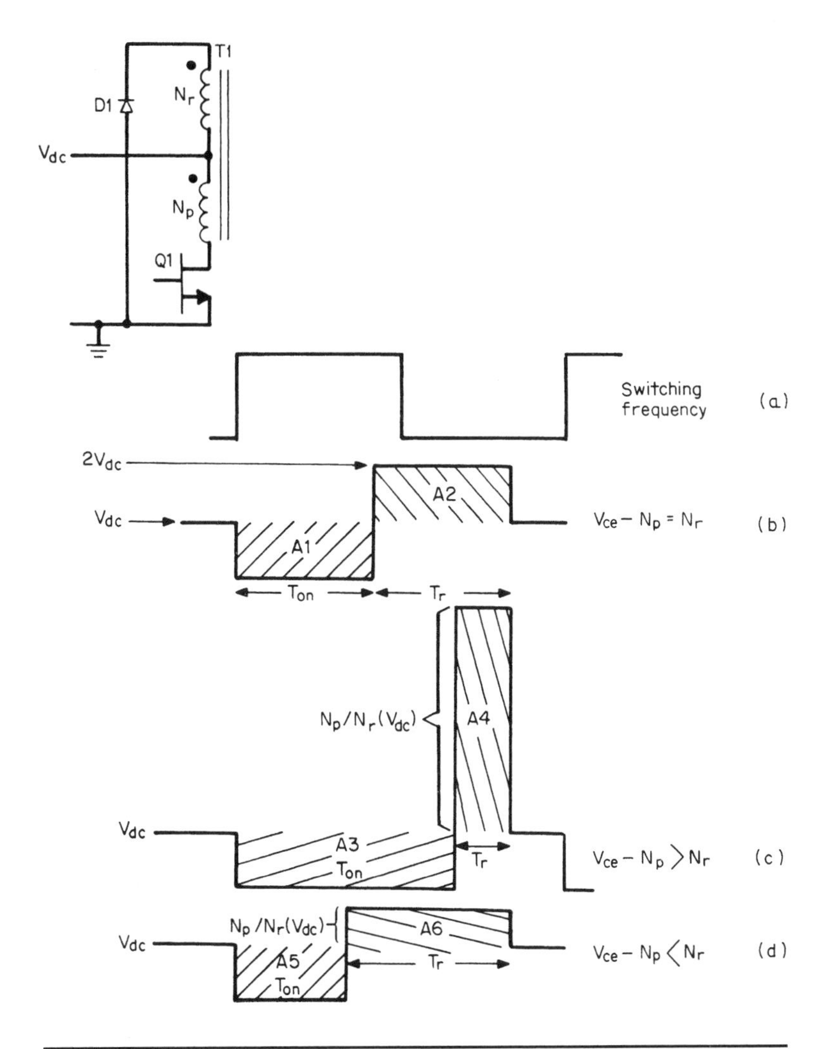

FIGURE 2.12 Forward converter–collector-to-emitter voltages for three N_p to N_r ratios. Note in all cases that reset volt-second product equals set volt-second product. (*a*) Switching frequency, (*b*) $N_p = N_r$, (*c*) $N_p > N_r$, (*d*) $N_p < N_r$.

smaller peak current stress has been traded for a longer voltage stress than in Figure 2.12*b*.

With N_r greater than N_p, the reset voltage is less than V_{dc}. Then if $T_{on} + T_r$ is still to equal $0.8T$, and the reset volt-seconds is to equal the set volt-seconds (area *A*5 = area *A*6 in Figure 2.12*d*), T_r must be longer and T_{on} must be shorter than $0.8T/2$, as the reset voltage is less than the set voltage. With T_{on} less than $0.8T/2$, the peak current must be higher for the same average current. Thus, in Figure 2.12*d*, a lesser voltage stress has been achieved at the cost of a higher peak current for the same output power as in Figure 2.12*b*. This can be seen quantitatively as

$$\text{Set } T_{on} + T_r = 0.8T; \quad \text{reset voltage} = V_r = \frac{N_p}{N_r} V_{dc} \tag{2.30}$$

For "on" volt-seconds equal to reset volt-seconds,

$$V_{dc} T_{on} = \frac{N_p}{N_r} V_{dc} T_r \tag{2.31}$$

Combining Eqs. 2.30 and 2.31,

$$\overline{T_{on}} = \frac{0.8T}{1 + N_r/N_p} \tag{2.32}$$

For 80% efficiency $P_{in} = 1.25P_o$ and P_{in} at $\underline{V_{dc}} = V_{dc}(I_{av}) = \underline{V_{dc}}\, I_{pft}(\overline{T_{on}})/T$ or $I_{pft} = 1.25(P_o/\underline{V_{dc}})(T/T_{on})$. Then from Eq. 2.32

$$I_{pft} = 1.56 \left(\frac{P_o}{\underline{V_{dc}}} \right) (1 + N_r/N_p) \tag{2.33}$$

and the maximum $Q1$ off-voltage stress $\overline{V_{ms}}$—exclusive of the leakage spike—is the maximum DC input voltage V_{dc} plus the reset voltage (voltage across N_p when the dot end of N_r is at ground). Thus

$$\overline{V_{ms}} = \overline{V_{dc}} + \frac{N_p}{N_r}(\overline{V_{dc}}) = \overline{V_{dc}}(1 + N_p/N_r) \tag{2.34}$$

Values of I_{pft} and V_{ms} calculated from Eqs. 2.33 and 2.34 are

N_r/N_p	I_{pft} **(from Eq. 2.33)**	V_{ms} **(from Eq. 2.34)**
0.6	$2.50(P_o/\underline{V_{dc}})$	$2.67\ \overline{V_{dc}}$ + leakage spike
0.8	$2.81(P_o/\underline{V_{dc}})$	$2.25\ \underline{V_{dc}}$ + ""
1.0	$3.12(P_o/\underline{V_{dc}})$	$2.00\ \underline{V_{dc}}$ + ""
1.2	$3.43(P_o/\underline{V_{dc}})$	$1.83\ \underline{V_{dc}}$ + ""
1.4	$3.74(P_o/\underline{V_{dc}})$	$1.71\ \underline{V_{dc}}$ + ""
1.6	$4.06(P_o/\underline{V_{dc}})$	$1.62\ \underline{V_{dc}}$ + ""

2.3.9 Forward Converter Magnetics

2.3.9.1 First-Quadrant Operation Only

The transformer core in the forward converter operates in the first quadrant of the hysteresis loop only. This can be seen in Figure 2.10.

When $Q1$ is "on," the dot end of $T1$ is positive with respect to the no-dot end, and the core is driven, say, in a positive direction on the hysteresis loop, and the magnetizing current ramps up linearly in the magnetizing inductance.

When $Q1$ turns "off," stored current in the magnetizing inductance reverses the polarity of voltages on all windings. The dot end of N_r goes negative until it is caught one diode drop below ground by catch diode $D1$. Now the magnetizing current that is stored in the magnetic core continues to flow. It simply transfers from N_p, where it had ramped upward during the $Q1$ "on" time, into N_r where it ramps back to zero during the "off" time. It flows out of the no-dot end of N_r into the positive end of the supply voltage V_{dc}, out of the negative end of V_{dc}, through $D1$, and back into N_r.

Since the dot end of N_r is positive with respect to its no-dot end during the $Q1$ "off" time, the magnetizing current I_d ramps linearly downward, as can be seen in Figure 2.10. When it has ramped down to zero (at the end of area $A2$ in Figure 2.10), there is no longer any stored energy in the magnetizing inductance and nothing to hold the dot end of N_r below the $D1$ cathode. The voltage at the dot end of N_r starts rising toward that at the $D1$ cathode. The voltage at the dot end of N_r starts rising toward V_{dc}, and that at the no-dot end of N_p ($Q1$ collector) starts falling from $2V_{dc}$ back down toward V_{dc}.

Thus operation on the hysteresis loop is centered about half the peak magnetizing current ($V_{dc}T_{on}/2L_m$). Nothing ever reverses the direction of the magnetizing current—it simply builds up linearly to a peak and relaxes back down linearly to zero.

This first-quadrant operation has some favorable and some unfavorable consequences. First, compared with a push-pull circuit, it halves the available output power from a given core. This can be seen from Faraday's law (see Eq. 1.17), which fixes the number of turns on the primary.

By solving Faraday's law for the number of primary turns, we get $N_p = E\,dt/A_e dB \times 10^{-8}$. If dB in the forward converter is limited to an excursion from zero to some B_{max}, instead of from $-B_{max}$ to $+B_{max}$ as in a push-pull topology, the number of primary turns for the forward converter will be twice that in each half primary of a push-pull operating from the same V_{dc}. Although the push-pull has two half primaries, each of which must support the same volt-second product as the forward converter primary, the push-pull provides two power pulses per period as compared with one for the forward converter. The end result is that a core used in a forward converter can process only half the output power available from the same core in a push-pull configuration.

However, the push-pull core at twice the output power will run somewhat warmer, as its flux excursion is twice that of the forward converter. Since core losses are proportional to the area of the hysteresis loop traversed, the push-pull core losses are twice that of the forward converter.

Yet total copper losses in both half primaries of a push-pull are no greater than that of a forward converter of half the output power, because the rms current in each push-pull half primary is equal to that in the forward converter primary. Since the number of turns in each push-pull half primary is half that of the forward converter primary of half the output power, they also have half the resistance. Thus total copper loss in a forward converter is equal to the total loss of the two half primaries in a push-pull of twice the output power.

2.3.9.2 Core Gapping in a Forward Converter

In Figure 2.3, we see the hysteresis loop of a ferrite core with no air gap. We see that at zero magnetizing force (0 Oe) there is a residual magnetic flux density of about $\pm$ 1000 G. This residual flux is referred to as *remanence.*

In a forward converter, if the core started at 0 Oe and hence at 1000 G, the maximum flux change in dB possible before the core is driven up into the curved part of the hysteresis loop is about 1000 G. It is desirable to stay off the curved part of the hysteresis loop, and hence the forward converter core with no air gap is restricted to a maximum dB of 1000 G. As shown earlier, the number of primary turns is inversely proportional to dB. Such a relatively small dB requires a relatively large number of primary turns. A large number of primary turns requires small wire size and hence decreases the current and power available from the transformer.

By introducing an air gap in the core, the hysteresis loop is tilted as shown in Figure 2.5, and magnetic remanence is reduced significantly. The hysteresis loop tilts over but still crosses the H (*coercive force*) axis with zero flux density at the same point. Coercive force for ferrites is seen to be about 0.2 Oe in Figure 2.3. An air gap of 2 to 4 mils will reduce remanence to about 200 G for most cores used at 200 to 500 W of output power. With remanence of 200 G, the dB before the core enters the curved part of the hysteresis loop is now about 1800 G, and fewer turns are permissible.

However, a penalty is paid in introducing an air gap. Figure 2.5 shows the slope of the hysteresis loop tilted over. The slope is dB/dH or core permeability, which has been decreased by adding the gap. Decreasing permeability decreases magnetizing inductance and increases magnetizing current ($I_m = V_{dc}T_{on}/L_m$). Magnetizing current contributes no output power to the load; it simply moves the operating point of the core around the hysteresis loop and contributes significant copper loss if it exceeds 10% of the primary load current.

2.3.9.3 Magnetizing Inductance with Gapped Core

Magnetizing inductance with a gapped core can be calculated as follows. Voltage across the magnetizing inductance is $L_m dI_m/dt$ and from Faraday's law:

$$V_{dc} = \frac{L_m dI_m}{dt} = \frac{N_p A_e dB}{dt} 10^{-8} \quad \text{or} \quad L_m = \frac{N_p A_e dB}{dI_m} 10^{-8} \tag{2.35}$$

where L_m = magnetizing inductance, H
N_p = number of primary turns
A_e = core area, cm^2
dB = core flux change, G
dI_m = change in magnetizing current, A

A fundamental law in magnetics is Ampere's law:

$$\int H \cdot dl = 0.4\pi NI$$

This states that if a line is drawn encircling a number of ampere turns NI, the dot product $H \cdot dl$ along that line is equal to $0.4\pi NI$. If the line is taken through the core parallel to the magnetic flux lines and across the gap, since H is uniform at a value H_i within the core and uniform at a value H_a within the gap, then

$$H_i l_i + H_a l_a = 0.4\pi N I_m \tag{2.36}$$

where H_i = magnetic field intensity in iron (ferrite), Oe
l_i = length of iron path, cm
H_a = magnetic field intensity in air gap, Oe
l_a = length of air gap, cm
I_m = magnetizing current, A

However, $H_i = B_i/u$, where B_i is the magnetic flux density in iron and u is the iron permeability; $H_a = B_a$ as the permeability of air is 1; and $B_a = B_i$ (flux density in iron = flux density in air) if fringing flux around the air gap is ignored. Then Eq. 2.36 can be written as

$$\frac{B_i}{u} l_i + B_i l_a = 0.4\pi N_p I_m \quad \text{or} \quad B_i = \frac{0.4\pi N I_m}{l_a + l_i/u} \tag{2.37}$$

Then $dB/dI_m = 0.4\pi N/(l_a + l_i/u)$, and substituting this into Eq. 2.35:

$$L_m = \frac{0.4\pi (N_p)^2 A_e \times 10^{-9}}{l_a + l_i/u} \tag{2.38}$$

Thus, introducing an air gap of length l_a to a core of iron path length l_i reduces the magnetizing inductance in the ratio of

$$\frac{L_{m\,(\text{with gap})}}{L_{m\,(\text{without gap})}} = \frac{l_i/u}{l_a + l_i/u} \tag{2.39}$$

It is instructive to consider a specific example. Take an international standard core such as the Ferroxcube 783E608-3C8. It has a magnetic path length of 9.7 cm and an effective permeability of 2300. Then if a 4-mil (= 0.0102-cm) gap were introduced into the magnetic path, from Eq. 2.39:

$$\begin{aligned} L_{m\,(\text{with gap})} &= \frac{9.7/2300}{0.0102 + 9.7/2300} L_{m\,(\text{without gap})} \\ &= 0.29 L_{m\,(\text{without gap})} \end{aligned}$$

A useful way of looking at a gapped core is to examine the denominator in Eq. 2.38. In most cases, u is so high that the term l_i/u is small compared with the air gap l_a, and the inductance is determined primarily by the length of the air gap.

2.3.10 Power Transformer Design Relations

2.3.10.1 Core Selection

As discussed in Section 2.2.9.1 on core selection for a push-pull transformer, the amount of power available from a core for a forward converter transformer is related to the same parameters—peak flux density, core iron and window areas, frequency, and coil current density in circular mils per rms ampere.

In Chapter 7, an equation will be derived giving the amount of available output power as a function of these parameters. This equation will be converted to a chart that permits selection of core size and operating frequency at a glance.

For the present, it is assumed that a core has been selected and that its iron and window areas are known.

2.3.10.2 Primary Turns Calculation

The number of primary turns is calculated from Faraday's law as given in Eq. 2.7. From Section 2.3.9.2, we see that in the forward converter with a gapped core, flux density moves from about 200 G to some higher value B_{max}.

In the push-pull topology as discussed in Section 2.2.9.4, this peak value will be set at 1600 G (for ferrites at low frequencies, where core losses are not a limiting factor). This avoids the problem of a much larger and more dangerous flux swing due to rapid changes in DC input voltage or load currents. Such rapid changes are not immediately compensated because the limited error-amplifier bandwidth can't correct the power transistor "on" time fast enough.

During this error-amplifier delay, the peak flux density can exceed the calculated normal steady-state value for a number of cycles. This can be tolerated if the normal peak flux density in the absence of a line or load transient is set to the low value of 1600 G. As discussed earlier, the excursion from approximately zero to 1600 G will take place in 80% of a half period to ensure that the core can be reset before the start of the next period (see Figure 2.12*b*).

Thus, the number of primary turns is set by Faraday's law at

$$N_p = \frac{(\underline{V_{dc}} - 1)(0.8T/2) \times 10^{+8}}{A_e dB} \tag{2.40}$$

where $\underline{V_{dc}}$ = minimum DC input, V
T = operating period, s
A_e = iron area, cm^2
dB = change in flux density, G

2.3.10.3 Secondary Turns Calculation

Secondary turns are calculated from Eqs. 2.25 to 2.27. In those relations, all values except the secondary turns are specified or already calculated. Thus (see Figure 2.10):

$$\underline{V_{dc}} = \text{minimum DC input, V}$$
$$T_{on} = \text{maximum "on" time, s}(= 0.8T/2)$$
$$N_m, N_{s1}, N_{s2} = \text{numbers of main and slave turns}$$
$$N_p = \text{number of primary turns}$$
$$V_d = \text{rectifier forward drop}$$

If the main output produces 5 V at high current as is often the case, a Schottky diode with forward drop of about 0.5 V is typically used. The slaves usually have higher output voltages that require the use of fast-recovery diodes with higher reverse-voltage ratings. Such diodes have forward drops of about 1.0 V over a large range of current.

2.3.10.4 Primary rms Current and Wire Size Selection

Primary equivalent flat-topped current is given by Eq. 2.28. That current flows for a maximum of 80% of a half period per period, so its maximum duty cycle is 0.4. Recalling that the rms value of a flat-topped pulse of amplitude I_p is $I_{\text{rms}} = I_p\sqrt{T_{\text{on}}/T}$, the rms primary current is

$$I_{\text{rms (primary)}} = \frac{3.12P_o}{V_{\text{dc}}}\sqrt{0.4} = \frac{1.97P_o}{V_{\text{dc}}} \tag{2.41}$$

If the wire size is chosen on the basis of 500 circular mils per rms ampere, the required number of circular mils is

$$\text{Circular mils needed} = \frac{500 \times 1.97P_o}{V_{\text{dc}}} = \frac{985P_o}{V_{\text{dc}}} \tag{2.42}$$

2.3.10.5 Secondary rms Current and Wire Size Selection

It is seen in Figure 2.11 that the secondary current has the characteristic shape of a ramp on a step. The pulse amplitude at the center of the ramp is equal to the average DC output current. Thus, the equivalent flat-topped secondary current pulse at V_{dc} (when its width is a maximum) has amplitude I_{dc}, width $0.8T/2$, and duty cycle $(0.8T/2)/T$ or 0.4. Then

$$I_{\text{rms(secondary)}} = I_{\text{dc}}\sqrt{0.4} = 0.632I_{\text{dc}} \tag{2.43}$$

and at 500 circular mils per rms ampere, the required number of circular mils for each secondary is

$$\text{Circular mils needed} = 500 \times 0.632I_{\text{dc}} = 316I_{\text{dc}} \tag{2.44}$$

2.3.10.6 Reset Winding rms Current and Wire Size Selection

The reset winding carries only magnetizing current, as can be seen by the dots in Figure 2.10. When $Q1$ is "on," diode $D1$ is reverse-biased, and no current flows in the reset winding. But magnetizing current builds up linearly in the power winding N_p. When $Q1$ turns "off," that magnetizing current must continue to flow. When $Q1$ current ceases, the current in the magnetizing inductance reverses all winding voltage polarities. When $D1$ clamps the dot end of N_r to ground, the magnetizing current transfers from N_p to N_r and continues flowing through the DC input voltage source V_{dc}, through $D1$, and back into N_r. Since the no-dot end of N_r is positive with respect to the dot end, the magnetizing current ramps downward to zero as seen in Figure 2.10.

The waveshape of this N_r current is the same as that of the magnetizing current that ramped upward when $Q1$ was "on," but it is reversed from left to right. Thus the peak of this triangle of current is $I_{p(\text{magnetizing})} = V_{\text{dc}}\overline{T_{\text{on}}}/L_{\text{mg}}$, where L_{mg} is the magnetizing inductance with an air gap as calculated from Eq. 2.39. The inductance without the gap is calculated from the ferrite catalog value of A_l, the inductance per 1000 turns. Since inductance is proportional to the square of the number of turns, inductance for n turns is $L_n = A_l(n/1000)^2$. The duration of this current triangle is $0.8T/2$ (the time required for the core to reset), and it comes at a duty cycle of 0.4.

It is known that the rms value of a repeating triangle waveform (no spacing between successive triangles) of peak amplitude I_p is $I_{\text{rms}} = I_p\sqrt{3}$. But this triangle comes at a duty cycle of 0.4, and hence its rms value is

$$I_{\text{rms}} = \frac{V_{\text{dc}}\overline{T_{\text{on}}}}{L_{\text{mg}}}\frac{\sqrt{0.4}}{\sqrt{3}}$$

$$= 0.365\frac{V_{\text{dc}}\overline{T_{\text{on}}}}{L_{\text{mg}}}$$

and at 500 circular mils per rms ampere, the required number of circular mils for the reset winding is

$$\text{Circular mils required} = 500 \times 0.365\frac{V_{\text{dc}}\overline{T_{\text{on}}}}{L_{\text{mg}}} \tag{2.45}$$

Most frequently, the magnetizing current is so small that the reset winding wire can be No. 30 AWG or smaller.

2.3.11 Output Filter Design Relations

The output filters $L1C1$, $L2C2$, and $L3C3$ average the voltage waveform at the rectifier cathodes. The inductor is selected to operate in continuous mode (see Section 1.3.6) at the minimum DC output current. The capacitor is selected to yield a specified minimum output ripple voltage.

2.3.11.1 Output Inductor Design

Recall from Section 1.3.6 that discontinuous mode condition occurs when the inductor current ramp drops to zero (see Figure 2.10). Since the DC output current is the value at the center of the ramp, discontinuous mode occurs at a minimum current I_{dc} equal to half the ramp amplitude dI as can be seen in Figure 2.10.

Now referring to Figure 2.11,

$$dI = 2\underline{I_{dc}} = \frac{(\underline{V_{rk}} - V_o)\overline{T_{on}}}{L1} \quad \text{or} \quad L1 = \frac{(\underline{V_{rk}} - V_o(\overline{T_{on}})}{2\underline{I_{dc}}}$$

But $V_o = \underline{V_{rk}}\overline{T_{on}}/T$. Then

$$L1 = \left(\frac{V_o T}{\overline{T_{on}}} - V_o\right)\frac{\overline{T_{on}}}{2\underline{I_{dc}}} = \frac{V_o(T/\overline{T_{on}} - 1)/\overline{T_{on}}}{2\underline{I_{dc}}}$$

But $\overline{T_{on}} = 0.8T/2$. Then

$$L1 = \frac{0.3V_o T}{\underline{I_{dc}}} \tag{2.46}$$

and if the minimum DC current $\underline{I_{dc}}$ is one-tenth the nominal output current I_{on}, then

$$L1 = \frac{3V_o T}{I_{on}} \tag{2.47}$$

2.3.11.2 Output Capacitor Design

It was seen in Section 1.3.7 that the output ripple is almost completely determined by the equivalent series resistance R_o of the filter capacitor. The peak-to-peak ripple amplitude is $V_{or} = R_o\, dI$, where dI is the peak-to-peak ripple current amplitude chosen by the selection of the ripple inductor as discussed earlier. Assuming that the average value of $R_o C_o$ for aluminum electrolytic capacitors over a large range of voltage and capacitance ratings is given by $R_o C_o = 65 \times 10^{-6}$ as in Section 1.3.7, then

$$C_o = 65 \times 10^{-6}/R_o = 65 \times 10^{-6}\frac{dI}{V_{or}} \tag{2.48}$$

where dI is in amperes and V_{or} is in volts for C_o in farads.

2.4 Double-Ended Forward Converter Topology

2.4.1 Basic Operation

Double-ended forward converter topology is shown in Figure 2.13. Although it has two transistors rather than one compared with the single-ended forward converter of Figure 2.10, it has a very significant

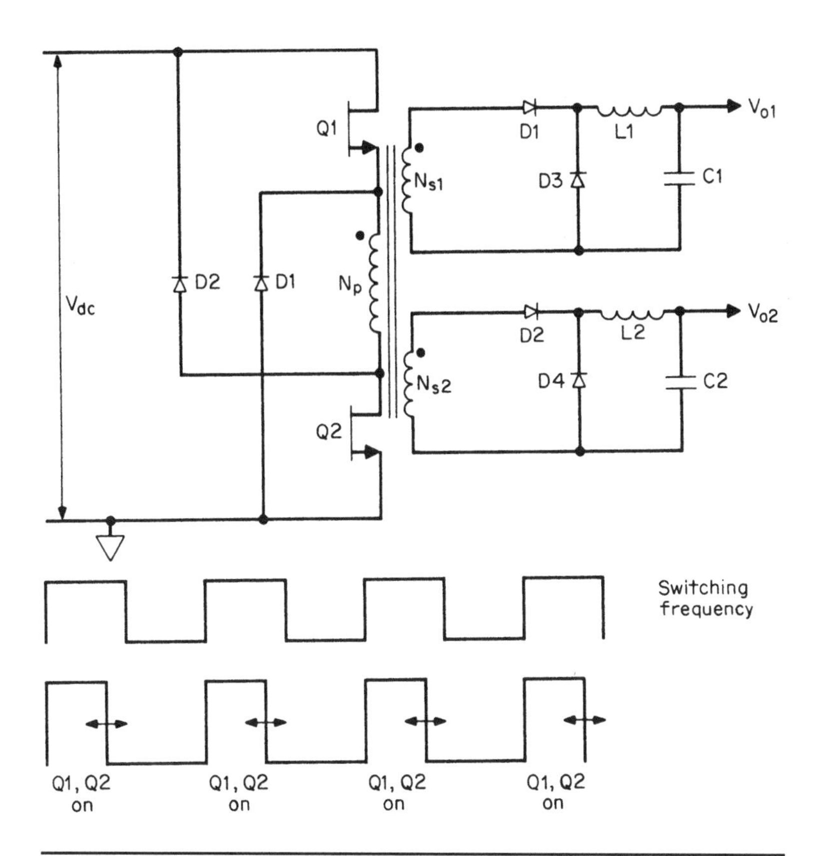

FIGURE 2.13 Double-ended forward converter. Transistors $Q1$ and $Q2$ are turned on and off simultaneously. Diodes $D1$ and $D2$ keep the maximum off-voltage stress on $Q1$, $Q2$ at a maximum of V_{dc} as contrasted with $2V_{dc}$ plus a leakage spike for the single-ended forward converter of Figure 2.10.

advantage. In the "off" state, both transistors are subjected to only the DC input voltage rather than twice that, as in the single-ended converter. Further, at turn "off," there is no leakage inductance spike.

It was pointed out in Section 2.3.7 that the off-voltage stress in the single-ended forward converter operating from a nominal 120-V AC line can be as high as 550 V when there is a 15% transient above a 10% steady-state high line and a 30% leakage spike.

Although a number of bipolar transistors have V_{cev} ratings up to 650 and even 850 V that can take that stress, it is far more reliable to use a double-ended forward converter with half the off-voltage stress. Reliability is of overriding importance in a power supply design, and in any weighing of reliability versus initial cost, the best and—in the long run—least expensive choice is reliability.

Further, for power supplies to be used in the European market where the AC voltage is 220 V (rectified DC voltage is nominally about 308 V), the single-ended forward converter is not usable at all because of the excessive voltage stress on the off transistor (see Eq. 2.29). The double-ended forward converter, the half bridge, and the full bridge (to be discussed in Chapter 3) are the only choices for equipment to be used in the European market.

The double-ended forward converter works as follows. In Figure 2.13, $Q1$ and $Q2$ are in series with the transformer primary. These transistors are turned on and off simultaneously. When they are "on," all primary and secondary dot ends are positive, and power is delivered to the loads. When they turn "off," current stored in the $T1$ magnetizing inductance reverses the voltage polarity of all windings. The negative-going dot end of N_p is caught at ground by diode $D1$, and the positive-going no-dot end of N_p is caught at V_{dc} by diode $D2$.

Thus the emitter of $Q1$ can never be more than V_{dc} below its collector, and the collector of $Q2$ can never be more than V_{dc} above its emitter. Leakage inductance spikes are clamped so that the maximum voltage stress on either transistor can never be more than the maximum DC input voltage.

The further significant advantage is that there is no leakage inductance energy to be dissipated. Any energy stored in the leakage inductance is not lost by dissipation in some resistive element or in the power transistors. Instead, energy stored in the leakage inductance during the "on" time is fed back into V_{dc} via $D1$ and $D2$ when the transistors turn "off." The leakage inductance current flows out of the no-dot end of N_p, through $D2$, into the positive end of V_{dc}, out of its negative end, and up through $D1$ back into the dot end of N_p.

Examination of Figure 2.13 reveals that the core is always reset in a time equal to the "on" time. The reverse polarity voltage across N_p when the transistors are "off" is equal to the forward polarity voltage across it when the transistors are "on." Thus the core will always be fully reset with a 20% safety margin before the start of a succeeding half cycle if the maximum "on" time is no greater than 80% of a half period. This is accomplished by choosing secondary turns so that the peak secondary voltage at minimum V_{dc} times the maximum duty cycle of 0.4 equals the desired output voltage (see Eq. 2.25).

2.4.1.1 Practical Output Power Limits

It should be noted that this topology still yields only one power pulse per period, just like the single-ended forward converter. Thus the power available from a specific core is pretty much the same for either the single- or double-ended configuration. As noted in Section 2.3.10.6, the reset winding in the single-ended circuit carries only magnetizing current during the power transistor "off" time. Since that current is small, the reset winding can be wound with very small wire. Thus, the absence of a reset winding in the double-ended circuit does not permit significantly larger power winding wire size and output power from a given core.

Because the maximum off transistor voltage stress cannot be greater than the maximum DC input voltage, however, the 200-W practical power limit for the single-ended forward converter discussed in Section 2.3.7 does not hold for the double-ended forward converter. With the reduced voltage stress, output powers of 400 to 500 W are obtainable, and transistors with the required voltage and current capability and adequate gain are available at low price.

Consider a double-ended forward converter operating from a nominal 120-V AC line with ± 10% tolerance and ± 15% allowance for transients on top of that. The maximum rectified DC voltage is 1.41 × 120 × 1.1 × 1.15 = 214 V, and the minimum rectified DC voltage is 1.41 × 120 ÷ 1.1÷1.15 = 134 V, and equivalent flat-topped primary current from Eq. 2.28 is $I_{pft} = 3.13P_o/\underline{V_{dc}}$, and for P_o = 400 W, $I_{pft} = 9.6$ A. This requirement can be satisfied quite easily, because both bipolar and MOSFET transistors with adequately high gain are available at low cost.

A double-ended forward converter with a voltage doubler from the 120-V AC line would be a better alternative (see Figure 3.1). This would double the voltage stress to 428 V but would halve the peak current to 4.8 A. With 4.8 A of primary current, RFI problems would be less severe. A bipolar transistor with a 400-V V_{ceo} rating could tolerate 428 V easily, with –1- to –5-V reverse bias at the instant of turn "off" (V_{cev} rating).

2.4.2 Design Relations and Transformer Design

2.4.2.1 Core Selection—Primary Turns and Wire Size

The transformer design for the double-ended forward converter proceeds exactly as for the single-ended converter. A core is selected from the aforementioned selection charts (to be presented in Chapter 7 on magnetics) for the required output power and operating frequency.

The number of primary turns is chosen from Faraday's law as in Eq. 2.40. There the minimum primary voltage is ($\underline{V_{dc}}$ – 2) as there are two transistors rather than one in series with the primary—but the transistor drops are insignificant since $\underline{V_{dc}}$ is usually 134 V (120 V AC). Maximum "on" time should be set at $0.8T/2$ and dB at 1600 G up to 50 kHz, or higher if not limited by core losses.

As mentioned for frequencies from 100 to 300 kHz, peak flux density may have to be set from about 1400 to 800 G, as core losses increase with frequency. But the exact peak flux density chosen depends on whether the newer, lower-loss materials are available. It also depends to some extent on transformer size—smaller cores can generally operate at higher flux density, because they have a larger ratio of radiating surface area to volume and hence can get rid of the heat they generate (which is proportional to volume) more easily.

Since there is only one current or power pulse per period, as in the single-ended forward converter, the primary current for a given output power and minimum DC input voltage is given by Eq. 2.28, and the primary wire size is chosen from Eq. 2.42.

2.4.2.2 Secondary Turns and Wire Size

Secondary turns are chosen exactly as in Sections 2.3.2 and 2.3.3 from Eqs. 2.25 to 2.27. Wire sizes are calculated as in Section 2.3.10.5 from Eq. 2.44.

2.4.2.3 Output Filter Design

The output inductor and capacitor magnitudes are calculated exactly as in Section 2.3.11 from Eqs. 2.46 to 2.48.

2.5 Interleaved Forward Converter Topology

2.5.1 Basic Operation—Merits, Drawbacks, and Output Power Limits

This topology is simply two identical single-ended forward converters operating on alternate half cycles with their secondary currents adding through rectifying "on" diodes. The topology is shown in Figure 2.14.

The advantage, of course, is that now there are two power pulses per period, as seen in Figure 2.14, reducing the ripple current; also each converter supplies only half the total output power.

Equivalent flat-topped peak transistor current is derived from Eq. 2.28 as $I_{pft} = 3.13\, P_{ot}/2\underline{V_{dc}}$ where P_{ot} is the total output power. This transistor current is half that of a single forward converter at the same total output power. Thus the expense of two transistors is offset by the lower peak current rating and lower cost than that of the higher current rating device.

Looking at it another way, two transistors of the same current rating used at the same peak current as one single-ended converter at a given output power in an interleaved converter would yield twice the output power of the single converter.

Also, since the intensity of EMI generated is proportional to the peak current, not to the number of current pulses, an interleaved converter of the same total output power as a single forward converter will generate less EMI.

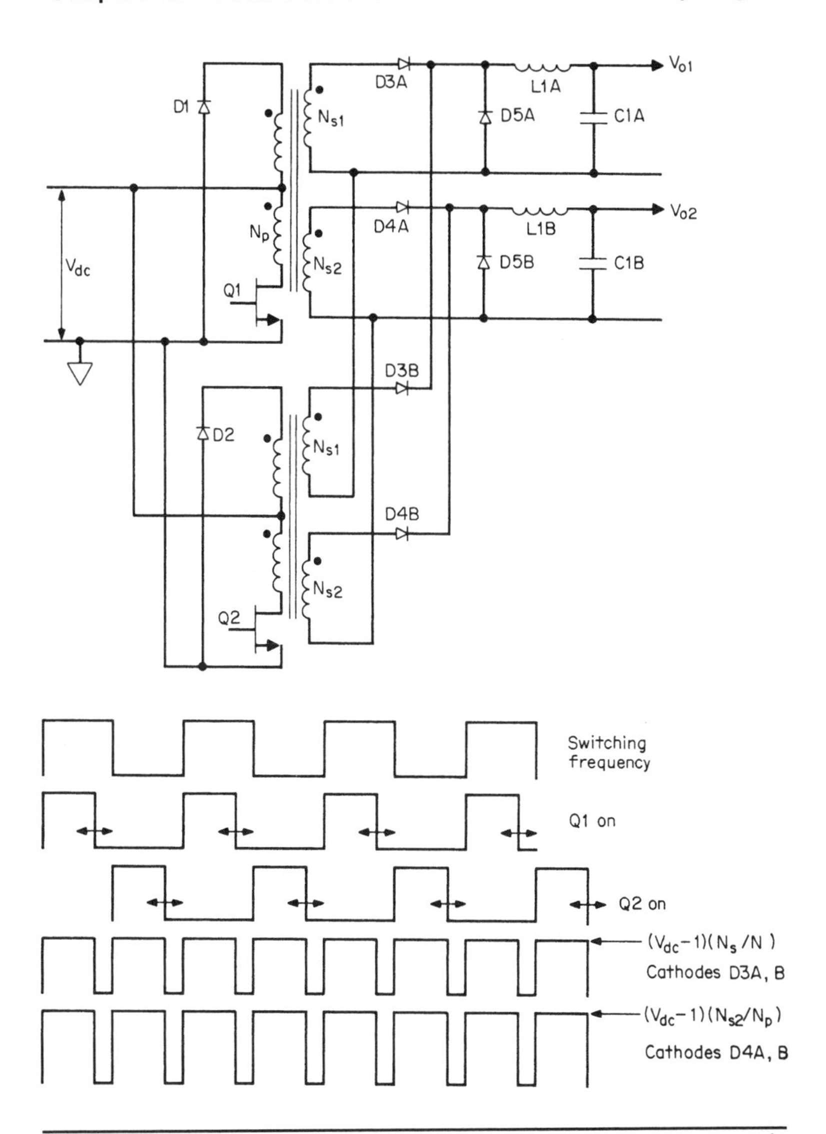

FIGURE 2.14 Interleaved forward converter. Interleaving the "on" times of *Q*1 and *Q*2 on alternate half cycles, and summing their secondary outputs, gives two power pulses per period but avoids the flux-imbalance problem of the push-pull topology.

If this topology is compared to a push-pull, it might be thought that the push-pull is preferable. Although both are two-transistor circuits, the two transformers in the interleaved forward converter are probably more expensive and occupy more space than a single large one in a push-pull circuit. But there is the ever-present uncertainty that the flux imbalance problem in the push-pull could appear under odd transient line and load conditions. The certainty that there is no flux imbalance in the interleaved forward converter is probably the best argument for its use.

There is one special, although not frequent, case where the interleaved forward converter is a much more desirable choice than a single forward converter of the same output power. This occurs when a DC output voltage is high—over about 200 V. In a single forward converter the peak reverse voltage experienced by the output free-wheeling diodes ($D5A$ or $D5B$) is twice that for an interleaved forward converter as the duty cycle in the latter is twice that in the former.

This is no problem when output voltages are low, as can be seen in Eq. 2.25. Transformer secondary turns are always selected (for the single forward converter) so that at minimum DC input, when the secondary voltage is at its minimum, the duty cycle T_{on}/T need not be more than 0.4 to yield the desired output voltage. Then for a DC output of 200 V, the peak reverse voltage experienced by the free-wheeling diode is 500 V. At the instant of power transistor turn "on," the free-wheeling diode has been carrying a large forward current and will suddenly be subjected to reverse voltage. If the diode has slow reverse recovery time, it will draw a large reverse current for a short time at 500-V reverse voltage and run dangerously hot.

Diodes with larger reverse voltage ratings generally have slower recovery times and can be a serious problem. The interleaved forward converter runs at twice the duty cycle and, for a 200 V-DC output, subjects the free-wheeling diode to only 250 V. This permits a lower voltage, faster-recovery diode with considerably lower dissipation.

2.5.2 Transformer Design Relations

2.5.2.1 Core Selection

The core for the two transformers will be selected from the aforementioned charts, to be presented in Chapter 7, but it will be chosen for half the total power output that each transformer must supply.

2.5.2.2 Primary Turns and Wire Size

The number of primary turns in the interleaved forward converter is still given by Eq. 2.40, as each converter's "on" time will still be $0.8T/2$ at minimum DC input. The core iron area A_e will be read from the catalogs for the selected core. Primary wire size will be chosen from Eq. 2.42 at half the total output power.

2.5.2.3 Secondary Turns and Wire Size

The number of secondary turns will be chosen from Eqs. 2.26 and 2.27, but therein the duty cycle will be 0.8 as there are two voltage pulses, each of duration $0.8T/2$ at V_{dc}. Wire size will still be chosen from Eq. 2.44, where I_{dc} is the actual DC output current that each secondary carries at a maximum duty cycle of 0.4.

2.5.3 Output Filter Design

2.5.3.1 Output Inductor Design

The output inductor sees two current pulses per period, exactly like the output inductor in the push-pull topology. These pulses have the same width, amplitude, and duty cycle as the push-pull inductor at the same DC output current. Hence the magnitude of the inductance is calculated from Eq. 2.20 as for the push-pull inductor.

2.5.3.2 Output Capacitor Design

Similarly, the output capacitor "doesn't know" whether it is filtering a full-wave secondary waveform from a push-pull topology or from an interleaved forward converter. Thus for the same inductor current ramp amplitude and permissible output ripple as the push-pull circuit, the capacitor is selected from Eq. 2.22.

Reference

1. K. Billings, *Switchmode Power Supply Handbook*, New York: McGraw-Hill, 1990.

CHAPTER 3

Half- and Full-Bridge Converter Topologies

3.1 Introduction

Half-bridge and full-bridge topologies stress their transistors to a voltage equal to the DC input voltage not to twice this value, as do the push-pull, single-ended, and interleaved forward converter topologies. Thus the bridge topologies are used mainly in offline converters where supply voltage would be more than the switching transistors could safely tolerate. Bridge topologies are almost always used where the normal AC input voltage is 220 V or higher, and frequently even for 120-V AC inputs.

An additional valuable feature of the bridge topologies is that primary leakage inductance spikes (Figures 2.1 and 2.10) are easily clamped to the DC supply bus and the energy stored in the leakage inductance is returned to the input instead of having to be dissipated in a resistive snubber element.

3.2 Half-Bridge Converter Topology

3.2.1 Basic Operation

Half-bridge converter topology is shown in Figure 3.1. Its major advantage is that, like the double-ended forward converter, it subjects the "off" transistor to only V_{dc} and not twice that value. Thus it is widely used in equipment intended for the European market, where the AC input voltage is 220 V.

First consider the input rectifier and filter in Figure 3.1. It is used universally when the equipment is to work from either 120-V AC American power or 220-V AC European power. The circuit always yields roughly 320-V rectified DC voltage, whether the input is 120 or

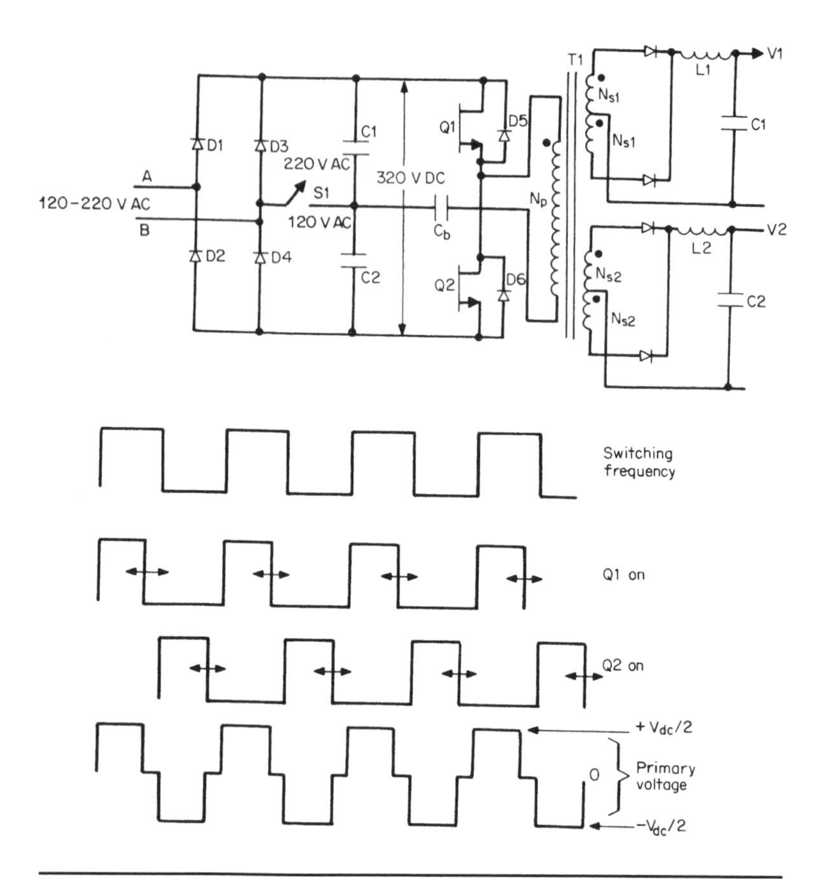

FIGURE 3.1 Half-bridge converter. One end of the power transformer primary is connected to the junction of filter capacitors $C1$, $C2$ via a small DC locking capacitor C_b. The other end is connected to the junction of $Q1$, $Q2$, which turn "on" and "off" on alternate half cycles. With $S1$ in the closed position, the circuit is a voltage doubler; in the open position, it is a full-wave rectifier. In either case, the rectified output is about 308 to 336 V_{dc}.

220 V AC. It does this when switch $S1$ is set to the open position for 220-V AC input, or to the closed position for 120-V AC input. The $S1$ component is normally not a switch; more often it is a wire link that is either installed for 120 V AC, or not for 220 V AC.

With the switch in the open 220-V AC position the circuit is a full-wave rectifier, with filter capacitors $C1$ and $C2$ in series. It produces a peak rectified DC voltage of about $(1.41 \times 220) - 2$ or 308 V. When the switch is in the closed 120-V AC position, the circuit acts as a voltage doubler. On a half cycle of the input voltage when A is positive relative to B, $C1$ is charged positively via $D1$ to a peak of $(1.41 \times 120) - 1$ or 168 V. On a half cycle when A is negative with respect to B, capacitor $C2$ is charged positively via $D2$ to 168 V. The total voltage across $C1$ and $C2$ in series is then 336 V. It can be seen in Figure 3.1 that with either transistor "on," the "off" transistor is subjected to the maximum DC input voltage and not twice that value.

Since the topology subjects the "off" transistor to only V_{dc} and not $2V_{dc}$, there are many inexpensive bipolar and MOSFET transistors that can support the nominal 336 DC V plus 15% upper maximum of 386 V. Thus the equipment can be used with either 120- or 220-V AC line inputs by making a simple switch or linkage change.

After Pressman *An automatic line voltage sensing and switching circuit that drives a relay or other device in the position of S1 is sometimes implemented. The added cost and circuit complexity is offset by making the switching action transparent to the end user of the equipment and by preventing the possible damaging error of running the supply at 220 V while connected for 120 V. ~ T.M.*

Assuming a nominal rectified DC voltage of 336 V, the topology works as follows: For the moment, ignore the small series blocking capacitor C_b. Assume the bottom end of N_p is connected to the junction of $C1$ and $C2$. Then if the leakages in $C1$, $C2$ are assumed to be equal, that point will be at half the rectified DC voltage, about 168 V. It is generally good practice to place equal bleeder resistors across $C1$ and $C2$ to equalize their voltage drops. Now $Q1$ and $Q2$ conduct on alternate half cycles. When $Q1$ is "on" and $Q2$ "off" (Figure 3.1), the dot end of N_p is 168 V positive with respect to its no-dot end, and the "off" stress on $Q2$ is only 336 V. When $Q2$ is "on" and $Q1$ "off," the dot end of N_p is 168 V negative with respect to its no-dot end and the emitter of $Q1$ is 336 V negative with respect to its collector.

This AC square-wave primary voltage produces full-wave square waveshapes on all secondaries—exactly like the secondary voltages in the push-pull topology. The selection of secondary voltages and wire sizes and the output inductor and capacitor proceed exactly as for the push-pull circuit.

3.2.2 Half-Bridge Magnetics

3.2.2.1 Selecting Maximum "On" Time, Magnetic Core, and Primary Turns

It can be seen in Figure 3.1, that if $Q1$ and $Q2$ are "on" simultaneously—even for a very short time—there is a short circuit across the supply voltage and the transistors will be destroyed. To make sure that this does not happen, the maximum $Q1$ or $Q2$ "on" time, which occurs at minimum DC supply voltage, will be set at 80% of a half period. The secondary turns will be chosen so that the desired output voltages are obtained with an "on" time of no more than $0.8T/2$. An "on"-time clamp will be provided to ensure that the "on" time can never be greater than $0.8T/2$ under fault or transient conditions.

The core is selected from the tables in Chapter 7 mentioned earlier. These tables give maximum available output power as a function of operating frequency, peak flux density, core and iron areas, and coil current density.

With a core selected and its iron area known, the number of primary turns is calculated from Faraday's law (Eq. 1.17) using the minimum primary voltage $(V_{dc}/2) - 1$, and the maximum "on" time of $0.8T/2$. Here, the flux excursion dB in the equation is twice the desired peak flux density (1600 G below 50 kHz, or less at higher frequency), because the half-bridge core operates in the first and third quadrants of its hysteresis loop—unlike the forward converter (Section 2.3.9), which operates in the first quadrant only.

3.2.2.2 The Relation Between Input Voltage, Primary Current, and Output Power

If we assume an efficiency of 80%, then

$$P_{in} = 1.25P_o$$

The input power at minimum supply voltage is the product of minimum primary voltage and average primary current at minimum DC input. At minimum DC input, the maximum "on" time in each half period will be set at $0.8T/2$ as discussed above, and the primary has two current pulses of width $0.8T/2$ per period T. At primary voltage $\underline{V_{dc}}/2$, the input power is $1.25P_o = (\underline{V_{dc}}/2)(I_{pft})(0.8T/T)$, where I_{pft} is the peak equivalent flat-topped primary current pulse. Then

$$I_{pft\,(half\ bridge)} = \frac{3.13P_0}{\underline{V_{dc}}} \tag{3.1}$$

3.2.2.3 Primary Wire Size Selection

Primary wire size must be much larger in a half bridge than in a push-pull circuit of the same output power. However, there are two half primaries in the push-pull, each of which has to support twice the voltage of the half-bridge primary when operated from the same supply voltage. Consequently, coil sizes for the two topologies are not much different. Half-bridge primary RMS current is

$$I_{rms} = I_{pft}\sqrt{0.8T/T}$$

and from Eq. 3.1

$$I_{\text{rms}} = \frac{2.79 P_o}{\underline{V_{\text{dc}}}} \tag{3.2}$$

At 500 circular mils per RMS ampere, the required number of circular mils is

$$\begin{aligned}\text{Circular mils needed} &= \frac{500 \times 2.79 P_o}{\underline{V_{\text{dc}}}} \\ &= \frac{1395 P_o}{\underline{V_{\text{dc}}}} \end{aligned} \tag{3.3}$$

3.2.2.4 Secondary Turns and Wire Size Selection

In the following treatment the number of secondary turns will be selected using Eqs. 2.1 to 2.3 for $\overline{T_{\text{on}}} = 0.8T/2$, and the term $V_{\text{dc}} - 1$ will be replaced by the minimum primary voltage, which is $(\underline{V_{\text{dc}}}/2) - 1$. The secondary RMS currents and wire sizes are calculated from Eqs. 2.13 and 2.14, exactly as for the full-wave secondaries of a push-pull circuit.

3.2.3 Output Filter Calculations

The output inductor and capacitor are selected using Eqs. 2.20 and 2.22 as in a push-pull circuit for the same inductor current ramp amplitude and desired output ripple voltage.

3.2.4 Blocking Capacitor to Avoid Flux Imbalance

To avoid the flux-imbalance problem discussed in connection with the push-pull circuit (Section 2.2.5), a small capacitor C_b is fitted in series with the primary as in Figure 3.1. Recall that flux imbalance occurs if the volt-second product across the primary while the core is set (moves in one direction along the hysteresis loop) differs from the volt-second product after it moves in the opposite direction.

Thus, if the junction of $C1$ and $C2$ is not at exactly half the supply voltage, the voltage across the primary when $Q1$ is "on" will differ from the voltage across it when $Q2$ is "on" and the core will walk up or down the hysteresis loop, eventually causing saturation and destroying the transistors.

This saturating effect comes about because there is an effective DC current bias in the primary. To avoid this DC bias, the blocking capacitor is placed in series in the primary. The capacitor value is selected

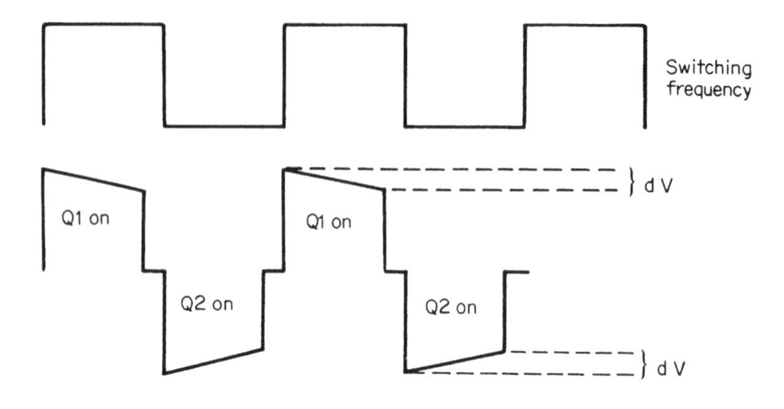

FIGURE 3.2 The small blocking capacitor C_b in series with the half-bridge primary (Figure 3.1) is needed to prevent flux imbalance if the junction of the filter capacitors is not at exactly the midpoint of the supply voltage. Primary current charges the capacitor, causing a droop in the primary voltage waveform. This droop should be kept to no more than 10%. (The droop in primary voltage, due to the offset charging of the blocking capacitor, is shown as dV.)

as follows. The capacitor charges up as the primary current I_{pft} flows into it, robbing voltage from the flat-topped primary pulse shown in Figure 3.2.

This DC offset robs volt-seconds from all secondary windings and forces a longer "on" time to achieve the desired output voltage. In general, it is desirable to keep the primary voltage pulses as flat-topped as possible.

In this example, we will assume a permissible droop of dV. The equivalent flat-topped current pulse that causes this droop is I_{pft} in Eq. 3.1. Then, because that current flows for $0.8T/2$, the required capacitor magnitude is simply

$$C_b = \frac{I_{pft} \times 0.8T/2}{dV} \tag{3.4}$$

Consider an example assuming a 150-W half bridge operating at 100 kHz from a nominal DC input of 320 V. At 15% low line, the DC input is 272 V and the primary voltage is $\pm 272/2$ or ± 136 V.

A tolerable droop in the flat-topped primary voltage pulse would be 10% or about 14 V.

Then from Eq. 3.1 for 150 W and $\underline{V_{dc}}$ of 272 V, $I_{pft} = 3.13 \times 150/272 = 1.73$ A, and from Eq. 3.4, $C_b = 1.73 \times 0.8 \times 5 \times 10^{-6}/14 = 0.49\ \mu F$. The capacitor must be a nonpolarized type.

3.2.5 Half-Bridge Leakage Inductance Problems

Leakage inductance spikes, which are so troublesome in the single-ended forward converter and push-pull topology, are easily avoided in the half bridge: they are clamped to V_{dc} by the clamping diodes $D5$, $D6$ across transistors $Q1$, $Q2$.

Assuming $Q1$ is "on," the load and magnetizing currents flow through it and through the primary leakage inductance of T1, the paralleled $T1$ magnetizing inductance, and the secondary load impedances that are reflected by their turn ratios squared into the primary. Then it flows through C_b into the $C1$, $C2$ junction. The dot end of N_p is positive with respect to its no-dot end.

When $Q1$ turns "off," the magnetizing inductance forces all winding polarities to reverse. The dot end of $T1$ starts to go negative by flyback action, and if this were to continue, it would put more than V_{dc} across $Q1$ and could damage it. Also, $Q2$ could be damaged by imposing a reverse voltage across it. However, the dot end of $T1$ is clamped by diode $D6$ to the supply rail V_{dc} and can go no more negative than the negative end of the supply.

Similarly, when $Q2$ is "on," it stores current in the magnetizing inductance, and the dot end of N_p is negative with respect to the no-dot end (which is close to $V_{dc}/2$). When $Q2$ turns "off," the magnetizing inductance reverses all winding polarities by flyback action and the dot end of N_p tries to go positive but is caught at V_{dc} by clamp diode $D5$. Thus the energy stored in the leakage inductance during the "on" time is returned to the supply rail V_{dc} via diodes $D5$, $D6$.

3.2.6 Double-Ended Forward Converter vs. Half Bridge

Both the half-bridge and double-ended forward converter (Figure 2.13) subject their respective "off" state transistors to only V_{dc} and not twice that. Thus, they are both candidates for the European market where the prime power is 220 V AC. Both methods have been used in such applications in enormous numbers, and it is instructive to consider the relative merits and drawbacks of each approach.

The most significant difference between the two approaches is that the half-bridge secondary provides full-wave output as compared with half-wave in the forward converter. Thus, the square-wave frequency in the half-bridge secondary is twice that in the forward converter, and hence, the output LC inductor and capacitor are smaller with the half bridge.

After Pressman *The term* frequency, *when applied to double-ended and single-ended converters, is not helpful. It is easier to consider secondary pulse* *repetition rate. If the pulse rate is the same for both types (conventionally, doubling the frequency of the single-ended case), the power throughput will be the same. It is just a matter of convention rather than a basic difference in power ratings. In the push-pull case, each positive and negative half cycle produces an output pulse resulting in two pulses per cycle (pulse frequency doubling). So simply producing two pulses from the single-ended topology in the same time period results in the same output.*

The real difference between the two is that the push-pull takes the flux in the core from a negative position on the BH loop to a positive position and, conversely, while the single-ended goes from zero to positive only. Potentially the push-pull has twice the flux range. However, above about 50 kHz, the p-p flux swing is limited by core loss to less than 200 mT typically, a flux swing that can be obtained easily from both the push-pull and single-ended topologies. ~K.B.

Peak secondary voltages are higher with the forward converter because the duty cycle is half that of the half bridge. This is significant only if DC output voltages are high—greater than 200 V, as discussed in Section 2.5.1.

There are twice as many turns on the forward converter primary as on the half bridge because the former must sustain the full supply voltage as compared with half that voltage in the half bridge. Having fewer turns on the half-bridge primary may reduce its winding cost and result in lower parasitic capacities.

After Pressman *Although there are less turns on the half bridge, the current is doubled and copper loss is proportional to I^2, so the wire must be twice the diameter for the same copper loss. ~K.B.*

One final marginal factor in favor of the half bridge is that the coil losses in the primary due to the proximity effect (Section 7.5.6.1) are slightly lower than in the forward converter.

Proximity effect losses are caused by eddy currents induced in one winding layer by currents in adjacent layers. Proximity losses increase rapidly with the number of winding layers, and the forward converter may have more layers. The half-bridge primary has half the turns of a double-ended forward converter primary of equal output power operating from the same DC supply voltage. However, this is balanced somewhat by the larger wire size required for the half bridge. Thus, the required number of circular mils for a forward converter primary is given by Eq. 2.42 as $985P_o/\underline{V_{dc}}$ and for a half bridge by Eq. 3.3 as $1395P_o/\underline{V_{dc}}$.

In a practical case, the lower proximity effect losses for the half bridge may be only a marginal advantage. Proximity effect losses will be discussed in more detail in Chapter 7.

3.2.7 Practical Output Power Limits in Half Bridge

Peak primary current and maximum transistor off-voltage stress determine the practical maximum available output power in the half bridge. This limit is about 400 to 500 W for a half bridge operating from 120-V AC input in the voltage-doubling mode, shown in Figure 3.1. It is equal to that required for the double-ended forward converter as discussed in Section 2.4.1.1 and which can be seen as follows: The peak equivalent flat-topped primary current is given by Eq. 3.1 as $I_{pft} = 3.13P_o/\underline{V_{dc}}$. For a ±10% steady-state tolerance and a 15% transient allowance on top of that, the maximum off-voltage stress is $V_{dc} = 1.41 \times 120 \times 2 \times 1.1 \times 1.15$ or 428 V. The minimum DC input voltage is $\underline{V_{dc}} = 1.41 \times 120 \times 2/1.1/1.15 = 268$ V.

Thus, for 500-W output, Eq. 3.1 gives the peak primary current as $I_{pft} = 3.13 \times 500/268 = 5.84$ A, and there are many transistor choices—either MOSFETs or bipolars—with 428-V, 6-A ratings. Bipolars must have a −1-V to −5-V reverse bias (to permit V_{cev} rating) at turn "off" to permit a safe "off" voltage of 428 V. Most adequately fast transistors at that current rating have a V_{ceo} rating of only 400 V.

The half bridge can be pushed to 1000-W output, but at the required 12-A rating, most available bipolar transistors with adequate speed have too low a gain. MOSFET transistors at the required current and voltage rating have too large an "on" drop and are too expensive for most commercial applications at the time of this writing.

Above 500 W, consider the full-bridge topology, a small modification of the half bridge but capable of twice the output power.

3.3 Full-Bridge Converter Topology

3.3.1 Basic Operation

The full-bridge converter topology is shown in Figure 3.3 with the same voltage-doubling full-wave bridge rectifying scheme as was shown for the half bridge (Section 3.2.1). It can be used as an offline converter from a 440-V AC line.

Its major advantage is that the voltage impressed across the primary is a square wave of $\pm V_{dc}$, instead of $\pm V_{dc}/2$ for the half bridge. Further, the maximum transistor off-voltage stress is only the maximum DC input voltage—just as for the half bridge. Thus, for transistors of the same peak current and voltage ratings, the full bridge is able to deliver twice the output power of the half bridge.

In the full bridge the transformer primary turns must be twice that of the half bridge as the primary winding must sustain twice the voltage. However, to get the same output power as a half bridge from the

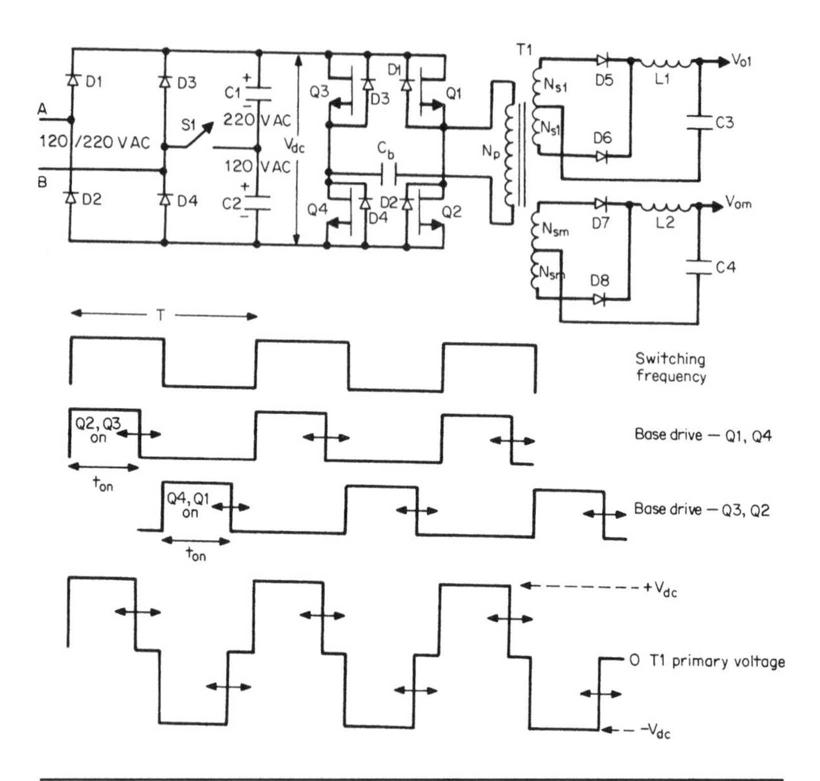

FIGURE 3.3 Full-bridge converter topology. Power transformer $T1$ is bridged between the junction of $Q1$, $Q2$ and $Q3$, $Q4$. Transistors $Q2$, $Q3$ are switched "on" simultaneously for an adjustable time during one half period; then transistors $Q4$, $Q1$ are simultaneously "on" for an equal time during the alternate half period. Transformer primary voltage is a square wave of $\pm V_{dc}$. This contrasts with the $\pm V_{dc}/2$ primary voltage in the half bridge and yields twice the available power.

same DC supply voltage, the peak and RMS currents are half that of the half bridge because the transformer primary supports twice the voltage as the half bridge. With twice the primary turns but half the RMS current, the full-bridge transformer size is identical to that of the half bridge at equal output powers. With a larger transformer, the full bridge can deliver twice the output of the half bridge with transistors of identical voltage and current ratings.

Figure 3.3 shows a master output, V_{om} and a single slave output, V_{o1}. The circuit works as follows. Diagonally opposite transistors ($Q2$ and $Q3$ or $Q4$ and $Q1$) are turned "on" simultaneously during alternate half cycles. Assuming that the "on" drop of the transistors is negligible, the transformer primary is thus driven with an alternating polarity square wave of amplitude V_{dc} and "on" time t_{on} determined by the feedback loop.

The feedback loop senses a fraction of V_{om}, and the pulse width modulator controls t_{on} so as to keep V_{om} constant against line and load changes. The slave outputs, as in all other topologies, are kept constant against AC line input changes, but only to within about 5 to 8% against load changes. If we assume a 1-V "on" drop in each switching transistor, 0.5-V forward drop in the master output Schottky rectifiers, and 1.0-V forward drops in the slave output rectifiers, we get

$$V_{om} = \left[(V_{dc} - 2)\frac{N_{sm}}{N_p} - 0.5\right]\frac{2t_{on}}{T} \tag{3.5a}$$

$$V_{om} \approx V_{dc}\frac{N_{sm}}{N_p}\frac{2t_{on}}{T} \tag{3.5b}$$

$$V_{o1} = \left[(V_{dc} - 2)\frac{N_{s1}}{N_p} - 1\right]\frac{2t_{on}}{T} \tag{3.6a}$$

$$V_{o1} \approx V_{dc}\frac{N_{s1}}{N_p}\frac{2t_{on}}{T} \tag{3.6b}$$

As in all pulse width modulated regulators, as V_{dc} goes up or down by a given percentage, the width modulator decreases or increases the "on" time by the same percentage so as to keep the product $(V_{dc})(t_{on})$ and, hence, the output voltages constant.

3.3.2 Full-Bridge Magnetics

3.3.2.1 Maximum "On" Time, Core, and Primary Turns Selection

In Figure 3.3, it can be seen that if two transistors that are vertically stacked above one another (Q3 and Q4, or Q1 and Q2) are turned "on" simultaneously, they would present a dead short-circuit across the DC supply bus and the transistors would fail. To ensure this does not happen, the maximum "on" time $\overline{t_{on}}$ will be chosen as 80% of a half period. This is "chosen" by selecting the turns ratios N_{sm}/N_p, N_{s1}/N_p, so that in those equations for $\underline{V_{dc}}$, with $\overline{t_{on}}$ equal to $0.8T/2$, the correct output voltages—V_{om}, V_{o1}—are obtained. The maximum "on" time occurs at minimum DC input voltage $\underline{V_{dc}}$—as can be seen in Eqs. 3.5*b* and 3.6*b*.

The magnetic core and operating frequency are chosen from the core-frequency selection chart in Chapter 7. With a core selected and its iron area A_e known, the number of primary turns N_p is chosen from Faraday's law (Eq. 1.17). In Eq. 1.17, E is the minimum primary voltage $(\underline{V_{dc}} - 2)$, and dB is the flux change desired in the time dt of $0.8T/2$. As discussed in Section 2.2.9.4, dB will be chosen as 3200 G

(−1600 to +1600 G) for frequencies up to 50 kHz and this will be reduced at higher frequencies because core losses increase.

3.3.2.2 Relation Between Input Voltage, Primary Current, and Output Power

Assume an efficiency of 80% from the primary input to the total output power. Then

$$P_o = 0.8P_{\text{in}} \quad \text{or} \quad P_{\text{in}} = 1.25P_o$$

At minimum DC input voltage $\underline{V_{\text{dc}}}$, on time per half period is $0.8T/2$, and duty cycle over a complete period is 0.8. Then neglecting the power transistor on drops, input power at $\underline{V_{\text{dc}}}$ is

$$P_{\text{in}} = \underline{V_{\text{dc}}}(0.8)I_{\text{pft}} = 1.25P_o$$

or

$$I_{\text{pft}} = \frac{1.56P_o}{\underline{V_{\text{dc}}}} \tag{3.7}$$

where I_{pft} is the equivalent primary flat-topped current as described in Section 2.2.10.1.

3.3.2.3 Primary Wire Size Selection

Current I_{pft} flows at a duty cycle of 0.8 so its RMS value is

$$I_{\text{rms}} = I_{\text{rms}}\sqrt{0.8}.$$

Then, from Eq. 3.7

$$I_{\text{rms}} = (1.56P_o/\underline{V_{\text{dc}}})\ \sqrt{0.8}$$

$$I_{\text{rms}} = \frac{1.40P_o}{\underline{V_{\text{dc}}}} \tag{3.8}$$

And at a current density of 500 circular mils per RMS ampere, the required number of circular mils is

$$\text{Circular mils needed} = \frac{500 \times 1.40P_o}{\underline{V_{\text{dc}}}} = \frac{700P_o}{V_{\text{dc}}} \tag{3.9}$$

3.3.2.4 Secondary Turns and Wire Size

The number of turns on each secondary is calculated from Eqs. 3.5*a* and 3.5*b*, where $\overline{t_{\text{on}}}$ is $0.8T/2$ for the specified minimum DC input

V_{dc}, N_p as calculated in Section 3.3.2.1, and all DC outputs are specified.

Secondary RMS currents and wire sizes are chosen exactly as for the push-pull secondaries as described in Section 2.2.10.3. The secondary RMS currents are given by Eq. 2.13 and the required circular mils for each half secondary is given by Eq. 2.14.

3.3.3 Output Filter Calculations

For the half-bridge and push-pull topologies that have full-wave output rectifiers, the output inductor and capacitors are calculated from Eqs. 2.20 and 2.22. Equation 2.20 specifies the output inductor for minimum DC output currents equal to one-tenth the nominal values. Equation 2.22 specifies the output capacitor for the specified peak-to-peak output ripple V_r and the selected peak-to-peak inductor current ripple amplitude.

3.3.4 Transformer Primary Blocking Capacitor

Figure 3.3 shows a small nonpolarized blocking capacitor C_b in series with the transformer. It is needed to avoid the flux-imbalance problem as discussed in Section 3.2.4.

Flux imbalance in the full bridge is less likely than in the half bridge, but still is possible. With bipolars, an "on" pair in one half cycle may have different storage times than the pair in the alternate half cycle. With MOSFETs, the "on" state voltage drops of the pairs for alternate half cycles may be unequal. In either case, if the volt-second product applied to the transformer primary in alternate half cycles is unequal, the core could walk off the center of the hysteresis loop, saturate the core, and destroy the transistors.

CHAPTER 4

Flyback Converter Topologies

Foreword

I find that many engineers and students have great difficulty with the design of flyback type converters. This is unfortunate because these topologies are very useful, and, in fact, they are not difficult to design.

The problem is not the intrinsic difficulty of the subject matter (or the ability of the student). The fault is related to the way the subject is traditionally taught.

Right from the start the normal term *flyback transformer* immediately projects the wrong mindset. Not unreasonably, designers set out to design a "flyback transformer" as if it were a real transformer. This is not the way to go.

We are all very familiar with transformers, very simple devices really—we put a voltage across a primary winding and we get a voltage on a secondary winding. The voltage ratio follows the turns ratio, irrespective of the output (or load) current. In other words, the transformer conserves the voltage transfer ratio (one volt per turn on the primary results in one volt per turn on the secondary. You want ten volts? Then use ten turns, very simple). However, notice an important property of transformers, the primary and secondary conduct at the same time. If current flows into the start of the primary winding it flows out of the start of the secondary winding at the same time.

Figure 4.1 shows the basic schematic of a flyback converter. Notice the when *Q*1 is "on" current flows into the primary winding of *T*1 but the secondary diodes are not conducting and there is no secondary current. When *Q*1 turns "off" the primary current stops, all winding voltages reverse by flyback action, and the output diodes and secondary windings now conduct current. So the primary and secondary windings in the flyback "transformer" conduct current at different times. This apparently minor difference dramatically changes the rules.

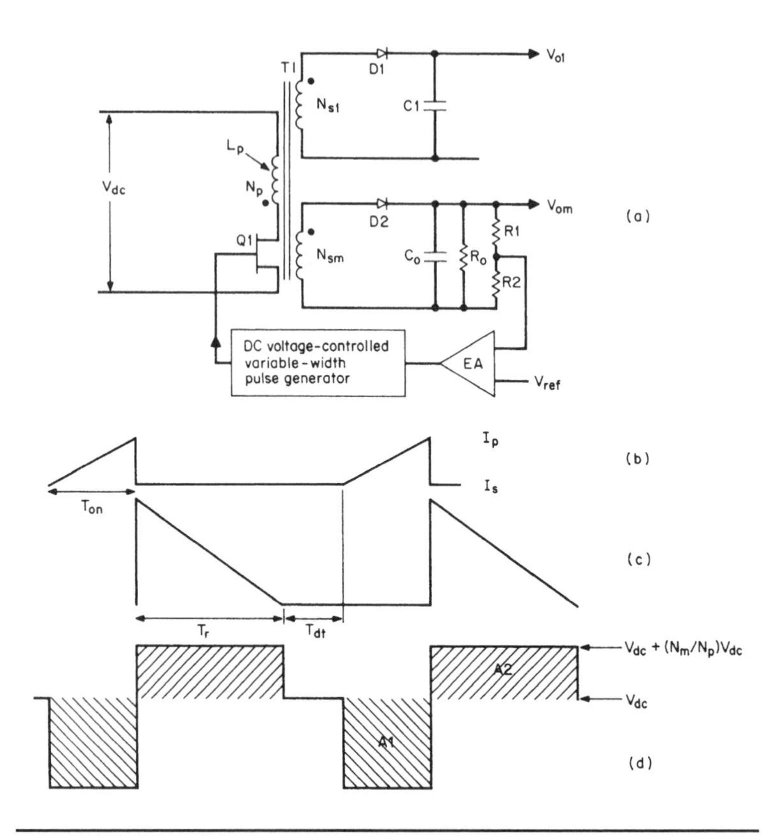

FIGURE 4.1 Basic flyback converter schematic. The action is as follows: When $Q1$ turns "on," all rectifier diodes become reverse-biased, and all output load currents are supplied from the output capacitors. T1 acts like a pure inductor and primary current builds up linearly in it to a peak I_p. When $Q1$ turns "off," all winding voltages reverse under flyback action, bringing the output diodes into conduction and the primary stored energy $1/2LI_p^2$ is delivered to the output to supply load current and replenish the charge on the output capacitors (the charge that they lost when $Q1$ was on). The circuit is discontinuous if the secondary current has decayed to zero before the start of the next turn "on" period of $Q1$.

Think about it! When $Q1$ is "on" only the primary winding is conducting (the other windings are not visible to the primary because they are not conducting). $Q1$ thinks it is driving an inductor. When $Q1$ turns "off" only the secondary windings conduct and now the primary winding cannot be seen by the secondaries (so now the secondaries think they are being driven by an inductor). So how does this change the rules? Well, functionally the so-called flyback "transformer" is really functioning as an inductor with several windings and follows the rules applicable to inductors.

The rules for an inductor with more than one winding are as follows: The primary to secondary ampere-turns ratios are conserved (not the voltage ratios, as was the case with a true transformer). For example, if the primary is, say, 100 turns and the current when $Q1$ turns "off" is 1 amp, then we have developed 100 ampere-turns in the primary. This must be conserved in the secondaries. With, say, a single secondary winding of 10 turns, the secondary current will be 10 amps (10T × 10A = 100 ampere-turns). In the same way, a single turn will develop 100 amps or 1000 secondary turns will develop 0.1 amps.

So where do we stand with regard to voltage? Well, to the first order, there is no correlation between primary and secondary voltages. The secondary voltage is simply a function of load. Consider the 10-turn 10-amp (100 ampere-turns) secondary winding example mentioned above. If we terminate the winding with a 1-ohm load, we will get 10 volts. What is more striking because the 10 amps must be conserved is that if we terminate it with 100 ohms, we will get 1000 volts! This is why the flyback topology is so useful for generating high voltages (don't try to open circuit this winding because it will destroy the semiconductors). With several secondary windings conducting at the same time, then the sum of all the secondary ampere-turns must be conserved.

So the lesson we learn here is that flyback "transformers" actually operate as inductors and must be designed as such. (In Chapter 7, I use the term *choke* instead of inductor because the core must support both DC and AC components of current.) If flyback "transformers" had originally been called by their correct functional name, "flyback chokes," then a lot of confusion could have been avoided.

We must not forget that voltage transformation is still taking place between primary and secondary windings even if they are not conducting at the same time. Taking the above example of 10 turns terminated in 100 ohms, the 1000 volts thus developed on this secondary winding will reflect back to the primary as 10,000 volts; this added to the supply of 100 volts will stress $Q1$ in its "off" state with 10,100 volts (where did I put that 11,000 volt transistor?). Hardly practical, but the theory holds.

So when designing flyback transformers keep the following key points in mind:

1. Remember you are not designing a transformer, you are designing a choke with additional windings.
2. The primary turns are selected to satisfy the AC voltage stress (volt-seconds) and the core AC saturation properties:

$$N_p = \frac{VT}{B\,Ae}$$

Where Np is minimum primary turns
V is the maximum primary DC voltage (volts)
T is the maximum "on" period for $Q1$ (microseconds)
B is the AC p-p flux swing (tesla) typically 200 mT for ferrite
Ae is the effective center pole area of the core (mm^2)

3. The secondary turns are optional. If you choose the same volts per turn on the secondary as was used for the primary, then the flyback voltage on $Q1$ will be twice the supply voltage.
4. When using a gapped ferrite core, the minimum core gap must be such that the core will not saturate for the sum of DC and AC magnetization current. More often the gap is chosen to satisfy the power transfer requirements. This normally results in a gap exceeding the minimum requirements. Remember the energy stored in the primary is

$$E(joules) = {}^1\!/_2 LI^2$$

Remember this is the maximum energy that can be transferred to the secondary, and then only in the discontinuous (complete energy transfer) mode. In the continuous mode, only part of this energy is transferred.

Note *Although reducing the inductance* L *may appear to reduce the stored energy, the current* I *increases in the same ratio as the inductance decreases. Since the* I *parameter is squared, the stored energy actually increases as* L *decreases.*

5. It is not recommended that you try to design for a defined inductance. It is better to let inductance be a dependant variable as changing the core gap or core material (permeability) will change the inductance.

Below in Chapter 4, Pressman follows the conventional "flyback transformer" approach, providing a very complete analysis. The reader may find it helpful to first read Chapter 7 in this book and Part 2, Chapters 1 and 2 in my book, shown as Reference 1 at the end of this chapter.

4.1 Introduction

All the topologies previously discussed (with the exception of the boost regulator Section 1.4 and the polarity inverter Section 1.5) deliver power to their loads during the period when the power transistor is turned "on."

However, the flyback topologies described in this chapter operate in a fundamentally different way. During the power transistor "on" time, they store energy in the power transformer. During this period, the load current is supplied from an output filter capacity only. When the power transistor turns "off," the energy stored in the power transformer is transferred to the load and to the output filter capacitor as it replaces the charge it lost when it alone was delivering load current.

The flyback has advantages and limitations, discussed in more detail later. A major advantage is that the output filter inductors normally required for all forward topologies are not required for flyback topologies because the transformer serves both functions. This is particularly valuable in low-cost multiple output power supplies yielding a significant saving in cost and space.

4.2 Basic Flyback Converter Schematic

The basic flyback converter topology together with typical current and voltage waveforms is shown in Figure 4.1. It is very widely used for low-cost applications in the power range from about 150 W down to less than 5 W. Its great initial attraction is immediately clear—it has no secondary output inductor, and the consequent saving in cost and volume is a significant advantage.

In Figure 4.1, flyback operation can be easily recognized from the position of the dots on the transformer primary and secondary (these dots show the starts of the windings). When $Q1$ is "on," the dot ends of all windings are negative with respect to their no-dot ends. Output rectifier diodes $D1$ and $D2$ are reverse-biased and all the output load currents are supplied from storage filter capacitors $C1$ and $C2$. These will be chosen as described below to deliver the load currents with the maximum specified ripple or droop in output voltages.

4.3 Operating Modes

There are two distinctly different operating modes for flyback converters: the continuous mode and the discontinuous mode. The waveforms, performance, and transfer functions are quite different for the two modes, and typical waveforms are shown in Figure 4.2. The value of primary inductance and the load current determine the mode of operation.

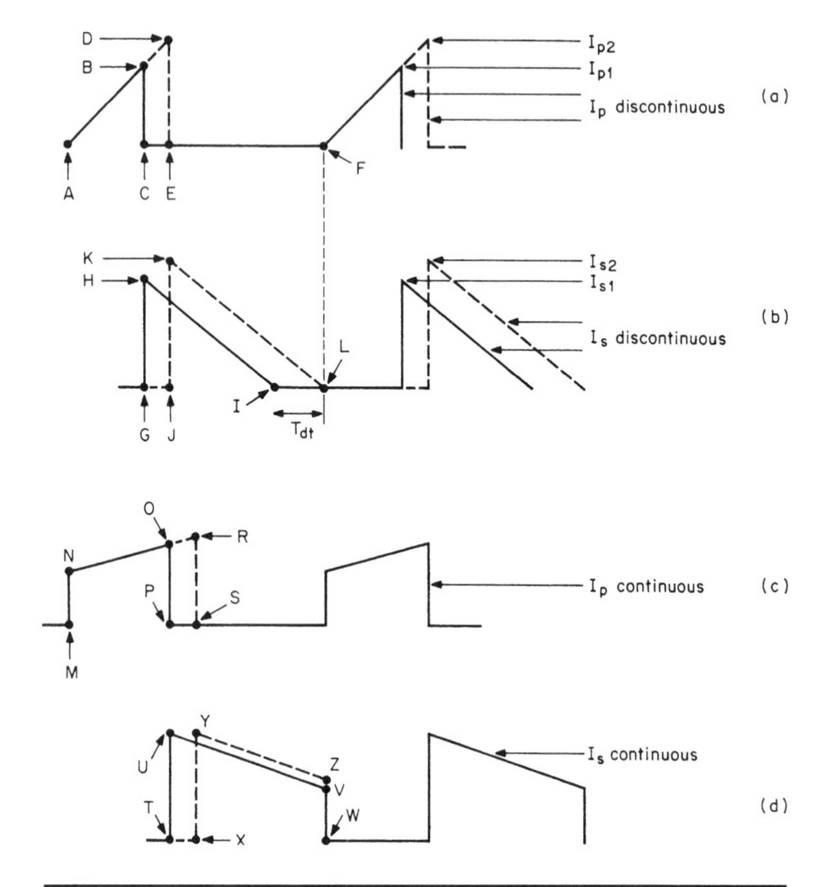

FIGURE 4.2 (*a* and *b*) Waveforms of a discontinuous-mode flyback at the point of transition to continuous-mode operation. Notice in the discontinuous mode, the current remains discontinuous (the transformer has periods of zero current) providing there is a dead time (T_{dt}) between the instant the secondary current reaches zero and the start of the next "on" period. (*c* and *d*) If the transformer is loaded beyond this point, some current remains in the transformer at the end of the "off" period and the next "on" period will have a sharp current step at its front end. This step is characteristic of the continuous mode of operation, as the secondary current no longer decays to zero at any part of the conduction period. There is a dramatic change in the transfer function at the point of entering continuous mode, and if the error-amplifier bandwidth has not been drastically reduced, the circuit will oscillate.

4.4 Discontinuous-Mode Operation

Figure 4.1 shows a master output and one slave output. As in all other topologies shown previously, a negative-feedback loop will be closed around the master output V_{om}. A fraction of V_{om} will be compared to a reference, and the error signal will control the "on" time of $Q1$ (the pulse width), so as to regulate the sampled output voltage equal to the reference voltage against line and load changes. Hence, the master output is fully regulated. However, the slaves will also be well regulated against line changes and somewhat less well against load changes because the secondary winding voltages tend to track the master voltage. As a result, the slave line and load regulation is better than for the previously discussed forward-type topologies.

During the $Q1$ "on" time, there is a fixed voltage across N_p and current in it ramps up linearly (Figure 4.1*b*) at a rate of $dI/dt = (V_{dc}-1)/L_p$, where L_p is the primary magnetizing inductance. At the end of the "on" time, the primary current has ramped up to $I_p = (V_{dc}-1)T_{on}/L_p$. This current represents a stored energy of

$$E = \frac{L_p(I_p)^2}{2} \tag{4.1}$$

where E is in joules
L_p is in henries
I_p is in amperes

Now when $Q1$ turns "off," the current in the magnetizing inductance forces a reversal of polarities on all windings. (This is called *flyback action*.) Assume, for the moment, that there are no slave windings and only the master secondary N_m. Since the current in an inductor cannot change instantaneously, at the instant of turn "off," the primary current transfers to the secondary at an amplitude $I_s = I_p(N_p/N_m)$.

After a number of cycles, the secondary DC voltage has built up to a magnitude (calculated below) of V_{om}. Now with $Q1$ "off," the dot end of N_m is positive with respect to its no-dot end and current flows out of it, but ramps down linearly (Figure 4.1*c*) at a rate $dI_s/dt = V_{om}/V_s$, where L_s is the secondary inductance. The discontinuous mode action is defined as follows.

If the secondary current has ramped down to zero before the start of the next $Q1$ "on" time, all the energy stored in the primary when $Q1$ was "on" has been delivered to the load and the circuit is said to be operating in the discontinuous mode.

Since an amount of energy E in joules delivered in a time T in seconds represents input power in watts, we can calculate the input power as follows: At the end of one period, power P drawn from V_{dc} is

$$P = \frac{1/2 L_p (I_p)^2}{T} W \tag{4.2a}$$

But $I_p = (V_{dc} - 1)T_{on}/L_p$. Then

$$P = \frac{[(V_{dc} - 1)T_{on}]^2}{2TL_p} \approx \frac{(V_{dc}T_{on})^2}{2TL_p} W \tag{4.2b}$$

As can be seen from Eq. 4.2*b*, the feedback loop maintains constant output voltage by keeping the product $V_{dc}T_{on}$ constant.

4.4.1 Relationship Between Output Voltage, Input Voltage, "On" Time, and Output Load

Let us assume an efficiency of 80%, then

$$\text{Input power} = 1.25 \text{ (output power)}$$
$$= \frac{1.25(V_o)^2}{R_o} = \frac{1/2(L_p I_p^2)}{T}$$

But $I_p = \underline{V_{dc}}\overline{T_{on}}/L_p$ since maximum "on" time $\overline{T_{on}}$ occurs at minimum supply voltage $\underline{V_{dc}}$, as can be seen from Eq. 4.2*b*.

Then $1.25\,(V_o)^2/R_o = 1/2 L_p \underline{V_{dc}}^2 \overline{T_{on}^2}/L_p^2 T$ or

$$V_o = \underline{V_{dc}}\overline{T_{on}}\sqrt{\frac{R_o}{2.5TL_p}} \tag{4.3}$$

Thus the feedback loop will regulate the output by decreasing T_{on} as V_{dc} or R_o goes up, increasing T_{on} as $V_{dc}R_o$ goes down.

4.4.2 Discontinuous-Mode to Continuous-Mode Transition

In Figures 4.2*a* and 4.2*b* the solid lines represent primary and secondary currents in the discontinuous mode. Primary current is a triangle starting from zero and rising to a level I_{p1} (point *B*) at the end of the power transistor "on" time.

At the instant of Q1 turn "off," the current I_{p1} established in the primary winding is transferred to the secondary so as to maintain the ampere-turns ratio. This current is dumped into the secondary capacitors and load during the "off" period. The secondary current ramps downward at a rate $dI_s/dt = (V_o + 1)/L_s$, where L_s is the secondary inductance, which is $(N_s/N_p)^2$ times the primary magnetizing inductance. This current reaches zero at time *I*, leaving a dead time T_{dt} before the start of the next turn "on" period at point *F*. All the current and hence energy stored in the primary during the previous "on" period has now been completely delivered to the load before the next turn "on." The average DC output current will be the average of the triangle *GHI* multiplied by its duty cycle of T_{off}/T.

Now, to remain in the discontinuous mode, there must be a dead time T_{dt} (Figure 4.2*b*) between the time the secondary current has dropped to zero and the start of the next power transistor "on" time. As more power is demanded (by decreasing R_o), T_{on} must increase to keep output voltage constant (see Eq. 4.3). As T_{on} increases (at constant V_{dc}), primary current slope remains constant and the peak current rises from *B* to *D* as shown in Figure 4.2*a*. Secondary peak current ($= I_p N_p/N_s$) increases from *H* to *K* in Figure 4.2*b* and starts later in time (from *G* to *J*).

Since the output voltage is kept constant by the feedback loop, the secondary slope V_o/L_s remains constant and the point at which the secondary current falls to zero moves closer to the start of the next turn "on." This reduces T_{dt} until a point *L* is reached where the secondary current has just fallen to zero at the instant of the next turn "on." This load current marks the end of the discontinuous mode. Notice that if the supply voltage falls, the "on" time T_{on} must increase as V_{dc} decreases to maintain constant output voltage and this will have the same effect.

Notice that as long as the circuit is in the discontinuous mode so that a dead time always remains, increasing the "on" time increases the area of the primary and secondary current triangle *GHI* up to the limit of the area *JKL*. Further, since the DC output current is the average of the secondary current triangle multiplied by its duty cycle, then during the very next "off" period following an increase in "on" time, more secondary current is immediately available to the load.

When the dead time has been lost, however, any further increase in load current demand will increase the "on" time and decrease the "off" time as the back end of the secondary current can no longer move to the right. The secondary current will start later than point *J* (Figure 4.2*b*) and from a higher point than *K*. Then at the start of the next "on" period (position *F* in Figure 4.2*a* or *L* in Figure 4.2*b*), there is still some current or energy left in the transformer.

Now the front end of the primary current will have a small step. The feedback loop tries to deliver the increased DC load current demand by keeping the "on" time later than point *J*. Now at each successive "off" time, the current remaining at the end of the "off" time and hence the current step at the start of the next "on" time increase.

Finally after many switching cycles, the front-end step of primary current and the back-end current at the end of the "off" time in Figure 4.2*d* are sufficiently high so that the area *XYZW* is somewhat larger than that sufficient to supply the output load current. Now the feedback loop starts to decrease the "on" time so that the primary trapezoid lasts from *M* to *P* and the secondary current trapezoid lasts from *T* to *W* (Figures 4.2*a* and 4.2*b*).

At this point, the volt-seconds across the transformer primary when the power transistor is "on" is equal to the "off" volt-seconds across it when the transistor is "off." For this condition, the transformer core is always reset to its original point on the hysteresis loop at the end of a full cycle. It is also the condition where the average or DC voltage across the primary is zero. This is an essential requirement, since the DC resistance in the primary is near zero and it is not possible to support a long-term DC voltage across zero resistance.

Once the continuous mode has been established, increased load current is supplied initially by an increase in "on" time (from *MP* to *MS* in Figure 4.2*c*). For fixed-frequency operation this results in a decrease in "off" time from *TW* to *XW* (Figure 4.2*d*) as the back end of the secondary current pulse cannot move further to the right in time because the dead time has vanished. Although the peak of the secondary current has increased somewhat (from point *U* to *Y*), the area lost in the decreased "off" time (*T* to *X*) is greater than the area gained in the slope change from *UV* to *YZ* in Figure 4.2*d*.

Thus, in the continuous mode, a sudden increase in DC output current initially causes a decrease in width and a smaller increase in height of the secondary current trapezoid. After many switching cycles, the average trapezoid height builds up and the width relaxes back to the point where the "on" volt-seconds again equals the "off" volt-seconds across the primary.

In addition, since the DC output voltage is proportional to the area of the secondary current trapezoid, the feedback loop, in attempting to keep the output voltage constant against an increased current demand, first drastically decreases the output voltage and then, after many switching cycles, corrects it by building up the amplitude of the secondary current trapezoid. This is the physical-circuits significance of the so-called right-half-plane-zero, which forces the drastic reduction in error-amplifier bandwidth to stabilize the feedback loop. The right-half-plane-zero will be discussed further in the chapter on loop stabilization.

After Pressman *In a fixed-frequency system, the immediate effect of increasing the "on" period (to increase primary and hence output current) will be to decrease the "off" period (the period for transfer of current to the output). Since the inductance of the transformer prevents rapid changes* *in current, the immediate effect of trying to increase current is to cause a short-term decrease in output current. (This is a transitory 180° phase shift between cause and effect). This short transitory phase shift is the cause of the right-half-plane-zero in the transfer function. It is a non-compensatable dynamic effect and forces the designer to provide a very low-frequency roll off in the control loop to maintain stability. Hence transient performance will not be good. The flyback converter in the continuous mode has a boost-like converter characteristic and any converter or combination of converters that have a boost-type characteristic will have the right-half-plane-zero problem. ~K.B.*

4.4.3 Continuous-Mode Flyback—Basic Operation

The flyback topology is widely used for high output voltages at relatively low power (≤5000 V at <15 W). It can also be used at powers of up to 150 W if DC supply voltages are high enough (≥160 V) so that primary currents are not excessive. The feature which makes it valuable for high output voltages is that it requires no output inductor. In forward converters, discussed above, output inductors become a troublesome problem at high output voltages because of the large voltages they have to sustain. Not requiring a high voltage free-wheeling diode is also a plus for the flyback in high voltage supplies.

After Pressman *A further advantage for high voltage applications is that relatively large voltages can be obtained with relatively fewer transformer turns. ~K.B.*

The flyback topology is attractive for multiple output supplies because the output voltages track one another for line and load changes, far better than they do in the forward-type converters described earlier. The absence of output inductors results in better tracking. As a result, flybacks are a frequent choice for supplies with many output voltages (up to 10 isolated outputs are not uncommon). The power can be in the range of 50 to 150 W.

Although they can be used from DC input voltages as low as 5 V, it is more usual to find them used for the usual rectified 160 VDC obtained from a 115-V AC power line input. By careful design of the turns ratios, they can also be used in universal line input applications ranging from the rectified output of 160 volts DC from 110 AC inputs up to the 320 V DC obtained from a 220-V AC power line, without the need for the voltage doubling–full-wave rectifying scheme (switch S1) shown in Figure 3.1.

The latter scheme, although very widely used, has the objectionable feature that to do the switching from 115 to 220 V AC, both ends of the switch in Figure 3.1 have to be accessible on the outside of the supply, which is a safety hazard. Or the supply must be opened to change the switch position. Both these alternatives have drawbacks. An alternative scheme not requiring switching will be discussed in Section 4.3.5.

Both modes have an identical circuit diagram, shown in Figure 4.1, and it is only the transformer's magnetizing inductance and output load current that determines the operating mode. It has been shown that with a given magnetizing inductance, a circuit that has been designed for the discontinuous mode will move into the continuous mode when the output load current is increased beyond a unique boundary. The mechanism for this and its consequence are discussed in more detail below.

The discontinuous mode (as shown in Figure 4.2*a*) does not have a front-end step in the primary current. At turn "off" (as shown in Figure 4.2*b*), the secondary current will be a decaying triangle that has ramped down to zero before the next turn "on." All the energy stored in the primary during the "on" period has been completely delivered to the secondary and thus to the load before the next turn "on."

In the continuous mode, however (as seen in Figure 4.2*c*), the primary current does have a front-end step and the characteristic of a rising current ramp following the step. During the "off" period of Q1 (Figure 4.2*d*), the secondary current has the shape of a decaying triangle sitting on a step with current still remaining in the secondary at the instant of the next turn "on" action. Clearly there is still some energy left in the transformer at the instant of the next turn "on."

The two modes have significantly different operating properties and usages. The discontinuous mode, which does not have a right-half-plane-zero in the transfer function, responds more rapidly to transient load changes with a lower transient output voltage spike.

A penalty is paid for this performance, in that the secondary peak current in the discontinuous mode can be between two and three times greater than that in the continuous mode. This is shown in Figures 4.2*b* and 4.2*d*. Secondary DC load current is the average of the current waveshapes in those figures. Also, assuming closely equal "off" times, it is obvious that the triangle in the discontinuous mode must have a much larger peak than the trapezoid of the continuous mode for the two waveshapes to have equal average values.

With larger peak secondary currents, the discontinuous mode has a larger transient output voltage spike at the instant of turn "off" (Section 4.3.4.1) and requires a larger *LC* spike filter to remove it.

Also, the larger secondary peak current at the start of turn "off" in the discontinuous mode causes a greater RFI problem. Even for moderate output powers, the very large initial spike of secondary current at the instant of turn "off" causes a much more severe noise spike on the output ground bus, because of the large *di/dt* into the output bus inductance.

After Pressman *A major advantage of the discontinuous mode is that the secondary rectifier diodes turn "off" under low current stress conditions. Also they are fully "off" before the next "on" edge of Q1. Hence the problem of diode reverse recovery is eliminated. This is a major advantage in high voltage applications as diode reverse recovery current spikes are difficult to eliminate and are a rich source of RFI. ~K.B.*

Due to the poor form factor, secondary RMS currents in the discontinuous mode can be much larger than those in the continuous mode. Hence, the discontinuous mode requires larger secondary wire size and output filter capacitors with larger ripple current ratings. The rectifier diodes will also run hotter in the discontinuous mode because of the larger secondary RMS currents. Further, the primary peak currents in the discontinuous mode are larger than those in the continuous mode. For the same output power, the triangle of Figure 4.2*a* must have a larger peak than the trapezoid of Figure 4.2*c*. The consequence is that the discontinuous mode with its larger peak primary current requires a power transistor of higher current rating and possibly higher cost. Also, the higher primary current at the turn "off" edge of *Q*1 results in a potential for greater RFI problems.

Despite all the disadvantages of the discontinuous mode, it is much more widely used than the continuous mode. This is so for two reasons. First, as mentioned above, the discontinuous mode, with an inherently smaller transformer magnetizing inductance, responds more quickly and with a lower transient output voltage spike to rapid changes in output load current or input voltage. Second, because of a unique characteristic of the continuous mode (its transfer function has a right-half-plane-zero, to be discussed in a later chapter on feedback loop stabilization), the error amplifier bandwidth must be drastically reduced to stabilize the feedback loop.

After Pressman *Modern power devices, such as the Power Integration's Top Switch range of products, have the "noisy" FET drain part of the chip isolated from the heat sink tab. This, together with integrated drive and control circuits, which further reduce radiating area, very much reduces the RFI problems normally associated with the discontinuous flyback topology. ~K.B.*

4.5 Design Relations and Sequential Design Steps

4.5.1 Step 1: Establish the Primary/Secondary Turns Ratio

For the most expedient design, there are a number of decisions that should be made in the following logical sequence.

First select a core size to meet the power requirements.

Next choose the primary/master secondary turns ratio N_p/N_{sm} to determine the maximum "off"-voltage stress $\overline{V_{ms}}$ on the power transistor in the absence of a leakage inductance spike as follows:

Neglecting the leakage spike, the maximum transistor voltage stress at maximum DC input $\overline{V_{dc}}$ and for a 1-V rectifier drop is

$$\overline{V_{ms}} = \overline{V_{dc}} + \frac{N_p}{N_{sm}}(V_o + 1) \tag{4.4}$$

where $\overline{V_{ms}}$ is chosen sufficiently low so that a leakage inductance spike of $0.3V_{dc}$ on top of that still leaves a safety margin of about 30% below the maximum pertinent transistor rating (V_{ceo}, V_{cer}, or V_{cev}).

4.5.2 Step 2: Ensure the Core Does Not Saturate and the Mode Remains Discontinuous

To ensure that the core does not drift up or down its hysteresis loop, the "on" volt-second product (*A*1 in Figure 4.1*d*) must equal the reset volt-second product (*A*2 in Figure 4.1*d*). Assume that the "on" drop of *Q*1 and the forward drop of the rectifier *D*2 are both 1 V:

$$(\underline{V_{dc}} - 1)\overline{T_{on}} = (V_o + 1)\frac{N_p}{N_{sm}}T_r \tag{4.5}$$

where T_r shown in Figure 4.1*c* is the reset time required for the secondary current to return to zero.

To ensure the circuit operates in the discontinuous mode, a dead time (T_{dt} in Figure 4.1*c*) is established so that the maximum "on" time $\overline{T_{on}}$, which occurs when V_{dc} is a minimum, plus the reset time T_r is only 80% of a full period. This leaves $0.2T$ margin against unexpected decreases in R_o, which according to Eq. 4.3 would force the feedback loop to increase T_{on} in order to keep V_o constant.

As for the boost regulator, which is also a flyback type (Sections 1.4.2 and 1.4.3), it was pointed out that if the error amplifier has been designed to keep the loop stable only in the discontinuous mode, it may break into oscillation if the circuit momentarily enters the continuous mode.

Increasing DC load current or decreasing V_{dc} causes the error amplifier to increase T_{on} in order to keep V_o constant (Eq. 4.3). This increased T_{on} eats into the dead time T_{dt}, and eventually the secondary current does not fall to zero by the start of the next $Q1$ "on" time. This is the start of the continuous mode, and if the error amplifier has not been designed with a drastically lower bandwidth than required for discontinuous mode, the circuit will oscillate. To ensure that the circuit remains discontinuous, the maximum "on" time that will generate the desired maximum output power is established:

$$\overline{T_{on}} + T_r + T_{dt} = T$$

or

$$\overline{T_{on}} + T_r = 0.8T \tag{4.6}$$

Now in Eqs. 4.5 and 4.6, there are two unknowns, as N_p/N_{sm} has been calculated from Eq. 4.4 for specified $\overline{V_{dc}}$ and $\overline{V_{ms}}$. Then from the last two relations

$$\overline{T_{on}} = \frac{(V_o + 1)(N_p/N_{sm})(0.8T)}{(\underline{V_{dc}} - 1) + (V_o + 1)(N_p/N_{sm})} \tag{4.7}$$

4.5.3 Step 3: Adjust the Primary Inductance Versus Minimum Output Resistance and DC Input Voltage

From Eq. 4.3, the primary inductance is

$$L_p = \frac{R_o}{2.5T}\left(\frac{\underline{V_{dc}}\,\overline{T_{on}}}{V_o}\right)^2 = \frac{(\underline{V_{dc}}\,\overline{T_{on}})^2}{2.5T\,\overline{P_o}} \tag{4.8}$$

4.5.4 Step 4: Check Transistor Peak Current and Maximum Voltage Stress

If the transistor is a bipolar type, it must have an acceptably high gain at the peak current operating current Ip. This is

$$I_p = \frac{\underline{V_{dc}}\,\overline{T_{on}}}{L_p} \tag{4.9}$$

where $\underline{V_{dc}}$ is specified and $\overline{T_{on}}$ is calculated from Eq. 4.7 and L_p is calculated from Eq. 4.8.

If $Q1$ is a MOSFET, it should have a peak current rating about 5 to 10 times the value calculated from Eq. 4.9 so that its "on"-state resistance is low enough to yield an acceptably low voltage drop and power loss.

4.5.5 Step 5: Check Primary RMS Current and Establish Wire Size

The primary current is a triangle of peak amplitude I_p (Eq. 4.9) at a maximum duration $\overline{T_{\text{on}}}$ out of every period T. Its RMS value (Section 2.2.10.6) is

$$I_{\text{rms(primary)}} = \frac{I_p}{\sqrt{3}}\sqrt{\frac{\overline{T_{\text{on}}}}{T}} \tag{4.10}$$

where I_p and $\overline{T_{\text{on}}}$ are as given by Eqs. 4.9 and 4.7.

At 500 circular mils per RMS ampere, the required number of circular mils is

$$\begin{aligned} \text{Circular mils required (primary)} &= 500\ I_{\text{rms(primary)}} \\ &= 500\frac{I_p}{\sqrt{3}}\sqrt{\frac{\overline{T_{\text{on}}}}{T}} \end{aligned} \tag{4.11}$$

4.5.6 Step 6: Check Secondary RMS Current and Select Wire Size

The secondary current is a triangle of peak amplitude $I_s = I_p(N_p/N_s)$ and duration T_r. Primary/secondary turns ratio N_p/N_s is given by Eq. 4.4 and $T_r = (T - T_{\text{on}})$. Secondary RMS current is then

$$I_{\text{rms(secondary)}} = \frac{I_p(N_p/N_s)}{\sqrt{3}}\sqrt{\frac{T_r}{T}} \tag{4.12}$$

At 500 circular mils per RMS ampere, the required number of circular mils is

$$\text{Secondary circular mils required} = 500 I_{\text{rms(secondary)}} \tag{4.13}$$

4.6 Design Example for a Discontinuous-Mode Flyback Converter

We will now look at a worked design example for a flyback converter with the following specifications:

V_o	5.0 V
$P_{o(\text{max})}$	50 W
$I_{o(\text{max})}$	10 A
$I_{o(\text{min})}$	1.0 A
$V_{\text{dc(max)}}$	60 V
$V_{\text{dc(min)}}$	38 V
Switching frequency	50 kHz

First, select the voltage rating of the transistor as this mainly determines the transformer turns ratio. Choose a device with a 200-V rating. In Eq. 4.4 choose the maximum stress V_{ms} on the transistor in the "off" state (excluding the leakage inductance spike) as 120 V. Then even with a 25% or 30-V leakage spike, this leaves a 50-V margin to the maximum voltage rating. Then from Eq. 4.4

$$120 = 60 + \frac{N_p}{N_{sm}}(V_o + 1) \quad \text{or} \quad \frac{N_p}{N_{sm}} = 10$$

Now choose maximum "on" time from Eq. 4.7:

$$\begin{aligned}\overline{T_{om}} &= \frac{(V_o + 1)(N_p/N_{sm})(0.8T)}{(V_{dc} - 1) + (V_o + 1)N_p/N_{sm}} \\ &= \frac{6 \times 10 \times 0.8 \times 20}{(38 - 1) + 6 \times 10} \\ &= 9.9\,\mu s\end{aligned}$$

From Eq. 4.8

$$\begin{aligned}L_p &= \frac{(V_{dc}\overline{T_{on}})^2}{2.5T\,P_o} \\ &= \frac{(38 \times 9.9 \times 10^{-6})^2}{2.5 \times 20 \times 10^{-6} \times 50} \\ &= 56.6\,\mu H\end{aligned}$$

From Eq. 4.9

$$\begin{aligned}I_p &= \frac{V_{dc}\overline{T_{on}}}{L_p} \\ &= \frac{38 \times 9.9 \times 10^{-6}}{56.6 \times 10^{-6}} \\ &= 6.6\text{ A}\end{aligned}$$

From Eq. 4.10, primary RMS current is

$$\begin{aligned}I_{rms(primary)} &= \frac{I_p}{\sqrt{3}}\sqrt{\frac{\overline{T_{on}}}{T}} \\ &= \frac{6.6}{\sqrt{3}} \times \sqrt{\frac{9.9}{20}} \\ &= 2.7\text{ A}\end{aligned}$$

From Eq. 4.11, primary circular mils requirement is

$$I_{(\text{primary circular mils})} = 500 \times 2.7 = 1350 \text{ circular mils}$$

This calls for a No. 19 wire of 1290 circular mils, which is close enough.

From Eq. 4.12, secondary RMS current is

$$I_{\text{rms(secondary)}} = \frac{I_p(N_p/N_s)}{\sqrt{3}}\sqrt{\frac{T_r}{T}}$$

But reset time T_r is

$$(0.8T - \overline{T_{\text{on}}}) = (16 - 9.9) = 6.1\,\mu\text{s}$$

Then

$$I_{\text{rms(secondary)}} = \frac{6.6 \times 10}{\sqrt{3}}\sqrt{\frac{6.1}{20}}$$
$$= 21 \text{ A}$$

We see that from Eq. 4.12, the required number of circular mils is 500 × 21 = 10,500. This calls for No. 10 wire, which is impractically large in diameter. A foil winding or a number of smaller diameter wires in parallel with an equal total circular-mil area would be used.

After Pressman *Contrary to popular belief, the wire size and leakage inductance in a flyback transformer are important design parameters. Multiple strands of wire in parallel are required, with the maximum wire size selected to minimize skin and proximity effects. Even though the secondary energy comes from the energy stored in the transformer core, leakage inductance must still be minimized to ensure good energy transfer from the primary winding to the secondary windings to reduce voltage spikes on Q1 at the "off" transition. This will reduce the amount of snubbing required on Q1 and reduce RFI. (See Chapter 7.) ~K.B.*

The output capacitor is chosen on the basis of specified peak-to-peak output voltage ripple as follows:

At maximum output current, the filter capacitor C_o carries the 10-A output current for all but the 6.1 μs reset period, or 13.9 μs. The voltage on this capacitor droops by $V = I(T - t_{\text{off}})/C_o$. Then for a voltage droop of 0.05 V

$$C_o = \frac{10 \times 13.9 \times 10^{-6}}{0.05}$$
$$= 2800\,\mu\text{F}$$

From Section 1.4.7, the average ESR of a 2800-μF aluminum electrolytic capacitor is

$$R_{esr} = 65 \times 10^{-6}/C_o = 0.023\ \Omega$$

At the instant of transistor turn "off," a peak secondary current of 66 amps flows through the above capacitor ESR, causing a thin spike of $66 \times 0.023 = 1.5$ V. This large-amplitude thin spike at transistor turn "off" is a universal problem with flybacks having a large N_p/N_s ratio. It is usually solved by using a larger filter capacitor than calculated as above (since R_{esr} is inversely proportional to C_o) and/or integrating away the thin spike with a small *LC* circuit.

After Pressman *Since the spike contains a large amount of high frequency components, combining several capacitors in parallel will reduce the spike amplitude significantly. Small ceramic and film caps are often used. ~K.B.*

Selecting a transformer core for a flyback topology circuit is significantly different than selecting for a forward converter. Remember in the flyback that when current flows in the primary, the secondary current is zero, and there is no current flow in the secondary to buck out the primary ampere turns as there is in the forward converters. Thus in the flyback, all the primary ampere turns tend to saturate the core.

In contrast, in non-flyback topologies, secondary load current flows when primary current flows and is in the direction (by Lenz's law) to cancel the ampere turns of the primary. It is only the primary magnetizing current that drives the core over its hysteresis loop and moves it toward saturation. That magnetizing current is kept a small fraction of the primary load current by providing a large magnetizing inductance and hence core saturation is not a basic a problem with non-flyback topologies.

Hence, flyback transformer cores must have some means of carrying large primary currents without saturating. This is done by choosing low permeability materials such as MPP (molybdenum permalloy powder) cores that have an inherent air gap or by using gapped ferrite cores (Section 2.3.9.3 and Chapter 7). This is discussed further in the following section.

4.6.1 Flyback Magnetics

Referring to Figure 4.1*a*, it is seen from the winding dots that when the transistor is "on" and current flows in the primary, no secondary currents flow. This is totally different from forward-type converters, in which current flows in the secondary when it flows in the primary. Thus in a forward-type converter, primary current flows into a dot end and secondary current flows out of a dot end.

Primary and secondary load ampere-turns then cancel each other out and do not move the core across its hysteresis loop. In the forward-type converters, it is only the magnetizing current that drives the core across the hysteresis loop and may potentially saturate it. But this magnetizing current is a small fraction (rarely >10%) of the total primary current.

In a flyback converter, however, the entire triangle of primary current shown in Figure 4.1*b* drives the core across the hysteresis loop as it is not canceled out by any secondary ampere turns. Thus, even at very low output power, an ungapped ferrite core would almost immediately saturate and destroy the transistor if nothing were done to prevent it.

To prevent core saturation in the flyback transformer, the core is gapped. The gapped core can be either of two types. It can be a solid ferrite core with a known air-gap length obtained by grinding down the center leg in EE or cup-type cores. The known gap length can also be obtained by inserting plastic shims between the two halves of an EE, cup, or UU core.

A more usual gapped core for flyback converters is the MPP or molypermalloy powder core. Such cores are made of a baked and hardened mix of magnetic powdered particles. These powdered particles are mixed in a slurry with a plastic resin binder and cast in the shape of a toroid. Each magnetic particle in the toroid is thus encapsulated within a resin envelope that behaves as a "distributed air gap" and acts to keep the core from saturating. The basic magnetic material that is ground up into a powder is Square Permalloy 80, an alloy of 79% nickel, 17% iron, and 4% molybdenum, made by Magnetics Inc. and Arnold Magnetics, among others.

The permeability of the resulting toroid is determined by controlling the concentration of magnetic particles in the slurry. Permeabilities are controlled to within ±5% over large temperature ranges and are available in discrete steps ranging from 14 to 550. Toroids with low permeability behave like gapped cores with large air gaps. They require a relatively large number of turns to yield a desired inductance but tolerate many ampere-turns before they saturate. Higher permeability cores require relatively fewer turns but saturate at a lower number of ampere-turns.

Such MPP cores are used not only for flyback transformers in which all the primary current is DC bias current. They are also used for forward converter output inductors where, as has been seen, a unique inductance is required at the large DC output current bias (Section 1.3.6).

4.6.2 Gapping Ferrite Cores to Avoid Saturation

Adding an air gap to a solid ferrite core achieves two results. First, it tilts the hysteresis loop as shown in Figure 2.5 and hence decreases its permeability, which must be known to select the number of turns for a desired inductance. Second, and more important, it increases the number of ampere turns it can tolerate before it saturates.

Core manufacturers often offer curves that permit calculation of the number of turns for a desired inductance and the number of ampere-turns at which saturation commences. Such curves are shown in Figure 4.3 and show A_{lg}, the inductance per 1000 turns with an air

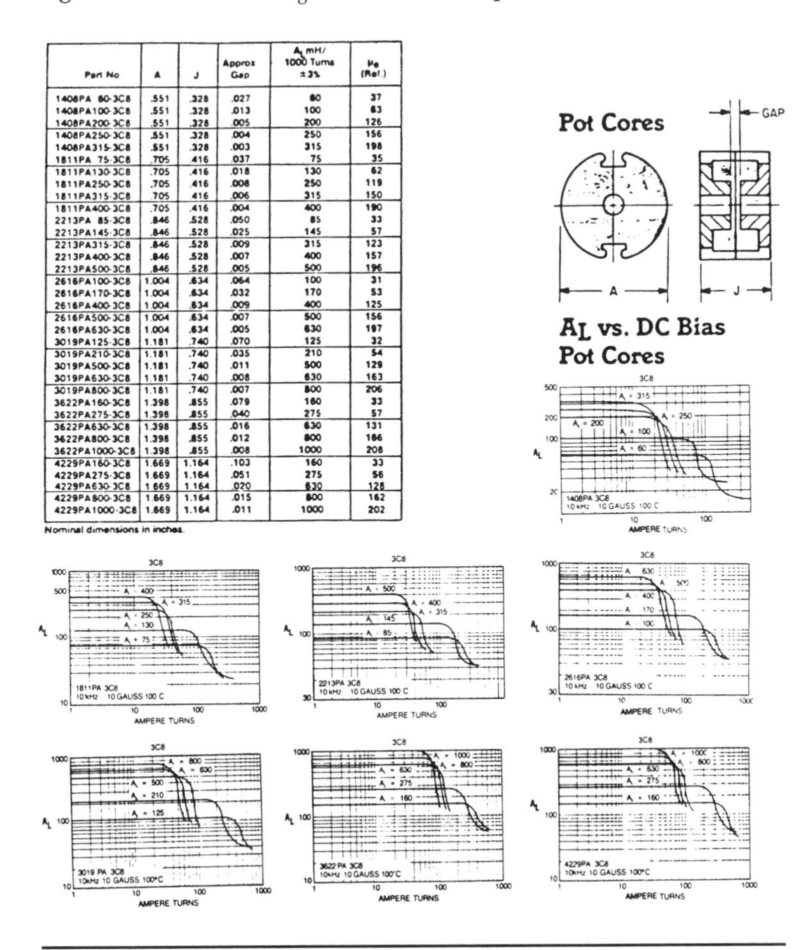

Part No	A	J	Approx Gap	A_L mH/ 1000 Turns ±3%	μ_e (Ref.)
1408PA 80-3C8	.551	.328	.027	80	37
1408PA100-3C8	.551	.328	.013	100	63
1408PA200-3C8	.551	.328	.005	200	126
1408PA250-3C8	.551	.328	.004	250	156
1408PA315-3C8	.551	.328	.003	315	198
1811PA 75-3C8	.705	.416	.037	75	35
1811PA130-3C8	.705	.416	.018	130	62
1811PA250-3C8	.705	.416	.008	250	119
1811PA315-3C8	.705	.416	.006	315	150
1811PA400-3C8	.705	.416	.004	400	190
2213PA 85-3C8	.846	.528	.050	85	33
2213PA145-3C8	.846	.528	.025	145	57
2213PA315-3C8	.846	.528	.009	315	123
2213PA400-3C8	.846	.528	.007	400	157
2213PA500-3C8	.846	.528	.005	500	196
2616PA100-3C8	1.004	.634	.064	100	31
2616PA170-3C8	1.004	.634	.032	170	53
2616PA400-3C8	1.004	.634	.009	400	125
2616PA500-3C8	1.004	.634	.007	500	156
2616PA630-3C8	1.004	.634	.005	630	197
3019PA125-3C8	1.181	.740	.070	125	32
3019PA210-3C8	1.181	.740	.035	210	54
3019PA500-3C8	1.181	.740	.011	500	129
3019PA630-3C8	1.181	.740	.008	630	163
3019PA800-3C8	1.181	.740	.007	800	206
3622PA160-3C8	1.398	.855	.079	160	33
3622PA275-3C8	1.398	.855	.040	275	57
3622PA630-3C8	1.398	.855	.016	630	131
3622PA800-3C8	1.398	.855	.012	800	166
3622PA1000-3C8	1.398	.855	.008	1000	208
4229PA160-3C8	1.669	1.164	.103	160	33
4229PA275-3C8	1.669	1.164	.051	275	56
4229PA630-3C8	1.669	1.164	.020	630	128
4229PA800-3C8	1.669	1.164	.015	800	162
4229PA1000-3C8	1.669	1.164	.011	1000	202

Nominal dimensions in inches.

FIGURE 4.3 Inductance per 1000 turns (A_{lg}) for various ferrite cores with various air gaps. Note the "cliff" points in ampere-turns where saturation commences. (*Courtesy Ferroxcube Corporation.*)

gap and the number of ampere-turns (NI_{sat}) where saturation starts to set in. Since inductance is proportional to the square of the number of turns, the number of turns N_l for any inductance L is calculated from

$$N_l = 1000\sqrt{\frac{L}{A_{lg}}} \tag{4.14}$$

Figure 4.3 shows A_{lg} curves for a number of different air gaps and the "cliff" point at which saturation starts. It can be seen that the larger the air gap, the lower the value of A_{lg} and the larger the number of ampere-turns at which saturation starts. If such curves were available for all cores at various air gaps, Eq. 4.14 would give the number of turns for any selected air gap from the value of A_{lg} read from the curve. The cliff point on the curve would tell whether, at those turns and for the specified primary current, the core had fallen over the saturation cliff.

Such curves, though, are not available for all cores and all air gaps. This is no problem, because A_{lg} can be calculated with reasonable accuracy from Eq. 2.39 using A_l with no gap, which is always given in the manufacturers' catalogs. The cliff point at which saturation starts can be calculated from Eq. 2.37 for any air gap. The cliff point corresponds to the flux density in iron B_i, where the core material itself starts bending over into saturation.

From Figure 2.3, it is seen that this is not a very sharp breaking point, but occurs around 2500 G for this ferrite material (Ferroxcube 3C8). Thus the cliff in ampere-turns is found by substituting 2500 G in Eq. 2.37. As noted in connection with Eq. 2.37, in the usual case, the air-gap length l_a is much larger than l_i/u as u is so large. Then the iron flux density as given by Eq. 2.37 is determined mainly by the air-gap length l_a.

4.6.3 Using Powdered Permalloy (MPP) Cores to Avoid Saturation

These toroidal cores are widely used and made by Magnetics Inc. (data in catalog MPP303S) and by Arnold Co. (data in catalog PC104G).

After Pressman *The term* transformer *in the phrase* flyback transformer *is a misnomer and is very misleading. For true transformer action to take place both primary and secondaries must conduct current at the same time. We are all aware that a true transformer conserves the primary to secondary voltage ratio (irrespective of current). In the flyback case the so-called transformer conserves the primary to secondary ampere-turns ratio (irrespective of voltage). This means it is really a "choke"—an inductor with a DC component of current and additional windings. I find it much easier and less confusing to design my flyback "transformers" from this perspective. The reader may also find it helpful to use this approach because the inductance* *becomes the dependant variable and can be easily adjusted to get the desired results. Chapter 7 deals with the design in this way.*

You will see from the following that although he does not mention it, Pressman is leading you in the direction of choke design. ~K.B.

The problem in designing a core of desired inductance at a specified maximum DC current bias is to select a core geometry and material permeability, such that the core does not saturate at the maximum ampere-turns to which it is subjected. There are a limited number of core geometries, each available in permeabilities ranging from 14 to 550. Selection procedures are described in the catalogs mentioned above, but the following has been found more direct and useful.

In the Magnetics Inc. catalog, one full page (Figure 4.4) is devoted to each size toroid, and for each size, its A_l value (inductance in millihenries per 1000 turns) is given for each discrete permeability. Figure 4.5, also from the Magnetics Inc. catalog, gives the falloff in permeability (or A_l value) for increasing magnetizing force in oersteds for core materials of the various available permeabilities. (Recall the oersted–ampere-turns relation in Eq. 2.6.)

A core geometry and permeability can be selected so that at the maximum DC current and the selected number of turns, the A_l and hence inductance has fallen off by any desired percentage given in Figure 4.5. Then at zero DC current, the inductance will be greater by that percentage. Such inductors or chokes are referred to as "swinging chokes" and in many applications are desirable. For example, if an inductor is permitted to swing a great deal, in an output filter, it can tolerate a very low minimum DC current before it goes discontinuous (Section 1.3.6). But this greatly complicates the feedback-loop stability design and, most often, the inductor in an output filter or transformer in a flyback will not be permitted to "swing" or vary very much between its zero and maximum current value.

Referring to Figure 4.4, it is seen that a core of this specific size is available in permeabilities ranging from 14 to 550. Cores with permeability above 125 have large values of A_l and hence require fewer turns for a specified inductance at zero DC current bias. But in Figure 4.5 it is seen that the higher-permeability cores saturate at increasingly lower ampere-turns of bias. Hence in power supply usage, where DC current biases are rarely under 1 A, cores of permeability greater than 125 are rarely used, and an inductance swing or change of 10% from zero to the maximum specified current is most often acceptable.

In Figure 4.5, it is seen that for a permeability dropoff or swing of 10%, core materials of permeabilities 14, 26, 60, and 125 can sustain maximum magnetizing forces of only 170, 95, 39, and 19 Oe, respectively. These maximum magnetizing forces in oersteds can be translated into maximum ampere-turns by Eq. 2.6 ($\overline{H} = 0.4\pi(\overline{NI})/l_m$), in

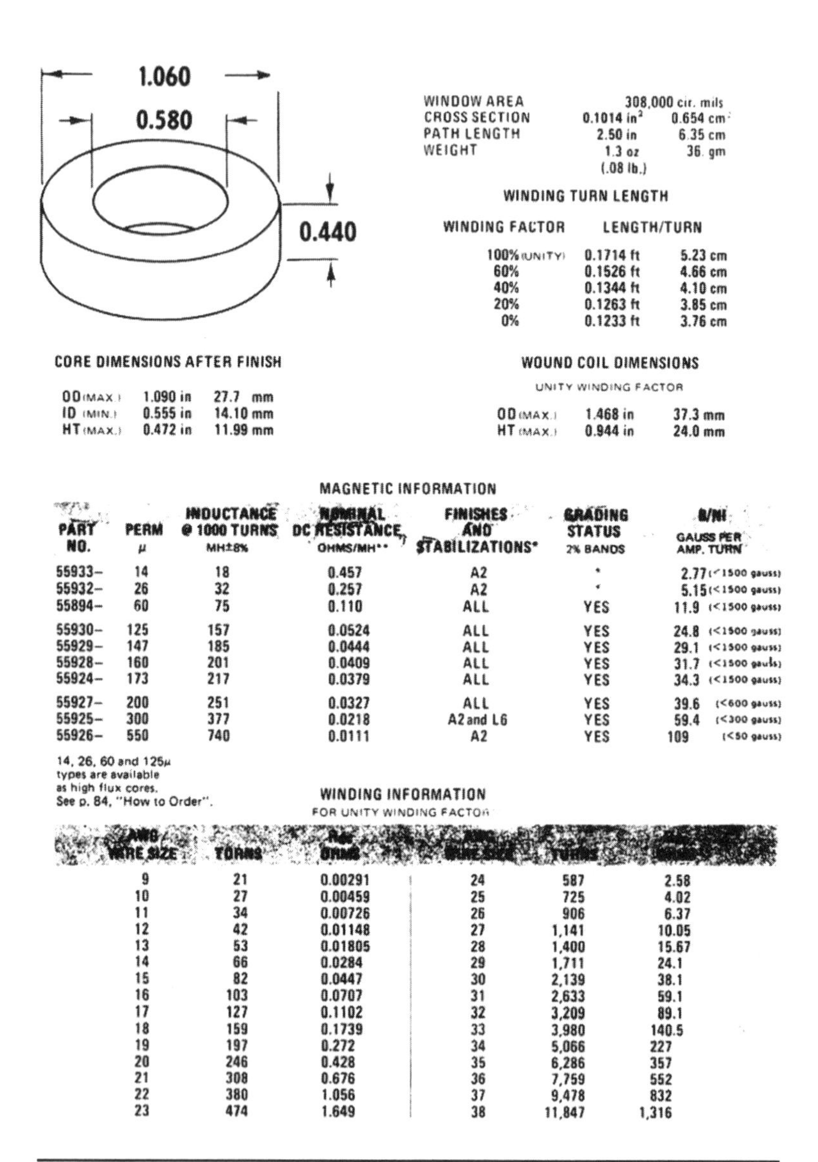

WINDOW AREA	308,000 cir. mils	
CROSS SECTION	0.1014 in^2	0.654 cm^2
PATH LENGTH	2.50 in	6.35 cm
WEIGHT	1.3 oz (.08 lb.)	36. gm

WINDING TURN LENGTH

WINDING FACTOR	LENGTH/TURN	
100% (UNITY)	0.1714 ft	5.23 cm
60%	0.1526 ft	4.66 cm
40%	0.1344 ft	4.10 cm
20%	0.1263 ft	3.85 cm
0%	0.1233 ft	3.76 cm

CORE DIMENSIONS AFTER FINISH

OD (MAX.)	1.090 in	27.7 mm
ID (MIN.)	0.555 in	14.10 mm
HT (MAX.)	0.472 in	11.99 mm

WOUND COIL DIMENSIONS

UNITY WINDING FACTOR

OD (MAX.)	1.468 in	37.3 mm
HT (MAX.)	0.944 in	24.0 mm

MAGNETIC INFORMATION

PART NO.	PERM μ	INDUCTANCE @ 1000 TURNS MH±8%	NOMINAL DC RESISTANCE OHMS/MH**	FINISHES AND STABILIZATIONS*	GRADING STATUS 2% BANDS	B/NI GAUSS PER AMP. TURN
55933–	14	18	0.457	A2	*	2.77 (<1500 gauss)
55932–	26	32	0.257	A2	*	5.15 (<1500 gauss)
55894–	60	75	0.110	ALL	YES	11.9 (<1500 gauss)
55930–	125	157	0.0524	ALL	YES	24.8 (<1500 gauss)
55929–	147	185	0.0444	ALL	YES	29.1 (<1500 gauss)
55928–	160	201	0.0409	ALL	YES	31.7 (<1500 gauss)
55924–	173	217	0.0379	ALL	YES	34.3 (<1500 gauss)
55927–	200	251	0.0327	ALL	YES	39.6 (<600 gauss)
55925–	300	377	0.0218	A2 and L6	YES	59.4 (<300 gauss)
55926–	550	740	0.0111	A2	YES	109 (<50 gauss)

14, 26, 60 and 125μ types are available as high flux cores. See p. 84, "How to Order".

WINDING INFORMATION

FOR UNITY WINDING FACTOR

AWG WIRE SIZE	TURNS	Rdc OHMS	AWG WIRE SIZE	TURNS	Rdc OHMS
9	21	0.00291	24	587	2.58
10	27	0.00459	25	725	4.02
11	34	0.00726	26	906	6.37
12	42	0.01148	27	1,141	10.05
13	53	0.01805	28	1,400	15.67
14	66	0.0284	29	1,711	24.1
15	82	0.0447	30	2,139	38.1
16	103	0.0707	31	2,633	59.1
17	127	0.1102	32	3,209	89.1
18	159	0.1739	33	3,980	140.5
19	197	0.272	34	5,066	227
20	246	0.428	35	6,286	357
21	308	0.676	36	7,759	552
22	380	1.056	37	9,478	832
23	474	1.649	38	11,847	1,316

FIGURE 4.4 A typical MPP core. With its large distributed air gap, it can tolerate a large DC current bias without saturating. It is available in a large range of different geometries. (*Courtesy Magnetics Inc.*)

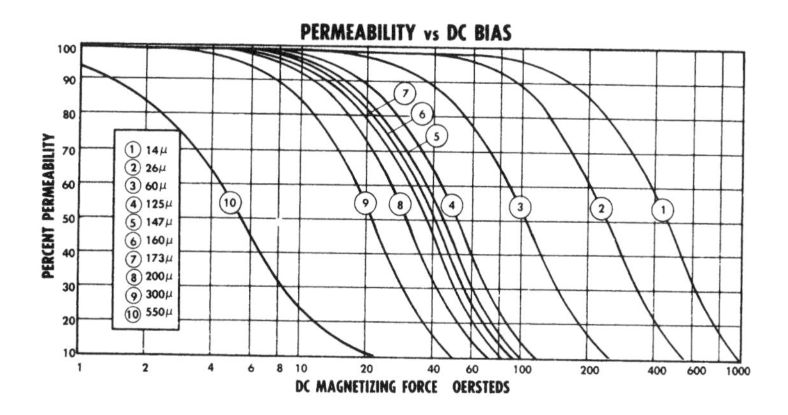

FIGURE 4.5 Falloff in permeability of A_1 for MPP cores of various permeabilities versus DC magnetizing force in oersteds. (*Courtesy Magnetics Inc.*)

which l_m is the magnetic path length in centimeters, given in Figure 4.3 for this particular core geometry as 6.35 cm.

From these maximum numbers of ampere-turns ($\overline{NI}$), beyond which inductance falls off more than 10%, the maximum number of turns ($\overline{N}$) is calculated for any peak current. From $\overline{N}$, the maximum inductance possible for any core at the specified peak current is calculated as $L_{max} = 0.9 A_1 (N_{max}/1000)^2$.

Tables 4.1, 4.2, and 4.3 show N_{max} and L_{max} for three often-used core geometries in permeabilities of 14, 26, 60, and 125 at peak currents of 1, 2, 3, 5, 10, 20, and 50 amperes. These tables permit core geometry and permeability selection at a glance without iterative calculations.

Table 4.1 is used in the following manner. Assume that this particular core has the acceptable geometry. The table is entered horizontally to the first peak current greater than specified value. At that peak current, move down vertically until the first inductance L_{max} greater than the desired value is reached. The core at that point is the only one which can yield the desired inductance with only a 10% swing. The number of turns N_d on that core for a desired inductance L_d within 5% is given by

$$N_d = 1000\sqrt{\frac{L_d}{0.95 A_l}}$$

where A_l is the value in column 3 in Table 4.1. If, moving vertically, no core can be found whose maximum inductance is greater than the desired value, the core with the next larger geometry (greater OD or greater height) must be used.

Magnetics Inc. core number	Permeability	A_l, mH per 1000 turns	Maximum $\overline{H}$ for 10% falloff in inductance	$\overline{NI}$ Maximum permissible ampere-turns corresponding to $\overline{H}$	Maximum permissible turns and inductance at those turns for a 10% inductance falloff at indicated peak currents							
						N_{max}/L_{max}						
Core	μ	A_l	H	NI	1A	2A	3A	5A	10A	20A	50A	I_p
55930	125	157	19	96	96	48	32	19	10	5	2	N_{max}
					1,382	339	145	56	15	3.5	0.6	L_{max}
55894	60	75	39	197	197	99	66	39	20	10	4	N_{max}
					2,620	662	294	103	27	7	1	L_{max}
55932	26	32	95	480	480	240	160	96	48	24	10	N_{max}
					6,635	1,659	737	265	66	17	3	L_{max}
55933	14	18	170	859	859	430	286	172	86	43	17	N_{max}
					11,954	2,995	1,325	479	120	30	5	L_{max}

Note: Magnetics Inc. MPP cores. All cores have outer diameter (OD) = 1.060 in, inner diameter (ID) = 0.58 in, height = 0.44 in, l_m = 6.35 cm. All inductances in microhenries.

TABLE 4.1 Maximum number of turns yielding maximum inductance for various peak currents I_p at maximum inductance falloff of 10% from zero current

Magnetics Inc. core number	Perme-ability	A_l, mH per 1000 turns	Maximum $\overline{H}$ for 10% falloff in inductance	$\overline{NI}$ Maximum permissible ampere-turns corresponding to $\overline{H}$	Maximum permissible turns and inductance at those turns for a 10% inductance falloff at indicated peak currents							
						N_{max}/L_{max}						
Core	μ	A_l	H	NI	1A	2A	3A	5A	10A	20A	50A	I_p
55206	125	68	19	77	77	39	26	15	8	4	2	N_{max}
					363	93	41	14	4	1	0.24	L_{max}
55848	60	32	39	158	158	79	53	32	16	8	3	N_{max}
					719	180	81	29	7	2	0.26	L_{max}
55208	26	14	95	385	385	193	128	77	39	19	8	N_{max}
					1,868	469	206	75	19	4.5	0.8	L_{max}
55209	14	7.8	170	689	689	345	230	138	69	34	14	N_{max}
					3,333	836	371	134	33	8	1.4	L_{max}

Note: Magnetics Inc. MPP cores: OD = 0.8 in, ID = 0.5 in, height = 0.25 in, l_m = 5.09 cm. All inductances in microhenries.

TABLE 4.2 Maximum number of turns and maximum inductance for various peak currents I_p at a maximum inductance falloff of 10% from zero current

Core	μ	A_l	$\overline{H}$	$\overline{NI}$	1A	2A	3A	5A	10A	20A	50A	I_p
55438	125	281	19	162	162	81	54	32	16	8	3	N_{max}
					6,637	1,659	737	259	65	16	2	L_{max}
55439	60	135	39	333	333	167	111	67	33	17	7	N_{max}
					13,473	3,389	1,497	545	132	35	6	L_{max}
55440	26	59	95	812	812	406	271	162	81	41	16	N_{max}
					35,011	8,753	3,900	1,394	348	89	14	L_{max}
55441	14	32	170	1454	1,454	727	485	291	145	73	29	N_{max}
					60,744	15,222	6,774	2,439	605	153	24	L_{max}

Note: Magnetics Inc. MPP cores: OD = 1.84 in, ID = 0.95 in, height = 0.71 in, l_m = 10.74 in. All inductances in microhenries.

TABLE 4.3 Maximum number of turns and maximum inductance for various peak currents I_p at a maximum inductance falloff of 10% from zero current

The core ID must be large enough to accommodate the number of turns of wire selected at the rate of 500 circular miles per RMS ampere, or the next larger size core must be used.

Tables 4.2 and 4.3 show similar data for smaller (OD = 0.80 in) and larger (OD = 1.84 in) families of cores. Similar charts can be generated for all the other available core sizes, but Tables 4.1 to 4.3 bracket about 90% of the possible designs for flyback transformers under 500 W or output inductors of up to 50 A.

A commonly used scheme for correcting the number of turns on a core when an initial selection has resulted in too large an inductance falloff should be noted. If, for an initially selected number of turns and a specified maximum current, the inductance or permeability falloff from Figure 4.5 is down by $P\%$, the number of turns is increased by $P\%$.

This moves the operating point further out by $P\%$, as the magnetizing force in oersteds is proportional to the number of turns. The core slides further down its saturation curve, and it might be thought that the inductance would fall off even more. But since inductance is proportional to the square of the number of turns, and magnetizing force is proportional only to the number of turns, the zero current inductance has been increased by $2P\%$ and magnetizing force has gone up only by $P\%$. The inductance is then correct at the specified maximum current. If the consequent swing is too large, a larger core must be used.

4.6.4 Flyback Disadvantages

Despite its many advantages, the flyback has the following drawbacks.

4.6.4.1 Large Output Voltage Spikes

At the end of the "on" time, the peak primary current is given by Eq. 4.9. Immediately after the end of the "on" time, that primary peak current, multiplied by the turns ratio N_p/N_s, is driven into the secondary where it decays linearly as shown in Figure 4.1*c*. In most cases, output voltages are low relative to input voltage, resulting in a large N_p/N_s ratio and a consequent large secondary current.

At the start of turn "off," the impedance looking into C_o is much lower than R_o (Figure 4.1) and almost all the large secondary current flows into C_o and its equivalent series resistor R_{esr}. This produces a large, thin output voltage spike, $I_p(N_p/N_s)R_{\text{esr}}$. The spike is generally less than 0.5 μs in width, as it is differentiated with a time constant of $R_{\text{esr}}C_o$.

Frequently a power supply specification calls for output voltage ripple only as an RMS or peak-to-peak fundamental value. Such a large, thin spike has a very low RMS value and, if a sufficiently large output filter capacitor is chosen, the supply can easily meet its RMS ripple specification but can have disastrously high, thin output spikes. It is common to see a 50-mV fundamental peak-to-peak output ripple with a 1-V thin spike sitting on top of it.

Thus, a small *LC* filter is almost always added after the main storage capacitor in flybacks. The *L* and *C* can be quite small as they have to filter out a spike generally less than 0.5 μs in width. The inductor is usually considerably smaller than the inductor in forward-type converters, but it still has to be stocked, and board space must be provided for it. Output voltage sensing for the error amplifier is taken before this *LC* filter.

4.6.4.2 Large Output Filter Capacitor and High Ripple Current Requirement

A filter capacitor for a flyback must be much larger than for a forward-type converter. In a forward converter, when the power transistor turns "off" (Figure 2.10), load current is supplied from the energy stored in both the filter inductor and capacitor. But in the flyback, that capacitor is necessarily larger because it is the stored energy in it alone that supplies current to the load during the transistor "on" time. Output ripple is determined mostly by the ESR of the filter capacitor (see Section 1.3.7). An initial selection of the filter capacitor is made on the basis of output ripple specification from Eq. 1.10.

Frequently, however, it is not the output ripple voltage requirement that determines the final choice of the filter capacitor. Ultimately it may be the ripple current rating of the capacitor selected initially on the basis of the output ripple voltage specification.

In a forward-type converter (as in a buck regulator), the capacitor ripple current is greatly limited by the output inductor in series with it (Section 1.3.6). In a flyback, however, the full DC load current flows from common through the capacitor during the transistor "on" time. During the transistor "off" time, a charge of equal ampere-second product must flow into the capacitor to replenish the charge it lost during the "on" time. Assuming, as in Figure 4.1, a sum of "on" time plus reset time of 80% of full period, the RMS ripple current in the capacitor is closely

$$I_{\text{rms}} = I_{\text{dc}}\sqrt{\frac{t_{\text{on}}}{T}} = I_{\text{dc}}\sqrt{0.8} = 0.89I_{\text{dc}} \qquad (4.15)$$

If the capacitor initially selected on the basis of output ripple voltage specifications did not also have the ripple current rating of Eq. 4.15, a larger capacitor or more units in parallel must be chosen.

4.7 Universal Input Flybacks for 120-V AC Through 220-V AC Operation

Here we consider universal or wide input range flyback topologies that do not have the auto ranging, voltage doubling methods previously described.

In Section 3.2.1, we considered a commonly used scheme that permitted operation from either a 120-V AC or 220-V AC line with minimal changes. As seen in Figure 3.1 at 120 V AC, switch S1 is thrown to the lower position, making the circuit into a voltage doubler that yields a rectified voltage of 336 V. With 220-V AC, S1 is thrown to the upper position and the circuit becomes a full-wave rectifier with $C1$ and $C2$ in series, yielding about 308 V. The converter is thus designed to always work from a rectified nominal input of 308 to 336 V DC by proper choice of the transformer turns ratio.

In some applications, it is preferable to eliminate the requirement of changing $S1$ from one position to the other in changing from 120- to 220-V AC operation. To change switch position without opening the power supply case, the switch must be accessible externally, and this is a safety hazard. The alternative is to change the switch internally, but this requires opening the power supply case to make the change, and this is a nuisance. Further, there is always the possibility that the switch is mistakenly thrown to the voltage doubling position when operated from 200 V AC. This, of course, would cause significant damage—the power transistor, rectifiers, and filter capacitors would be destroyed.

An alternative is the universal line voltage unit that does not require switching and can tolerate the full range of line inputs from 115 to 220 V AC. The rectified 115 V input will be 160 V DC and the 220 V AC will be 310 V DC.

A flyback converter, designed with a small primary/secondary turns ratio, can ensure that the "off"-voltage stress at high AC input does not overstress the power transistor.

The maximum "on" time $\overline{T_{on}}$ at the minimum value of the 220-V AC input is calculated from the corresponding minimum rectified DC input as in Eq. 4.7 and the rest of the magnetics design can proceed as shown in the text following Eq. 4.7. The minimum "on" time occurs at the maximum value of the 220-V AC input. Since the feedback loop keeps the product of $V_{dc}T_{on}$ constant (Eq. 4.3), minimum "on" time is $\underline{T_{on}} = \overline{T_{on}}(\underline{V_{dc}}/\overline{V_{dc}})$ where $\underline{V_{dc}}$ and $\overline{V_{dc}}$ correspond to the minimum and maximum values of the 220-V AC line.

The maximum "on" time with 115-V AC input is still given by Eq. 4.7 and will be greater than with 220 V, as the term $\underline{V_{dc}} - 1$ is smaller. But the primary inductance L_p given by Eq. 4.8, which is proportional to the product $V_{dc}T_{on}$, is still the same as that product is kept constant by the feedback loop. So long as the transistor can operate with the minimum "on" time calculated for the maximum DC corresponding to high AC input, there is no problem. With bipolar transistors operating at a high frequency, transistor storage time could prevent operation at too low an "on" time. An example will clarify this.

Eq. 4.4 gives the maximum "off" stress in terms of the maximum DC input voltage, the output voltage, and the N_p/N_s turns ratio. Assume in that equation that $\overline{V_{ms}}$ is 500 V; many bipolar transistors can safely sustain that voltage with a negative base bias at turn "off" (V_{cev} rating). At 220 V AC, the nominal V_{dc} is 310 V. Assume that the maximum at high line with a worst-case transient is 375 V. Then for a 5-V output, Eq. 4.4 gives a turns ratio of 21.

Now assume that minimum DC supply voltage is 80% of nominal. Assume a switching frequency of 50 kHz (period T of 20 μs). Maximum "on" time is calculated from Eq. 4.7 at the minimum DC input corresponding to minimum AC input of 0.8 × 115 or 92 V AC. For the corresponding DC input of 1.41 × 92 or about 128 V, maximum "on" time calculated from Eq. 4.7 is 7.96 μs.

Minimum "on" time occurs at maximum input voltage. Assuming a 20% high line, the maximum DC input is $1.2 \times 220 \times 1.41 = 372$ V. Since the feedback loop keeps the product of $V_{dc}T_{on}$ constant (Eq. 4.3), "on" time at the 20% high line of 264 V AC is (128/372)(7.96) or 2.74 μs. The circuit can thus cope with either a 20% low AC line input of 92 V AC from a nominal 115 V AC, or a 20% high AC input of 264 V AC from the nominal 220-V AC line by readjusting its "on" time from 7.96 to 2.74 μs.

If this were attempted at higher switching frequencies, the minimum "on" time at a 220-V AC line would become so low as to prohibit the use of bipolar transistors, which could have 0.5- to 1.0-μs storage time. The upper-limit switching frequency at which the above scheme can be used with bipolar transistors is about 100 kHz.

It is instructive to complete the above design. Assume an output power of 150 W at 5-V output. Then $R_o = 0.167\ \Omega$ and the primary inductance from Eq. 4.8 is

$$L_p = \left(\frac{0.167}{2.5 \times 20 \times 10^{-6}}\right)\left(\frac{128 \times 7.96 \times 10^{-6}}{5}\right)^2$$

$$= 139\ \mu\text{H}$$

and the peak primary current from Eq. 4.9 is

$$I_p = \frac{128 \times 7.96 \times 10^{-6}}{139 \times 10^{-6}} = 7.33\ \text{A}$$

There are many reasonably priced bipolar transistors with a V_{cev} rating above 500 V having adequate gain at 7.33 A.

Table 4.1 shows that the 55932 MPP core can tolerate a maximum of 480 ampere-turns, beyond which its inductance will fall off by more than 10% (at 5 A, column 9 shows that the maximum turns is 96 for a maximum inductance of 265 μH). For maximum ampere-turns, the inductance is $32{,}000 \times 0.9(66/1000)^2 = 125\mu$ H. If (as discussed in Section 4.2.3.2) 10% more turns are added, the inductance at 7.33 A will increase by 10% to 138 μH, but at zero current, the inductance will "swing" up to 20% above that.

If the 20% inductance swing is undesirable, the lower permeability core 55933 of Table 4.1 can be used. Table 4.1 shows that the maximum ampere-turns stress is 859. For 7.33 A, the maximum number of turns is 859/7.33 or 117. The maximum inductance for a swing of only 10% is $(0.117)^2 \times 18000 \times 0.9$ or 222 μH. For the desired 139 μh, the required turns are $1000\sqrt{0.139/18 \times 0.95} = 90$.

Thus a design not requiring voltage doubling/full-wave rectifier switching when operation is changed from 115 to 220 V AC is possible. But this subjects the power transistor to a leakage inductance spike at turn "off" of about 500 V. The lower reliability of this scheme must be weighed against the use of a double-ended forward converter or half bridge—both of which subject the "off" transistor to only the maximum DC input (375 V in the preceding example) with no leakage spike. Of course, for 115/220-V AC operation, the rectifier switching of Figure 3.1 must be accepted.

After Pressman *Modern FETs (for example, the Power Integrations "Top Switch" devices) very much simplify the design of universal input flyback type supplies, which are now an accepted and standard topology for lower power applications. Very good application notes are available for these devices. ~K.B.*

4.8 Design Relations—Continuous-Mode Flybacks

4.8.1 The Relation Between Output Voltage and "On" Time

Look once again at Figure 4.1. When the transistor Q1 is "on," the voltage across the primary is close to $V_{dc}-1$ with the dot end negative with respect to the no-dot end, and the core is driven—say, up the hysteresis loop. When the transistor turns "off," the magnetizing current reverses the polarity of all voltages in order to remain constant. The primary and secondary are driven positive, but the secondary is caught and clamped to $V_{om}+1$ by $D2$—assuming a 1-V forward drop.

This reflects across to the primary as a voltage $(N_p/N_s)(V_{om}+1)$, with the dot end now positive with respect to the no-dot end. All the current that was flowing in the primary (I_{PO} in Figure 4.2*c*) now transfers to the secondary as I_{TU} in Figure 4.2*d*. The initial magnitude of the secondary current I_{TU} is equal to the final primary current at the end of the "on" time (I_{PO}) times the turns ratio N_p/N_s. Since the dot end of the secondary is now positive with respect to the no-dot end, the secondary current ramps downward with the slope *UV* in Figure 4.2*d*.

Since the primary is assumed to have zero DC resistance, it cannot sustain a DC voltage averaged over many cycles. Thus in the steady state, the volt-second product across it when the transistor is "on" must equal that across it when the transistor is "off"—i.e., the voltage across the primary averaged over a full cycle must equal zero. This is equivalent to saying the core's downward excursion on the *BH* loop during the "off" time is exactly equal to the upward excursion during the "on" time. Then

$$(\underline{V_{dc}}-1)\overline{t_{on}} = (V_{om}+1)\frac{N_p}{N_s}t_{off}$$

or

$$V_{om} = \left[(\underline{V_{dc}}-1)\frac{N_s}{N_p}\frac{\overline{t_{on}}}{t_{off}}\right]-1 \tag{4.16}$$

and since there is no dead time in continuous mode, $\overline{t_{on}}+t_{off}=T$, and

$$V_{om} = \left[\frac{(\underline{V_{dc}}-1)(N_s/N_p)(\overline{t_{on}}/T)}{1-\overline{t_{on}}/T}\right]-1 \tag{4.17a}$$

$$= \left[\frac{(\underline{V_{dc}}-1)(N_s/N_p)}{(T/\overline{t_{on}})-1}\right]-1 \tag{4.17b}$$

The feedback loop regulates against DC input voltage changes by decreasing t_{on} as V_{dc} increases, or increasing t_{on} as V_{dc} decreases.

4.8.2 Input, Output Current-Power Relations

In Figure 4.6, the output power is equal to the output voltage times the average of the secondary current pulses. For I_{csr} equal to the current at the center of the ramp in the secondary current pulse

$$\begin{aligned} P_o &= V_o I_{csr}\frac{t_{off}}{T} \\ &= V_o I_{csr}(1-\overline{t_{on}}/T) \end{aligned} \tag{4.18}$$

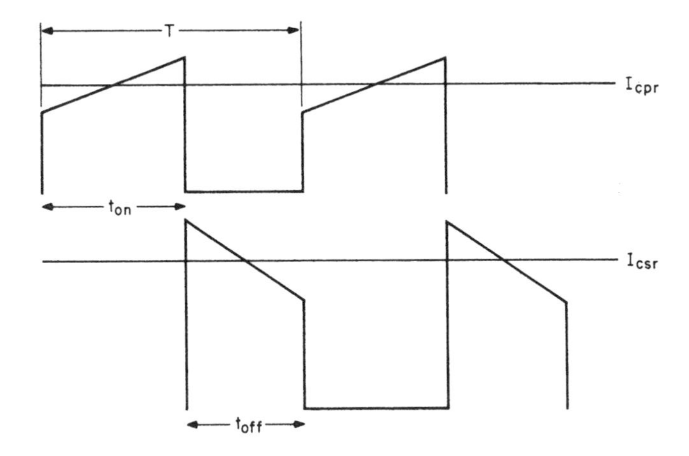

FIGURE 4.6 Real-time relation between the primary and secondary current waveforms in a continuous-mode flyback converter. Current is delivered to the output capacitor only during the "off" period of Q1. At a fixed DC input voltage, t_{on} and t_{off} remain constant. Output load current changes are accommodated by the feedback loop by changing the magnitude of the current at the center of the primary current ramp I_{cpr}, which results in a change at the center of the secondary current ramp (I_{csr}). This occurs over many switching cycles by temporary increases in "on" time until the average current pulse amplitudes build up and then relax to the new steady-state values of t_{on} and t_{off}.

or

$$I_{csr} = \frac{P_o}{V_o(1 - \overline{t_{on}}/T)} \tag{4.19}$$

In Eqs. 4.18 and 4.19, $\overline{t_{on}}/T$ is given by Eq. 4.17 for specified values of V_{om} and V_{dc}, and turns ratio N_s/N_p from Eq. 4.4, which was chosen for acceptably low maximum "off"-voltage stress at maximum DC input.

Further, for an assumed efficiency of 80%, $P_o = 0.8P_{in}$ and I_{cpr} is equal to the current at the center of the ramp in the primary current pulse:

$$P_{in} = 1.25 \quad P_o = V_{dc} I_{cpr} \frac{\overline{t_{on}}}{T}$$

or

$$I_{cpr} = \frac{1.25P_o}{(V_{dc})(\overline{t_{on}}/T)} \tag{4.20}$$

After Pressman *In the continuous mode, the duty cycle is defined by the voltage ratio. Changes in load current try to reflect into the primary, but for transient load changes, the transformer inductance limits the rate of change of current. Hence the first and immediate effect of a transient load increase is to cause a decrease in output voltage, resulting in an increase in the "on" period of Q1 (to increase the primary current). But this results in a further drop in output voltage because there is an immediate decrease in the energy-transferring "off" period (the secondary conducting period). It takes many cycles before the new higher current conditions are established, at which point the duty cycle returns to its original value. This is a dynamic effect intrinsic to the topology and cannot be compensated by the control loop. In terms of control theory, this translates to a right-half-plane-zero. ~K.B.*

4.8.3 Ramp Amplitudes for Continuous Mode at Minimum DC Input

It has been shown that the threshold of continuous-mode operation occurs when there is just the beginning of a step at the front end of the primary current ramp. Referring to Figure 4.6, the step appears when the current at the center of the primary ramp I_{cpr} just exceeds half the ramp amplitude dI_p. That value of $I_{cpr}(\underline{I_{cpr}})$ is then the minimum value at which the circuit is still in the continuous mode. From Eq. 4.20, I_{cpr} is proportional to output power and hence for the minimum output power $\underline{P_o}$ corresponding to $\underline{I_{cpr}}$

$$I_{cpr} = \frac{dI_p}{2} = \frac{1.25\underline{P_o}}{(\underline{V_{dc}})(\overline{t_{on}}/T)}$$

or

$$dI_p = \frac{2.5\underline{P_o}}{(\underline{V_{dc}})(\overline{t_{on}}/T)} \tag{4.21}$$

In Eq. 4.21, $\overline{t_{on}}$ is taken from Eq. 4.17 at the corresponding value minimum of $V_{dc}(\underline{V_{dc}})$. The slope of the ramp dI_p is given by $dI_p = (\underline{V_{dc}} - 1)\overline{t_{on}}/L_p$, where L_p is the primary magnetizing inductance. Then

$$\begin{aligned} L_p &= \frac{(\underline{V_{dc}} - 1)\overline{t_{on}}}{dI_p} \\ &= \frac{(\underline{V_{dc}} - 1)(\underline{V_{dc}})(\overline{t_{on}})^2}{2.5\underline{P_o}T} \end{aligned} \tag{4.22}$$

Here again, $\underline{P_o}$ is the minimum specified value of output power and $\overline{t_{on}}$ is the maximum "on" time calculated from Eq. 4.17 at the minimum specified DC input voltage $\underline{V_{dc}}$.

4.8.4 Discontinuous- and Continuous-Mode Flyback Design Example

It is instructive to compare discontinuous- and continuous-mode flyback designs at the same output power levels and input voltages. The magnitudes of the currents and primary inductances will be revealing.

Assume a 50-W, 5-V output flyback converter operating at 50 kHz from a telephone industry prime power source (38 V DC minimum, 60 V maximum). Assume a minimum output power of one-tenth the nominal, or 5 W.

Consider first a discontinuous-mode flyback. Choosing a bipolar transistor with a 150-V V_{ceo} rating is very conservative, because it is not necessary to rely on the V_{cer} or V_{cev} ratings that permit larger voltages. Then in Eq. 4.4, assume that the maximum "off"-voltage stress $\overline{V_{ms}}$ without a leakage spike is 114 V, which permits a 36-V leakage spike before the V_{ceo} limit is reached. Then Eq. 4.4 gives $N_p/N_s = (114 - 60)/6 = 9$.

Eq. 4.7 gives the maximum "on" time as

$$\overline{t_{on}} = 6 \times 9 \times 0.8 \frac{20 \times 10^{-6}}{37 + 6 \times 9}$$

$$= 9.49\ \mu s$$

and primary inductance for $R_o = 5/10 = 0.5\ \Omega$ from Eq. 4.8 is

$$L_p = \frac{0.5}{2.5 \times 20^{-6}} \left(\frac{38 \times 9.49}{5} \right)^2 \times 10^{-12}$$

$$= 52\ \mu H$$

Peak primary current from Eq. 4.9 is

$$I_p = \frac{38 \times 9.49 \times 10^{-6}}{52 \times 10^{-6}}$$

$$= 6.9\ A$$

and the start of the secondary current triangle is

$$I_{s(peak)} = (N_p/N_s) I_p = 9 \times 6.9 = 62\ A$$

Recall that in the discontinuous flyback, the reset time Tr—the time for the secondary current to decay back to zero—plus the maximum "on" time is equal to $0.8T$ (Eq. 4.6). Reset time is then $T_r = (0.8 \times 20) - 9.49 = 6.5\ \mu s$, and the average value of the secondary current triangle (which should equal the DC output current) is

$$I \text{ (secondary average)} = \frac{I_{s(\text{peak})}}{2}\frac{T_r}{T}$$

$$= \left(\frac{62}{2}\right)\frac{6.5}{20} = 10 \text{ A}$$

which is the DC output current.

Now consider a continuous-mode flyback for the same frequency, input voltages, output power, output voltage, and the same N_p/N_s ratio of 9. From Eq. 4.17*b*, calculate $\overline{t_{on}}/T$ for $V_{dc} = 38\ V$ as

$$5 = \left[\frac{(37/9)(\overline{t_{on}}/T}{1 - \overline{t_{on}}/T}\right] - 1$$

or $\overline{t_{on}}/T = 0.5934$ and $\overline{t_{on}} = 11.87\ \mu\text{s}$, $t_{off} = 8.13\ \mu\text{s}$ and from Eq. 4.19

$$I_{csr} = \frac{50}{(5)(1 - 0.5934)} = 24.59 \text{ A}$$

and the average of the secondary current pulse, which should equal the DC output current, is

$$I\text{(secondary average)} = I_{csr}(t_{off}/T) = 24.59 \times 8.13/20 = 10.0 \text{ A}$$

which checks. From Eq. 4.20, $I_{cpr} = 1.25 \times 50/(38)(11.86/20) = 2.77$ A.

From Eq. 4.22, for the minimum input power of 5 W at the minimum DC input voltage of 38 V, $L_p = 37 \times 38(11.86)^2 \times 10^{-12}/2.5 \times 5 \times 20 \times 10^{-6} = 791\ \mu\text{H}$.

The contrast between the discontinuous and continuous modes will now be clear from the following table, which compares the required primary inductances, and primary and secondary currents at minimum DC input of 38 V.

	Discontinuous	Continuous
Primary inductance, μH	52	791
Primary peak current, A	6.9	2.77
Secondary peak current, A	62.0	24.6
On time, μs	9.49	11.86
Off time, μs	6.5	8.13

The lower primary current and especially the secondary current for the continuous mode are certainly an advantage, but the much larger primary inductance that slows up response to load current changes, and the right-half-plane-zero that requires a very low error-amplifier bandwidth to achieve loop stabilization, can make the continuous mode a less desirable choice in applications that require good transient load response. In fixed-load applications this is not a problem.

4.9 Interleaved Flybacks

An interleaved flyback topology is shown in Figure 4.7. It consists of two or more discontinuous-mode flybacks whose power transistors

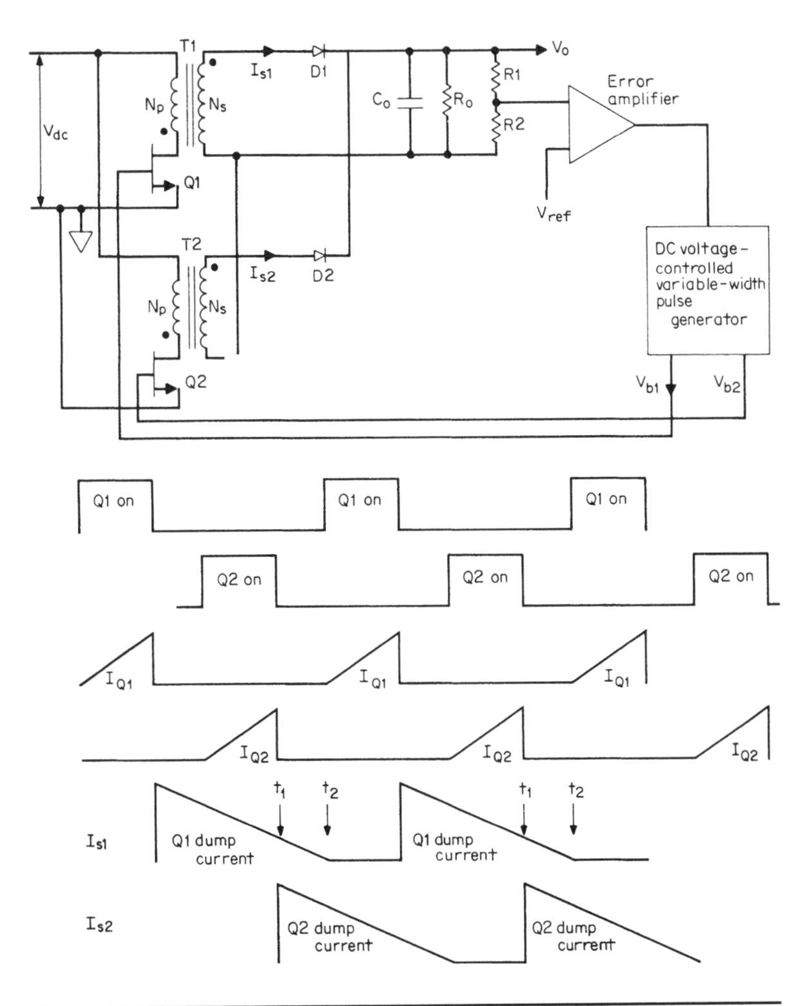

FIGURE 4.7 Interleaving two discontinuous-mode flybacks on alternative half cycles to reduce peak currents. Output powers of up to 300 W are possible with reasonably low peak currents.

are turned "on" at alternate half cycles and whose secondary currents are summed through their rectifying diodes.

It can be used at power levels up to 300 W, limited mainly by the high-peak primary and especially secondary currents. Although that power level can be obtained with a single continuous-mode flyback with reasonable currents, it may be better to accept the greater cost and volume of two or more interleaved discontinuous-mode flybacks. Both input and output ripple currents are much smaller and of higher frequency. Increasing the number of elements with suitable phase shift between drive pulses will further reduce the ripple current. Further, the discontinuous mode's faster response to load current changes, greater error-amplifier bandwidth, and the elimination of the right-half-plane-zero loop stabilization problem may make this a preferred choice.

A single discontinuous-mode flyback at the 300-W level is impractical because of the very high peak primary and secondary currents, as can be seen from Eqs. 4.2, 4.7, and 4.8.

At a lower power of 150 W, a single forward converter is very likely a better choice than the two interleaved flybacks because of the considerably lower secondary peak current of the forward converter. The interleaved flyback has been shown here for the sake of completeness and for its possible use at lower power levels when many (over five) outputs are required.

4.9.1 Summation of Secondary Currents in Interleaved Flybacks

The magnetics design of each flyback in an interleaved flyback proceeds exactly as for a single flyback at half the power level, because the secondary currents add into the output through their "ORing" rectifier diodes.

Even when both secondary diodes dump current simultaneously (as from t_1 to t_2), there is no possibility that one diode can back-bias the other and supply all the load current. This can happen if one attempts to sum the currents of two low-impedance voltage sources. If one of the low-impedance voltage sources has a slightly higher open-circuit voltage or a lower forward-drop OR diode, it will back-bias the other diode and supply all the load current by itself. This can over-dissipate the diode or the transistor supplying that diode.

Looking back into the secondary of a flyback, however, there is a high-impedance current source, which is the secondary inductance. Thus the current dumped into the common load by either diode is unaffected by the other diode simultaneously supplying load current.

4.10 Double-Ended (Two Transistor) Discontinuous-Mode Flyback

4.10.1 Area of Application

The topology is shown in Figure 4.8*a*. Its major advantage is that, using the scheme of the double-ended forward converter of Figure 2.13, its power transistors in the "off" state are subjected to only the maximum DC input voltage. This is a significant advantage over the single-ended forward converter of Figure 4.1, where the maximum "off"-voltage stress is the maximum DC input voltage plus the reflected secondary voltage $(N_p/N_s)(V_o + 1)$ plus a leakage inductance spike that may be as high as one-third of the DC input voltage.

4.10.2 Basic Operation

The lower "off"-voltage stress comes about in the same way as for the double-ended forward converter of Figure 2.13. Power transistors $Q1$, $Q2$ are turned "on" simultaneously. When they are "on," the dot end of the secondary is negative, $D3$ is reverse-biased, and no secondary current flows. The primary is then just an inductor, and current in it ramps up linearly at a rate of $dI_1/dt = V_{\rm dc}/(L_m + L_l)$, where L_m and L_l are the primary magnetizing and leakage inductances, respectively. When $Q1$ and $Q2$ turn "off," as in the previous flybacks, all primary and secondary voltages reverse polarity, $D3$ becomes forward-biased, and the stored energy in $L_m = 1/2L_m(I_1)^2$ is delivered to the load.

As shown previously, the "on" or set volt-second product across the primary must equal the "off" or reset volt-second product. At the instant of turn "off," the bottom end of L_l attempts to go far positive but is clamped to the positive end of $V_{\rm dc}$. The top end of L_m attempts to go far negative but is clamped to the negative end of $V_{\rm dc}$. Thus the maximum voltage stress at either $Q1$ or $Q2$ can never be more than $V_{\rm dc}$.

The actual resetting voltage V_r across the magnetizing inductance L_m during the "off" time is given by the voltage reflected from the secondary $(N_p/N_s)(V_o + V_{D3})$. The voltage across L_m and L_l in series is the DC supply voltage, and hence, as seen in Figure 4.8*b*, the voltage across the leakage inductance L_l is $V_l = (V_{\rm dc} - V_r)$.

The division of the $V_{\rm dc}$ supply voltage across L_m and L_l in series during the "off" time is a very important point in the circuit design and establishes the transformer turns ratio N_p/N_s as discussed below.

The price paid for this advantage is, of course, the requirement for two transistors and the two clamp diodes, $D1$, $D2$.

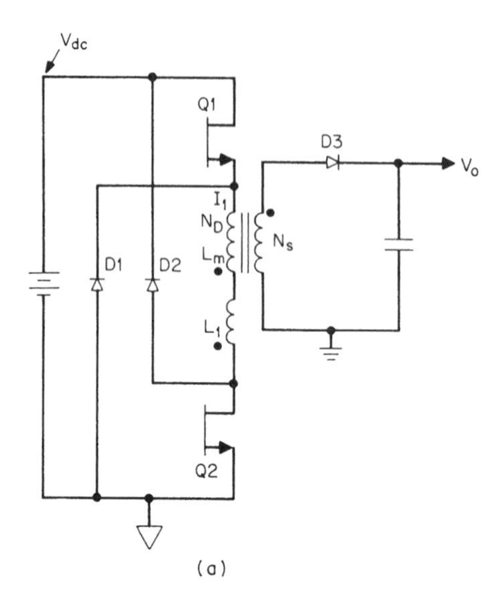

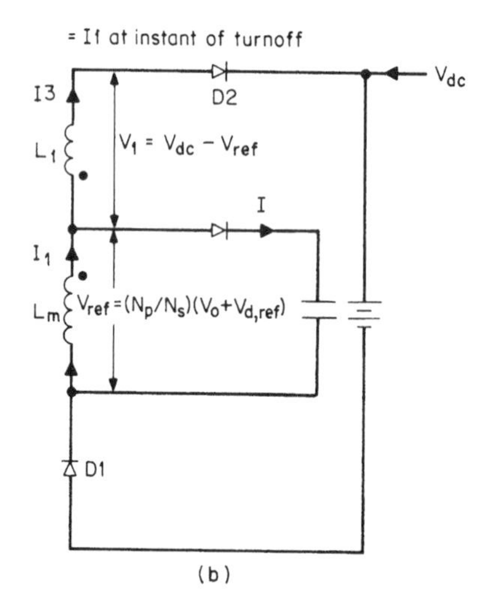

FIGURE 4.8 Circuit during $Q1$ and $Q2$ "off" time. Current I_1, stored in L_m during $Q1$, $Q2$ "on" time, also flows through leakage inductance L_l. During the "off" time, energy stored in L_m must be delivered to the secondary load as reflected into the primary across L_m. But I_1 also flows through L_l, and during the "off" time, the energy it represents $({}^1/_2 L_l I^2)$ is returned to the input source V_{dc} through diodes $D1$, $D2$. This robs energy that should have been delivered to the output load and continues to rob energy until I_1, the leakage inductance current, falls to zero. To minimize the time for I_1 in L_l to fall to zero, V_l is made significantly large by keeping the reflected voltage $V_r (= N_p/N_s)(V_o + V_{D3})$ low by setting a low N_p/N_s turns ratio. A usual value for V_r is two-thirds of the minimum V_{dc}, leaving one-third for V_l.

4.10.3 Leakage Inductance Effect in Double-Ended Flyback

Figure 4.8*b* shows the circuit during the $Q1$, $Q2$ "off" time. The voltage across L_m and L_l in series is clamped to V_{dc} through diodes $D1$, $D2$ The voltage V_r across the magnetizing inductance is clamped against the reflected secondary voltage and equals $(N_p/N_s)(Vo + V_{D3})$. The voltage across L_l is then $V_l = V_{dc} - V_r$.

At the instant of turn "off," the same current I_1 flows in L_m and $L_l(I_3 = I_1$ at instant of turn "off"). That current in L_l flows through diodes $D1$, $D2$ and returns its stored energy to the supply source V_{dc}. The L_l current decays at a rate of $dI_1/dt = V_l/L_l$ as shown in Figure 4.9*a* as slope AC or AD. The current in L_m (initially also equal to I_1) decays at a rate V_r/L_m and is shown in Figure 4.9*a* as slope AB.

The current actually delivering power to the load is I_2—the difference between the currents in L_m and L_l. This is shown as current RST in Figure 4.9*b* if the L_1 current slope is AC of Figure 4.9*a*. The larger area current UVW in Figure 4.9*c* results if the L_1 current slope is faster, as AD of Figure 4.9*a*. It should be evident in Figures 4.9*b* and 4.9*c* that so long as current still flows in leakage inductance L_l, through $D1$ and $D2$ back into the supply source, all the current available in L_m does

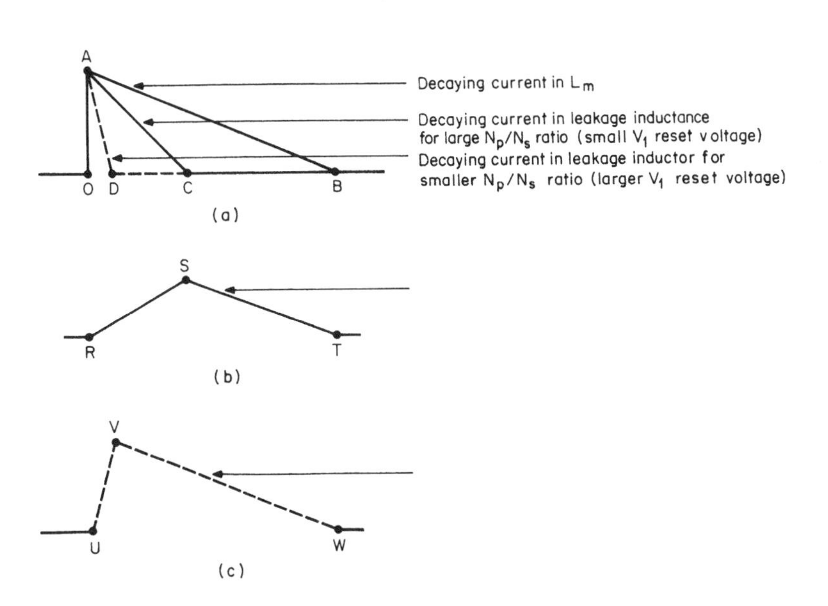

FIGURE 4.9 (*a*) Currents in magnetizing and leakage inductances in double-ended flyback. (*b*) Current into reflected load impedance for large N_p/N_s ratio. AB–AC of Figure 4.9*a*. (*c*) Current into reflected load impedance for smaller N_p/N_s ratio. AB–AD of Figure 4.9*a*.

not flow into the reflected load but is partly diverted back into the supply.

It can thus be seen from Figures 4.9*b* and 4.9*c* that to maximize the transfer of L_m current to the reflected load and to avoid a delay in the transfer of current to the load, the slope of the leakage inductance current decay should be maximized (slope *AD* rather than *AC* in Figure 4.9*a*). Or in magnetics–power supply jargon, the leakage inductance current should be rapidly reset to zero.

Since the rate of decay of the leakage inductance current is V_l/L_l and $V_l = V_{dc} - (N_p/N_s)(V_o + V_{D3})$, choosing lower values of N_p/N_s increases V_l and hastens leakage current reset. A usual value for the reflected voltage $(N_p/N_s)(V_o + V_{D3})$ is two-thirds of V_{dc}, leaving one-third for V_l. Too low a value for V_r will require a longer time to reset the magnetizing inductance, rob from the available *Q*1, *Q*2 "on" time, and decrease the available output power.

Once N_p/N_s has been fixed to yield $V_l = V_{dc}/3$, the maximum "on" time for discontinuous operation is calculated from Eq. 4.7, L_m is calculated from Eq. 4.8 and I_p from Eq. 4.9, just as for the single-ended flyback.

References

1. Billings, K., *Switchmode Power Supply Handbook,* McGraw-Hill, New York, 1989.
2. Chryssis, G., *High Frequency Switching Power Supplies,* 2nd Ed., McGraw-Hill, New York, 1989, pp. 122–131.
3. Dixon, L., "The Effects of Leakage Inductance on Multi-output Flyback Circuits," *Unitrode Power Supply Design Seminar Handbook,* Unitrode Corp., Lexington, Mass., 1988.
4. Patel, R., D. Reilly, and R. Adair, "150 Watt Flyback Regulator," *Unitrode Power Supply Design Seminar Handbook,* Unitrode Corp., Lexington, Mass., 1988.

CHAPTER 5

Current-Mode and Current-Fed Topologies

5.1 Introduction

In this chapter, current-mode[1–7] and current-fed[9–20] topologies are grouped into one family, despite their very significant differences, because they both rely on controlling input current and output voltage. However, they do this in quite different ways.

5.1.1 Current-Mode Control

Current-mode control (Figure 5.3) has two control loops: a slow outer loop (via $R1$,$R2$ and error amp EA), which senses DC output voltage and delivers a control voltage (V_{eao}), to a much faster inner current control loop (via $R1$, V_i, and the pulse width modulator PWM). R_i senses peak transistor currents (the peak choke current) and keeps the peak current constant on a pulse-by-pulse basis. The end result is that it solves the magnetic flux imbalance problem in the current-mode version of the push-pull topology and restores push-pull as a viable approach in applications where the uncertainty of other solutions to flux imbalance is a drawback (Section 2.2.8). Further, the constant power transistor current pulses simplify the feedback-loop design.

After Pressman *Because the converter in this example is a forward type, the secondary current reflects back into the primary. By sensing the current in the common return of* Q1 *and* Q2*, the inner current control loop effectively is looking at the current flow in the output choke* L_0*. The fast inner loop maintains the peak current in* L_0 *constant on a pulse-by-pulse basis, changing only slowly in response to voltage adjustments. In this way, the peak output current in* L_0 *is the controlled parameter. This takes* L_0 *out of the small signal transfer function of the outer loop, allowing faster response in the closed loop system. At the same time, because current is the controlled parameter, current* *limit and short circuit protection are intrinsic in the topology. Further, since current is controlled on a pulse-by-pulse basis, any tendency for current imbalance in Q1 and Q1 is eliminated and staircase saturation of T1 is no longer a possibility. Finally, the effect of any changes in supply voltage is automatically eliminated from the peak output current in* L_o*, so that line regulation is automatically better. ~K.B.*

5.1.2 Current-Fed Topology

A current-fed topology derives its input current from an input inductor (choke) as shown in Figure 5.9. In this example, the top end of a push-pull forward converter transformer gets its supply from input inductor *L*1. Thus the power train is driven from the high impedance current source (the input inductor *L*1) rather than the low impedance of a rectifier filter capacitor or perhaps the low-source impedance of a source battery. This higher source impedance helps to solve the flux imbalance problem in *T*1 and offers other significant advantages.

5.2 Current-Mode Control

In all the voltage-mode topologies discussed so far, output voltage alone is the controlled parameter. In those circuits, regulation against load current changes occurs because current changes cause small output voltage changes that are sensed by a voltage-monitoring error amplifier, which then corrects the power transistor "on" time to maintain output voltage constant. Output current itself is not monitored directly.

In the 1980s, the new topology *current-mode control* appeared, in which both voltage and current were monitored. The scheme had been known previously, but was not widely used as it required discrete circuit components to implement it. When a new Unitrode™ pulse-width-modulating (PWM) chip—the UC1846—appeared, with all the features needed to implement current-mode control, the advantages of the technique were quickly recognized and it was widely adopted.

After Pressman *As of 2008, many similar current-mode control ICs are now available. Unitrode is now part of Texas Instruments. ~K.B.*

Where two 180° out-of-phase width-modulated drive signals are required as in the push-pull, half-bridge, full-bridge, interleaved forward converter, or flyback, the UC1846 can be used to implement current-mode control. A lower-cost, single-ended PWM controller, the UC1842, is currently available to implement current mode in single-ended circuits such as forward converters, flybacks, and buck regulators.

5.2.1 Current-Mode Control Advantages

5.2.1.1 Avoidance of Flux Imbalance in Push-Pull Converters

Flux imbalance was discussed in Section 2.2.5. It occurs in a push-pull converter when the transformer core operates asymmetrically about the origin of its hysteresis loop. The consequence is that the core moves up toward saturation and one transistor draws more current during its "on" time than does the opposite transistor (Figure 2.4*c*).

As the core drifts further off center of the origin, it goes deeply into saturation and may destroy the power transistor. A number of ways to cope with flux imbalance have been described in Section 2.2.8. These schemes work, but under unusual line or load transient conditions and especially at higher output powers, there is never complete certainty that flux imbalance cannot occur.

Current-mode monitors current on a pulse-by-pulse basis and forces alternate pulses to have equal peak amplitudes by correcting each transistor's "on" time so that current amplitudes must be equal. This puts push-pull back into the running in any proposed new design and is a valuable contribution to the repertoire of possible topologies. For example, if a forward converter with no flux imbalance problem were chosen to be certain of no flux imbalance in the absence of current mode, a severe penalty would be paid.

Eq. 2.28 shows the peak primary current in a forward converter is $3.13(P_o/\underline{V_{dc}})$. But Eq. 2.9 shows it is only half that or $1.56(P_o/\underline{V_{dc}})$ for the push-pull. At low output powers, it is not a serious drawback to use the forward converter with twice the peak current of a push-pull at equal output power, especially since the forward converter has only one transistor. But at higher output power, twice the peak primary current in a forward converter than in a push-pull becomes prohibitive.

The push-pull is a very attractive choice for telephone industry power supplies where the maximum DC input voltage is specified as only 38–60 V. Having it in its current-mode version with a certainty that flux imbalance cannot exist is very valuable.

5.2.1.2 Fast Correction Against Line Voltage Changes Without Error Amplifier Delay (Voltage Feed-Forward)

It is inherent in the details of how current mode works that a line voltage change immediately causes a change in power transistor "on" time. This change is corrected without having to wait for an output voltage change to be sensed after a relatively long delay by a conventional voltage error amplifier. The details of how this comes about will be discussed below.

5.2.1.3 Ease and Simplicity of Feedback-Loop Stabilization

All the topologies discussed above with the exception of flybacks have an output *LC* filter. An *LC* filter has a maximum possible phase shift of 180° not far above its resonant frequency of $f_o = \frac{1}{2\pi\sqrt{LC}}$, and gain between input and output falls very rapidly with increasing frequency. As frequency increases, the impedance of the series *L* arm increases and that of the shunt arm decreases.

This possible large phase shift and rapid change of gain with frequency complicates feedback-loop design. More important, the elements around the error amplifier required to stabilize the loop are more complex and can cause problems with rapid changes in input voltage or output current.

In a small-signal analysis of the current-mode outer voltage loop, however, which calculates gain and phase shift to consider the possibility of oscillation, the output inductor does not appear even though it is physically in series with the output shunt capacitor. So for small signal changes, the voltage loop behaves as if the inductor were not there.

The circuit behaves as if there were a constant current feeding the parallel combination of the output capacitor and the output load resistor. Such a network can yield only 90° rather than 180° of phase shift, and the gain between input and output falls half as rapidly as for a true *LC* filter (–20 dB per decade rather than –40 dB per decade). This simplifies feedback-loop design, simplifies the circuitry around the error amplifier required for stabilization, and avoids problems arising from rapid line or load changes. The details of why this is so will be discussed below.

5.2.1.4 Paralleling Outputs

A number of current-mode power supplies may be operated in parallel, each with an equal share of the total load current. This is achieved by sensing current in each supply with equal current sensing resistors, which convert transistor peak current pulses to voltage pulses. These are compared in a voltage comparator to a common error-amplifier output voltage, which forces peak current-sensing voltages and hence peak currents in the parallel supplies to be equal.

5.2.1.5 Improved Load Current Regulation

Current mode has better load current regulation than voltage mode. The improvement is not as great as that in voltage regulation, however, which is greatly enhanced by the feed-forward characteristic inherent in current mode. The improved load current regulation comes about because of the greater error-amplifier bandwidth possible in current mode.

5.3 Current-Mode vs. Voltage-Mode Control Circuits

To understand the differences and advantages of current mode over voltage mode, it is essential first to see how voltage-mode control circuitry works. The basic elements of a typical voltage-mode, PWM control circuit are shown in Figure 5.1. That block diagram shows most of the elements of the SG1524, the first of many integrated-circuit control chips that have revolutionized the switching power supply industry. The SG1524, originally made by Silicon General Corporation, is now manufactured by many other companies and in improved versions such as the UC1524A (Unitrode) and SG1524B (Silicon General).

5.3.1 Voltage-Mode Control Circuitry

In Figure 5.1, an oscillator generates a 3-V sawtooth V_{st}. The DC voltage at the triangle base is about 0.5 V and, at the peak, about 3.5 V. The period of the sawtooth is set by external discrete components R_t and C_t and is approximately equal to $T = R_t C_t$.

An error amplifier compares a fraction of the output voltage KV_o to a voltage reference V_{ref} and produces an error voltage V_{ea}. V_{ea} is compared to the sawtooth V_{st} in a voltage comparator (PWM). Note that the fraction of the output KV_o is fed to the inverting input of the error amplifier so that when V_o goes up, the error-amplifier output V_{ea} goes down.

In the PWM voltage comparator, the sawtooth is fed to the noninverting input and V_{ea} is fed to the inverting input. Thus the PWM output is a negative-going pulse of variable width. The pulse is negative for the entire time the sawtooth is below the DC level of the error-amplifier output V_{ea} or from t_1 to t_2. As the DC output voltage goes—say—slightly positive, KV_o goes slightly positive, and V_{ea} goes negative and closer to the bottom of the sawtooth. Thus the duration of the negative-going pulse V_{pwm} decreases.

The duration of this negative-going pulse is the duration of the power transistor "on" time. Further, since in all the voltage-mode topologies discussed above, the DC output voltage is proportional to the power transistor "on" time, decreasing the "on" time brings the DC output voltage back down by negative-feedback loop action. The duration of the negative pulse V_{pwm} increases as the output DC voltage decreases.

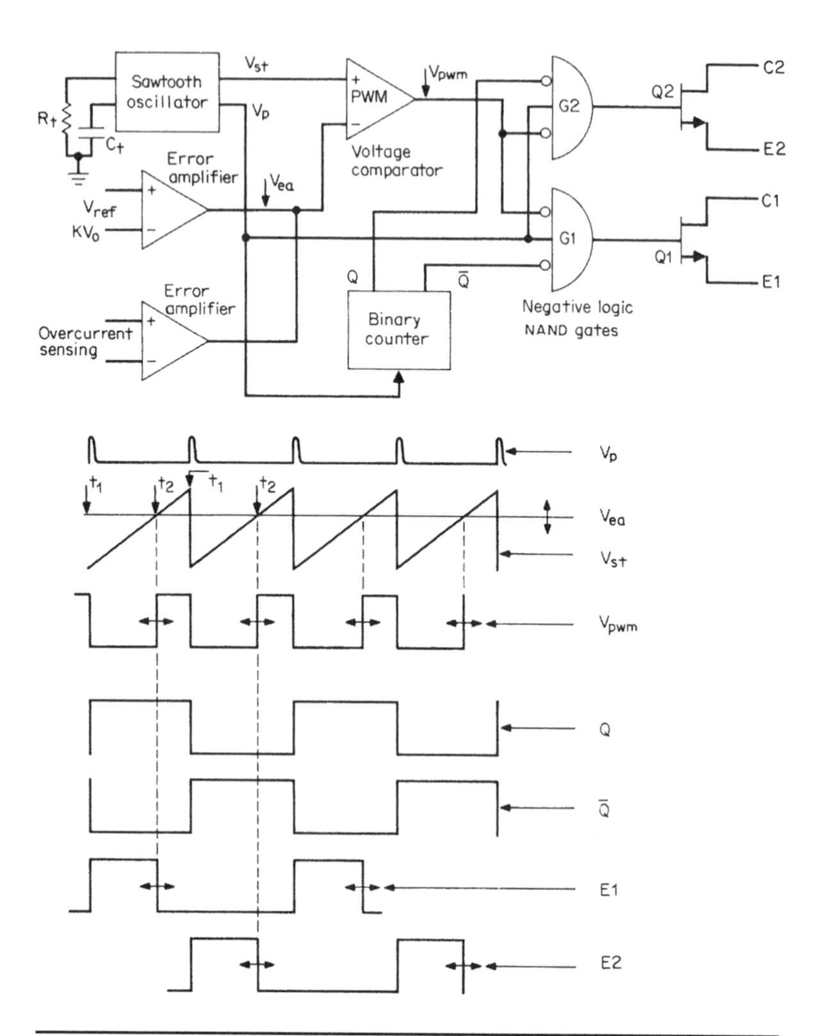

FIGURE 5.1 A basic voltage-mode PWM controller. The output voltage is sensed directly by the error amplifier. Regulation against load current changes occurs only after the current changes cause small output voltage changes. The current-limit amplifier operates to shut down the supply only when a maximum current limit is exceeded. Transistor "on" time is from start of sawtooth until the sawtooth crosses V_{ea}.

The UC1524 is designed primarily for push-pull-type topologies, so the single negative-going pulse of adjustable width, coming once per sawtooth period, must be converted to two 180° out-of-phase pulses of the same width. This is done with the binary counter and negative logic NAND gates G_1 and G_2. A positive-going pulse V_p occurring at the end of each sawtooth is taken from the sawtooth oscillator and used to trigger the binary counter.

Outputs from the binary counter Q and $\bar{Q}$ are then out-of-phase square waves at half the sawtooth frequency. When they are negative, these square waves steer negative V_{pwm} pulses alternately through negative logic NAND G_1 and G_2. These gates produce a positive output only for the duration of time that the inputs are negative. Thus the bases (and emitters) of output transistors $Q1$ and $Q2$ are positive only on alternate half cycles and only for the same duration as the V_{pwm} negative pulses.

The "on" time of the power transistors must correspond to the time the V_{pwm} pulse is negative for the complete circuit to have negative feedback, since KV_o is connected to the inverting terminal of the error amplifier. Thus if the power transistors are of the NPN type, they must be fed from the emitters of $Q1$, $Q2$, or if of the PNP type, from the collectors. If current amplifiers are interposed between the bases of the output transistors and $Q1$, $Q2$, polarities must be such that $Q1$, $Q2$ are "on" when the output transistors are "on."

The narrow positive pulse V_p is fed directly into gates $G1$, $G2$. This forces both gate outputs to be "low" simultaneously for the duration of V_p, and both output transistors to be "off" for that duration. This ensures that if the pulse width of V_{pwm} ever approached a full half period, both power transistors could never be "on" simultaneously at the end of the half period. In a push-pull topology, if both transistors are simultaneously "on" even for a short time, they are subjected to both high current and the full supply voltage and could be destroyed.

This, then, is a voltage-mode circuit. Power transistor or output current is not sensed directly. The power transistors are turned "on" at the beginning of a half period and turned "off" when the sawtooth V_{st} crosses the DC level of the error-amplifier output, which is a measure of output voltage only.

The complete details of the SG1524 are shown in Figure 5.2*a*. The negative logic NAND gates $G1$, $G2$ of Figure 5.1 are shown in Figure 5.2*a* as positive logic NOR gates. These perform the same function for requiring all "lows" to make a "high" and are identical to any one "high" forcing a "low."

In Figure 5.2*a*, when pin 10 goes "high," the associated transistor collector goes "low" and brings the error-amplifier output (pin 9) down to the base of the sawtooth. This reduces output transistor "on" times to zero and shuts down the supply. In the current limit comparator, if pin 4 is 200 mV more positive than pin 5, the error-amplifier output is also brought down to ground (there is an internal phase inversion, not shown) and the supply is shut down. Pins 4 and 5 are bridged across a current-sensing resistor in series with the current being monitored. If current is to be limited to I_m, the resistor is selected as $R_s = 0.2/I_m$.

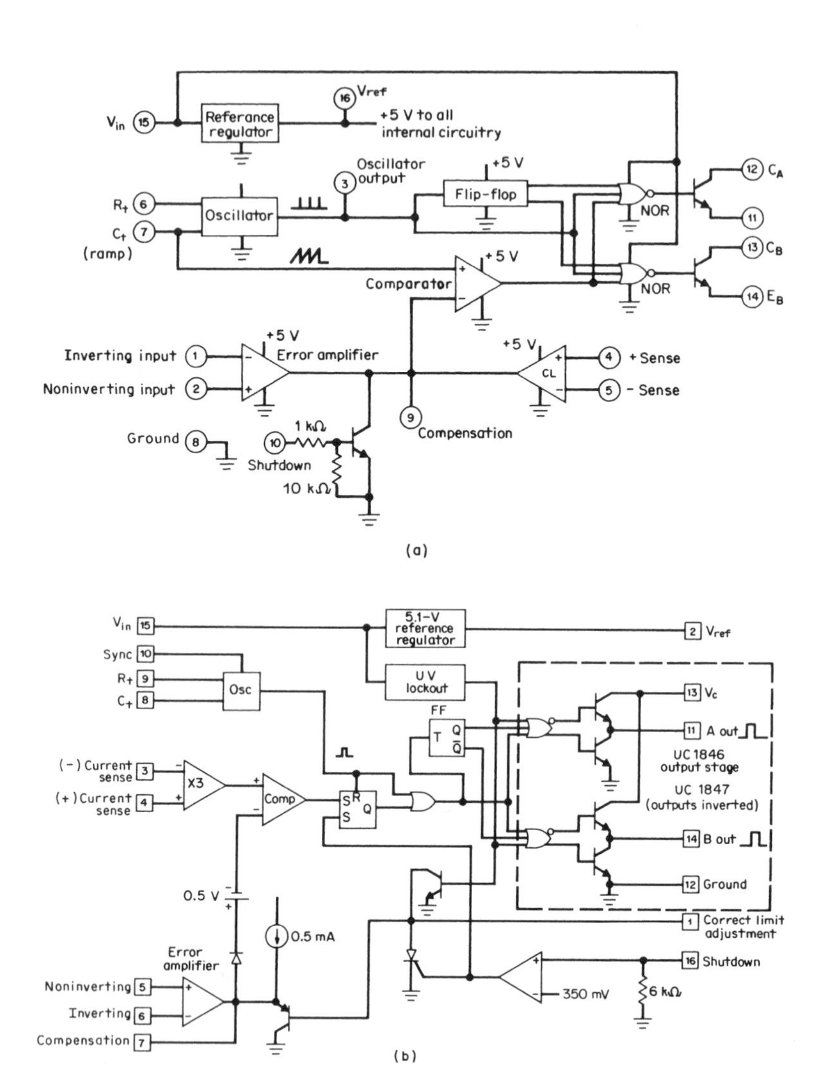

FIGURE 5.2 (*a*) PWM chip SG1524, the first integrated-circuit pulse-width-modulating control chip. (*Courtesy Silicon General Corp.*) (*b*) PWM chip UC1846, Unitrode's first integrated-circuit current-mode control chip. (*Courtesy Unitrode Corp.*)

5.3.2 Current-Mode Control Circuitry

Circuitry of the first integrated-circuit current-mode control chip (Unitrode UC1846) is shown in Figure 5.2*b*. Figure 5.3 shows its basic elements controlling a push-pull converter.

Note in Figure 5.3 that there are two feedback loops—an outer loop consisting of output voltage sensor (EA) and an inner loop comprising

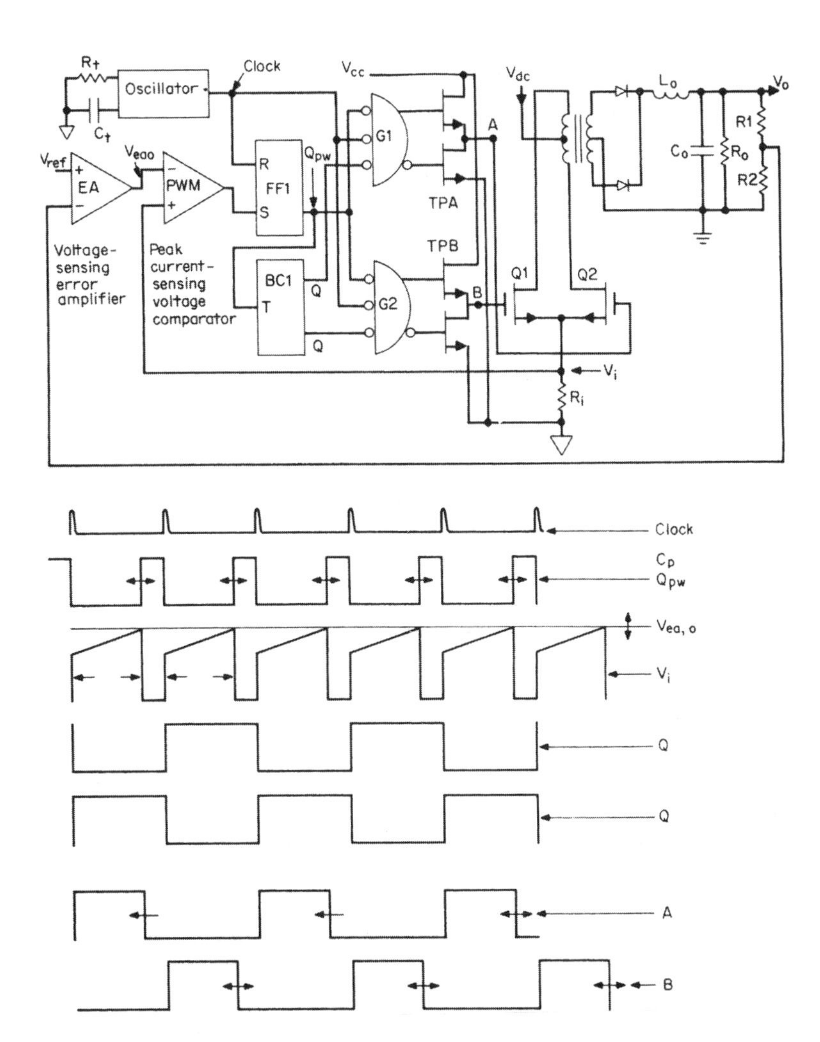

FIGURE 5.3 Current-mode controller UC1846, driving a push-pull MOSFET converter. Transistors are turned "on" alternately at each clock pulse. They are turned "off" when the peak voltage across the common current-sensing resistor equals the output voltage of the voltage-sensing error amplifier. PWM forces all Q1, Q2 current pulses to have equal peak amplitudes.

primary peak current sensor (PWM) and current-sensing resistor R_i which converts ramp-on-a-step transistor currents to ramp-on-a-step voltages.

Line and load current changes are regulated by varying power transistor "on" time. "On" time is determined by both the voltage-sensing error-amplifier output V_{eao} and the PWM voltage comparator, which compares V_{eao} to the ramp-on-a-step voltage at the top of the current-sensing resistor R_i.

Because the secondaries all have output inductors, the secondary currents have the characteristic ramp-on-a-step shape. These reflect as identical-shaped currents, which are smaller by the N_s/N_p ratio, in the primary and the output transistors. Those currents flowing in the common emitters through R_i produce the ramp-on-a-step voltage waveshape V_i. Power transistor "on" time is then determined as follows: An internal oscillator, whose period is set by external discrete components R_t, C_t, generates narrow clock pulses C_p. The oscillator period is approximately $0.9R_tC_t$. At every clock pulse, feed-forward *FF*1 is reset, causing its output Q_{pw} to go "low." The duration of the "low" time at Q_{pw}, it will soon be seen, is the duration of the "high" time at either of the chip outputs *A* or *B* and, hence, the duration of the power transistor "on" times.

When the PWM voltage comparator output goes "high," *FF*1 is set, thus terminating the Q_{pw} "low" and hence the "high" time at *A* or *B*, and turns "off" the power transistor which had been "on." Thus the instant at which the PWM comparator output goes "high" determines the end of the "on" time.

The PWM comparator compares the ramp-on-a-step current-sensing voltage V_i to the output of the voltage error-amplifier EA. Hence when the peak of V_i equals V_{eao}, the PWM output goes positive and sets *FF*1, Q_{pw} goes "high," and whichever of *A* or *B* had been "high" goes "low." The power transistor that had just been "on" is now turned "off."

A "low" output from *FF*1 occurs once per clock period. It starts "low" at every clock pulse and goes back "high" when the PWM noninverting input equals the DC level of the EA output. Most frequently, power transistors *Q*1, *Q*2 will be N types, which require positive-going signals for turn "on." Thus these equal-duration negative-going pulses are steered alternately through negative logic NAND gates *G*1 and *G*2, becoming 180° out-of-phase, positive-going pulses at the chip outputs *A* and *B*.

Chip output stages *TPA* and *TPB* are "totem poles." When the bottom transistor of a totem pole is "on," the top one is "off" and vice versa. Output nodes *A* and *B* have very low output impedance. When the bottom transistor is "on," it can "sink" (absorb inward-directed current) 100 mA continuous and 400 mA during the "high"-to-"low" transition. When the top transistor is "on," it can "source" (emit outward-directed current) 100 mA continuous and 400 mA during the "low"-to-"high" transition.

Steering is done by binary counter *BC*1, which is triggered once per clock pulse on the leading edge of the pulse. The negative-going *Q* pulses steer the negative Q_{PW} pulses alternately through negative logic NAND gates *G*1, *G*2. The chip outputs *A* and *B* are 180° out-of-phase positive pulses whose duration is the same as that of the negative pulses Q_{PW}.

Note that Q_{PW} is positive from the end of the "on" time until the start of the next turn "on." This forces the bubble outputs of *G*1, *G*2 "high" and brings points *A* and *B* both "low." This "low" at both power transistor inputs during the dead time between the turn "off" of one transistor and the turn "on" of the other is a valuable feature. It presents a low impedance at the "off"-voltage level and prevents noise pickup from turning the power transistors "on" spuriously. While the bubble outputs of *G*1, *G*2 are both "high," their no-bubble outputs are both "low," and thus turn "off" the upper transistors of the totem poles *TPA* and *TPB* and avoid over-dissipating them.

It can be seen also that the narrow positive clock pulse is fed as a third input to NAND gates *G*1, *G*2. This makes bubble outputs from *G*1, *G*2 "high" and outputs *A*, *B* simultaneously "low" for the duration of the clock pulse. This guarantees that under fault conditions, if the controller attempts a full half period "on" time (Q_{PW} "low" and either *A* or *B* "high" for a full half period), there will be a dead time between the end of one "on" time and the start of the opposite "on" time. Thus the power transistors can't conduct simultaneously.

5.4 Detailed Explanation of Current-Mode Advantages

5.4.1 Line Voltage Regulation

Consider how the controller regulates against line voltage changes. Assume that line voltage (and hence V_{dc}) goes up. As V_{dc} goes up, the peak controlled secondary voltage will go up and after a delay in L_o, V_o will eventually go up. Since secondary DC voltages are proportional to secondary winding peak voltages and power transistor "on" time, the "on" time must decrease because the peak secondary voltage has increased. Then, after a delay through the error amplifier, V_{eao} will go down and, in the PWM comparator, the ramp in V_i will become equal to the lowered value of V_{eao} earlier in time. Thus, "on" time will be decreased and the output voltage will be brought back down.

If this were the only mechanism to correct against line voltage changes, however, the correction would be slow due to the delays in

L_o and the error amplifier, but there is a shortcut around those delays. As V_{dc} goes up, the peak voltage at the input to the output inductor V_{sp} increases, the slope of inductor current DI_s / dt increases, and hence the slope of the ramp of V_i increases. Now the faster ramp equals V_{eao} earlier in time, and the "on" time is shortened without having to wait for V_{eao} to move down and shorten the "on" time. Output voltage transients resulting from input voltage transients are smaller in amplitude and shorter in duration because of this feed-forward characteristic.

5.4.2 Elimination of Flux Imbalance

Consider the waveform V_i in Figure 5.3. It is taken from the current-sensing resistor R_i and is hence proportional to power transistor currents. The "on" time ends when the peak of the ramp in V_i equals the output voltage of the error amplifier V_{eao}. It can be seen in Figure 5.3 that peak currents on alternate half cycles cannot be unequal as in Figure 2.4*b* and 2.4*c* because the error-amplifier output V_{eao} is essentially horizontal and cannot change significantly within one cycle because of limited EA bandwidth.

If the transformer core got slightly off center and started walking up into saturation on one side, the voltage V_i would become slightly concave upward close to the end of that "on" time. It would then equal V_{eao} earlier and terminate that "on" time sooner. Flux increase in that half cycle would then cease, and in the next half cycle, since the opposite transistor would not have a foreshortened "on" time, the core flux would be brought back down and away from saturation.

Since the peaks of the voltage ramps in Figure 5.3 (V_i) are equal, peak currents on alternate half cycles must be equal. Thus the inequality of alternate currents and flux imbalance shown in Figure 2.4*b* are not possible.

5.4.3 Simplified Loop Stabilization from Elimination of Output Inductor in Small-Signal Analysis

Refer to Figure 5.3. In a small-signal analysis to determine whether the outer voltage loop is stable, it is assumed that the loop is opened at some point and a small sinusoidal signal of variable frequency is inserted at the input side of the break. The gain and phase shift versus frequency are calculated through all the loop elements starting from the input side of the break, around to the same point at the output side of the loop break. By tailoring the error-amplifier gain and phase shift properly in relation to the other elements in the open loop (primarily the output *LC* filter), the closed loop is made stable.

The variable frequency is often inserted at the input to the error amplifier. In Chapter 12 on feedback loop stability analysis, it will be shown how gain and phase shift through the error amplifier may be calculated and tailored to achieve the desired results.

Considering Figure 5.3, the concept of gain and phase shift of a sinusoidal signal from the error-amplifier output to the input of the *LC* filter may not be obvious. Of primary importance is the fact that the highest frequency to which the loop will respond significantly is well below the switching frequency of the converter. The error-amplifier output V_{eao} is, therefore, a slowly changing or essentially DC voltage that, when it equals the peak of the ramp-on-a-step pulse sequence V_i, results in a sequence of negative-going pulses at Q_{pw} whose duration depends on V_{eao}. The Q_{pw} negative pulses result in a sequence of positive-going pulses at the input to the *LC* filter.

It may seem puzzling to speak of gain and phase shift of sinusoidal signals in view of this odd operation of converting a voltage level to a sequence of pulses at the switching frequency. The situation may be clarified as follows.

If there is a sinusoidal signal at the error-amplifier input, it is amplified and phase-shifted at the EA output. Thus V_{eao} is sinusoidally amplitude modulated at that frequency. The Q_{pw} negative pulses are similarly pulse width modulated at that frequency. So are the "on" times of the positive-going pulses at the output rectifiers pulse width modulated at that frequency. Hence, the voltage at the output rectifier cathodes, which is proportional to the pulse widths, when averaged over a time long compared to the switching period, is simply amplitude modulated at the same frequency as was inserted at the error-amplifier input.

So long as the modulation period is long compared to the switching period, the modulation operation is a sinusoid-to-pulse width-to-sinusoid converter. The gain of this modulation operation will be discussed further in the chapter on feedback loop stability.

In the converter of Figure 5.3, there remains only the problem of calculating the gain and phase shift versus frequency for the sinusoid through the *LC* filter. A sine wave voltage at the rectifier cathodes will be phase shifted 90° by the *LC* filter at the resonant frequency $\frac{1}{2\pi\sqrt{LC}}$ and 180° at frequencies above that, and gain from input to output will fall at –40 dB/decade above resonance.

In current mode, however, the PWM comparator forces the output at the rectifier cathodes to be a sequence of width-modulated constant-current pulses—not voltage pulses. Thus at the input to the *LC* filter, the averaged waveform is a constant-current, not a constant-voltage, sinusoid.

With a constant-current sinusoid, the filter inductor cannot act to change phase. The circuit behaves, in this small-signal analysis, as if the inductor were not present. Thus after the rectifier cathodes, the gain and phase shift correspond to that of a constant-current sinusoid flowing into the parallel combination of the output capacitor and load resistor. Such a circuit can yield a maximum phase shift of only 90°, and a gain-versus-frequency characteristic that falls at –20 dB per decade, rather than –40 dB.

Chapter 12 on feedback stability analysis will show that this greatly simplifies the error-amplifier design, yields greater bandwidth, and improves the response of the closed-loop circuit to step changes in load current and line voltage. For now, Figure 5.4*a* and 5.4*b* show a comparison of the error-amplifier feedback networks required to stabilize a voltage-mode circuit (Figure 5.4*a*) and a current-mode circuit (Figure 5.4*b*).

After Pressman *Notice the inductor is only taken out of the loop for small signal changes (it is still there in fact). For larger transient changes, the inductor will still limit the slew rate and cannot be ignored for large changes (where the control amplifiers bottom or top out at the limit of their range). ~K.B.*

5.4.4 Load Current Regulation

In Figure 5.3, the V_i voltage waveform is proportional to power transistor currents, which are related to controlled secondary current by the transformer turns ratio.

At a DC input voltage V_{dc}, the peak secondary voltage is $V_{sp} = V_{dc}(N_s/N_p)$. For an "on" time of t_{on} in each transistor, the DC output voltage is $V_o = V_{sp}(2t_{on}/T)$—just as for a voltage-mode push-pull circuit. The "on" time starts at the clock pulse, as shown in Figure 5.3, and ends when the V_i ramp equals the voltage error-amplifier output.

If the DC voltage goes up as described, initially the V_i ramp rate increases and shortens the "on" time as it reaches the original V_{eao} level earlier in time. This yields a fast correction for a step change in input voltage and the "on" time remains shorter as required by the preceding relation for the increase in peak secondary voltage.

The mechanism for load current regulation, though, is different. For a fast step increase—say—in DC load current, the DC output voltage drops momentarily somewhat because the *LC* output filter has a surge impedance of approximately $\sqrt{LC}$. After the delay in the error amplifier, V_{eao} moves up an amount determined by the EA gain.

Now V_i must ramp longer and hence higher in amplitude for it to reach equality with the higher V_{eao}. The secondary peak current and hence the output inductor current are thus larger in amplitude. The up-slope of the inductor current lasts longer and eats somewhat into the dead time before the opposite transistor turns "on."

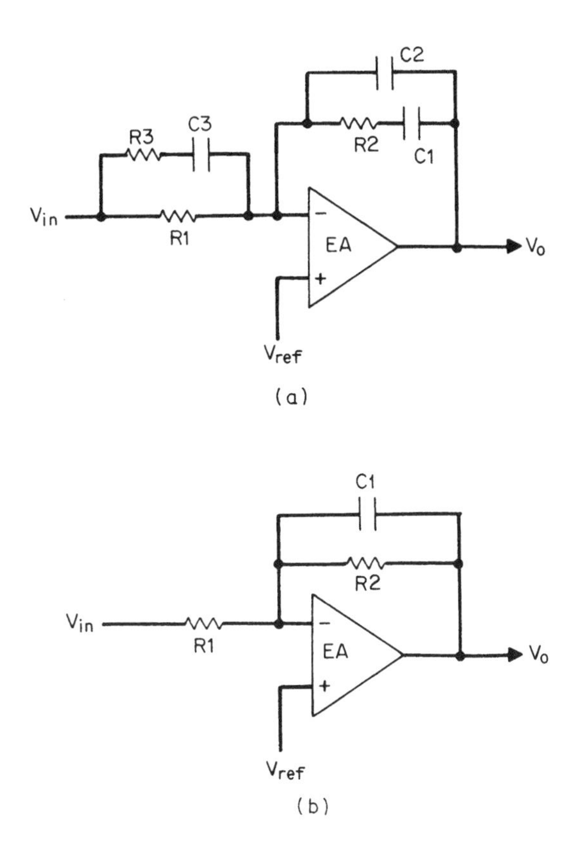

FIGURE 5.4 (*a*) Typical compensating network for a voltage-mode power supply. The complex input-feedback network in voltage mode is necessary because the output inductor with the filter capacitor together yields a 180° phase shift and a –40 dB/decade gain versus frequency characteristic, which make loop stabilization more difficult. (*b*) Typical compensation network for a current-mode power supply. In current mode, the source driving the output inductor is an effective "current source." The output inductor does not contribute to phase shift. The circuit acts at its output as if there were a constant current driving the parallel combination of the output filter capacitor and the output load resistance. Such a network yields a maximum 90° phase shift and a –20 dB/decade gain versus frequency characteristic. This permits the simpler input-feedback network for loop stabilization. It also copes much more easily with large-amplitude load and line changes.

With a shorter dead time, when the opposite transistor turns "on" at the beginning of the dead time, the current remaining in the inductor will be greater than it had been in the previous cycle. Thus the front-end step in each current pulse represented by V_i will be greater than that in the previous cycle.

This process continues for a number of switching cycles, until the step part of the ramp-on-a-step current waveform builds up sufficiently to supply the increased demand for DC load current. As this current builds up, the DC output voltage gradually builds back up and V_{eao} relaxes back down, returning the "on" time to its original value. The time to respond to a change in DC load current is thus seen to be dependent on the size of the output inductor, since a smaller value permits more rapid current changes. The response time also depends on the bandwidth of the error amplifier.

5.5 Current-Mode Deficiencies and Limitations

5.5.1 Constant Peak Current vs. Average Output Current Ratio Problem[1-4]

Current mode controls the peak transistor currents (and hence the peak output inductor/choke currents) constant at a level needed to supply the required mean DC load current to give the mean DC output voltage dictated by the voltage error amplifier, as shown in Figure 5.3.

The DC load current is the average of the output inductor current so that keeping the peak transistor current constant, and hence the peak output inductor current constant, does not keep the average inductor current and hence output current constant. Because of this, in the unmodified current-mode scheme described thus far, changes in the DC input voltage will cause momentary changes in the DC output voltage. After a short delay the output voltage change will be corrected by the voltage error amplifier in the outer feedback loop, as this is the loop that ultimately sets output voltage.

After Pressman *This is now referred to as the "peak to average current ratio" effect. The problem stems from the fact that maintaining the peak inductor current constant does not maintain the average output current constant, because duty cycle changes change the average value but not the peak value. This can become a problem for wide duty cycle changes, leading to subharmonic instability. It is corrected by ramp compensation (Section 5.5.3). ~K.B.*

However, the inner loop, in keeping peak inductor current constant, does not supply the correct average inductor current and output voltage changes again. The effect is an oscillation that commences at every change in input voltage and that may continue for some time. The mechanism can be better understood from an examination of the up- and down-slopes of the output inductor currents in Figure 5.5.[3]

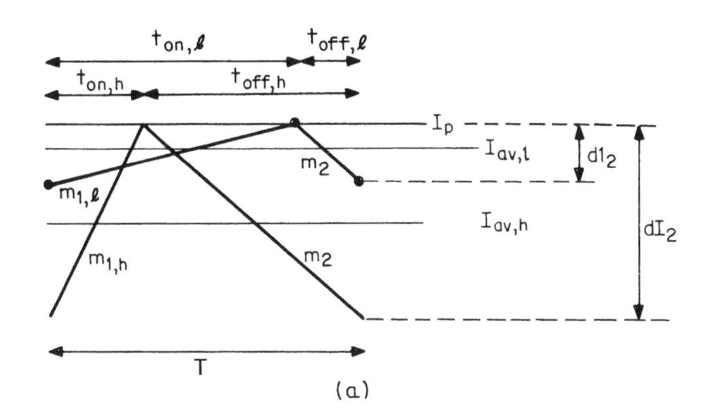

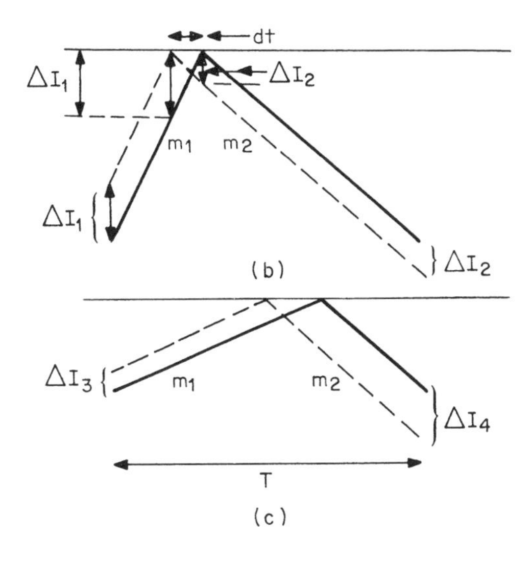

FIGURE 5.5 Problems in current mode. (*a*) Output inductor currents at high and low input voltages. In current mode, peak inductor currents are constant. At low DC input, t_{on} is maximum, yielding average inductor current I_{avl}. At high DC input, "on" time decreases to keep output voltage constant. But average inductor current I_{avh} is lower at high DC input. Since output voltage is proportional to average—not peak—inductor current, this causes oscillation when input voltage is changed. Slope m_2 is inductor current down-slope, which is not affected by loop action and is constant. Slope m_{1l} is inductor current up-slope at low line; m_{1h} is inductor current up-slope at high line. (*b*) For a duty cycle less than 50%, an initial inductor current disturbance I_1 results in smaller I_2 disturbances in successive cycles until the disturbances die out. (*c*) For a duty cycle greater than 50%, an initial inductor current disturbance I_3 results in larger I_4 disturbances in successive cycles. The disturbances grow and then decay, resulting in an oscillation.

Figure 5.5*a* shows the up- and down-slopes of the output inductor current for two different DC input voltages in current mode. Slope m_2 is the down-slope $= dI_1/dt = V_o/L_o$. It is seen to be constant for the two different DC input voltages. At the high input voltage, "on" time is short at $t_{on,h}$ and at the lower DC input, "on" time is longer at $t_{on,l}$.

The peak inductor currents are constant because the power transistor peak currents are kept constant by the PWM comparator (see Figure 5.3). The DC voltage input V_{eao} to that comparator is constant since the outer feedback loop is keeping V_o constant. The constant V_{eao} then keeps V_i peaks constant, and hence transistor and output inductor peak currents are constant.

In Figure 5.5*a*, in the steady state, the current change in the output inductor during an "on" time is equal and opposite to that during an "off" time. If this were not so, there would be a DC voltage across the inductor, and since it is assumed that the inductor has negligible resistance, it cannot support DC voltage.

It can be seen in Figure 5.5*a* that the average inductor current at low DC input is higher than it is at high DC input voltage. This can be seen quantitatively as

$$
\begin{aligned}
I_{av} &= I_p - \frac{dI_2}{2} \\
&= I_p - \left(\frac{m_2 t_{off}}{2}\right) \\
&= I_p - \left[\frac{m_2(T - t_{on})}{2}\right] \\
&= I_p - \left(\frac{m_2 T}{2}\right) + \left(\frac{m_2 t_{on}}{2}\right)
\end{aligned}
\tag{5.1}
$$

Since the voltage feedback loop keeps the product of $V_{dc}t_{on}$ constant, at lower DC input voltage when the "on" time is higher, the average output inductor current I_{av} is higher, as can be seen from Eq. 5.1 and Figure 5.5*a*.

Further, since the DC output voltage is proportional to the average and not the peak inductor current, as DC input goes down, DC output voltage will go up. DC output voltage will then be corrected by the outer feedback loop and a seesaw action or oscillation will occur.

This phenomenon does not occur in voltage-mode control, in which only DC output voltage is controlled. Also, since DC output voltage is proportional to average and not peak inductor current, keeping output voltage constant maintains average inductor current constant.

5.5.2 Response to an Output Inductor Current Disturbance

A second problem that gives rise to oscillation in current mode is shown in Figure 5.5*b* and 5.5*c*. In Figure 5.5*b*, it is seen that at a fixed DC input voltage, if for some reason there is an initial current disturbance ΔI_1, after a first down-slope the current will be displaced by an amount ΔI_2.

Further, if the duty cycle is less than 50% ($m_2 < m_1$), as in Figure 5.5*b*, the output disturbance ΔI_2 will be less than the input disturbance ΔI_1, and after a few cycles, the disturbance will die out. If the duty cycle is greater than 50% ($m_2 > m_1$) as in Figure 5.5*c*, the output disturbance ΔI_4 after one cycle is greater than the input disturbance ΔI_3. This can be seen quantitatively from Figure 5.5*b* as follows. For a small current displacement ΔI_1, the current reaches the original peak value earlier in time by an amount dt where $dt = \Delta I_1/m_1$.

On the inductor down-slope, at the end of the "on" time, the current is lower than its original value by an amount ΔI_2 where

$$\Delta I_2 = m_2 dt = \Delta I_1 \frac{m_2}{m_1} \tag{5.2}$$

Now with m_2 greater than m_1, the disturbances will continue to grow but eventually decay, giving rise to an oscillation.

5.5.3 Slope Compensation to Correct Problems in Current Mode[1–4]

Both current-mode problems mentioned above can be corrected as shown in Figure 5.6, in which the original, unmodified output of the error amplifier is shown as the horizontal voltage level *OP*. The "slope compensation" scheme for correcting the preceding problems consists of adding a negative voltage slope of magnitude m to the output of the error amplifier. By proper selection of m in a manner discussed below, the output inductor average DC current can be made independent of the power transistor "on" time. This corrects the problems indicated by both Eqs. 5.1 and 5.2.

In Figure 5.6, the up-slope m_1 and down-slope m_2 of output inductor current are shown. Recall that in current mode, the power transistor "on" time starts at every clock pulse and ends at the instant the output of the PWM comparator reaches equality with the output of the voltage error-amplifier as shown in Figure 5.3. In slope compensation, a negative voltage slope of magnitude $m = dV_{ea}/dt$ starting at clock time is added to the error-amplifier output. The magnitude of m is calculated thus: In Figure 5.6, the error-amplifier output at any time

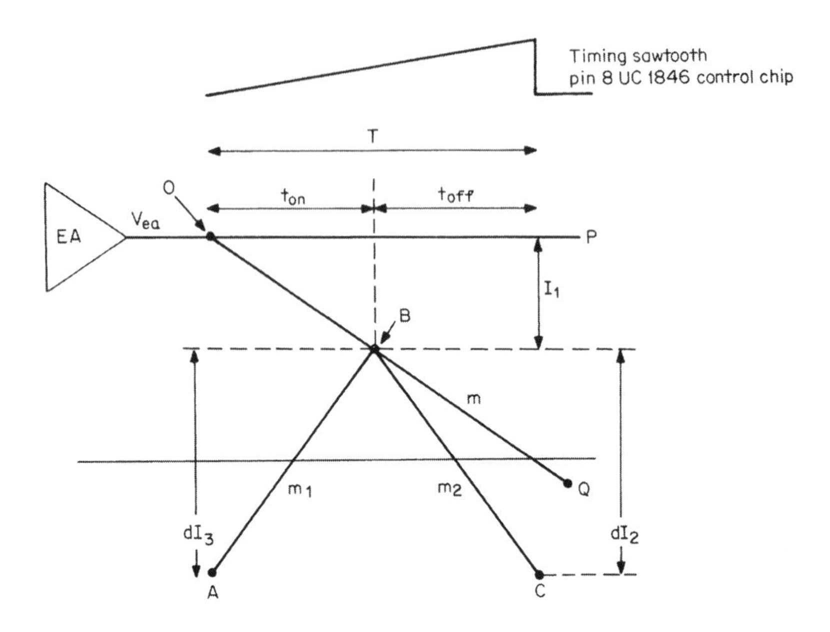

FIGURE 5.6 Slope compensation. By adding a negative voltage slope of magnitude $m = N_s/N_p(R_i)(m_2/2)$ to the error-amplifier output (Figure 5.3), the two problems shown in Figure 5.5 are corrected.

t_{on} after a clock pulse is

$$V_{ea} = V_{eao} - mt_{on} \tag{5.3}$$

where V_{eao} is the error-amplifier output at clock time. The peak voltage V_i across the primary current-sensing resistor R_i in Figure 5.3 is

$$V_i = I_{pp}R_i = I_{sp}\frac{N_s}{N_p}R_i$$

in which I_{pp} and I_{sp} are the primary and secondary peak currents, respectively. But $I_{sp} = I_{sa} + dI_2/2$, where I_{sa} is the average secondary or average output inductor current and dI_2 in Figure 5.6 is the inductor current change during the "off" time ($= m_2t_{off}$). Then

$$I_{sp} = I_{sa} + \frac{m_2t_{off}}{2}$$

$$= I_{sa} + \frac{m_2}{2}(T - t_{on})$$

So

$$V_i = \frac{N_s}{N_p}R_i\left[I_{sa} + \frac{m_2}{2}(T - t_{on})\right] \tag{5.4}$$

Equating Eqs. 5.3 and 5.4, which is what the PWM comparator does, we obtain

$$\frac{N_s}{N_p}R_i I_{sa} = V_{eao} + t_{on}\left(\frac{N_s}{N_p}R_i\frac{m_2}{2} - m\right) - \left(\frac{N_s}{N_p}R_i\frac{m_2}{2}T\right)$$

It can be seen in this relation that if

$$\frac{N_s}{N_p}R_i\frac{m_2}{2} = m = \frac{dV_{ea}}{dt} \tag{5.5}$$

then the coefficient of the t_{on} term is zero and the average output inductor current is independent of the "on" time. This then corrects the above two problems arising from the fact that without compensation, current mode maintains the peak, and not the average, output inductor current constant.

5.5.4 Slope (Ramp) Compensation with a Positive-Going Ramp Voltage[3]

In the previous section it was shown that if a negative ramp of magnitude given by Eq. 5.5 is added to the error-amplifier output, the two current-mode problems described above are corrected.

The same effect is obtained by adding a positive-going ramp to the output of the current-sensing resistor V_i (Figure 5.3) and leaving the error-amplifier output voltage V_{eao} (Figure 5.3) unmodified. Adding a positive ramp to V_i is simpler and is the more usual approach. That adding the appropriate positive ramp to V_i also makes the average output inductor current independent of "on" time can be shown as follows: A ramp voltage of slope dV/dt will be added to the voltage V_i of Figure 5.3, and the resultant voltage will be compared in the PWM to the error-amplifier output V_{eao} of that figure. When the PWM finds equality of those voltages, its output terminates the "on" time. Then $V_i + dV/dt = V_{eao}$. Substitute V_i from Eq. 5.4:

$$\frac{N_s}{N_p}R_i\left[I_{sa} + \frac{m_2}{2}(T - t_{on})\right] + \frac{dV}{dt}t_{on} = V_{eao}$$

Then

$$\frac{N_s}{N_p}R_i I_{sa} + \frac{N_s}{N_p}R_i\frac{m_2}{2}T + t_{on}\left(\frac{dV}{dt} - \frac{N_s}{N_p}R_i\frac{m_2}{2}\right) = V_{eao}$$

From the above, it is seen that if the slope dV/dt of the voltage added to V_i is equal to $(N_s/N_p)R_i m_2/2$, the terms involving t_{on} in the preceding relation vanish and the secondary average voltage I_{sa} is independent of the "on" time. Note that $m_2(= V_o/L_o)$ is the current down-slope of the output inductor as defined earlier.

5.5.5 Implementing Slope Compensation[3]

In the UC1846 chip, a positive-going ramp starting at every clock pulse is available across the timing capacitor (pin 8 in Figure 5.2*b*). The voltage at that pin is

$$V_{\text{osc}} = \frac{\Delta V}{\Delta t} t_{\text{on}} \tag{5.6}$$

where $\Delta V = 1.8$ V and $\Delta t = 0.45 R_t C_t$.

As seen in Figure 5.7, a fraction of that voltage, whose slope is $\Delta V/\Delta t$, is added to V_i (the voltage across the current-sensing resistor).

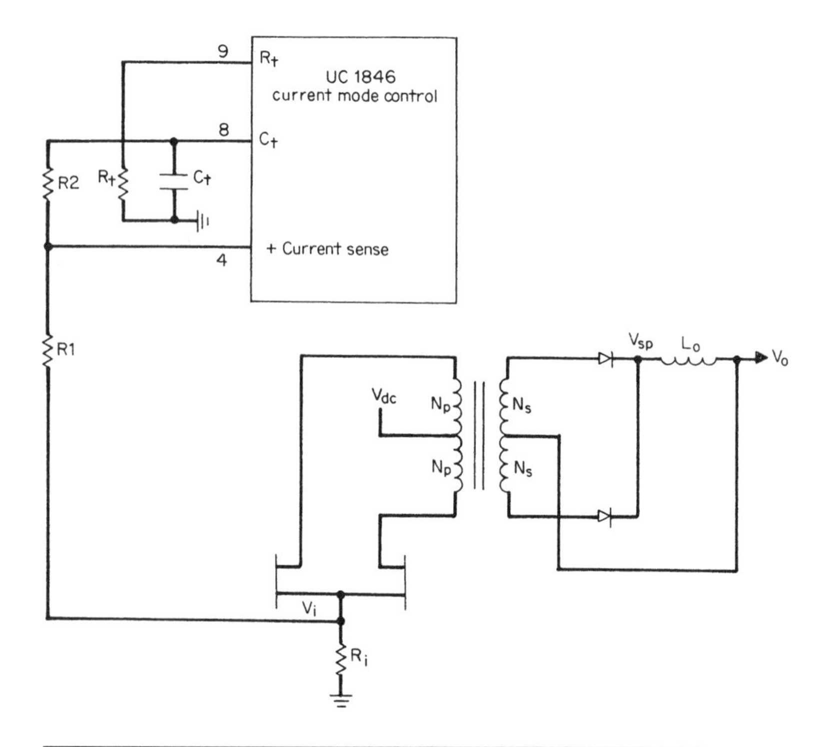

FIGURE 5.7 Slope compensation in the UC1846 current-mode control chip. A positive ramp voltage is taken from the timing capacitor, scaled by resistors $R1$, $R2$ and added to the voltage on the current resistor R_i. By choosing R_1, R_2 to make the slope of the voltage added to V_i equal to half the down-slope of the output inductor current reflected into the primary and multiplied by R_i, the average output inductor current is rendered independent of power transistor "on" times.

That slope is set to $(N_s/N_p)R_i(m_2/2)$ by resistors $R1$, $R2$. Thus in Figure 5.7, since R_i is much less than $R1$, the voltage delivered to the current-sensing terminal (pin 4) is

$$V_i + \frac{R1}{R1+R2}V_{\text{osc}} = V_i + \frac{R1}{R1+R2}\frac{\Delta V}{\Delta t}t_{\text{on}} \tag{5.7}$$

and setting the slope of that added voltage equal to $(N_s/N_p)R_i m_2/2$, we obtain

$$\frac{R1}{R1+R2} = \frac{(N_s/N_p)(R_i)(m_2/2)}{\Delta V/\Delta t} \tag{5.8}$$

in which $\Delta V/\Delta t = 1.8/(0.45R_tC_t)$.

Since $R1 + R2$ drains current from the timing capacitor, they change operating frequency. Then either $R1 + R2$ is made large enough so that the frequency change is small, or a buffer amplifier is interposed between pin 8 and the resistors. Usually $R1$ is preselected and $R2$ is calculated from Eq. 5.8.

After Pressman *With large values of inductance* L_o *or at higher frequencies, the slope on the current waveform (Figure 5.5) as it approaches the point of transition to the "off" state can approach zero. Hence any small noise spike can cause early or late switching resulting in jitter and noise in the output. In effect, the gain of the fast current control loop becomes very high. Close attention to layout and using a non-inductive current-sensing resistor for* R_i *or a DCCT may help. But in many cases the solution requires a reduction in inductance resulting in an increase in high frequency ripple current. ~K.B.*

5.6 Comparing the Properties of Voltage-Fed and Current-Fed Topologies

5.6.1 Introduction and Definitions

All topologies discussed thus far have been of the voltage-fed type. *Voltage-fed* implies that the source impedance of whatever drives the topology is low and hence there is no way of limiting the current drawn from it during unusual conditions at power switch turn "on" or turn "off," or under various fault conditions in the topology.

There are various ways of implementing "current limiting" with additional circuitry, which senses an over-current condition and takes some kind of corrective action such as narrowing the controller's switching pulse width or stopping it completely. But all such schemes are not instantaneous; they involve a delay over a number of switching cycles during which there can be excessive dissipation in either the power transistors or output rectifiers and dangerous voltage or current spiking. Thus such over-current sensing schemes are of no help in the case of high transient currents at the instant of the power switch turn "on" and turn "off."

The low-source impedance in voltage-fed topologies is that of the filter capacitor in offline converters or of the battery in battery-powered converters. In compound schemes that use a buck regulator to preregulate the rectified DC voltage of the AC line rectifier, it is the very low-output impedance of the buck regulator itself.

In current-fed topologies, the high instantaneous impedance of an inductor is interposed between the power source and the topology itself. This provides a number of significant advantages, especially in high power supplies (> 1000 W), high output voltage supplies (> 200 V), and multi-output supplies where close tracking between slaves and a master output voltage is required.

Advantages of the current-fed technique can be appreciated by examining the usual shortcomings of high-power, high-output-voltage, and multi-output voltage-fed topologies.

5.6.2 Deficiencies of Voltage-Fed, Pulse-Width-Modulated Full-Wave Bridge[9]

Figure 5.8 shows a conventional voltage-fed full bridge—the usual choice for a switching supply at 1000-W output. At higher output powers, high output voltages, or multiple output voltages, it has the following significant shortcomings.

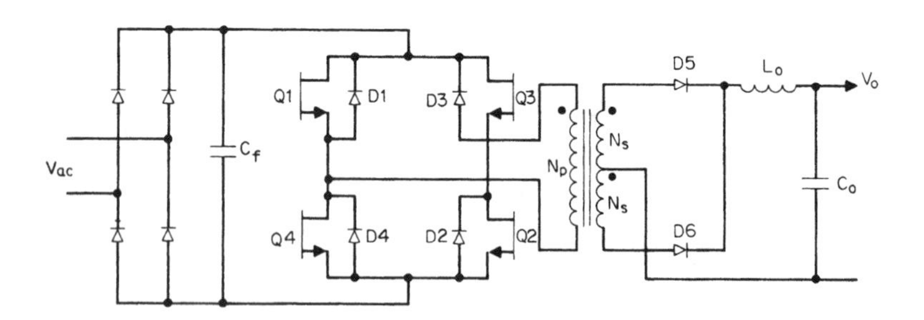

FIGURE 5.8 A conventional voltage-fed full bridge, often used for higher output powers typically 1000 W or more. The low-source impedance of the filter capacitor C_f and the need for the output inductor L_o are significant drawbacks for output powers over 1000 W and output voltages over a few hundred volts. Further, in a multi-output power supply, the requirement for an output inductor at each output makes the topology expensive in cost and space.

5.6.2.1 Output Inductor Problems in Voltage-Fed, Pulse-Width-Modulated Full-Wave Bridge

For high-output voltages, the size and cost of the output inductor L_o (or inductors in a multi-output supply) becomes prohibitive as can be seen from the following. The inductor is selected to prevent going into the discontinuous mode or running dry at the specified minimum DC load current (Sections 1.3.6 and 2.2.14.1). For a minimum DC load current of one-tenth the nominal I_{on}, Eq. 2.20 gives the magnitude of the inductor as $L_o = 0.5V_oT/I_{on}$.

Now consider a 2000-W supply at $V_o = 200$ V, $I_{o(\text{nominal})} = 10$ A, and a minimum DC output current of 1 A. To minimize the size of the output inductor, T should be minimized, and a switching frequency of 50 kHz might be considered. At 50 kHz, for $V_o = 200$ V, $I_{on} = 10$ A, Eq. 2.20 yields L_o of 200 μH.

The inductor must carry the nominal current of 10 A without saturating. Inductors capable of carrying large DC bias currents without saturating are discussed in a later chapter and are made either with gapped ferrite or powdered iron toroidal cores. A 200-μH 10-A inductor using a powdered iron toroid would have a diameter about 2.5 in and a height about 1.0 in.

Although this is not a prohibitive size for a single-output 2-kW supply, a supply with many outputs, higher output voltage, or higher output power, the size and cost of many large inductors would be a serious drawback. For high-output voltages (> 1000 V), even at low-output currents, the output inductor is far more troublesome because of the large number of turns required to support the high voltage across the inductor. This high voltage—especially during the dead time when cathodes of $D5$, $D6$ of Figure 5.8 are both "low"—can produce corona and arcing.

A further problem with a topology requiring output inductors, as shown in Figure 5.8, is the poor cross regulation or change in output voltage of a slave when current changes in the master (Section 2.2.2). The output inductors in both the master and slave must be large enough to prevent discontinuous mode operation and large-output voltage changes at minimum load currents.

The current-fed topology (Figure 5.10) discussed below avoids many of the above problems, as it does not require multiple output inductors. It uses a single input inductor $L1$ in place of the individual output inductors, and is positioned before the high frequency switching bridge circuit and after line rectification and storage capacitors. Thus DC output voltages are the peak rather than the average of the transformer secondary voltages. Voltage regulation is achieved by pulse-width modulation of the bridge, or as in Figure 5.10, by a buck regulator transistor switch $Q5$ ahead of the $L1$ inductor.

5.6.2.2 Turn "On" Transient Problems in Voltage-Fed, Pulse-Width-Modulated Full-Wave Bridge[9]

In Figure 5.8, diagonally opposite transistors are simultaneously "on" during alternate half cycles. The maximum "on" time of each pair is designed to be less than 80% of a half period. This ensures a $0.2T/2$ dead time between the turn "off" of one transistor pair and the turn "on" of the other. This dead time is essential, for if the "on" time of alternate pairs overlapped by even a fraction of a microsecond, there would be a dead short circuit across the filter capacitor, and with nothing to limit current flow, the transistors would fail immediately.

During the dead time, all four transistors are "off," the anodes of output rectifiers *D*5, *D*6 are at zero volts, and the voltage at the input end of filter inductor L_o has swung down to keep the current constant. The input end of L_o is clamped at one diode drop below ground by *D*5, *D*6, which act as free-wheeling diodes. The current that had been flowing in L_o before the dead time (roughly equal to the DC output current) continues to flow in the same direction. It flows out through the ground terminal into the secondary center tap, where it divides equally with half flowing through each of *D*5 and *D*6 and back into the input end of L_o.

At the start of the next half cycle when, say, *Q*1, *Q*2 turn "on," the no-dot end of the *T*1 primary is high and the no-dot end of the *T*1 secondary (anode of *D*6) attempts to go high. But the cathode of *D*6 is looking into the cathode of *D*5, which is still conducting half the DC output current. Until *D*6 supplies a current equal to and canceling the *D*5 forward current, it is looking into the low impedance of a conducting diode ($\sim$10 Ω).

This low secondary impedance reflects as a low impedance across the primary. But this low impedance is in series with the transformer's leakage inductance, which limits the primary current during the time required to cancel the *D*5 free-wheeling current. Because of the high-impedance current-limiting effect of the leakage inductance, transistors *Q*1 and *Q*2 remain in saturation until the *D*5 free-wheeling current is canceled.

When the *D*5 current is canceled, it still has a low impedance because of its reverse-recovery time, which may range from 35 ns (ultra-fast-recovery type) to 200 ns (fast-recovery type). For a reverse-recovery time of t_r, supply voltage of V_{cc}, and transformer primary leakage inductance of L_l, the primary current overshoots to $V_{cc}t_r/L_l$. This overshoot current can pull the transistors out of saturation and either damage or destroy them.

Finally, when the output rectifier recovers abruptly, there is a damped oscillatory ring at its cathode. The first positive half cycle of this ring can more than double the reverse voltage stress on the diode and possibly destroy it. Even in lower power supplies, it is often necessary to put series RC snubbers across the rectifiers to damp the oscillation. The penalty paid for this is, of course, dissipation in the resistors.

5.6.2.3 Turn "Off" Transient Problems in Voltage-Fed, Pulse-Width-Modulated Full-Wave Bridge[9]

In Figure 5.8, there is a spike of high power dissipation at turn "off" as a result of the instantaneous overlap of falling current and rising voltage across the "off"-turning transistors.

Consider that $Q3$ and $Q4$ are "on" and have received turn "off" signals at their bases. As $Q3$, $Q4$ commence turning "off," current stored in the leakage and magnetizing inductance of $T1$ force a polarity reversal across the primary. The bottom end of $T1$ primary goes immediately positive and is clamped via $D1$ to the positive rail at the top of C_f. The top end of $T1$ primary goes immediately negative and is clamped via $D2$ to the negative rail at the bottom end of C_f. Now voltages across $Q3$ and $Q4$ are clamped at V_{cc} so long as diodes $D1$, $D2$ conduct. There are no leakage inductance voltage spikes across $Q3$, $Q4$ as in push-pull or single-ended forward converter topologies. Energy stored in the leakage inductance is returned without dissipation to the input capacitor C_f.

However, while the voltage across $Q3$, $Q4$ is held at V_{cc}, the current in these two transistors falls linearly to zero in a time t_f determined by their reverse base drives. This overlap of a fixed-voltage V_{cc} and a current falling linearly from a value I_p results in dissipation averaged over a full period T of

$$PD = V_{cc}\frac{I_p}{2}\frac{t_f}{T} \tag{5.9}$$

It is instructive to calculate this dissipation for, say, a 2-kW supply operating at 50 kHz from a nominal V_{cc} of 336 V (typical V_{cc} for an offline inverter operating from a 120-V AC line in the voltage-doubling mode as in Section 3.1.1). Assume a minimum V_{cc} of 0.9 (336 V) or 302 V. Then from Eq. 3.7, the peak current is

$$I_p = \frac{1.56P_o}{\underline{V_{dc}}} = 1.56\frac{2000}{302} = 10.3\text{ A}$$

A bipolar transistor at this current has a fall time of perhaps 0.3 μs. Since peak currents are independent of DC input voltage, calculate overlap dissipation from Eq. 5.9 at a high line of $1.1 \times 336 = 370$ V. For the dissipation in either $Q3$ or $Q4$, Eq. 5.9 gives $PD = V_{cc}(I_p/2)(t_f/T) = 370(10.3/2)(0.3/20) = 28.5$ W, and for the four transistors in the bridge, total overlap losses would be 114 W.

It is of interest to calculate the dissipation per transistor during the "on" time. This is $V_{ce(sat)} I_c T_{on}/T$ and for a typical $V_{ce(sat)}$ of 1.0 V and an "on" duty cycle of 0.4 is only $1 \times 10.3 \times 0.4$ or 4.1 W.

Even though the 28.5 W of overlap dissipation per transistor can be reduced with four load- and line-shaping "snubbers" (to be discussed in a later chapter), these snubbers reduce transistor losses only by diverting them to the snubber resistors with no improvement in efficiency. It will be shown that in the current-fed topology, only two snubbers will be required, reducing transistor overlap dissipation to a negligible value. The price paid for this is the dissipation in each of the two snubber resistors of somewhat more than that in the voltage-fed full bridge.

5.6.2.4 Flux-Imbalance Problem in Voltage-Fed, Pulse-Width-Modulated Full-Wave Bridge

Flux imbalance, or operation not centered about the origin of the transformer's *BH* loop, was discussed in Section 2.2.5 in connection with the push-pull and in Section 3.2.4 for the half bridge. It arises because of unequal volt-second products applied to the transformer primary on alternate half cycles. As the core drifts farther and farther off center on the *BH* loop, it can move into saturation where it is unable to sustain the supply voltage and destroy the transistor.

Flux imbalance can also arise in the conventional full-wave bridge because of a volt-second imbalance on alternate half cycles. This can come about with bipolar transistors because of unequal storage times on alternate half cycles or with MOSEFT transistors because of unequal MOSEFT "on"-voltage drops. The solution for the full-wave bridge is to place a DC blocking capacitor in series with the primary. This prevents a DC current bias in the primary and forces operation to be centered about the *BH* loop origin. The size of such a DC blocking capacitor is calculated as in Section 3.2.4 for the half bridge.

The current-fed circuits, discussed below, do not require DC blocking capacitors, providing another advantage over voltage-fed circuits. This is still an advantage despite the relatively small size and cost of such blocking capacitors.

5.6.3 Buck Voltage-Fed Full-Wave Bridge Topology—Basic Operation

This topology is shown in Figure 5.9. It avoids many of the deficiencies of the voltage-fed pulse-width-modulated full-wave bridge in high-voltage, high-power, multi-output supplies.

Consider first how it works. There is a buck regulator preceding a square-wave inverter, which has only capacitors after the secondary

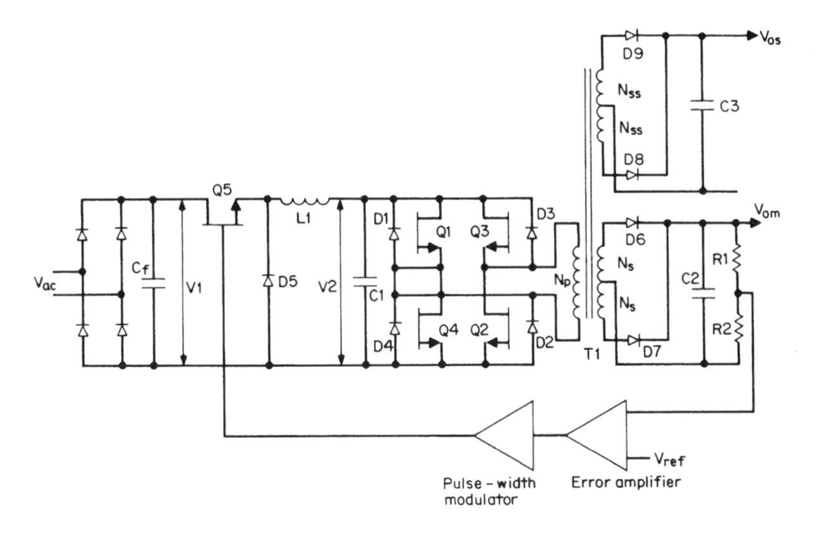

FIGURE 5.9 Buck voltage-fed full bridge. The buck regulator preceding the full bridge eliminates the output inductors in a multi-output supply, but the low-source impedance of the buck capacitor and the low-output impedance of the buck regulator still leave many drawbacks to this approach. *Q*5 is pulse-width-modulated, but *Q*1 to *Q*4 are operated at a fixed "on" time at about 90% of a half period to avoid simultaneous conduction. The output filters *C*2, *C*3 are peak rather than averaging rectifiers. Practical output powers of about 2 kW to 5 kW are realizable.

rectifying diodes. Thus the DC output voltage at the filter capacitor is the peak of the secondary voltage less the negligible rectifying diode drop. Neglecting also the inverter transistor "on" drop, the DC output voltage is $V_o = V_2(N_s/N_p)$, where V_2 is the output of the buck regulator. The inverter transistors are not pulse-width-modulated. They are operated at a fixed "on" time—roughly 90% of a half period to avoid simultaneous conduction in the two transistors positioned vertically one above another. Diagonally opposite transistors are switched "on" and "off" simultaneously.

Feedback is taken from one of the secondary outputs (usually the output with highest current or tightest output voltage tolerance) and used to pulse-width-modulate the buck transistor *Q*5. This bucks down the rectified, unregulated DC voltage V_1 to a DC value V_2, which is usually selected to be about 25% lower than the lowest rectified voltage V_1 corresponding to the lowest specified AC input voltage. The turns ratio N_s/N_p is then chosen so that for this value of V_2, the correct master output voltage $V_{\text{om}} = V_2(N_s/N_p)$ is obtained.

The feedback loop, in keeping V_{om} constant against line and load changes, then keeps V_2 constant (neglecting relatively constant rectifier diode drops) at $V_2 = V_o(N_p/N_s)$. Additional secondaries, rectifier diodes, and peak-rectifying filter capacitors can be added for slave outputs.

Alternatively, feedback can be taken from $C1$ to keep V_2 constant. From V_2 to the outputs, the circuit is open-loop. But all output voltages are still quite insensitive to line and load changes because they change only slightly with forward drops in diode rectifiers and "on" drops of the transistors, which change only slightly with output currents. Thus the output voltages are all largely proportional to V_2.

Taking feedback from V_2 results in somewhat less constant output voltage, but avoids the problem of transmitting a pulse-width-modulated control voltage pulse across the boundary from output to input common. If an error amplifier is located on output common with a pulse-width modulator on input common, it avoids the problem of transmitting the amplified DC error voltage across the output-input boundary. Such a scheme usually involves the use of an optocoupler, which has wide tolerances in gain and is not too reliable a device.

5.6.4 Buck Voltage-Fed Full-Wave Bridge Advantages

5.6.4.1 Elimination of Output Inductors

The first obvious advantage of the topology for a multi-output supply is that it replaces many output inductors with a single input inductor with consequent savings in cost and space.

Since there are no output inductors in either the master or slaves, there is no problem with large output voltage changes that result from operating the inductors in discontinuous mode (Sections 1.3.6 and 2.2.4). Slave output voltages track the master over a large range of output currents, within about ±2%, rather than the ±6 to ±8% with output inductors in continuous mode, or substantially more in discontinuous mode.

After Pressman *Providing the outputs share a common return, this problem can also be solved in the multiple output inductor case by using the coupled inductor approach. Here a single inductor has a winding for each output wound on a single core. The transformer type coupling between the windings also eliminates many of the problems shown above.*[1] *~K.B.*

The input inductor is designed to operate in continuous mode at any current above the minimum. Since it is unlikely that all outputs are at minimum current simultaneously, this indicates a higher total minimum current and a smaller input inductor (Section 1.3.6).

Further, even if the input inductor goes discontinuous, the master output voltage will remain substantially constant, but with somewhat more output ripple and somewhat poorer load regulation. The feedback loop will keep the main output voltage constant even in discontinuous mode through large decreases in "on" time of the buck transistor (Figure 1.6*a*).

Further, since the slave outputs are clamped to the main output in the ratio of their respective turns ratios, slaves will also remain constant against large line and load changes.

Elimination of output inductors, with the many turns required to sustain high AC voltages for high voltage DC outputs, makes 2000- to 3000-V outputs easily feasible. Higher output voltages—15,000 to 30,000 V—at relatively low-output currents as for cathode-ray tubes, or high-voltage high-current outputs as for traveling-wave tubes, are easily obtained by conventional diode-capacitor voltage multipliers after the secondaries.[8]

5.6.4.2 Elimination of Bridge Transistor Turn "On" Transients

With respect to the full-wave pulse-width-modulated bridge of Figure 5.8, Section 5.6.2.2 discussed turn "on" transient current stresses in the bridge transistors (*Q*1 to *Q*4), and excessive voltage stress in the rectifying diodes (*D*5, *D*6).

It was pointed out in Section 5.6.2.2 that these stresses arose because the rectifier diodes were also acting as free-wheeling diodes. At the instant of turn "on" of one diagonally opposite pair (say *Q*1, *Q*2), *D*6 was still conducting as a free-wheeling diode. Until the forward current in *D*6 was canceled, the impedance seen by *Q*1, *Q*2 was the leakage inductance of *T*1 in series with the low forward impedance of *D*6 reflected into the primary.

Subsequently, when *Q*1, *Q*2 forced a current into the primary sufficient to cancel the *D*6 forward current, there was still a low impedance reflected into the primary because of reverse recovery time in *D*6. This caused a large primary current overshoot that overstressed *Q*1, *Q*2. At the end of the recovery time, when the large secondary current overshoot terminated, it caused an oscillation and excessive voltage stress on *D*6.

This current overstress on the bridge transistors and voltage overstress on the output rectifiers does not occur with the buck voltage-fed topology of Figure 5.9. The inverter transistors are operated with a dead time ($\sim 0.1T/2$) between the turn "off" of one pair of transistors and the turn "on" of the other pair.

During this dead time when none of the bridge transistors are "on," no current flows in the output rectifiers and output load current is supplied from the filter capacitors alone. Thus at the start of the next half period, the "on"-turning rectifier diode is not loaded down with a conducting free-wheeling diode as in Figure 5.8. The opposite diode has long since ceased conducting; thus there is no current overstress in the bridge transistors, no recovery time problem in the rectifier diodes, and hence no overvoltage stress in them.

5.6.4.3 Decrease of Bridge Transistor Turn "Off" Dissipation

In Section 5.6.2.3, it was calculated that for a 2000-W supply operating from a nominal input of 120-V AC in the input voltage-doubling mode, the bridge dissipation is 28.5 W at maximum AC input for each of the four transistors in the voltage-fed, pulse-width-modulated bridge circuit of Figure 5.8.

In the buck voltage-fed, full-wave bridge (Figure 5.9), this dissipation is somewhat less. This is so because, even at maximum AC input, the "off"-turning bridge transistors are subjected to bucked-down voltage V_2 (Figure 5.9) of about 0.75 times the minimum rectified voltage as discussed in Section 5.6.3. For the minimum rectified DC of 302 V (Section 5.6.2.3), this is 0.75×302 or 227 V. This compares favorably to the 370 V DC at maximum AC input as calculated in Section 5.6.2.3.

The peak current from the bucked-down 227 V will not differ much from the 10.3 A calculated in Section 5.6.2.3. Thus assume a total efficiency of 80%, as for the circuit of Figure 5.8. Assume that half the losses are in the bridge and half in the buck regulator of Figure 5.9. Then for a bridge efficiency of 90%, its input power is 2000/0.9 or 2222 W. With a preregulated input, the bridge transistors can operate at 90% duty cycle without concern about simultaneous conduction. Input power is then $0.9 I_p V_{\text{dc}} = 2222$ W. For V_{dc} of 227 V as above, this yields I_p of 10.8 A. Calculating bridge transistor dissipation as in Section 5.6.2.3 for a current fall time t_f of 0.3 μs out of a period T of 20 μs, dissipation per transistor is $(I_p/2)(V_{\text{dc}})(t_f/T) = (10.8/2)227 \times 0.3/20 = 18.4$ W. This is 74 W for the entire bridge as compared to 114 W for the circuit of Figure 5.8 as calculated in Section 5.6.2.3.

5.6.4.4 Flux-Imbalance Problem in Bridge Transformer

This problem is still the same as in the topology of Figure 5.8. A volt-second unbalance can occur because of unequal storage times for bipolar bridge transistors or because of unequal "on" voltages for MOSFET transistors. The solution for both the Figure 5.8 and Figure 5.9 topologies is to insert a DC blocking capacitor in series with the transformer primary.

5.6.5 Drawbacks in Buck Voltage-Fed Full-Wave Bridge[9,10]

Despite the advantages over the pulse-width-modulated full-wave inverter bridge, the buck voltage-fed full-wave bridge has a number of significant drawbacks.

First, there are the added cost, volume, and power dissipation of the buck transistor $Q5$ (Figure 5.9) and the cost and volume of the buck LC filter ($L1$, $C1$). The added cost and volume of these elements is partly compensated by the saving of an inductor at each output. The added dissipations of the buck regulator $Q5$ and the free-wheeling diode $D5$ are most often a small percentage of the total losses for a $\geq$ 2000-W power supply.

Second, there are turn "on" and turn "off" transient losses in the buck transistor, which can be greater than its DC conduction losses. These can be reduced in the transistor by diverting them to passive elements in snubbers. But the losses, cost, and required space of the snubbers is still a drawback. Turn "on"–turn "off" snubbers will be discussed in the later section on the buck current-fed full-wave bridge.

The turn "off" transient losses in the bridge transistors, although less than for the pulse-width-modulated bridge of Figure 5.8, still remain significant. (See the discussion in Section 5.6.4.3.)

Finally, under conditions of unusually long storage time at high temperature and low load or low line, at the turn "on" of one transistor pair, the opposite pair may still be "on." With the low-source impedance of the buck filter capacitor and the momentary short circuit across the supply bus, this will cause immediate failure of at least one, and possibly all, of the bridge transistors.

5.6.6 Buck Current-Fed Full-Wave Bridge Topology—Basic Operation[9,10]

This topology is shown in Figure 5.10.[6] It has no output inductors and is exactly like the buck voltage-fed full-wave bridge of Figure 5.9 with the exception that there is no buck filter capacitor $C1$. Instead, there is a virtual capacitor $C1V$, which is the sum of all the secondary filter capacitors reflected by the squares of their respective turns ratios into the $T1$ primary. The filtering by this virtual capacitor $C1V$ is exactly the same as that of a real capacitor of equal magnitude.

Thus, by replacing all the output inductors of the pulse-width-modulated full-wave bridge of Figure 5.8 with a single primary side inductor as in Figure 5.9, all the advantages described in Section 5.6.2.1 for the Figure 5.9 circuit are also obtained for the circuit of Figure 5.10.

Bridge transistors $Q1$ to $Q4$ are not pulse-width-modulated, as they were in Figure 5.8. In this topology, diagonally opposite transistors are

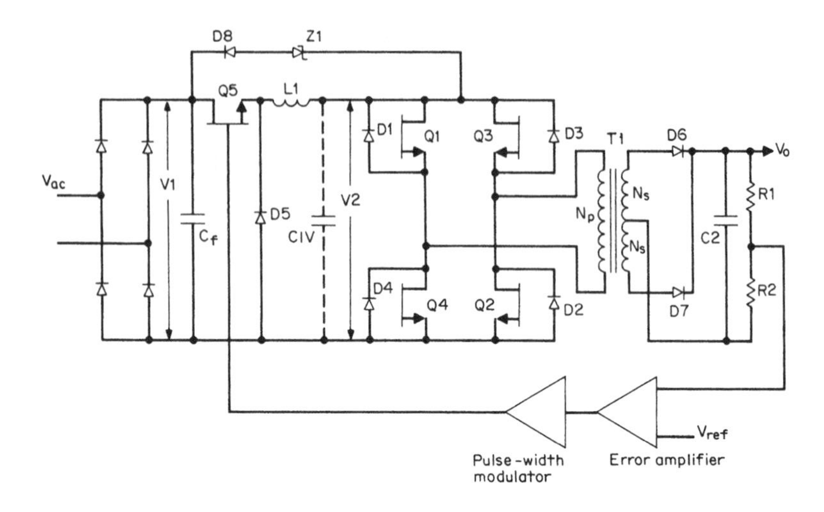

FIGURE 5.10 Buck current-fed full-wave bridge. The buck filter capacitor $C1$ is omitted. There is a virtual capacitor $C1V$ there—it is the sum of all the output capacitors of the master and slaves reflected into the primary. Diagonally opposite transistors are turned "on" simultaneously. By causing the "off"-turning and "on"-turning pair to overlap in the "on" state for a short time (~ 1 μs), significant advantages are obtained. During the overlap of the "off"- and "on"-turning pairs, the high impedance looking into $L1$ (with $C1$ missing) forces all input and output nodes of the bridge to collapse to zero volts. It is the high impedance looking back into $L1$ that gives the source driving the bridge the characteristic of a constant current generator. $Z1$, $D8$ constitute an upper clamp to limit $V2$ when the previously "on" transistors turn "off."

simultaneously "on" during alternate half cycles without the normal "off dead time" between the turn "off" of one pair and the turn "on" of the next pair, as was required for the voltage-fed circuit of Figure 5.9. Each pair in Figure 5.10 is kept "on" deliberately for slightly more than a half period, either by depending on the storage times of slow bipolar transistors or by delaying the turn "off" time by ~1 μs or so when using faster bipolar or MOSFET devices. Output voltage regulation is achieved by pulse-width-modulating the "on" time of the buck transistor $Q5$ as was done for the buck voltage-fed circuit of Figure 5.9.

Significant advantages accrue from the physical removal of buck filter capacitor $C1$ of Figure 5.9 and the deliberate overlapping "on" times of alternate transistor pairs. These advantages are described as follows.

5.6.6.1 Alleviation of Turn "On"–Turn "Off" Transient Problems in Buck Current-Fed Bridge[9,10]

For the pulse-width-modulated full-wave bridge of Figure 5.8, Section 5.6.2.2 described excessive current, power dissipation stresses in the bridge transistors, and voltage stresses in the output rectifier diodes at the instant of turn "on." Such stresses do not occur in the current-fed circuit of Figure 5.10 because of the overlapping "on" times of alternate transistor pairs and the high impedance seen looking back into $L1$ with no filter capacitor physically present at that node.

This can be seen from Figures 5.11 and 5.12. Consider in these figures that $Q3$, $Q4$ had been "on" and $Q1$, $Q2$ commence turning "on" at T_1. Transistors $Q3$, $Q4$ remain "on" until T_2 (Figure 5.12), resulting in an overlap time of $T_2 - T_1$. At T_1, as $Q1$, $Q2$ come "on," a dead short circuit appears at the output of $L1$, and since the impedance looking into $L1$ is high, the voltage V_2 collapses to zero (Figure 5.12*c*). $L1$ is a large inductor and current in it must remain constant at its initial value I_L. Thus as current in $Q1$, $Q2$ rises from zero toward I_L (Figure 5.12*f* and 5.12*g*), current in $Q3$, $Q4$ falls from I_L toward zero (Figure 5.12*d* and 5.12*e*).

Note, the rising current in $Q1$, $Q2$ occurs with zero voltage at V_2, so there is also zero voltage between nodes A and B in Figure 5.11. Hence, there is no voltage across $Q1$, $Q2$ as their current rises, and there is no dissipation in them. At some later time T_3, currents in $Q1$, $Q2$ have risen to $I_L/2$ and currents in $Q3$, $Q4$ have fallen from I_L to $I_L/2$, thus summing to the constant current I_L from inductor $L1$.

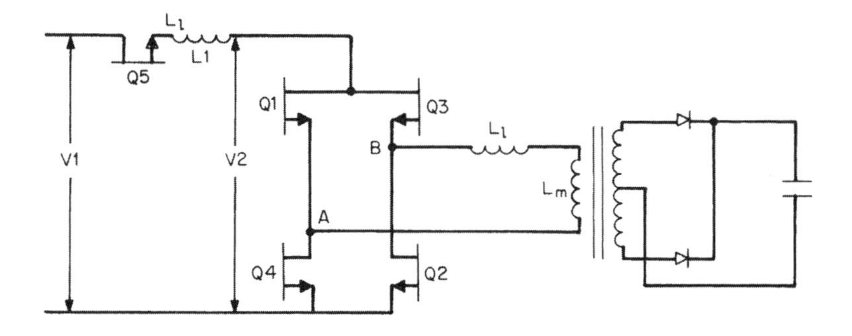

FIGURE 5.11 During the overlap, when all four transistors are "on," the voltage $V2$ and that across nodes A, B collapse to zero. Energy stored in leakage inductance L_l is fed to the load via the transformer instead of being dissipated in a snubber resistor or being returned to the input bus as in conventional circuits. Hence, there is no turn "on" transient dissipation in the bridge transistors or overvoltage stress in output rectifiers.

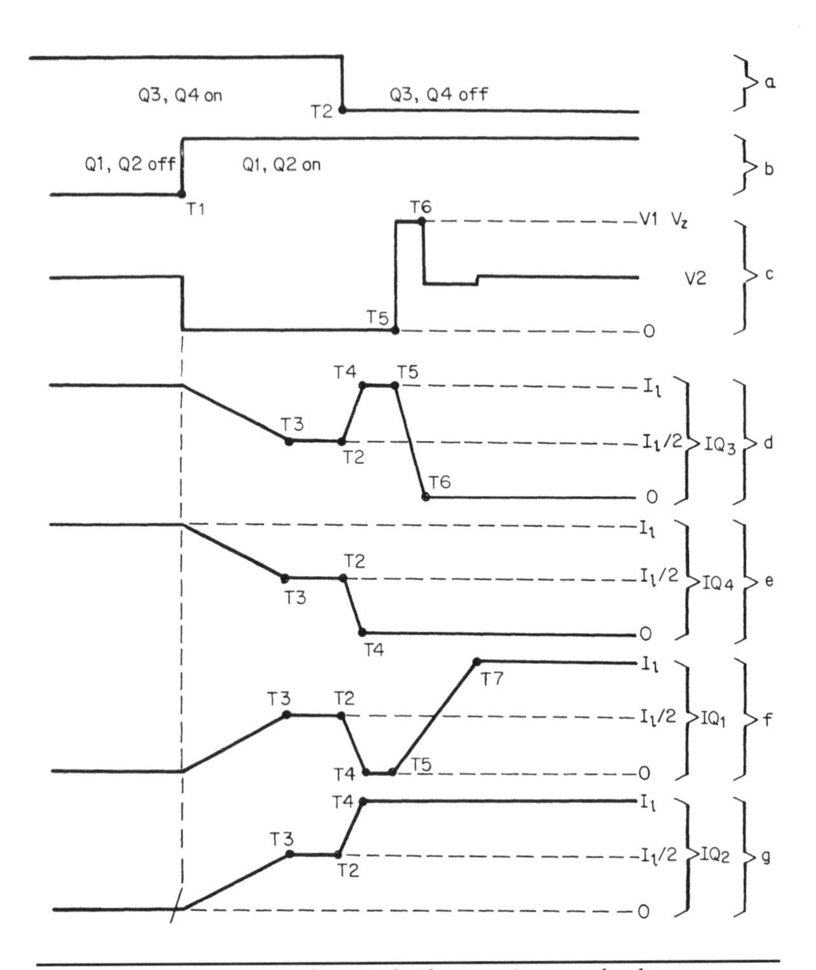

FIGURE 5.12 Current waveforms in bridge transistors and voltages at bridge input during the overlapping "on" times of all four bridge transistors in buck current-fed topology.

Assume, as a worst-case scenario, that *Q*3 is slower than *Q*4, and *Q*4 turns "off" first. Note that at T_2, when *Q*3 and *Q*4 are commanded "off," the voltage V_2 is zero, so *Q*4 turns "off" with zero voltage across it and little turn "off" dissipation. As I_{Q4} falls from $I_L/2$ toward zero (T_2 to T_4), I_{Q2} rises from $I_L/2$ toward I_L to maintain the constant I_L demanded by *L*1. As I_{Q2} rises from $I_L/2$ toward I_L, I_{Q3} rises from $I_l/2$ to I_L to supply I_{Q2}. Again, since *L*1 demands a constant current I_L, as I_{Q3} rises toward I_L, I_{Q1} falls from $I_L/2$ to zero at T_4.

During the time T_1 to T_4, while V_2 is zero volts, the voltage across the transformer primary (*A* to *B* in Figure 5.11) will also fall. Current

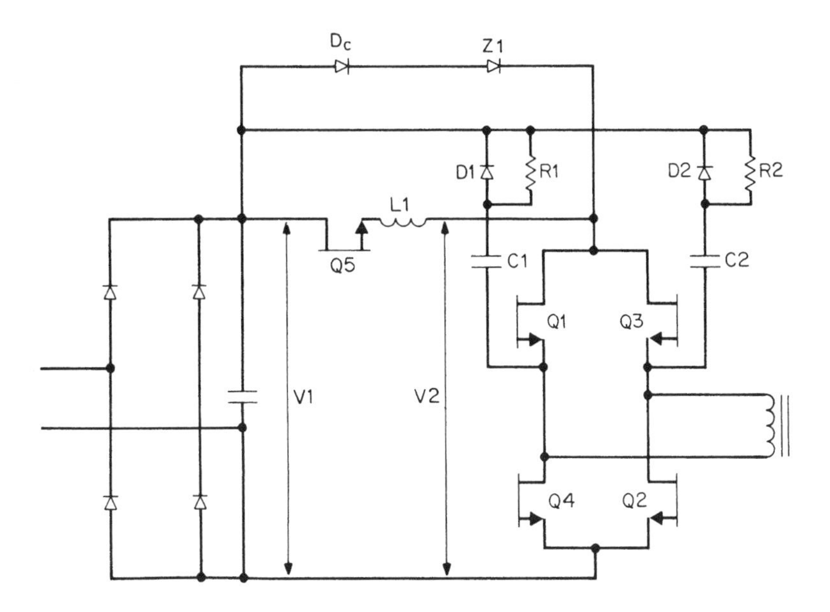

FIGURE 5.13 The buck current-fed bridge. In this circuit only two turn "off" snubbers ($R1$, $C1$, $D1$ and $R2$, $C2$, $D2$) are required. An upper voltage clamp ($Z1$, D_c) is required to limit $V2$ when the last of the "off"-turning transistors turns "off."

had been stored in the transformer leakage inductance L_L while $Q3$, $Q4$ were "on." As voltage A to B collapses, the voltage across the primary leakage inductance reverses to keep the current constant. Thus the leakage inductance acts like a generator and delivers this stored energy through the transformer to the secondary load instead of returning it to the input supply bus or to dissipative snubbers as in conventional circuits.

At a later time T_5, the slower transistor $Q3$ starts turning "off." As current in it falls from I_L to zero (Figure 5.12*d*), current I_{Q1} tries to rise from zero to I_L to maintain the constant current I_L demanded by $L1$. But I_{Q1} rise time is limited by the transformer leakage inductance (Figure 5.12*f*). Since I_{Q3} fall time is generally greater than I_{Q1} rise time, voltage V_2 will overshoot its quiescent value and must be clamped to avoid overstressing $Q3$, as its emitter is now clamped to ground by the conducting $Q2$. The clamping is done by a zener diode $Z1$, as shown in Figures 5.10 and 5.13.

Voltage overshoot of V_2 during the slower transistor ($Q3$) turn "off" time results in somewhat more dissipation in it than in the circuit of the conventional pulse-width-modulated bridge (Figure 5.8). This dissipation is $(V_1 + V_z)(I_L/2)(T_6 - T_5)/T$ for Figure 5.10, but only

$V_1(I_L/2)(T_6 - T_5)/T$ for Figure 5.8. In Figure 5.8, there are four transistors that have relatively high turn "off" dissipation. In Figures 5.10 and 5.13, only the two transistors with slow turn "off" time have high dissipation. As discussed above, the faster transistor suffers no dissipation at turn "off" as it turns "off" at zero voltage; and at turn "on," all transistors have negligible dissipation, because the transformer leakage inductance is in series with them, so they turn "on" at zero voltage.

The increased dissipation of the two transistors at turn "off" can be diverted from the transistors to resistors by adding the snubbing networks $R1$, $C1$, $D1$ and $R2$, $C2$, $D2$ of Figure 5.13. Design of such turn "off" snubbing circuits will be discussed in the later chapter on snubbers.

5.6.6.2 Absence of Simultaneous Conduction Problem in the Buck Current-Fed Bridge

In the buck voltage-fed bridge of Figure 5.9, care must be taken to avoid simultaneous conduction in transistors positioned vertically above one another ($Q1$, $Q4$ or $Q3$, $Q2$). Such simultaneous conduction comprises a short circuit across $C1$. Since $C1$ has a low impedance, it can supply large currents without its output (V_2) dropping very much. Thus the bridge transistors could be subjected to simultaneous high voltage and high current, and one or more would immediately fail.

Even if a dead time between the turn "off" of one transistor pair and the turn "on" of the other is designed in to avoid simultaneous conduction, it still may occur under various odd circumstances, such as high temperature and/or high load conditions when transistor storage time may be much lower than data sheets indicate or low input voltage (in the absence of maximum "on" time clamp or undervoltage lockout) as the feedback loop increases "on" time to maintain constant output voltage.

But in the buck current-fed bridge, simultaneous conduction is actually essential to its operation and the inductor limits the current, hence it provides the advantages discussed above. Further, in the buck current-fed bridge, since the "on" time is slightly more than a half period for each transistor pair, the peak current is less than in the buck voltage-fed bridge, whose maximum "on" time is usually set at 90% of a half period to avoid simultaneous conduction.

5.6.6.3 Turn "On" Problems in Buck Transistor of Buck Current- or Buck Voltage-Fed Bridge[10]

The buck transistor in either the voltage- or current-fed bridge suffers from a large spike of power dissipation at the instant both of turn "on" and turn "off," as can be seen in Figure 5.14.

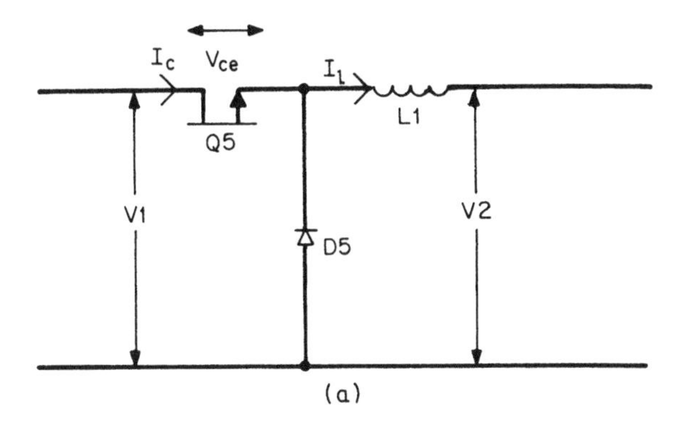

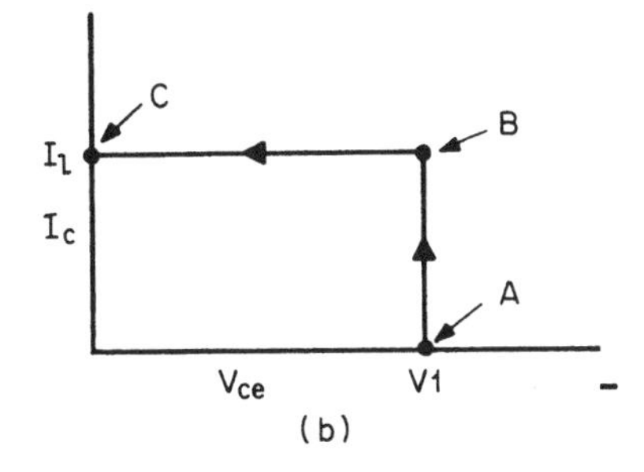

FIGURE 5.14 (*a*) The buck transistor in the buck current- or voltage-fed topology has a very unfavorable voltage-current locus at the instant of turn "on." It operates throughout its current rise time at the full input voltage *V*1 until the forward current in free-wheeling diode *D*5 has been canceled. This generates a large spike of dissipation at turn "on." (*b*) I_c vs. V_{ce} locus during turn "on" of buck transistor *Q*5. Voltage V_{ce} remains constant at *V*1 until the current in *Q*5 has risen to I_l (*A* to *B*) and canceled the forward current I_l in free-wheeling diode *D*5. Then, if capacitance at the *Q*5 emitter is low and *D*5 has a fast recovery time, it moves very rapidly to its "on" voltage of about 1 V (*B* to *C*).

After Pressman *Because* L1 *forces a constant current to flow in* Q5 *as it turns "off,"* Q5 *is subject to both an increasing voltage and a constant current until the emitter voltage drops below zero, when* D5 *conducts and the* L1 *current commutates from* Q5 *to* D5. *The peak power occurs at half voltage, when* $P_P = V1/2 \times I_L$. *Faster switching devices will reduce the average power loss, but cannot reduce the peak power unless an alternative path is provided for the L1 current during the turn "off" edge of* Q5. *~K.B.*

Consider first the instantaneous voltage and current of *Q*5 during the turn "on" interval. The locus of rising current and falling voltage during that interval is shown in Figure 5.14*b*. Just prior to *Q*5 turning "on," free-wheeling diode *D*5 is conducting and supplying inductor current I_L. As *Q*5 commences turning "on," its collector is at V_1, its emitter is at one diode (*D*5) drop below common. The emitter does not move up from common until the current in *Q*5 has risen from zero to I_L and canceled the *D*5 forward current.

Thus, during the current rise time to t_r, the $I_c - V_{ce}$ locus is from points *A* to *B*. During t_r, the average current supplied by *Q*5 is $I_L/2$ and the voltage across it is V_1. Once current in *Q*5 has risen to I_L, assuming negligible capacitance at the *Q*5 emitter node and fast recovery time in *D*5, the voltage across *Q*5 rapidly drops to zero along the path *B* to *C*. If there is one turn "on" of duration t_r in a period *T*, the dissipation in *Q*5, averaged over *T*, is

$$PD_{\text{turnon}} = V_1 \frac{I_L t_r}{2T} \tag{5.10}$$

It is of interest to calculate this dissipation for a 2000-W buck current-fed bridge operating from the rectified 220-V AC line. Nominal rectified DC voltage (V_1) is about 300 V, minimum is 270 V, and maximum is 330 V. Assume that the bucked-down DC voltage V_2 is 25% below the minimum V_1 or about 200 V.

Further, assume the bridge inverter operates at 80% efficiency, giving an input power of 2500 W. This power comes from a V_1 of 200 V, and hence the average current in *L*1 is 12.5 A. Assume that *L*1 is large enough so that the ripple current in I_L can be neglected.

Then, for an assumed 0.3-μs current rise time (easily achieved with modern bipolar transistors) and a *Q*5 switching frequency of 50 kHz, turn "on" dissipation at maximum AC input voltage is (from Eq. 5.10)

$$PD = 330\left(\frac{12.5}{2}\right)\left(\frac{0.3}{20}\right) = 31\text{ W}$$

Note in this calculation that the effect of poor recovery time in *D*5 has been neglected. This has been discussed in Section 5.6.2.2 in connection with the poor recovery time of output rectifiers of the bridge inverter. This problem can be far more serious for the free-wheeling diode of the buck regulator, for *D*5 must have a much higher voltage rating—at least 400 V for the maximum V_1 of 330 V—and high-voltage diodes have poorer recovery times than lower-voltage ones. Thus the *Q*5 current can considerably overshoot the peak of 12.5 A that *D*5 had been carrying. Further, the oscillatory ring after the recovery time, discussed in Section 5.6.2.2, can cause a serious voltage overstress in free-wheeling diode *D*5.

Turn "on" dissipation in *Q*5 and voltage overstress of *D*5 can be eliminated with the turn "on" snubber of Figure 5.15.

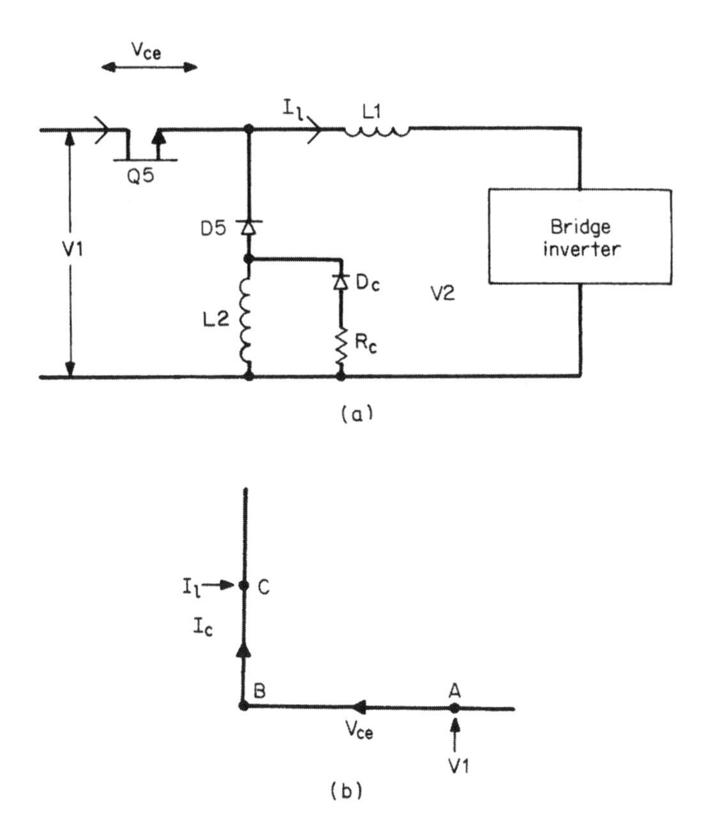

FIGURE 5.15 (*a*) Turn "on" snubber—$L2$, D_c, R_c—eliminates turn "on" dissipation in $Q5$, but at the price of an equal dissipation in R_c. When $Q5$ commences turning "on," $L2$ drives the $Q5$ emitter voltage up to within 1 V of its collector. As $Q5$ current rises toward I_l, the current in $L2$, which has been stored in it by $L1$ during the $Q5$ "off" time, decreases to zero. Thus the voltage across $Q5$ during its turn "on" time is about 1 V rather than $V1$. During the next $Q5$ "off" time, $L2$ must be charged to a current I_l without permitting too large a drop across it. Resistor R_c limits the voltage across $L2$ during its charging time. (*b*) $Q5$'s locus of falling voltage (A to B) and rising current (B to C) during turn "on," with the snubber of Figure 5.15*a*.

5.6.6.4 Buck Transistor Turn "On" Snubber—Basic Operation

The turn "on" snubber of Figure 5.15*a* does not reduce circuit dissipation. Power is diverted from the vulnerable semiconductor $Q5$, where it is a potential failure hazard, to the passive resistor R_s, which can far more easily survive the heat. It works as follows. An inductor $L2$ is added in series with the free-wheeling diode $D5$. While $Q5$ is "off," the inductor load current I_L flows out of the bottom of the bridge transistors, up through the bottom end of $L2$, through free-wheeling diode $D5$, and back into the front end of $L1$. This causes the top end of $L2$ to be slightly more negative than its bottom end.

As $Q5$ commences turning "on," it starts delivering current into the cathode of $D5$ to cancel its forward current. This current flows down into $L2$, opposing the load current it is carrying. Since the current in an inductor ($L2$) cannot change instantaneously, the voltage polarity across it reverses instantaneously to maintain constant current.

The voltage at the top end of $L2$ rises, pushing the free-wheeling diode cathode up with it until it meets the "on"-turning $Q5$ emitter voltage. The $Q5$ emitter is forced up to within $V_{ce(sat)}$ of its collector, and now $Q5$ continues increasing its current, but at a V_{ce} voltage of about 1 V rather than the V_1 voltage of 370 V it had to sustain in the absence of $L2$.

When the $Q5$ current has risen to I_L (in a time t_r), the forward current in $D5$ has been canceled and $Q5$ continues to supply the load current I_L demanded by $L1$. Since the voltage across $Q5$ during the rise time t_r is only 1 V, its dissipation is negligible. Further, because of the high impedance of $L2$ in series with the $D5$ anode, there is negligible recovery time current in $D5$. The current-voltage locus of $Q5$ during the turn "on" time is shown in Figure 5.15*b*.

5.6.6.5 Selection of Buck Turn "On" Snubber Components

For the preceding sequence of events to proceed as described, the current in $L2$ must be equal to the load current I_L at the start of $Q5$ turn "on" and must have decayed back down to zero in the time t_r that current from $Q5$ has risen to I_L. Since the voltage across $L2$ during t_r is clamped to V_1, the magnitude of $L2$ is calculated from

$$L2 = \frac{V_1 t_r}{I_L} \tag{5.11}$$

For the above example, V_1 was a maximum of 330 V, t_r was 0.3 μs, and I_L was 12.5 A. From Eq. 5.11, this yields $L2 = 330 \times 0.3/12.5 = 7.9\,\mu H$.

The purpose of R_c, D_c in Figure 5.15*a* is to ensure that at the start of $Q5$ turn "on," current in $L2$ truly is equal to I_L and that it has reached that value without overstressing $Q5$.

Consider, for the moment, that R_c, D_c were not present. As $Q5$ turned "off," since current in $L1$ cannot change instantaneously, the input end of $L1$ goes immediately negative to keep current constant. If $L2$ were not present, $D5$ would clamp the front end of $L1$ (and hence the $Q5$ emitter) at common, and permit a voltage of only V1 across $Q5$. But with $L2$ present, the impedance looking out of the $D5$ anode is the high instantaneous impedance of $L2$. As $Q5$ turned "off," I_L would be drawn through $L2$, pulling its top end far negative. This would put a large negative voltage at the $Q5$ emitter, and with its collector at V_1 (370 V in this case), it would immediately fail.

Thus R_c and D_c are shunted around $L2$ to provide a path for I_L at the instant $Q5$ turns "off," and R_c is selected low enough so that the voltage drop across it at a current I_L plus V_1 is a voltage stress that $Q5$ can safely take. Thus

$$VQ_{5(\max)} = V_1 + R_c I_L \tag{5.12}$$

In the preceding example, $V_{1(\max)}$ was 330 V. Assume that $Q5$ had a V_{ceo} rating of 450 V. With a t1- to t5-V reverse bias at its base at the instant of turn "off," it could safely sustain the V_{cev} rating of 650 V. Then to provide a margin of safety, select R_c so that $V_{Q5(\max)}$ is only 450 V. Then from Eq. 5.12, $450 = 330 + R_c \times 12.5$ or $R_c = 9.6\Omega$.

5.6.6.6 Dissipation in Buck Transistor Snubber Resistor

Examination of Figure 5.15*a* shows that essentially the constant current I_L is charging the parallel combination of R_c and $L2$. The Thevenin equivalent of this is a voltage source of magnitude $I_L R_c$ charging a series combination of R_c and $L2$. It is well known that in charging a series inductor L to a current I_p or energy $\frac{1}{2}L(I_p)^2$, an equal amount of energy is delivered to the charging resistor. If L is charged to I_p once per period T, the dissipation in the resistor is $\frac{1}{2}L(I_p)^2/T$.

In the preceding example where $T = 2\ \mu\text{s}$, $L = 7.9\ \mu\text{H}$, and $I_p = 12.5$ A

$$PD_{\text{snubber resistor}} = \frac{(1/2)(7.9)(12.5)^2}{20} = 31\ W$$

Thus, as mentioned above, this snubber has not reduced circuit dissipation; it has only diverted it from the transistor Q5 to the snubbing resistor.

5.6.6.7 Snubbing Inductor Charging Time

The snubbing inductor must be fully charged to I_L during the "off" time of the buck transistor. The charging time constant is L/R, which in the above example is $7.9/6.4 = 1.23\ \mu\text{s}$. The inductor is 95% fully charged in three time constants or 3.7 μs.

In the preceding example, switching period T was 20 μs. To buck down the input of 330 V to the preregulated 200 V, "on" time is $T_{\text{on}} = 20(200/330) = 12\ \mu\text{s}$. This leaves a $Q5$ "off" time of 8 μs, which is sufficient, as the snubbing inductor is 95% fully charged in 3.7 μs.

5.6.6.8 Lossless Turn "On" Snubber for Buck Transistor[10,21,22]

Losses in the snubbing resistor of Figure 5.15*a* can be avoided with the circuit of Figure 5.16. Here, a small transformer $T2$ is added. Its primary turns N_p and gap are selected so that at a current I_L, its inductance is the same as $L2$ of Figure 5.15*a*. The polarities at the primary and secondary are as shown by the dots.

When $Q5$ turns "off," the front end of $L1$ goes negative to keep I_L constant. I_L flows through $D5$ and N_p, producing a negative voltage V_n at the dot end of N_p and voltage stress across $Q5$ of $V_1 + V_n$. Voltage V_n is chosen so that $V_1 + V_n$ is a voltage that $Q5$ can safely sustain. To maintain the voltage across N_p at V_n when $Q5$ has turned "off," the turns ratio N_s/N_p is selected equal to V_1/V_n. When $Q5$ turns "off," as the dot end of N_p goes down to V_n, the no-dot end of N_s goes positive and is clamped to V_1, holding the voltage across N_p to V_n.

Prior to $Q5$ turn "on," $L1$ current flows through N_p, $D5$. As $Q5$ commences turning "on," its emitter looks into the high impedance of N_p and immediately rises to within one volt of its collector. Thus, current in $Q5$ rises with only one volt across it, and its dissipation is negligible. All the energy stored in N_p when $Q5$ was "off," is returned via $L1$ to the load with no dissipation. $Q5$ turn "off" dissipation can be minimized with a turn "off" snubber (Chapter 11).

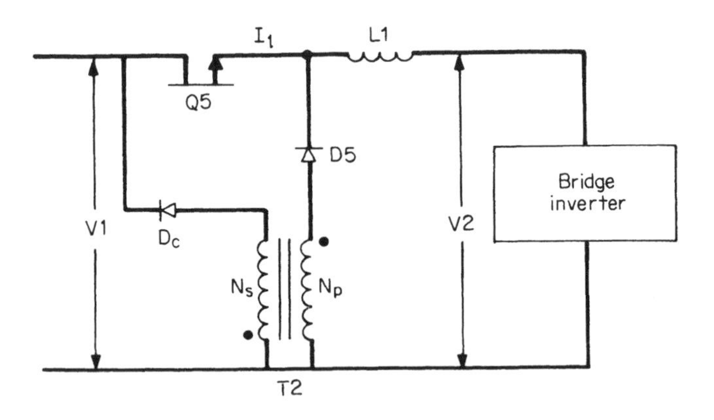

FIGURE 5.16 Non-dissipative turn "on" snubber. When $Q5$ turns "off," $L1$ stores a current I_l in N_p of $T1$. The negative voltage at the dot end of N_p during this charging time is fixed by the turns ratio N_s/N_p. If the top end of N_p is to be permitted to go to only V_n negative when $Q5$ turns "off," the voltage stress on $Q5$ is $V_1 + V_n$. When the dot end of N_p has gone negative to V_n, the no-dot end of N_s has been driven up to $V1$, D_c clamps to $V1$, clamping the voltage across N_p to the preselected V_n. Thus N_s/N_p is chosen as $V1/V_n$. The charging of N_p is not limited by a resistor as in Figure 5.14, so there is no snubber dissipation.

5.6.6.9 Design Decisions In Buck Current-Fed Bridge

The first decision to be made on the buck current-fed bridge is when to use it. It is primarily a high-output-power, high-output-voltage topology.

In terms of cost, efficiency, and required space, it is a good choice for output powers in the range of 1 to 10 or possibly 20 kW. For high-output voltages—above about 200 V—and above about 5 A output current—the absence of output inductors makes it a good choice. For output powers above 1 kW, the added dissipation, volume, and cost of the buck transistor is not a significant increase above what is required in a competing topology such as a pulse-width-modulated full-wave bridge.

It is an especially good choice for a multi-output supply consisting of one or more high-output voltages (5000 to 30,000 V). In such applications, the absence of output inductors permits the use of capacitor-diode voltage multiplier chains.[8,13] Also, the absence of output inductors in the associated lower-output voltages partly compensates for the cost and volume of the buck transistor and its output inductor.

The next design decision is the selection of the bucked-down voltage (V_2 of Figure 5.10). This is chosen at about 25% below the lowest ripple trough of V_1 (Figure 5.10) at the lowest specified AC input. Inductor $L1$ is chosen for continuous operation at the calculated minimum inductor current I_L corresponding to the minimum total output power at the preselected value of V_2. It is chosen as in Section 1.3.6 for a conventional buck regulator.

The output capacitors are not chosen to provide storage or reduce ripple directly at the output, because the overlapping conduction of bridge transistors minimizes this requirement. Rather they are chosen so that when reflected into the primary, the equivalent series resistance R_{esr} of all reflected capacitors is sufficiently low as to minimize ripple at V_2. Recall from Section 1.3.7, in calculating the magnitude of the output capacitor, it was pointed out that output ripple in a buck regulator V_{br} is given by

$$V_{br} = \Delta I \, R_{esr}$$

in which ΔI is the peak-to-peak ripple current in the buck inductor and is usually set at twice the minimum DC current in it so that the inductor is on the threshold of discontinuous operation at its minimum DC current. Minimum DC current in this case is the current at minimum specified output power at the preselected value of V_2. Thus with R_{esr} selected so as to yield the desired ripple at V_2, ripple at each secondary is

$$V_{sr} = V_{br} \frac{N_s}{N_p}$$

There is an interesting contrast in comparing a current- to a voltage-fed bridge at the same bucked-down voltage (V_2 of Figures 5.9 and 5.10).

For the voltage-fed bridge, a maximum "on" time of 80% of a half period must be established to ensure that there is no simultaneous conduction in the two transistors positioned vertically one above another. With the low impedance looking back into the buck regulator of the voltage-fed circuit, such simultaneous conduction would subject the bridge transistors to high voltage and high current and destroy one or more of them.

In the current-fed circuit, such slightly overlapping simultaneous conduction is essential to its operation and "on" time of alternate transistor pairs is slightly more than a full half period at any DC input voltage. In addition, since the "on" time of a voltage-fed bridge (Figure 5.9) is only 80% of a half period, its peak current must be 20% greater than that of the current-fed bridge at the same output power.

It should also be noted that the number of primary turns as calculated from Faraday's law (Eq. 2.7) must be 20% greater in the current-fed bridge, since the "on" time is 20% greater for a flux change equal to that in a voltage-fed bridge at the same V_2.

5.6.6.10 Operating Frequencies—Buck and Bridge Transistors

The buck transistor is usually synchronized to and operates at twice the square-wave switching frequency of the bridge transistors. Recall that it alone is pulse-width-modulated, and that the bridge devices are operated at a 50-percent duty cycle with a slightly overlapping "on" time.

Frequently, however, the scheme of Figure 5.17*a* with two buck transistors ($Q5A$ and $Q5B$) is used to reduce dissipation. They are synchronized to the bridge transistor frequency and are turned "on" and pulse-width-modulated on alternate half cycles of the bridge square-wave frequency. Thus the DC and switching losses are shared between two transistors with a resulting increase in reliability.

5.6.6.11 Buck Current-Fed Push-Pull Topology

The buck current-fed circuit can also be used to drive a push-pull circuit as in Figure 5.18 with the consequent saving of two transistors over the buck current-fed bridge. Most of the advantages of the buck current-fed bridge are realized and the only disadvantage is that the push-pull circuit power transistors have greater voltage stress. This voltage stress is twice V_2, rather than V_2 as in the bridge circuit. But V_2 is the pre-regulated and bucked-down input voltage—usually only 75% of the minimum V_1 input. This is usually about the same as the maximum DC input of a competing topology—like the pulse-width-modulated full-wave bridge (Figure 5.8).

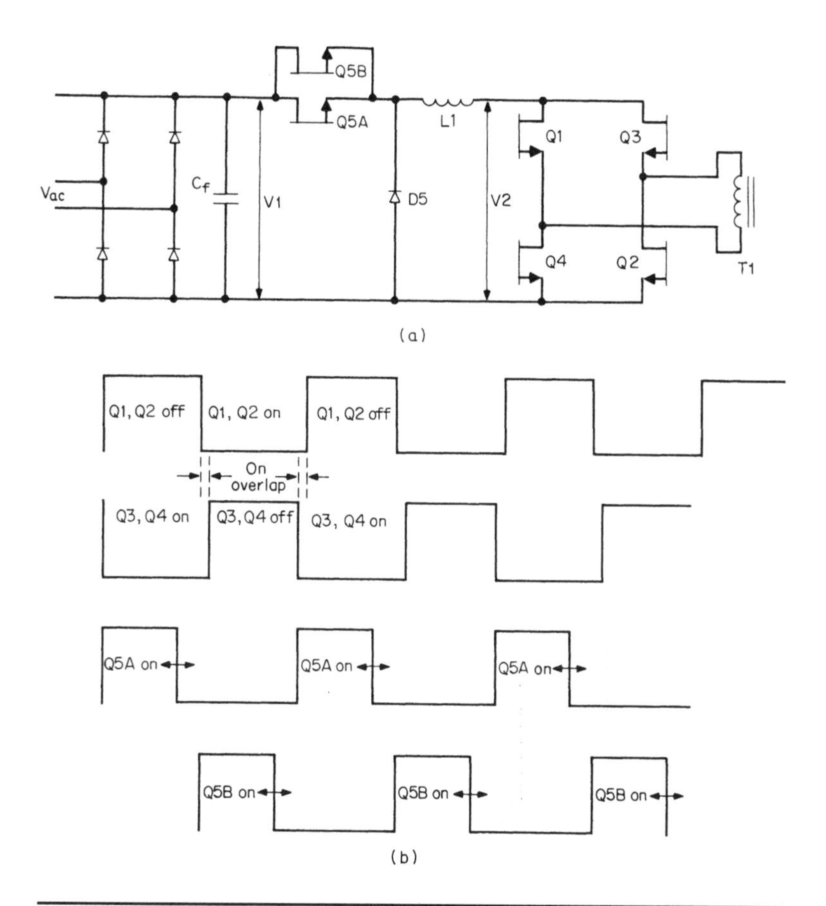

FIGURE 5.17 (*a*) Buck transistor *Q*5 can be a single transistor operating at twice the frequency of the bridge transistors and synchronized to them, or more usually, it is two synchronized transistors that are both pulse-width-modulated and are "on" during alternate half periods of the bridge transistors. (*b*) To reduce dissipation in the buck transistor, it is usually implemented as two transistors, each synchronized to the bridge transistors and operated at the same square-wave frequency as the bridge devices. Transistors *Q*5*A*, *Q*5*B* are pulse-width-modulated. Bridge transistors are not and are operated with a small "on" overlap time.

However, the major advantages of the current-fed technique—no output inductors and no possibility of flux imbalance—still exist.

The topology can be used to greatest advantage in supplies of 2 to 5 kW, especially if there are multiple outputs or at least one high-voltage output.

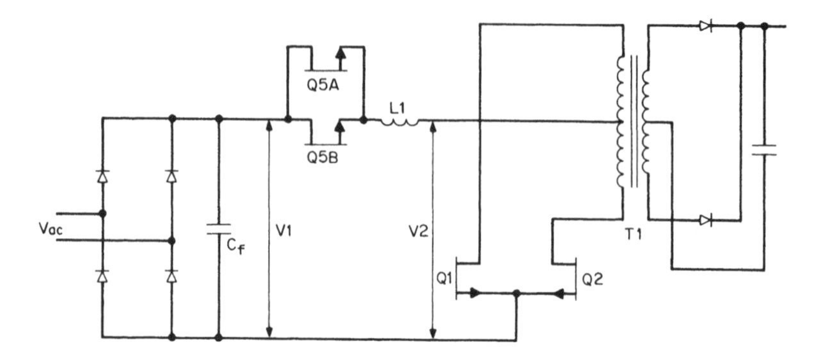

FIGURE 5.18 The current-fed topology can also be implemented as a buck push-pull circuit. As in the buck current-fed bridge, the capacitor after the buck inductor *L*1 is omitted, and *Q*1, *Q*2 are operated with a deliberately overlapping "on" time. Only buck transistors *Q*5*A*, *Q*5*B* are pulse-width-modulated. Output inductors are not used. All the advantages of the buck current-fed bridge are retained. Although "off"-voltage stress is twice *V*2 (plus a leakage spike) instead of *V*2 as in the bridge, it is still significantly less than twice *V*1 because *V*2 is bucked down to about 75% of the minimum value of *V*1. This circuit is used at lower power levels than the buck current-fed bridge and offers the savings of two transistors.

5.6.7 Flyback Current-Fed Push-Pull Topology (Weinberg Circuit [23])

This topology[1,23] is shown in Figure 5.19. Effectively it has a flyback transformer in series with a push-pull inverter. It has many of the valuable attributes of the buck current-fed push-pull topology (Figure 5.18), and since it requires no pulse-width-modulated input transistor (*Q*5), it has lower dissipation, cost, and volume, and greater reliability.

It might be puzzling at first glance to see how the output voltage is regulated against line and load changes, since there is no *LC* voltage-averaging filter at the output. The diode-capacitor at the output is a peak, rather than an averaging, circuit. The answer is that the averaging or regulating is done at the push-pull center tap to keep V_{ct} relatively constant. The output voltage (or voltages) is (are) kept constant by pulse-width-modulating the *Q*1, *Q*2 "on" time. Output voltage is simply $(N_s/N_p)V_{ct}$ and a feedback loop sensing V_o controls the *Q*1, *Q*2 "on" times to keep V_{ct} at the correct value to maintain V_o constant. The relation between the *Q*1, *Q*2 "on" times and output voltage is shown below.

The circuit retains the major advantage of the current-fed technique—a single-input inductor but no output inductors, which makes it a good choice for a multi-output supply with one or more

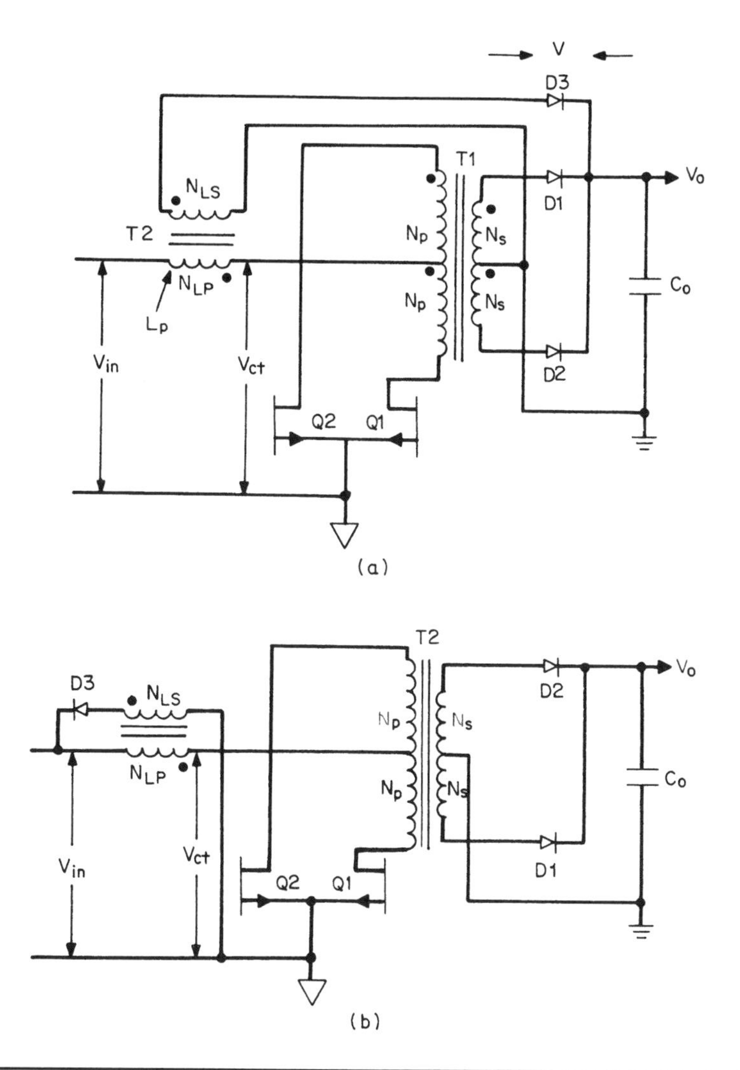

FIGURE 5.19 (*a*) Flyback current-fed push-pull topology (Weinberg circuit[23]). This is essentially a flyback transformer in series with a pulse-width-modulated push-pull inverter. It is used primarily as a multi-output supply with one or more high-voltage outputs, as it requires no output inductors and only the one input flyback transformer *T*2. The high impedance seen looking back into the primary of *T*2 makes it a "current-fed" topology, with all the advantages shown in Figure 5.18. Here, the *T*2 secondary is shown clamped to V_o. Transistors *Q*1, *Q*2 may be operated either with a "dead time" between "on" times or with overlapping "on" times. Its advantage over Figure 5.18 is that it requires no additional input switching transistors. The usual output power level is 1 to 2 kW. (*b*) Shows the same circuit as Figure 5.19*a*, but with the flyback secondary clamped to V_{in}. This results in less input current ripple but more output voltage ripple.

high-voltage outputs. Further, because of the high-source impedance of the flyback transformer primary $L1$, the usual flux-imbalance problem of voltage-fed push-pulls does not result in transformer saturation and consequent transistor failure. Its major usage is at the 1- to 2-kW power level.

Two circuit configurations of the flyback current-fed push-pull topology are shown in Figure 5.19*a* and 5.19*b*. Figure 5.19*a* shows the flyback secondary returned to the output voltage through diode *D*3; in Figure 5.19*b*, the diode is returned to the input voltage. When the diode is returned to V_o, output ripple voltage is minimized; when it is returned to V_{in}, input ripple current is minimized. Consider first the configuration of Figure 5.19*a*, where the diode is returned to the output.

The configuration of Figure 5.19*a* can operate in two significantly different modes. In the first mode, *Q*1 and *Q*2 are never permitted to have overlapping "on" times at any DC input voltage. In the second mode, *Q*1 and *Q*2 may have overlapping "on" times throughout the entire range of specified DC input voltage. The circuit can also be set up to shift between the two modes under control of the feedback loop as the input voltage varies.

It will be shown below that in the non-overlapping mode, power is delivered to the secondaries at a center tap voltage V_{ct} lower than the DC input voltage (buck-like operation) and in the overlapping mode, power is delivered to the secondaries at a center tap voltage V_{ct} higher than the DC input voltage (boost-like operation). Since V_{ct} is relatively low in the non-overlapping mode, *Q*1, *Q*2 currents are relatively high for a given output power. But with the lower V_{ct} voltage, "off"-voltage stress in *Q*1, *Q*2 is relatively low. In the overlapping mode, since V_{ct} is higher than V_{in}, *Q*1, *Q*2 currents are lower for a given output power but "off"-voltage stress in *Q*1, *Q*2 is higher than that for the non-overlapping mode.

The circuit is usually designed not to remain in one mode throughout the full range of input voltages. Rather, it is designed to operate in the overlapping mode with an "on" duty cycle T_{on}/T greater than 0.5, and in the non-overlapping mode with T_{on}/T less than 0.5 as the DC input voltage shifts from its minimum to its maximum specified values. This permits proper operation throughout a larger range of DC input voltages than if operation remained within one mode throughout the entire range of DC input voltage.

5.6.7.1 Absence of Flux-Imbalance Problem in Flyback Current-Fed Push-Pull Topology

Flux imbalance is not a serious problem in this topology because of the high-impedance current-fed source that feeds the push-pull transformer center tap.

The current-fed nature of the circuit arises from the flyback transformer, which is in series with the push-pull center tap. The high impedance looking back from the push-pull center tap is the magnetizing inductance of the flyback primary.

In a conventional voltage-fed push-pull inverter, unequal volt-second products across the two half primaries cause the flux-imbalance problem (Section 2.2.5). The transformer core moves off center of its hysteresis loop and toward saturation. Because of the low impedance of a voltage source, current to the push-pull center tap is unlimited and the voltage at that point (V_{ct}) remains high. The core then moves further into saturation, where its impedance eventually vanishes and transistor currents increase drastically. With high current and voltage, the transistors will fail.

With the high impedance looking back into the dot end of N_{LP} as shown in Figure 5.19, however, as the push-pull core moves into saturation drawing more current, the high current causes a voltage drop at V_{ct}. This reduces the volt-second product on the half primary, which is moving toward saturation, and prevents complete core saturation.

Thus the high source impedance of N_{LP} does not fully prevent core saturation. In the worst case, it keeps the core close to the knee of the *BH* loop, which is sufficient to keep transistor currents from rising to disastrous levels. The major drawback of push-pull circuit flux imbalance is thus not a problem with this inverter.

5.6.7.2 Decreased Push-Pull Transistor Current in Flyback Current-Fed Topology

In a conventional pulse-width-modulated push-pull, driven at the center tap from a low-impedance voltage source, it is essential to avoid simultaneous conduction in the transistors by providing a dead time of about 20% of a half period between turn "off" of one transistor and turn "on" of the other. This results in higher peak transistor current for the same output power, since output power is proportional to average transistor current.

This dead time is essential in the voltage-fed push-pull, for if $Q1$, $Q2$ were simultaneously "on," the half primaries could not sustain voltage. Then, the transistor collectors would rise to the supply voltage, which would remain high, and with high voltage and high current, the transistors would fail.

In the current-fed circuit, there is no problem if both transistors are simultaneously "on" under transient or fault conditions, when the DC input voltage is momentarily lower than specified or with storage times greater than specified, because of the high impedance looking back into the dot end of N_{LP}. Should both transistors turn "on" briefly at the same time, V_{ct} simply drops to zero and current drawn from the input source is limited by the impedance of the input inductor.

Thus even in the "non-overlap" mode, no dead time need be provided between the turn "off" of one transistor and the turn "on" of the other. If there is a momentary overlap because of storage time, V_{ct} simply collapses to zero and no harm results. Hence by eliminating the 20% dead time, the peak current required for a given output power is decreased by 20% at the same value of V_{ct}. Further, as discussed above, in the overlap mode, overlapping "on" time need not be the small amount arising from transistor storage time but can be a deliberately large fraction of a half period.

5.6.7.3 Non-Overlapping Mode in Flyback Current-Fed Push-Pull Topology—Basic Operation

The circuit operation can be understood from examination of the significant voltage and current waveforms shown in Figure 5.20.

Operation will be explained on the assumption that the "on" potentials of transistors $Q1$, $Q2$ are negligibly small and can be neglected, considering their actual "on" drop of about 1 V would complicate the design equations and hamper understanding of important circuit behavior. Also, the forward drops V_d of diodes $D1$, $D2$, $D3$ are assumed equal.

In Figure 5.19*a*, when either $Q1$ or $Q2$ is "on," the voltage across the corresponding half secondary is clamped to $V_o + V_d$. Then the voltage at the push-pull center tap V_{ct} is clamped to $(N_p/N_s)(V_o + V_d)$, as can be seen in Figure 5.20*d*. The ratio N_p/N_s is chosen so that V_{ct} is 25% lower than the bottom of the input ripple trough at the lowest specified value of V_{in}.

Thus when either transistor is "on," the dot end of N_{LP} is negative with respect to its no-dot end and current flows through to the push-pull center tap, which is clamped to the voltage V_{ct}. The waveshapes of the currents that flow are shown in Figure 5.20*g* and 5.20*h*. These currents have the ramp-on-a-step waveshape characteristic of any buck regulator operating in the continuous conduction mode as discussed in Section 1.3.2.

When the "on" transistor turns "off," the dot end of N_{LP} goes positive to maintain the L_p current constant. The dot end of N_{LS} also goes positive until $D3$ is forward-biased and clamps to V_o. The ratio N_{LP}/N_{LS} is set equal to N_p/N_s of the push-pull transformer (hereafter these ratios will be designated as N). The reflected voltage on the flyback primary is $N(V_o + V_d)$. Thus, when either transistor turns "off," V_{ct} rises to $V_{dc} + N(V_o + V_d)$ and stays there until the opposite transistor turns "on," as shown in Figure 5.20*d*. The waveshape of Figure 5.20*d* permits calculation of the relation between the output voltage and "on" time as follows.

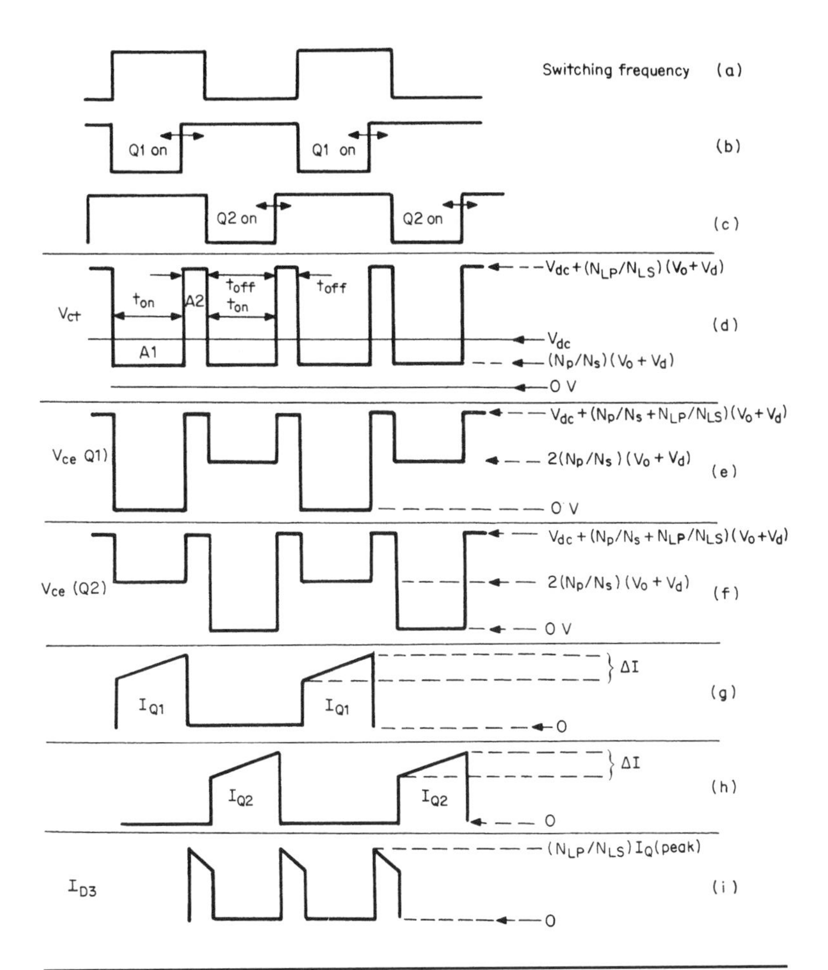

FIGURE 5.20 The key voltage and current waveforms in flyback current-fed topology when operating in non-overlapping conduction mode. Power is delivered to the load only when either $Q1$ or $Q2$ is "on." Power is delivered at a supply voltage of $(N_p/N_s)(V_o + V_d)$, which is less than V_{dc}.

5.6.7.4 Output Voltage vs. "On" Time in Non-Overlapping Mode of Flyback Current-Fed Push-Pull Topology

In Figure 5.20*d* it can be seen that during t_{on}, V_{ct} is $N(V_o + V_d)$ and during t_{off}, it is $V_{dc} + N(V_o + V_d)$. The average of voltage V_{ct} must be equal to V_{dc}, the DC voltage at the front end of L_p. This is because L_p is assumed to have negligible DC resistance, so the voltage across it averaged over a full or half period must equal zero. Another way of expressing this is that in Figure 5.20*d*, the volt-second area $A1$ must equal the volt-second area $A2$:

$$A1 = [V_{dc} - N(V_o + V_d)]t_{on}$$

and

$$V_{dc}t_{on} - NV_o t_{on} - NV_d t_{on} = NV_o t_{off} + NV_d t_{off}$$

or

$$NV_o(t_{on} + t_{off}) = V_{dc}t_{on} - NV_d(t_{on} + t_{off}) \quad \text{but} \quad t_{on} + t_{off} = \frac{T}{2}$$

Then

$$V_o = \left(\frac{2V_{dc}t_{on}}{NT}\right) - V_d$$

or

$$V_o = \left(2V_{dc}\frac{N_s}{N_p}\right)\frac{t_{on}}{T} - V_d \tag{5.13}$$

Thus the feedback loop regulates V_o by width-modulating t_{on}, just as in all previous circuits, to keep the product $V_{dc}t_{on}$ constant.

5.6.7.5 Output Voltage Ripple and Input Current Ripple in Non-Overlapping Mode

Choosing N_{LP}/N_{LS} equal to N_p/N_s in Figure 5.19*a* results in negligible V_o ripple, as can be seen in Figure 5.21. The voltages delivered to the anodes of *D*1, *D*2, and *D*3 are all equal in amplitude. The currents delivered through D1, D2 are NI_{Q1} and NI_{Q2}, which are equal. Further, since $N_{LP}/N_{LS} = N_p/N_s$, during t_{off}, the current delivered through *D*3 is also NI_{Q1}. Thus, there is no gap in time during which C_o must supply or absorb current. The total load current is at all times being supplied through *D*1, *D*2, or *D*3 and C_o serves no energy storage function.

Output ripple is the product of the secondary ripple current amplitude ΔI_s times the equivalent series resistance R_{esr} of C_o. Also, $\Delta I_s = N\Delta I_p$ where ΔI_p is set to twice the minimum current at the center of the *Q*1, *Q*2 ramps at minimum output power; ΔI_p is set to the desired value as discussed in Section 1.3.6 by choosing L_p sufficiently large that it does not go into the discontinuous conduction mode above minimum output power. Then C_o is chosen to minimize R_{esr} as discussed in Section 1.3.7.

There may be very narrow (<1 μs) voltage spikes at each turn "on"/turn "off" transition, as can be seen in Figure 5.21. These occur if the voltage fall time at the anode of *D*1, *D*2, or *D*3 is slightly faster

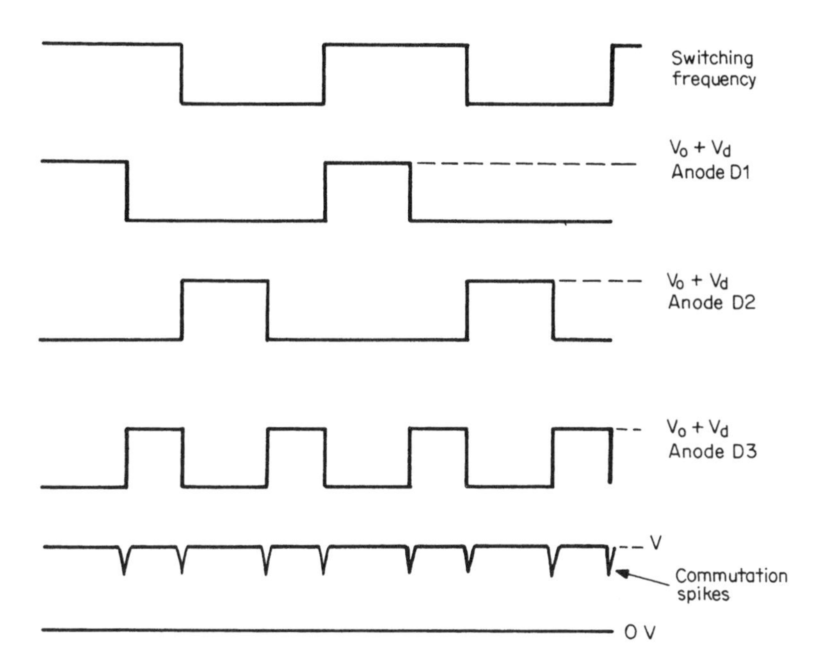

FIGURE 5.21 In Figure 5.19*a*, peak voltages applied to the anodes of *D*1, *D*2, *D*3 are all equal if N_p/N_s is chosen equal to N_{LP}/N_{LS}, but if fall time at one anode is slightly faster than the rise time of the next "on"-coming anode, there will be a narrow commutation spike at the output.

than the voltage rise time of the next "on"-turning diode. Such spikes are easily eliminated with small *LC* integrators.

Current drawn from V_{in} is discontinuous. As seen in Figure 5.20*i*, although flyback secondary current flows during the "off" time, making output current continuous, input current falls to zero during each "off" time. Discontinuous input current requires the addition of a space-consuming RF1 input filter to keep large transient current off the input lines. By returning the flyback secondary and *D*3 to the input as in Figure 5.19*b*, input current will never fall to zero but will ramp up and down with the amplitude shown in Figure 5.20*g* and 5.20*h*. This will reduce the size of the required RFI input filter greatly or may even make it unnecessary.

5.6.7.6 Output Stage and Transformer Design Example—Non-Overlapping Mode

It is instructive to run through a typical design example of the output stage and transformer for the flyback current-fed push-pull topology operating in the non-overlapping mode.

Only a single master secondary will be considered. Additional secondaries may be added, and their required turns will be related to that of the master in the ratios of their output voltages. Slave secondary output voltages will track the master to within about 2%—much more closely than is possible in supplies with output *LC* filters.

The design example will assume the following conditions:

Output power	2000 W
Output voltage	48 V
Efficiency	80%
Switching frequency	50 kHz ($T = 20$ μs)
Diode voltage drops	1 V
DC input voltage (from 115-V, ±15% AC line)	Maximum, 184 V; nominal, 160 V; minimum, 136 V

The first decision is to select V_{ct} during the "on" time. As in Section 5.6.7.3, V_{ct} is set at 75% of the minimum DC input voltage, or $0.75\times 136 = 102$ V.

The turns ratio is now selected to yield 102 V during the "on" time. From Figure 5.20*d*, V_{ct} during the "on" time is

$$V_{ct} = \frac{N_p}{N_s}(V_o + V_d) \quad \text{or} \quad \frac{N_p}{N_s} = N = \frac{102}{48+1} = 2$$

Transistor current amplitudes I_{Q1}, I_{Q2} and "on" times are calculated to permit prediction of primary and secondary RMS currents and hence wire sizes. From the "on" times and preselected V_{ct}, the number of primary turns will be calculated from Faraday's law once a transformer core area is selected for the specified output power.

From Eq. 5.13

$$\frac{t_{on}}{T} = \frac{(V_o + V_d)(N_p/N_s)}{2V_{dc}}$$

From this, Table 5.1 can be constructed.

When either transistor is "on," it delivers a current whose waveshape is shown in Figure 5.20*g* and 5.20*h*. This current is delivered to the push-pull center tap at V_{ct} of 102 V. When the transistor turns

V_{dc}, V	t_{on}/T	t_{on}, μs
200	0.245	4.9
184	0.266	5.3
160	0.306	6.12
136	0.360	7.2

TABLE 5.1

"off," the current shown in Figure 5.20i is delivered via $D3$ to the secondary load. This current (NI_Q) is delivered during the "off" time at a voltage $V_o + V_d$ but is equivalent to current I_Q delivered at a voltage $N(V_o + V_d)$. Effectively, this current is being delivered at 100% duty cycle at a voltage of $N(V_o + V_d)$ or at 102 V.

Assuming an efficiency of 80% downstream from the push-pull center tap, input power at the center tap is 2000/0.8 or 2500 W. Average current into the center tap is then 2500/102 = 24.5 A. This is very close to the current at the center of the current ramp in Figure 5.20g and 5.20h.

Current in each push-pull half primary is thus approximated by an equivalent flat-topped pulse whose amplitude I_{pk} is 24.5 A and whose duration is given in Table 5.1. The RMS value of this is $I_{pk}\sqrt{t_{on}/T}$. Since the RMS input current is a maximum at minimum $V_{dc} = 136$ V, the RMS current in each half secondary is $I_{rms} = 24.5\sqrt{0.36} = 14.7$A. At a current density of 500 circular mils per RMS ampere, the required number of circular mils for each half primary is $500 \times 14.7 = 7350$ circular mils.

The transformer core will be selected from charts in the coming section on magnetics design, which has been discussed in Section 2.2.9.1. Jumping ahead, these charts will show that a Ferroxcube EC70 core (an international standard type) with an area of 2.79 cm^2 can deliver 2536 W at 48 kHz, and can be used.

The number of turns per half primary is calculated from Faraday's law (Eq. 1.17) at the maximum "on" time (7.2 μs in Table 5.1) and at a primary voltage of 102 V. Losses are quite small using Ferroxcube core material type 3F3 at 50 kHz, in the order of 60 mW/cm^3 at a peak flux density of 1600 G. For a core volume of 40.1 cm^3, total core losses are only 2.4 W. This is low enough so that copper losses of even twice that much will still leave the transformer at a safely low temperature. From Faraday's law, $N_p = V_p t_{on} \times 10^{r8}/A_e \Delta B = 102(7.2 \times 10^{t6})10^{r8}/2.79 \times 3200 = 8$ turns, and for $N_p/N_s = 2$, each half secondary has 4 turns.

Finally, the secondary wire size must be calculated. Each half secondary delivers the characteristic ramp-on-a-step waveform shown in Figure 5.20g and 5.20h. The current at the center of the ramp is the DC output current. The maximum pulse width occurs at minimum DC input of 136 V and is seen in Table 5.1 to be 7.2 μs. To calculate RMS secondary current, the pulse can be approximated by a rectangular pulse of amplitude I_{dc} and pulse width 7.2 μs. For an output power of 2000 W, the DC output current is 2000/48 = 41.6 A. The RMS value of this rectangular current pulse of 41.6 A, 7.2-μs pulse width, once per 20 μs is $41.6\sqrt{7.2/20} = 25$ A.

At a current density of 500 circular mils per RMS ampere, the required wire area for the half secondaries is $500 \times 25 = 12,500$ circular mils. For such a large area, the secondary would most likely be wound with metal foil of thickness and width to yield the required circular mil area. The primary with a required 7350-circular-mil area would be wound with a number of paralleled small diameter wires.

5.6.7.7 Flyback Transformer for Design Example of Section 5.6.7.6

The preceding design was based on the inductor L_p operating in the continuous mode (Section 1.3.6). It was shown there that discontinuous mode operation occurs when ΔI, the peak-to-peak ramp amplitude, is less than twice the minimum DC current, which is the current in the center of the ramp at minimum output power.

In the design example above, assume that minimum output power is one-tenth the nominal output power. At nominal output power, current at the center of the ramp was calculated above as 25 A. Then current at the center of the ramp at minimum output power is 2.5 A, and the peak-to-peak ramp amplitude (ΔI of Figure 5.20g and 5.20h) is 5.0 A. But $\Delta I = V_L t_{\text{on}}/L_p$, where V_L is the voltage across L_p during t_{on}, and from Table 5.1 at $V_{\text{dc}} = 136$ V, $t_{\text{on}} = 7.2\,\mu\text{s}$. Then

$$L_p = \frac{(136 - 102)(7.2 \times 10^{-6})}{5.0} = 49\,\mu\text{H}$$

Wire size for the flyback secondary must be calculated at high DC input, for it is then that its current pulse width t_{off} is greatest (Table 5.1). Further, as above, the equivalent flat-topped pulse amplitude is I_{dc}. Then from Table 5.1, its maximum width is $10-5.3 = 4.7\,\mu\text{s}$. Maximum RMS current in the flyback secondary is then

$$\begin{aligned} I_{\text{rms}(T2\text{ secondary})} &= 41.6\sqrt{\frac{t_{\text{off}}}{0.5T}} \\ &= 41.6\sqrt{\frac{4.7}{10}} \\ &= 28.5\text{ A} \end{aligned}$$

At 500 circular mills per RMS ampere, that winding requires a circular-mil area of $500 \times 28.5 = 14{,}260$ circular mils.

Wire size for the flyback primary is calculated at minimum DC input voltage, for it is then that primary current has greatest pulse width and hence largest RMS value. Then from Table 5.1

$$I_{\text{rms(primary)}} = 25\sqrt{\frac{t_{\text{on}}}{0.5T}} = 25\sqrt{\frac{7.2}{10}} = 21.2\text{ A}$$

At 500 circular mils per RMS ampere, it requires $500 \times 21.2 = 10,600$ circular mils.

Both primary and secondary would most likely be wound with metal foil rather than round wire for those large required areas.

Since the flyback transformer has no secondary current flowing when primary current flows, all the primary current drives the core toward saturation. To maintain the required 49-μH primary inductance at 25 A of primary current, the core must be a gapped ferrite, powdered Permalloy, or powdered iron type (Section 4.3.3).

5.6.7.8 Overlapping Mode in Flyback Current-Fed Push-Pull Topology—Basic Operation[14]

In the non-overlapping mode of Figure 5.20 ($t_{on}/T < 0.5$), it is difficult to accommodate a large ratio of maximum to minimum DC input voltage. Since the maximum "on" time is $0.5T$ at minimum DC, then at high DC inputs, "on" time will be a small fraction of a period and may approach 1 to 3 μs at a 100- to 50-kHz switching rate. But bipolar transistors with their appreciable storage times cannot work reliably down to such low "on" times.

By operating with overlapping "on" times ($T_{on}/T > 0.5$), as in Figure 5.22, however, a much larger range of maximum to minimum DC input voltages is possible.

The usual integrated-circuit pulse-width-modulating chips cannot be used for the overlapping mode because their two 180° out-of-phase outputs have a maximum duty cycle D of only 0.5. A number of schemes using several discrete integrated-circuit packages and capable of a duty cycle from 0 to 100% have been described in the literature.[14,19]

Overlapping-mode operation is achieved using the same circuit as shown in Figure 5.19, by proper choice of the turns ratios N_p/N_s and N_{LP}/N_{LS} (hereafter designated N_1, N_2, respectively). Circuit operation will be described for the scheme of Figure 5.19*a*, where diode $D3$ is returned to the output voltage rather than the input. It will be recalled that this connection minimizes output voltage ripple rather than input current ripple.

The circuit operation can be understood from the waveforms of Figure 5.22 and the circuit of Figure 5.19*a*. For the overlapping mode, $Q1$ and $Q2$ are simultaneously "on" during T_1 intervals, and only one of these is "on" during T_{off} intervals (t_2 to t_3 when $Q2$ is "off," and t_4 to t_5 when $Q1$ is "off"). Power is delivered to the load only during the T_{off} times.

When both transistors are "on," the half primaries cannot support voltage, and the push-pull center tap voltage falls to zero as seen in Figure 5.22*d*. The full input voltage V_{dc} is applied across the flyback primary inductance L_p, in which current ramps up linearly at a rate $dI/dT = V_{dc}/L_p$. The division of this current between $Q1$ and $Q2$ is roughly even, and is seen as the upward-going ramps from t_3 to t_4 and t_5 to t_6. During T_1, $D3$ is reverse-biased, and there is no voltage across

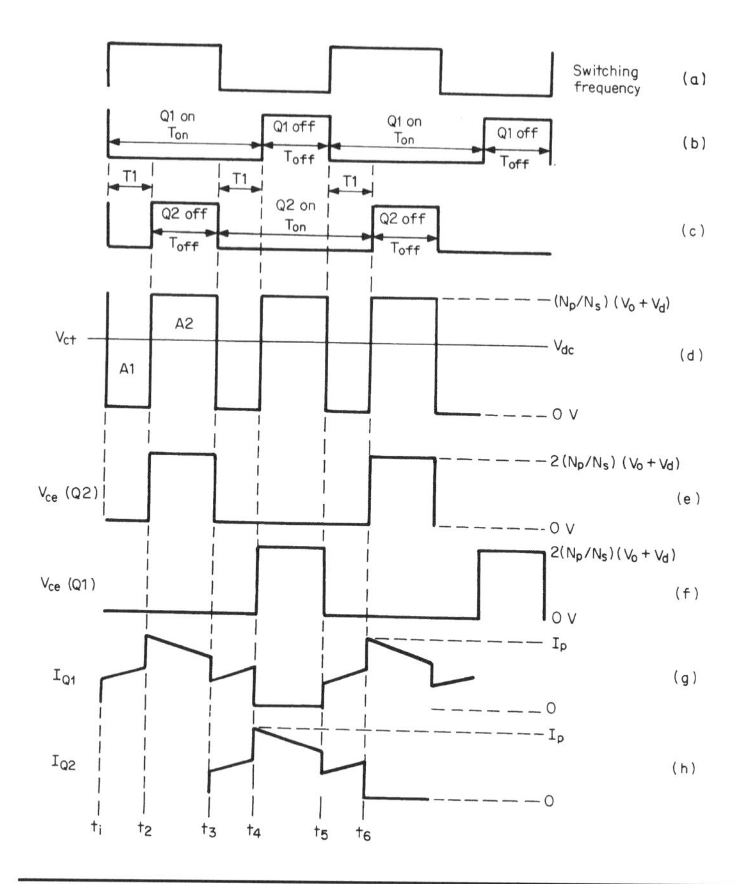

FIGURE 5.22 Typical voltage and current waveforms in flyback current-fed topology when operating in overlapping mode ($T_{on} > T_{off}$). This mode permits a much larger range of input voltages. Power is delivered to the load only in the interval when one transistor is "on" and one is "off." It is delivered at voltage V_{ct}, which is higher than the DC input voltage (boost operation). In the non-overlapping mode of Figure 5.20, it is delivered at a voltage lower than the DC input (buck operation).

the push-pull secondaries. Hence all the output power is supplied from the output filter capacitor C_o during T_1 intervals.

When $Q2$ turns "off" at t_2, $Q1$ is still "on." Now the $Q1$ half primary can support voltage, and V_{ct} begins to rise, as does the $D1$ anode voltage. The $D1$ anode rises until its cathode reaches V_o, the secondary is clamped to $V_o + V_d$, and V_{ct} is clamped to $N_1(V_o + V_d)$, as seen in Figure 5.22*d*.

The turns ratio $N_{LP}/N_{LS}(= N_2)$ is chosen large enough so that when one transistor is on and one "off" (T_{off} intervals), with the maximum voltage across N_{LP}, the voltage across N_{LS} is insufficient to forward-bias $D3$. This permits V_{ct} to be clamped to $N_1(V_o + V_d)$ during T_{off}. Some of the current stored in L_p during T_1 intervals and current from V_{dc} is delivered via the push-pull primary to the load at a voltage $N_1(V_o + V_d)$. It will soon be seen that $D3$ is forward-biased and delivers load power at some sufficiently higher DC input voltage.

The current in $Q1$ at the instant t_2 is equal to the sum of the $Q1$ and $Q2$ currents at the instant just prior to $Q2$ turn "off," since the current in L_p cannot change instantly. During t_2 to t_3, the current ramps downward (Figure 5.22*g*), because N_1 will be chosen high enough that $N_1(V_o + V_d)$ is greater than V_{dc}. With the dot end of L_p positive with respect to its no-dot end, current in it and in $Q1$ ramps downward.

When $Q2$ turns "on" at t_3, again both transistors are "on" and their half primaries cannot support voltage; V_{ct} again drops to zero and remains there until t_4, when $Q1$ turns "off" (Figure 5.22*d*). From t_4 to t_5, V_{ct} is again clamped to $N_1(V_o + V_d)$.

From Figure 5.22*d*, the relation between output/input voltages and "on" time can be calculated as follows.

5.6.7.9 Output/Input Voltages vs. "On" Time in Overlapping Mode

Refer to Figure 5.19*a*. When both transistors were "on," the dot end of L_p was negative with respect to the no-dot end. When one transistor turned "off" (during T_{off}), the polarity across L_p reversed to keep current in it constant. The voltage at the dot end of L_p rose until it was clamped to $N_1(V_o + V_d)$ by the clamping action at the secondary.

Now since L_p has negligible DC resistance, it cannot support a DC voltage. Thus the voltage across it, averaged over a full or half period, must equal zero. Since the input end of L_p is at V_{dc}, so must the output end be averaged over a half period. Another way of stating this is that in Figure 5.22*d*, the area $A1$ must equal area $A2$. Or

$$\begin{aligned} V_{dc}T_1 &= [N_1(V_o + V_d) - V_{dc}]T_{off} \\ &= N_1 V_o T_{off} + N_1 V_d T_{off} - V_{dc} T_{off} \\ V_o N_1 T_{off} &= V_{dc}(T_1 + T_{off}) - N_1 V_d T_{off} \end{aligned}$$

Since $T_1 + T_{off} = T/2$

$$V_o = \left(\frac{V_{dc}T}{2N_1 T_{off}}\right) - V_d \quad \text{and} \quad T_{off} = T - T_{on}$$

Then for $D = T_{on}/T$

$$V_o = \left[\frac{V_{dc}}{2N_1(1 - D)}\right] - V_d \tag{5.14a}$$

and from Eq. 5.14*a*, the duty cycle for any DC input is

$$D = \frac{2N_1(V_o + V_d) - V_{dc}}{2N_1(V_o + V_d)} \tag{5.14b}$$

5.6.7.10 Turns Ratio Selection in Overlapping Mode

Equation 5.14*a* gives the relation between output/input voltages and "on" time for the overlapping mode for a preselected choice of push-pull turns ratio N_1. A good choice for N_1 is the value calculated from Eq. 5.14*a*, which makes $D = 0.5$ at the nominal input voltage V_{dcn}. Then for all DC input voltages less than V_{dcn}, there will be overlapping "on" times ($D > 0.5$) and output voltage versus "on" time is given by Eq. 5.14*a* for that calculated N_1.

For input voltages greater than V_{dcn}, D is less than 0.5, there is no overlapping "on" time, and Eq. 5.14*a* no longer holds. The output voltage versus "on" time relation will now involve N_2. It did not involve N_2 for D greater than 0.5 in Eq. 5.14*a* because N_2 had been made large enough that during T_{off}, $D3$ was reverse-biased and the peak voltage at V_{ct} involved only N_1 (see Figure 5.22*d*).

Thus, the first choice is the N_1 from Eq. 5.14*a*, which makes $D = 0.5$ for nominal input voltage V_{dcn}. From Eq. 5.14*a*

$$N_1 = \frac{V_{dcn}}{2(V_o + V_d)(1 - 0.5)} = \frac{V_{dcn}}{V_o + V_d} \tag{5.15}$$

Next $N_2 (= N_{LP}/N_{LS})$ must be selected so that during T_{off} in Figure 5.22*d*, the maximum voltage across N_{LS} does not forward-bias $D3$. The maximum N_{LS} voltage occurs at the maximum voltage across N_{LP}, which is a maximum when V_{dc} is a minimum (see Figure 5.22*d*). Maximum flyback secondary voltage is then $[N_1(V_o + V_d) - V_{dc(min)}]/N_2$. Further, since the $D3$ cathode is at V_o, in order for $D3$ not to be forward-biased:

$$\frac{N_1(V_o + V_d) - V_{dc(min)}}{N_2} < V_o + V_d$$

or

$$N_2 > \frac{[N_1(V_o + V_d) - V_{dc(min)}]}{V_o + V_d} \tag{5.16a}$$

To avoid problems arising from push-pull transformer leakage inductance spikes, N_2 is usually selected to be twice this minimum value.[14] Thus

$$N_2 = \frac{2[N_1(V_o + V_d) - V_{dc(min)}]}{V_o + V_d} \tag{5.16b}$$

5.6.7.11 Output/Input Voltages vs. "On" Time for Overlap-Mode Design at High DC Input Voltages, with Forced Non-Overlap Operation

With N_1 selected from Eq. 5.15, and N_2 from 5.16*b*, when V_{dc} is less than the nominal, the relation between output voltage and "on" time is given by Eq. 5.14*a*. At nominal input V_{dcn}, $D = T_{on}/T$ is 0.5, and at DC input voltages greater than V_{dcn}, D is less than 0.5 and there is no overlapping "on" time. Waveforms for this input voltage range are shown in Figure 5.23.

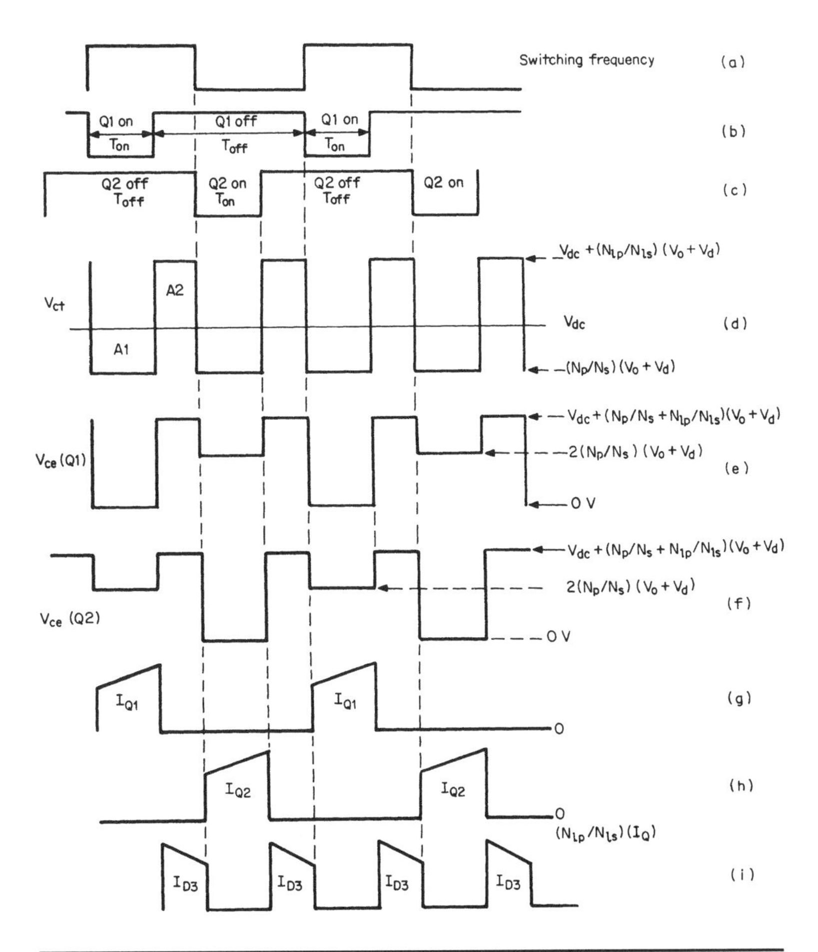

FIGURE 5.23 Circuit of Figure 5.20 in overlap mode when the DC input voltage has risen sufficiently to force it into non-overlap mode. There is a smooth transition between overlap and non-overlap modes if turns ratios are chosen correctly.

For the conditions of Figure 5.23, whenever $Q1$ or $Q2$ is "on," the secondaries are clamped to $(V_o + V_d)$ and the center tap is clamped to $N_1(V_o + V_d)$, where N_1 is calculated from Eq. 5.15. When either transistor turns "off," the dot end of L_p rises to keep current in it constant. As the dot end of L_p rises, so does the dot end of N_{LS} until it clamps to $V_o + V_d$ via $D3$. This clamps the dot end of L_p (or V_{ct}) to $V_{dc} + N_2(V_o + V_d)$, as seen in Figure 5.23*d*.

Again in Figure 5.23, since the DC voltage averaged over a half cycle must equal zero, the area $A1$ must equal area $A2$. Or

$$[V_{dc} - N_1(V_o + V_d)]T_{on} = [N_2(V_o + V_d)][(T/2) - T_{on}]$$

From this, for $T_{on}/T = D$,

$$V_o = \frac{V_{dc}D - N_2 V_d(0.5 - D) - N_1 V_d D}{N_2(0.5 - D) + N_1 D} \tag{5.17a}$$

In Eq. 5.17a, since the diode forward drops V_d are about 1 V, the last two terms in the numerator are small compared to $V_{dc}D$ and can be neglected. The equation can then be rewritten as

$$V_o = \frac{V_{dc}D}{N_2(0.5 - D) + N_1 D} \tag{5.17b}$$

And from Eq. 5.17*b*, the duty cycle at any DC input is

$$D = \frac{0.5 V_o N_2}{V_{dc} - V_o(N_1 - N_2)} \tag{5.18}$$

When designing for overlap mode, N_1 is calculated from Eq. 5.15 and N_2 from 5.16*b*. At DC input voltages less than nominal V_{dcn}, the feedback loop sets the duty cycle in accordance with Eq. 5.14*a* to maintain V_o constant. This duty cycle will be greater than 0.5.

When input voltage has risen to V_{dcn}, the duty cycle has decreased to 0.5 to keep V_o at the same value. When DC input voltage has risen above V_{dcn}, the feedback loop sets the duty cycle in accordance with Eq. 5.17*b* to maintain the output constant. This duty cycle will now be less than 0.5.

The transition from $D > 0.5$ to $D < 0.5$ will be smooth and continuous as V_{dc} rises through V_{dcn}. A much larger range of DC input voltage can now be tolerated than if the design were restricted entirely to non-overlap mode, as in Section 5.6.7.6.

5.6.7.12 Design Example—Overlap Mode

Using the overlap mode design, it is instructive to calculate "on" times for a range of DC input voltages. This will be done for the design example of Section 5.6.7.6, which restricts operation to non-overlap mode. Recall in that design example that V_o was 48 V, nominal input voltage V_{dcn} was 160 V, minimum input voltage $V_{dc(min)}$ was 100 V, switching frequency was 50 kHz, and P_o was 2000 W. From Eq. 5.15

$$N_1 = \frac{V_{dcn}}{V_o + V_d} = \frac{160}{48 + 1} = 3.27$$

and from Eq. 5.16*b*

$$\begin{aligned} N_2 &= \frac{2[(3.2)(49) - 100]}{49} \\ &= 2\left(3.27 - \frac{100}{49}\right) = 2.46 \end{aligned}$$

and for $V_{dc} < V_{dcn}$, from Eq. 5.14*b*

$$\begin{aligned} D &= \frac{[2N_1(V_o + V_d) - V_{dc}]}{2N_1(V_0 + V_d)} \\ &= 1 - \left(\frac{V_{dc}}{2 \times 3.27 \times 49}\right) \\ &= 1 - \left(\frac{V_{dc}}{320.5}\right) \end{aligned} \tag{5.19}$$

and for $V_{dc} > V_{dcn}$, from Eq. 5.17*b*

$$\begin{aligned} D &= \frac{0.5 V_o N_2}{V_{dc} - V_o(N_1 - N_2)} \\ &= \frac{0.5 \times 48 \times 2.46}{V_{dc} - 48(3.27 - 2.46)} \\ &= \frac{59}{V_{dc} - 38.9} \end{aligned} \tag{5.20}$$

and from Eqs. 5.19 to 5.22, we can construct Table 5.2.

V_{dc}, V	D	T_{on}, μs	T_{off}, μs	I_p, A (Eq. 5.21)	I_{rms}, A (Eq. 5.22)
50	0.840	16.9	3.1	50.2	24.6
100	0.688	13.8	6.2	25.2	15.1
136	0.576	11.5	8.5	18.3	12.2
160	0.500	10.0	10.0	15.6	11.0
175	0.433	8.67	11.3	13.8	
185	0.404	8.08	11.9	13.1	
200	0.366	7.32	12.7	12.3	

TABLE 5.2

Comparing Table 5.2 to Table 5.1, in which operation is restricted to non-overlap mode, it is seen that allowing both overlap and non-overlap modes permits a larger range of DC input voltages and larger "on" times at high input voltages. This permits the use of bipolar transistors, which don't operate reliably with short "on" times close to their storage times.

5.6.7.13 Voltages, Currents, and Wire Size Selection for Overlap Mode

Transistor currents and transformer RMS currents can be calculated from the waveforms of Figures 5.22 and 5.23. Wire sizes will be selected from the RMS currents at the rate of 500 circular mils per RMS ampere.

First, consider operation at V_{dc} less than nominal, so there will be overlapping conduction with the waveforms of Figure 5.22. Assume an efficiency of 80% as in the design example of Section 5.6.7.6. Input power is then $P_o/0.8 = 2000/0.8 = 2500$ W. Note that whether V_{dc} is above or below nominal, power is supplied to the load through the push-pull transformer at a center tap voltage of $N_1(V_o + V_d)$. For V_{dc} less than nominal (Figure 5.22*d*), the center tap voltage is boosted up to $N_1(V_o + V_d)$. For supply voltages greater than nominal, center tap voltage is bucked down (Figure 5.23*d*) to the same value. In this design example, $N_1(V_o + V_d) = 3.27(48 + 1) = 160$ V.

The equivalent flat-topped current pulse I_p into the push-pull center tap will be calculated. This is close to the current at the center of the ramp in Figure 5.22*g*. Power into the center tap is

$$P_{in} = 2500 = 160 I_p \frac{2T_{off}}{T} \quad \text{or} \quad I_p = \frac{156}{T_{off}} \tag{5.21}$$

Peak currents for supply voltages less than nominal are calculated from Eq. 5.23 and shown in Table 5.2. If $Q1$, $Q2$ are bipolar transistors, the base drive current must be adequate to saturate them at that peak current. The transistors are chosen for a maximum collector-emitter voltage of $V_{dc(max)} + (N_1 + N_2)(V_o + V_d)$ from Figure 5.23*e* and 5.23*f*, since this is greater than $2N_1(V_o + V_d)$ of Figure 5.22*e* and 5.22*f*. Allowance should be made for a leakage inductance spike.

Although power is delivered to the load only during the two "off" times per period (Figure 5.22*g* and 5.22*h*), each half secondary carries I_p during one "off" time but also $I_p/2$ during the two T_1 times per period. Note in Figure 5.22, $T_1 = (T/2) - T_{off}$. The RMS current carried by each half secondary is

$$I_{rms} = I_p \left(\frac{T_{off}}{T}\right)^{1/2} + \left(\frac{I_p}{2}\right) \left(\frac{T - 2T_{off}}{T}\right)^{1/2} \tag{5.22}$$

These RMS currents are shown in Table 5.2. Wire size for each half secondary will be selected at the rate of 500 circular mils per RMS ampere, and it is seen in Table 5.2 that maximum RMS current occurs at minimum DC input. RMS currents for supply voltages above V_{dcn} are lower than those below V_{dcn}, so the RMS currents of Table 5.2 dictate the wire sizes.

The flyback transformer secondary carries the pulses I_{D3} shown in Figure 5.23. As DC voltage goes up, the transistor "on" times decrease toward zero, and the I_{D3} pulses widen until they reach a full half period each. All the output load current is then fully supplied by flyback action from the flyback secondary. Since the center of the ramp of the *D*3 pulses is the DC output current, the flyback secondary winding should be sized for the worst-case condition: to carry the DC output current at 100-percent duty cycle.

Finally, wire size for the flyback primary must be chosen. Table 5.2 gives the RMS currents per half primary at supply voltages of less than nominal. Since the flyback primary carries the currents of both half primaries, its RMS current is twice that shown in the table.

Examination of Figure 5.22 shows that the astonishingly high currents at low DC input should be expected, for in Figure 5.22, as the supply voltage goes lower, the T_{off} times become shorter. Since power is delivered to the load only during the T_{off} times when voltage exists at the push-pull center tap, the very short T_{off} times demand high peak and RMS currents to supply the output power.

References

Current Mode:

1. B. Holland, "A New Integrated Circuit for Current Mode Control," *Proceedings Powercon 10*, 1983.
2. W. W. Burns and A. K. Ohri, "Improving Off Line Converter Performance with Current Mode Control," *Proceedings Powercon 10*, 1983.
3. "Current Mode Control of Switching Power Supplies," Unitrode Power Supply Design Seminar Manual SEM 400, 1988, Unitrode Corp., Lexington, MA.
4. T. K. Phelps, "Coping with Current Mode Regulators," *Power Control and Intelligent Motion* (*PCIM Magazine*), April 1986.
5. C. W. Deisch, "Simple Switching Control Method Changes Power Converter into a Current Source," 1978 IEEE.
6. R. D. Middlebrook, "Modelling Current Programmed Regulators," *APEC Conference Proceedings*, March 1987.
7. G. Fritz, "UC3842 Provides Low Cost Current Control," Unitrode Corporation Application Note U-100, Unitrode Corp., Lexington, MA.

Current Fed:

8. A. I. Pressman, *Switching and Linear Power Supply, Power Converter Design*, p. 146, Switchtronix Press, Waban, MA, 1977.

9. E. T. Calkin and B. H. Hamilton, "A Conceptually New Approach for Regulated DC to DC Converters Employing Transistor Switching and Pulse Width Control," *IEEE Transactions on Industry Applications*, 1A: 12, July 1986.
10. E. T. Calkin and B. H. Hamilton, "Circuit Techniques for Improving the Switching Loci of Transistor Switches in Switching Regulators," *IEEE Transactions on Industry Applications*, 1A: 12, July 1986.
11. K. Tomaschewski, "Design of a 1.5 kW Multiple Output Current Fed Converter Operating at 100 kHz," *Proceedings Powercon 9*, 1982.
12. B. F. Farber, D. S. Goldin, C. Siegert, and F. Gourash, "A High Power TWT Power Processing System," *PESC Record*, 1974.
13. R. J. Froelich, B. F. Schmidt, and D. L. Shaw, "Design of an 87 Per Cent Efficient HVPS Using Current Mode Control," *Proceedings Powercon 10*, 1983.
14. V. J. Thottuvelil, T. G. Wilson, and H. A. Owen, "Analysis and Design of a Push Pull Current Fed Converter," *IEEE Proceedings*, 1981.
15. J. Lindena, "The Current Fed Inverter—A New Approach and a Comparison with the Voltage Fed Inverter," *Proceedings 20th Annual Power Sources Conference*, pp. 207–210, 1966.
16. P. W. Clarke, "Converter Regulation by Controlled Conduction Overlap," U.S. Patent 3,938,024, issued Feb. 10, 1976.
17. B. Israelson, J. Martin, C. Reeve, and A. Scown, "A 2.5 kV High Reliability, TWT Power Supply: Design Techniques for High Efficiency and Low Ripple," *PESC Record*, 1977.
18. J. Biess and D. Cronin, "Power Processing Module for Military Digital Power Sub System," *PESC Record*, 1977.
19. R. Redl and N. Sokal, "Push Pull Current Fed, Multiple Output DC/DC Power Converter with Only One Inductor and with 0 to 100% Switch Duty Ratio," *IEEE Proceedings*, 1980.
20. R. Redl and N. Sokal, "Push Pull, Multiple Output, Wide Input Range DC/DC Converter—Operation at Duty Cycle Ratio Below 50%," *IEEE Proceedings*, 1981.
21. L. G. Meares, "Improved Non-Dissipative Snubber Design for Buck Regulator and Current Fed Inverter," *Proceedings Powercon 9*, 1982.
22. E. Whitcomb, "Designing Non-Dissipative Snubber for Switched Mode Converters," *Proceedings Powercon 6*, 1979.
23. A. H. Weinberg, "A Boost Regulator with a New Energy Transfer Principle," *Proceedings of the Spacecraft Power Conditioning Electronics Seminar*, European Space Research Organization Publication Sp-103, September 1974.
24. Rudolf P. Severns and Gordon (Ed) Bloom, *Modern DC-to-DC Switchmode Power Converter Circuits*, Van Nostrand Reinhold Company, 1985.
25. Keith Billings, *Switchmode Power Supply Handbook*, McGraw-Hill, New York, 1989.
26. Wm. T. McLyman, *Transformer and Inductor Design Handbook*, Marcel Dekker Inc., New York, 1978.
27. Wm. T. McLyman, *Magnetic Core Selection for Transformers and Inductors*, Marcel Dekker Inc., New York, 1982.

CHAPTER 6

Miscellaneous Topologies

6.1 SCR Resonant Topologies—Introduction

The silicon controlled rectifier (SCR) has been used in DC/AC inverters and DC/DC power supplies for over 25 years.[1,2] They are used because they are available with higher voltage and current ratings, and at lower cost than bipolar or MOSFET transistors. Because SCRs are normally higher voltage and current rated, they are used primarily for supplies of over 1000 W. A significant feature of an SCR for high-power inverters is that it does not suffer from secondary breakdown, the most frequent failure mode of transistors.

The SCR is a solid-state switch that is easily turned "on" by a narrow pulse at its gate input terminal; it then latches and stays "on" after the input is removed. Having been turned "on," it must now be turned "off" at some point. This is not so easy, as it cannot be done from the gate. There are many schemes for turning an SCR "off" or "commutating it off." Essentially, all these schemes involve reducing its "on" current to zero by diverting the current to an alternate path for a minimum turn "off" period t_q. SCR turn "off" will be discussed below.

After Pressman *A gate controlled switch, or gate turn off (GTO), a device similar to the SCR, can be turned "on" and "off" from the gate, providing the correct operating conditions are maintained. ~K.B.*

A major problem with early SCR type DC/AC or DC/DC supplies is that they could not operate reliably at switching frequencies much over 8 to 10 kHz. This was because even the fastest inverter-type SCRs available at that time did not have a reliable high-impedance "off" state until about 10 to 20 μs after they had been commutated "off" and their internal currents had dropped to zero, due to the recombination time in the substrate. Thus they could not be subjected to high voltage stress until the 10 to 20 μs recombination time had elapsed, after their currents had been reduced to zero.

Further, early SCRs could not tolerate a large *dV/dt* across their output terminals even after the recombination time had elapsed. Most were specified at a maximum rate of change of output voltage of 200 V/μs, and at a *dV/dt* faster than that, they would spontaneously turn back "on" again, independent of the input control voltage.

Early inverter-type SCRs also could not tolerate a large rate of change of output current *dI/dt* at the instant of turn "on." Most were specified in the range of 100 to 400 A/μs. At *dI/dt* faster than specified, average junction temperatures would rise, local hot spots would develop on the chips, and the SCRs would either fail immediately or degrade to the failure point in a short period of time.

With switching frequencies thus limited to 10 kHz, transformers, inductors, and capacitors were still relatively large, which made the overall size of a DC/AC or DC/DC converter too large in many applications. Further, switching frequencies of 10 kHz and under are in the middle of the audio range, and the audible noise emitted from such converters made them unacceptable in an office or even factory environment. To be acceptable in such environments, switching frequencies must be above the highest audible frequency of about 20 kHz.

About 1977, RCA developed the asymmetrical silicon controlled rectifier (ASCR), which solved most of these problems and made possible DC/AC and DC/DC converters operating up to 40 or 50 kHz.

Conventional SCRs can sustain (or block) reverse voltages across their output terminals equal to their forward-voltage blocking capability. But in a large number of SCR circuits, reverse voltage at the output terminals is clamped to one or two diode drops, or to a maximum of about 2 V, making large reverse-voltage blocking capability unnecessary. By making certain changes on the SCR chip, RCA was able to achieve turn "off" times t_q of 4 μs. (t_q is the time after SCR forward current has dropped to zero until the full-rated forward voltage can be reapplied.) The price paid for this reduction in t_q is that the reverse voltage blocking capability is reduced to 7 V, but this is more than adequate in many inverter circuits.

Thus with t_q times of 4 μs, this RCA device (S7310) made inverters at switching frequencies of 40 to 50 kHz possible in a host of circuit configurations. The S7310 had many other very useful features. Its *dV/dt* and *dI/dt* ratings were 3000 V/μs and 2000 A/μs with only a 1-V negative bias on the input terminal. Compare this with the 20 V/μs and 400 A/μs for conventional SCRs. Further, the device was available in voltage ratings of 800 V and RMS current ratings of 40 A. These advances in voltage, current, and t_q ratings made inverters and

DC/DC power supplies possible with output power ratings of 4000 W using only two ASCRs in a 40-kHz half-bridge circuit configuration.

In its first few years on the market the S7310 sold for about $5. No other transistor circuit topology could even approach 4000 W with only two switching devices costing $10. Unfortunately, the S7310 is no longer manufactured by RCA, but equivalent asymmetrical SCRs are made by other manufacturers. A similar Marconi ASCR, type ACR25U, has a blocking time t_q of 4 μs, with voltage ratings of up to 1200 V, and RMS current ratings of 40 A.

6.2 SCR and ASCR Basics

The SCR symbol is shown in Figure 6.1. Its input terminal 1 is designated the *gate*, terminal 2 is the *anode*, and 3 is the *cathode*. When it is "on," current flows from the anode to the cathode. When it is "off," the maximum voltage it can sustain or block from anode to cathode is designated V_{DRM}. ASCR types are available with V_{DRM} voltage ratings ranging from 400 to 1200 V.

Once turned "on," the anode current is determined by the supply voltage and load impedance from anode to supply source. The anode-to-cathode voltage versus anode current characteristics are given in the data sheets for a specific device. For the Marconi ACR25U, a 40A RMS device, anode-to-cathode voltage at 100-A anode current is typically 2.2 V (see Figure 6.2).

The reason for this long anode voltage fall time is that it takes a long time for the anode current carriers to spread uniformly throughout the chip area. Initially, the current carriers are concentrated in only a small fraction of the chip area and instantaneous anode-to-cathode resistance is high, causing a high instantaneous "on" voltage. After a time, the current carriers spread uniformly throughout the chip, and the "on" voltage drops to the quiescent level given in Figure 6.2.

Thus, most of the dissipation in the SCR occurs during the turn "on" time. This dissipation is the integral $\int I_a V_a dt$. In most SCR circuits, currents have the shape of a half sinusoid rather than a square wave, which is helpful. As seen in Figure 6.5, if anode current pulses were square waves, the front edge of the current pulse would flow at an

FIGURE 6.1 Silicon controlled rectifier and ASCR symbol.

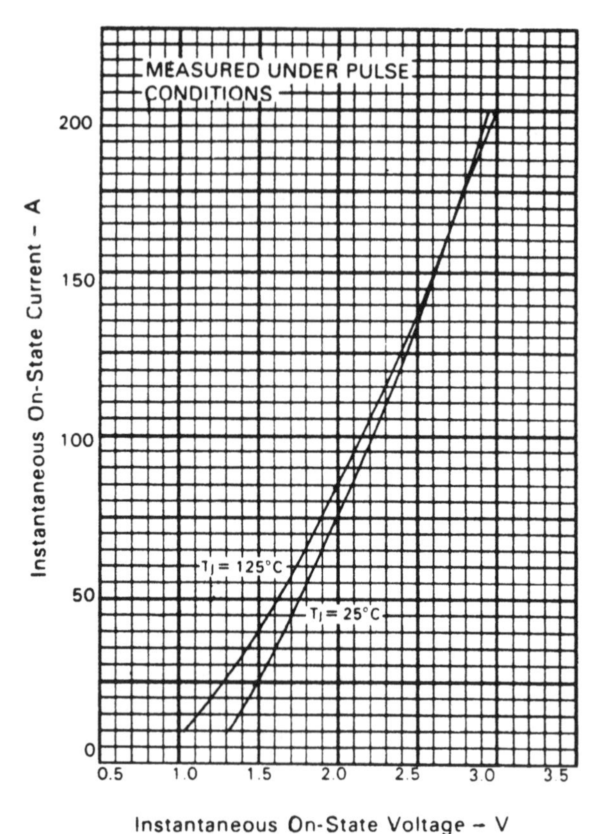

FIGURE 6.2 Anode current versus anode voltage.

anode voltage in the vicinity of 25 V, and dissipation would be high. Figure 6.5 also shows that if anode current pulses are half sinusoidal, their base width should be longer than 2.5 μs to avoid an "on" anode potential greater than 5 V throughout the entire half sinusoid.

Gate pulse duration should be greater than 400 ns for 100 A of anode current. The gate-to-cathode voltage during the duration of the gate current pulse is shown in Figure 6.4 and is in the range of 0.9 to 3 V for a large range of gate currents. Once turned "on," the anode will latch "on" and stay conducting after the gate turn "on" pulse is gone. The anode "on" potential ranges from 1.2 to 2.2 V for an anode current range of 20 to 100 A, as seen in Figure 6.2.

The device is turned "on" by a gate-to-cathode current pulse whose amplitude and duration are not well defined in the data sheets. Figure 6.3 shows anode current delay and rise time to 100 A as a function of

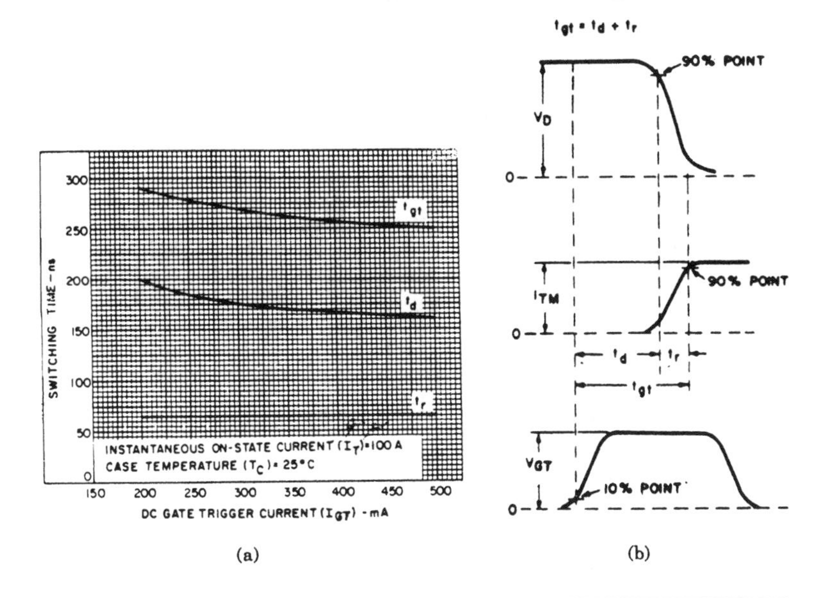

FIGURE 6.3 (*a*) Switching times. Typical switching times t_{gt}, t_d, t_r, versus gate trigger current. (*b*) Relationship among "off"-state voltage, "on"-state current, and gate trigger voltage showing reference points for definition of turn "on" time t_{gr}. (*Note:* Figure 6.3*a* and 6.3*b* illustrate the original RCA S7310—a type of SCR similar to the Marconi type ACR25U.)

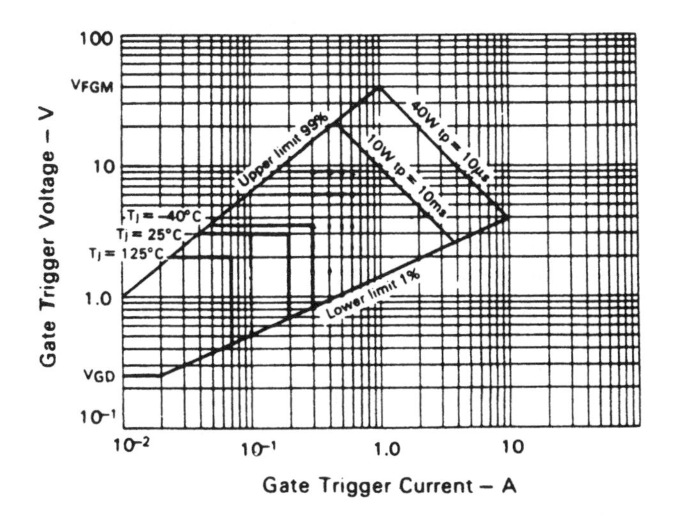

FIGURE 6.4 Gate voltage versus current.

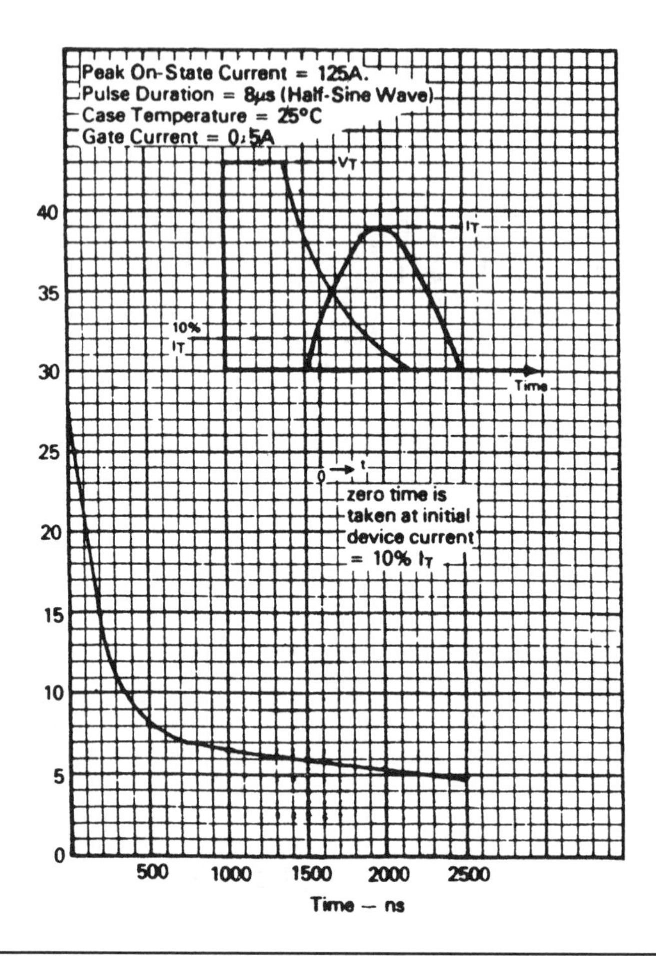

FIGURE 6.5 Anode-to-cathode voltage fall time for Marconi ASCR type ACR25U.

gate trigger current. Gate current pulse width also determines current rise time to some extent, but this is seldom given. Typically for the ACR25U, gate current should be in the range of 90 to 200 mA for an anode current of 100 A.

Anode current rise time shown in Figure 6.3 is not as important for an SCR as is anode-to-cathode voltage fall time. This is obvious from Figure 6.5, which shows that even with a 500-mA gate current pulse, with a half-sinusoid 8-μs anode current pulse 125 A in amplitude, the anode-to-cathode voltage has fallen to only 5 V in 2.5 μs. This is still twice the anode-to-cathode quiescent voltage at that current shown in Figure 6.2. Figures 6.6 and 6.7 show maximum dV/dt and t_q for the Marconi ACR25U.

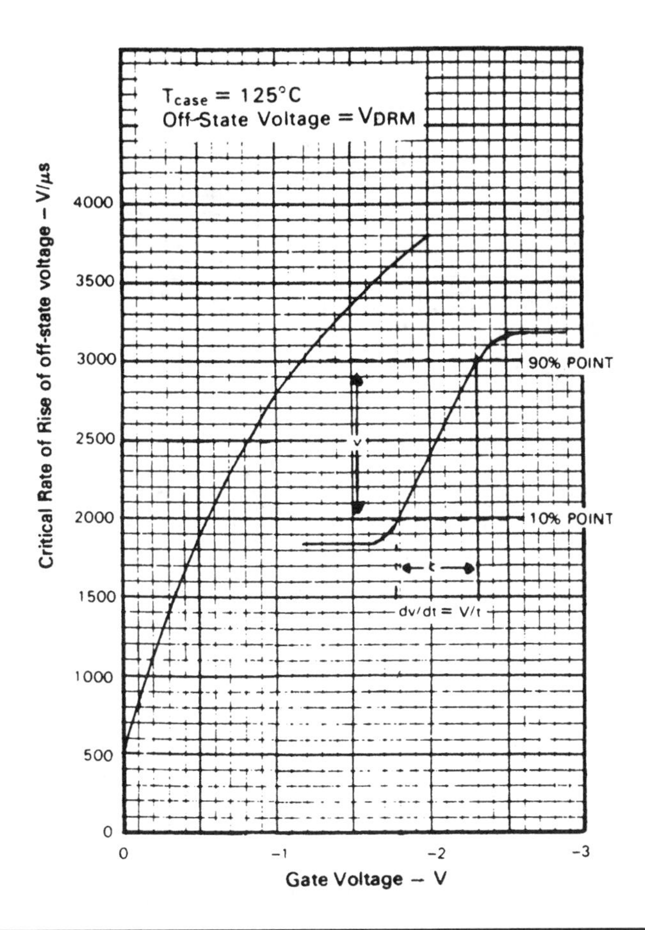

FIGURE 6.6 Marconi ASCR type ACR25U characteristics. Minimum linear critical rate of rise of "off"-state voltage versus gate voltage.

6.3 SCR Turn "Off" by Resonant Sinusoidal Anode Current—Single-Ended Resonant Inverter Topology

It was pointed out above that an SCR is easily turned "on" with a narrow pulse but stays latched "on" after the pulse has gone. To turn it "off," anode current must be reduced to zero for a time equal to at least the specified t_q time of the device. Further, after the t_q time, the reapplied anode voltage rise-time rate must be less than the specified dV/dt rating of the SCR.

All this is easily achieved by forcing the SCR anode current to be sinusoidal in shape, and this offers other significant advantages as

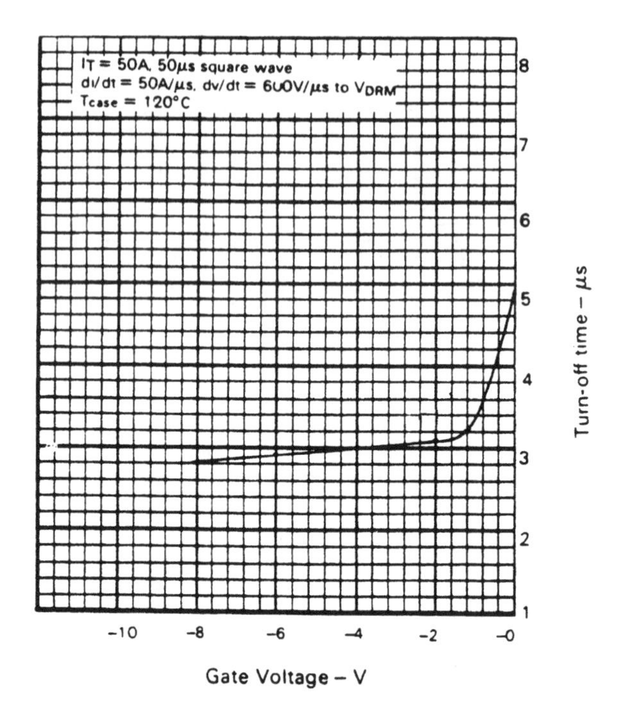

FIGURE 6.7 Marconi ASCR type ACR25U characteristics. Typical circuit commutated turn "off" time versus gate voltage at turn "off."

well. The basic scheme and its advantages are most easily described with a typical single-ended SCR resonant converter such as that shown in Figure 6.8.[3–8]

The SCR, an inductor L, and a capacitor C are arranged in series. Before the SCR is fired, capacitor C is charged to some positive voltage through the larger constant-current inductor L_c. When the SCR is triggered "on" with a narrow gate pulse, the equivalent circuit is that of a switch closure applying a step waveform to a series resonant LC circuit. Current in the circuit is shocked into a resonant "ring" whose period is $t_r = 2\pi\sqrt{LC}$.

Current increases in the SCR sinusoidally, goes through its first negative peak, and decreases sinusoidally to zero at the end of a half period ($= \pi\sqrt{LC}$). As the sine wave of current in the SCR reaches zero at t_1, it reverses direction and flows sinusoidally for the next half cycle through the anti-parallel diode $D1$. During this half cycle of diode conduction time T_d, the SCR is clamped with a reverse voltage of about 1 V by $D1$. This maximum reverse voltage is safely below the 7- to 10-V reverse-voltage specification of the asymmetrical SCR.

If T_d is greater than the specified t_q time of the device, the SCR has safely extinguished itself at the end of T_d without the need for any external "commutation" circuitry, and forward voltage may be safely reapplied. The half sinusoids of current through the SCR and anti-parallel diode both provide power to the load resistor R_o.

After t_2, when the $D1$ current has fallen back to zero, both $Q1$ and $D1$ are safely "off" and the constant current from L_c commences charging C back up with its left end positive. During the interval t_3 to t_2, the change in voltage on C corresponds to that fraction of its stored energy ($CV^2/2$) that is equal to the energy delivered to the load the next time $Q1$ is triggered. After a time t_t (the triggering period), $Q1$ can be triggered "on" again and the cycle repeats.

As the load is increased (resistor R_o is decreased), the amplitude of the first half cycle increases and its duration increases somewhat. This decreases the duration of the second half cycle T_d, and care must be taken that the load is not increased to the point where T_d is shorter than t_q, the SCR turn "off" time, or the SCR will not turn "off" successfully.

By choosing the triggering period t_t in the range of 1.5 to 2 times the resonant period t_r at minimum line input and maximum load, the "off" time $t_t - t_r$ shown in Figure 6.8 is not too large and the output across R_o is a fairly distortion-free sine wave over a large range of load resistance. Capacitance across R_o can reduce the distortion due to the time gap $t_t - t_r$. The circuit can thus be used as a DC/AC converter. As the supply voltage V_{dc} is increased, output power increases since the sine-wave peak amplitudes increase. To maintain constant output voltage as V_{dc} or R_o goes up, the triggering frequency can be decreased (increases t_t). This maintains a roughly constant peak AC voltage at the output, although distortion increases as the time gap ($t_t - t_r$) increases.

The circuit is more useful as DC/DC converter with regulated and isolated output as in Figure 6.9, where R_o is replaced by the primary of a power transformer with rectifying diodes and a capacitor filter at the secondary. Line and load changes are regulated by varying the triggering frequency f_t.

As the line voltage or load resistance increases the sine-wave peaks in Figure 6.8 increase, but their base widths remain roughly constant and equal to $\pi\sqrt{LC}$. To regulate against line and load changes, the rectified DC output is sensed with a voltage error amplifier that alters the switching frequency to maintain constant output. If output voltage rises because of an increase in line voltage or a decrease in DC load current, the switching frequency is decreased so as to take less power per unit time from the input. Similarly, a decrease in output is corrected by an increase in switching frequency.

This method of regulating the output voltage—by varying the switching frequency—is common to most resonant power supplies,

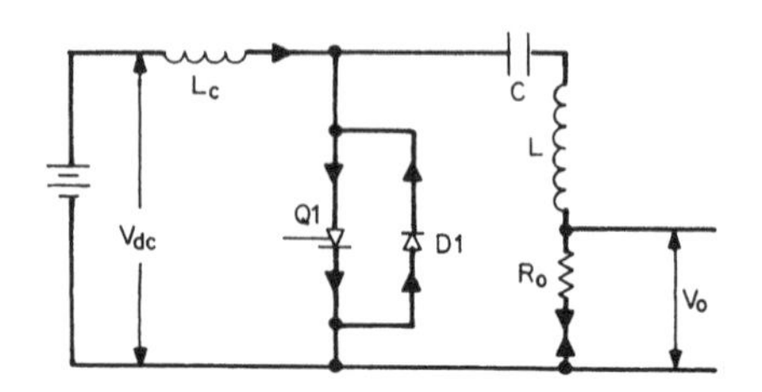

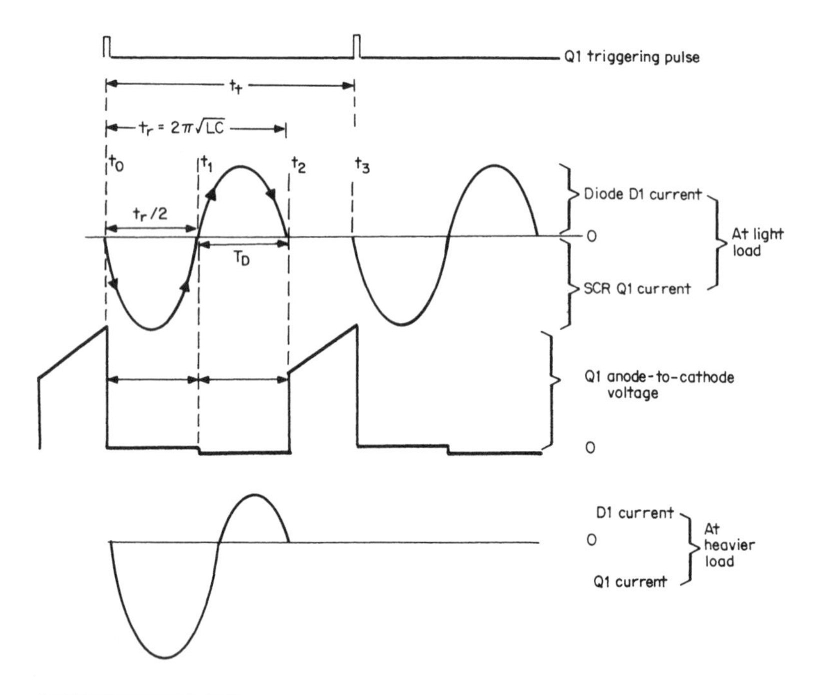

FIGURE 6.8 A single-ended SCR resonant converter. Inductor L_c charges C to a voltage higher than V_{dc}. When $Q1$ is fired, a sinusoidal current flows through $Q1$, delivering power into R_o. At t_1, this sinusoidal current reverses direction and flows through $D1$, delivering power to R_o. If T_d is greater than t_q time of Q_1, the SCR self-extinguishes. During $(t_t - t_r)L_c$ recharges C and the cycle repeats.

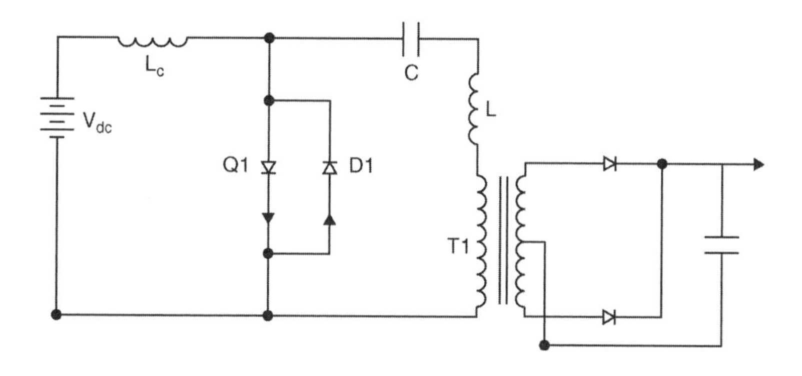

FIGURE 6.9 A transformer-coupled, series-loaded, single-ended SCR resonant converter.

because the output current or voltage pulses are constant in width. Nonresonant topologies are usually operated at a constant switching frequency and regulated by varying pulse width to control the DC output voltage.

In constant-frequency schemes, the power supply switching frequency is generally synchronized to the horizontal line rate in an associated display terminal or to the system's clock rate. This makes it easier to tolerate any RFI noise pickup on the display screen and lessens the possibility of computer logic errors due to noise pickup.

Since this advantage is lost in variable-frequency resonant topologies, they are not acceptable in many applications. Although it can be argued that RFI noise pickup, if it exists, is more troublesome with variable-frequency switching supplies, it is less likely to exist. The sinusoidal currents in resonant supplies have much lower *di/dt* than the square-wave currents of fixed-frequency, adjustable pulse width supplies, so they emit less RFI.

The SCR supply was originally made resonant to ensure turn "off" of the SCR at the zero crossing of the current sinusoid. This added a further significant advantage. It was seen in Section 1.2.4 that with square waves of current, most of the losses in the switching device occur at turn "off" as a result of the overlap of falling current and rising voltage. But with sinusoidal currents, turn "off" occurs at zero voltage across the device and these losses are almost nonexistent. However, the turn "on" losses due to the relatively slow voltage fall time (Section 6.1 and Figure 6.5) can be high. If the sine-wave base width $\pi\sqrt{LC}$ is greater than about 8 μs, these losses are also not excessive (Figure 6.5).

The DC/DC converter of Figure 6.9, used with a single 800-V, 45-A RMS SCR, can generate 1-kW of output power.[4] An inductor is not needed in the secondary output, because the SCR and diode currents shown in Figure 6.8 are constant-current pulses whose magnitude is close to the voltage applied across the series *LC* elements divided by $\sqrt{L/C}$. These constant currents reflect into the secondary and flow into the output resistor, producing constant-voltage pulses of similar waveshape. The filter capacitor alone averages the pulses to obtain a constant ripple-free output voltage. Not requiring an output inductor, the circuit can be used as a high voltage supply.

A quantitative design example will be presented after the following discussion of two widely used SCR resonant-bridge DC/DC inverters.

6.4 SCR Resonant Bridge Topologies—Introduction

The resonant half bridge (Figure 6.10) and full bridge (Figure 6.11) are the most useful SCR circuits. The half bridge, with two 800-V 45-A RMS SCRs (Marconi ACR25UO8LG), can deliver up to 4 kW of AC or rectified DC power from a rectified 220-V AC line. A 1200-V version of the device can generate up to 8 kW in a full-bridge circuit. Full-bridge operation is much the same as the half-bridge except that SCR voltage stresses are twice and current levels are half those of the half bridge for equal output power. Hence only the half bridge will be discussed here in detail.

The half bridge can be operated *series-loaded* as in Figure 6.10, with the secondary load reflected via transformer $T1$ in series with a series resonant circuit ($C3$ with the series combination of $L3$ and $L1$ when $Q1$ is "on" or with the series combination of $L3$ and $L2$ when $Q2$ is "on").

The secondary load in the series-loaded circuit, when reflected into the primary, must not appear as too high an impedance or the resonant circuit Q will be low and the "on" SCR may not be safely commutated "off" by the above-described resonant current reversal. No output inductor is required in the series-loaded configuration, so it can be used for either high or low output DC voltages. The series-loaded circuit can safely tolerate an output short circuit, since the normal load impedance in series with the resonating LC is already small compared to the impedance of the LC elements. As discussed below, however, there is a problem when the output load is open-circuited.

Alternatively, the half bridge can be shunt-loaded as in Figure 6.12. Here, the output load reflected into the $T1$ primary is connected across the resonating capacitor $C3$. In this case, the output load reflected across $C3$ must not be too low or the resonant circuit Q will be so low as to prevent resonant turn "off" of the SCR. Thus, this configuration

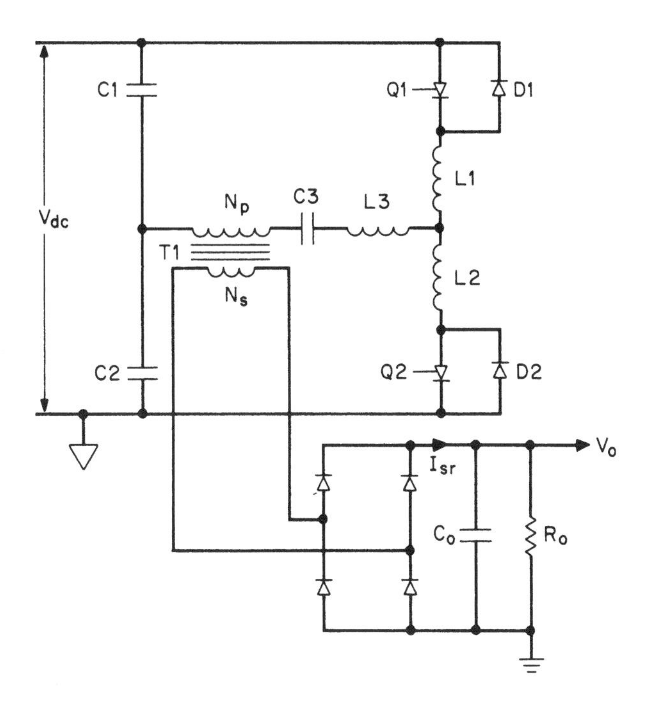

FIGURE 6.10 Series-loaded SCR resonant half bridge. SCRs $Q1$ and $Q2$ are fired on alternate half cycles. Capacitor $C3$ resonates with $L3$ and $L1$ when $Q1$ is "on" and with $L3$ and $L2$ when $Q2$ is "on." After the firing of an SCR, its current goes through a half sinusoid, reverses direction, and flows through the associated anti-parallel diode. If the duration of the diode current is greater than the t_q time of the SCR, the SCR self-extinguishes.

can easily tolerate an open circuit at the output, but not a short circuit.

The series-loaded configuration is analyzed as a current source driving the $T1$ primary whereas the shunt-loaded circuit is better analyzed as a voltage source. The shunt-loaded circuit does require secondary output inductors when DC output is required, although they may be omitted if large output voltage ripple can be tolerated. For a DC/DC converter, the series-loaded circuit is a better choice.

6.4.1 Series-Loaded SCR Half-Bridge Resonant Converter—Basic Operation[9,10]

The series-loaded SCR resonant half-bridge circuit is shown in Figure 6.10, and its significant waveforms in Figure 6.13. In Figure 6.10, when $Q1$ is triggered "on," the equivalent circuit is that of a

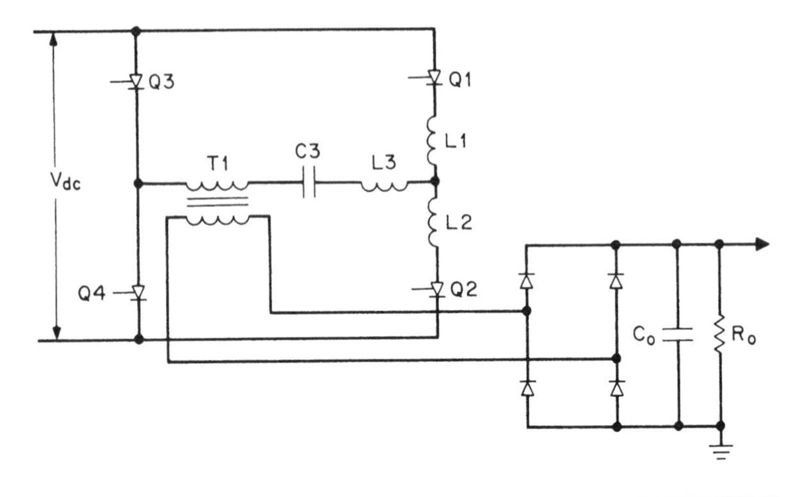

FIGURE 6.11 SCR resonant full bridge. This can deliver twice the output power of the half bridge.

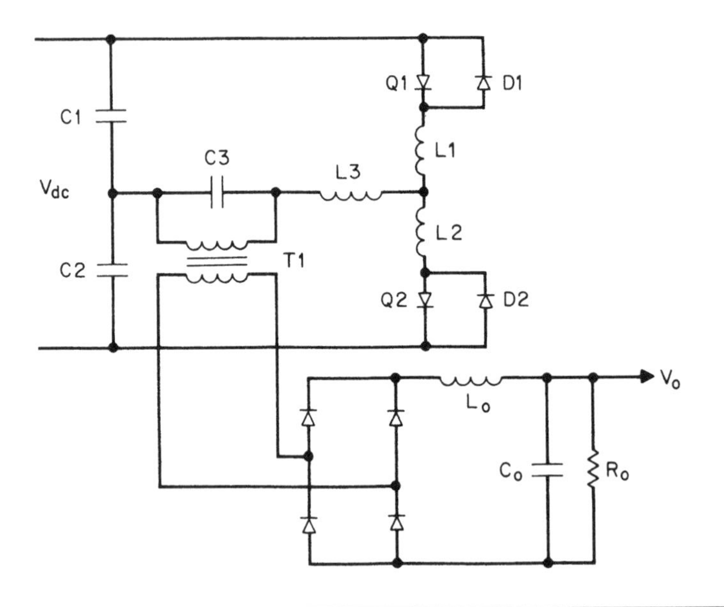

FIGURE 6.12 A shunt-loaded SCR resonant half bridge. Power is taken in shunt across the resonant capacitor. The load resistance reflected into the primary from the secondary must be high so as not to excessively lower the resonant circuit Q, and prevent successful turn "off" of the SCR.

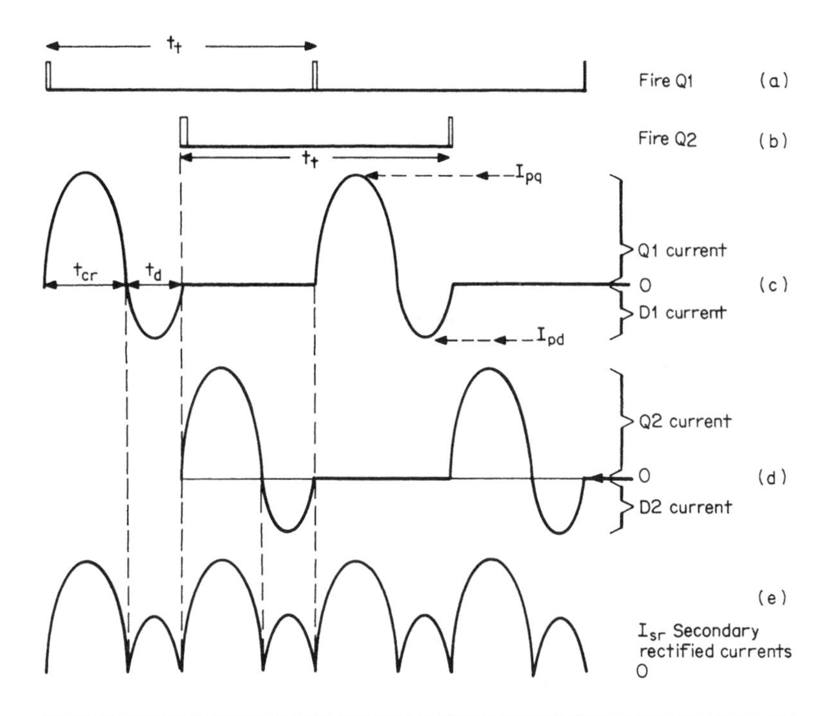

FIGURE 6.13 Significant currents for series-loaded SCR half bridge of Figure 6.10 at minimum DC input voltage and maximum load output. As line voltage or load resistance increases, the feedback loop decreases triggering frequency so as to space the $Q1$, $A2$ sine waves farther apart and maintain the average output current and voltage constant.

voltage step of magnitude $Vdc/2$ applied to a series combination of $L3 + L1$ resonating with $C3$. That resonant circuit is series-loaded with the $T1$ secondary resistance reflected into the primary, which is shunted by the $T1$ magnetizing inductance.

If the Q of the equivalent circuit is sufficiently high, current in it is shocked into a sinusoidal "ring" as shown in Figure 6.13*a*. During the first half cycle, a half sinusoid current pulse flows through $Q1$. At the end of that half cycle at t_1, the current reverses and continues flowing through the anti-parallel diode $D1$. From t_1 to t_2, current in the SCR is zero, and if that time is greater than the specified t_q time of the device, the SCR extinguishes itself and can safely sustain forward voltage again.

The resonant period is $t_r = 2\pi\sqrt{(L_1 + L_3)C_3}$. At light load, the SCR and diode conduction times (t_{cr} and t_d, respectively) are almost equal. As the secondary load increases (R_o decreases), t_{cr} and I_{pq} increase and t_d, I_{pd} decrease. Load power must not be increased beyond the point where t_d is less than the maximum specified value of t_q, or the SCR will not turn "off" successfully.

With *Q*1 safely "off" at the next half cycle of the triggering period t_t, *Q*2 is fired. The resonant period is now $2\pi\sqrt{(L_2 + L_3)C_3}$. Current waveforms of *Q*2, *D*2 are similar to those of *Q*1, *D*1 and are shown in Figure 6.13*b*. The *Q*1, *D*1 and *Q*2, *D*2 current waveforms, shown in Figure 6.13*c*, are multiplied in *T*1 by the turns ratio N_p/N_s, rectified, and summed by the output diodes.

The DC output voltage is the average of the Figure 6.13*e* current waveform, multiplied by the secondary load resistor R_o. The output filter capacitor averages the output current waveforms to yield a constant, ripple-free DC output voltage, without requiring an output inductor. After t_t the entire cycle repeats.

The timing relations shown in Figure 6.13 hold at minimum DC input voltage and minimum output load resistance (maximum output power). The values of resonating inductance and capacitance are chosen so that the current amplitudes and spacings shown in Figure 6.13 yield the correct average output current at the desired output voltage, at minimum line and maximum load. Calculations to achieve this are shown below. As line and load change, a feedback-loop that senses output voltage adjusts SCR triggering frequency f_t to keep output voltage constant.

As DC input voltage and hence the peak currents in Figure 6.13 increase, the feedback loop decreases f_t to maintain constant secondary average current and hence constant output voltage. Further, at a fixed DC input voltage and with fixed *L* and *C*, peak currents (Figure 6.13*c*, *d*) are constant, so as R_o goes up, the feedback loop decreases f_t to maintain constant average output voltage. Secondary current waveshape with DC input or output load higher than minimum is shown in Figure 6.14.

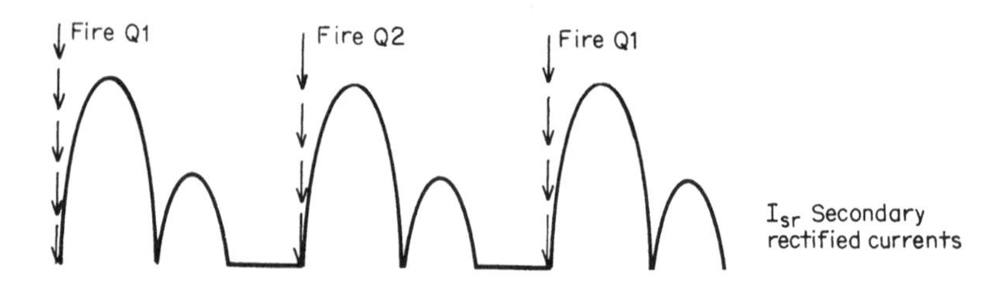

FIGURE 6.14 Bridge output current I_{sr} at higher input voltage. Peak current is higher, forcing feedback loop to decrease triggering frequency to maintain constant output voltage.

6.4.2 Design Calculations—Series-Loaded SCR Half-Bridge Resonant Converter[9,10]

The circuit configuration is shown in Figure 6.10, and its significant waveforms in Figures 6.13 and 6.14. The discussion herein is based on a paper by D. Chambers.[9]

The first choice to be made is the resonant frequency. Assume the use of the Marconi ASCR type ACR25. Figure 6.7 shows its typical turn "off" time t_q as 5 μs, from an "on" current of 50 A with a gate bias of 0 V. Assume the worst case is 20% higher or 6 μs. Also assume desired operation at minimum DC input and maximum output power at the edge of the continuous mode (as in Figure 6.13 with no gap between the zero crossing of the diode current and the start of the opposite SCR current). Then the absolute minimum resonant period as in Figure 6.13 would be 12 μs, corresponding to a resonant frequency of 83 kHz.

However, it was noted in Section 6.3 that at high output power, the diode conduction time t_d shortens to a not easily calculable value. If it is less than t_q, the SCR might not turn "off" successfully. Thus there should be more margin in the t_d time.

Further, Figure 6.5 shows that the anode-to-cathode "on" voltage does not fall very quickly to its quiescent value of 2 to 3 V. To keep the high anode-to-cathode voltage time to a small fraction of SCR conduction time t_{cr}, the resonant half period should be increased to at least four times 2.5 μs. As seen in Figure 6.5, this puts the anode-to-cathode voltage at the peak of the sinusoidal anode current at about 3 V, and is a reasonable compromise. Thus the resonant period is chosen as 20 μs ($f_r = 50$ kHz). Or

$$T_r = 2\pi\sqrt{(L_3 + L_1)(C_3)} = 20 \times 10^{-6} \tag{6.1}$$

Next the peak voltage on the $T1$ primary (Figure 6.10) must be determined. Following the suggestion in the Chambers paper,[9] this will be chosen as 60% of the minimum voltage across one of the bridge capacitors or

$$\text{Minimum primary voltage} = V_{\text{p(min)}} = \frac{0.6\ V_{\text{dc(min)}}}{2} \tag{6.2}$$

Assuming a bridge output rectifier with a 1-V drop across each rectifier diode, this fixes the $T1$ turns ratio at

$$\frac{N_p}{N_s} = \frac{0.6V_{\text{dc(min)}}}{2(V_o + 2)} \tag{6.3}$$

At minimum line input and maximum output current, the secondary currents are as shown in Figure 6.13*e* with no time gaps between the termination of current in one anti-parallel diode and turn "on" of current in the opposite SCR. Assume to a close approximation, that even at maximum DC current, SCR and diode currents have equal widths of a half period. Also assume—as in the Chambers paper[9]—that the diode peak current is one-fourth that of the SCR. Then the average of these Figure 6.13*e* SCR plus diode currents is

$$I_{\text{secondary average}} = I_{o(\text{dc})} = \frac{2I_{\text{ps}}}{\pi}\frac{T}{2T} + \frac{2I_{\text{ps}}}{4\pi}\frac{T}{2T}$$
$$= \frac{1.25 I_{\text{ps}}}{\pi}$$

where I_{ps} is the peak primary SCR current after reflection into the secondary. Then

$$I_{\text{ps}} = 0.8\pi I_{o(\text{dc})} \tag{6.4}$$

$$I_{\text{pp}} = I_{\text{ps}}\frac{N_s}{N_p} = 0.8\pi I_{o(\text{dc})}\frac{N_s}{N_p} \tag{6.5}$$

where I_{pp} is the peak primary SCR current.

In Figure 6.10, V_{ap}, the voltage applied to the series resonant elements when, say, *Q*1 turns "on" is the voltage across the bridge capacitor *C*1 plus the transformer voltage peak of 0.6 $V_{\text{dc(min)}}/2$. Or

$$V_{\text{ap}} = \frac{V_{\text{dc(min)}}}{2} + \frac{0.6V_{\text{dc(min)}}}{2} = 0.8\ V_{\text{dc(min)}} \tag{6.6}$$

It can be shown that, to a close approximation, when a step voltage V_{ap} is applied to a series *LC* circuit, the peak amplitude of the first resonant current pulse is

$$I_{\text{pp}} = \frac{V_{\text{ap}}}{\sqrt{L/C}} \tag{6.7}$$

In this case, where $L = (L_1 + L_3)$, $C = C_3$:

$$\sqrt{(L_1 + L_3)/C_3} = \frac{V_{\text{ap}}}{I_{\text{pp}}}$$

or

$$\sqrt{(L_1 + L_3)/C_3} = \frac{0.8V_{\text{dc(min)}}}{0.8\pi I_{o[\text{dc(min)}]}(N_s/N_p)}$$
$$= \frac{V_{\text{dc(min)}}(N_p/N_s)}{\pi I_{o(\text{dc})}} \tag{6.8}$$

The resonating elements $(L_3 + L_1) = (L_3 + L_2)$ and C_3 are fixed for specified values of $V_{\text{dc(min)}}$ and maximum output current $I_{o(\text{dc})}$ using Eq. 6.1, which gives their product, and Eq. 6.8, which gives their ratio. The transformer turns ratio is fixed from Eq. 6.3.

The ratio of L_3/L_1 is chosen to minimize "off"-voltage stress on the SCRs. A smaller ratio produces less "off" stress and less dV/dt stress. The precise ratio is best determined empirically. Inductance L_3 comprises the transformer primary leakage inductance plus some external inductance. It is best not to rely on the leakage inductance alone as it varies widely, and that would result in large variability in the resonant period.

6.4.3 Design Example—Series-Loaded SCR Half-Bridge Resonant Converter

The preceding relations will now be used in a design example. Assume the circuit of Figure 6.10 with the following specifications:

Output power	2000 W
Output voltage	48 V
Output current $I_{o(dc)}$	41.7 A
Input DC voltage, nominal	310 V
Input DC voltage, maximum	370 V
Input DC voltage, minimum	270 V

Then from Eq. 6.3 $N_p/N_s = 0.6 \times 270/2(48+2) = 1.62$ and from Eq. 6.1

$$2\pi\sqrt{(L_3+L_1)C_3} = 20 \times 10^{-6}$$

or

$$\sqrt{(L_3+L_1)C_3} = 3.18 \times 10^{-6} \tag{6.1a}$$

From Eq. 6.8

$$\begin{aligned}\sqrt{\frac{(L_3+L_1)}{C_3}} &= \frac{V_{dc(min)}(N_p/N_s)}{\pi I_{o(dc)}} \\ &= \frac{270 \times 1.62}{\pi \times 41.7} \\ &= 3.34\end{aligned} \tag{6.8a}$$

From Eqs. 6.1a and 6.8a $C_3 = 0.95\ \mu$F and $L3 + L1 = 10.6\ \mu$H and from Eqs. 6.6 and 6.7

$$\begin{aligned}I_{pp} &= \frac{V_{ap}}{\sqrt{(L_3+L_1)/C_3}} \\ &= \frac{0.8 \times 270}{3.34} = 64.7A\end{aligned}$$

Since the maximum duty cycle of this peak SCR current is $t_r/2t_{t(\min)} = 0.25$ (Figure 6.13*a* and 6.13*c*), the SCR RMS current is $I_{\mathrm{rms(SCR)}} = 64.7 \times \sqrt{0.25}/\sqrt{2} = 22.9$ A. This is well within the maximum 40-A RMS capability of the Marconi ACR25U. Also, as assumed above, the anti-parallel diode peak current is one-fourth of the SCR current or $34.7/4 = 16.2$ A. With a 1.62 turns ratio, the peak rectifier diode currents will be 104.8 and 26.2 A, corresponding to the peak SCR and anti-parallel diode currents. With such high rectifier diode currents, a full-wave rectifier with one series diode rather than a bridge with two diodes is preferable.

6.4.4 Shunt-Loaded SCR Half-Bridge Resonant Converter[6,12]

The ancestor of most practical SCR resonant bridge power supplies is a shunt-loaded resonant half bridge used as a DC to AC inverter, which is described in Neville Mapham's classic paper.[6] The circuit was essentially that of Figure 6.12, with a resistive load at the transformer secondary instead of the rectifier and output filter.

Mapham's paper described no attempt to regulate the output, but showed that over a 10/1 range of output load currents with no feedback, the output AC voltage was constant to within 1% and generated a relatively distortion-free sine wave. A detailed and elegant computer analysis showed that the peak sine-wave output could be regulated against input line changes by changing the SCR triggering frequency.

The computer analysis is presented in terms of normalized relations $R_o/\sqrt{L/C}, I_{\mathrm{scr}}\sqrt{L/C}, I_{\mathrm{diode}}\sqrt{L/C}/E, V_o/E$, which permits a simplified design of the circuit with a resistance loaded secondary. Despite the LC filter at the output, it provides a good guide to that circuit.

In the shunt-loaded configuration of Figure 6.12, output regulation is achieved by varying the SCR triggering frequency. Output voltage waveshape at the cathodes of the rectifier in Figure 6.12 is similar to the current waveshape of Figures 6.13 and 6.14 for the series-loaded circuit. The output *LC* filter averages the voltage waveshape for the shunt-loaded circuit, whereas the output capacitor alone is needed to average the current waveshape for the series-loaded case.

Since the shunt-loaded circuit requires an output inductor, it has no advantage over the series-loaded configuration. The shunt circuit has only been touched on here to bring attention to the Mapham paper with its useful computer analysis, which is basic to a full understanding of the series-loaded circuit. The shunt-loaded circuit has been used with MOSFETs as a much higher frequency DC/DC converter.[13]

6.4.5 Single-Ended SCR Resonant Converter Topology Design[3,5]

The single-ended SCR resonant converter (Figure 6.8) was discussed in qualitative terms in Section 6.3. A more quantitative analysis containing some simplifying approximations is presented here, but it is not accurate enough to permit a workable design. A more rigorous discussion would require computer analysis of the circuit.

The circuit is shown again in Figure 6.15 in a more useful configuration, with the load coupled to the series resonant circuit through

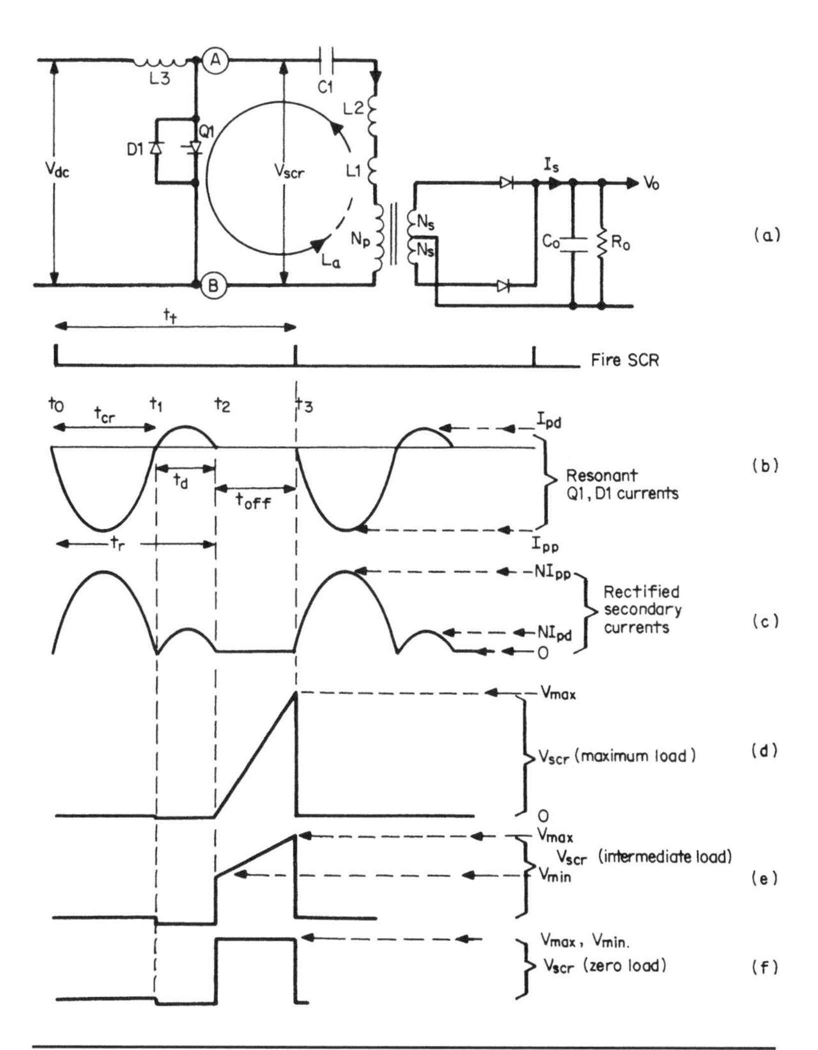

FIGURE 6.15 Single-ended SCR resonant converter.

an isolating transformer. The circuit is a series-loaded circuit with the secondary load reflected into the primary in series with the resonating elements (C_1 and $L_1 + L_2$). The resonating inductance is the leakage inductance L_1 of $T1$ plus another discrete inductance L_2 to make up the total inductance required for the desired resonant period. If L_2 is relatively large compared to L_1, the unavoidable variation in L_1 is not significant.

Although in series with the resonating elements, the $T1$ magnetizing inductance does not affect the resonant period or circuit operation in most circumstances, as it is shunted by the secondary load resistance reflected into the primary. With an open-circuited secondary, however, the high impedance of the magnetizing inductance can kill the Q of the resonant circuit and prevent SCR turn "off." By gapping the $T1$ core, the circuit can operate successfully with large values of R_o.

Recall the basic circuit operation. Assume that the left-hand side of $C1$ has been charged up to some DC voltage V_{max} higher than V_{dc} (as in the boost regulator of Section 1.4.1). When $Q1$ is fired with a narrow trigger pulse, $Q1$ turns "on" and will stay "on" until its current falls to zero. With $Q1$ "on," a voltage step of amplitude V_{max} is applied to all the series elements to the right of $Q1$. This causes a half sine wave of current of duration t_{cr} to flow through $Q1$ in loop L_a. When the $Q1$ primary current falls to zero at t, it reverses and continues flowing as a half sine wave of duration t_d through $D1$.

At low loads, the durations of the $Q1$ and $D1$ half sine waves (t_{cr} and t_d) are close to $\pi\sqrt{(L_1 + L_2)C_1}$. At higher loads, t_{cr} increases somewhat and t_d decreases. If t_d at its minimum is longer than the t_q time of the SCR, the SCR will automatically extinguish, and can safely sustain forward voltage without falsely refiring. During the "off" time $t_{off} = (t_3 - t_2)$, current in $L3$, which is chosen at least 20 times $(L_1 + L_2)$, charges $C1$ back to its original voltage V_{max}, and $Q1$ can be fired again. The cycle repeats with a triggering period

$$t_t = t_r + t_{off} = (2\pi\sqrt{(L_1 + L_2)C_1}) + \mathrm{toff}$$

The SCR and diode primary currents produce similarly shaped sinusoids in the secondary, $N(= N_s/N_p)$ times as large. They are rectified and summed by rectifier diodes $D3$, $D4$ as shown in Figure 6.15*c*. The average of these secondary current pulses is produced by C_o without an output inductor, and is equal to the DC output current.

The output voltage is regulated by varying the triggering period to maintain a constant average output current. As the DC input voltage or output resistance R_o increase, t_t is increased by a feedback loop to maintain constant output voltage. As the input voltage or output resistance decrease, t_t is decreased.

6.4.5.1 Minimum Trigger Period Selection

The peak SCR and diode currents (I_{pcr}, I_{pd}) of Figure 6.15*b* are determined by the peak voltage to which the resonating capacitor $C1$ is charged. These peak currents must be known in order to fix the triggering period and hence the average output current and voltage.

SCR voltage V_{scr} waveshapes are shown in Figure 6.15*d*–*f* for maximum, intermediate, and zero output power. At the instant of an SCR trigger pulse, $C1$ has been charged up to a voltage V_{max} and hence has stored energy $C1\,(V_{max})^2/2$. At t_2, some of this energy has been delivered to the load by the SCR and diode half sine waves.

At intermediate powers, there still is some energy left in the capacitor. When the diode current has fallen back to zero at t_2, the left side of $C1$ is unclamped and current from $L3$ starts charging $C1$ up. With some voltage still across $C1$, V_{scr} first steps up by the amount of that voltage V_{min} and then starts rising more slowly as shown in Figure 6.15*e*.

At maximum load, all the stored energy has been delivered to the load by the end of an "off" time at t_0, and hence at the start of the next "off" time at t_3, there is no remaining voltage on $C1$ and hence no front-end step as shown in Figure 6.15*d*.

At zero DC load, all the energy stored in $C1$ at the end of an "off" time still remains at the start of the next "off" time. Hence V_{max} is equal to V_{min} as in Figure 6.15*f*. From the waveshapes in Figure 6.15*d* to 6.15*f*, the minimum triggering period t_t can be established. Since the inductor $L3$ cannot support a DC voltage, the average voltage at its output end must equal that at its input end, which is V_{dc}. Thus

$$V_{dc} = \frac{V_{max} + V_{min}}{2}\frac{t_{off}}{t_t}$$

or

$$V_{max} + V_{min} = \frac{2V_{dc}t_t}{t_{off}} \tag{6.9}$$

And from Figure 6.15*d*, at maximum load $V_{min} = 0$ and

$$V_{max} = \frac{2V_{dc}t_t}{t_{off}}$$

From Figure 6.15*b*, $t_{off} = t_t - t_r$, so

$$V_{max} = \frac{2V_{dc}}{1 - t_r/t_t} \tag{6.10}$$

From Eq. 6.10, V_{max} is calculated for various ratios of t_r/t_t for an off-line converter. Maximum power occurs at minimum AC line input and maximum DC output current, since those are the conditions for

t_r/t_t	V_{max} (at $V_{dc} = 138$ V)	V_{max}, V
0.7	6.6 V_{dc}	911
0.6	5.0 V_{dc}	690
0.5	4.0 V_{dc}	552
0.4	3.3 V_{dc}	455
0.3	2.9 V_{dc}	393

TABLE 6.1

maximum SCR current. Assume that nominal and minimum line inputs are 115 and 98 V AC, respectively, giving approximately 160 and 138 V of rectified DC. For the minimum V_{dc} of 138 V from Eq. 6.10, we can construct Table 6.1.

From Table 6.1, a good compromise choice for t_r/t_t is 0.6. This still permits using a reasonably inexpensive 800-V SCR. For $t_r = 16$ μs, the resonant half period is 8 μs, which allows the "on"-turning SCR to spend most of its "on" time at a low anode-to-cathode voltage (Figure 6.5). Then t_t, the minimum trigger period, is 26.6 μs (maximum trigger frequency is 38 kHz). In regulating down to lower output power, the minimum trigger frequency will be about one-third of that or about 13 kHz—not too far down into the audible range.

6.4.5.2 Peak SCR Current Choice and *LC* Component Selection

The peak SCR and diode *D*1 currents are shown in Figure 6.15*b*. Like the half bridge (Section 6.4.2), assume that the peak diode current is one-fourth the SCR current. Rectified secondary currents are as shown in Figure 6.15*c*. The average of those currents for a *T*1 turns ratio *N* is

$$\begin{aligned} I_{s(av)} &= \frac{2I_{pp}N}{\pi}\frac{t_r}{2t_t} + \frac{2I_{pp}N}{4\pi}\frac{t_r}{2t_t} \\ &= 1.25 I_{pp} N \frac{t_r}{\pi t_t} \end{aligned} \tag{6.11}$$

in which I_{pp} is the peak primary current (Figure 6.15*b*). This is equal to the average or DC output current at minimum line input and maximum current output. The output voltage at minimum output load resistance R_o is

$$V_o = 1.25 N I_{pp} R_o \frac{t_r}{\pi t_t} \tag{6.12}$$

The transformer turns ratio must be chosen and the magnitudes of the resonant *LC* components selected to yield the peak resonant currents given above.

As in Section 6.4.2 per the suggestion in the Chambers paper,[9] select the $T1$ primary voltage to be 60% of the voltage across the entire loop (points A to B in Figure 6.15*a*). This is the $0.6V_{max}$ of Eq. 6.10, which is clamped through the turns ratio N against the output voltage plus one diode rectifier drop. Then

$$N = \frac{0.6V_{max}}{V_o + 1} \tag{6.13}$$

When the SCR is fired, the voltage applied to the series resonant circuit elements is $V_{ap} = V_{max} + 0.6V_{max} = 1.6V_{max}$ and the peak amplitude of the half-sine-wave SCR current pulse is

$$\begin{aligned} I_{pp} &= \frac{V_{ap}}{\sqrt{(L_1 + L_2)/C_1}} \\ &= \frac{V_{max}}{\sqrt{(L_1 + L_2)/C_1}} \end{aligned} \tag{6.14}$$

Since maximum DC output current in Eq. 6.11 is specified, and all other terms in that relation are known (Eq. 6.13 and Table 6.1), Eq. 6.14 gives the ratio $(L_1 + L_2)/C1$. Also since the resonant period t_r was chosen in Section 6.4.5.1 as 16 μs, we obtain

$$t_r = 2\pi\sqrt{(L_1 + L_2)C_1} = 16 \times 10^{-6} \tag{6.15}$$

Between Eqs. 6.14 and 6.15, there are two unknowns and two equations and so both $C1$ and $(L_1 + L_2)$ are determined.

6.4.5.3 Design Example

Design a single-ended SCR resonant converter with the following specifications:

Output power	1000 W
Output voltage	48 V
Output current	20.8 A
AC input, nominal	115 V AC RMS
AC input, minimum	98 V AC RMS
Rectified DC, nominal	160 V
Rectified DC, minimum	138 V

From Table 6.1, chose $t_r/t_t = 0.6$, giving $V_{max} = 690$ V. From Eq. 6.13

$$\begin{aligned} N &= \frac{0.6V_{max}}{V_o + 1} \\ &= \frac{0.6 \times 690}{49} \\ &= 8.44 \end{aligned}$$

From Eq. 6.11

$$I_{s(\mathrm{av})} = I_{o(\mathrm{dc})} = 20.8 = 1.25 I_{\mathrm{pp}} N \frac{t_r}{\pi t_t}$$
$$= 1.25 I_{\mathrm{pp}} \times 8.44 \left(\frac{0.6}{\pi}\right)$$

or $I_{\mathrm{pp}} = 10.3$ A. Then, from Eq. 6.14

$$I_{\mathrm{pp}} = 10.3 = \frac{1.6 V_{\max}}{\sqrt{L_1 + L_2)/C_1}}$$
$$= \frac{1.6 \times 690}{\sqrt{(L_1 + L_2)/C_1}}$$

or

$$\sqrt{\frac{(L_1 + L_2)}{C_1}} = 107.8 \tag{6.14a}$$

From Eq. 6.15

$$t_r = 16 \times 10^{-6} = 2\pi \sqrt{(L_1 + L_2) C_1}$$

or

$$\sqrt{(L_1 + L_2) C_1} = 2.55 \times 10^{-6} \tag{6.15a}$$

and from Eqs. 6.14*a* and 6.15a, $C_1 = 0.024\ \mu\mathrm{F}$, $L_1 + L_2 = 275\ \mu\mathrm{H}$.

A lower value for t_r/t_t would yield lower maximum SCR voltage stress (Table 6.1) and possibly greater reliability. This would result in larger t_t (lower trigger frequency), and at low output power, the resulting t_t (Eq. 6.12) would bring the trigger frequency down far into the audible frequency range.

6.5 Cuk Converter Topology—Introduction[14-16]

In its specialized area of application, this is a very imaginative and valuable topology. Its major advantage is that both input and output ripple currents are continuous, i.e., there is no time gap where the ripple current falls to zero. In contrast, the buck regulator of Figure 1.4 has continuous output current if L_o is made sufficiently large (Figure 1.4*f*), but input current is discontinuous (Figure 1.4*d*). In the boost regulator (Figures 1.10 and 1.11), input current is continuous (Figure 1.11, I_{L1}) but the output current through the rectifier diode is discontinuous (Figure 1.11, I_{D1}).

In applications where very low input and output noise are essential, it is important to have the input and output ripple currents ramp up and down without switching to zero as in Figure 1.4*f*. At the inputs of most topologies (forward converters, push-pulls, bucks, flybacks, and bridges), this is usually done by adding an RFI input filter, but this adds cost and space.

6.5.1 Cuk Converter–Basic Operation

Both input and output current are continuous in the Cuk converter (Figure 6.16*a*) as seen in Figure 16.6*e* and 6.16*f*. By making $L1$ and $L2$ sufficiently large, the amplitude of the current ramps can be made extremely small. As discussed below, by winding $L1$ and $L2$ on the same core, the ripple amplitude can be reduced to zero.

The circuit in its basic form is shown in Figure 6.16*a* with common input and output DC return. Input and output returns can be DC isolated by adding a transformer, as will be discussed below.

In its basic non-isolated version, the circuit works as follows (see Figure 6.16*a*). When $Q1$ turns "on," $V1$ goes steeply negative to approximately zero volts. Since the voltage across a capacitor cannot change instantaneously, $V2$ goes negative an equal amount, reverse-biasing $D1$ as seen in Figure 6.16*c* and 6.16*d*. With V_{dc} across $L1$, its current ramps up linearly, adding to its stored energy. Before $Q1$ turned "on," the left-hand end of $C1$ was charged up to voltage V_p, and its right-hand end was clamped to common via D1. The stored energy in C1 was $C_1 V_p^2/2$.

When $Q1$ turns "on," $C1$ acts a battery delivering current down through $Q1$, up through R_o, and back through $L2$ into the right-hand end of $C1$. Thus the stored energy in $C1$ and in $L2$ delivers power into R_o, and charges the output filter capacitor to a negative voltage $-V_o$.

When $Q1$ turns "off," $V1$ goes positive to some voltage V_p, and V_2 follows it up but gets clamped to common by $D1$ as discussed above. Now with the left-hand end of $L2$ at common and the right-hand end at –Vo, the current in $L2$ flows down through $D1$, up through Ro, and back into the right-hand end of $L2$.

When Q1 is "on," current in $L1$ ramps up at a rate $dI/dt = V_{dc}/L1$ (Figure 6.16*e*). Since $V2$ has gone down by the same amount as $V1$ (V_p), the left-hand end of $L2$ is at $-V_p$ and current in $L2$ ramps up at a rate $dI/dt = (V_o - V_p)/L2$ as seen in Figure 6.16*f*.

When $Q1$ is "off," $V1$ has risen to a voltage V_p that is higher than V_{dc}, and current in $L1$ ramps downward at a rate $dI/dt = (V_p - V_{dc})/L1$ (Figure 6.16*e*).

With $Q1$ "off," the left-hand end of $L2$ is clamped to common through $D1$ and with its right-hand end at $-V_o$, current in $L2$ ramps downward at a rate $di/dt = V_o/L2$ as in Figure 6.16*f*. If $L1$, $L2$ are

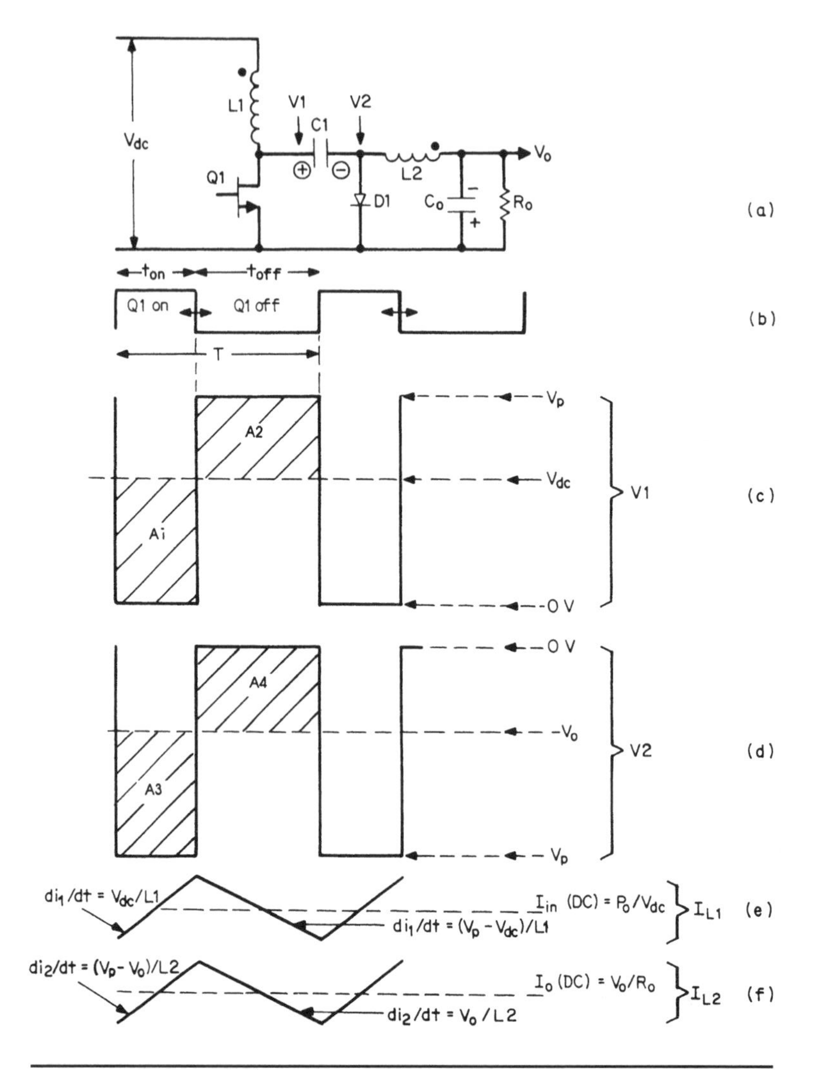

FIGURE 6.16 Basic Cuk converter with input and output not isolated, and significant voltage and currents.

made large enough, these ramp currents average to a non-zero DC level.

6.5.2 Relation Between Output and Input Voltages, and *Q*1 "on" Time

Since *L*1 has close to zero DC resistance, it cannot support a DC voltage. Hence the voltage at its bottom end (*V*1), averaged over one cycle, must equal the DC voltage at its top end (V_{dc}). This is equivalent to stating that in Figure 6.16*c*, the area *A*1 in volt-seconds is equal to the area *A*2. Thus

$$V_p \frac{t_{off}}{T} = V_{dc} \tag{6.16a}$$

or

$$V_p = V_{dc} \frac{T}{t_{off}} \tag{6.16b}$$

Since the voltage change at *V*2 equals the voltage change at *V*1, the bottom end of the *V*2 voltage is at $-V_p$ during t_{on}. During t_{off}, the top end of *V*2 is clamped close to 0 *V* by diode *D*1.

Similarly, since *L*2 cannot support a DC voltage, the average over one cycle at its left-hand end (*V*2) must equal the DC voltage at its right-hand end ($-V_o$). This is equivalent to stating that in Figure 6.16*d*, the area *A*3 in volt-seconds is equal to the area *A*4. Thus

$$V_p \frac{t_{off}}{T} = V_o \tag{6.17a}$$

or

$$V_p = V_o \frac{T}{t_{on}} \tag{6.17b}$$

Equating relations 6.16*b* and 6.17*b* yields

$$V_o = \frac{V_{dc} t_{on}}{t_{off}} \tag{6.18}$$

It is seen from Eq. 6.18 that the magnitude of the DC output voltage can be less than, equal to, or greater than the DC input voltage depending on the ratio t_{on}/t_{off}.

6.5.3 Rates of Change of Current in *L*1, *L*2

It is interesting to note that for $L_1 = L_2$, the upslopes of current in *L*1 and *L*2 during t_{on} are equal and their downslopes during t_{off} are also equal. This is shown in Figure 6.16*e* and 6.16*f*. It is this fact that makes it possible to reduce current ripple at the input completely to zero and makes this Cuk converter an ultra-low-noise circuit.

Equality of upslope and downslope currents can be shown as follows. During t_{on}, the upslope of *L*1 current is

$$\frac{+di_1}{dt} = \frac{V_{dc}}{L_1}$$

and again during t_{on}, the upslope of $L2$ current is

$$\frac{+di_2}{dt} = \frac{V_p - V_o}{L_2}$$

From Eqs. 6.16*a* and 6.18

$$\frac{+di_2}{dt} = \frac{1}{L_2}\left(\frac{V_{dc}T}{t_{off}} - \frac{V_{dc}t_{on}}{t_{off}}\right)$$

$$= \frac{V_{dc}}{L_2 t_{off}}(T - t_{on})$$

$$= \frac{V_{dc}}{L_2}$$

and for $L_1 = L_2$ the upslopes of current in the inductors are equal during t_{on}.

Now consider the downslopes of current during t_{off}. In $L2$, $-di_2/dt = V_o/L2$, and from Eq. 6.18

$$\frac{-di_2}{dt} = \frac{V_{dc}}{L_2}\frac{t_{on}}{t_{off}} \tag{6.19a}$$

and in $L1$, $-di_1/dt = (V_p - V_{dc})L_1$. But from Eq. 6.16*b*

$$\frac{-di_1}{dt} = \frac{(V_{dc}T/t_{off}) - V_{dc}}{L_2}$$

$$= \frac{V_{dc}}{L_2}\frac{T - t_{off}}{t_{off}}$$

$$= \frac{V_{dc}}{L_2}\frac{t_{on}}{t_{off}} \tag{6.19b}$$

Thus for $L1 = L2$, the downslopes of the inductor currents during t_{off} are equal, and of magnitude given by Eq. 6.19*a* and 6.19*b*. This is seen in Figure 6.16*e* and 6.16*f*.

6.5.4 Reducing Input Ripple Currents to Zero

The current ramps at the input and output can be reduced to zero yielding pure DC in those lines, if $L1$ and $L2$ are wound on the same core with the polarities as shown in Figure 6.17.

If $L1$, $L2$ are wound on the same core as in Figure 6.17, the assembly is a transformer. During the "on" time, current flows from V_{dc} into the dot end of $L1$ in an upgoing ramp as seen in Figure 6.16*e*. But in $L2$, its dot end is positive with respect to its no-dot end, and by transformer

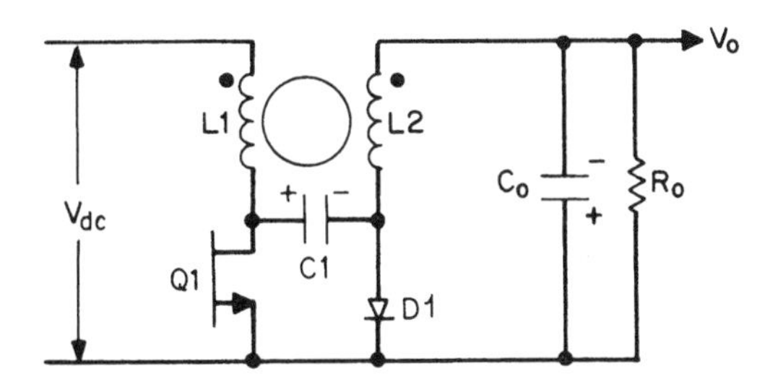

FIGURE 6.17 If $L1$, $L2$ are wound on the same core, input and output ripple currents are very close to zero. The ramp currents into the primary are bucked out by an equal and oppositely directed ramp rate reflected into the primary from the secondary. A similar cancellation occurs in the secondary.

action, it forces a voltage across $L1$, making its dot end positive with respect to its no-dot end. This forces a "secondary" current from the dot end of $L1$ back into V_{dc}. This current is a positive-going ramp of the same upslope as the upslope of current from V_{dc}.

Since these two upslopes are of equal magnitude (as demonstrated in Section 6.5.3) but with the currents flowing in opposite directions, the net current change during t_{on} is zero; i.e., current flow in V_{dc} is pure DC. That current is $P_o E / V_{dc}$, where P_o is the output power and E is the efficiency.

A similar line of reasoning demonstrates that if the coupling between $L1$ and $L2$ is 100%, the current in the load is also pure DC with no ripple component. That current is V_o / R_o.

Further, since it has been shown that the current downslopes in $L1$ and $L2$ during t_{off} are equal, similar reasoning indicates that there are also no ripple currents in the input or output lines during t_{off}.

6.5.5 Isolated Outputs in the Cuk Converter

In most instances, output returns must be DC-isolated from input returns. This can be done with the addition of a 1/1 isolating transformer as shown in Figure 6.18. Output voltage is still determined by the ratio t_{on}/t_{off}, and the output polarity can be either positive or negative, depending on which end of the secondary circuit is common.

Thus, although the Cuk converter of Figure 6.18 is a very clever way of producing pure DC input and output currents, the requirement for two pieces of magnetics (the $L1$,$L2$ inductor core and $T1$) is a high price to pay for the advantage.

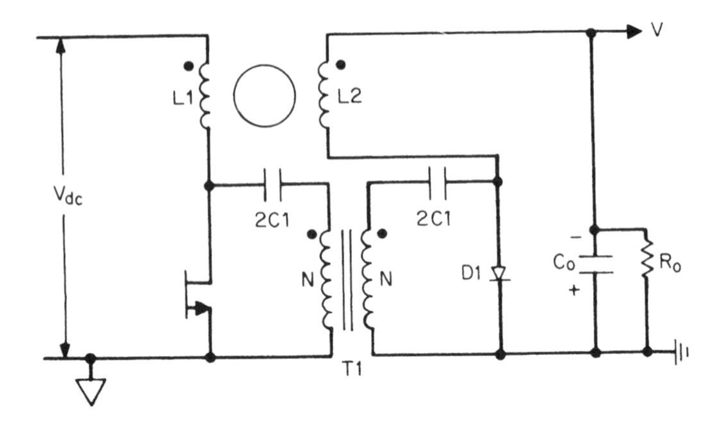

FIGURE 6.18 By adding a 1/1 isolating transformer, output return is DC-isolated from input return. Either a positive or negative output is now possible depending on which end of the secondary is common.

6.6 Low Output Power "Housekeeping" or "Auxiliary" Topologies—Introduction[15-17]

These are not strictly "topologies" having a broad range of uses; rather, they are specialized circuits for unique applications. Since they generate an output voltage that serves a vital function in any switching power supply design, they are discussed here as separate topologies.

All the topologies discussed thus far require a low power (1 to 3 W) supply of about 10- to 45-V output. It is used to feed the usual pulse-width-modulating (PWM) chip for the main "power train," and power the logic and sensing circuits that perform various housekeeping functions. Such housekeeping functions may include overcurrent sensing, overvoltage sensing and protection, remote signaling, and correction of turn "on" and turn "off" sequencing for each output in a multi-output supply.

These housekeeping supplies need not always be regulated, since the usual loads can tolerate a relatively large range of supply voltage (± 15% maximum). But reliability is improved, and more predictable operation of the main power train results, if the housekeeping supply is regulated—usually ±2% is adequate. The important objectives for these supplies is that they be low in parts count and cost, and occupy only a small fraction of the space that the main power train occupies, with all its outputs.

6.6.1 Housekeeping Power Supply—on Output or Input Common?

In any new design, an initial major decision must be made as to whether the housekeeping power supply with the PWM chip that it powers should be located on output common, which in most cases is DC-isolated from input common. The main switching power transistors are located on input common—one end of the rectified AC line in off-line converters, or at one end of the DC prime power source in battery-operated DC/DC converters.

To regulate the output voltage, a DC error amplifier must be located on output common to sense output voltage, compare it to a reference voltage, and produce an amplified error voltage. This error voltage is the difference between the reference and a fraction of the output voltage. The amplified error voltage is then used to control width of the pulses that drive the main power transistor or transistors, which are located on input common. A typical example of this is shown in Figure 2.1.

Since output and input common are DC-isolated, and may be tens or hundreds of volts apart, the width-modulated pulse cannot be DC-coupled to the power transistor.

Thus if the error amplifier and pulse-width modulator are on output common (usually in a PWM chip), the width-modulated pulse is transferred across the output–input barrier, often by a pulse transformer. It is the function of the housekeeping supply, whose input power comes from the prime power source at input common, to produce the usual 10 to 15 V for the 1 to 3 W of housekeeping power referenced to output common.

Such a housekeeping supply is also often used when the PWM chip is located on input common. Although power for the chip can be derived from an auxiliary winding on the main transformer when the main power transistor is being driven, if the drive is shut down, (e.g., for overvoltage or overcurrent reasons), that power goes away and it is no longer possible to energize remote indicators. Further, on shutdown, as voltage from the auxiliary winding goes away, supply voltage for the PWM chip decays. A race condition leading to excessive pulse width can occur and may cause failures.

In general, it is far more reliable to have a housekeeping supply that is always present, instead of deriving ("bootstrapping") it from an auxiliary winding on the main power transformer.

An alternative method of transmitting a measure of the output voltage across the output–input barrier is to width-modulate the power transistor on input common via an optical coupler. This also requires a housekeeping supply on input common if bootstrapping from an auxiliary winding on the main power transformer is considered undesirable. Some schemes for implementing these housekeeping supplies are discussed below.

6.6.2 Housekeeping Supply Alternatives

In an attempt to minimize the parts count of housekeeping supplies, many designers have resorted to single- or two-transistor transformer coupled, self-oscillating circuits. This saves the space and cost of a PWM chip or a stable multivibrator for generating the required AC drive frequency to some kind of a driven converter. Further, by using a transformer coupled feedback oscillator, adding a separate winding on the transformer provides output power referenced to any desired DC voltage. The feedback from a collector to a base winding that keeps the circuit oscillating also provides sufficient drive to deliver the required power from the output winding.

Such self-oscillating housekeeping converters are very well covered in Keith Billings' handbook.[16]

These self-oscillating converters appear at first glance to be very attractive because of their simple circuits and low parts counts. Without additional circuitry, most do not produce regulated DC output voltage and have various other shortcomings. Adding circuitry to regulate the output and overcome any other shortcomings increases internal dissipation, parts count, and complexity.

At some point it becomes debatable whether these self-oscillating converters are a better choice than a conventional single-transistor, low-power, low-component-count driven converter such as a flyback, fed from its own PWM chip.

With the increasingly lower price of PWM chips, no need for an output inductor, and regulating output voltage by sensing a slave winding voltage on input common, the flyback is a viable alternative to a self-oscillating circuit.

Nevertheless, two of the most frequently considered self-oscillating types—the Royer and Jensen oscillating converters—are considered below and compared to a simple flyback.

6.6.3 Specific Housekeeping Supply Block Diagrams

Figures 6.19 to 6.21 show the block diagrams of three reasonable approaches to a housekeeping supply where the error amplifier and pulse width modulator are in a PWM chip on output common.

6.6.3.1 Housekeeping Supply for AC Prime Power

Figure 6.19 shows the simplest and most frequently used scheme when the prime power is AC. A small (usually 2- to 6-W) 50/60-Hz

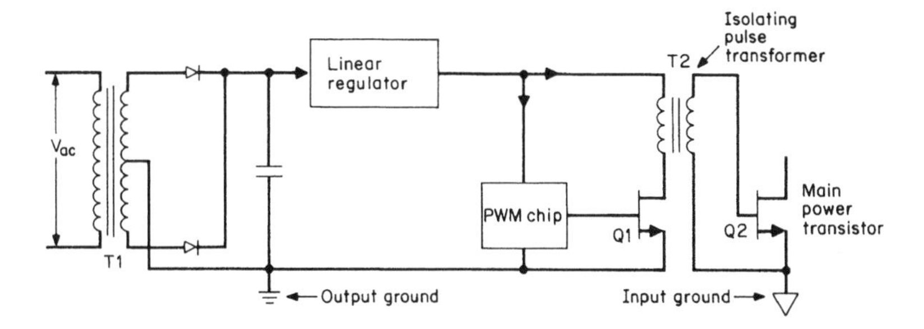

FIGURE 6.19 For an off-line converter, the simplest housekeeping supply is a small (2-W) 60-Hz isolating transformer with its secondary generating a rectified 15 V, referenced to output common. This is followed by an inexpensive linear regulator (12-V output) that is referenced to output.

transformer is powered from the AC input and has its secondary referenced to output common. Such transformers are available from a large number of manufacturers and have tapped primaries so that they can be fed from either 115 or 220 V AC—either 50 or 60 Hz.

Typical sizes are 1.88 × 1.56 × 0.85 in for a 6VA unit or 1.88 × 1.56 × 0.65 in for a 2VA unit. They come with a large range of standard secondary voltages and usually have two secondaries that can be wired in series for a full-wave center-tapped output rectifier or in parallel for a bridge output rectifier. The secondary is rectified and filtered with a capacitor input filter, and the rectified DC is returned to output or input common as desired.

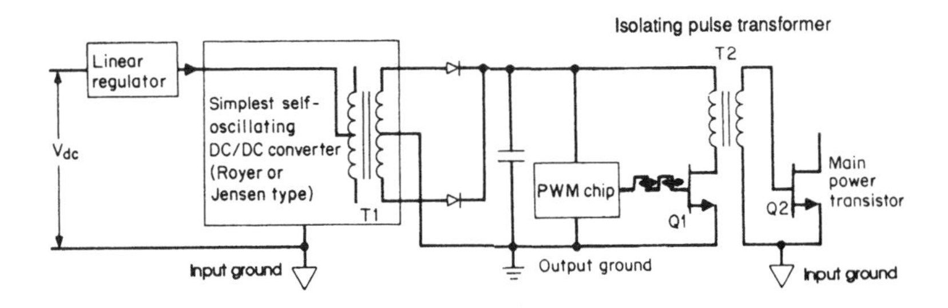

FIGURE 6.20 A housekeeping supply for DC input voltage. When AC voltage is not present to provide a rectified DC at output common, a simple magnetically coupled feedback oscillator fed from the DC at input common is used as a DC/DC converter to provide output voltage to a PWM chip on output common. The housekeeping supply output is proportional to the input voltage. A regulator may be unnecessary if the main PWM chip can tolerate ±10% input voltage variation.

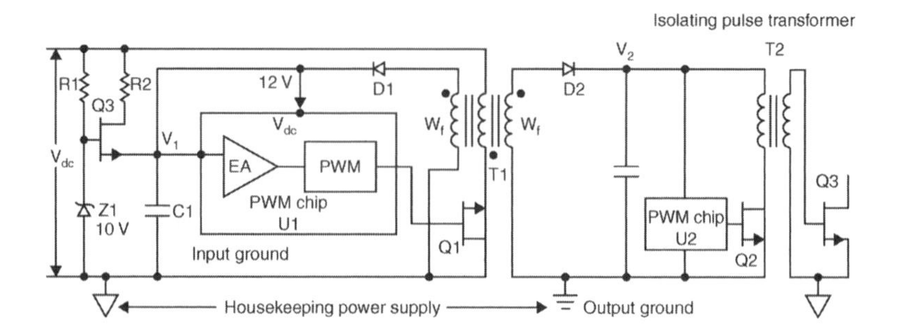

FIGURE 6.21 A minimum-parts-count flyback as a housekeeping supply. Although this approach may have more parts than Figure 6.20, it generates a regulated output *V*2 without requiring any linear regulators, and at the 2- to 3-W power level, may require less input power than that shown in Figures 6.19 and 6.20.

The rectified DC will have the same tolerance as the AC input—usually ±10%. Since most PWM chips can accept DC inputs ranging from 8 to 40 V, it is not essential to regulate the output. But in general, safer and more predictable performance results if the output is regulated. Thus, the transformer secondary voltage frequently is chosen to yield a rectified DC voltage of about 3 V above the desired regulated DC voltage and an inexpensive integrated-circuit linear regulator in a TO-220 package is added in series after the filter capacitor as shown in Figure 6.19. The configuration is usually designed to yield a regulated ± 12-V output and achieves an efficiency of about 55% at 3-W output at a 10% high-line input.

6.6.3.2 Oscillator-Type Housekeeping Supply for AC Prime Power

When the prime input power is DC, there is no AC voltage easily available to produce a rectified DC at output common. Figure 6.20 shows a configuration often used in this case.[17] A simple magnetically coupled feedback oscillator fed from the DC input produces high-frequency square-wave output in a secondary referenced to output common.

The secondary is rectified and filtered with a capacitor input filter, and the resultant DC is returned to output common. Because of the high-frequency square wave, the filter capacitor after secondary rectification is far smaller than for the 60-Hz rectifier-filter of Figure 6.19.

The rectified DC output in such a scheme is most often proportional to DC input voltage. Thus, if DC output voltage variation of about ±10% is acceptable, this is a very efficient and low-component-count scheme if an efficient oscillator is available. There are a number of oscillator configurations that can achieve efficiencies of 75 to 80% at an output power level of 3 W.[16] One particularly useful oscillator, the Royer circuit, will be discussed below.

If regulated DC output is required, the oscillator is preceded by an integrated-circuit linear regulator as shown in Figure 6.20. Adding the linear preregulator would drop the worst case total efficiency down to about 44%, assuming a telephone industry supply where the maximum DC input is 60 V and the linear preregulator drops that to 35 V—just below the minimum specification for a telephone industry supply.

By replacing the linear regulator of Figure 6.20 with a simple buck regulator at a small increase in parts count and cost, efficiency can be brought up to 70 and possibly 75% at the 3-W output power level.

6.6.3.3 Flyback-Type Housekeeping Supplies for DC Prime Power

With the added cost and component count needed to produce regulated DC in the scheme of Figure 6.20, simple self-oscillator schemes begin to lose their attractiveness. Figure 6.21 shows a third alternative for the housekeeping supply. It is a simply flyback driven from one of the many inexpensive current PWM chips. It is powered from the DC prime power at input common and a secondary winding *W*1 delivers DC voltage referenced to output common.

In Figure 6.21, the PWM chip is powered via emitter-follower *Q*3 at initial turn "on." The *Q*3 output voltage is one base-emitter voltage drop less than the 10-V zener diode *Z*1. The resulting 9-V output at the *Q*3 emitter is enough to power the PWM chip (*U*1) that commences driving flyback transistor *Q*1 and delivers, via *W*1, output power to the main PWM chip (*U*2) on output common.

As *U*1 is now powered via *Q*3, a bootstrap winding W_f on the flyback transformer starts generating output voltage. The number of turns on W_f is chosen to produce about 12 *V* at filter capacitor *C*1, which is higher than the 9 V at that point when it is being fed from the *Q*3 emitter. With the emitter of *Q*3 at 12 V and its base at 10 V, its base is reversed-biased and it turns "off." The auxiliary PWM chip on input common now continues to be powered from W_f, and the main PWM chip on output common continues to be powered via *W*1.

Resistor *R*2 is chosen small enough to deliver the current required by *U*1 during the initial turn "on" interval. Dissipation in R2 is negligible as that interval lasts only a few tens of microseconds. Resistor *R*1 carries current continuously but dissipates very little power as it carries only the *Q*3 base current, which needs to be only about 1 mA to supply the initial startup current of about 10 to 20 mA for *U*1.

The error amplifier in U1 senses its own bootstrapped supply voltage V_1 and keeps it constant as a "master." The output voltage V_2 from

$W1$ is a "slave" and is also quite constant (within 1 to 2%) as slaves track the master very well in a flyback topology.

6.6.4 Royer Oscillator Housekeeping Supply—Basic Operation[17,18]

This configuration is shown in Figure 6.22*a*. It was one of the earliest applications of transistors to power electronics and was conceived in 1955, only a few years after transistors were invented.[1]

It was used to generate square-wave AC and with rectification, as a DC/DC converter up to a power level of a few hundred watts. In its original form, it had two significant drawbacks that limited its usefulness and in some cases made it unreliable, but with the addition of three small changes and with more modern components, it has become a valuable circuit at power levels as low as 10 W and up to 300 W.

The original basic Royer oscillator shown in Figure 6.22*a* works as follows. It is a push-pull circuit with positive feedback from the collectors to the base windings to keep it oscillating. The positive feedback can be seen from the dots on the collector and base windings.

Assume that Q1 is "on" and is in saturation. The no-dot end of the primary N_{p1} is positive, and hence the no-dot end of the base winding N_{b1} is also positive. Voltage across N_{p1} is V_{dc} (assuming negligible V_{ce} drop). N_{p1} delivers output current to the load via N_{s1} and also enough current to the $Q1$ base via N_{b1} and $R1$ to keep $Q1$ "on" and in saturation at the maximum current reflected into the primary by the minimum R_o.

The $T1$ transformer core is made of material with a square hysteresis loop, as seen in Figure 6.22*b*. Assume that when $Q1$ turned "on" initially, the core was at point C on its hysteresis loop. With a voltage V_{dc} across N_{p1}, the rate of change of flux density in the core is given by Faraday's law as

$$\frac{dB}{dt} = \frac{Vdc \times 10^{+8}}{N_{p1} A_e} \tag{6.20}$$

The core moves up the hysteresis loop from negative saturation $-B_s$ to positive saturation $+B_s$ along the path CDE. The time required for this is given by Eq. 6.20 as

$$\begin{aligned} T1 = \frac{T}{2} &= \frac{dBN_{p1} A_e \times 10^{-8}}{V_{dc}} \\ &= \frac{2B_s N_{p1} A_e \times 10^{-8}}{V_{dc}} \end{aligned} \tag{6.21}$$

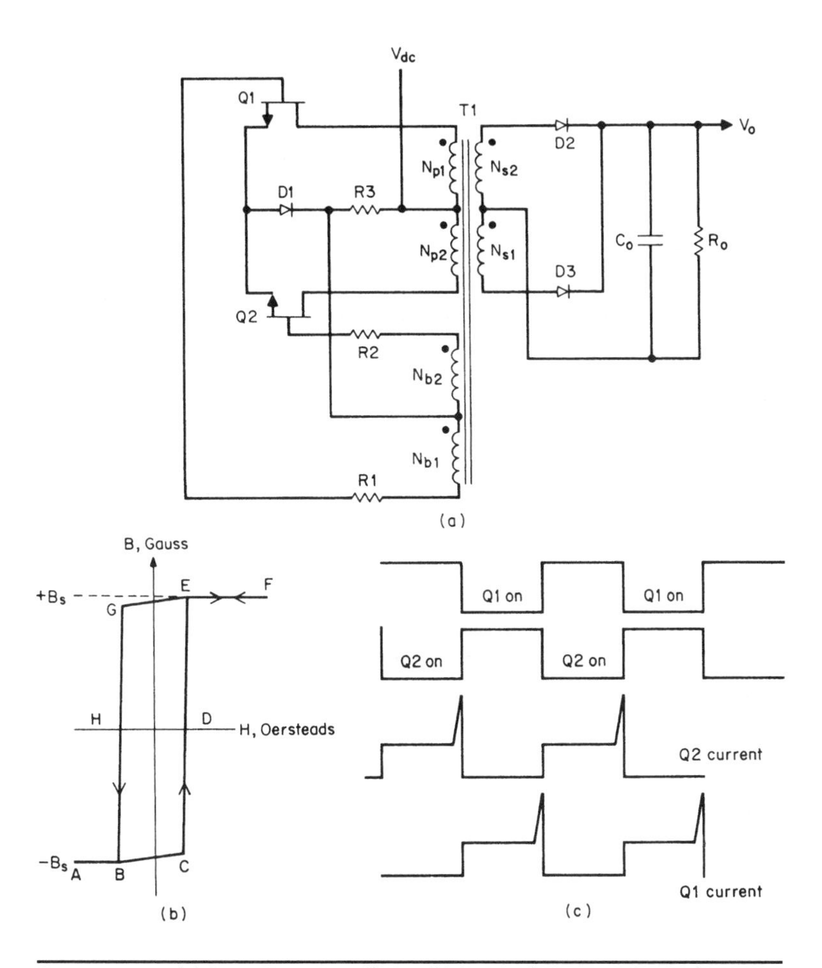

FIGURE 6.22 (*a*) Basic Royer oscillator. (*b*) Square hysteresis loop of *T*1 core. (*c*) Characteristic high current spikes at end of "on" time. These spikes are a major drawback in Royer oscillators. As long as the core is on the vertical part of its hysteresis loop, the positive feedback from N_p to N_b windings keeps a transistor "on" and in saturation. When the core has moved to either the top or the bottom of its hysteresis loop, coupling between the collector and base windings immediately drops to zero as the core permeability in such a square loop material is unity. The "on" transistor's base voltage and current drop to zero, and its collector voltage starts to rise. Some small residual air coupling couples this rising collector voltage into the opposite base, and by positive feedback, the opposite transistor turns "on." In one half period, the core moves along the path *CDEF*, then in the next half period along the path *FEGHBA*.

When the core has reached point E, it is saturated, its permeability is close to unity, and coupling between the $Q1$ collector and base windings suddenly drops to zero. The $Q1$ base current quickly drops to zero and $Q1$ collector voltage starts rising.

A small residual air coupling from the dot end of N_{p1} to the dot end of N_{b2} starts turning $Q2$ "on." As it commences turning "on," positive feedback from the N_{p2} to the N_{b2} winding speeds up the turn "on" process until $Q2$ is fully "on." When $Q1$ was "on," the no-dot end of N_{p1} was positive and the core moved up the hysteresis loop. With $Q2$ "on," the dot end of N_{p2} is positive and the core is driven back down the hysteresis loop along the path EGHBA.

It requires the same $T/2$ given by Eq. 6.21 to move back down the hysteresis loop. The preceding cycles repeat and the circuit oscillates at a frequency given by

$$F = \frac{1}{T} = \frac{V_{dc} \times 10^{+8}}{4B_s N_p A_e} \tag{6.22}$$

In Figure 6.22*a*, the function of resistor $R3$ is to start the circuit oscillating. When V_{dc} is first applied, neither $Q1$ nor $Q2$ is "on" and the above cycles cannot commence. Current from V_{dc} flows down through $R3$ to the base windings center tap, the half base windings, the base resistor, and then the bases, and the cycle can now start.

In general, the transistor with the highest gain will be the one to turn "on" first. Once the circuit is oscillating, base current flows from the base winding, through its base resistor, its base, out of the transistor emitter, through $D1$, and back into the base winding center tap.

The circuit of Figure 6.22*a* shows only one of many possible base drive configurations. The base resistors shown serve to limit base current that may be excessive at high temperature and cause long transistor storage delay. Collector current may be limited with emitter resistors. Baker clamps (to be discussed in a later chapter on bipolar base drives) may be used to make the circuit less sensitive to load changes, production spread in transistor gain, and temperature.

6.6.4.1 Royer Oscillator Drawbacks[17]

The basic Royer oscillator has two major drawbacks, but these can be corrected by quite simple means. The effect of the first drawback can be seen in Figure 6.22*c* as an ultra-high-current spike at the end of a transistor "on" time. This spike may last only 1 to 2 μs but may be three to five times the current prior to the spike.

The spike occurs at a collector voltage about equal to the supply voltage and thus adds significantly to the transistor dissipation. Since it comes at simultaneously high current and voltage, it may exceed the safe operating area (SOA) boundary and cause failure by "secondary breakdown" even if the average dissipation is low.

The spike is inherent to the very nature of the Royer oscillator, and it can be explained as follows. During the time—say—$Q1$ is "on," $Q2$ has a reverse bias and is held "off" (observe the dots at the base windings). When the core has moved up—say, to the top of its hysteresis loop—it saturates, the windings can no longer support voltage, and the $Q1$ collector voltage rises.

However, its base voltage does not immediately go negative to turn "off" collector current. It goes negative only after the stored base charges drain away and $Q2$ has turned "on" sufficiently to produce a solidly negative voltage at the no-dot end of N_{p2} and hence at the no-dot end of N_{b1}. During this delay between core saturation at the end of one "on" time and the flopover to the opposite transistor turn "on," the "off"-turning transistor operates with high collector voltage and a high-current spike and may fail.

The second drawback is really a partial cause of the first. It is the long delay between core saturation on one side and turn "on" of the opposite transistor. During this delay, voltage at the base of the "off"-turning transistor hangs on at about 0.5 V, and drifts slowly negative before being pulled down abruptly by the opposite transistor turning solidly "on."

While the "off"-turning base is drifting slowly down from the 0.5-V level, the transistor is still partially "on" at a high collector-to-emitter voltage. While the base hangs on thus, the partially "on" transistor will often oscillate at a very high frequency. This can easily be corrected with small capacitors—empirically chosen between 100 and 500 pF—cross-coupled from the collectors to the opposite bases.

There is a further drawback in that the oscillator square-wave frequency is directly proportional to supply voltage (Eq. 6.22). There are many systems-related objections to a variable-frequency switching power supply. They all relate to the fact that any RFI generated will cover a wider and more continuous frequency spectrum with a variable frequency, as opposed to a fixed-frequency switching power supply.

Figure 6.23 shows the schematic and critical waveforms for a typical Royer oscillator DC/DC converter for 2.4-W output operating from 38 V DC—the minimum specified input for a telephone industry power supply.

The Royer oscillator is clearly low in parts count, but the waveform on the bottom-right of Figure 6.23 shows the aforementioned spikes at the end of turn "on" and also in this case at the start of turn "on." Collector voltages are shown in the waveform on the left. The low efficiency of only 50.6% is a consequence of the dissipation due to the current spikes at turn "on" and turn "off."

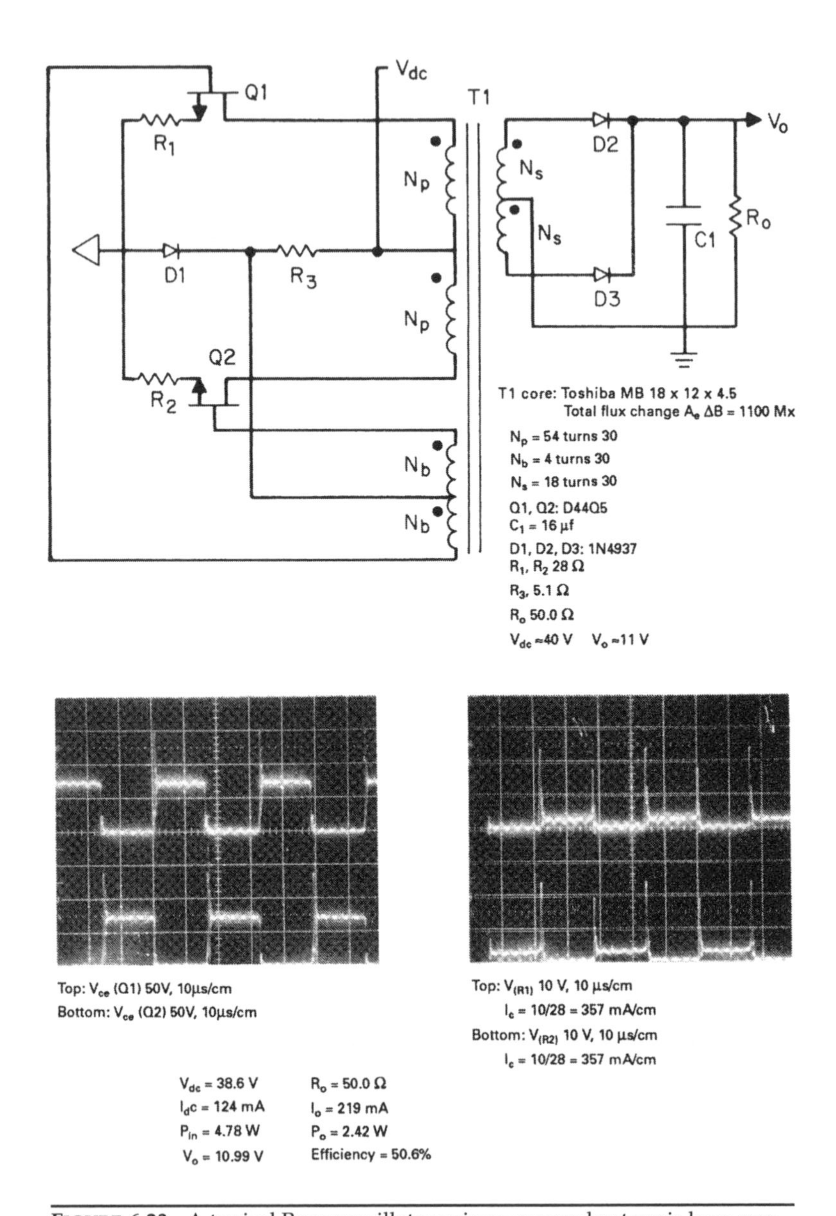

FIGURE 6.23 A typical Royer oscillator using a square hysteresis loop core, frequently used as a low-power "housekeeping supply" to power a PWM chip on output common, with its own power derived from the power source on input common. The high-current spikes at the end of "on" time and often at start of turn "on" make it unreliable and undesirable despite its low parts count.

6.6.4.2 Current-Fed Royer Oscillator[19]

By simply adding an inductor in series with Royer transformer center tap, the aforementioned current spikes at turn "off" and turn "on" are eliminated and efficiency is greatly increased.

The addition of the series inductor makes the circuit constant-current-fed as opposed to voltage-fed, and achieves all the advantages of current-fed topologies discussed in Section 5.6. The series inductor helps in the following way. When the core has saturated on one side, the associated transistor commences having a large current spike with a large di/dt. Since the current in an inductor cannot change instantaneously, the voltage at the transformer center tap drops down to common and the collector current is limited to the value it had just prior to core saturation.

The start current from *R*3 turns "on" the opposite transistor and both transistors are "on" simultaneously for at least the duration of the storage time in the "off"-turning transistor. This transistor turns "off" at zero collector-to-emitter voltage—the condition for minimum transient turn "off" dissipation (Section 2.2.12.1). The "on"-turning transistor turns "on" at zero collector-to-emitter voltage, which also minimizes transient turn "on" losses.

The benefits of this current-fed Royer can be seen in Figure 6.24. There the Royer circuit of Figure 6.23 was fed from an adjustable voltage power supply through a series 630-μH inductor (50 turns on a 1408-3C8 ferrite core with a total 2-mil air gap).

The transistor currents shown in Figure 6.24 show no sign of an end of "on"-time spike. The numerical data of Figure 6.24 are summarized in Table 6.2.

It is seen from Table 6.2 that efficiency averages about 71% with a constant load over the 38- to 60-V range of telephone industry specifications for power supplies. This compares favorably with the 50.6% efficiency for the same Royer and 49.8-Ω load resistor, but without the series input inductor (Figure 6.23).

The voltage drop down to zero at the transformer center tap due to the input inductor is clearly seen in Figures 6.24 and 6.25.

If output voltage variations of 11 to 18 V for input changes of 38 to 60 V were acceptable, the unregulated, current-fed Royer DC/DC converter would be a good choice because of its very low parts count.

6.6.4.3 Buck Preregulated Current-Fed Royer Converter

In many applications, regulated output voltage is required. Output voltage regulation can be achieved with very little more complexity and cost by preceding the Royer with a buck regulator as in Figures 6.26 and 6.27. Since buck regulators can quite easily be built with efficiencies of 90%, total efficiency does not suffer too much even though

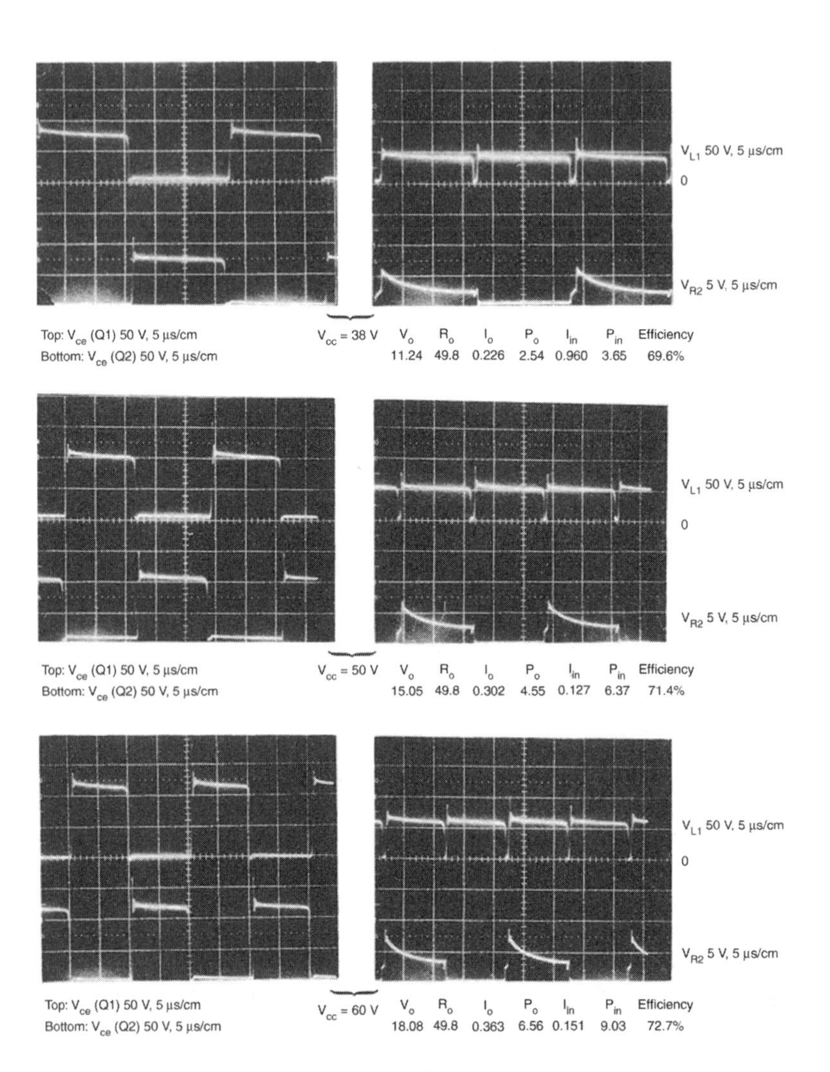

FIGURE 6.24 Waveform in a current-fed Royer oscillator. By adding an inductor in series between V_{cc} and the transformer center tap, the high-current spikes at the start and end of the transistor "on" time (Figure 6.22*c*) are eliminated and efficiency improves greatly. This occurs because the center tap voltage drops to zero when both transistors are simultaneously "on" for a brief instant at each transition. Waveforms are for the circuit of Figure 6.22*a* with an inductor of 630 μH in series with V_{cc} at 38, 50, and 60 V.

$V_{dc(in)}$, V	$I_{dc(in)}$, mA	P_{in}, W	V_{out}, V	R_o, Ω	P_{out}, W	Efficiency, %
38.0	96	3.65	11.24	49.8	2.54	69.6
50.0	127	6.37	15.05	49.8	4.55	71.4
60.0	151	9.03	18.08	49.8	6.56	72.7

TABLE 6.2

the power is handled twice—in the buck and in the Royer. Figure 6.27 shows the composite efficiency ranges from 57.9 to 69.5% over an input voltage range of 38 to 60 V and an output power range of 2.3 to 5.7 W.

In Figure 6.27, the buck regulator loop senses the output of the buck itself, keeping it constant and running the Royer open loop. This is often good enough, since the Royer output voltage is constant for constant input voltage and has quite good open-loop load regulation.

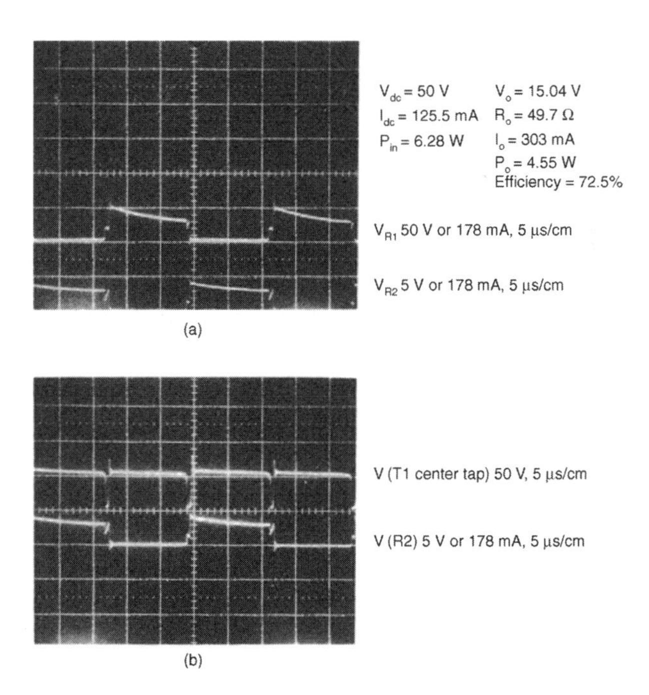

FIGURE 6.25 (*a*) Voltage across the emitter resistors in Figure 6.22 with a 1630 μH inductor in series with *T*1 center tap. (*b*) Circuit as in Figure 6.24, showing *T*1 center tap voltage dropping to zero at each transmission.

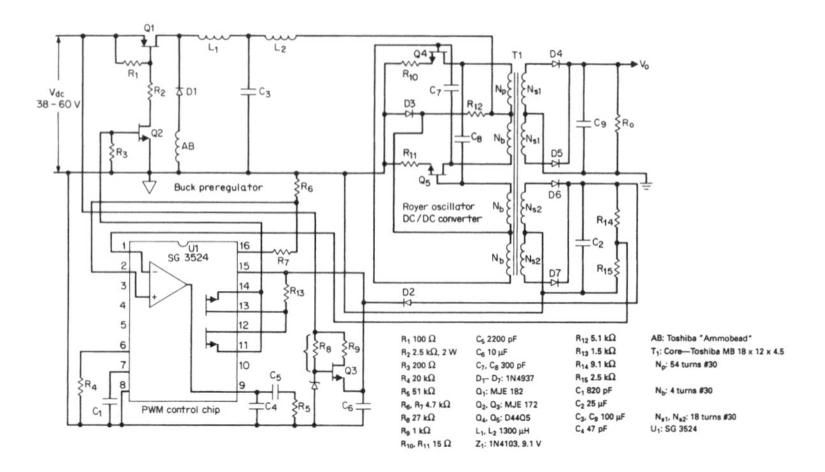

FIGURE 6.26 A buck regulator driving a current-fed Royer DC/DC converter for constant V_o load. Feedback can be taken from the buck output, with the Royer DC/DC-converter operated open loop. This yields regulation of better than 0.5% with an input change from 38 to 60 V, but load regulation of only ±5%. For better load regulation, feedback is taken from the bootstrapped slave output, which is referenced to input common as shown.

In Figure 6.27 it is seen that output voltage change over the above-mentioned line and load changes was only from 9.79 to 10.74 V—adequate for a housekeeping power supply.

If better load regulation is desired, the error amplifier in the buck can sense a bootstrapped slave secondary off the main power transformer as described in Section 6.6.3.3 and Figure 6.21.

The circuit details for the data of Figure 6.27 are shown in Figure 6.26.

6.6.4.4 Square Hysteresis Loop Materials for Royer Oscillators

The transformer core for a Royer oscillator must have a square hysteresis loop. If the loop is not square, turn "on" flipover from one transistor to the other will be sluggish, and in the worst case may not occur.

The "on" transistor may push the core to the top of the hysteresis loop and hang up there, delivering sufficient base drive to keep itself "on," yet not turning the opposite transistor "on." If this occurs, the partially "on" transistor will fail in a few tens of microseconds.

Most ferrite core materials do not have a sufficiently square hysteresis loop, but there are various other materials that do. The earliest material was an alloy of 79% nickel, 17% iron, and 4% molybdenum available from a number of manufacturers under various trade names.

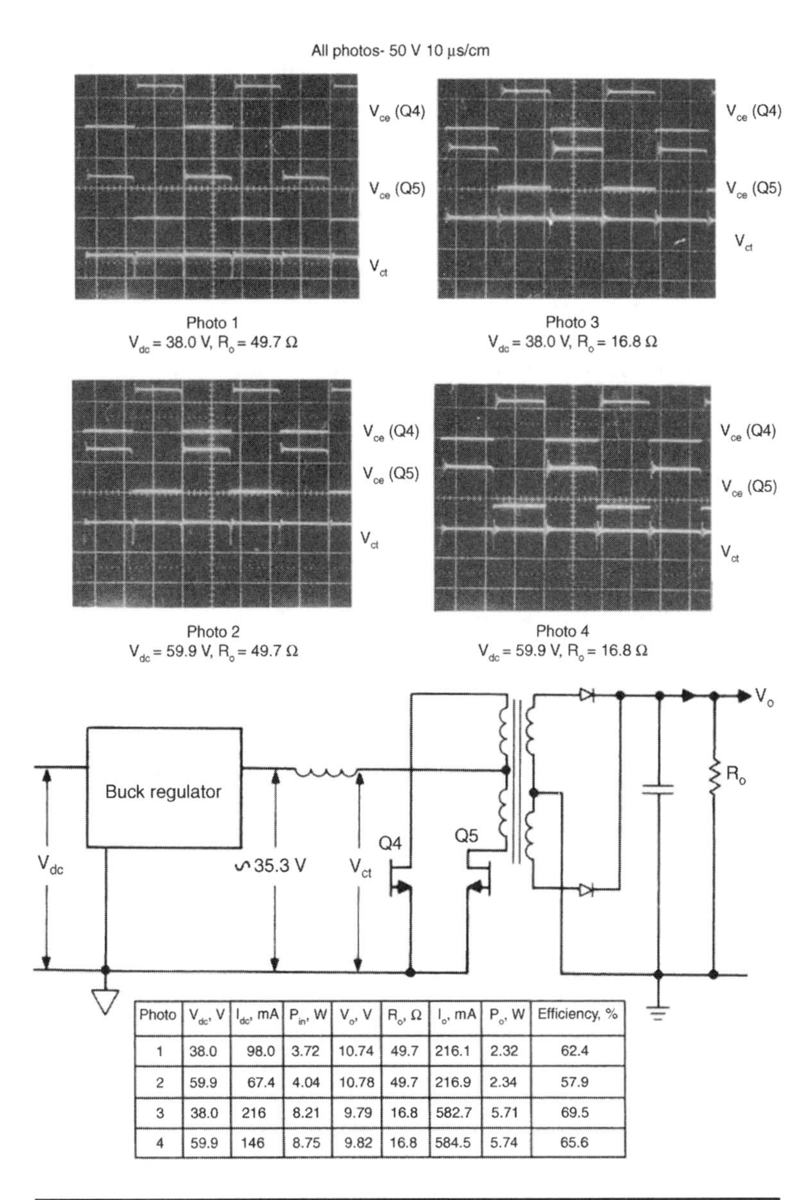

Photo	V_{dc}, V	I_{dc}, mA	P_{in}, W	V_o, V	R_o, Ω	I_o, mA	P_o, W	Efficiency, %
1	38.0	98.0	3.72	10.74	49.7	216.1	2.32	62.4
2	59.9	67.4	4.04	10.78	49.7	216.9	2.34	57.9
3	38.0	216	8.21	9.79	16.8	582.7	5.71	69.5
4	59.9	146	8.75	9.82	16.8	584.5	5.74	65.6

FIGURE 6.27 Waveforms and data on the current-fed Royer of Figure 6.26 driven from a buck regulator. Feedback is from the buck output.

Magnetics Inc. has probably the largest selection of standard core sizes made from its material called Square Permalloy 80. Other manufacturers' trade names for roughly the same material are 4-79 Permalloy, Square Mu 79, and Square Permalloy. The material has a saturation flux density ranging between 6600 and 8200 G.

The material is produced in a thin tape, wound into a toroidal core, and then encased in aluminum or a nonmetallic case. The tape is available in 1- or ½-mil thickness. Core losses increase rapidly with frequency, and just as with power transformers, in which higher frequencies require thinner laminations to minimize losses, the ½-mil tape cores should be used beyond 50 kHz. Beyond 100 kHz, losses in even the ½-mil cores become prohibitive.

In the 1980s, "amorphous" magnetic material with low losses at high frequencies for flux swings between $+B_S$ and $-B_S$ became available. It permits building Royer oscillators of up to 200 kHz with acceptably low losses and core temperature rise.

After Pressman *Be careful when designing single transformer Royer self-oscillating circuits using the very square loop "amorphous" magnetic material. Oscillation requires flyback action from the core and some of these materials have a flux remnants value (Br) very near the saturation value, so the core will latch in the saturated state and will not oscillate. ~K.B.*

Amorphous cores are manufactured in the United States under the trade name Metglas by Allied Corporation and Magnetics Inc., and by Toshiba under the trade name Amorphous-MB. These core materials have saturation flux densities ranging between 5700 and 6200 G.

Since they are made of thin tapes, all of these square-loop cores have relatively low iron area compared to ferrite cores. Thus at the same frequency, they require more turns than would a ferrite core, if a square-loop ferrite core were available.

However, the number of turns for a Royer is no problem with the small-area tape-wound cores, as can be seen from Eq. 6.22, which shows that the required number of turns is inversely proportional to saturation flux density, frequency, and iron area. Although the iron area of the tape-wound cores is small, their saturation flux density is close to twice that of any available square hysteresis loop ferrite. Since Royers with tape-wound cores can be built at 50 to 200 kHz where fewer turns are necessary, there is no problem with excessive turns.

Nevertheless, if a ferrite core is desired because of its larger iron area, there are a few sources of a square hysteresis loop ferrite core. One such is material Type 83 from the Fair-rite Corporation (Wallkill, New York). It has a saturation flux density of 4000 G, but its losses operating at ±4000 G are sufficiently high that the maximum operating

Core material	Saturation flux density, G	Core losses* 50 kHz	W/cm^3 100 kHz
Toshiba MB	6000	0.49	1.54
Metglas 2714A	6000	0.62	1.72
Square Permalloy 80 (1/2 mil)	7800	0.98	2.26
Square Permalloy 80 (1 mil)	7800	4.2	9.6
Fair-rite Type 83	4000	4.0†	30.0

*For flux excursions between positive and negative saturation
†1 W/cm^3 at 25 kHz

TABLE 6.3

frequency is 50 kHz. Available square hysteresis core materials are listed in Table 6.3.

It should be noted that most of these tape-wound cores have relatively small radiating surface area and hence high thermal resistance (in the range of 40 to 100°C/W). Unless bound to a heat sink, total losses should be kept under 1 W.

Hysteresis loops for the above materials are shown in Figure 6.28.

6.6.4.5 Future Potential for Current-Fed Royer and Buck Preregulated Current-Fed Royer

It may seem surprising in a text on modern power supply design to devote much space to the Royer circuit, which was cast aside 30 years ago, but the Royer, operated in the current-fed mode, with small collector-to-opposite-base flipover capacitors, and with the new low-loss amorphous cores, is very attractive in many applications.

If line regulation is not required, it is extremely low in parts count (Figures 6.23 and 6.24). Its major fields of application are where prime input is low-voltage DC—e.g., 48 V for telephone industry supplies, 28 V for aircraft supplies, and 12 or 24 V for automotive supplies.

With the new available cores, they can generate up to 200 or possibly 300 W. Since they require no output inductors, they can easily generate high voltage—with a multi-turn secondary or an output voltage multiplier. If regulated output voltage is required, they can be preceded by a high efficiency buck regulator (Figures 6.26 and 6.27).

It appears at this writing that in the coming years, there will be widespread renewed interest in the current-fed Royer oscillator DC/DC converter.

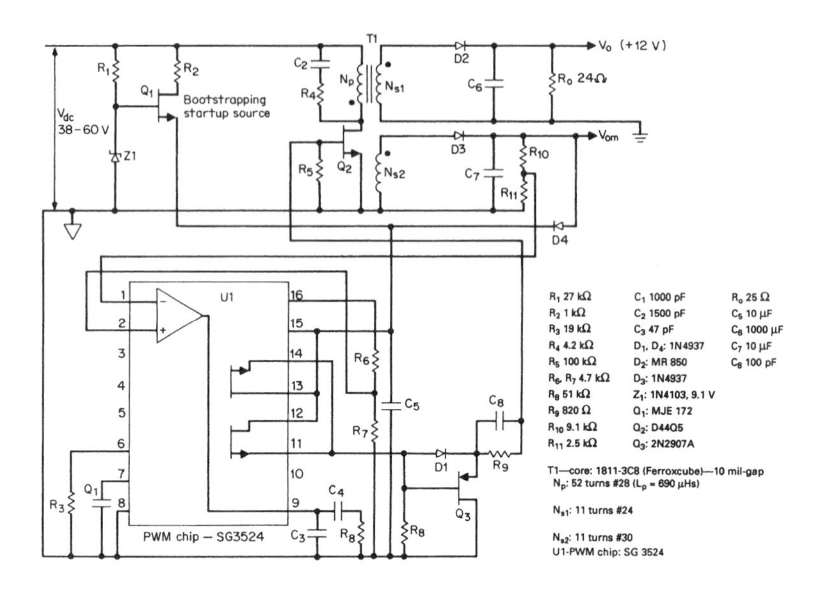

FIGURE 6.28 A low-power flyback housekeeping power supply with isolated output V_o.

6.6.5 Minimum-Parts-Count Flyback as a Housekeeping Supply

The low-power flyback scheme of Figure 6.21 as a housekeeping supply is detailed in Figure 6.28.

The circuit was designed as discontinuous-mode flyback from the design relations presented in Chapter 4. It was designed for 6 W of output power at a switching frequency of 50 kHz from a supply voltage of 38 to 60 V—the usual range for a telephone industry power supply. The circuit as is can easily deliver twice the output power without overstressing any components. Beyond 6 W at less than 38 V input, it will enter the continuous mode and oscillate unless the feedback loop is changed (Section 4.3).

In Figure 6.28, the regulated output is a master secondary V_{om} referred to input common. The housekeeping output V_o is a slave returned to output common where it can drive the PWM chip for the main power supply. As discussed in Section 4.2, because flybacks have no output inductors, slaves track the master very closely. Thus regulating V_{om} on input common keeps V_o on output common sufficiently constant for a housekeeping supply.

In Figure 6.28, $Q1$ supplies voltage to the PWM chip during startup. After the supply is up and delivering its output voltages, V_{om} takes over and supplies the chip via diode $D4$. The voltage at the $D4$ cathode is about 11 V. Since the base of *Q*1 is kept at 9.1 V by *Z*1, it is biased "off" as soon as its emitter rises to about 9 V via *D*4.

Significant waveforms and performance data for the circuit are shown in Figure 6.29*a*, 6.29*b*, and 6.29*c* for input voltages of 38, 50, and 60 V, respectively. Efficiencies are about 70%. This is not spectacular, and no effort has been made to optimize circuit efficiency.

It was intended here only to show the significant classic waveforms of an actual operating, discontinuous-mode flyback. Figure 6.29*a*, for

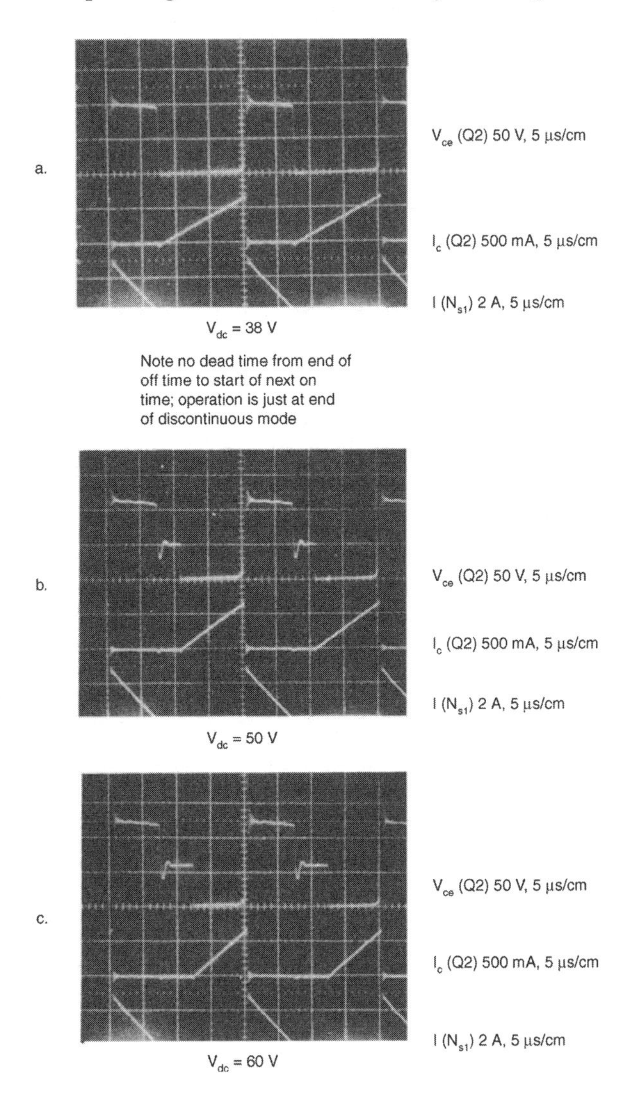

FIGURE 6.29 Significant waveforms for low-power flyback of Figure 6.28.

a DC input voltage of 38 V, shows the circuit to be just at the threshold of discontinuous mode (Section 4.4.1 and Figure 4.8). Primary current starts ramping up with no dead time the instant after the previous secondary current has ramped down to zero. Figure 6.29*b* and 6.29*c* show the same waveforms at 50 and 60 V, respectively, in which there is now a dead time between the instant secondary current has ramped down to zero and the start of the next turn "on."

The flyback has fewer components than the buck-current-fed Royer of Figure 6.26, but it is seen in Figure 6.29 that at 6 W of output power, peak secondary current for the flyback is 3 A. This compares unfavorably to 0.36 A for the buck-current-fed Royer (Figure 6.25*a*). The higher flyback secondary current can produce a greater RFI problem that requires a larger output filter, and possibly a small *LC* filter after the main filter capacitor to eliminate the output spike at the instant of turn "off," due to ESR in the main capacitor.

6.6.6 Buck Regulator with DC-Isolated Output as a Housekeeping Supply

Figure 1.9 shows another possible inexpensive, low-parts-count scheme for generating a DC-isolated power supply. It is described in Section 1.3.8. Care must be taken in this scheme that the current drawn from the secondary is not sufficient to cause the primary current to go into the discontinuous mode, or regulation will suffer.

References

1. B. D. Bedford and R. G. Hoft, *Principles of Inverter Circuits,* Wiley, New York, 1964.
2. *General Electric SCR Manual,* 6th ed., General Electric Co., Auburn, NY, 1979.
3. I. Martin, *Operating Characteristics of Self-Commutated Sinewave SCR Inverters,* RCA Application Note AN-6745, RCA, Somerville, NJ, 1978.
4. I. Martin, *Regulating the SCR Inverter Power Supply,* RCA Application Note AN-6856, RCA, Somerville, NJ, 1980.
5. Z. F. Chang, *Application of ASCR in 40 kHz Sine Wave Converter,* RCA Application Note ST-6867, RCA, Somerville, NJ, 1980.
6. N. Mapham, "An SCR Inverter with Good Regulation and Sine Wave Output," *IEEE Transactions on Industry and General Applications,* IGA-3 (2), 1967.
7. N. Mapham, "Low Cost Ultrasonic Frequency Inverter Using Single SCR," *IEEE Transactions on Industry and General Applications,* IGA-3 (5), 1967.
8. I. Martin, "Application of ASCR's to High Frequency Inverters," *Proceedings Powercon 4,* May 1977.
9. D. Chambers, "Designing High Power SCR Resonant Converters for Very High Frequency Operation," *Proceedings Powercon 9,* 1982.
10. D. Chambers, "A 30 kW Series Resonant X Ray Generator," *Powertechnics Magazine,* January 1986.
11. K. Check, "Designing Improved High Frequency DC/DC Converters with a New Resonant Thyristor Technique," Intel Corporation, Hillsboro, OR.

12. See Reference 1, Chapters 8 and 10.
13. D. Amin, "Applying Sinewave Power Switching Techniques to the Design of High Frequency Off-Line Converters," *Proceedings Powercon 7*, 1980.
14. S. Cuk and D. Middlebrook, *Advances in Switched Mode Power Converters,* Vols. 1, 2, Teslaco, Pasadena, CA, 1981.
15. G. Chryssis, *High Frequency Switching Power Supplies*, 2d ed., McGraw-Hill, New York, 1989.
16. K. Billings, *Switchmode Power Supply Handbook,* McGraw-Hill, New York, 1990.
17. A. Pressman, *Switching and Linear Power Supply, Power Converter Design,* Switchtronix Press, Waban, MA, 1977.
18. G. H. Royer, "A Switching Transistor AC to DC Converter," *AIEE Transactions,* July 1955.
19. D. V. Jones "A Current Sourced Inverter with Saturating Output Transformer," *IEEE Proceedings*, 1981.

PART 2

Magnetics and Circuit Design

CHAPTER 7

Transformers and Magnetic Design

7.1 Introduction

In Part 1, we considered the characteristics of most of the more frequently used topologies in sufficient depth to allow us to make a suitable choice of topologies that best meets the power supply specifications. Frequently, the topology is selected to minimize the power transistor's off-voltage stress at high line and the peak current stress at maximum output power. Other considerations would be to minimize parts count, cost, and required volume of the complete supply. Minimizing potential RFI problems is also a frequent factor in choice of topology and working frequency.

After a topology is selected, the next major decisions are to select an operating frequency and minimum transformer core size which yields the specified maximum output power. To make the frequency and transformer core selection, it is necessary to know the numerical relations between desired output power and transformer parameters such as transformer size, area of core, core window or bobbin winding area, peak flux density, operating frequency, and coil current density. In the following sections, equations giving these relations will be derived for the most frequently used topologies.

The above relations will be used in equation form to select a transformer core and operating frequency. One method is to estimate the required core size and frequency. We can then calculate the approximate power available from the selected core, frequency, and the remaining parameters. This initial estimate can then be easily corrected if the desired power is not available. Since all the parameters are interrelated, such interactive calculations may have to be done several dozen times before a satisfactory combination of the parameters is found, a somewhat cumbersome procedure.

TIP *Reference 17 provides a selection nomogram on page 3.68 allowing the optimum core size to be selected directly from the required power and selected operating frequency. ~K.B.*

A better method is obtained by putting the equations into a chart. The following charts show frequency increasing (in multiples of 8 kHz to the right) in vertical columns and specific core sizes from various manufacturers in horizontal rows with available output power (calculated from the equations) at the column-row intersections.

The cores are arranged in horizontal rows of increasing output power. Thus, at a glance one can choose an operating frequency and move vertically through the rows until the first core of sufficient power is found. Alternatively, if a specific core whose dimensions fit the available space is chosen, one can move horizontally through columns of increasing frequency to find the desired output power.

The charts shown below are for various core geometries from four major core manufacturers.

Core losses versus frequency and peak flux density are shown for widely used core materials from various core manufacturers. The available ferrite core geometries and their usage are discussed. Core and copper loss calculations are presented. A significant contributor to copper losses—proximity effect—is described. Transformer temperature rise calculation from the sum of core and copper losses is demonstrated.

7.2 Transformer Core Materials and Geometries, and Peak Flux Density Selection

7.2.1 Ferrite Core Losses versus Frequency and Flux Density for Widely Used Core Materials[17-20]

Most switching power supply transformers are made with ferrite cores. Ferrites are ceramic ferromagnetic materials having a crystalline structure consisting of mixtures of iron oxide with either manganese or zinc oxide. Their eddy current losses are negligible, as their electrical resistivities are very high. Core losses comprise mainly hysteresis losses which are low enough to permit use of some materials up to a frequency of 1 MHz. Ferrite cores are available from many manufacturers (such as Ferroxcube-Philips, Magnetics Inc., Ceramic Magnetics Inc., Ferrite International, and Fair-rite, among others) and some overseas manufacturers (TDK, Siemens, Thomson-CSF, Tokin).

Each manufacturer has a number of different mixes of the various oxides, processed in various ways to achieve different advantages. Some materials are tailored to yield the minimum core loss point at higher frequency (>100 kHz), to shift the minimum core loss temperature point to a higher value (90°C), or to achieve minimum core losses at the most usual combination of high frequency and peak flux density. However, the DC hysteresis loops of most vendors' ferrites that are intended for switching power transformer applications are quite similar. At 100°C, they are within 10% of complete saturation in the region of 3000 to 3200 G, have a coercive force of 0.10 to 0.15 Oe, and have a residual flux density of 900 to 1200 G.

The major factors affecting material selection are summarized in curves of core loss (usually expressed in milliwatts per cubic centimeter) versus frequency and peak flux density. A typical curve of these data, plus the DC hysteresis loop for the Ferroxcube-Philips high-frequency material 3F3, is shown in Figure 7.1. Core losses for some widely used materials are given in Table 7.1.

Losses in Table 7.1, taken from core manufacturer data sheets and, although it is seldom pointed out, are for bipolar magnetic circuits in which the flux excursion extends into the first and third quadrants of the hysteresis loop (push-pulls, half and full bridges). Forward converters and flybacks operate in the first quadrant only.

Since ferrite core losses are hysteresis losses only and these losses are proportional to the area of the hysteresis loop, it might be thought that unipolar magnetic circuits, which traverse only half of the hysteresis loop, would have half the core losses of bipolar circuits at the same peak flux density.

There is considerable difference of opinion among manufacturers on this. Some say unipolar circuit losses are one-fourth the quoted and measured values for bipolar circuits at the same peak flux density. They reason that if a unipolar circuit swings from 0 to B_{max} gauss, it is equivalent to a bipolar circuit swinging around a mean value of $B_{max}/2$, with a peak excursion of $B_{max}/2$. Further, since core losses are roughly proportional to the square of the peak flux excursion in a bipolar circuit, halving the peak flux excursion reduces losses by a factor of 4.

TDK offers a curve showing that unipolar circuit losses for a zero to B_{max} excursion are a factor K_{fc} times as great as losses in a bipolar circuit swinging from $-B_{max}$ to $+B_{max}$. The factor K_{fc} is shown as being frequency-dependent. It is 0.39 at 20 kHz, 0.35 at 60 kHz, and 0.34 at 100 kHz.

A conservative approach is to accept the argument that losing half the area of a hysteresis loop (for unipolar circuits) should reduce losses measured for a bipolar circuit by a factor of 2, so it will be assumed

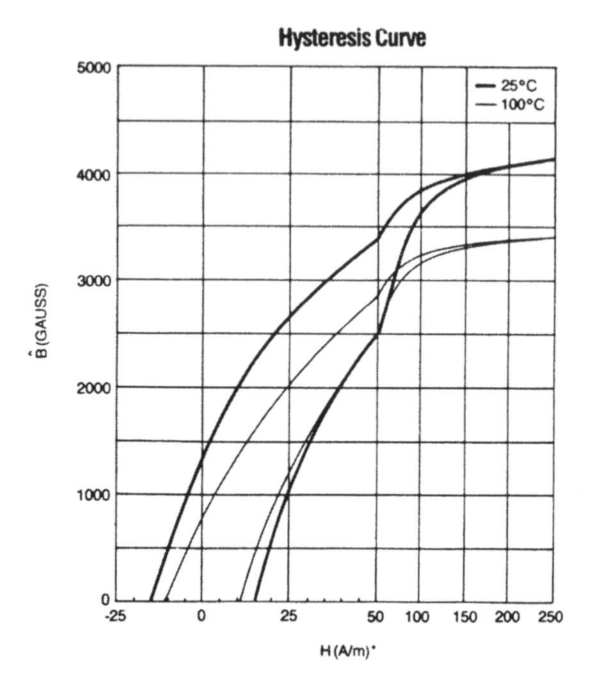

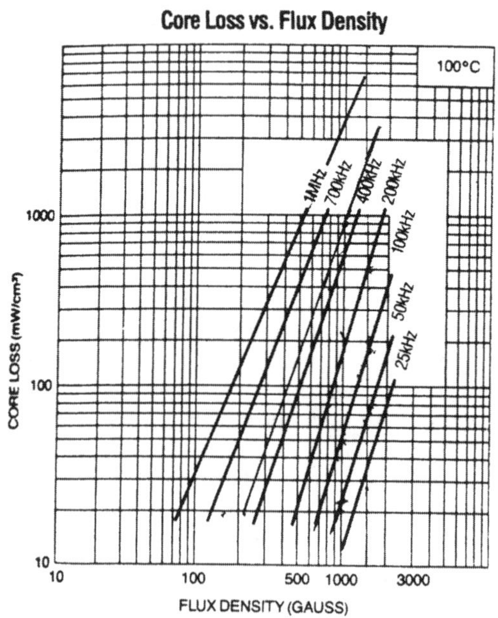

FIGURE 7.1 Significant characteristics of 3F3—a high-frequency, low-loss core material (*Courtesy of Ferroxcube-Philips Corp.*)

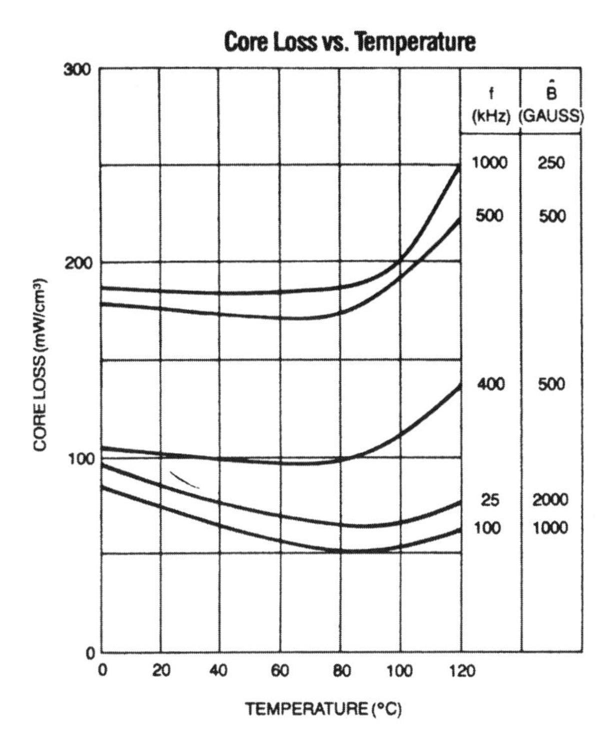

FIGURE 7.1 *Continued.*

herein that unipolar circuits at the same peak flux density as listed in Table 7.1 will have losses half those shown in the table.

7.2.2 Ferrite Core Geometries

Ferrite cores are manufactured in a relatively small number of geometric shapes and varying dimensions within the shapes. The shapes and dimensions of four core manufacturers' products are described in their catalogs.[1–4]

Many of the core shapes and dimensions in these catalogs are international standards and are available from various manufacturers in their proprietary core materials. Cores which are international standards are listed in publications from the Magnetic Material Producers Association (MMPA),[5,6] and in IEC publications from the American National Standards Institute.[7]

The core geometries shown in Figure 7.2 are pot or cup cores, RM cores, EE cores, PQ cores, UU or UI cores. The pot core is shown in Figure 7.2*e*. It is used mostly at power levels up to 125 W, and usually in

Frequency, kHz	Material	Core loss, mW/cm³ for various peak flux densities, G					
		1600	1400	1200	1000	800	600
20	Ferroxcube 3C8	85	60	40	25	15	
	Ferroxcube 3C85	82	25	18	13	10	
	Ferroxcube 3F3	28	20	12	9	5	
	Magnetics Inc.-R	20	12	7	5	3	
	Magnetics Inc.-P	40	18	13	8	5	
	TDK-H7C1	60	40	30	20	10	
	TDK-H7C4	45	29	18	10		
	Siemens N27	50			24		
50	Ferroxcube 3C8	270	190	130	80	47	22
	Ferroxcube 3C85	80	65	40	30	18	9
	Ferroxcube 3F3	70	50	30	22	12	5
	Magnetics Inc.-R	75	55	28	20	11	5
	Magnetics Inc.-P	147	85	57	40	20	9
	TDK-H7C1	160	90	60	45	25	20
	TDK-H7C4	100	65	40	28	20	
	Seimens N27	144			96		
100	Ferroxcube 3C8	850	600	400	250	140	65
	Ferroxcube 3C85	260	160	100	80	48	30
	Ferroxcube 3F3	180	120	70	55	30	14
	Magnetics Inc.-R	250	150	85	70	35	16
	Magnetics Inc.-P	340	181	136	96	57	23
	TDK-H7C1	500	300	200	140	75	35
	TDK-H7C4	300	180	100	70	50	
	Seimens-N27	480			200		
	Siemens-N47				190		

TABLE 7.1 Core Losses at 100°C for Some Materials at Various Frequencies and Peak Flux Densities

Frequency, kHz	Material	Core loss, mW/cm^3 for various peak flux densities, G					
		1600	1400	1200	1000	800	600
200	Ferroxcube 3C8				700	400	190
	Ferroxcube 3C85	700	500	350	300	180	75
	Ferroxcube 3F3	600	360	250	180	85	40
	Magnetics Inc.-R	650	450	280	200	100	45
	Magnetics Inc.-P	850	567	340	227	136	68
	TDK-H7C1	1400	900	500	400	200	100
	TDK-H7C4	800	500	300	200	100	45
	Seimens-N27	960			480		
	Siemens-N47				480		
500	Ferroxcube 3C85				1800	950	500
	Ferroxcube 3F3		1800	1200	900	500	280
	Magnetics Inc.-R		2200	1300	1100	700	400
	Magnetics Inc.-P		4500	3200	1800	1100	570
	TDK-H7F						100
	TDK-H7C4		2800	1800	1200	980	320
1000	Ferroxcube 3C85						2000
	Ferroxcube 3F3				3500	2500	1200
	Magnetics Inc.-R				5000	3000	1500
	Magnetics Inc.-P						6200

Note: Data are for bipolar magnetic circuits (first- and third-quadrant operation). For unipolar circuits (forward converter, flyback), divide flux density by 2.

TABLE 7.1 Core Losses at 100°C for Some Materials at Various Frequencies and Peak Flux Densities (*Continued*)

DC/DC converters. Its major advantage is that the coil on the bobbin around the center post is almost entirely enclosed by ferrite material. This decreases its radiating magnetic field and hence is used when EMI or RFI problems must be minimized.

The major disadvantage of the pot core is the narrow slot in the ferrite through which the coil leads exit. This makes it difficult to use at high input or output currents requiring large wire diameter, or in multi-output supplies with many wires exiting.

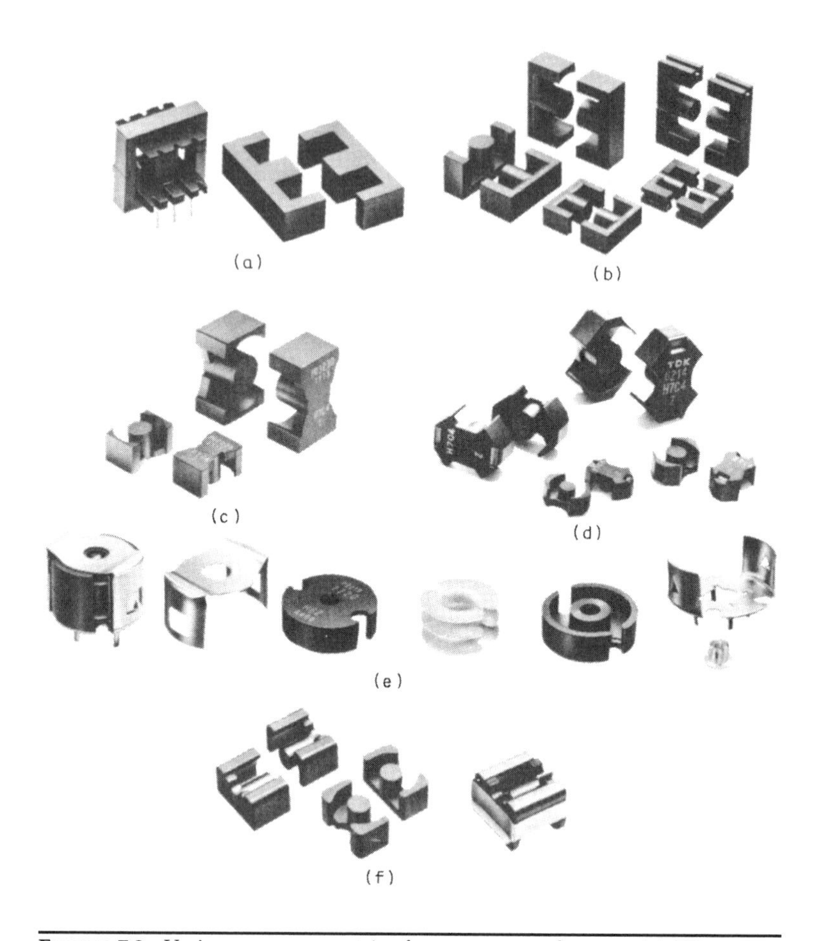

FIGURE 7.2 Various core geometries for power transformers: (*a*) EE cores; (*b*) EC and ETD cores; (*c*) PQ cores; (*d*) RM cores; (*e*) pot cores; (*f*) LP cores (*Courtesy of TDK Corp.*)

It is also not a good choice for a high-voltage supply, even at low power. Leads carrying a high voltage may arc because of the close spacing in the narrow exit notch in the ferrite.

Many pot cores are available with gaps of various sizes in the center leg so that they may carry a DC bias current without saturating. This permits their use as output inductors in buck regulators (Section 1.3.6), forward converters (Sections 2.3.9.2, 2.3.9.3), push-pulls (Section 2.2.8.1), and flybacks (Section 4.3.1).

Most often, when a core is available with a gapped center leg, the manufacturer gives its A_l value (inductance in millihenries per

1000 turns) and the cliff point in ampere turns at which it falls over its saturation cliff.

If a gapped core is required, it is more cost effective—and from a performance viewpoint, preferable—to use a core with a gapped center leg rather than using ungapped core halves separated with the proper thickness of plastic shims. Shimming core halves will not yield reproducible A_l values over time, temperature, and production spread. Also, gapping the outer leg will increase EMI.

The most widely used are EE cores (Figure 7.2*a*) because there is ample room for coil leads entering and leaving the bobbin. Since the coil is not fully surrounded by ferrite, it does produce a larger EMI-RFI field. However, airflow around the windings is unimpeded, which therefore run cooler. EE cores are available with either a square or round center leg. Round-center-leg cores (EC or ETD types; Figure 7.2*b*) have a small advantage in that the mean length of a turn is about 11% shorter than for a square-legged core of equal center-leg area. Coil resistance is thus about 11% less for equal numbers of turns, and copper loss and temperature rise are somewhat lower.

There is a large range of EE core sizes, and depending on frequency and peak flux density, they can deliver output powers from under 5 W up to 5 or possibly 10 kW. By using two square-center-leg EE cores side by side, the core area is doubled, requiring half the number of turns for the same voltages, peak flux density, and frequency (Faraday's law; Eq. 2.7). This doubles the power available from a single core, and may result in a smaller transformer than using a single core of the next larger size.

The RM or "square" core, a compromise between a pot and an EE core, is shown in Figure 7.2*d*. It is effectively a pot core with a much wider notch cut out of the ferrite. It is thus easier to bring larger diameter or many wires in and out of the coil, so this core is usable for much higher output power levels and for multi-output transformers. The larger ferrite notch also provides easier access for convection air currents than in a pot core, which results in smaller temperature rise.

Because the coil is not as fully surrounded by ferrite as in a pot core, it causes more EMI-RFI radiation. Yet, because there is more ferrite surrounding the coil than in an EE core, its EMI radiation is less than that for an EE of equal output power.

RM cores are available with or without a center-leg hole. The center-leg hole is used for mounting with a bolt or in frequency-sensitive applications. By inserting an adjustable ferrite "tuning rod" into the center hole, the A_l value may be adjusted by as much as 30%. Although this tuning feature is not usable in power transformers because of increased energy losses, it is usable for frequency-sensitive filters.

The geometry of the PQ core (Figure 7.2*c*; Magnetics Inc. and TDK) is such that it provides an optimum ratio of volume to radiating surface and coil winding area. Since core losses are proportional to core volume, and heat radiation capability is proportional to radiating surface area, these cores have a minimized temperature rise for a given output power. Further, since the PQ core volume-to-coil-winding-area ratio is optimized, the volume is minimized for a given output power.

LP cores (Figure 7.2*f*; TDK) are specifically designed for low-profile transformers. They have long center legs, which minimize leakage inductance.

UU or UI cores (not shown in Figure 7.2) are used mainly for high-voltage or ultra-high-power applications. They are rarely used at power levels under 1 kW. Their large window area compared to an EE core of equal core area permits much larger wire sizes or many more turns. But their much larger magnetic path length does not yield as close primary-secondary coupling as in an EE core, and results in larger leakage reactances.

7.2.3 Peak Flux Density Selection

As discussed in Section 2.2.9.3, the number of transformer primary turns will be calculated from Faraday's law (Eq. 1.17, 2.7) for a preselected peak flux density B_{max}. It is seen in Eq. 1.17, that the larger the flux excursion (larger value of B_{max}), the fewer the primary turns so the larger the permissible wire size, and therefore the greater the available output power.

There are two limitations to peak flux density in ferrite cores. The first is core losses and the resulting core temperature rise. Core losses in most ferrite materials are proportional to the 2.7th power of the peak flux density, so high peak flux densities cannot be permitted, especially at higher frequencies. Most ferrites—even the lossiest ones—have such low losses at 25 kHz and below, that core losses are not a limiting factor at those frequencies (see Table 7.1). At these low frequencies, peak flux excursion may possibly extend far up into the curved area of the *BH* hysteresis loop. However, care must be taken that the core does not move so far into saturation that the primary current increases uncontrollably, or the power transistor will be destroyed. Ferrite core losses also increase roughly as the 1.7th power of the switching frequency. Thus at higher frequencies, for the more lossy materials (Table 7.1), attempting a high peak flux density to minimize the number of turns results in such high losses that temperature rise will be excessive. Core losses are equal to the loss factor in milliwatts per cubic centimeter, times the core volume in cubic centimeters.

Hence at 50 kHz and above, less lossy (somewhat more expensive) core material must be used, or peak flux density must be reduced. Reducing the peak flux density requires increasing the number of primary turns (Eq. 2.7) and hence requires smaller wire size for the same core-bobbin winding area. With smaller wire sizes, primary and secondary currents are smaller and available output power is decreased. Thus at high frequencies (> ~ 50 kHz) the least lossy core material must be used, and peak flux density must be chosen sufficiently low that total core and copper losses result in an acceptably low temperature rise. Temperature rise calculation for the sum of core plus coil losses will be demonstrated in Section 7.4.

Even at low frequencies, however, where core losses are not a limiting factor, peak flux density cannot be permitted to move excessively high up on the *BH* loop to minimize the number of primary turns. It is seen in Figure 2.3 that the *BH* loop is still roughly linear up to 2000 G (this is near the end of the linear portion for most ferrite materials). Exceeding this flux density will increase the magnetizing current near the end of the transistor "on" time and will unnecessarily increase coil and transistor losses. But designing (choosing N_p in Eq. 2.7) for a maximum of 2000 G for most ferrites can be risky. For fast transient line or load steps, if the feedback error amplifier is not fast enough for a few switching cycles, peak flux density may move up to hard saturation (> 3200 G at 100°C) and destroy the power transistor. This is discussed in more detail in Section 2.2.9.4.

Thus, even below 50 kHz where ferrite core loss is not a limiting factor, peak flux density will be chosen at 1600 G in all designs herein. At higher frequencies, where even the least lossy material at 1600 G results in excessive power losses, peak flux density will be reduced. The "available output power tables" to be developed below will show how output power may be calculated easily for these reduced peak flux densities.

7.3 Maximum Core Output Power, Peak Flux Density, Core and Bobbin Areas, and Coil Currency Density

7.3.1 Derivation of Output Power Relations for Converter Topology

Refer to Figure 7.3 for the forward converter topology. The following output power relation will be based on the following assumptions:

1. Efficiency of the *power train*—the ratio of input power to the sum of all output powers, neglecting control circuit dissipation—is 80%.
2. The *space factor SF*, the fraction of total bobbin winding area occupied by current-conducting metal, is 0.4. The area is occupied by

the primary and secondary wires covered with insulation, layer insulations, any RFI or Faraday shields, and empty space. SF is typically in the range 0.4 to 0.6 in transformer design because of the many factors contributing to waste (area not used to conduct current) of the total bobbin winding area. One significant factor is that turns within a coil layer often are widely spaced to make all bobbin layers of equal width, to improve magnetic coupling between layers and reduce leakage inductances. Also, adherence to European safety specifications (VDE) at this time requires leaving a 4-mm gap between each end of a layer and the ends of the bobbin. There is also the thickness of the insulation layers. VDE specifications generally required three layers of 1-mil-thick insulating material between layers; if secondaries are to be sandwiched between halves of the primary (often done to reduce proximity-effect copper losses), that sacrifices 6 mils of bobbin height. Finally, there is the practical problem that it is difficult to safely assemble the core and bobbin if the bobbin height is fully utilized.

3. Primary current waveshape is as shown in Figure 7.3. At minimum V_{dc} input, "on" time is a maximum at $0.8T/2$ (Sections 2.3.2, 2.3.5). Primary current has the ramp-on-a-step waveshape because of inductors at each secondary output. The ramp swings ±10% about the center value I_{pft}. The primary current waveshape can be accurately approximated by a rectangular pulse of peak amplitude I_{pft} of duty cycle $0.8T/2T$ or 0.4 (Section 2.3.5). Then for minimum $V_{dc} = V_{dc(min)}$

$$
\begin{aligned}
P_o &= 2.8P_{in} = 0.8V_{dc(min)}\left[I_{av} \text{ at } V_{dc(min)}\right] \\
&= 0.8V_{dc(min)}(I_{pft})\frac{0.8T}{2T} \\
&= 0.32V_{dc(min)}I_{pft}
\end{aligned}
\tag{7.1}
$$

But the RMS of a rectangular waveform of amplitude I_{pft}, duty cycle of 0.4, is $I_{rms} = I_{pft}\sqrt{0.4}$ or $I_{pft} = 1.58I_{rms}$. Then

$$
\begin{aligned}
P_o &= 0.32V_{dc(min)}(1.58I_{rms}) \\
&= 0.506V_{dc(min)}I_{rms}
\end{aligned}
\tag{7.2}
$$

From Faraday's law

$$V_p = N_p A_e \frac{\Delta B}{\Delta T} \times 10^{-8}$$

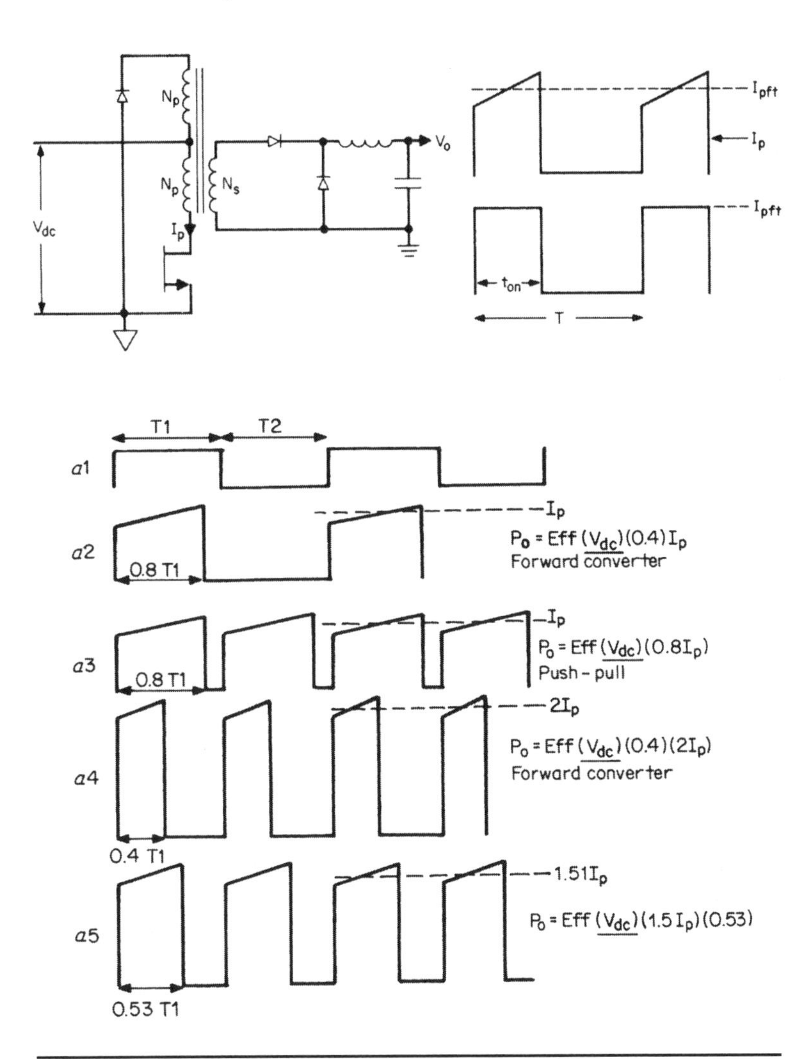

FIGURE 7.3 Forwarded converter topology and primary current waveshape I_p. Equivalent flat-topped current waveshape I_{pft} is used to calculate output power relation to $B_{\max}$, frequency, A_e, A_b, and D_{cma}. Turns ratio N_s/N_p is chosen to yield $t_{\text{on}} = \text{g}0.8T/2$ at minimum V_{dc} for specified V_o.

where V_p = primary volts ($\approx V_{\text{dc}}$)
N_p = number of primary turns
A_e = core area, cm^2
ΔB = flux density change, G (0 to $B_{\max}$)
ΔT = time, s for this flux change = $0.4T$

At $V_{dc(min)}$, $\Delta B/\Delta T = B_{max}/0.4T$. Then for $f = 1/T$ from Eq. 7.2

$$P_o = \frac{0.506 I_{rms} N_p A_e B_{max} f}{0.4} \times 10^{-8}$$
$$= 1.265\, N_p B_{max} A_e f \times 10^{-8} (I_{rms}) \tag{7.3}$$

Assume that primary and all secondaries operate at the same current density D_{cma} in circular mils per RMS ampere. Bobbin area occupied by the reset winding is negligible as it carries only magnetizing current, and it is usually smaller than No. 30 AWG.

Let A_b = bobbin winding area, in^2
A_p = primary winding area, in^2
A_s = secondary winding area (total of all secondaries), in^2
A_{ti} = area of one turn of primary power winding, in^2

Then for a space factor SF of 0.4 and $A_p = A_s$

$$A_p = 0.20 A_b = N_p A_{ti}$$

or

$$A_{ti} = \frac{0.2 A_b}{N_p} \tag{7.4}$$

Current density D_{cma} in circular mils per RMS ampere is

$$D_{cms} = \frac{A_{tcm}}{I_{rms}}$$

in which A_{tcm} is primary wire area in circular mils. Then

$$I_{rms} = \frac{A_{tcm}}{D_{cma}} \tag{7.5}$$

Area in square inches equals area in circular mils times $(\pi/4)\, 10^{-6}$, so

$$A_{tcm} = \frac{4 A_{ti} \times 10^{+6}}{\pi} = \frac{4(0.20 A_b) 10^{+9}}{\pi N_p}$$

From Eq. 7.5

$$I_{rms} = \frac{0.8 A_b \times 10^{+6}}{\pi N_p D_{cma}} \tag{7.6}$$

and putting Eq. 7.6 into Eq. 7.3

$$P_o = \left(1.265 N_p B_{max} A_e f \times 10^{-8}\right) \frac{0.8 A_b \times 10^{+6}}{\pi N_p D_{cma}}$$
$$= \frac{0.00322\, B_{max} f A_e A_b}{D_{cma}}$$

A_b is in square inches. If it is to be expressed in square centimeters, divide by 6.45. Then

$$P_o = \frac{0.00050\, B_{\text{max}} f\, A_e\, A_b}{D_{\text{cma}}} \tag{7.7}$$

where P_o is in watts for B_{max} in gauss, A_e and A_b are in square centimeters, f is in hertz, and D_{cma} in circular mils per RMS ampere.

7.3.2 Derivation of Output Power Relations for Push-Pull Topology

For the same assumptions as in Section 7.3.1, $P_o = 0.8P_{\text{in}} = 0.8V_{\text{dc(min)}}$ $[I_{\text{av}}$ at $V_{\text{dc(min)}}]$. In a push-pull, at $V_{\text{dc(min)}}$, each transistor is "on" a maximum of $0.8T/2$ within its half period. For two such pulses per period, the total duty cycle of current drawn from $V_{\text{dc(min)}}$ is 0.8, and during its "on" time, each transistor and half primary carries an equivalent flat-topped current pulse of amplitude I_{pft}. Output power is

$$\begin{aligned} P_o &= 0.8V_{\text{dc(min)}}\,(0.8I_{\text{pft}}) \\ &= 0.64V_{\text{dc(min)}}\,I_{\text{pft}} \end{aligned} \tag{7.8}$$

But each half primary carries current at a duty cycle of only 0.4, so RMS current in each half primary is $I_{\text{rms}} = I_{\text{pft}}\sqrt{0.4}$ or $I_{\text{pft}} = 1.58I_{\text{rms}}$. So

$$\begin{aligned} P_o &= 0.64V_{\text{dc(min)}}\,(1.58I_{\text{rms}}) \\ &= 1.01V_{\text{dc(min)}}\,I_{\text{rms}} \end{aligned} \tag{7.9}$$

Again for $SF = 0.4$, with half the total coil area devoted to the primary and half to the secondary, and each winding operating at a current density of D_{cma} circular mils per RMS ampere, we obtain

$$A_p = 0.20V_b = 2N_pA$$

or

$$A_{\text{ti}} = \frac{0.1A_b}{N_p} \tag{7.10}$$

in which N_p = number of primary turns
A_p = core area, cm^2
A_{ti} = area of a single turn of primary wire, in^2

$$D_{\text{cma}} = \frac{A_{\text{tcm}}}{I_{\text{rms}}}$$

$A_{\rm tcm}$ = wire area in circular mils and $I_{\rm rms}$ = RMS current per half primary

$$I_{\rm rms} = \frac{A_{\rm tcm}}{D_{\rm cma}} \tag{7.11}$$

$A_{\rm ti} = A_{\rm tcm}\,(\pi/4)10^{-6}$. Putting this into Eq. 7.10, we obtain

$$A_{\rm tcm} = 0.1273\frac{A_b}{N_p}10^{+6}$$

and putting this into Eq. 7.11

$$I_{\rm rms} = 0.1273\frac{A_b}{N_p D_{\rm cma}}10^{+6}$$

and putting this into Eq. 7.9

$$P_o = 1.01\,V_{\rm dc(min)}\frac{0.1273 A_b}{N_p D_{\rm cma}}10^{+6}$$

$$= 0.129\frac{V_{\rm dc(min)}\,A_b}{N_p D_{\rm cma}}10^{+6} \tag{7.12}$$

and finally from Faraday's law

$$V_{\rm primary\ min} \approx V_{\rm primary\ min} = \left(\frac{N_p A_e \Delta B}{\Delta T}\right)10^{-8}$$

In a push-pull, flux swing is $2B_{\rm max}$ in a time $0.4T$ at $V_{\rm dc(min)}$. Put this into Eq. 7.12:

$$P_o = 0.129(N_p A_e)\frac{2B_{\rm max}}{0.4T}\,\frac{A_b}{N_p D_{\rm cma}}10^{-2}$$

$$= \frac{0.00645 B_{\rm max} f\ A_e A_b}{D_{\rm cma}}$$

Again A_b is in square inches. If it is expressed in square centimeters, divide this last relation by 6.45.

$$P_o = \frac{0.0010 B_{\rm max} f\ A_e A_b}{D_{\rm cma}} \tag{7.13}$$

where P_o is in watts for $B_{\rm max}$ in gauss, A_e and A_b are in square centimeters, f is in hertz, and $D_{\rm cma}$ is in circular mils per RMS ampere. This result—specifically, that the power available from a given core in a push-pull topology is twice that for the same core in a forward converter topology—might have been foreseen.

In the push-pull, each transformer half must sustain the same voltage as a forward converter fed from the same supply voltage. But in the push-pull, the available flux change is $2B_{max}$ as compared to B_{max} in the forward converter. Thus from Faraday's law, the number of turns on a half primary of the push-pull is half that in the forward converter for the same B_{max}. But since there are two half secondaries in the push-pull, the total number of turns in the push-pull is equal to the number of turns in the forward converter for equal V_{dc} and B_{max} in both circuits (neglecting the insignificant space occupied by the reset winding in the forward converter).

In the push-pull, however, half the total output power is delivered through each half secondary. Thus, for equal output powers in a push-pull and a forward converter, the peak and RMS currents in each push-pull transformer half is half that in the forward converter (compare Eqs. 2.11 to 2.28 and 2.9 to 2.41). Thus the required circular-mil wire area and hence the required bobbin winding space for each half push-pull are half that for a forward converter of equal output power. So for equal bobbin winding space, the push-pull can deliver twice the power of a forward converter from the same core, as indicated by Eqs. 7.17 and 7.13.

7.3.2.1 Core and Copper Losses in Push-Pull, Forward Converter Topologies

Comparing Eqs. 7.7 and 7.13, it is seen that the push-pull topology can yield twice the output power of a forward converter using the same-sized (same $A_e A_b$ product) core.

There is a slight penalty paid in the push-pull transformer: at twice the forward converter output power, it will run warmer than in the forward converter. Doubling forward converter output power by going to a push-pull does double the core losses, but copper losses remain unchanged. This can be seen as follows: In the push-pull (Figure 7.3*a*3), each half primary must sustain the same voltage as the primary of the forward converter fed from the same supply voltage. In the forward converter, the flux density changes from zero to some preselected value B_{max} in a time $0.8T1$ (Figure 7.3*a*2). In the push-pull, the flux density change is from $-B_{max}$ to B_{max} or $2B_{max}$ in the same time $0.8T1$.

In Faraday's law, the number of primary turns is directly proportional to the applied voltage and inversely proportional to the flux density change. Thus, the number of turns in each half primary of the push-pull is half that of the forward converter primary for equal peak flux densities.

For the push-pull to have twice the output power of the original forward converter, the peak current in each half period must be equal to that of the forward converter. Since the duty cycle of current in each half primary of the push-pull is equal to that of the forward converter of half the output power, RMS current in each push-pull half primary is equal to that of the forward converter primary.

Thus, the forward converter primary and each push-pull half primary will use equal wire sizes (equal numbers of circular mils) since the wire size will be selected on the basis of 500 circular mils per RMS ampere, and since the number of half primary turns is half that of the full primary in the forward converter of half the output power, the two half primaries of the push-pull occupy a volume equal to that of the forward converter.

The push-pull half primary has half the turns of, and RMS current equal to, the forward converter primary, and hence has half the I^2R copper loss of the full-forward converter primary. Thus, total copper loss for the two half-primaries of the push-pull equals that of the forward converter of half the output power.

However, core losses for the push-pull will be twice that for the forward converter. In the push-pull, the flux density change is from $-B_{\text{max}}$ to B_{max}, but in the forward converter only from zero to B_{max}. Core losses are proportional to the area of the hysteresis loop transversed and the frequency.

The push-pull, operating over the first and third quadrant of the hysteresis loop, will thus have twice the losses of the forward converter operating over only the first quadrant of the hysteresis loop (see Section 7.2.1). With the newer low-loss core material, the increased core losses will not become the limiting factor in transformer temperature rise at frequencies below about 30 kHz.

7.3.2.2 Doubling Output Power from a Given Core Without Resorting to a Push-Pull Topology

In the previous section, it was pointed out that for the same core, a push-pull topology at the lower frequencies can yield twice the output power of a forward converter. The only penalty paid is a doubling of the core losses; copper losses remain unchanged.

A push-pull with its two transistors, however, has added cost and space drawbacks. Also, the push-pull transformer is harder to wind, and hence more expensive, as three wires must be taken out of each winding as compared to two for a forward converter. Also, there is the everpresent possibility of flux imbalance (Section 2.2.5) if current mode is not used.

If it is desired to double the output power of a given forward converter without increasing its core size, the following is an alternative to a push-pull (see Figure 7.3*a*).

Below about 30kHz the push-pull yields twice the output of a forward converter from a given core because there are two current pulses per period rather than one for the forward converter. Thus, an alternative to the push-pull is to retain the forward converter topology, and have it give one current pulse, whose amplitude is twice that of the push-pull, per half period of the push-pull. This can be seen in Figure 7.3*a*4.

This is another way of saying that the forward converter frequency is doubled (forward converter frequency is defined as the inverse of the time required for a complete traversal of the hysteresis loop back to its starting point). Equation 7.7 states the available output power from a core is directly proportional to frequency.

Thus, going to twice the original forward converter frequency would unquestionably yield twice the output power from the core, but it would result in somewhat more than twice (close to three times) the core losses. Core losses are roughly proportional to the 1.7th power of the frequency (Table 7.1 and Section 7.2.3).

After Pressman *Doubling the pulse repetition rate in the forward converter will not necessarily increase core loss compared with the equivalent push-pull example, providing the peak flux density remains the same. Remember the peak-to-peak flux change in the push-pull is twice the change found in the forward converter, so a single excursion of the push-pull has the same loss as two pulses of the forward converter. If you prefer, the area of the B/H loop for a single pulse of the forward converter is half that of a single pulse of the push-pull converter.*

Copper losses in a forward converter at twice the frequency and peak primary current, and hence twice the output power, would remain unchanged. This will be discussed below.

If doubling the frequency and peak primary-current amplitude of a forward converter doubles its output power and costs only a threefold increase in core loses, this may be a viable alternative to a push-pull at the original frequency. The push-pull at the original frequency also achieves double output power, but at a cost of only doubling the core losses.

Doubling the forward converter frequency and its peak current amplitude is practical only at original frequencies below 50 to 80 kHz. At higher original frequencies, the increased AC switching and snubber losses (see Chapter 11) at the doubled frequency would result in poor efficiency.

In Figure 7.3*a*4, the original forward converter frequency and peak primary current were doubled to achieve a doubling of output power. But, this has its own drawbacks. First, the higher peak current will cause more severe RFI problems, and it may force the selection of a higher current, more expensive transistor.

If the doubled peak primary current is unattractive, an alternative might be to increase the maximum transistor "on" time. In Figure 7.3*a*4, it was chosen as $0.8T1/2$ so as to ensure that the transformer core can be fully reset with an equal reset time and a guaranteed "dead" time before the start of the next turn "on" (Section 2.3.2). This is achieved by setting the ratio N_r/N_p (transformer reset winding to power winding turns ratio) equal to 1.0. (See Figure 7.3.)

In Section 2.3.8, it was shown that making N_r/N_p less than 1.0 permits a larger "on" time at a cost of a larger reset voltage and higher transistor off-voltage stress. In Figure 7.3*a*5, an N_r/N_p ratio of 0.5 was thus chosen. This yielded a larger maximum "on" time of $0.53T1$ and a peak current of $1.51I_p$ as compared to $0.4T1$ and a peak current of $2I_p$ when N_r/N_p is 1.0. The penalty is that the peak transistor off-voltage stress is $3V_{dc}$ rather than $2V_{dc}$ for the case of $N_r/N_p = 1.0$.

As discussed in Section 2.3.8, setting the N_r/N_p ratio less than 1.0 decreases transistor peak current stress, but increases its peak voltage stress. Ratios of N_r/N_p less than 0.5 generally lead to unacceptably high off-voltage stress.

As stated above, going to twice the forward converter frequency and peak transistor current doubles the output power but does not increase copper losses. This can be seen below as follows:

Figure No.	Frequency	I_p	"On" duty cycle	I_{rms}	N	Wire area	Wire resistance	$(I_{rms})^2R$	P_0
7.3*a*2	F1	I_p	0.4	$0.632\,I_p$	N	$A1$	$R1$	$(I_{rms})^2R$	P_0
7.3*a*4	2F1	$2I_p$	0.4	$1.264I_p$	$0.5N$	$2A1$	$0.25R1$	$(I_{rms})^2R$	$2P_0$

Since the double-frequency forward converter has twice the RMS current, it will have twice the wire area of the original converter of half the output power. Since it has half the number of primary turns, its resistance is one-fourth the resistance of the original forward converter. With twice the RMS current, its I^2R losses are equal to that of the original forward converter of half the output power.

7.3.3 Derivation of Output Power Relations for Half Bridge Topology

The half bridge is shown in Figure 3.1. Again assume that minimum DC input voltage is $V_{dc(min)}$. Maximum "on" time per transistor is $0.8T/2$ and occurs at $V_{dc(min)}$. Thus

Efficiency = 80%
A_e, A_b = core, bobbin area, cm^2
A_{bi} = bobbin area, in^2
A_p = primary area, in^2
SF = 0.4, primary and total secondary areas equal
D_{cma} = current density, circular mils/RMS A (all windings operate at same current density)
A_{ti} = wire area, in^2
A_{tcm} = wire area, circular mils
N_p = number of primary turns
I_{pft} = equivalent flat-topped primary current pulse

As I_{pft} comes at a duty cycle of 0.8, its RMS value is

$$I_{rms} = I_{pft}\sqrt{0.8} = 0.894 I_{pft} \quad \text{or} \quad I_{pft} = 1.12 I_{rms}$$

Then

$$\begin{aligned} P_o &= 0.8 P_{in} = 0.8 \frac{V_{dc(min)}}{2} \left[I \text{ average at } V_{dc(min)} \right] \\ &= 0.4 V_{dc(min)} 0.8 I_{pft} \\ &= 0.32 V_{dc(min)} I_{pft} \\ &= 0.358 V_{dc(min)} I_{rms} \end{aligned} \tag{7.14}$$

and

$$\begin{aligned} A_p &= 0.2 A_{bi} = N_p A_{ti} \\ A_{ti} &= \frac{0.2 A_{bi}}{N_p} \\ A_{ti} &= A_{tcm} (\pi/4) 10^{-6} \end{aligned}$$

So

$$A_{tcm} = 0.255 \left(\frac{A_{bi}}{N_p} \right) 10^{+6} \tag{7.15}$$

$$\begin{aligned} I_{rms} &= \frac{A_{tcm}}{D_{cma}} \\ &= 0.255 \frac{A_{bi}}{N_p D_{cma}} 10^{+6} \end{aligned} \tag{7.16}$$

Putting Eq. 7.16 into Eq. 7.14 we obtain

$$P_o = 0.0913 \frac{V_{dc(min)} A_{bi}}{N_p D_{cma}} 10^{+6} \tag{7.17}$$

From Faraday's law, since $V_{dc(min)}/2$ is applied to the primary

$$V_{p(min)} = \frac{V_{dc(min)}}{2} = N_p A_e \frac{\Delta B}{\Delta T} 10^{-8}$$

where ΔB is $2B_{max}$ and ΔT is $0.4T$. Then

$$V_{dc(min)} = 10 N_p f A_e B_{max} 10^{-8}$$

Putting this into Eq. 7.17, we have

$$P_o = \frac{0.00913 B_{max} f A_e A_{bi}}{D_{cma}}$$

and dividing by 6.45 for the bobbin area in square centimeters:

$$P_o = \frac{0.0014 B_{\text{max}} f A_e A_b}{D_{\text{cma}}} \qquad (7.18)$$

in which P_o is in watts, B_{max} is in gauss, f is in hertz, and A_e and A_b are in square centimeters.

7.3.4 Output Power Relations in Full Bridge Topology

A given core used in a full bridge topology can yield no more output power than the same core used in a half bridge. The full bridge can deliver twice the output power of the half bridge, but it requires a larger core to do so. This comes about as follows. A full bridge primary must sustain twice the supply voltage of the half bridge and hence must have twice the number of primary turns (Faraday's law and Section 3.3.2.1).

If the same fraction of the total bobbin area as in the half bridge is to be utilized for the primary, the wire area must be halved. If the wire area is halved, operating the same current density (circular mils per RMS ampere), the permissible RMS current must be halved. The full bridge transformer core, operating at twice the primary voltage and half the current of a half bridge, delivers the same output power as the half bridge with the same core.

A full bridge primary, operating at twice the voltage and half the current as a half bridge, delivers the same power, but at the same primary current as a half bridge, it delivers twice the output power. However, the full bridge requires a larger core winding area and hence a larger core to contain twice the number of turns at the same current density as a half bridge.

7.3.5 Conversion of Output Power Equations into Charts Permitting Core and Operating Frequency Selection at a Glance

Equations 7.7, 7.13, and 7.18 are valuable in selecting a core and operating frequency for a desired output power. As discussed in Section 7.1, using them requires a number of time-consuming iterative calculations.

The charts of Table 7.2*a* and 7.2*b* avoid such calculations by showing the available output power as calculated from these equations for a peak flux density B_{max} of 1600 G and a coil current density D_{cma} of 500 circular mils per RMS.

Core	A_e, cm^2	A_b, cm^2	A_eA_b, cm^4	Output power in watts at									Volume, cm^3
				20 kHz	24 kHz	48 kHz	72 kHz	96 kHz	150 kHz	200 kHz	250 kHz	300 kHz	
EE Cores, Ferroxcube-Philips													
814E250	0.202	0.171	0.035	1.1	1.3	2.7	4.0	5.3	8.3	11.1	13.8	16.6	0.57
813E187	0.225	0.329	0.074	2.4	2.8	5.7	8.5	11.4	17.8	23.7	29.6	35.5	0.89
813E343	0.412	0.359	0.148	4.7	5.7	11.4	17.0	22.7	35.5	47.3	59.2	71.0	1.64
812E250	0.395	0.581	0.229	7.3	8.8	17.6	26.4	35.3	55.1	73.4	91.8	110.2	1.93
782E272	0.577	0.968	0.559	17.9	21.4	42.9	64.3	85.8	134.0	178.7	223.4	268.1	3.79
E375	0.810	1.149	0.931	29.8	35.7	71.5	107.2	143.0	223.4	297.8	372.3	446.7	5.64
E21	1.490	1.213	1.807	57.8	69.4	138.8	208.2	277.6	433.8	578.4	722.9	867.5	11.50
783E608	1.810	1.781	3.224	103.2	123.8	247.6	371.4	495.1	773.7	1031.6	1289.4	1547.3	17.80
783E776	2.330	1.810	4.217	135.0	161.9	323.9	485.8	647.8	1012.2	1349.5	1686.9	2024.3	22.90
E625	2.340	1.370	3.206	102.6	123.1	246.2	369.3	492.4	769.4	1025.9	1282.3	1538.8	20.80
E55	3.530	2.800	9.884	316.3	379.5	759.1	1138.6	1518.2	2372.2	3162.9	3953.6	4744.3	43.50
E75	3.380	2.160	7.301	233.6	280.4	560.7	841.1	1121.4	1752.2	2336.3	2920.3	3504.4	36.00
EC Cores, Ferroxcube-Philips													
EC35	0.843	0.968	0.816	26.1	31.3	62.7	94.0	125.3	195.8	261.1	326.4	391.7	6.53
EC41	1.210	1.350	1.634	52.3	62.7	125.5	188.2	250.9	392.0	522.7	653.4	784.1	10.80
EC52	1.800	2.130	3.834	122.7	147.2	294.5	441.7	588.9	920.2	1226.9	1533.6	1840.3	18.80
EC70	2.790	4.770	13.308	425.9	511.0	1022.1	1533.1	2044.2	3194.0	4258.7	5323.3	6388.0	40.10

TABLE 7.2a Maximum Available Output Power in Forward Converter Topology

Core	A_e, cm^2	A_b, cm^2	A_eA_b, cm^4	Output power in watts at									Volume, cm^3
				20 kHz	24 kHz	48 kHz	72 kHz	96 kHz	150 kHz	200 kHz	250 kHz	300 kHz	
ETD Cores, Ferroxcube-Philips													
ETD 29	0.760	0.903	0.686	22.0	26.4	52.7	79.1	105.4	164.7	219.6	274.5	329.4	5.50
ETD 34	0.971	1.220	1.185	37.9	45.5	91.0	136.5	182.0	284.3	379.1	473.8	568.6	7.64
ETD 39	1.250	1.740	2.175	69.6	83.5	167.0	250.6	334.1	522.0	696.0	870.0	1044.0	11.50
ETD 44	1.740	2.130	3.706	118.6	142.3	284.6	427.0	569.3	889.5	1186.0	1482.5	1779.0	18.00
ETD 49	2.110	2.710	5.718	183.0	219.6	439.2	658.7	878.3	1372.3	1829.8	2287.2	2744.7	24.20
Pot Cores, Ferroxcube-Philips													
704	0.070	0.022	0.002	0.0	0.1	0.1	0.2	0.2	0.4	0.5	0.6	0.7	0.07
905	0.101	0.034	0.003	0.1	0.1	0.3	0.4	0.5	0.8	1.1	1.4	1.6	0.13
1107	0.167	0.054	0.009	0.3	0.3	0.7	1.0	1.4	2.2	2.9	3.6	4.3	0.25
1408	0.251	0.097	0.024	0.8	0.9	1.9	2.8	3.7	5.8	7.8	9.7	11.7	0.50
1811	0.433	0.187	0.081	2.6	3.1	6.2	9.3	12.4	19.4	25.9	32.4	38.9	1.12
2213	0.635	0.297	0.189	6.0	7.2	14.5	21.7	29.0	45.3	60.4	75.4	90.5	2.00
2616	0.948	0.407	0.386	12.3	14.8	29.6	44.4	59.3	92.6	123.5	154.3	185.2	3.53
3019	1.380	0.587	0.810	25.9	31.1	62.2	93.3	124.4	194.4	259.2	324.0	388.8	6.19
3622	2.20	0.774	1.563	50.0	60.0	120.1	180.1	240.2	375.2	500.3	625.4	750.5	10.70
4229	2.660	1.400	3.724	119.2	143.0	286.0	429.0	572.0	893.8	1191.6	1489.6	1787.5	18.20

TABLE 7.2a Maximum Available Output Power in Forward Converter Topology (*Continued*)

Core	A_e, cm^2	A_b, cm^2	A_eA_b, cm^4	Output power in watts at									Volume, cm^3
				20 kHz	24 kHz	48 kHz	72 kHz	96 kHz	150 kHz	200 kHz	250 kHz	300 kHz	
RM Cores, Ferroxcube-Philips													
RM5	0.250	0.095	0.024	0.8	0.9	1.8	2.7	3.6	5.7	7.6	9.5	11.4	0.45
RM6	0.370	0.155	0.057	1.8	2.2	4.4	6.6	8.8	13.8	18.4	22.9	27.5	0.80
RM8	0.630	0.310	0.195	6.2	7.5	15.0	22.5	30.0	46.9	62.5	78.1	93.7	1.85
RM10	0.970	0.426	0.413	13.2	15.9	31.7	47.6	63.5	99.2	132.2	165.3	198.3	3.47
RM12	1.460	0.774	1.130	36.2	43.4	86.8	130.2	173.6	271.2	361.6	452.0	542.4	8.34
RM14	1.980	1.100	2.178	69.7	83.6	167.3	250.9	334.5	522.7	697.0	871.2	1045.4	13.19
PQ Cores, Magnetics Inc.													
42016	0.620	0.256	0.159	5.1	6.1	12.2	18.3	24.4	38.1	50.8	63.5	76.2	2.31
42020	0.620	0.384	0.238	7.6	9.1	18.3	27.4	36.6	57.1	76.2	95.2	114.3	2.79
42620	1.190	0.322	0.383	12.3	14.7	29.4	44.1	58.9	92.0	122.6	153.3	183.9	5.49
42625	1.180	0.502	0.592	19.0	22.7	45.5	68.2	91.0	142.2	189.6	236.9	284.3	6.53
43220	1.700	0.470	0.799	25.6	30.7	61.4	92.0	122.7	191.8	255.7	319.6	383.5	9.42
43230	1.610	0.994	1.600	51.2	61.5	122.9	184.4	245.8	384.1	512.1	640.1	768.2	11.97
43535	1.960	1.590	3.116	99.7	119.7	239.3	359.0	478.7	747.9	997.2	1246.6	1495.9	17.26
44040	2.010	2.490	5.005	160.2	192.2	384.4	576.6	768.8	1201.2	1601.6	2002.0	2402.4	20.45

Note: From Eq. 7.7, $P_o = 0.00050 B_{max} f A_e A_b / D_{cma}$, where P_o is in watts, B_{max} in gauss, A_e and A_b in square centimeters, f in hertz, D_{cma} in circular mils per rms ampere, bobbin winding space factor = 40 percent. For B_{max} = 1600 G. For other B_{max}, multiply by $B_{max}/1600$. Fr D_{cma} = 500 circular mils/rms ampere. For other D_{cma}, multiply by $500/D_{cma}$. For push-pull topology, multiply powers by a factor of 2.

TABLE 7.2a Maximum Available Output Power in Forward Converter Topology (*Continued*)

Core	A_e, cm^2	A_b, cm^2	A_eA_b, cm^4	Output power in watts at									Volume, cm^3
				20 kHz	24 kHz	48 kHz	72 kHz	96 kHz	150 kHz	200 kHz	250 kHz	300 kHz	
EE Cores, Ferroxcube-Philips													
814E250	0.202	0.171	0.035	3.1	3.7	7.4	11.2	14.9	23.2	30.9	38.7	46.4	0.57
813E187	0.225	0.329	0.074	6.6	8.0	15.9	23.9	31.8	49.7	66.3	82.9	99.5	0.89
813E343	0.412	0.359	0.148	13.3	16.0	31.8	47.8	63.6	99.4	132.5	165.7	198.8	1.64
812E250	0.395	0.229	20.6	24.8	49.3	74.1	98.7	154.2	154.2	205.6	257.0	308.4	1.93
782E272	0.577	0.968	0.559	50.0	60.3	120.1	180.4	240.2	375.3	500.4	625.6	750.7	3.79
E375	0.810	1.149	0.931	83.4	100.5	200.1	300.6	400.2	6254	833.9	1042.4	1250.8	5.64
E21	1.490	1.213	1.807	161.9	195.2	388.6	583.8	777.2	1214.6	1619.4	2024.3	2429.1	11.50
783E608	1.810	1.781	3.224	288.8	348.1	693.1	1041.2	1386.2	2166.2	2888.4	3610.4	4332.5	17.80
783E776	2.330	1.810	4.217	377.9	455.5	906.7	136.2	1813.4	2834.0	3778.7	4723.4	5668.1	22.90
E625	2.340	1.370	3.206	287.2	346.2	689.2	1035.5	1378.5	2154.3	2872.4	3590.4	4308.6	20.80
E55	3.530	2.800	9.884	885.6	1067.5	2125.1	3192.5	4250.1	6642.0	8856.1	11070.1	13284.1	43.50
E75	3.380	2.160	7.301	654.2	788.5	1569.7	2358.2	3139.3	4906.1	6541.5	8176.9	9812.3	36.00
EC Cores, Ferroxcube-Philips													
EC35	0.843	0.968	0.816	73.1	88.1	175.4	263.6	350.9	548.4	731.2	913.9	1096.7	6.53
EC41	1.210	1.350	1.634	146.4	176.4	351.2	527.6	702.4	1097.7	1463.6	1829.5	2195.4	10.80
EC52	1.800	2.130	3.834	343.5	414.1	824.3	1238.4	1648.6	2576.4	3435.3	4294.1	5152.9	18.80
EC70	2.790	4.770	13.308	1192.4	1437.3	2861.3	4298.6	5722.6	8943.2	11924.2	14905.3	17886.4	40.10

TABLE 7.2b Maximum Available Output Power in Half or Full Bridge Topology

Core	A_e, cm^2	A_b, cm^2	$A_e A_b$, cm^4	Output power in watts at 20 kHz	24 kHz	48 kHz	72 kHz	96 kHz	150 kHz	200 kHz	250 kHz	300 kHz	Volume, cm^3
ETD Cores, Ferroxcube-Philips													
ETD 29	0.760	0.903	0.686	61.5	74.1	147.6	221.7	295.1	461.2	614.9	768.6	922.4	5.50
ETD 34	0.971	1.220	1.185	106.1	127.9	254.7	382.6	509.4	796.1	1061.4	1326.8	1592.1	7.64
ETD 39	1.250	1.740	2.175	194.9	234.9	467.6	702.5	935.3	1461.6	1948.8	2436.0	2923.2	11.50
ETD 44	1.740	2.130	3.706	332.1	400.3	796.8	1197.1	1593.7	2490.6	3320.8	4150.9	4981.1	18.00
ETD 49	2.110	2.710	5.718	512.3	617.6	1229.4	1846.9	2458.9	3842.6	5123.4	6404.3	7685.1	24.20
Pot Cores, Ferroxcube-Philips													
704	0.070	0.022	0.002	0.1	0.2	0.3	0.5	0.7	1.0	1.4	1.7	2.1	0.07
905	0.101	0.034	0.003	0.3	0.4	0.7	1.1	1.5	2.3	3.1	3.8	4.6	0.13
704	0.070	0.022	0.002	0.1	0.2	0.3	0.5	0.7	1.0	1.4	1.7	2.1	0.07
905	0.101	0.034	0.003	0.3	0.4	0.7	1.1	1.5	2.3	3.1	3.8	4.6	0.13
1107	0.167	0.054	0.009	0.8	1.0	1.9	2.9	3.9	6.1	8.1	10.1	12.1	0.25
1408	0.251	0.097	0.024	2.2	2.6	5.2	7.8	10.4	16.3	21.8	27.2	32.7	0.50
1811	0.433	0.187	0.081	7.3	8.7	17.4	26.2	34.8	54.4	72.6	90.7	108.8	1.12
2213	0.635	0.297	0.189	16.9	20.4	40.5	60.9	81.9	126.7	169.0	211.2	253.5	2.00
2616	0.948	0.407	0.386	34.6	41.7	83.0	124.6	165.9	259.3	345.7	432.1	518.6	3.53
3019	1.380	0.587	0.810	72.6	87.5	174.2	261.6	348.3	544.4	725.8	907.2	1088.7	6.19
3622	2.020	0.774	1.563	140.1	168.9	336.1	505.0	672.3	1050.7	1400.9	1751.1	2101.3	10.70
4229	2.660	1.400	3.724	333.7	402.2	800.7	1202.9	1601.3	2502.5	3336.7	4170.9	5005.1	18.20

TABLE 7.2b Maximum Available Output Power in Half or Full Bridge Topology (*Continued*)

Core	A_e, cm^2	A_b, cm^2	$A_e A_b$, cm^4	Output power in watts at									Volume, cm^3
				20 kHz	24 kHz	48 kHz	72 kHz	96 kHz	150 kHz	200 kHz	250 kHz	300 kHz	
RM Cores, Ferroxcube-Philips													
RM5	0.250	0.095	0.024	2.1	2.6	5.1	7.7	10.2	16.0	21.3	26.6	31.9	0.45
RM6	0.370	0.155	0.057	5.1	6.2	12.3	18.5	24.7	38.5	51.4	64.2	77.1	0.80
RM8	0.630	0.310	0.195	17.5	21.1	42.0	63.1	84.0	131.2	175.0	218.7	262.5	1.85
RM10	0.970	0.426	0.413	37.0	44.6	88.8	133.5	177.7	277.7	370.2	462.8	555.4	3.47
RM12	1.460	0.774	1.130	101.3	122.0	243.0	365.0	485.9	759.4	1012.5	1265.6	1518.8	8.34
RM14	1.980	1.100	2.178	195.1	235.2	468.3	703.5	936.5	1463.6	1951.5	2439.4	2927.2	13.19
PQ Cores, Magnetics Inc.													
42016	0.620	0.256	0.159	14.2	17.1	34.1	51.3	68.2	106.7	142.2	177.8	213.3	2.31
42020	0.620	0.384	0.238	21.3	25.7	51.2	76.9	102.4	160.0	213.3	266.6	320.0	2.79
42620	1.190	0.322	0.383	34.3	41.4	82.4	123.8	164.8	257.5	343.3	429.2	515.0	5.49
42625	1.180	0.502	0.592	53.1	64.0	127.4	191.3	254.7	398.1	530.8	663.4	796.1	6.53
43.220	1.700	0.470	0.799	71.6	86.3	171.8	258.1	343.6	536.9	715.9	894.9	1073.9	9.42
43230	1.610	0.994	1.600	143.4	172.8	344.1	516.9	688.1	1075.4	1433.9	1792.4	2150.9	11.97
43535	1.960	1.590	3.116	279.2	336.6	670.0	1006.6	1340.1	2094.2	2792.3	3490.4	4188.4	17.26
44040	2.010	2.490	5.005	448.4	540.5	1076.1	1616.6	2152.1	3363.3	4484.4	5605.5	6726.6	20.45

Note: From Eq. 7.18, $P_o = 0.0014 B_{max} f A_e A_b / D_{cma}$, where P_o is in watts, B_{max} in gauss, A_e and A_b in square centimeters, f in hertz, D_{cma} in circular mils per rms ampere, bobbin winding space factor = 40 percent. For B_{max} = 1600 G. For other B_{max}, multiply by $B_{max}/1600$. Fr D_{cma} = 500 circular mils/rms ampere. For other D_{cma}, multiply by $500/D_{cma}$.

TABLE 7.2b Maximum Available Output Power in Half or Full Bridge Topology (*Continued*)

The reason for the selection of 1600 G is discussed in Sections 7.2.3 and 2.2.9.4. At frequencies above about 50 kHz, excessive core losses for some of the more lossy materials may dictate a lower value $B_{max,l}$ for the peak flux density. The charts make a rapid calculation of the available power simple. If the lowered available power shown in the charts at the lowered peak flux density is multiplied by $(B_{max,l}/1600)$, it yields the power at the actual B_{max}.

The selection of $D_{cma} = 500$ circular mils per RMS ampere is a common compromise in transformer design. A higher density (lower value of D_{cma}) would result in more copper losses, and a lower density would unnecessarily increase the coil size. Current densities down to 300 circular mils per RMS ampere are acceptable, but densities below this should definitely be avoided.

Actually, the choice of D_{cma} specifies only the DC wire resistance. In subsequent sections skin and proximity effects will be discussed. These effects produce eddy currents in the wires, cause the currents to flow in only a fraction of the wire area, and hence may make the effective wire resistance considerably higher than the values shown in wire tables for wires of a specified circular-mil area. Nevertheless, choosing a current density of 500 circular mils per RMS ampere is a good starting point.

The charts of Table 7.2 are used as follows. First, choose a topology which yields the best combination of power transistor off-voltage and peak-current stress. Another topology selection criterion is to minimize the cost of components.

Note that the cores are arranged vertically in order of increasing $A_e A_b$ product and hence increasing output power capability. If familiarity or experience dictates or suggests a particular operating frequency, that vertical frequency column is entered. Now move vertically downward and choose the first core whose output power is at least the specified maximum power.

If a specific core is chosen first because it fits the available space, go to that core and move horizontally to the right to the first frequency which yields at least the specified maximum output power.

If a desired core does not yield the required output power at a selected frequency in—say—a forward converter topology, a push-pull might be considered. The push-pull (voltage- or current-mode) topology with the same core offers twice the output power at the same frequency. If voltage-mode push-pull is selected, all the precautions relating to flux imbalance (Section 2.2.8) should be kept in mind.

Thus, by moving upward to smaller cores and to the right to higher frequencies, an optimum core-frequency combination can be found. For a given output power, at higher frequencies, the core gets smaller but core losses and transformer temperature rise, and transistor switching losses increase.

7.3.5.1 Peak Flux Density Selection at Higher Frequencies

Care should be taken in the use of Tables 7.2*a* and 7.2*b*. The powers shown are available only if operation at a peak flux density of 1600 G at the selected frequency does not cause excessive temperature rise. At frequencies in the range of 20 to 50 kHz, core losses are so low that temperature rise at a peak flux density of 1600 G is negligibly small, even for the most lossy materials of Table 7.1.

However, core losses increase roughly as the 1.6th power of the frequency and the 2.7th power of the peak flux density. Thus at frequencies above about 50 kHz, peak flux density may have to be reduced below 1600 G by increasing the number of primary turns to keep the transformer temperature rise acceptably low.

In general, smaller cores can more easily tolerate a higher peak flux density at high frequencies than larger cores. This is so because core losses are proportional to volume, but core cooling is proportional to radiating surface area. Thus, as a core gets larger, its volume increases faster than its surface area, and the internal heat generated increases more rapidly than the surface area which cools it.

A specific example can easily demonstrate this. Consider the Ferroxcube-Philips E55 core in Table 7.2*b*, which shows that if operation at 200 kHz and 1600 G were possible, it would be capable of 8856 W of output power in a half bridge topology. From Table 7.1 for 3C85 material, its losses are 700 mW/cm^3 at 1600 G and 200 kHz. For its volume of 43.5 cm^3, its dissipation is $0.7(43.5) = 30.5$ W. Coil losses (considered in the following section) probably equal this.

Consider a smaller core, the 813E343. From Table 7.2*b*, its output power capability at 1600 G, 200 kHz is 133 W in a half bridge. For a volume of 1.64 cm^3 and the same 700 mW/cm^3, its core losses are only 1.15 W. Thus, neglecting coil losses, the 813E343 with a volume of 1.64 cm^3 and core losses of 1.15 W would run at a far lower temperature than the E55 core with a volume of 43.5 cm^3 and 30.5 W of core losses.

Calculation of actual transformer temperature rise due to core plus coil losses will be demonstrated in the following section.

For the larger cores, the powers shown in Table 7.2*a* and 7.2*b* might not be obtainable at frequencies above 50 kHz as operation at 1600 G may result in excessive temperature rise. Peak flux density B_{max} would then have to be reduced to somewhere in the range 1400 to 800 G. Actual output powers are then those shown in Table 7.2*a* and 7.2*b* multiplied by $B_{max}/1600$.

Table 7.2*a* and 7.2*b* show output powers for the two major American core manufacturers. Many of their cores are interchangeable in their geometries and A_e values, but they are made from proprietary core materials that have different core losses in milliwatts per cubic centimeter. Core interchangeability (with regard to geometry and A_e only) and corresponding type numbers are shown in

Table 7.3 for Ferroxcube-Philips, Magnetics Inc., and TDK. Manufacturers' catalogs[1–4] also show a variety of core accessories—bobbins, and assembly and mounting hardware.

Each manufacturer produces these cores in their own proprietary core materials that have unique core loss characteristics in milliwatts per cubic centimeter.

7.4 Transformer Temperature Rise Calculations[8]

Transformer temperature rise above the ambient-air environment depends on total core plus coil (copper) losses, and radiating surface area. Forced air flowing past the transformer can lower temperature rise considerably, depending on the airflow rate in cubic feet per minute.

There is no way of calculating transformer temperature rise analytically with great accuracy. It can be estimated within about 10°C with some empirical curves based on the concept of thermal resistance of a radiating surface area. Recall that the definition of thermal resistance R_t of a heat sink is the temperature rise (usually in degrees Celsius) per watt of dissipation. Then temperature rise dT for a power dissipation P is simply $dT = PR_t$.

TIP *Reasonably accurate temperature rise predictions are now possible using the calculated total transformer loss (the sum of copper and core losses) and information related to effective wound component area and cooling methods. See Reference 18. ~K.B.*

Some core manufacturers list R_t for their various cores, implying that R_t multiplied by the total core plus copper losses yields the temperature rise of the outer surface of the core. An educated guess of typically 10 to 15°C is often assumed for the temperature rise of the internal hot spot (usually the core center leg) above the core's outer surface.

Temperature rise is dependent not only on the radiating surface area but also on the total dissipation. The greater the power dissipation from a radiating surface, the greater the temperature differential between the surface and the ambient air, and the more easily the surface area loses its heat or the lower the thermal resistance.

Transformer temperature rise will be estimated herein[8] as if the transformer's total outer surface area (2 × *width* × *height* 2 × *width* × *thickness* 2 × *height* × *thickness*) were the radiating area of an equivalent heat sink. The thermal resistance of this equivalent heat sink will be modified by the total dissipation (total of core plus copper losses).

Ferroxcube-Philips	Magnetics Inc.	TDK
	EE Cores	
814E250	41205	
813E187	41808	EE19
813E343		
812E250		
782E272		
E375	43515	
E21	44317	
783E608		EE4/42/15
783E776		
E625	44721	
E55		EE55/55/21
E75	45724	
	EC Cores	
EC35	43517	EC35
EC41	44119	EC41
EC52	45224	EC52
EC70	47035	EC70
	ETD Cores	
ETD29		
ETD34	43434	ETD34
ETD39	43939	ETD39
ETD44	44444	ETD44
ETD49	44949	ETD49
	Pot Cores	
704	40704	P7/4
905	40905	P9/5
1107	41107	P11/17
1408	41408	P14/8

TABLE 7.3 Core Type Numbers for Geometrically Interchangeable Cores

Ferroxcube-Philips	Magnetics Inc.	TDK
	Pot Cores (*Continued*)	
1811	41811	P14/8
2213	42213	P22/13
2616	42616	P26/16
3019	43019	P30/19
3622	43622	P36/22
4229	44229	P42/29
	RM Cores	
RM4	41110	RM4
RM5	41510	RM5
RM6	41812	RM6
RM7		RM7
RM8	42316	RM8
RM10	42819	RM10
RM12	43723	RM12
RM14		RM14
	PQ Cores	
	42016	PQ20/16
	42020	PQ20/20
	42620	PQ2620
	42625	PQ26/25
	43220	PQ32/20
	43230	PQ32/30
	43535	PQ32/30
	44040	PQ40/40
		PQ50/50

TABLE 7.3 Core Type Numbers for Geometrically Interchangeable Cores (*Continued*)

An empirical curve of heat-sink thermal resistance versus total surface area is shown in Figure 7.4*a*. It is the average of a large number of heat sinks of different sizes and shapes from different heat-sink manufacturers. The curve is the thermal resistance at a 1-W power level, and is a straight line on a log-log graph.

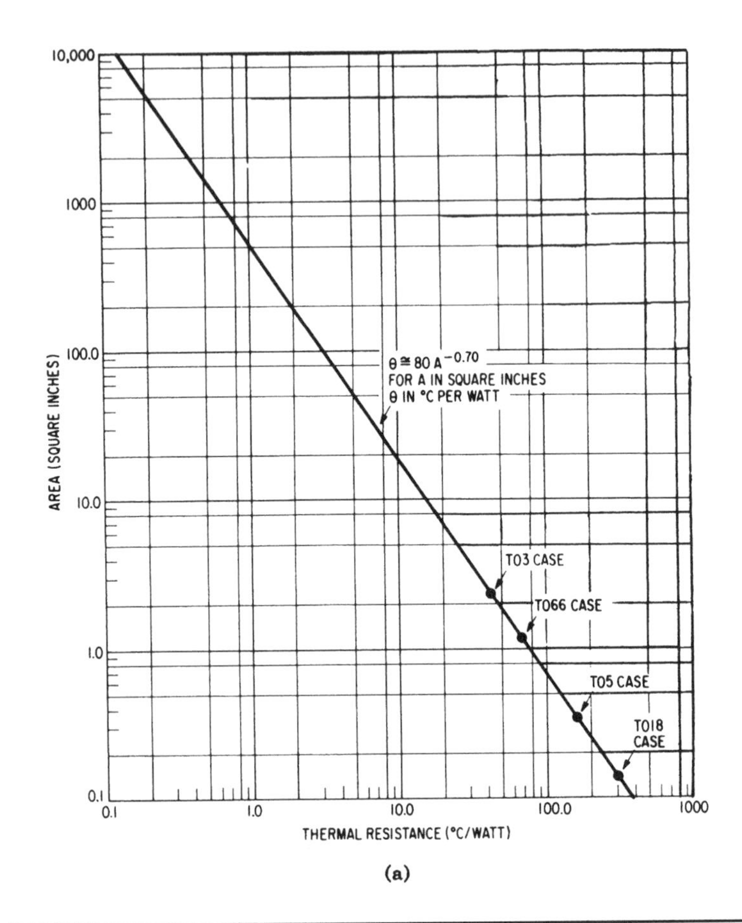

FIGURE 7.4 Calculating transformer temperature rise from its equivalent heat-sink area (total core area of both faces plus edge area). (*a*) Thermal resistance versus total heat-sink area. Total area means area of both sides of a flat plate, or both sides of all fins plus the back of a finned heat sink. Curve is at a power dissipation of 1 W. Use multiplying factor of Figure 7.4*b* for other power levels. (*b*) Normalized thermal resistance versus power dissipation in a heat sink. (*c*) Heat-sink temperature rise versus power dissipation for various heat-sink areas. From 7.19: $T = 80A^{-0.70}P^{0.85}$. Figure 7.4*b* is represented analytically by $K_1 = P^{-0.15}$. Combining Figure 7.4*a* $(= 80A^{-0.70})$ and 7.4*b* gives the temperature rise for any transformer power dissipation and radiating surface areas as $T = 80A^{-0.70}P^{0.85}$.

Although the thermal resistance of a finned heat sink depends somewhat on fin shape and spacing, and if the surface is blackened or aluminized, these are second-order effects. To a close approximation, the thermal resistance of a heat sink depends almost entirely on its radiating surface area only.

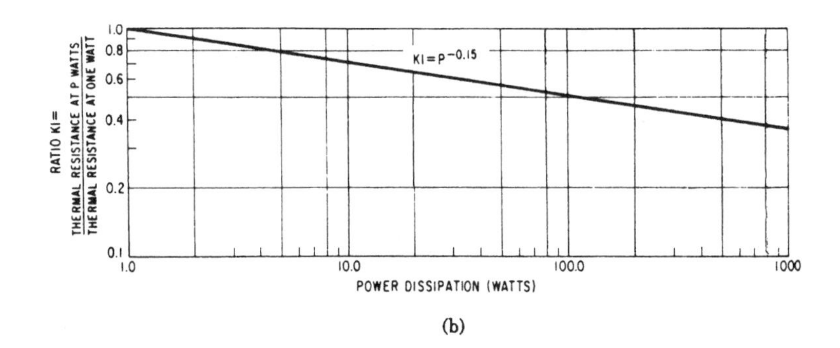

(b)

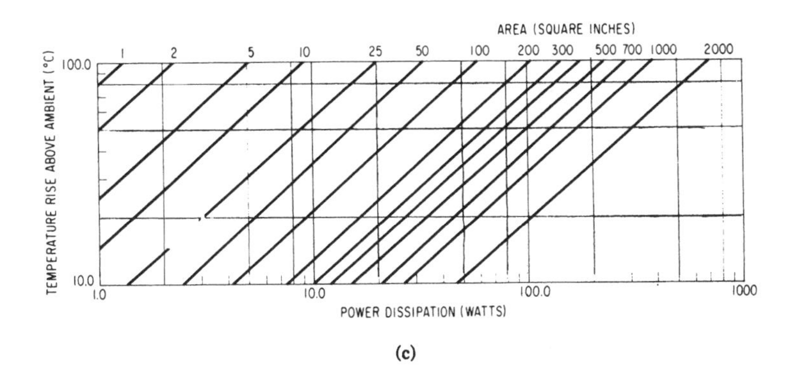

(c)

FIGURE 7.4 *(Continued)*

Also from various heat-sink manufacturers' catalogs, an average empirical curve is shown in Figure 7.4*b*, which gives the variation of thermal resistance with power dissipation.

From Figure 7.4*a* and 7.4*b*, the more directly useful curves of Figure 7.4*c* are derived. Figure 7.4*c* gives the temperature rise above ambient for various heat-sink areas (diagonal lines) and power dissipation. A transformer's outer surface temperature rise will hereafter be read from Figure 7.4*c* for the sum of its core plus copper losses and total radiating surface area as defined above. It is interesting to read from Figure 7.4*c* the temperature rise of the two cores discussed in Section 7.3.5.1. There it was calculated that an E55 of 43.5 cm^3 volume, operated at 1600 G and 200 kHz, dissipated 30.5 W. Its radiating surface area as defined above is 16.5 in^2. From Figure 7.4*c*, neglecting copper losses entirely, at 30.5 W of core losses its temperature rise is 185°C.

The smaller 813E343 core of 1.64-cm^3 volume, also operated at 1600 G and 200 kHz, has 1.15 W of core losses. Its radiating surface area, calculated as above, is 1.90 in^2. From Figure 7.4*c*, neglecting

Core	Radiating surface area, in^2	Thermal resistance, °C/W	
		Measured by manufacturer	Calculated from Figure 7.4*a*
EC35	5.68	18.5	23.7
EC41	7.80	16.5	19.0
EC52	10.8	11.0	12.6
EC70	22.0	7.5	9.2

TABLE 7.4 Core Thermal Resistance

copper losses, its temperature rise is only 57°C. It is thus verified that it is easier for smaller cores to deliver the powers shown in Figure 7.2*a* and 7.2*b* at 1600 G and high frequency.

It is of interest to compare the thermal resistance of some cores as measured by the manufacturer and as calculated ($R_t = 80A^{-0.70}$) from Figure 7.4*a* (see Table 7.4).

7.5 Transformer Copper Losses

7.5.1 Introduction

In Section 7.3 it was stated that wire size for all windings would be chosen to yield a current density of 500 circular mils per RMS ampere. It was assumed there that copper losses would be calculated as $(I_{rms})^2 R_{dc}$, where R_{dr} is the winding's DC resistance as calculated from its length and resistance in ohms per foot as read from the wire tables for the selected wire size. It was also assumed that I_{rms} is the RMS current as calculated from its waveshape (Sections 2.2.10.2, 2.3.10.4).

There are two effects—*skin* and *proximity* effects—which can cause the winding losses to be significantly greater than $(I_{rms})^2 R_{dr}$.

Both skin and proximity effects arise from eddy currents, which are induced by the varying magnetic fields in the coil. Skin effect is caused by eddy currents induced in a wire by the magnetic field of the current carried by the wire itself. Proximity effect is caused by eddy currents induced in wires by magnetic fields of currents in adjacent wires or adjacent layers of the coil.

Skin effect causes current in a wire to flow only in a layer on the surface of the wire. The depth of this skin or annular conducting area is inversely proportional to the square root of the frequency. Thus, as frequency increases, a progressively larger part of solid wire area is lost, increasing the AC resistance and hence copper losses.

One might not expect skin effect to increase wire resistance significantly at low frequencies, since the skin depth is 17.9 mils at 25 kHz for example.

However, currents in conventional switching supplies have rectangular waveshapes, whose high-frequency Fourier components comprise a considerable proportion of the total energy. Thus high AC resistance at these high-frequency harmonics is a concern even in a 25-kHz rectangular current waveform.

Skin effect will be discussed quantitatively in the following sections.

Proximity effect, caused by eddy currents induced by varying magnetic fields from adjacent conductors or adjacent coil layers, can cause considerably more copper loss than skin effect.

Proximity-effect losses can be especially high in multi-layer coils. This is so partly because the induced eddy currents crowd the net current into a small fraction of the copper wire area, increasing its resistance. What makes proximity effect more serious is that these induced eddy currents can be many times greater than the net current flowing in the individual wire or wire layers. This will be discussed quantitatively in the following sections.

7.5.2 Skin Effect

Skin effect[9–17] had been known, and equations had been derived for skin depth versus frequency, as far back as 1915.[9] The means by which induced eddy currents cause current to crowd into the thin outer skin of a conductor can be seen in Figure 7.5, which shows a section of a round conductor sliced through a diameter. It carries its main current in the direction of OA. If not for skin effect, the current would be uniformly distributed throughout the wire volume.

All current flowing in the direction OA is encircled by magnetic flux lines normal to OA. Consider a thin filament of current flowing along the axis OA. By Fleming's right-hand rule, its magnetic flux lines are in the direction shown by the arrow in the figure—from 1 to 2 to 3 and around back to 1.

Consider two flat loops (X and Y) within the wire. They are on a wire diameter and extend the full length of the wire. These loops are symmetrically displaced to either side of the wire axis. The magnetic flux lines flow up through loop X (shown by the dots in the center of the loop), and back down through loop Y (shown by the crosses in the center of loop Y).

By Faraday's law, when a varying magnetic field flows through an area, a current is induced in a line encircling that area. By Lenz's law, the polarity of the magnetic field induced by the eddy current flow opposes that which caused the eddy current.

Thus eddy currents will be induced in loops X and Y that flow in the directions shown. By the right-hand rule, the current in loop X flows

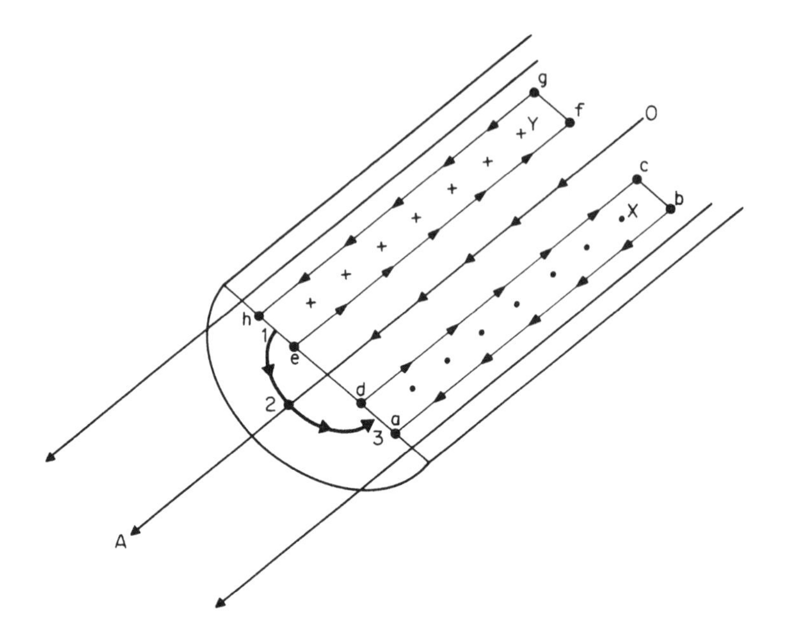

FIGURE 7.5 Eddy currents in a round wire cause skin effect—current canceling in the center of the wire, and its crowing into the outer skin. The magnetic field of the current in the wire induces voltages in loops such as *abcd* and *efgh*. The polarity of these voltages causes eddy currents to flow around the boundaries of the loops. The direction of the eddy currents is opposite to the main current flow on the inside of the loops (d to c and e to f), and in the same direction as the main current flow on the outside of the loops (b to a and g to h). The consequence is the canceling of current flow on the inside of the wire and its concentration in a skin on its outside.

clockwise in the direction d to c to a to d. That direction of current flow causes a magnetic field to go down through the center of the loop in opposition to the field from the main filament of current along OA. Similarly, in loop Y, the eddy current flows counterclockwise (*efgh* and back to e) so as to cause a magnetic field to come up through the center of the loop in opposition to the magnetic field of the main current filament along OA.

Note that the eddy currents along arms *dc* and *ef* are in a direction opposite to that of the main current filament OA, and tend to cancel it. Further, eddy currents along arms *ab* and *gh* (along the outer skin of the wire) are in the same direction as the main current, and tend to reinforce it.

Thus the net current—the sum of the eddy currents and the main current which caused them—is canceled at the center of the wire and crowded into the outer skin. Thus at high frequency, the total current-carrying area is less than the full wire area and the AC resistance is greater than the DC resistance by an amount determined by the skin thickness.

7.5.3 Skin Effect—Quantitative Relations

Skin depth is defined as the distance below the surface where the current density has fallen to $1/e$, or 37%, of its value at the surface. The relation between skin depth and frequency has been derived by many sources[9] and for copper wire at 70°C is

$$S = \frac{2837}{\sqrt{f}} \tag{7.19}$$

where S is the skin depth in mils and f is frequency in hertz.

Table 7.5 shows skin depth for copper wire at 70°C at various frequencies as calculated from Eq. 7.19.

Consider conductors of circular cross section. The relationship among a DC resistance R_{dc}, an AC resistance due to skin effect R_{ac},

Frequency, kHz	Skin depth, mils*
25	17.9
50	12.7
75	10.4
100	8.97
125	8.02
150	7.32
175	6.78
200	6.34
225	5.98
250	5.67
300	5.18
400	4.49
500	4.01

* From Eq. 7.19. Skin depth $S = 2837/\sqrt{f}$; S in mils for f in hertz.

TABLE 7.5 Skin Depth in Copper Wire at 70°C

and a resistance change ΔR due to skin effect is

$$R_{ac} = R_{dc} + \Delta R = R_{dc}\left(1 + \frac{\Delta R}{R_{dc}}\right) = R_{dc}(1 + f)$$

or

$$\frac{R_{ac}}{R_{dc}} = 1 + f \tag{7.20}$$

From the skin depth relation of Eq. 7.19, the ratio R_{ac}/R_{dc} can be calculated for any wire size at any frequency. Since the resistances are inversely proportional to wire conducting area, and the conducting area of the wire is the annular ring whose inner radius is $(r - S)$, for any skin depth S, wire radius r, and diameter d

$$\begin{aligned}\frac{R_{ac}}{R_{dc}} &= \frac{\pi r^2}{\pi r^2 - \pi(r - S)^2}\\ &= \frac{(r/S)^2}{(r/S)^2 - (r - S)^2/S^2}\\ &= \frac{(r/S)^2}{(r/S)^2 - (r/S - 1)^2}\\ &= \frac{(d/2S)^2}{(d/2S)^2 - (d/2S - 1)^2}\end{aligned} \tag{7.21}$$

Eq. 7.21 indicates that the wire's AC-to-DC resistance $R_{ac}/R_{dc} = (1/f)$ is dependent only on the ratio of wire diameter to skin depth. Figure 7.6 plots R_{ac}/R_{dc} against the ratio d/S from Eq. 7.21.

7.5.4 AC/DC Resistance Ratio for Various Wire Sizes at Various Frequencies

Because of skin effect, the AC-to-DC resistance ratio of round wire is dependent on the ratio of the wire diameter to skin depth (Eq. 7.21). Further, since skin depth is inversely proportional to the square root of frequency, different-sized wires have different AC-to-DC resistance ratios, and these ratios increase with frequency.

Table 7.6 shows this for all even-numbered wire sizes at 25, 50, 100, and 200 kHz. In this table, d/S (wire diameter/skin depth ratio) is calculated from the maximum bare wire diameter as given in the wire tables and skin depth is calculated from Eq. 7.19 (Table 7.5). From these d/S ratios, R_{ac}/R_{dc} is calculated from Eq. 7.21 or read from Figure 7.6.

It is apparent from Table 7.6 that large-diameter wires have a large AC/DC resistance ratio, which increases greatly with frequency. Thus No. 14 wire has a diameter 64.7 mils and a skin depth of 17.9 mils at 25 kHz (Table 7.5). This yields a d/S ratio of 3.6, and from Figure 7.6,

AWG	Diameter d, mils	25 kHz			50 kHz			100 kHz			200 kHz		
		Skin depth S, mils	d/S	R_{ac}/R_{dc}	Skin depth S, mils	d/S	R_{ac}/R_{dc}	Skin depth S, mils	d/S	R_{ac}/R_{dc}	Skin depth S, mils	d/S	R_{ac}/R_{dc}
12	81.6	17.9	4.56	1.45	12.7	6.43	1.85	8.97	9.10	2.55	6.34	12.87	3.50
14	64.7	17.9	3.61	1.30	12.7	5.09	1.54	8.97	7.21	2.00	6.34	10.21	2.90
16	51.3	17.9	2.87	1.10	12.7	4.04	1.25	8.97	5.72	1.70	6.34	8.09	2.30
18	40.7	17.9	2.27	1.05	12.7	3.20	1.15	8.97	4.54	1.40	6.34	6.42	1.85
20	32.3	17.9	1.80	1.00	12.7	2.54	1.05	8.97	3.60	1.25	6.34	5.09	1.54
22	25.6	17.9	1.43	1.00	12.7	2.02	1.00	8.97	2.85	1.10	6.34	4.04	1.30
24	20.3	17.9	1.13	1.00	12.7	1.60	1.00	8.97	2.26	1.04	6.34	3.20	1.15
26	16.1	17.9	0.90	1.00	12.7	1.27	1.00	8.97	1.79	1.00	6.34	2.54	1.05
28	12.7	17.9	0.71	1.00	12.7	1.00	1.00	8.97	1.42	1.00	6.34	2.00	1.00
30	10.1	17.9	0.56	1.00	12.7	0.80	1.00	8.97	1.13	1.00	6.34	1.59	1.00
32	8.1	17.9	0.45	1.00	12.7	0.64	1.00	8.97	0.90	1.00	6.34	1.28	1.00
34	6.4	17.9	0.36	1.00	12.7	0.50	1.00	8.97	0.71	1.00	6.34	1.01	1.00

Note. Skin depths are taken from Table 7.5; R_{ac}/R_{dc} from Eq. 7.21.

TABLE 7.6 AC/DC Resistance Ratios Due to Skin Effect

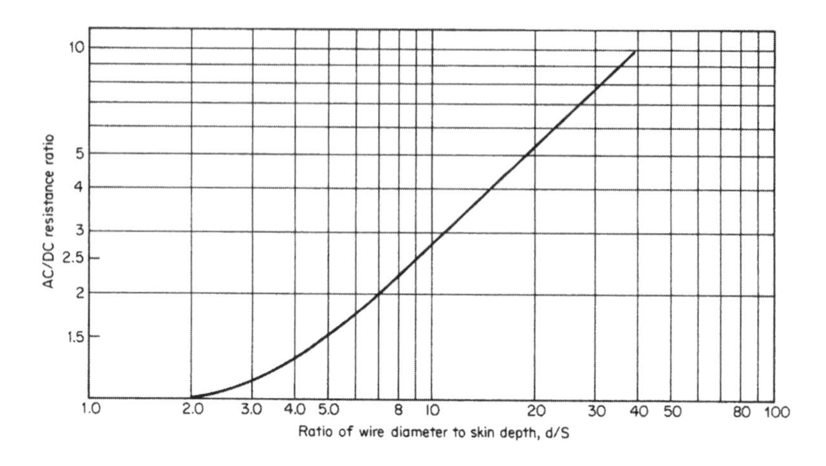

FIGURE 7.6 AC-to-DC resistance ratio for round wires versus ratio of wire diameter to skin depth (d/S). *(From Eq. 7.21.)*

its AC/DC resistance ratio is already 1.25. But the same wire at 200 kHz has a skin depth of 6.34 mils and a d/S ratio of 10.2—and from Figure 7.6, its AC/DC resistance ratio is 3.3!

However, Table 7.6 should not be misinterpreted. Although Figure 7.6 shows that R_{ac}/R_{dc} increases as wire diameter increases (d/S increases), R_{ac} actually decreases as wire diameter increases, and larger wire sizes will yield lower copper losses. This is because R_{dc} is inversely proportional to d^2, and decreases more rapidly than R_{ac} decreases as a result of increasing d. This is because R_{ac} is inversely proportional to the area of the annular skin, whose depth is S. Thus as d increases, R_{ac} decreases.

Large-diameter wire is much too lossy to use at high frequencies. Rather than using a single large-diameter wire, a number of parallel smaller-diameter wires with the same total circular-mil area can be used. This increases the total area of conducting annular skin zones, and can be seen as follows. If, say, two parallel wires are to have a total circular mil area of a single larger wire, the diameter of the two smaller wires must be $D/\sqrt{2}$, in which D is the diameter of the single original wire.

For skin depth S, the annular skin of the single original wire has total area πDS. Because skin depth is not related to wire diameter, only to frequency (Eq. 7.20), the total area of the skins of the two smaller wires is $2\left[\pi(D/\sqrt{2})S\right] = \pi\sqrt{2}DS$; and the conducting area of the two skins with two smaller wires is larger by a factor of $\sqrt{2}$, or an additional 41% over that of a single wire with the same skin depth and equal circular-mil area.

This fact gave rise to the invention of Litz wire,[16] which consists of a number of individual, fine, insulated wires or strands woven together in a bundle. In this way, on average moving over the length of the bundle, each strand spends equal times at all positions in the bundle, near the center and at its periphery. This minimizes both skin and proximity effects.[12]

Litz wire is about 5% more expensive than solid wire, but this is not as much of a drawback as the difficulty of handling it in a production environment. Care must be taken that all the fine strands (usually from 28 to 50 AWG) are soldered together at each end. It is reported that if some of the strands are broken or for some reason not connected at both ends, losses increase significantly. Also, other effects such as audible noise or vibration can occur.

General practice is to avoid Litz wire at switching frequencies up to 50 kHz. It is occasionally used at 100 kHz, and its use should be weighed against the use of up to four parallel small diameter wires.

Some appreciation of the tradeoff is obtained from Table 7.6, where it is seen that AC resistance of No. 18 wire is only 5% greater than its DC resistance at 25 kHz because of skin effect. This is not too significant. At 50 kHz, R_{ac} is 15% higher and at 100 and 200 kHz, it is 40 and 85% higher, respectively.

These numbers do not take into account that in most switching supply topologies, currents have rectangular waveshapes in which much of the energy is in the harmonics. When the losses in the harmonics of the current square wave are considered, the AC/DC resistance ratios of Table 7.6 will be seen to increase more rapidly with frequency. This will be taken up in the following section.

For high currents (usually above 15 to 20 A in secondaries), thin copper foil is often used, rather than Litz wire or multiple strands of solid wire. The foil is cut to the bobbin width (or less if VDE safety specifications must be observed). Foil thickness is typically chosen about 37% greater than the skin depth at the fundamental switching frequency. The foil can be covered with a 1-mil layer of plastic (Mylar) and wrapped as a ribbon around the bobbin for the required number of turns.

7.5.5 Skin Effect with Rectangular Current Waveshapes[14]

The ratio of AC-to-DC resistance is strongly dependent on the wire diameter/skin depth ratio (Figure 7.6), but skin depth is dependent on frequency (Eq. 7.19). In most switching power supply topologies, current waveshapes are rectangular with much of the energy residing in the harmonics. The question thus arises at what frequency to calculate skin depth. Venkatramen has rigorously analyzed this issue.[14]

A simplifying approximation has been made herein to permit estimation of the ratio of AC to DC resistance, and hence calculation of copper losses.

It is assumed that the majority of the energy in the square current waveshape resides in the first three harmonics. Skin depth S is then calculated from Eq. 7.19 for each of the three harmonics of the most usual switching frequencies: 25, 50, 100, and 200 kHz.

The average skin depth S_{av} is found for those switching frequencies. From this average skin depth, the average ratio d/S_{av} is calculated for a range of even-numbered wire sizes. This d/S_{av} is then used in reading R_{ac}/R_{dc} from Figure 7.6. The results are shown in Table 7.7.

Depending on how much of the square-wave energy is contained in harmonics above the third, Table 7.6 might give an optimistic estimate of skin effect losses for square current waveforms. Table 7.8 presents a comparison of R_{ac}/R_{dc} for No. 18 wire as read from Tables 7.6 and 7.7.

7.5.6 Proximity Effect

Proximity effect[11–15] is caused by alternating magnetic fields arising from currents in adjacent wires, adjacent turns of the same wire, and more seriously, adjacent winding layers in a multi-layer coil.

It is more serious than skin effect because the latter increases copper losses only by restricting the conducting area of the wire to a thin skin on its surface, but it does not change the magnitude of the currents flowing—only the current density at the wire surfaces. In contrast, in proximity effect, eddy currents caused by magnetic fields of currents in adjacent coil layers increase exponentially in amplitude as the number of coil layers increases.

7.5.6.1 Mechanism of Proximity Effect

Figure 7.7 shows how proximity effect comes about. There, currents are shown flowing in opposite directions (AA' and BB') in two parallel conductors. For simplicity, the conductors are shown as having a thin rectangular cross section and are closely spaced. The conductors could just as well be round wires or flat layers of closely spaced round wires such as adjacent layers in a transformer coil.

By Faraday's law, this varying magnetic field flowing through the area of loop 5678 induces a voltage in series with any line bounding the area of the loop. By Lenz's Law, the direction of this induced voltage produces a current flow in the area boundary such that its magnetic field is in the direction opposite to that of the magnetic field which induced the current flow.

Thus the current flow is counterclockwise in loop 5678. It is seen that on the bottom of the loop, the current flow is in the same direction (7 to 8) as the main current in the upper conductor (B to B'), and

Wire no.	Diameter d, mils	25 kHz Skin depth S, mils	25 kHz d/S	25 kHz R_{ac}/R_{dc}	50 kHz Skin depth S, mils	50 kHz d/S	50 kHz R_{ac}/R_{dc}	100 kHz Skin depth S, mils	100 kHz d/S	100 kHz R_{ac}/R_{dc}	200 kHz Skin depth S, mils	200 kHz d/S	200 kHz R_{ac}/R_{dc}
12	81.6	13.2	6.18	1.85	9.66	8.45	2.40	6.83	11.95	3.30	4.83	16.89	4.50
14	64.7	13.2	4.90	1.50	9.66	6.70	1.90	6.83	9.47	2.65	4.83	13.40	3.70
16	51.3	13.2	3.89	1.25	9.66	5.31	1.59	6.83	7.51	2.12	4.83	10.62	2.90
18	40.7	13.2	3.08	1.13	9.66	4.21	1.35	6.83	5.96	1.75	4.83	8.43	2.36
20	32.3	13.2	2.45	1.05	9.66	3.34	1.17	6.83	4.73	1.45	4.83	6.69	1.90
22	25.6	13.2	1.94	1.00	9.66	2.65	1.07	6.83	3.75	1.25	4.83	5.30	1.56
24	20.3	13.2	1.54	1.00	9.66	2.10	1.01	6.83	2.97	1.12	4.83	4.20	1.35
26	16.1	13.2	1.22	1.00	9.66	1.67	1.00	6.83	2.36	1.04	4.83	3.33	1.17
28	12.7	13.2	0.96	1.00	9.66	1.31	1.00	6.83	1.86	1.00	4.83	2.63	1.07
30	10.1	13.2	0.77	1.00	9.66	1.05	1.00	6.83	1.48	1.00	4.83	2.09	1.01
32	8.1	13.2	0.61	1.00	9.66	0.84	1.00	6.83	1.19	1.00	4.83	1.68	1.00
34	6.4	13.2	0.48	1.00	9.66	0.66	1.00	6.83	0.94	1.00	4.83	1.33	1.00

Note. This is a simplifying approximation. It is assumed that most energy resides in the first three harmonics of the square-wave fundamental frequency. Average skin depth for the square-wave current is then taken as the average of the first three harmonics of each fundamental as read in Table 7.5. From these average skin depths d/S is calculated, and from this R_{ac}/R_{dc} is read from Figure 7.6.

TABLE 7.7 Skin Effect AC/DC Resistance Ratios for Square-Wave Currents at Four Commonly Used Switching Frequencies

Frequency, kHz	R_{ac}/R_{dc} (Table 7.6) sine-wave currents	R_{ac}/R_{dc} (Table 7.7) square-wave currents
25	1.05	1.13
50	1.15	1.35
100	1.40	1.75
200	1.85	2.36

TABLE 7.8 AC-to-DC Resistance for No. 18 Wire

reinforces that current. On the top edge of the loop, current flow is opposite (5 to 6) to the main current and tends to cancel it. This occurs in all loops parallel to 5678 throughout the conductor width.

The result is an eddy current flowing along the full length of the bottom surface of the upper conductor in the direction 7 to 8, and returning along the upper surface of the upper conductor where it is canceled by the main current.

A similar analysis shows an eddy current flows along the full length of the upper surface of the bottom conductor in the direction to reinforce the main current. In the bottom surface of the lower conductor, it is in the direction opposite to the main current flow and tends to cancel it.

Thus currents in the two conductors are confined to a thin skin on the conductor surfaces that face one another. The depth of the skin is related to frequency, as in skin effect.

7.5.6.2 Proximity Effect Between Adjacent Layers in a Transformer Coil

Current in the individual wires in a layer of a transformer coil flow parallel to one another, and in the same direction. The current in a layer can then be considered to flow in a thin rectangular sheet whose height is the wire diameter and whose width is that of the coil. Thus there will be induced eddy currents that flow the full length of the winding. They will flow in thin skins on the interfaces between adjacent coil layers, just as described in the previous section for proximity effect in two adjacent flat conductors.

However, it is very significant that the amplitude of these eddy currents increases exponentially with the number of layers. It is this that makes proximity current effect much more serious than skin effect.

A widely referenced, classic paper by Dowell[13] analyzes proximity effect in transformers and derives curves showing the ratio of AC- to-DC resistance R_{ac}/R_{dc} as a function of the number of winding layers, and the ratio of wire diameter to skin depth. A detailed summary of Dowell's results is beyond the scope of this text, but it is well covered

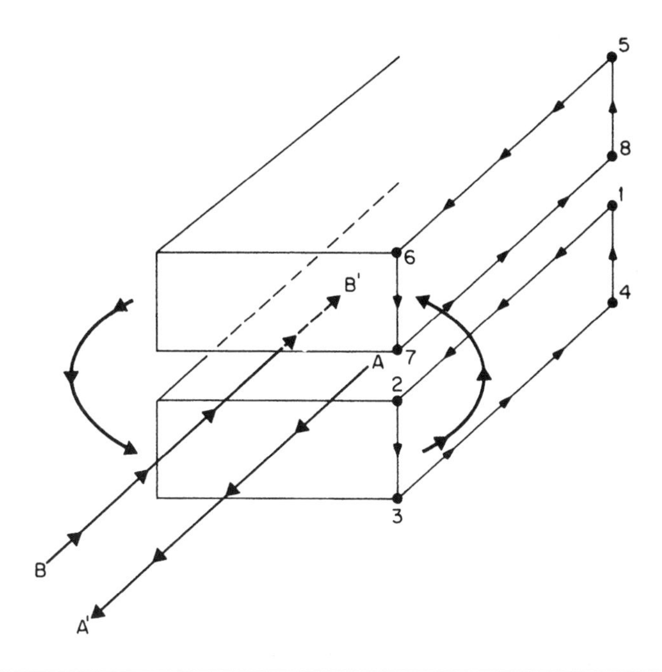

FIGURE 7.7 Magnetic field of currents in the lower conductor induces voltage in adjacent upper conductor. The resultant eddy current shown in this diagram flows along the full length of the wires on their top and bottom surfaces. On the top surface of the bottom conductor, the eddy current is in the same direction and reinforces the main current flow (AA'). On the bottom surface of the bottom conductor, the eddy current is in the direction opposite to that of the main current flow and cancels it. On the bottom surface of the upper conductor, the eddy current is in the same direction as that of the main current and reinforces it. On the top surface of the top conductor, the direction of the eddy current is opposite to that of the main current and cancels it. The consequence is that current in each conductor is confined to thin skins in the surfaces facing each other. The bottom conductor is surrounded by a magnetic field, which is shown coming out of its edge 1234, passing into the edge of the upper conductor, out of the opposite edge, and returning back into the far edge of the lower conductor. By Fleming's right-hand rule, the direction of the magnetic field is into edge 5678 of the upper conductor.

by Snelling.[12] A good discussion of Dowell's curves, showing physically why R_{ac}/R_{dc} increases exponentially with the number of layers, is given by Dixon.[11]

Herein, Dowell's curves will be presented, and a discussion will be given of their use and significance on the basis of Dixon's treatment.

Figure 7.8*a* shows an EE core with three primary layers. Each layer can be considered as a single sheet carrying a current $I = NI_t$, in which

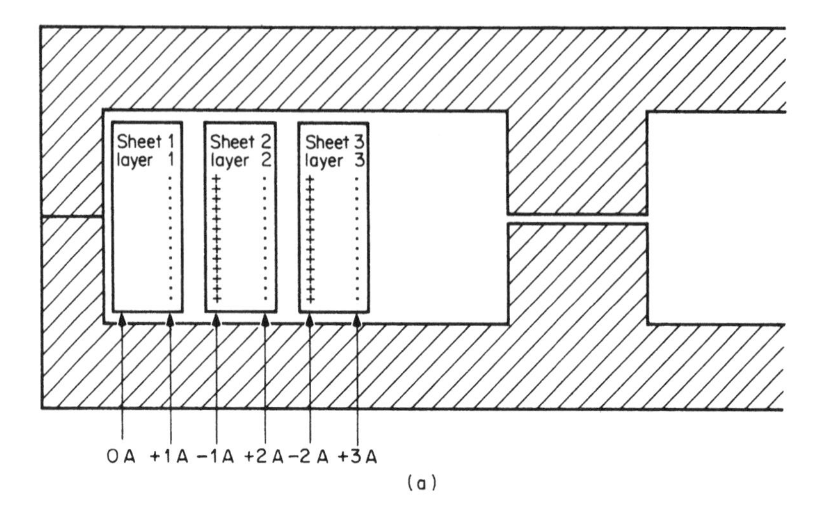

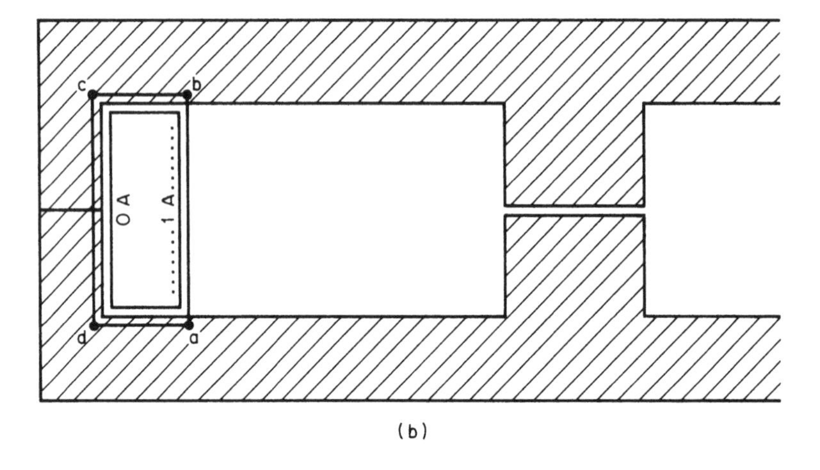

FIGURE 7.8 (*a*) Exponential buildup of surface eddy currents in a multi-layer coil. (*b*) Current in the first layer is confined to a thin skin on its surface facing away from the ferrite material as dictated by Ampere's law.

N is the number of turns in the layer and I_t is the current per turn. Now recall Ampere's law which states that $rH\,dl = 0.4\pi I$, or the line integral of $H\,dl$ around any closed loop, is equal to $0.4\pi I$, where I is the total current enclosed by the loop. This is the magnetic equivalent of Ohm's law, which states that the applied voltage to a closed loop is equal to the sum of all the voltage drops around that loop.

If the line integral is taken around the loop *abcd* in Figure 7.8*b*, the magnetic reluctance (magnetic analog of resistance) along the path *bcda* is low, since the path is in ferrite material which has high permeability. Thus most of the magnetic field intensity appears along the path *ab* which lies between sheets 1 and 2 and almost none of it lies along the leftmost surface of sheet 1. Since it is the magnetic field intensity along the surfaces which cause the skin currents to flow, all the current I that is carried by sheet 2 flows on its rightmost surface in, say, the plus direction (indicated by the dots) and no current flows on its leftmost surface.

Now consider the currents in sheet 2 (Figure 7.8*a*), and let us assume for this discussion that all winding currents are 1 A. Proximity effect as described for Figure 7.7 will cause eddy currents to flow on its left- and rightmost surfaces to a depth equal to the skin depth for that frequency. The magnetic field intensity cannot penetrate more than a skin depth below the right-hand surface of sheet 1 or the left-hand surface of sheet 2.

If the integral $rH\,dl$ is taken around the loop *efgh* (through the centers of sheets 1 and 2), since there is zero field intensity along that path, the net current enclosed by that path must be zero by Ampere's law. Further, since the current on the rightmost surface of sheet 1 is 1 A in the plus direction, the current in the left-hand skin of sheet 2 must be 1 A, but in the minus direction (indicated by crosses).

However, the net current in each of the three sheets is 1 A. Hence, with a –1 current in the left skin of sheet 2, the current in its right-hand skin must be 2 A.

In a similar argument, the current in the left-hand skin of sheet 3 is –2 A, forcing the current in its high-hand skin to be 3 A.

It can be seen from this intuitive reasoning that proximity effect causes eddy currents in the skins of a multi-layer coil to increase exponentially with the number of layers. The Dowell[13] analysis covered in the next section verifies this quantitatively.

7.5.6.3 Proximity Effect AC/DC Resistance Ratios from Dowell Curves

Dowell's analysis[13] yields the widely referenced curves of Figure 7.9. They show the ratio of AC/DC resistance ($F_R = R_{ac}/R_{dc}$) versus a factor

$$\frac{h\sqrt{F_l}}{\Delta}$$

in which h = effective round wire height = 0.866 (wire diameter d) = $0.866d$

Δ = skin depth (from Table 7.5)

F_l = copper layer factor $= N_l d/w$ (where N_l = number of turns per layer, w = layer width, d = wire diameter; note $F_l = 1$ for foil)

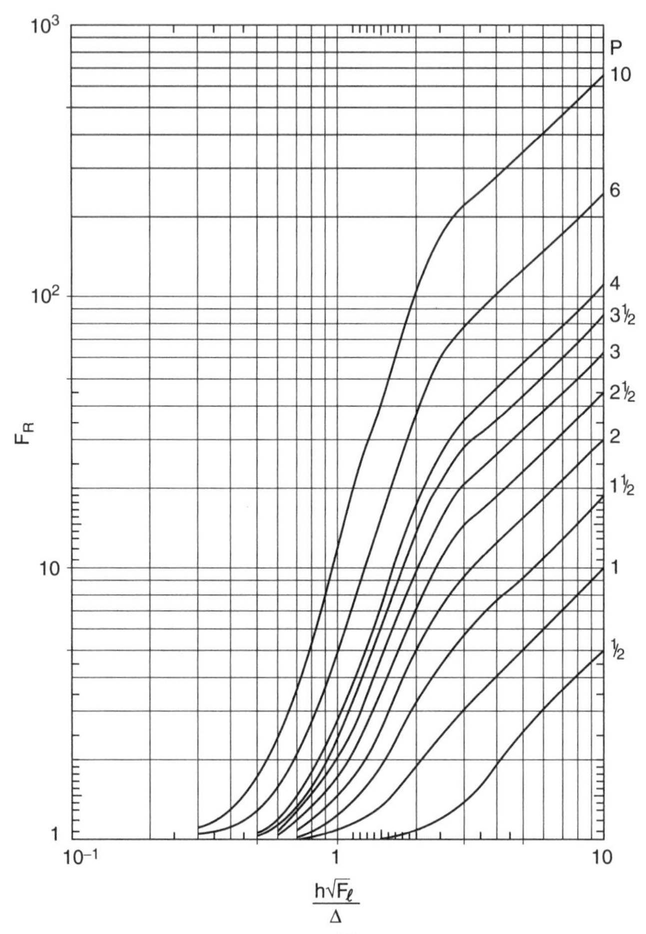

FIGURE 7.9 Ratio of AC to DC resistance due to proximity effect. The ratio is given for a number of different values of a variable p, which is the number of coil layers per portion. A "portion" is defined as a region where the low-frequency magnetomotive force ($rH\,dl = 0.4\pi NI$) ranges from zero to a peak, and back to zero. This "portion"—often misinterpreted—is clarified thus. Consider that the primary and secondary are both multi-layer windings, stacked on the bobbin with the primary layers innermost, followed by the secondary layers on top as in Figure 7.10*a*. Moving outward from the center leg of the core, the magnetomotive force ($rH\,dl = 0.4\pi NI$) increases linearly as shown in Figure 7.10*a*. (*From Dowell, Ref. 1.*)

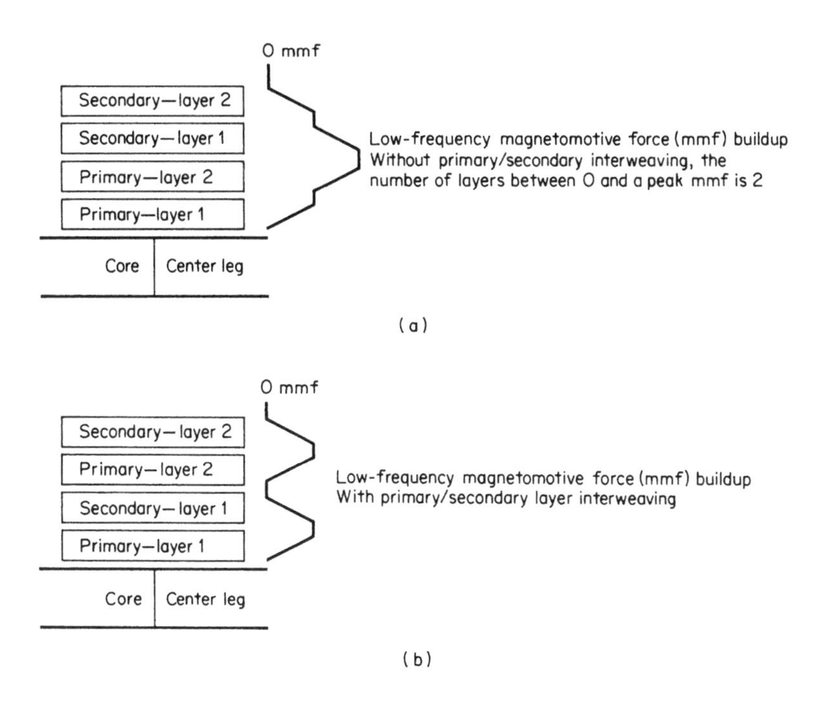

FIGURE 7.10 (*a*) Low-frequency magnetomotive force buildup in two-layer primary, secondary when primary and secondary layers are not interleaved. A "portion" is defined as the region between zero and a peak in magnetomotive force. Here there are two layers per portion. See Figure 7.9 for significance of "layers per portion." (*b*) By interleaving primary and secondary layers, the number of layers per portion has decreased to 1 and the AC/DC resistance ratio has decreased significantly (Figure 7.9).

As the line integral is taken over an increasing distance out from the innermost primary layer, it encloses more ampere turns. Then at the secondary-primary interface, $rH\ dl$ has reached a peak and starts falling linearly. In a conventional transformer (unlike a flyback), the secondary ampere turns are always simultaneously in the direction opposite to that of the primary ampere turns. Stated differently, when current in a primary flows into a "dot" end, secondary current flows out of the dot end.

When the line integral is taken over the last secondary layer, $rH\ dl$ has fallen back to zero. This is just another way of stating "the total secondary ampere turns bucks out the primary ampere turns"—except for the small primary magnetizing current.

Thus a "portion" is a region between zero and a peak magnetizing force, and for two secondary and two primary layers, sequenced as in Figure 7.10*a*, the number of layers per portion p is 2. In Figure 7.9, for—say—a $\left(h\sqrt{F_l}\right)/\Delta$ratio of 4, R_{ac}/R_{dc} is about 13!

The number of ampere turns in each half secondary in Figure 7.10*a* is equal to half the total ampere turns of the primary. If the two primary and secondary layers are interleaved as in Figure7.10*b*, the low-frequency magnetomotive forces are as shown. Now the number of layers between zero and a peak of magnetomotive force is only 1. For the same $\left(h\sqrt{F_l}\right)/\Delta$ratio of 4 (from Figure 7.9), the ratio R_{ac}/R_{dc} per portion is now only 4 instead of 13! Thus the total AC resistance of either the primary or secondary is only 4 instead of 13 times its DC resistance.

Note in Figure 7.9 that the number of layers per portion is shown going down to ½. The significance of ½ layer per portion can be seen in Figure 7.11. There it is seen that if the secondary consists of only one layer, the point at which the low-frequency magnetomotive force comes back to zero is halfway through the thickness of the secondary layer. Figure 7.9 shows that for the same $\left(h\sqrt{F_l}\right)/\Delta$ ratio of 4, R_{ac}/R_{dc} is 2 instead of 4 for the case of Figure 7.10*b*.

Figure 7.9 is very valuable in selecting a primary wire size or a secondary foil thickness at a rate other than the previously quoted "500 circular mils per RMS ampere." That choice usually leads to large values of h/Δ at high frequencies and, as seen in Figure 7.9, to very large values of R_{ac}/R_{dc}.

It is often preferable to choose a smaller wire diameter or foil thickness, yielding a $\left(h\sqrt{F_l}\right)/\Delta$ in the region of, say, 1.5. Of course, this would increase R_{dc}, but the smaller ratio of R_{ac}/R_{dc} may yield a lower R_{ac} and lower copper losses.

When interleaving windings in a push-pull circuit with two primary layers and two secondary layers, the simultaneously conducting half primary and secondary should be adjacent to each other as in Figure 7.12*a*. Placing the nonconducting secondary adjacent to the conducting primary (Figure 7.12*b*) would induce eddy currents in it even when it is nonconducting. Placing the nonconducting secondary outside the conducting one places it in a region where the half primary and half secondary ampere turns cancel during conduction time during each half cycle. Then, the line integral *rH dl* in that region is zero, magnetomotive force is zero, and hence no eddy currents flow in it during the conduction time of the opposite half secondary.

Note that in a flyback circuit, primary and secondary currents are not simultaneous. Thus, interleaving primary and secondary windings does not reduce proximity effect in flybacks. That can be done only by keeping the number of layers to a minimum and using finer wire than obtained from the rule of "500 circular mils per RMS ampere." Although that increases DC resistance, it decreases R_{ac}/R_{dc}, from Figure 7.9.

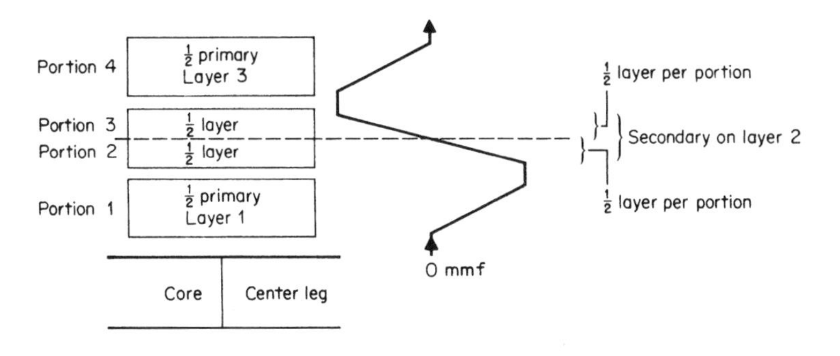

FIGURE 7.11 Half layers per portion (see Figure 7.9). With a single-layer secondary sandwiched in between two half primaries, the ampere terms of a half primary are bucked out by half the current in the secondary. In the definition of "layers per portion" of Figure 7.9, each half of the secondary operates at one-half layer per portion.

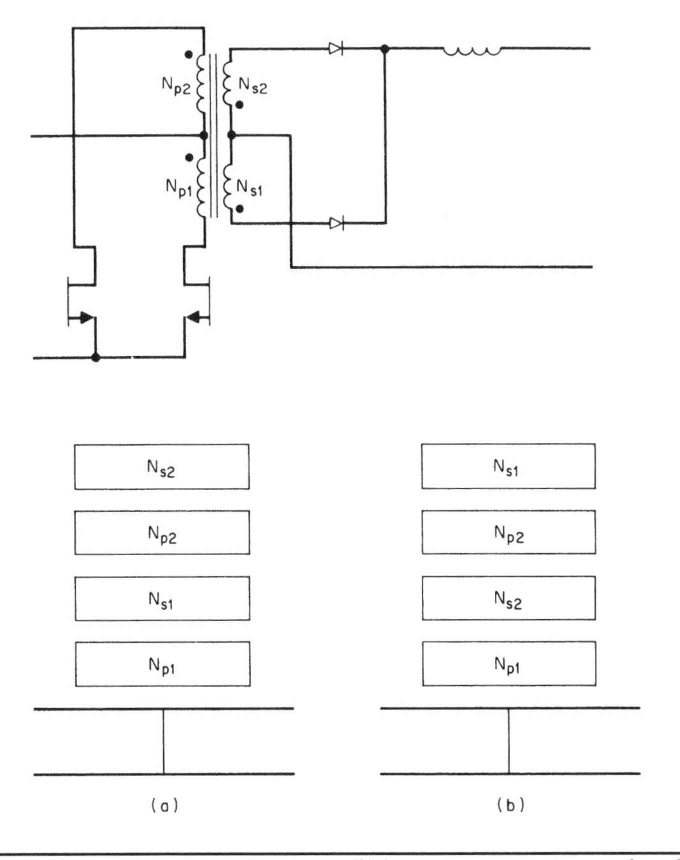

FIGURE 7.12 Correct (*a*) and incorrect (*b*) layer sequencing in a push-pull transformer. The sequencing of Figure 2.12*b* will produce significantly more eddy current losses than that of Figure 7.12*a* because of proximity effect.

7.6 Introduction: Inductor and Magnetics Design Using the Area Product Method

In previous sections, the late Mr. Pressman refers to any wound component with the property of inductance as "inductors." At this point, we will break from this convention to introduce a different term: "choke." I will use this traditional and somewhat neglected term for wound components that carry significant DC bias currents with relatively small AC ripple currents and voltages. A good example of a choke application would be seen in the low pass output filter of a typical switching supply.

The term "inductor" in this chapter will be limited to wound components which carry alternating currents and voltages, but are not required to support any significant DC bias current, while the term "choke" will be used for "inductors" that carry a significant DC bias current.

The reason for using the two discrete terms will become clear as we proceed. It will be shown that the design process for inductors is quite different from that used for chokes. Further, the materials used for the two components can also be quite different. Chokes tend to be much more tedious to design, due to the large number of interactive and divergent variables that need to be reconciled by the designer. As a result, the choke design process is intrinsically iterative. Although the various charts shown here, and provided by the core manufacturer, help to reduce the amount of iteration required for optimum design, they do not completely eliminate it.

The design approach used in the following chapters will depend on the application. Due to the many divergent variables, the final design tends to be a compromise, with emphasis being placed on the parameters that the designer decides are most important in a particular application. This could be any of the following: minimum cost, minimum size, minimum loss, maximum current, and maximum inductance. Since the optimum conditions for these basic requirements are different, a tradeoff is forced on the designer, and the design challenge is to obtain the best compromise for the intended application.

In this chapter, I use figures, charts, nomograms, and tables to establish the values of the unknown variables. Although this may not appear to be as precise as the formulae used in previous sections, it is realistic, because manufacturing spreads are such that approximate results are the realistic norm in the design of inductors and chokes. The charts show general trends for a few typical examples, rather than absolute values. Manufacturers are constantly improving their materials, so for best results the designer should refer to latest data provided by the manufacturer for actual design applications.

Nomograms, charts, and tables yield fast solutions, and the trend in a variable is more easily seen. Engineers that have less experience in this design area will find that the visual approach to the subject used here quickly provides a good understanding of the essential design parameters. I also make extensive use of the formulae and charts developed by Colonel Wm. T McLyman.[18,19] He did a great service to the industry with his many years of measurement and research on magnetic materials for the Jet Propulsion Laboratory. Finally, the designer will find that the "Area Product" figure of merit favored by the "Colonel" (this is his name, not his rank—smart parents!) is a very useful design tool. A brief explanation follows.

7.6.1 The Area Product Figure of Merit

In the following sections we will look at inductor and choke design examples and will make extensive use of a figure of merit developed by Colonel Wm. T. McLyman, called the area product (AP). This is a very powerful design tool, and can be used to indicate many properties of the core. It greatly simplifies the design process.

The area product of a core is simply the product of the center pole area multiplied by the area of the available winding window, that is, the area available for the copper wire and insulation. When both areas are measured in square centimeters, the product is in centimeters to the fourth power, and it is simply a figure of merit.

It has been shown[17,18,19] that the area product is a good indicator of the power rating of the core for transformer applications, but it can also be used to select an optimum core size in a choke design. It is a versatile parameter because it can also predict other key parameters, such as surface area, temperature rise, turns, and inductance.

The AP is now quoted by many core manufacturers. However, when not shown, it may be calculated quite easily from the core dimensions as follows:

In general

$$AP = A_e A_w$$

in which

AP = area product (cm^4)
A_w = core winding window area (cm^2) *(Use one window of an E core)*
A_e = effective area of the center pole (cm^2)

The area product figure of merit will be used extensively in the following choke design sections; you will find it is a pivotal design parameter.

Those who prefer a more in-depth derivation of the various magnetic equations, charts, and nomograms used in the following sections will find them shown more fully in the references at the end of this chapter. Engineers who fully master all the theoretical and practical requirements for optimum design of the various wound components used in switchmode supplies will find that they have developed rare and valuable design skills.

7.6.2 Inductor Design

We will start by looking at some typical switchmode power supply inductors. Inductors are a little easier to understand and design, so it is a good place to start. This approach leads us towards the more complex iterative design process required for the design of chokes in Sections 7.7, 7.8, 7.9, and 7.10.

In switchmode applications, inductors (wound components that do not support any DC bias current) will normally be limited to the following types:

- Low power signal-level inductors with no DC current component
- Common-mode line filter inductors (these are special dual-wound inductors that carry large but balanced line frequency currents)
- Series-mode line filter inductors (these are inductors that carry large unbalanced line frequency currents)
- Rod core inductors (small inductors wound on ferrite or iron powder rods)

7.6.3 Low Power Signal-Level Inductors

Here we consider inductors that are normally used in signal applications. They are not required to support any DC current, or even very much AC current or voltage stress. They are used in signal-level tuning and filter applications. The design of such inductors is relatively straightforward. For high frequency applications, the core material will normally be ferrite. The inductance and required turns is obtained directly from the A_l value provided by the manufacturer for the selected core. In signal applications the power is small and the copper loss is not a problem, so large numbers of turns can be used if large inductances are required.

The A_l value, provided by the core manufacturer, should provide the inductance of a single turn on the core, and includes the core properties and the effect of any air gap that the manufacturer may have provided. Remember that with any wound component, the inductance increases as N^2, so the inductance of the finished winding will be

$$L_n = N^2 A_{l1}$$

TIP *Take care, because the manufacturer may provide A_l values for a single turn (A_{l1}) or for many turns (A_{ln}), where n is typically 100 turns or even 1000 turns. To avoid errors, I prefer to normalize the A_l value down to a single turn A_{l1} where*

$$A_{l1} = A_{ln}/N^2$$

~K.B.

In power supply applications, signal-level inductors have limited use, so their design will not be covered in any more detail here; simply calculate the inductance from the above equation.

7.6.4 Line Filter Inductors

Line filter inductors are found in low-pass RFI filters used at the input of a switching power supply. Here, their function is to minimize the conduction of high-frequency RFI electrical noise back into the supply lines. The RFI typically comes from electrical noise generated by switching devices in the power supply.

Figure 7.13*a* shows a basic schematic of a typical line filter, often used to satisfy FCC conducted-mode RFI noise rejection limits in direct-off-line switchmode supplies. The filter circuit has a balanced common-mode inductor $L1$, with two identical windings and common-mode decoupling capacitors $C1$ and $C2$. This is followed by a series-mode inductor $L2$ with series-mode decoupling capacitors $C4$ and $C3$.

7.6.4.1 Common-Mode Line Filter Inductors

For the common-mode inductor $L1$ (a, b), we require the maximum inductance consistent with reasonable size and cost, so we choose the highest permeability core materials (typically >5,000 μ), because high permeability provides more inductance per turn. Low core loss would normally be considered an advantage in any design, but it is not essential in this application as the high frequency AC stress is normally very low so that in this type of inductor, core loss should not be a problem.

Common-mode inductors are a special case because although they may support DC currents and low frequency AC currents, the DC and low frequency magnetizing stress is bucked out by the contra winding arrangement.

7.6.4.2 Toroidal Core Common-Mode Line Filter Inductors

Figure 7.13*b* shows a typical common-mode inductor design, with two separate windings on a high permeability toroidal ferrite core.

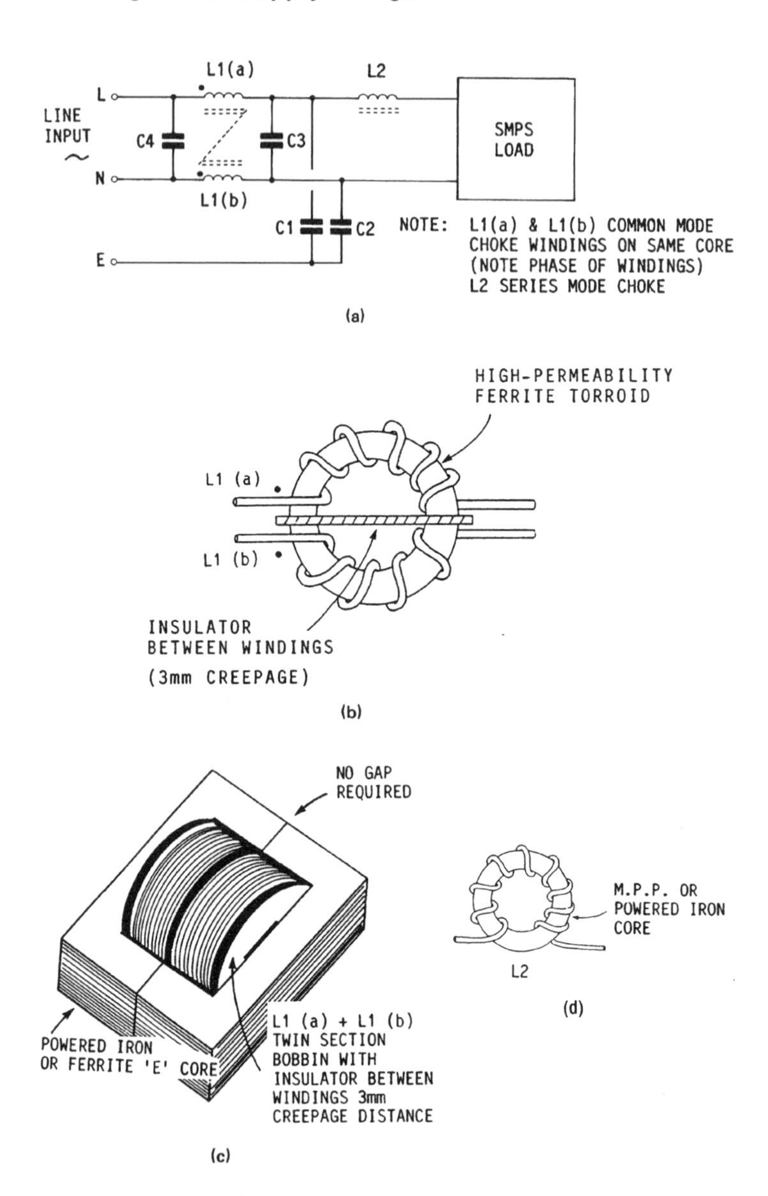

FIGURE 7.13 (*a*) A typical RFI line filter circuit, showing common-mode and series-mode filter elements. Such circuits are often used to reduce the conduction of RFI interference currents from the switching elements in the SMPS into the input supply lines. (*b*) An example of a common-mode line filter inductor, wound on a high permeability toroidal core. (*c*) An example of a common-mode line filter inductor wound on a two section bobbin E core. (*d*) An example of a series-mode line filter inductor, wound on a high loss iron powder toroidal core.

These two windings form inductors *L*1a and *L*1b, being wound on a single core to form a tightly coupled dual-wound common-mode line-filter inductor. Insulating material is located between the windings, and from the windings to core, to meet safety agency insulation and creepage spacing requirements.

Notice that this inductor has two isolated windings, with exactly the same number of turns on each winding. The windings are tightly coupled and connected into the circuit in such a way that the two windings are in series anti-phase for the low-frequency series-mode 60-Hz line currents. (Notice that when the normal 60-Hz line current flows into the start of the top winding, it flows out of the start of the lower winding, and vice versa.) Hence, the magnetic field that results from the 60-Hz series-mode AC line current (or the DC supply currents, in DC converters) will cancel to zero in the core. With the two windings connected in this way, the only inductance presented to the 60-Hz supply current is the leakage inductance between the two windings, and on a toroidal core this will be very small, so we can see that the inductor is effectively transparent to the normal series-mode supply currents.

With this anti-phase connection, the low-frequency AC (or DC) series-mode supply current will not contribute to core saturation. Hence, a very high permeability core material may be used without concern for saturation, and without the need for a core air gap. Toroidal cores do not have any core gap and often have the highest permeability, so a large inductance can be obtained with only a few turns.

For the common-mode noise, however (high-frequency noise currents or voltages which appear on both supply lines at the same time with respect to the ground plane, terminal E), the two windings are in parallel and are in phase, and a very high inductance is presented between the power supply noise source and line input terminals L and N. As a result, the majority of any common-mode noise currents from the switching devices in the power supply are bypassed to the ground plane by capacitors *C*1 and *C*2. This arrangement effectively prevents any significant common-mode RFI currents from being conducted back into the input supply lines.

The design approach for the common-mode inductor is very straightforward. Select a high permeability toroidal core of convenient size, and wind it with two single layer windings as shown in Figure 7.13*b*, using a wire gauge selected for the maximum RMS supply current. The current density can be quite high (700 to 1000 amps/cm^2) because the core loss is negligible and the open single layer winding will cool very effectively. Although multiple layers can be used, this is not recommended as the increased inter-winding capacitance will decrease the self-resonant frequency, reducing the effective high-frequency noise rejection ratio.

The effective inductance can be calculated from the turns and the A_l value for the chosen core. Realize the two windings are effectively in parallel for common-mode noise rejection conditions, and the effective turns are those of only one winding. Hence

$$L_n = N^2 A_{l1}$$

TIP *At the prototype stage, the actual inductance typically required from the common-mode inductor is not known, as the magnitude of the RFI problem depends on many factors which have probably not been determined at this stage. The final design is best determined by measurement of the actual conducted RFI currents using a spectrum analyzer, after everything has been built to its final standard. The test should include any chassis or box, and all heat sinks and switching devices with the intended mounting hardware. If more attenuation is required at this stage, then capacitors C1 and C2 can be adjusted within limits.*

If more inductance is required, then a larger core must be used. Supplies designed for patient-connected medical applications have very stringent limits on ground return currents, which confine the maximum value of the decoupling capacitors C1 and C2. The designer should expect to use much larger common-mode RFI chokes in such applications. ~K.B.

7.6.4.3 E Core Common-Mode Line Filter Inductors

E cores can also be used for common-mode line filter inductors. Figure 7.13*c* shows a typical example. Generally, E cores are easier to wind and have lower manufacturing costs. The main disadvantages of E cores are more variable inductance, lower core permeability, and larger inter-winding capacitance.

TIP *For the following reasons, the inductance of an E core is generally much more variable than that of a toroidal core. All E cores are made in two parts. The mating of these two parts is never perfect, so a small air gap is inevitable. With high permeability core material, this gap (although very small) has a significant effect on the total assembled permeability. Further, any contamination in the gap will result in large variations. The permeability of a 5000 perm material may be reduced by as much as 60% in an E core form, so the inductance will be lower in the same ratio. However, cost is a powerful incentive, and E cores are very often used for this application, in spite of the above limitations. Some manufacturers supply cores with the mating surfaces ground optically flat, and such cores will retain much higher permeability providing they are assembled in clean conditions. Typically, such cores are supplied in matched pairs, which must be kept together for the best results. ~K.B.*

7.6.5 Design Example: Common-Mode 60 Hz Line Filter

In the following design example, a very simple and expedient approach will be taken. An E core is selected to fit a convenient size and cost requirement. For prototypes this is not an unreasonable approach, because the filtering needs are not fully established at this stage. Hence, it will be assumed that we simply want to obtain the maximum common-mode inductance possible from the selected high-permeability ferrite E core, while at the same time limiting the temperature rise to 30°C.

In common-mode line filters, the high frequency noise voltages are quite small and the core loss will be negligible, so we consider only copper loss in our temperature rise calculations. Typically, a two section bobbin is used to provide good isolation between the two windings. The bobbin sections are completely filled with wire, allowing for insulation, using two identical but separate windings.

In this example, we must choose a wire gauge such that the copper loss at full load current will result in a temperature rise of 30°C or less. This design approach provides the maximum number of turns and hence the maximum inductance that can be obtained from the chosen core and temperature rise. The resulting pile wound bobbin will have good low frequency noise rejection, but the inter-winding capacitance may be quite large, and we will see that the high frequency attenuation may be compromised to some extent.

TIP *A large inter-winding capacitance will result in a low self-resonant frequency, and the high-frequency noise components may effectively bypass the inductor. However, it will be shown later that the higher-frequency components can be more effectively blocked by the series-mode inductor shown as L2 in Figure 7.13b. We will see that L2 normally has a very high self-resonant frequency. ~K.B.*

7.6.5.1 Step 1: Select Core Size and Establish Area Product

In general, select an E core that meets the cost and mechanical size requirements, and obtain its area product (AP) value from the manufacturer's data. See Table 7.9, or you can calculate the AP as follows:

TIP *The area product is the product of the core area (A_e) and the usable winding window area (A_{wb}), both in cm^2. (Include only the area of one window of the E core.) If a bobbin is to be used, for conservative design, take the internal winding area of the bobbin rather than the core, as shown in Figure 7.16. Include both sections of the bobbin. ~K.B.*

	AWG Winding Data (Copper, Wire, Heavy Insulation)						
AWG	Diameter, copper, cm	Area, copper, cm^2	Diameter, insulation, cm	Area, insulation, cm^2	fl/cm 20°C	fl/cm 100°C	A for 450 A/cm^2
10	.259	.052620	.273	.058572	.000033	.000044	23.679
11	.231	.041729	.244	.046738	.000041	.000055	18.778
12	.205	.033092	.218	.037309	.000052	.000070	14.892
13	.183	.026243	.195	.029793	.000066	.000088	11.809
14	.163	.020811	.174	.023800	.000083	.000111	9.365
15	.145	.016504	.156	.019021	.000104	.000140	7.427
16	.129	.013088	.139	.015207	.000132	.000176	5.890
17	.115	.010379	.124	.012164	.000166	.000222	4.671
18	.102	.008231	.111	.009735	.000209	.000280	3.704
19	.091	.006527	.100	.007794	.000264	.000353	2.937
20	.081	.005176	.089	.006244	.000333	.000445	2.329
21	.072	.004105	.080	.005004	.000420	.000561	1.847
22	.064	.003255	.071	.004013	.000530	.000708	1.465
23	.057	.002582	.064	.003221	.000668	.000892	1.162
24	.051	.002047	.057	.002586	.000842	.001125	.921
25	.045	.001624	.051	.002078	.001062	.001419	.731
26	.040	.001287	.046	.001671	.001339	.001789	.579
27	.036	.001021	.041	.001344	.001689	.002256	.459
28	.032	.000810	.037	.001083	.002129	.002845	.364
29	.029	.000642	.033	.000872	.002685	.003587	.289
30	.025	.000509	.030	.000704	.003386	.004523	.229
31	.023	.000404	.027	.000568	.004269	.005704	.182
32	.020	.000320	.024	.000459	.005384	.007192	.144
33	.018	.000254	.022	.000371	.006789	.009070	.114

TABLE 7.9 Magnet Wire Table for AWG 10 Through 41, Showing Current Ratings for Choke and Transformer Designs at a Typical Current Density of 450 amps/cm^2 (Notice, increasing or decreasing the AWG value by three steps changes the area of the copper by a factor of two. For example, two wires of 18 AWG have the same area as one wire of 15 AWG.)

	AWG Winding Data (Copper, Wire, Heavy Insulation)						
AWG	Diameter, copper, cm	Area, copper, cm^2	Diameter, insulation, cm	Area, insulation, cm^2	fl/cm 20°C	fl/cm 100°C	A for 450 A/cm^2
34	.016	.000201	.020	.000300	.008560	.011437	.091
35	.014	.000160	.018	.000243	.010795	.014422	.072
36	.013	.000127	.016	.000197	.013612	.018186	.057
37	.011	.000100	.014	.000160	.017165	.022932	.045
38	.010	.000080	.013	.000130	.021644	.028917	.036
39	.009	.000063	.012	.000106	.027293	.036464	.028
40	.008	.000050	.010	.000086	.034417	.045981	.023
41	.007	.000040	.009	.000070	.043399	.057982	.018

TABLE 7.9 Magnet Wire Table for AWG 10 Through 41, Showing Current Ratings for Choke and Transformer Designs at a Typical Current Density of 450 amps/cm^2 (Notice, increasing or decreasing the AWG value by three steps changes the area of the copper by a factor of two. For example, two wires of 18 AWG have the same area as one wire of 15 AWG.) *(Continued)*

The area product (AP) is defined as

$$AP = A_e A_{wb} (cm^4)$$

in which

A_e is the area of the core (cm^2)
A_{wb} is the winding area of the bobbin (cm^2)

7.6.5.2 Step 2: Establish Thermal Resistance and Internal Dissipation Limit

In general, with the value of *AP* established above, enter Figure 7.14 at the lower edge and project up to the required temperature rise line. At the intercept with the temperature rise line, project left to get the predicted thermal resistance (R_{th}) for the fully wound finished inductor on the left scale. R_{th} is given in °C/watt, assuming an ambient temperature of 25°C.

With R_{th}, we can calculate the permitted winding dissipation (W_{cu}) that just gives the specified 30°C temperature rise (ΔT) as follows:

$$W_{cu} = \Delta T / R_{th} (\text{watts})$$

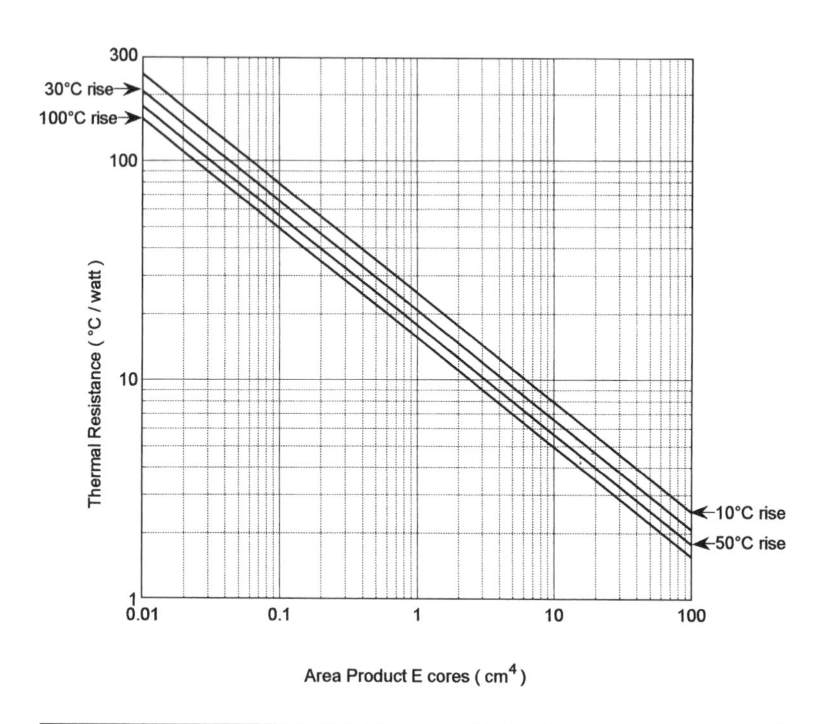

FIGURE 7.14 Link between the area product and thermal resistance for fully wound standard E cores, with temperature rise above an ambient of 25°C in free air, as a parameter.

For this example, we assume an EC35 E core has been chosen, with an effective area product of 1.3 cm^4. A bobbin will be used, and this reduces the effective area product to 1.1 cm^4. Entering the lower scale of Figure 7.14 with $AP = 1.1$ cm^4 and projecting up to the 30°C line, a thermal resistance (R_{th}) of 20°C/watt is predicted on the left scale. In this nomogram, the R_{th} intercept is for a temperature rise (ΔT) of 30°C above an ambient of 25°C. Hence, assuming zero core loss, the maximum permitted copper loss in the winding (W_{cu}) will be

$$W_{cu} = \Delta T / R_{th} = 30/20 = 1.5 \text{ watts}$$

7.6.5.3 Step 3: Establish Winding Resistance

We can now calculate the maximum permitted winding resistance R_w at the working 60-Hz AC current of 5 A RMS, that will generate a copper loss of 1.5 watts as follows:

We have W_{cu} copper loss of 1.5 watts, and with current of 5 A RMS; hence,

$$R_{wp} = W_{cu}/I^2 = 1.5/25 = 60 \text{ milliohms (or 0.16 ohms)}$$

Having established the resistance of a fully wound bobbin, we can establish the wire gauge and the number of turns to just fill the bobbin. (In this example the winding is split into two windings of 30 milliohms each.) From the effective turns and A_l value we can then establish the inductance.

7.6.5.4 Step 4: Establish Turns and Wire Gauge from the Nomogram Shown in Figure 7.15

Many manufacturers provide data giving the resistance of a fully wound bobbin using various wire gauges. However, in this example, we will get the wire gauge and turns from Figure 7.15 as follows:

With the resistance (0.06 Ω), enter the top horizontal scale of resistance and project down to the upper (positively sloping) "resistance and turns" line for the EC35 E core, as shown in the example. The intersection with the EC35 line is projected left to give the number of turns—56 turns in this example. From the same point, project right to the intersection with the (negatively sloping) "wire gauge and turns" line for the EC35 core. This intersection is then projected down to the lower scale as shown to give the wire gauge—approximately 17 AWG in this example.

TIP *For ease of winding, more strands of a smaller wire may be preferred. Table 7.9 provides the resistance for a range of magnet wire from 10 to 41 AWG. Notice that adding 3 to the AWG number will give a wire of half the cross sectional area, so two strands of 20 AWG will have the same copper area as one of 17 AWG. This relationship is maintained throughout the AWG table. ~K.B.*

For resistance values of less than 50 milliohms, enter the nomogram from the bottom scale of resistance and project up to the lower group of "resistance and turns" lines. In a common-mode inductor, the winding is split into two equal parts. Hence the bobbin would be wound with two windings of 28 turns of 17 AWG, or two strands of 20 AWG, per winding.

7.6.5.5 Step 5: Calculating Turns and Wire Gauge

If preferred, the wire gauge and the number of turns can be calculated from first principles as follows.

For this example, we will consider only one winding occupying half of the twin section bobbin, as shown in Figure 7.16. Allowing space at the top of the bobbin for insulation material, the usable window area A_w for one side is 30 mm^2. When round magnet wire is used, the packing factor is typically 60%, allowing for the insulation of the wire and the fact that the round wire does not completely fill the cross

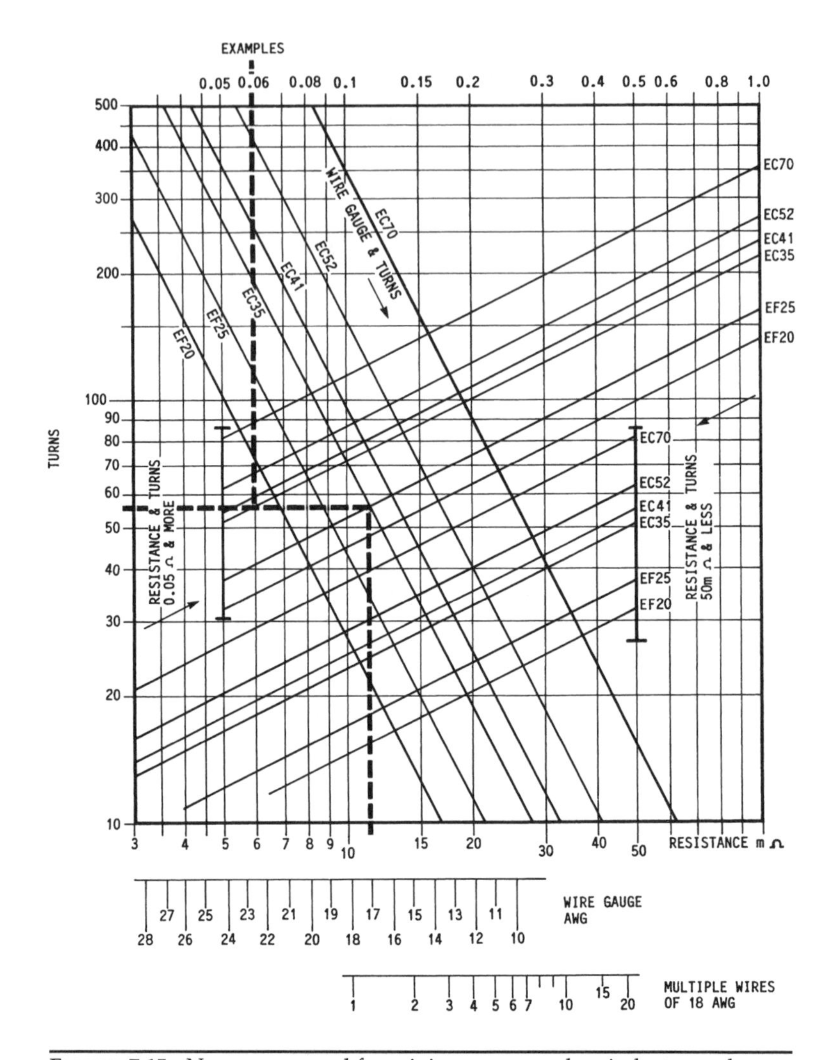

FIGURE 7.15 Nomogram used for minimum copper loss inductor and choke designs. This nomogram can be used to quickly establish the number of turns and optimum wire size for a fully wound E core bobbin.

sectional area with solid copper. Hence the effective usable area A_{cu} for solid copper is

$$A_{cu} = 0.6A_w = 0.6(30) = 18\ \text{mm}^2(0.18\ \text{cm}^2)$$

The mean diameter of the bobbin is 1.6 cm, so the mean length per turn (MLT) is 5.02 cm.

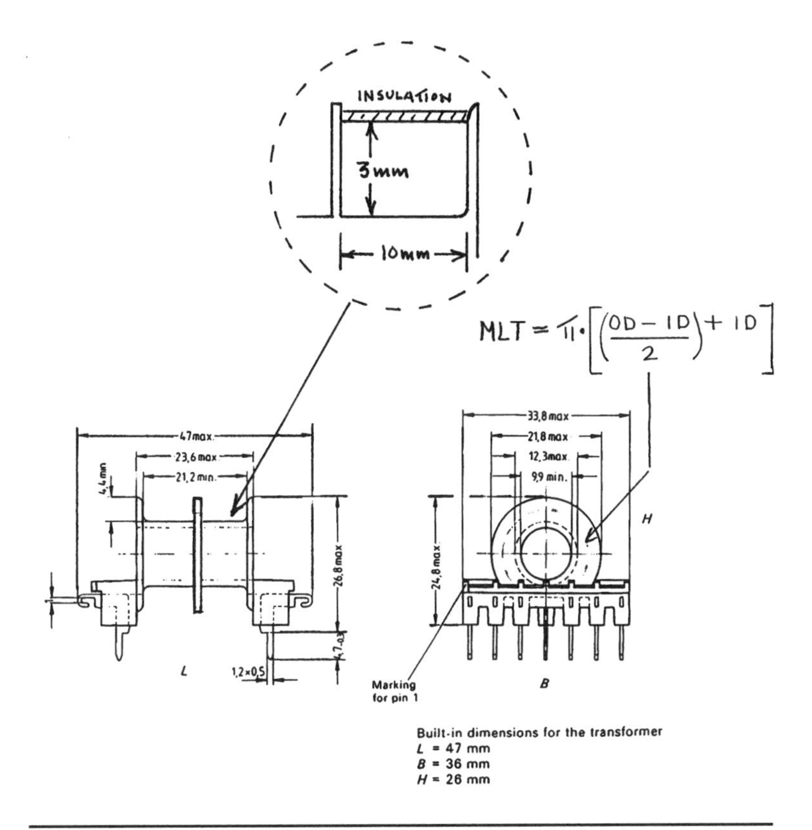

FIGURE 7.16 A_{wb}, the effective area of the winding window for a two-flange bobbin, decreases as a result of the space taken up by the bobbin material and insulation, which reduces the effective area product. The mean length per turn (MLT) is used to establish the length of wire and hence the resistance of a fully wound bobbin.

We can now calculate the theoretical resistance R_x of a single turn of solid copper that will fully occupy the available window area of one side of the bobbin R_{cu} using the nominal bulk resistivity of copper as follows:

The bulk resistivity of solid copper at 70°C is $\rho = 1.9\ \mu\Omega$ cm.

Hence

$$R_x = \frac{\rho MLT}{A_w} = \frac{1.9 \times 10^{-6}\Omega\ \text{cm}(5.02\ \text{cm})}{0.18\ \text{cm}^2} = 53\ \mu\Omega$$

We have shown above that to limit the temperature rise to 30°C, the total resistance of the winding R_w for a full bobbin is limited to 60 mΩ, or 30 mΩ for each half.

The resistance of a single turn of solid copper in the half section is 53 μΩ and the required resistance of the complete half winding is 30 mΩ. We can now calculate the number of turns required to produce a winding of 30 mΩ for one side of the bobbin as follows:

$$N = (R_w/R_x)^{1/2} = (30\ \text{m}\Omega/53\ \mu\Omega)^{1/2} = 24\ \text{turns}$$

TIP *The squared term comes from the fact that bobbin area must remain fully wound, so each time you double the turns you must half the area of copper and this will double the resistance per turn, so the resistance increases as* N^2. *~K.B.*

Having established the number of turns, we can now establish the area of copper available for the wire that will just fit in the available space using that number of turns as follows:

$$A_{\text{cuw}} = A_{\text{cu}}/N = 0.18/24 = 0.0075\ \text{cm}^2$$

From Table 7.9, we see that a wire between 18 and 19 AWG has this area.

We can check the result. Choosing the larger 18 AWG wire, we can calculate the resistance of the half winding R_{cu} from the total length of the winding, and the Ω/cm for 18 AWG shown in the table:

$$R_{\text{cu}} = N\,(MLT)\left(\frac{\Omega}{\text{cm}}\ of\ 18\ \text{AWG}\right)$$

$$= 24\,(5.02\ \text{cm})\left(0.00024\frac{\Omega}{\text{cm}}\right) = 29\ \text{m}\Omega$$

The final result (29 mΩ) for each half winding (58 mΩ total) is near the value obtained from the nomogram used previously.

7.6.6 Series-Mode Line Filter Inductors

In Figure 7.13*a*, $L2$ is in series with the 60-Hz line input supply. Its function is to offer as much impedance as possible to series-mode RFI currents. These noise currents flow from the supply to say the L terminal of the input through the SMPS load, and return via the N terminal to the supply, or the converse.

We have shown that due to the phasing of the common-mode inductor $L1$ it does not provide any inductance to the series-mode current. Hence a separate inductor $L2$ is normally required for the series-mode currents. Even though $L2$ may not carry a DC current in this position, it does carry a large peak line frequency (60 Hz) current, and a high forcing voltage exists across the inductor. Also the duration of the current pulse at the peak of the AC waveform is very long compared with the SMPS switching frequency. Hence, the inductor can be considered to be driven from a constant-current, line frequency source. The peak line current has a saturating effect similar to that of DC, and in fact in a DC/DC converter it is DC. The inductance is too small to have any significant effect on the 60 Hz current or any DC current.

TIP *With capacitive input rectifier circuits, it may be difficult to calculate the peak current in* L2. *Consider a typical off-line capacitive input rectifier circuit. The input capacitor is normally quite large and a large current pulse flows on the peak of the applied AC voltage waveform while the input rectifier diodes conduct. In this application it would be essential to ensure that* L2 *does not saturate during this current pulse. As a result,* L2 *must be designed to carry at least the peak line current without saturating. The peak current depends upon a number of ill-defined variables. These include the line source impedance, circuit resistance, input capacitor ESR, and total loop inductance. Therefore, it is often better to simply measure the current and calculate the peak flux density in the core. A 30% safety margin should be provided to allow for component and line impedance variations. Power factor corrected systems have a much lower and well-defined current. ~K.B.*

To prevent $L2$ saturating, it may be necessary to use a gapped ferrite core or a low-permeability iron powder core. If the peak current is known, the design of the series-mode inductor may proceed in the same way as in the choke designs shown later. In such designs, use the peak forced AC current in place of the DC current shown in the calculations. In DC/DC converters use the maximum DC current. Hence, in general, the design of the series-mode input inductor $L2$ should follow the same approach as that used for choke design. (See Section 7.7.)

If the inductance of $L2$ is to be less than 50 μH (in many cases this will be sufficient), the simple rod core inductor described below can be used, which has the advantages that it is simple, low cost, and will not saturate.

7.6.6.1 Ferrite and Iron Powder Rod Core Inductors

For small low-inductance applications in the range from 5 to 50 μH, the designer should consider using simple open-ended ferrite or iron powder rod cores, bobbin cores and spools, or axial lead ferrite beads.

By careful attention to minimizing the inter-winding capacitance (for example, by using spaced windings and insulating the wire from the rod former), the self-resonant frequency of RFI inductors wound on open-ended rods can be made very high. We now consider the design of a rod core inductor for $L2$.

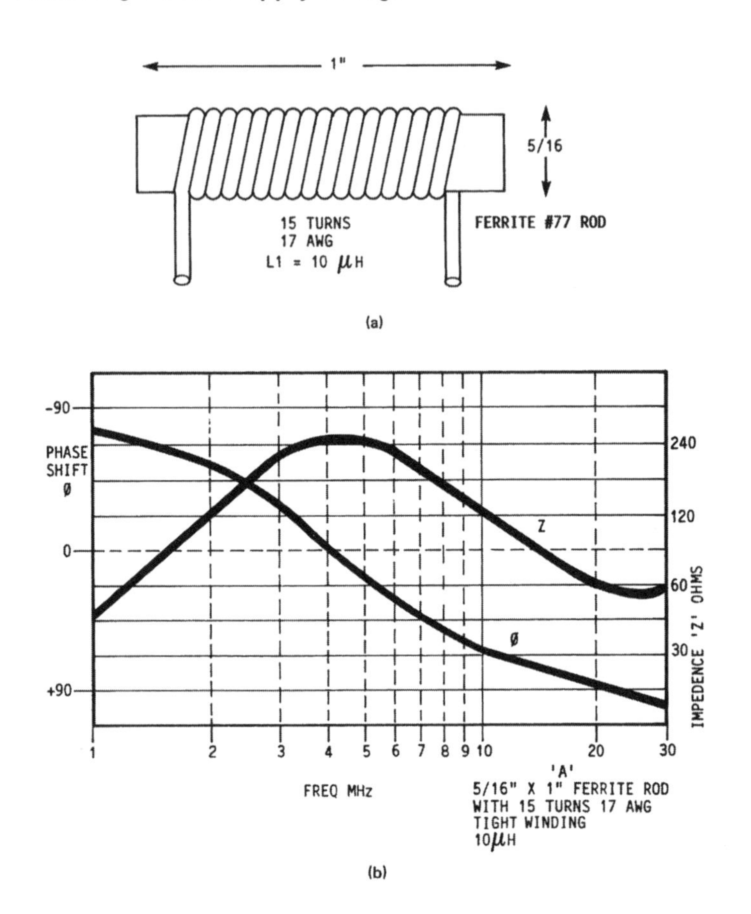

FIGURE 7.17 (*a*) An example of a rod core inductor/choke. (*b*) An impedance and phase plot, showing how the impedance and phase shift change with respect to frequency, for a rod core inductor/choke with a tight winding. Notice the maximum impedance is at 4 MHz. (*c*) The impedance and phase shift of the same rod core inductor/choke with a low capacitance, spaced winding. Notice the maximum impedance is now at 6 MHz, improving the high frequency attenuation.

An example of a ferrite rod inductor is shown in Figure 7.17*a*. These simple inductors, when used in RFI filters together with low-ESR capacitors, can be very effective in reducing high-frequency noise spikes. In many cases, the high frequency AC current is much smaller than the mean 60 Hz or DC current, so the high-frequency magnetic radiation from open-ended rods, spools, or bobbins (normally the most objectionable parameter for this type of inductor) is acceptably small and should not present an EMI problem.

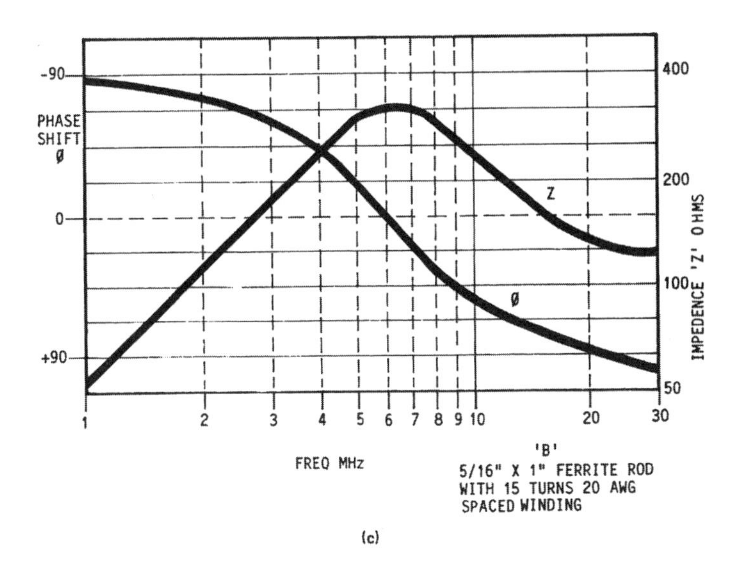

FIGURE 7.17 *Continued.*

Normal ferrite materials may be used for rod inductors, since the long external air path prevents saturation of the core, even with high permeability materials. The next section shows how a rod core inductor design can be optimized to have the maximum impedance at a high frequency.

7.6.6.2 High-Frequency Performance of Rod Core Inductors

Notice that with any wound component, the self inductance of the winding is effectively in parallel with the inter-winding capacitance, and forms a parallel L-C circuit. Hence, a small inter-winding capacitance will result in a high self-resonant frequency.

Above the resonant frequency, the inductor starts to look more like a capacitor and the impedance is lower, so noise components of sufficiently high frequency bypass the inductor. To get the best high-frequency attenuation, the inter-winding capacitance should be minimized.

Figure 7.17*a* shows a one inch long, 5/16″ diameter ferrite rod inductor, wound with 15 turns of closely packed 17 AWG wire. Figure 7.17*b* shows a plot of the phase shift and impedance as a function of frequency for this design. Notice that the phase shift is zero and the impedance is maximum at the self-resonant frequency of 4 MHz. At higher or lower frequencies, the impedance is lower, as you would expect from a parallel resonant circuit.

The second impedance plot in Figure 7.17*c* shows the improvement obtained by reducing the inter-winding capacitance. This plot was obtained from the same inductor sample, after spacing the windings and insulating them from the rod with 10-mil Mylar tape. In the second sample, 15 turns of 20-gauge wire was used with a space between each turn. The plot shows that the reduction in inter-winding capacitance has increased the impedance and shifted the self-resonant frequency to 6.5 MHz. This will increase the range of the noise spectrum that is rejected in the final application.

7.6.6.3 Calculating Inductance of Rod Core Inductors

Figure 7.18 shows the effective permeability of rod core inductors with respect to the initial permeability of the core material, with the geometric ratio L/d as a parameter. Notice the inductance depends more on the geometry of the winding than on the permeability of the core material.

For most practical applications, in which the length-to-diameter ratio is in the order of 3:1 or greater, the initial permeability of the core material does not significantly affect the effective permeability of the finished product. Hence, the inductance is not very dependent on the core permeability in most practical applications. The chart is for iron powder, but it can also be applied to high-permeability ferrite rod chokes with little error.

With the effective permeability established, the equations shown in Figure 7.18 allow the inductance to be calculated according to the construction and geometric ratio. The external, intrinsically large air gap in rod core inductors prevents the saturation of high-permeability ferrite rods, even when the DC current is very large, so these inductors satisfy the DC bias current requirements for chokes, and are sometimes referred to as RF chokes.

The wire gauge should be chosen for acceptable dissipation and temperature rise at the working current; a current density of 600 to 1000 A/cm^2 is acceptable. Iron powder rods are also suitable for this application. The lower permeability of these materials is not much of a disadvantage, as the large air gap swamps the initial permeability of the core material. The core loss is normally very low as the high-frequency flux swing is quite small. The increased core loss of the iron powder material improves the attenuation.

This completes the section on "inductors," and we will now look at the design of "chokes." We will find that the design of chokes is quite different, as they present many more variables.

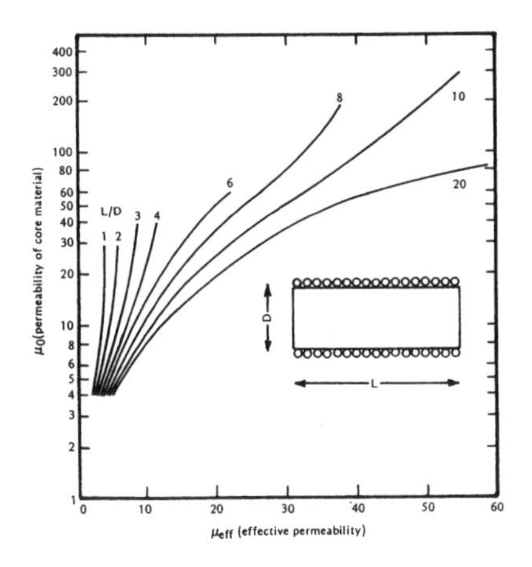

Single-Layer Coil

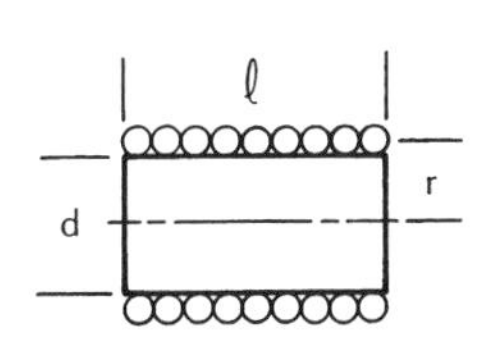

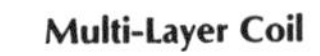

Multi-Layer Coil

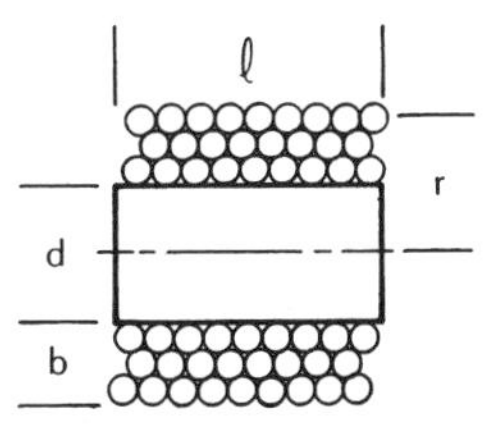

Single-layer coil:

$$L = \frac{\mu eff\,(rN)^2}{9r + 10\ell}$$

or

$$N = \frac{1}{r}\left[\frac{L(9r + 10\ell)}{\mu eff}\right]^{1/2}$$

Multi-layer coil:

$$L = \frac{0.8\,(\mu eff)(rN)^2}{6r + 9\ell + 10b}$$

or

$$N = \frac{1}{r}\left[\frac{L(6r + 9\ell + 10b)}{(0.8)(\mu eff)}\right]^{1/2}$$

WHERE:

L = Inductance (microhenries)

μeff = Effective permeability of core (see graph below)

N = Number of turns

r = radius of coil (inches)

d = diameter of core (inches)

ℓ = length of coil/core (inches)

b = coil build (inches)

FIGURE 7.18 This chart shows how the effective permeability of rod core chokes changes as a function of the material initial permeability, with the ratio of length to diameter as a parameter. (*Courtesy of Micrometals Inc.*)

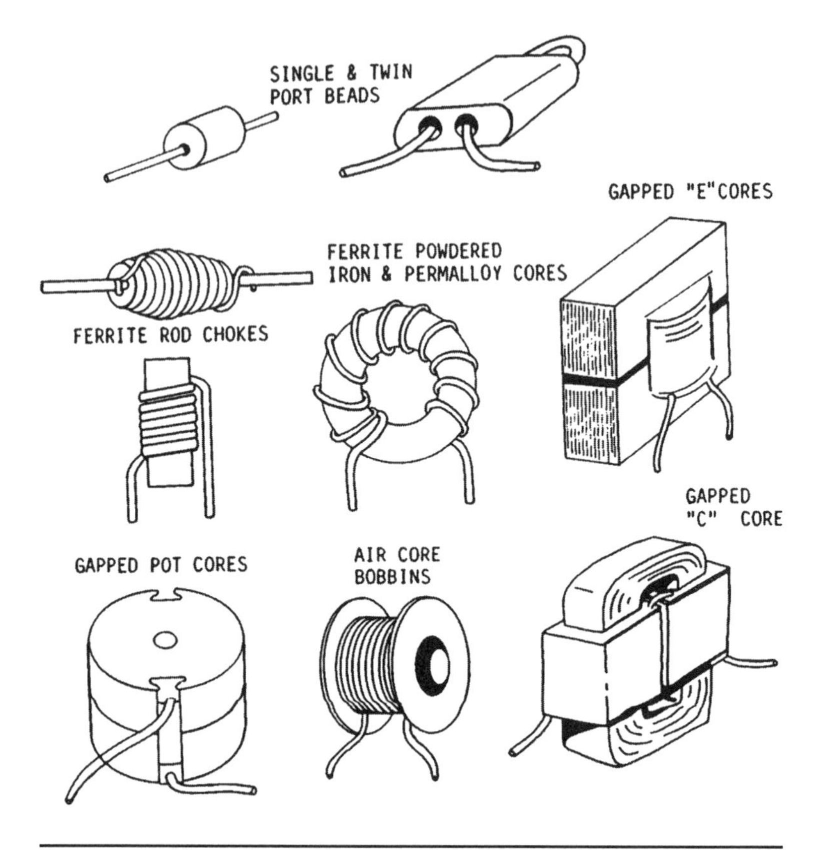

FIGURE 7.19 Examples of inductors and chokes used in switch-mode buck, boost, and low-pass filter applications and in flyback "transformer" designs.

7.7 Magnetics: Introduction to Chokes—Inductors with Large DC Bias Current

Chokes (inductors that carry a large component of DC current) are used extensively in switchmode supplies. Chokes range from small ferrite beads, used to profile the drive currents of switching transistors or diodes, up to the large high-current chokes used in power output filters. Some typical examples of switchmode chokes are shown in Figure 7.19.

A good working knowledge of the design procedure for chokes is essential for the best results. The power supply engineer needs to develop considerable skill in the choice of core type, material, design, and size, and winding design if the most cost-effective chokes are to be designed. The subject is very diverse; there are no ideal design methods, because there are no ideal magnetic materials. It is a matter of matching the selection to the application, which improves with the acquisition of design experience.

In this chapter we will limit our discussion to gapped ferrite E cores, powder E cores, and toroidal powder cores, because they are the types most often used in conventional high-frequency choke applications.

7.7.1 Equations, Units, and Charts

In previous chapters, the late Mr. Pressman makes extensive use of equations to explain the design of wound components. In this chapter, I adopt a more visual approach, using the B/H magnetization loop and various charts, nomograms, and tables to obtain the required solutions.

Since most engineers are designing to meet specific applications, they are most often concerned with real cores, that is, cores of well defined dimensions, so I prefer to use flux density B as a design parameter, rather than total magnetic flux Φ, where

$$B(tesla) = \Phi/A_e$$

A_e is the effective core area in mm^2.

In many examples, I have modified the equations dimensionally and rationalized the units to yield the most convenient solutions. I also make extensive use of the formulae developed by Colonel Wm. T McLyman, who did a great service to the industry with many years of measurement and research on magnetic materials for the Jet Propulsion Laboratory. For those engineers that prefer a more in-depth analysis, including the derivation of the various magnetic equations and nomograms used here, and the supporting formulae, this may be found in references 17, 18, and 19, listed at the end of the chapter.

In previous sections, we found that the design of inductors was quite straightforward, because there was little or no DC bias current. To better understand the severe limitations imposed on the design of chokes, when a large DC bias current is present, we need to examine the B/H loop once again with particular attention to the saturation properties of some typical core materials, and the effect of an air gap.

7.7.2 Magnetization Characteristics (B/H Loop) with DC Bias Current

Figure 7.20 shows the top quadrant of a typical B/H loop, for a low permeability iron powder core and a ferrite core material, with and without an air gap. To the first order, the horizontal axis H (oersteds) is

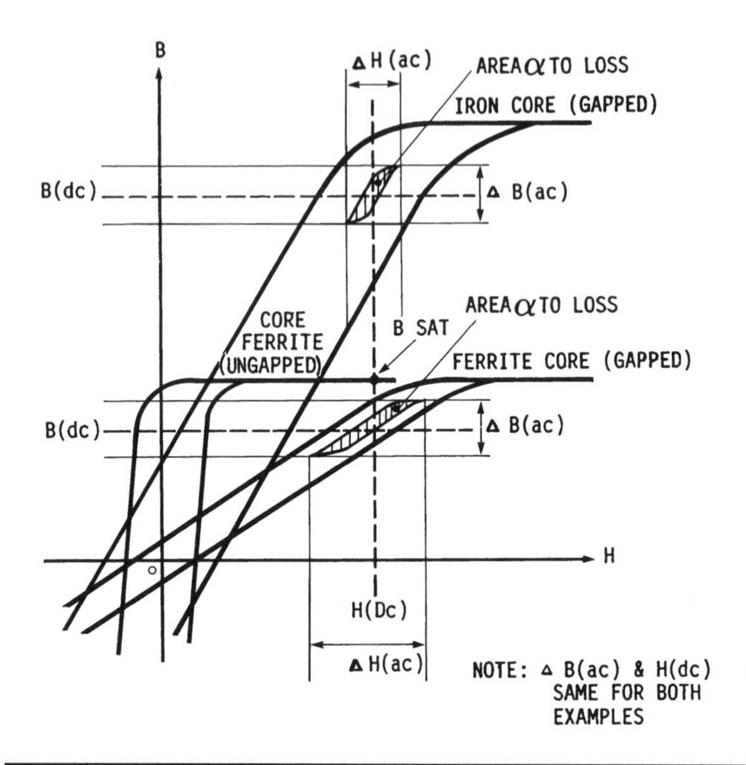

FIGURE 7.20 An example of the first quadrant of a B/H loop for gapped and non-gapped ferrite cores, and non-gapped powder cores in single-ended (not push-pull) applications. The figure shows how the mean flux density (B_{dc}) is a function of the magnetization force (H), resulting from the mean DC bias current in the winding and the permeability or air gap used in the choke core. It also shows the minor B/H loops responsible for the core loss, and how the flux density swing (ΔB) is a function of the AC stress only.

proportional to the DC bias current, while the vertical axis B is the core flux density in tesla. (1 tesla = 10,000 gauss, or 10 kG, so 1 millitesla = 10 gauss (1 mT = 10 G))

Notice that for an arbitrary value of DC bias current, resulting in a magnetization value H_{dc} shown by the vertical dotted line, that the ungapped ferrite core is completely saturated. Also notice that the slope of the B/H loop for the ferrite at saturation is zero (horizontal). This means that the effective permeability of the ungapped ferrite core at this value of H is zero, so the choke will have near zero inductance. Clearly, an ungapped ferrite core is not much use as a choke with this value of DC bias current.

The iron powder core, with a much smaller initial permeability (lower slope), is not saturated at the same value of H_{dc}. The mean slope of the B/H loop around H_{dc} is still significant and the choke will have significant inductance. Notice that the same ferrite core, but now with an air gap, is no longer saturated at H_{dc} either, and so the gapped ferrite core retains some inductance.

For the same magnetizing force (H_{dc}), the value of B_{dc} (the mean induced flux density due to the DC bias current) is higher in the iron core example. This means the iron powder core can support a larger range of ΔB (flux density swing) without saturating, so the powder core has an ability to support more applied AC volt seconds. It stores more energy and can accommodate a larger range of ripple voltage and ripple current.

The area of the B/H loop for the iron powder core is much larger than the gapped ferrite core, so the iron powder core has a propensity for more core loss. However, higher core loss is not necessarily inevitable, because in the final design, the actual working loss depends on the flux excursion ΔB (flux density swing) and the working frequency. Hence, where possible, iron powder cores can and should be used, as they generally cost less.

We will now look at the parameters controlling the core magnetization force H_{dc}.

7.7.3 Magnetizing Force H_{dc}

To better understand the magnetizing force H we turn to the B/H loop again. The horizontal scale is the magnetizing force H.[17,18,19] The general equation for H in SI units is

$$H = \frac{0.4\pi NI}{\ell}$$

where H = magnetizing force in oersteds
N = turns
ℓ = the length of the magnetic path around the core (cm)
I = the DC bias current in the winding (amps)

In the finished choke, N and ℓ are defined, so that $H \propto I$; that is, the magnetizing force H_{dc} is proportional to the bias current I_{dc}.

So here we see the first unavoidable compromise. With a particular core material, defined size, and turns, the larger the DC bias current, the lower the slope the B/H loop must have to prevent saturation. To do this we must select a core material with a lower permeability, or increase the length of the core air gap, which will also lower the effective permeability. With a lower permeability, the inductance per turn will be lower. Hence, there is an inevitable tradeoff between the ability to support DC bias current and the magnitude of the inductance that can be achieved—increasing one reduces the other.

7.7.4 Methods of Increasing Choke Inductance or Bias Current Rating

What must we do to increase the inductance and/or bias current in a particular design? Increasing the turns may help, providing we are not near saturation, because inductance increases as N^2 while H increases in proportion to N. If we are near saturation, however, increasing the turns will not do it, because H will increase in the same way as increasing the current. This will force the core deeper into saturation, so that an even lower permeability material must be used, and we are on a path of diminishing returns.

Consider the inductance formula[17,18,19] as follows:

$$L = \frac{N A_e \Delta B}{\Delta I}$$

where L = inductance (henrys)
N = turns
A_e = area of the center pole (mm^2)
ΔI = a small change in bias current (amps)
ΔB = the corresponding change in flux density (teslas)

The slope of the B/H loop at the working point (the working permeability of the chosen core) is proportional to $\Delta B/\Delta I$ and this is defined in any particular design, hence the remaining variables are N and A_e. So

$$L \propto N A_g$$

N cannot be increased very much without causing DC saturation, so the only way to increase the inductance is to increase A_e (the area of the center pole). So with a particular core material, the only way to get more inductance, or accommodate a larger DC bias current, is to use a larger core. We will see later that the size of the core is defined by the energy storage number W where

$$W = \tfrac{1}{2} L I^2 \; (joules)$$

In a similar way, we can show that increasing the slope of the ratio $\Delta B/\Delta I$ or the working permeability will also provide more inductance, but we see from Figure 7.20 that this requires a core material with a larger saturating flux density. Even if a suitable material exists, it normally has more core loss and this loss may become the limiting factor.

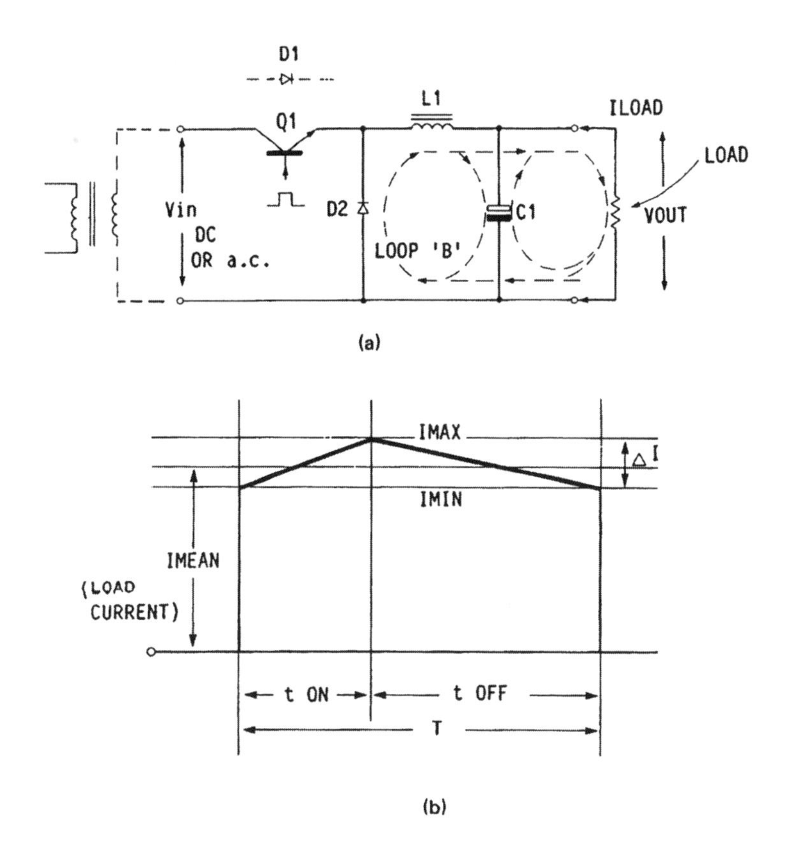

FIGURE 7.21 (*a*) A buck regulator power section example, showing the choke $L1$ and the essential power elements $Q1$, $D2$, and $C1$, together with the output current loops. (*b*) A typical current waveform for continuous-mode operation.

7.7.5 Flux Density Swing ΔB

To complete the picture we should now look at the vertical axis of the B/H loop (the flux density B). However, before we do this, we will consider a choke application to see how B is related to the working AC stress conditions.

Figure 7.21*a* shows a schematic of the power section of a typical buck regulator with a choke shown at $L1$. The expected choke current waveform for continuous-mode operation is shown at 7.21*b*. Notice there is a triangular ripple current centered on a mean DC bias current, which is the mean DC load current.

Under steady state conditions, this current waveform is established by the switching action of $Q1$, the action of the choke $L1$, the diode $D1$, capacitor $C1$, and the load as follows.

Under steady state conditions, when $Q1$ turns "on," the supply voltage is impressed on the left side of the choke $L1$ while the right side has the constant DC output voltage on $C1$, which is lower than the supply voltage. There is a forward voltage of $V_{L1} = V_{in} - V_{out}$ across $L1$, and the current in $L1$ will increase linearly as defined by $di/dt = V_{L1}/L1$.

When $Q1$ turns "off," the left side of $L1$ clamps to near zero via $D2$ by flyback action, and $L1$ has a reverse voltage of V_{out} plus a diode drop, and the current decreases linearly.

We characterize this action in this example as continuous operation because the current in the choke never goes to zero. The mean inductor current is the load current. We will refer to the mean load current as the DC bias current for the design of the choke.

From the B/H loop (Figure 7.20), we can see that the mean value of the flux density B_{dc} is defined by the magnetizing force H_{dc} and the slope and shape of the B/H loop. We will now consider the AC conditions, that is, the action of the applied ripple voltage as the transistor switches "on-off."

For the AC conditions, it is more convenient to enter the B/H loop from the left on the B scale, starting at the value of B_{dc} that has been defined by the mean DC bias current. Centered on this value of B_{dc}, the flux density in the core must increase during the "on" time such that the rate of change of flux density (the flux linkages within the winding) offsets the applied AC voltage as follows:

$$\Delta B = \frac{V_{L1} t}{N A_e} \text{ (teslas)}$$

Where ΔB = the flux density change (teslas)
V_{L1} = the forward voltage across choke $L1$ (volts)
t = the period the voltage is applied ($Q1$ "on" time in μ sec)
N = turns
A_e = area of the center pole of the choke core (mm^2)

In a similar way, during the "off" period of $Q1$, with V_{L1} being negative, the flux density change will have a negative slope, returning the current and the core flux to the same values each time $Q1$ just turns "on" again.

TIP *Notice in the above equation, since* N *and* A_e *are constant in a particular design, the flux density swing* ΔB *is defined by the applied volt seconds* (Vt)*. Hence, the change or "swing" in flux* ΔB *is defined and remains constant. It is not a function of the core material, core gap, or permeability. In other words, the core has nothing to do with the required change in flux* ΔB*; this change must take place to offset the applied volt seconds.*

Hence, for the AC voltage stress, we enter the B/H loop from the left on the vertical axis B *at* B_{dc}*, and impose on this an additional change in flux* ΔB*, defined by the applied volt seconds and centered on* B_{dc}*. The slope of the B/H loop (the permeability) defines the corresponding change in* H*, and hence, the ripple current associated with the applied ripple voltage. If the core is near saturation, the flux density cannot increase any further, there will be no "back emf" to offset the applied volt seconds, and the choke will look like a short circuit at that point. The current will increase rapidly, limited only by the resistance of the winding. Hence, the choke design must have a sufficient saturation margin to accommodate both the flux density* B_{dc} *developed by the applied DC bias current, and the additional imposed increase* ΔB/2 *required by the AC voltage. ~K.B.*

We can now study the complete action of the buck regulator and its interaction with the B/H loop, and see the tradeoff between the ferrite and powder materials. A clear understanding of the next three paragraphs will allow the designer to know immediately the implication of core material selection on the choke design, simply by looking at the B/H loops for the materials.

Look again at the B/H loop in Figure 7.20. Start by entering the H axis with H_{dc}, project up (dotted line) to the upper iron powder B/H loop, and then left to the mean working point B_{dc}. This is the mean working flux density for the powder core, caused by the DC bias current in the winding. Notice for the lower permeability gapped ferrite core, the working point B_{dc} is lower for the same mean current.

We now impose on this B_{dc} working point the AC component ΔB (the change or swing in flux density caused by the applied square wave switching voltage as $Q1$ turns "on" and "off"). The core now sweeps out the minor B/H loop shown. As the flux swings, the value of H changes by ΔH(AC) as shown. This translates to a change in current, producing the triangular ripple current shown in Figure 7.21*b*. If the B/H loop is curved, the ripple current will show the same curvature. If the core approaches saturation, the B/H loop slope rapidly decreases, resulting in a rapid increase in H, which translates to a rapid increase in current. Hence, impending saturation is clearly seen as a sudden increase (spike) of current near the positive peak of the current ripple waveform.

Notice that the same flux density swing imposed on the lower B_{dc} working point of the gapped ferrite core results in a much larger swing in ΔH because the slope of the B/H loop is lower. This results in a larger ripple current for the same applied ripple voltage (this waveform is not shown). In other words, the inductance of the gapped ferrite core is lower than the inductance of the powder core at the same working point, and as a result the ripple current is greater. Although the width of the minor B/H loop is greater for the ferrite core, the area of the loop is in fact much lower, as is the core loss. So just by looking at the B/H loop we can predict the performance of the choke material, and decide if it has the characteristics we want in our design.

Let us now look at a second tradeoff. We can choose a lower loss, gapped ferrite core material, but the lower saturation flux density results in less inductance and more ripple current. Hence with the same size core, at the same DC bias current, we get less inductance from the gapped ferrite core. We will see later that the energy storage number for the core ($\frac{1}{2}LI^2$) links the core size to the inductance and DC bias current rating.

7.7.6 Air Gap Function

At this point, I ask the reader's indulgence for the following very basic and detailed explanation of the function of the air gap. I include this here because this function is often poorly understood and causes much confusion. If you are well versed in the subject, you may prefer to go directly to Section 7.7.7.

We can see from the B/H loop (Figure 7.20) and the following equation that for the DC current component, the mean value of B_{dc} is a function of H_{dc}, which is linked to the DC bias current as follows:

$$H = \frac{0.4\pi N \hat{I}}{\ell} \text{oersteds}$$ [17,18,19]

So $H_{dc} \propto I_{dc}$ and this is applied to the horizontal scale. The DC induced flux density (B_{dc}) is the dependent variable, defined by the magnetization characteristic. Hence, the DC induced flux density B_{dc} will change for any change in core permeability (B/H loop slope), air gap, or DC bias current.

However, for the AC conditions, the change in B (ΔB) is quite independent of the above factors being defined by the following formula:

$$\Delta B = \frac{\Delta Vt}{NA_e}(tesla)$$

For AC conditions, with N and A_e being constant, we can see that the AC induced flux change (ΔB) is only a function of the externally applied volt seconds (Vt), and is not influenced by the core material. Hence, the AC induced change (ΔB) is applied to the vertical axis B, and is proportional to the applied AC conditions. Essentially, ΔB rides on top of the working point H_{dc} set up by the DC bias current. The AC flux swing ΔB is defined by the need to offset the externally applied volt seconds. Permeability does not figure in the above AC equation, so changing the mean DC current, permeability, or air gap does not change the required peak-to-peak flux density swing ΔB.

However, changing the permeability will change the slope of the B/H loop, so it changes the link between ΔB and ΔH. This translates to a change in ripple current. In other words, changing the core permeability changes the inductance of the choke and hence the ripple current.

Changing the permeability of the material or the air gap does not change the AC fluxing ΔB, and it does not change the AC saturation flux density. However, by reducing the B/H loop slope, reducing permeability, and reducing inductance, the flux developed by the DC bias current decreases, and this can prevent saturation otherwise caused by the sum of the DC and AC fluxing components. This is clearly shown in Figure 7.20, for the ungapped and gapped ferrite core B/H loops.

In summary, an air gap or permeability change changes the flux density caused by the DC component of current, but does not change the flux density swing required by the AC component.

7.7.7 Temperature Rise

The final, and perhaps most important, limiting factor in choke design is temperature rise. In general, the copper loss in chokes is greater than in inductors because the larger DC bias current contributes to the I^2R loss. As a result, in most applications the core loss is less than the copper loss, and in fact it can be quite small in some cases. However, the temperature rise is a function of the total loss, and when core loss is significant it must be included.

Temperature rise is a function of many variables, including choke location, air flow, and the effects of any surrounding components that may contribute to the temperature rise. The various charts and nomograms used here assume free air conditions, so the surface area and radiation properties control the temperature rise. The charts assume 45% convection cooling, and 55% radiation at 0.95 emissivity. In any event, the temperature of a choke should always be checked finally in the working prototype, where the layout and general thermal design introduce additional "difficult to determine" thermal effects. We will now look at some material properties.

7.8 Magnetics Design: Materials for Chokes—Introduction

In this chapter, we look at the more important properties of various materials available for use in the design of chokes.

Normally, the core material is chosen to best satisfy the parameters that the designer considers to be most important in a particular design. These include operating frequency, the ratio of DC bias current to

AC ripple current, required inductance, temperature rise, saturation margin, cost considerations, and any special mechanical requirements. Because these requirements are divergent, the choice will normally be the best compromise, since no single magnetic material will be found, irrespective of price, to satisfy all these needs at the same time.

The saturating effect of the DC bias current makes the design of chokes more difficult, and severely limits our ability to obtain large inductance values. Hence, choke design is always a compromise that balances core material, core size, core loss, copper loss, current rating, inductance, and temperature rise. All these interdependent and divergent variables must be reconciled in the final design.

There are simply no ideal core material selections; it depends on the application and the many variables, including the skill and preferences of the designer. The bottom line is that a core that provides millihenrys of inductance as a pure inductor may provide only microhenrys of inductance when configured as a choke.

7.8.1 Choke Materials for Low AC Stress Applications

In some cases, the choice of core material is relatively straightforward. When the AC ripple current or frequency is quite low (for example, in series-mode 60 Hz line input filters), the core loss is unlikely to be a major factor. In such applications, a low cost, high permeability, high saturation flux density material will be chosen. This might be iron powder, or even gapped silicon iron transformer laminations. Such materials have the advantage of high saturation flux density, high permeability, and low cost. As a result of the higher permeability, fewer turns are required to obtain the required inductance, and the core will remain out of saturation with a larger DC bias current. The reduced turns will result in lower copper losses. Core loss would not normally be a problem in such applications.

7.8.2 Choke Materials for High AC Stress Applications

At the other end of the range, where the operating frequencies and AC ripple currents are much larger, core losses will need to be considered, and lower loss materials will be required. These include the various lower permeability, lower loss materials, such as powdered iron, Molypermalloy MPP, Kool Mμ®, and gapped ferrite. All these materials have low permeability, and more turns are required to obtain the required inductance, so both the copper and core losses will be greater.

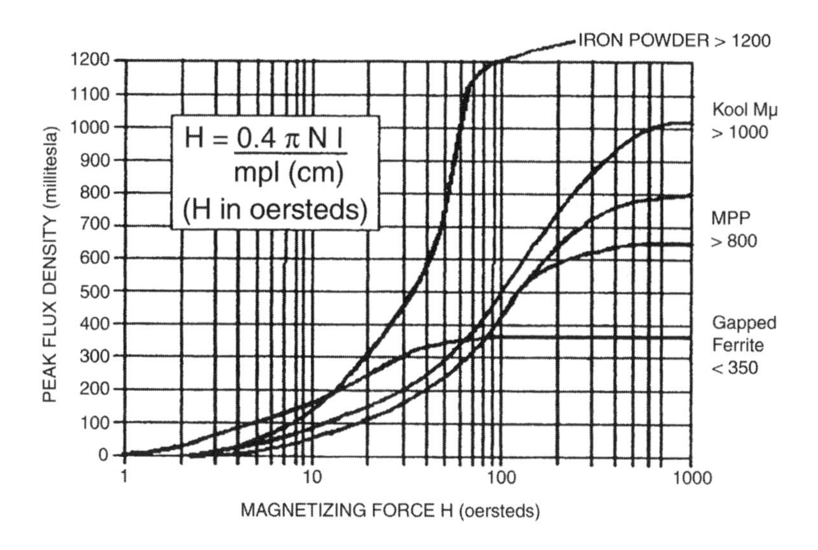

FIGURE 7.22 A general picture showing typical core saturation characteristics for iron powder, Kool Mμ, MPP, and gapped ferrite materials.

7.8.3 Choke Materials for Mid-Range Applications

Between these two extremes, the best choice of core geometry and material is not so obvious. There is a tradeoff between core loss and copper loss. It would help at this stage to look at the basic properties of some of the materials available to us, and compare some of the essential characteristics; we will start by looking at saturation properties.

7.8.4 Core Material Saturation Characteristics

Figure 7.22 shows the saturation characteristics of some typical core materials. The horizontal scale H (oersteds) is proportional to the DC bias current and the turns, and the vertical scale shows, in very general terms, the median flux density that may be expected against H_{dc} for each type of material. In the ferrite example, an air gap has been introduced to give a permeability of about 60. The important parameter to notice in this chart is the magnitude of the flux density at which a particular material saturates. This tends to be material specific and does not change greatly with the range of permeabilities offered for each type of material. Contrary to popular belief, a gap does not change

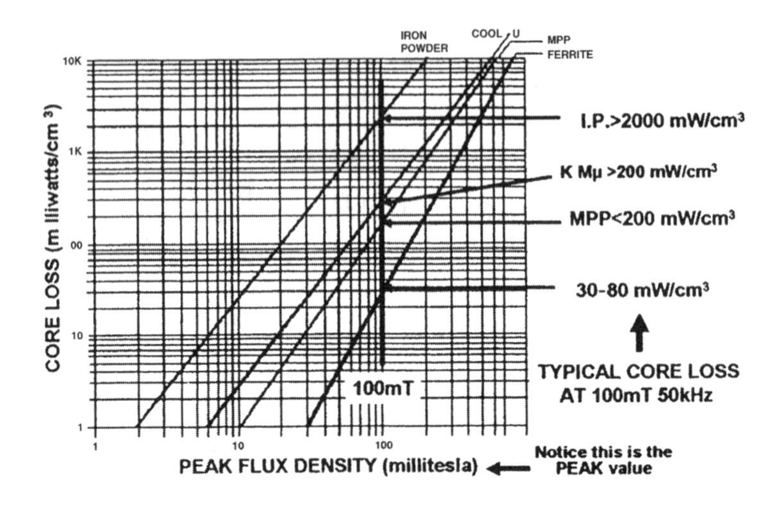

FIGURE 7.23 A general picture showing typical core loss characteristics for iron powder, Kool Mμ, MPP, and gapped ferrite materials.

saturation flux density. The materials arranged in ascending order of saturation flux density follow:

1) All ferrite materials saturate near 0.35 tesla.
2) MPP materials saturate in the range 0.65 to 0.8 tesla.
3) Kool Mμ saturates near 1.0 tesla.
4) Iron powder saturates above 1.2 tesla.

To convert tesla to gauss, multiply by 10^4: $1\ T = 10^4\ G$.

Figure 7.22 shows the first parameter that would be of interest to the designer in the core selection process. Clearly, we would choose the highest saturation material if there were no other limitations, but we must now consider core loss.

7.8.5 Core Material Loss Characteristics

For the same materials considered above, Figure 7.23 shows core loss due to the AC fluxing component ΔB. In this example, we assume a peak AC fluxing of 100 milliteslas, or 200 mT peak-peak, at a frequency of 50 kHz. Note that the chart is drawn on a log/log scale, so differences are much greater at the top of the chart than at the bottom.

Core loss is a function of AC conditions, that is, the flux density swing and frequency. Notice for the AC conditions shown in the chart, the material losses in ascending order are as follows:

1) For the gapped ferrite, it is near 30 mW/cm^3.
2) For the MPP, it is typically three times greater, at about 100 mW/cm^3.

3) Kool Mμ losses are nearly six times greater, at 200 mW/cm^3.
4) Iron powder losses are sixty-five times greater, above 2000 mW/cm^3.

We see an enormous spread in core loss, ranging from 30 mW/cm^3 up to 2000 mW/cm^3, a spread in loss of 65 to 1. These results are very generalized and are intended for guidance only, but the trend is very clear.

TIP *The core loss examples in Figure 7.24 are taken at an AC fluxing* (B_{ac}) *of 100 milliteslas peak. The manufacturers usually assume push-pull operation in their core loss specifications, so the loss nomogram shown assumes a peak-peak AC swing* (ΔB) *of 200 mT. The contribution from the DC fluxing* (B_{dc}) *is not included because, in the first order, the DC fluxing does not contribute to the core loss. ~K.B.*

The natural preference for a material with high saturation flux density is very much at odds with the corresponding core loss. These are just two of the many divergent variables that must be reconciled by the designer of chokes. The temptation to choose the highest saturation material is clearly in conflict with the need to reduce the core loss. Hence, with the limitations of existing magnetic materials, the final selection will always be a compromise.

7.8.6 Material Saturation Characteristics

A core must be able to support the maximum DC bias current, plus any over-current condition, plus the AC fluxing, without saturating. So the shape of the saturation characteristic (the curvature of the B/H loop) is an important selection factor. This parameter indicates the ability of the chosen core to support the total magnetizing force without premature saturation.

Figure 7.24 shows the saturation characteristics of gapped ferrite compared with various powder materials. The gapped ferrite material maintains a relatively constant permeability as the magnetizing force increases. This means that the inductance changes little with increasing load current, but it then saturates quite suddenly. When you design chokes using gapped ferrite material, make sure that saturation will not occur during maximum current operation. A good safety margin must be provided by selecting an adequate air gap.

7.8.7 Material Permeability Parameters

Having selected the most suitable material for the application, we should now look at the variations in the performance of the selected family of materials under various DC and AC working conditions.

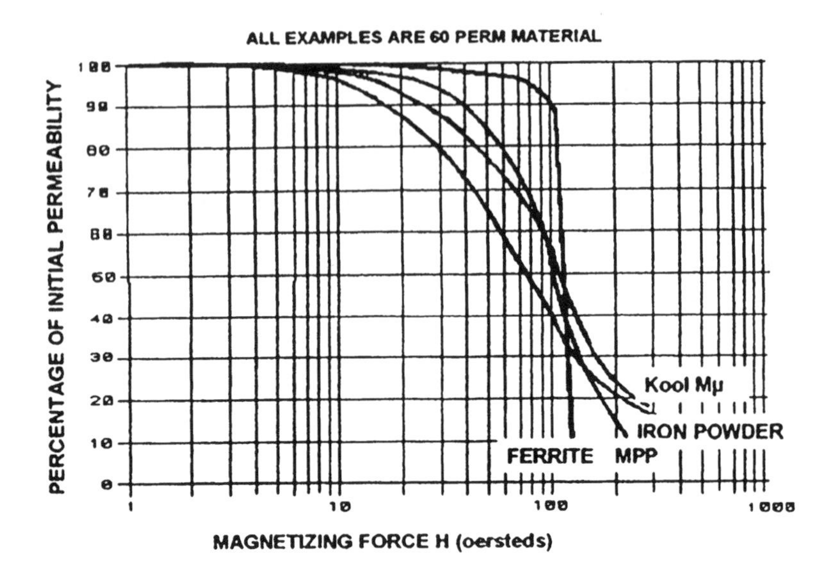

FIGURE 7.24 A general picture showing typical core saturation characteristics, as a function of DC magnetizing force for iron powder, Kool Mμ, MPP, and gapped ferrite materials.

In general, we will find that each material family provides a wide range of permeabilities. However, with ferrite cores, the working permeability is controlled by the thickness of the air gap. In choke applications the air gap is normally quite large, so the initial permeability of the ferrite material plays only a small part in the effective permeability of the gapped core. Neglecting fringing effects, ferrite core loss does not change significantly with air gap.

Powder materials are available in a large range of permeabilities that are controlled by the manufacturer during the mixing of the various powders and nonmagnetic binders. Hence the effective air gap is distributed throughout the bulk of the material and, in general, the higher permeability materials have higher core losses.

Figure 7.25 shows typical magnetizing characteristics for Kool Mμ materials, used in toroids, ranging in permeability from 26 μ to 125 μ. The higher permeability materials approach saturation at a lower magnetizing force. For example, at 100 oersteds, the 90 μ material shows only 25% of its initial permeability, whereas the lower permeability 26 μ material still shows over 80%. The difference is caused by the shape of the magnetizing B/H loop, rather than a change in the intrinsic saturation value. The flux density for saturation tends to be

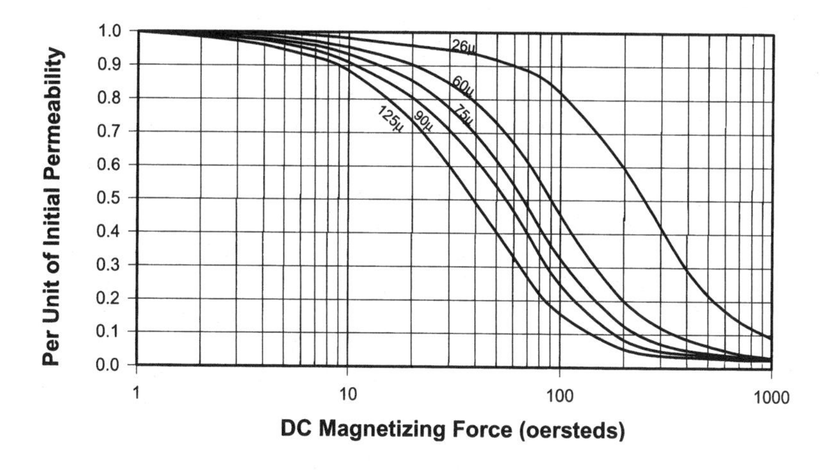

FIGURE 7.25 A chart showing the magnetization characteristics for Kool Mμ powder materials used in toroids, indicating how the core permeability decreases as the magnetization force (DC bias current) increases. There is a separate chart for E cores. (*Courtesy of Magnetics Inc.*)

close for all permeabilities of Kool Mμ materials. Iron powder shows similar properties.

In the above example, the difference between the 90 μ and the 26 μ materials is not as large as it might seem, because 80% of 25 μ is an effective 20 μ, whereas 25% of 90 μ is an effective 22.5 μ. So at a magnetizing force 100 oersteds, for the same number of turns, the working inductance will be nearly the same with either core material.

A parameter to notice is the curvature of the characteristic. This shows how much the permeability changes or "swings" with the value of magnetizing force. We can see that the working permeability of the 90 μ material (and hence the effective inductance) changes at a much lower DC bias current than that of the 26 μ material. Other powder materials show similar characteristics. This change, or "swing," in permeability can be used to advantage in the design of swinging chokes.

7.8.8 Material Cost

Cost is always a major selection criterion. Since this is a variable, we will only compare relative costs. The historically highest cost materials have been MPP and ferrite, because raw material and manufacturing costs are high. These are followed by the various Kool Mμ materials where price tends to be more variable. Iron powder is normally the lowest cost material.

Iron powder material has been known to age more rapidly than other materials at high temperature, and you should check the properties of the latest materials for this limitation and be sure that your design is well within the temperature limits. (There are variations among the different manufacturers.)

7.8.9 Establishing Optimum Core Size and Shape

In any design, the first step is the initial choice of core size and configuration. This can be quite confusing, because a bewildering range of core topologies and core sizes exist, and it can be difficult to decide on the optimum size and shape for a particular application.

The selection of the type of core is a little more straightforward. Here I have limited our selection to E cores or toroidal cores, although any shape may be used if a suitable core can be found. All of the previous materials are available in toroidal form. Iron powder cores and Kool Mμ cores are also available as E cores and building blocks. At this time, MPP cores are only available in toroidal form, because the material is difficult to work. Ferrite cores are available in many shapes.

The shape and type of the core is often a matter of the designer's preference, and any special mechanical requirements of the design. Toroidal cores can be more difficult to wind, and in any event are not suitable for gapped ferrite designs, since toroids cannot easily be gapped. E cores and blocks are more suitable for high current applications, where copper strip windings are often used.

7.8.10 Conclusions on Core Material Selection

We can conclude from the above that, in some cases, the selection of core material is quite straightforward. If the DC bias current is much larger than the AC stress, or the working frequency is low, then a natural selection would be the higher saturating flux density materials such as iron powder, to get less turns. In such examples, the core loss is less important, because the copper loss is likely to exceed the core loss by a large margin.

At the other end of the scale, the applied AC voltage stress is high, the working frequency is high, or the inductance is high and the DC bias current is small. The core loss is going to be the limiting factor, and clearly a lower loss powder material or gapped ferrite would be chosen. Between these two extremes, many other factors will control the choice, and similar results can be obtained from several quite different selections.

7.9 Magnetics: Choke Design Examples

7.9.1 Choke Design Example: Gapped Ferrite E Core

In the design example shown below, we consider the design approach for a choke using a gapped ferrite E core. We start by choosing a core size. This is where the area product[17,18,19] concept comes to our aid; the area product (AP) of a core provides a figure of merit for selecting core size and many other parameters. Although various methods can be used to establish the optimum core size, in the following examples we will use the AP approach.

We start by designing the choke $L1$ for the buck regulator shown in Figure 7.21, using a gapped ferrite E core. We will use a nomogram developed for gapped ferrite E cores to establish the AP of the core and hence the core size. This method provides a fast, simple, and effective solution to the general design requirement. It yields a typical "middle of the road" design that can be adjusted easily to meet specific requirements.

The nomogram[17] in Figure 7.26 has been developed for gapped ferrite E cores. It shows the area product, and hence the core size, as a function of load current, with the required inductance as a parameter. Further, the AP links the inductance to the copper loss, and thus the temperature rise.

The nomogram assumes an ambient of 20°C, and a 30°C temperature rise above ambient, a maximum flux density (B_m) of 250 mT, and a copper packing factor of 0.6, meaning that only 60% of the available winding window is occupied by copper. These are all typical values for this type of design.

In this nomogram, the AP is based on the product of core pole area and useful bobbin window area (rather than core window area). Where bobbins are to be used, this is a more conservative approach, since the AP can be reduced considerably by the window space used for the bobbin and insulating material. (Alternatively, you can adjust the packing factor to allow for the bobbin.) The following design example shows how this nomogram is used. We will assume the following electrical specification for the buck regulator.

Buck regulator specification:

Buck regulator specification
Input voltage = 25 V
Output voltage = 5 V
Maximum output current = 10 A
Frequency = 25 kHz
Maximum ripple current = 20% max (2 A peak-peak)
Maximum temperature rise = 30°C above ambient

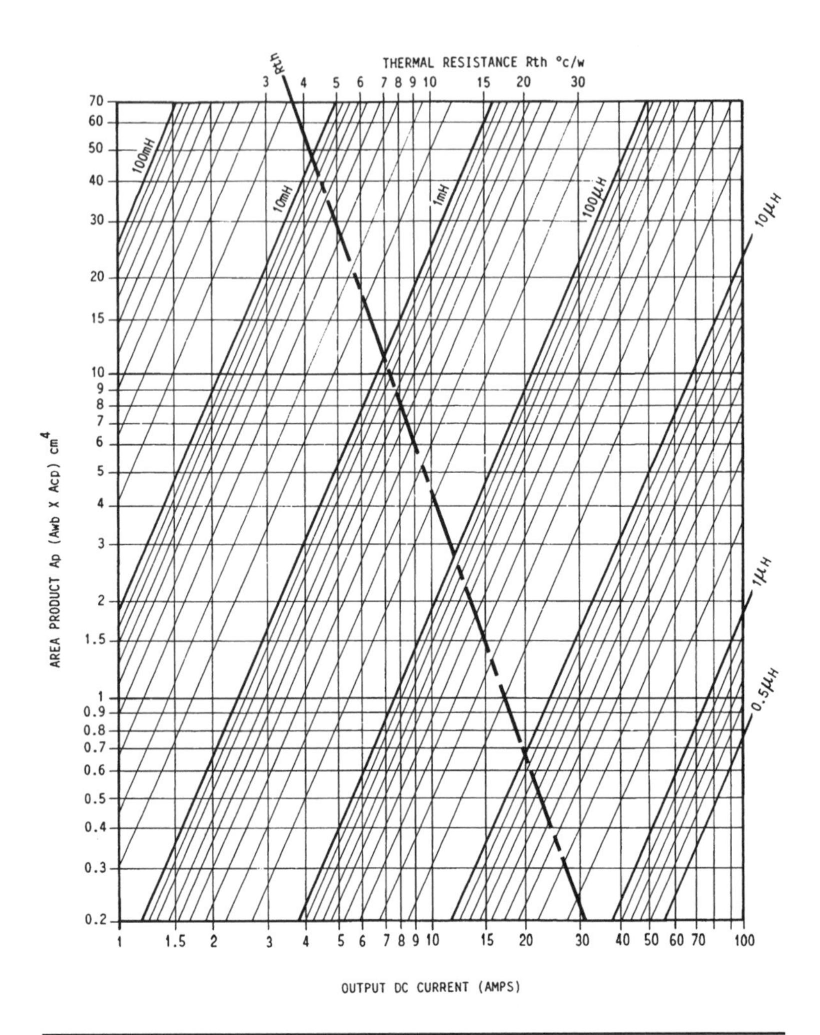

FIGURE 7.26 This nomogram can be used for gapped ferrite E core choke design. It links the area product (and hence core size) to the mean choke current, with inductance as a parameter.[17]

7.9.2 Step 1: Establish Inductance for 20% Ripple Current

The choke inductance has not yet been specified. We will now establish the inductance required for 20% ripple current, as follows.

Consider the current waveform shown in Figure 7.21*b*. This is the waveform for full load and maximum input voltage. The 20% ripple current waveform is centered on the mean DC current of 10 amps. We can establish the required inductance from the slope of the current waveform as follows (if less ripple current is required then simply draw the lower value, but remember the inductance must then be larger):

Frequency $f = 25$ kHz. Hence total cycle period $T = 1/f = 40\ \mu s$
Output voltage = 5 V
Input voltage = 25 V
We can calculate "on" time of $Q1$ (t_{on}):

For a buck regulator, the duty ratio D is t_{on}/T, and in steady state conditions, this is the same as the voltage ratio V_{out}/V_{in}:

$$t_{on} = \frac{T V_{out}}{V_{in}} = \frac{40(5)}{25} = 8\ \mu s$$

and

$$t_{off} = T - t_{on} = 40 - 8 = 32\ \mu s$$

By inspection of Figure 7.21*a*, we can see under steady state conditions that during the "off" period of $Q1$, the current shown in loop B is established by flyback action. The diode $D1$, being forward-biased during this period, takes the left side of $L1$ negative by about 0.6 volts while the right side remains at +5 volts, being maintained near this value by the energy stored in the large capacitor $C1$. Since in a closed loop control system, the output voltage is maintained constant and the choke voltage is therefore constant during the "off" period of $Q1$, we will use this period to establish the inductance.

The voltage across the choke $L1$ during this flyback action is the output voltage plus a diode drop. That is, 5 + 0.6 = 5.6 volts, and we can now calculate the inductance as follows:

During the "off" period the current decays linearly at a rate defined by

$$V_{L1} = \frac{L\,\Delta I}{\Delta t} = 5.6 \text{ volts}$$

Hence

$$L = \frac{V_{L1}\,\Delta t}{\Delta I} = \frac{5.6(32 \times 10^{-6})}{2} = 87\ \mu H$$

7.9.3 Step 2: Establish Area Product (AP)

We will use the nomogram shown in Figure 7.26 to find the AP for the required core and, hence, the core size, using the specified current and calculated inductance as follows:

Enter the bottom of the nomogram with the required current of 10 A, and project up to meet the nearest required inductance of 90 μH, this yields (to the left) an AP of approximately 1.5. This value falls between the EC35 core (AP = 1.3) and the EC41 core (AP = 2.4). Since core sizes change in large increments, absolute values of AP and hence temperature rise are not always possible.

The larger EC4I core is chosen in this example.

7.9.4 Step 3: Calculate Minimum Turns

The minimum number of turns that may be wound on a core to give the required inductance, without exceeding the flux density of 250 mT, is given by the following equation:

$$N_{\min} = \frac{L I_{\max} 10^4}{B_{\max} A_e}$$

where $N_{\min}$ = minimum turns
L = required inductance (henrys)
$I_{\max}$ = maximum current (amps)
$B_{\max}$ = maximum flux density (teslas)
A_e = center pole area (mm^2)

In this example

$$N_{\min} = \frac{90 \times 10^{-6}(11)10^4}{250 \times 10^{-3}(106 \times 10^{-2})} = 37 \text{ turns}$$

7.9.5 Step 4: Calculate Core Gap

In this example, a ferrite E core is to be used, and to prevent the core saturating for the DC current conditions, an air gap is required. The initial permeability of the ferrite core material is much greater than the permeability of the gapped core. Hence, we can assume that most of the reluctance is in the air gap.

TIP *We have a magnetic path around the core consisting mainly of a ferrite core material with a permeability between 2000 and 6000. This is in series with an air gap with a permeability of only one. Even though the gap length is much smaller than the core length, its very low permeability swamps any effect the core has, and we can neglect the core permeability in our calculations. This is the normal situation when gapped ferrite is used for chokes. ~K.B.*

The approximate air gap length ℓ_g (neglecting fringe effects) is given by the following equation:

$$\ell_g = \frac{\mu_r \mu_0 N^2 A_e 10^2}{L}$$

where ℓ_g = total air gap (mm)
$\mu_o = 4\pi \times 10^{-7}$
$\mu_r = 1$ (the relative permeability of air)
N = turns
A_e = the effective area of center pole (cm^2)
L = inductance (henrys)

In this example

$$\ell_g = \frac{4\pi \times 10^{-1}(37^2)(106 \times 10^{-2})(10^{-1})}{90 \times 10^{-6}} = 2 \text{ mm (0.078 inches)}$$

This is the total length of the air gap required in the core, and if possible, this should be confined to the center pole to minimize external magnetic radiation. With this type of choke, however, the ripple current component is normally small, and the gap may extend right across the core (a butt gap), and the resulting magnetic radiation will not be excessive. A butt gap is half the total, or 1 mm in this example, because the gap is split into two parts of 1 mm each—the center pole and the outer legs, totaling 2 mm.

TIP *If preferred, the majority of any remaining external magnetic field may be effectively reduced by fitting a copper screen right around the finished part in the area of the air gaps as shown in Figure 7.27. With an EC core, the area of the center pole is less than the sum of the outer legs, and if the gap extends*

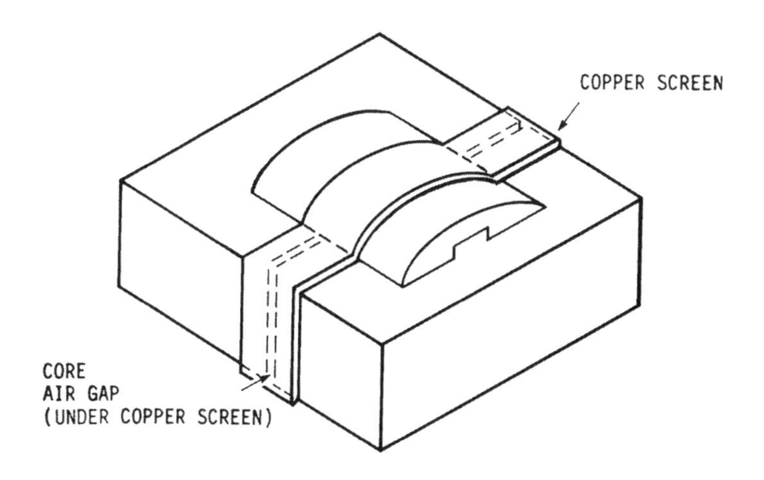

FIGURE 7.27 A copper screen may be fitted around the outside of a gapped ferrite E core choke to reduce EMI radiation and fringing at the air gap.[17]

right across the core, then the effective leg gap will be reduced by the ratio of pole to leg areas. In any event, because of the neglected core permeability and fringe effects, some adjustment of the air gap may be necessary to obtain optimum results.

When the AC stress is large, fringing at the air gap will increase the eddy current and skin effects in the wire near the gap which can cause local hot spots in the winding. In this case, extending the gap right across the core will help to reduce the hot spot effects. ~K.B.

7.9.6 Step 5: Establish Optimum Wire Size

In the design of chokes, the criteria for selecting the wire size differs from that used for transformers, due to the larger DC current component in chokes. This means that the copper loss normally exceeds the core loss by a large margin. Since the ripple current is often quite small, skin and proximity effects are not normally as big a problem as they are in transformer designs. Hence, for minimum copper loss, the wire size should be maximized. To do this, the available winding space should be completely filled with the required number of turns using a wire size that will completely fill the bobbin. (An exception to this is power factor correction chokes in which the high frequency ripple current is significant and skin effects should not be neglected.)

TIP *Information on the gauge and number of turns for a fully wound bobbin, together with the winding resistance, is often provided by the bobbin or core manufacturers. ~K.B.*

The nomogram in Figure 7.28 shows the relation between the area product and the number of turns that you would expect to get on a fully wound bobbin for wire gauges between 10 AWG and 28 AWG. It applies to standard E cores.

To use the nomogram shown in Figure 7.28, enter from the left with the area product for the chosen core (1.6 cm^4 in this example) and also on the lower scale, enter with the turns (37 turns in this example). At the intercept, we see from the diagonal lines that a wire gauge between 14 AWG and 16 AWG is indicated.

TIP *To make it easier to wind, you may prefer to use several strands of a thinner wire, as this will also improve the packing factor and reduce skin effects. In general, going down three gauges provides a wire with 50% of the cross sectional area. Hence, two strands of 18 AWG would have the same copper area as one strand of 15 AWG. This relationship is maintained throughout the AWG table, so four strands of 21 AWG could be used. ~K.B.*

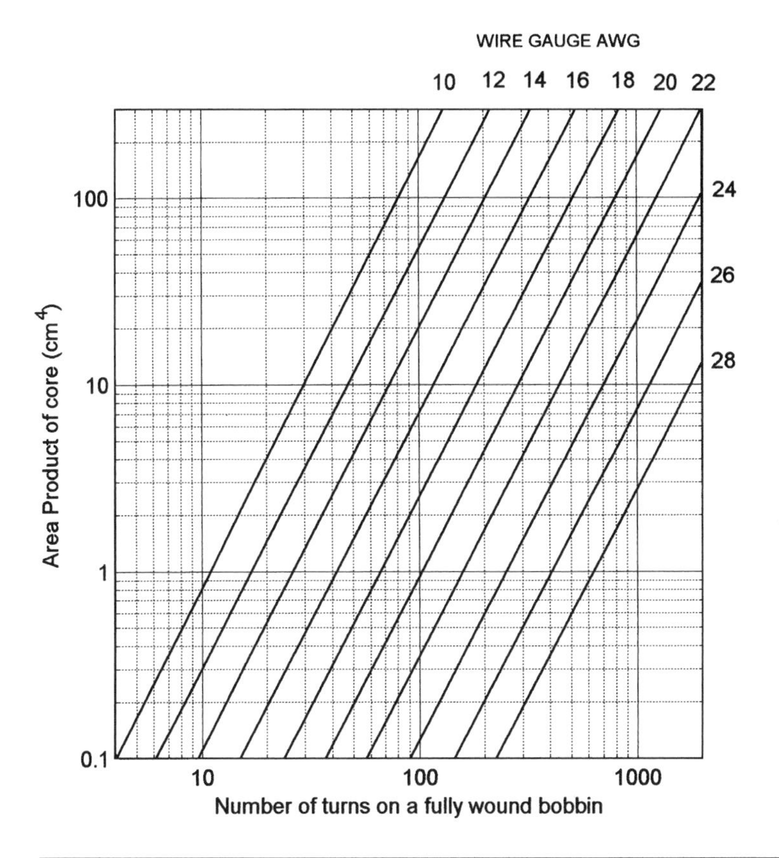

FIGURE 7.28 A nomogram used for E core choke designs, showing the link among area product, turns, and wire size for fully wound bobbins.[17]

7.9.7 Step 6: Calculating Optimum Wire Size

If preferred, the wire size for a fully wound bobbin can be calculated from basic principles as follows:

$$d = \frac{[A_w K_u]^{1/2}}{N}$$

where d = wire diameter, mm

A_w = total winding window area, mm^2

K_u = winding packing factor

N = turns

In this example,

$$d = 138 \text{ mm}^2 \text{ (EC41)}$$
$$K_u = 0.6 \text{ (for round wire)}$$
$$N = 37$$

Hence,

$$d = \frac{[138 \times 0.6]^{1/2}}{37} = 1.5 \text{ mm}$$

indicating a wire size of 15 AWG from Table 7.9.

7.9.8 Step 7: Calculate Winding Resistance

The DC resistance of the wound choke can be obtained from the bobbin manufacturer's information, or it may be calculated using the mean diameter of the wound bobbin, the turns, and the wire size. In any event, it should be measured after the choke is wound, as winding stress and packing factors will depend on the winding technique, and these will affect the final overall resistance. Remember, the resistance of copper will increase approximately 0.43%/°C above its value at 20°C. This makes the effective resistance 34% higher at 100°C, and the designer should allow for this when calculating the working resistance and copper loss.

The length of the winding, and hence the resistance, may be established from basic principles using the mean diameter of the bobbin and the number of turns, as follows:

Mean diameter of EC41 bobbin $d_b = 2$ cm

The mean length per turn (MLT) is πd_b

Total length of wire:

$$\ell_w = MLT\,N = \pi(2)37 = 233 \text{ cm}$$

From Table 7.9, the resistance of 14 AWG wire is between 83 mΩ/cm at 20°C and 110 mΩ/cm at 100°C, giving a total wound resistance (R_c) between 19.3 mΩ and 25.8 mΩ in this example.

7.9.9 Step 8: Establish Power Loss

Typically, the ripple current is small so the skin and proximity effects are negligible. Hence, the mean DC current and DC resistance can be used with little error in the power loss calculations. To the first order, the copper power loss is given by $I^2 R_c$ so the power loss in the finished choke will be

$$P = I^2 R_c \text{ watts}$$

In this example, the current is 10 A, and R_c is between 19.3 and 25.8 milliohms so $I^2 R_c$ is between 1.9 and 2.6 watts. Hence, the power loss in the copper (W_{cu}) will be between 1.9 and 2.6 W, depending on the working temperature.

7.9.10 Step 9: Predict Temperature Rise—Area Product Method

The temperature rise depends on the total power loss (core loss plus copper loss), the surface area, the emissivity of the core, and the air flow in the final application. In the interest of simplicity, we will assume free air conditions and neglect number of second-order effects, as they result in only a small error in the final predicted temperature rise.

In any event, the temperature of the choke should always be checked in the working prototype, where the layout and general thermal design will introduce additional "difficult to determine" thermal effects. It has been shown[17,18,19] that the "scrapless" E-core geometry allows the surface area of the final wound core to be related to its area product.

The nomogram in Figure 7.29 has two functions: 1) it shows the surface area of the wound E core as a function of area product (top and left scales and the dashed diagonal AP line), and 2) it also shows the predicted temperature rise as a function of dissipation, with surface area as a parameter (lower scale and solid diagonal temperature rise lines). This nomogram will be used to predict the temperature rise of the EC41 core when the total maximum wound component loss is 2.6 W.

The area product of the EC41 core from Table 7.10 is 2.4 cm^4. When a bobbin used the window area is reduced, and the area product is also reduced to near 1.6, we enter the nomogram at the top with an AP of 1.6 and project down to the intercept with the AP line (the dashed diagonal line). This intercept provides the surface area on the left scale (42 cm^2 in this example).

Enter the nomogram again on the bottom scale with the total dissipation (2.6 watts) and project up to intercept the horizontal surface area line (42 cm^2). The nearest diagonal solid lines predict the temperature rise. By interpolating between the lines we get a prediction of near 40°C rise above ambient in this example. This is more than the intended 30°C, but we would expect a higher temperature rise because we used a bobbin, reducing the area of the winding window.

Figure 7.14 can also be used to establish the temperature rise as described in Section 7.6.

7.9.11 Step 10: Check Core Loss

The temperature rise calculations in the preceding area product design approach assumed that the ferrite core loss would be negligible. You

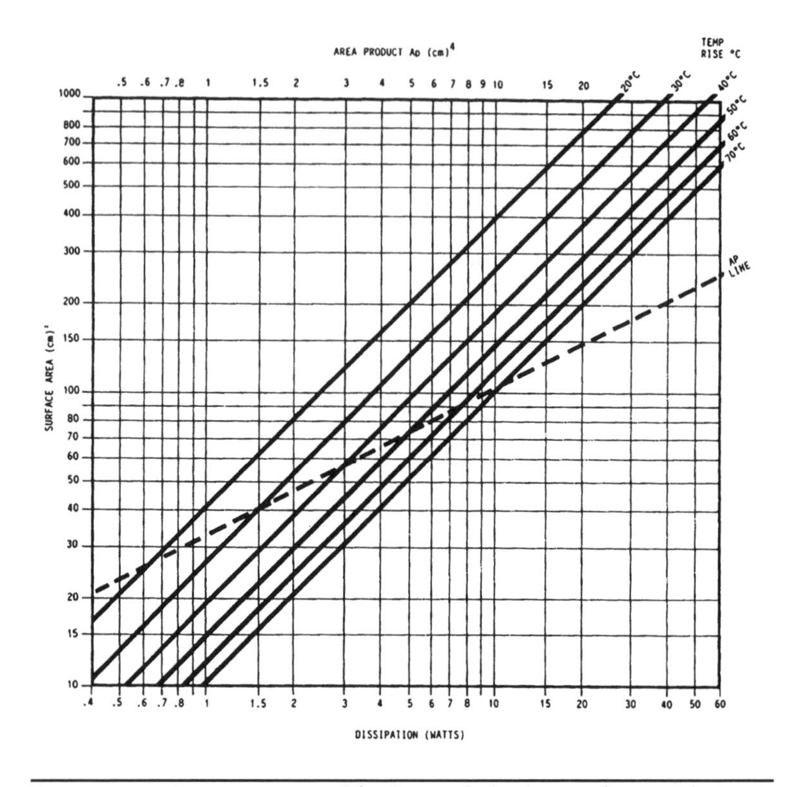

FIGURE 7.29 A nomogram used for E-core choke design, showing the surface area as a function of area product and linking the internal dissipation with temperature rise.[17]

will find in general that this is a fair assumption for gapped ferrite cores. However, we will now examine and verify this. The core loss may be calculated as follows.

The core loss is made up of eddy-current and hysteresis losses, both of which increase with frequency and AC flux excursion. The loss factor depends on the material and is provided in the manufacturers' material specifications.

TIP *The manufacturers supply core loss information related to peak flux density assuming push-pull operation. Hence, the published graphs assume a symmetrical flux density excursion of about zero, and the indicated* B_{max} *is the peak value, which is half the push-pull peak-to-peak flux density swing. Hence, when calculating the core loss for buck and boost chokes and flyback applications that use only the first quadrant of the B/H loop, the loss obtained from the manufacturers' loss diagrams should be divided by 2 when*

Core Type	Core Size cm	A_e cm^2	A_{WB} cm^2	AP cm^4	MPL cm	MLT cm	Volume cm^3
E 100	100/27	7.38	9.75	72	27.4	14.8	202
E 80	80/20	3.92	10.2	40	18.4	11.9	72.3
F 11	72/19	3.68	5.44	20	13.7	11.5	50.3
Din 5525	55/25	4.20	3.15	13.2	12.3	8.9	52.0
Din 5521	55/21	3.53	3.15	11.12	12.4	8.5	44.0
E 60	60/16	2.48	3.51	8.7	11.0	9.0	27.2
E 175	56/19	3.37	2.08	7.0	10.7	8.5	36.0
Din 4220	42/20	2.33	2.18	5.0	9.7	8.4	22.7
Din 4215	42/15	1.78	2.18	3.9	9.7	7.5	17.3
E 1625	47/15	2.34	1.64	3.83	8.9	6.5	20.8
E core	42/9	1.07	2.24	2.40	9.8	5.8	10.5
E 121	40/12	1.49	1.33	1.98	7.7	6.1	11.5
E 1375	34/9	0.87	1.31	1.14	6.9	5.2	5.6
E 2627	31/9	0.83	0.85	0.70	6.2	4.6	5.1
Din 307	30/7	0.60	0.99	0.59	6.7	4.0	4.0
E 2425	25/6	0.74	0.60	0.45	7.3	3.8	3.0
			EC CORES				
EC 35	34/9	0.84	1.55	1.3	7.74	5	6.5
EC 41	40/11	1.21	2.0	2.4	8.93	6	10.8
EC 52	52/13	1.80	3.0	5.4	10.5	7.3	18.8
EC 70	70/16	2.79	6.38	17.8	14.4	9.5	40.1

A_e = Effective area of center pole (cm^2); A_{wb} = Effective area of bobbin winding window (cm^2); AP = Area Product (cm^4); MPL = Magnetic path length around core (cm); MLT = Mean wire length per turn (cm^2); and Volume = Volume of core (cm^3)

TABLE 7.10 A Small Selection of Standard Ferrite E Cores, Showing the Essential Parameters for Choke Design (See manufacturer's data for a comprehensive listing and full data.)

entering the diagrams with the peak flux density B_{max}. *Alternatively, enter with* $B_{max}/2$. *~K.B.*

In the preceding example, the AC flux density excursion is given by

$$\Delta B_{ac} = \frac{V_{L1} t_{off}}{N A_e}$$

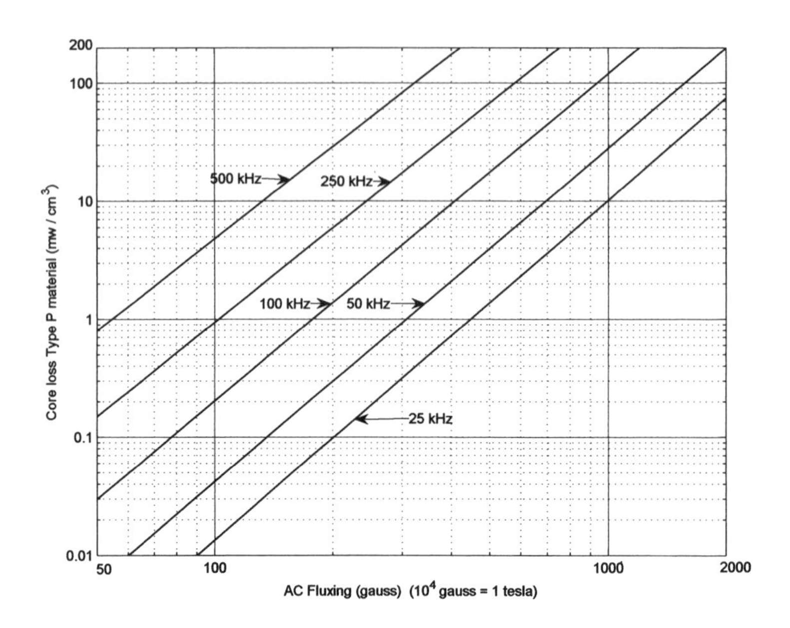

FIGURE 7.30 Core loss for Magnetics® type P ferrite material, as a function of peak AC flux density, with working frequency as a parameter. See manufacturers' data for latest information.
Note: Manufacturers assume push-pull operation, so the peak-peak flux swing (ΔB) is twice the peak value. When using these charts for single-ended, first-quadrant applications, such as buck regulator chokes, enter with ΔB and divide the indicated loss by two.

where ΔB_{ac} = AC flux density swing (teslas)
V_{L1} = the inductor voltage (volts)
t_{off} = Q1 "off" period (μs)
N = turns
A_e = effective area of core pole (mm^2)

For the above example

$V_{L1} = 5.6$ V
$t_{off} = 32$ μs
$N = 37$
$A_e = 71\ mm^2$

Hence

$$\Delta B_{ac} = \frac{5.6(32)}{37(71)} = 68\ mT(680\ \text{G})$$

$$B_{peak}(\text{for loss chart}) = \Delta B_{ac}/2 = 340\ \text{G}$$

With a typical ferrite material at a flux density of 340 gauss and a frequency of 20 kHz, the core loss (from Figure 7.30) will be less than 2 mW/cm^3. The EC41 has a volume of 10.8 cm^3, giving a total core loss of less than 22 mW, a negligible loss. Hence, ferrite core loss will normally be insignificant, except for high-frequency and large ripple current applications, and our above assumption was indeed valid.

Because it is intrinsically higher, it may not be possible to neglect the core loss of a powdered iron core for the same application in which a ferrite core yielded negligible core loss. Therefore, you should always calculate the actual core loss of iron powder materials, and if significant, add it to the copper loss to establish the temperature rise. Remember, this is a free air prediction so the proximity of other components and the air flow in the final design may affect this value. Measurements in the finished product may indicate the need to modify the final design.

7.10 Magnetics: Choke Designs Using Powder Core Materials—Introduction

In place of gapped ferrite, we can use materials with intrinsically lower permeabilities that do not require an air gap. Here, we compare the essential properties of various powder core materials to see how they may be used in choke designs.

We have seen in the previous gapped ferrite choke design example that low permeability is essential to prevent the large DC bias current from saturating the core. With ferrite cores, the low permeability was obtained by introducing an air gap into the magnetic path. The initial permeability of ferrite materials may range from 1000 μ to 5000 μ. We have seen that for choke use, the air gap must significantly reduce this permeability, and a typical range of permeability between 10 μ to 500 μ should be expected, depending on the application.

As an alternative to gapped ferrite, we can use one of the various low permeability powder core materials. Powder cores are constructed from finely divided ferromagnetic dust, compressed under high pressure into cores of various forms and sizes. The magnetic material is bonded together by a nonmagnetic carrier in such a way that each particle is spaced from its neighbors by nonmagnetic, electrically insulating material to reduce eddy current effects. As a result, the effective "air" gap is distributed throughout the body of the material. This distributed gap significantly lowers the intrinsic permeability, and cores are typically available in a range of initial permeabilities from 10 μ to 500 μ.

Because the permeability is defined by the manufacturing process, rather than an adjustable air gap, the powder cores are available only in discrete permeability steps. The higher saturation flux density of powder material over ferrite, together with the lower permeability, makes the energy storage capability of powder cores higher than gapped ferrite cores, so that slightly smaller chokes are possible using these materials, providing core loss is reasonably low.

A further advantage of the distributed gap is the elimination of the sudden discontinuity in the magnetic path associated with the large discrete air gaps used in the gapped ferrite choke designs described in the previous chapters. This gives a more uniform radiated magnetic field, with the advantage that hot spots associated with fringing at the gap of gapped ferrite designs are reduced. Although an air gap may be used to further lower the effective permeability of powder E cores, this is rarely done because a lower permeability, lower loss powder material would normally be a better choice.

7.10.1 Factors Controlling Choice of Powder Core Material

Many types of powder materials are available, but for the switchmode choke applications that we consider here, we limit the range to the more popular types, including iron powder, Molypermalloy powder (MPP), and Kool Mμ. These materials are available in toroidal, E core, C core, and block forms. We have seen in Section 7.8 that the core material is chosen to satisfy several divergent performance parameters. These include operating frequency, core loss, saturating flux density, ratio of DC bias current to AC ripple current, required inductance, range of current, temperature rise, and any special mechanical requirements.

Why are these parameters divergent? Well, with the materials available at this time, there is an inevitable tradeoff among various parameters, and improving one typically depreciates another. For example, choosing a low permeability material to reduce the core loss inevitably increases the copper loss. So to make an optimum choice, it is necessary to evaluate the relative performance of the various materials and select according to the most important application requirements. To remind us what these material parameters are, we will review some of the more important factors covered in Section 7.8.

7.10.2 Powder Core Saturation Properties

Before we proceed to a specific powder core design, we need to look more closely at the various materials available to us and compare the essential performance properties. We should look again at Figure 7.22.

This shows some typical saturation characteristics for gapped ferrite, iron powder, MPP, and Kool Mμ materials.

The horizontal scale H (oersteds) is proportional to the product of DC bias current or mean load current and the turns, and the vertical scale B (milliteslas) shows in general terms the median flux density that may be expected for H_{dc} for each type of material.

At this point, the most important parameter to consider in our design is the flux density at which particular families of materials saturate. Remember that in the ferrite example, an air gap was introduced to give a permeability near 60 μ. So let us review this.

TIP *Contrary to popular belief, the air gap in a ferrite core does not change the saturation flux density B, it changes only the magnetizing force* H *required to cause saturation. ~K.B.*

From Figure 7.22, the typical saturation flux density for each material, shown in ascending order, is as follows:

1) Ferrite, 0.35 T
2) MPP, 0.65 to 0.8 T
3) Kool Mμ, nearly 1.0 T
4) Iron powder, above 1.2 T
5) High flux (not shown) , 1.5 T

To convert from tesla to gauss, multiply by 10^4.

Clearly, with all else being equal, we would prefer the higher saturation material because it would give more inductance and lower copper loss. However, we must now consider material loss, so we look again at Figure 7.23.

7.10.3 Powder Core Material Loss Properties

Figure 7.23 shows typical core loss for the materials considered above. We see the core loss caused by the AC fluxing component ΔB and the AC stress applied to the choke, which is a function of the flux density swing and the frequency. In this example, it is for AC fluxing of 200 milliteslas peak-peak at 50 kHz. We see that typical loss in ascending order is as follows:

1) Ferrite is near 30 mW/cm^3.
2) MPP is typically three times greater, near 100 mW/cm^3.
3) Kool Mμ is nearly six times ferrite, near 200 mW/cm^3.
4) Iron powder is sixty-five times ferrite, above 2000 mW/cm^3.

We see that the ideal selection for minimum core loss is completely reversed to the optimum selection for maximum saturation properties. The two choices are in direct conflict, and we must make the best compromise selection.

So how should we proceed from here? Fortunately, some selections are quite straightforward, so we will now look more closely at both extremes of the design range. At one end, we have low AC stress conditions and hence low core loss, resulting in copper loss–limited designs, and at the other extreme, we have high AC stress conditions giving core loss–limited designs.

We can get a better picture of the relative performance of the various materials for a wide range of AC stress conditions by plotting a small selection of typical materials on the same loss chart.

Figure 7.31 shows material loss as a function of AC fluxing (ΔB) for various powder materials and type P ferrite at a working frequency

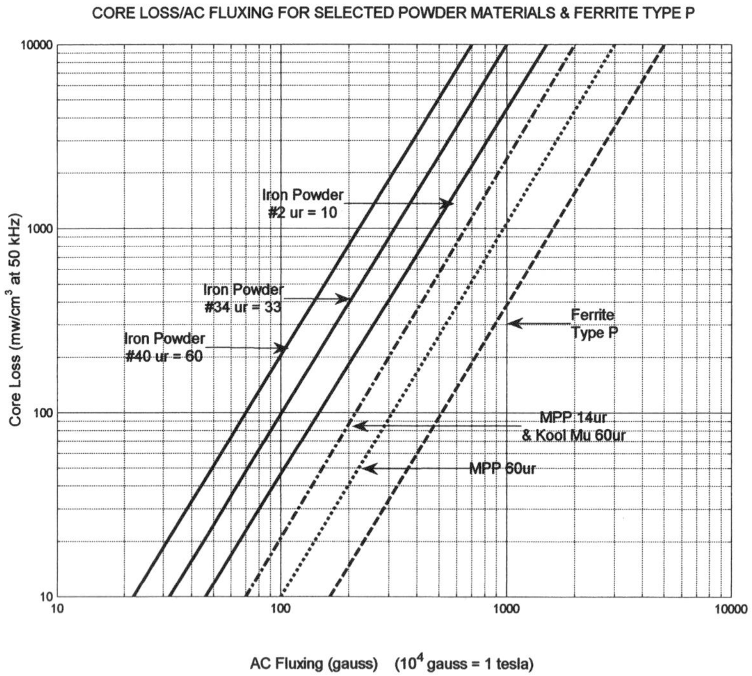

FIGURE 7.31 Loss as a function of AC fluxing at 50 kHz for iron powder, MPP, Kool Mμ, and type P ferrite materials. Within the same family, the higher permeability materials have higher loss. Not all materials follow this trend, so for specific materials the designer should always refer to the manufacturers' data.

of 50 kHz. The chart provides a direct comparison of the general loss characteristics of the various materials available to us.

In general, the higher permeability powder materials have considerably greater loss, although there are some exceptions. There are considerable variations even within a given material family, depending on the working conditions. However, if we consider the core loss at an AC fluxing of 700 gauss peak or 1400 gauss peak-peak at 50 kHz, we see the following losses, in ascending order:

Material	Relative Permeability μr	Loss (mW/cm^3)
Ferrite type P	2500	100
MPP 14 μr and Kool Mμ	60	500
MPP 60 μr	60	1100
Iron powder #2	10	2100
Iron powder #34	33	5000
Iron powder #60	60	10,000

For these materials, we see a spread in loss of two decades. At this time, gapped ferrite provides the lowest core loss, and the smallest change in core loss with change in permeability. This is because the permeability for ferrite is defined mainly by the air gap, whereas the core loss is defined by the selection of ferrite material. The various powder and MPP cores display a much greater change in core loss with change in permeability, because the material structure changes. Iron powder material shows the largest loss and the largest overall variation, while Kool Mμ shows a much smaller loss and a small spread (not shown here).

Take care, because some specific selections (not shown here) may not conform to the general rule, so you should always check the latest manufacturer's data for specific materials.

7.10.4 Copper Loss-Limited Choke Designs for Low AC Stress

In some cases, the selection of core material is quite straightforward. For example, where the AC ripple current and/or frequency is low, such as in a series-mode 60-Hz line input RFI filter choke, the core loss will never be a major factor and the highest permeability iron powder, or even gapped laminated silicon iron material, would be chosen. These materials have the advantage of high saturation flux density (B_{sat}), high permeability, and low cost.

As a result of the higher permeability, fewer turns are needed to obtain the required inductance, resulting in lower copper losses. The higher saturation value means the core will remain below saturation with a larger DC bias current. The low-frequency and/or low-ripple current means the core loss is not a significant problem in such applications, and the copper loss is likely to exceed the core loss even with the much higher core loss materials such as iron powder. Hence, these are called copper loss–limited designs. The main design thrust is to minimize the winding resistance. For such applications, the higher permeability iron powder materials would be a natural selection, but be sure to calculate the core losses in the final design just to be sure. We will look at a specific example of a copper loss limited design using a Kool Mμ powder core in Section 7.11.

7.10.5 Core Loss–Limited Choke Designs for High AC Stress

At the other end of the spectrum, where the operating frequencies and/or AC stress currents are much greater, the flux density swing is much greater and core losses will predominate, so this determines the choice of material.

Typical examples of chokes for high AC stress conditions are high-voltage, high-frequency switching regulators, and active boost type power factor correction chokes. For such applications, lower permeability, loss, and saturation flux materials such as powdered iron, Molypermalloy, Kool Mμ, and gapped ferrite are preferred. As a result of the lower permeability, more turns will be required to obtain the required inductance, and the copper loss will be greater.

In core loss–limited designs, even with the lowest loss materials, both the copper loss and core loss will likely be greater than the previous low AC stress examples. Hence, such designs are referred to as core loss limited, as the core loss remains the dominant factor. This will push the material selection toward the much lower loss MPP, Kool Mμ, and gapped ferrite materials. Again, the choice is not obvious, and it is necessary to calculate the actual copper and core loss to be sure you have made the optimum choice of material for the intended application. We look at core loss–limited design examples in Section 7.12.

7.10.6 Choke Designs for Medium AC Stress

Between the two extremes shown above, many factors control the choice of core type and core material. The optimum choice is not clear and similar results can be obtained from different materials and designs. It becomes necessary to calculate the relative core and copper loss for the chosen design, and adjust it to get the optimum result iteratively. There are well-written computer design programs that are ideal for this iterative optimization process, but make sure the program includes all the known variables.

Kool Mμ material can make the design process a little easier, because the core loss is reasonably low and remains reasonably constant throughout the range of permeabilities, reducing the effect of one of the variables. Further, the cost of this material is intrinsically less than the high nickel content MPP materials. The lower permeability iron powder materials should always be considered, providing the AC stress is not too great, since they may be satisfactory and are the cheapest.

7.10.7 Core Material Saturation Properties

Another important selection factor is the ability of a core to support the DC magnetizing force without saturation. That is, can it support the working DC bias current, plus the AC component, plus a reasonable over-current condition? Figure 7.24 shows the saturation characteristics of gapped ferrite compared with various powder materials. Notice that the ferrite saturates quite suddenly. Hence, it is important to ensure that saturation does not occur with a reasonable over-current when designing chokes using gapped ferrite cores. A good safety margin should be provided by selecting an adequate air gap.

Powder cores have a much more progressive drop in permeability as the current increases, so the inductance "swings," but some minimum inductance is still maintained even under large transient over-current conditions, so powder cores provide a much better over-current safety margin.

The curvature in the magnetizing characteristic of the various powder materials shows that the permeability changes significantly with magnetizing force H, which is proportional to load current, and this "permeability swing" can be used to design "swinging chokes."

7.10.8 Core Geometry

Selection of core geometry is a little easier. Here, we have limited our selection to E cores, toroids, C cores, and blocks. At this time, all of the powder materials are available in toroidal form. Iron powder and Kool Mμ cores are also available as E cores, C cores, and blocks. Ferrite cores are available in many forms. At the time of going to press, MPP cores are available only in toroidal form because the material is difficult to work.

TIP *I understand at least one manufacturer is planning to make MPP E cores in the future. Due to the low loss of this material you may want to consider these when they become available. ~K.B.*

E cores, C cores, and blocks have the advantage of ease of winding when large inductances and, hence, many turns are required. A typical application is a common-mode RFI filter. Powder material building blocks can be assembled in many forms and allow custom designs. They are particularly suitable for larger current applications, where the copper strip windings are normally used.

7.10.9 Material Cost

Cost varies over time so we will only compare relative costs. Historically the highest cost material has been MPP, because the raw material costs are very high. MPP is 79% nickel, and since this is a limited resource material, the costs are likely to remain high. This is closely followed by the various ferrite materials. Traditionally the lowest cost material has been the various iron powders. Kool Mμ does not contain nickel, and the price should be quite low, but it tends to be more variable, being driven more by market forces and manufacturing cost than by raw material cost. E cores, C cores, and blocks tend to be more expensive to manufacture, but have lower winding costs.

Powder materials are available in a large range of permeabilities that are controlled by the manufacturer while mixing the various powders and nonmagnetic binders. Hence, the effective air gap is distributed throughout the bulk of the material. When ferrite material is used for choke designs, however, the final permeability is controlled by the thickness of the air gap. The air gap normally accounts for the majority of magnetic circuit reluctance, so the initial permeability of the ferrite material plays only a small part in the effective permeability of the gapped core. Hence, there is little advantage in choosing the more costly, low loss, high permeability ferrite materials for gapped choke applications.

TIP *Iron powder materials have been known to deteriorate more rapidly than the other powder materials at temperatures above 90°C. The aging process is associated with the properties of the binder, so improvements in this technology are possible. If you intend to use this material, you should check the application temperature and look at the most recent high temperature aging properties of the chosen material. There are performance differences among the various manufacturers. ~K.B.*

We will now look at some specific powder core choke design examples.

7.11 Choke Design Example: Copper Loss Limited Using Kool Mμ Powder Toroid

7.11.1 Introduction

In this section, we will look at a copper loss–limited choke design, using a toroidal Kool Mμ powder core. The same methods are also suitable for the design of chokes using other low permeability materials such as iron powder and molybdenum permalloy in both toroidal and E core forms.

Copper loss–limited designs have the majority of the total power losses confined to the copper wire used to wind the choke. Due to the large ratio of mean DC load current to AC ripple current, most choke designs for switchmode applications are copper loss limited. This is particularly true when using gapped ferrite, Kool Mμ, or MPP powder, because with these materials, the core loss is often significantly lower than the copper loss. In fact, in such designs we can normally neglect core loss in the initial design process.

The first choice facing the designer is the selection of a suitable core size. The size of the chosen core must satisfy the mean load current, inductance, and temperature rise requirements. Several design methods can be used. Manufacturers often provide charts allowing the selection of core size related to the energy storage ability of the core. Other methods involve the use of charts or nomograms linking the required current and inductance to core size and other parameters such as temperature rise, the number of turns, and wire size.

In this example, we will use the area product method, because this method is universal and can be applied to any core material, size, and shape, providing suitable charts are available for each type.

7.11.2 Selecting Core Size by Energy Storage and Area Product Methods

At this point, we introduce a new choke design parameter—the "energy storage number ($\frac{1}{2}LI^2$)." Some choke design methods start by establishing the energy storage number for the choke. For this, we need to define the mean DC load current and the inductance required for the intended application. We can then calculate the energy storage number as follows:

$$W = \tfrac{1}{2}\, L1^2$$

where W is the energy storage number (stored energy in millijoules, or milliwatt-seconds)
L is the inductance (millihenrys)
I is the mean load current (amps)

It will be clear that in complete energy transfer systems such as discontinuous flyback converters, the energy storage ability of the core has a direct bearing on the energy transferred per cycle. In such systems, energy is stored in the core at the start of a power cycle and transferred to the output at the end of each power cycle, so it is easy to see a direct link between the energy storage ability of the core and output power.

For chokes in continuous-mode operation, such as those used in the output LC filters of switchmode power supplies, the connection between energy storage and choke size is not so obvious. Most of the energy stored in the core is a result of the mean DC current component and this energy remains in the core from one cycle to the next. However, the copper loss is related to I^2R, and the number of turns, and hence resistance R, is related to the inductance L. It has been shown[17,18,19] that the energy storage criterion is related to core size even in continuous-mode operation.

Having established the energy storage number, we can link this to the area product for toroidal cores using Figure 7.32. This chart also includes the predicted temperature rise as a parameter. Hence, from this chart, we can select an area product and then a core size to provide

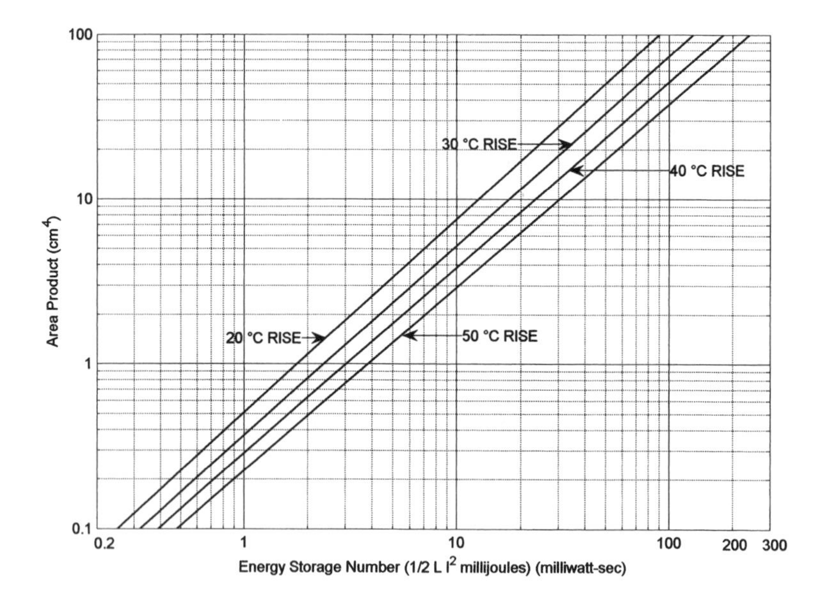

FIGURE 7.32 A chart for toroidal cores using Kool Mμ materials, showing the link between energy storage ($\frac{1}{2}LI^2$) and area product, with temperature rise as a parameter. This chart provides the area product and hence core size for toroidal choke designs.

both the energy storage number and a predicted temperature rise, by using the appropriate diagonal temperature line. The chart covers a range from 20 to 60°C. Other core types such as E cores require a slightly different chart because the ratio of surface area to area product is different.

7.11.3 Copper Loss-Limited Choke Design Example

For this design example, we will use a Kool Mμ toroidal core to design the choke $L1$ in the buck regulator example shown in Figure 7.21.

We will assume the following design parameters:

1) Mean load current 10 amps.
2) Required inductance 1.2 mH
3) Temperature rise limited to 40°C

7.11.3.1 Step 1: Calculate Energy Storage Number

Energy storage $W = \frac{1}{2} LI^2$

Hence $W = \frac{1}{2} (1.2 \times 10^{-3}) \times 10^2 = 60$ millijoules

7.11.3.2 Step 2: Establish Area Product and Select Core Size

With the energy storage number and Figure 7.32, we can establish the area product and hence the core size.

We enter this chart on the lower scale with the calculated energy storage number, and project up to the diagonal area product line meeting the temperature rise requirements. From this intercept, the area product is indicated on the left scale. With this area product, we can select a core size from the area product values provided by the core manufacturer or from Table 7.11. Alternatively, we can calculate the area product of a selected core from the window area and pole area as shown in previous sections.

For this example, entering Figure 7.32 on the lower scale with an energy storage number of 60 mJ and projecting up to the 40°C rise diagonal line, we see on the left scale that an area product of 28 cm^4 is indicated.

From the manufacturer's data for Kool Mμ toroidal cores shown in Table 7.11, we see the nearest larger core is 77868, with an area product of 31.8 cm^4, so we select this.

7.11.3.3 Step 3: Calculate Initial Turns

To establish the turns, initially we use the published permeability for the chosen core and core material. In this example, with this core size,

Core #	Core Size cm	A_e cm^2	A_W cm^2	AP cm^4	MPL cm	MLT cm	Volume cm^3	A_l #26	A_l #60
77908	79/17	2.27	18	40.8	20	7.5	45.3	37	
77868	79/14	1.77	18	31.8	20	6.9	34.7	30	
77110	58/15	1.44	9.5	13.7	14.3	6.2	20.7	33	75
77716	52/14	1.25	7.5	9.38	12.7	5.8	15.9	32	73
77090	47/16	1.34	6.1	8.19	11.6	5.9	15.6	37	86
77076	37/11	0.68	3.6	2.47	9.0	4.3	6.1	24	56
77071	34/11	0.67	2.9	1.97	8.1	4.3	5.5	28	61
77894	28/12	0.65	1.6	1.02	6.35	4.1	4.1	32	75
77351	24/10	0.39	1.5	0.58	5.88	3.34	2.3	22	51
77206	21/7	0.23	1.1	0.26	5.09	2.64	1.2	14	32
77120	17/7	0.19	0.7	0.14	4.11	2.44	0.79	15	35

A_e = Cross sectional area of core (cm^2); A_w = Total area of winding window (cm^2); AP = Area product (cm^4); MPL = Mean length of magnetic path (cm); MLT = Mean length of turn (40% fill factor) (cm); Volume = Volume of core (cm^3); A_l#26 = Inductance factor for #26 material (mH for 1000 turns) and A_l#60 = Inductance factor for #60 material (mH for 1000 turns)

TABLE 7.11 Essential Magnetic Parameters for a Small Selection of Kool Mμ Toroidal Cores for Choke Design Applications (see manufacturers' data for a more comprehensive listing)

we have only a single choice of core permeability, the #26 mix. From Table 7.11, we see that this material has an A_l value of 30 mH/1000^2 = 30×10^{-9}, and the initial turns may be calculated as follows:

In general, $L = N^2 A_l$

Therefore

$$N = \sqrt{\frac{L}{Al}} = \sqrt{\frac{1.2 \times 10^{-3}}{30 \times 10^{-9}}} = 200 \text{ turns}$$

At this point, we should look at Figure 7.25. This chart shows the relative permeability of Kool Mμ material with respect to the DC magnetizing force H_{dc}. It is an indication of the curvature of the B/H loop. Due to this curvature, the initial permeability falls as the magnetization force increases, and the turns calculation becomes an iterative process. (Until the turns are calculated, we do not know the value of H, so we do not know the relative permeability, and until the final permeability is known, we cannot finalize the turns calculation. If you have the equation for the curve, you can go directly to the final value.)

7.11.3.4 Step 4: Calculate DC Magnetizing Force

We can calculate the initial value of H_{dc} as follows:

$$H_{dc} = \frac{0.4\pi NI}{MPL}$$

where H_{dc} = magnetizing force (oersteds)
N = initial turns
I = DC current (amps)
MPL = magnetic path length (cm)

From Table 7.11, the MPL for the 77868 toroidal core is 20 cm, so the initial H_{dc} is

$$H_{dc} = \frac{0.4\pi(200)10}{20} = 126 \text{ oersteds}$$

7.11.3.5 Step 5: Establish New Relative Permeability and Adjust Turns

From Figure 7.25, the relative permeability for the #26 material is now only 85% of its initial value, so the new A value will be $30 \times 0.85 = 25.5$ and the turns must be increased as follows:

$$N = \sqrt{\frac{L}{A_l}} = \sqrt{\frac{1.2 \times 10^{-3}}{25.5 \times 10^{-6}}} = 69 \text{ turns}$$

We will round this off to 70 turns. For more accurate results, you can repeat the above iterative process to home in on the final, more accurate value.

7.11.3.6 Step 6: Establish Wire Size

Since we believe this will be a copper loss–limited design (meaning that the copper loss will greatly exceed the core loss), we will design for minimum winding resistance by using the maximum gauge of wire that will conveniently fit in the available window area. Skin and proximity effects are likely to be quite small, because the AC ripple current is small compared with the mean DC current. Hence, we would normally use the largest wire gauge that will fit (however, for ease of winding, multiple strands making the same area may be used).

For toroids, the normal fill factor using round wire and allowing room for the winding shuttle is 40%, and we will use this value to establish the wire size. From Table 7.11, the window area (Aw) for the 77868 core is 18 cm.[2] Hence 40% of the window area provides an effective usable copper area (*Awcu*) of 7.2 cm^2. With 70 turns the area of a single copper wire will be

$$Awcu = 7.2 = 10.3 \text{ mm}^2$$

$$N = 70$$

From the winding table, Table 7.9, we see that the nearest wire size is 17 AWG with an area of 12.2 mm^2 so we choose this.

TIP *The 40% fill factor used for toroids allows 30% of the inner diameter to be free of wire to allow room for the winding machine shuttle, so some flexibility on wire size is possible. ~K.B.*

7.11.3.7 Step 7: Establish Copper Loss

To calculate the copper loss, we need to know the resistance of the winding. From Table 7.10, we see that the mean winding length for a 40% fill factor on the 77868 core is 6.9 cm. Hence the total length of the winding will be

$$6.9 \times 70 = 483 \text{ cm (4.83 meters)}$$

17 AWG wire (or multiple strands of the same total area) has a nominal resistance of 0.01657 Ω/meter to give a total winding resistance (*Rcu*) of $4.83 \times 0.01657 = 0.08$ ohms.

Hence, the copper losses (I^2 *Rcu*) will be $10^2 \times 0.08 = 8$ watts.

7.11.3.8 Step 8: Check Temperature Rise by Energy Density Method

From Figure 7.32, we originally chose an area product and hence core size to give a temperature rise of not more than 40°C; we can now check this selection as follows:

The temperature rise of the finished choke depends upon the total losses and the effective surface area of the wound component.

Table 7.11 shows the surface area of the 77383 core with a 40% fill is 203 cm^2. The copper loss is 8 watts, giving a thermal energy density of 0.039 watts per cm^2 at the surface.

Entering Figure 7.33 from the left with an energy density of 0.039 watts per cm^2, we see from the intercept with the 25°C ambient line that a temperature rise of 31°C above ambient is predicted, starting from an ambient temperature of 25°C. (The lower temperature rise in this example is a result of choosing the larger core and larger wire.)

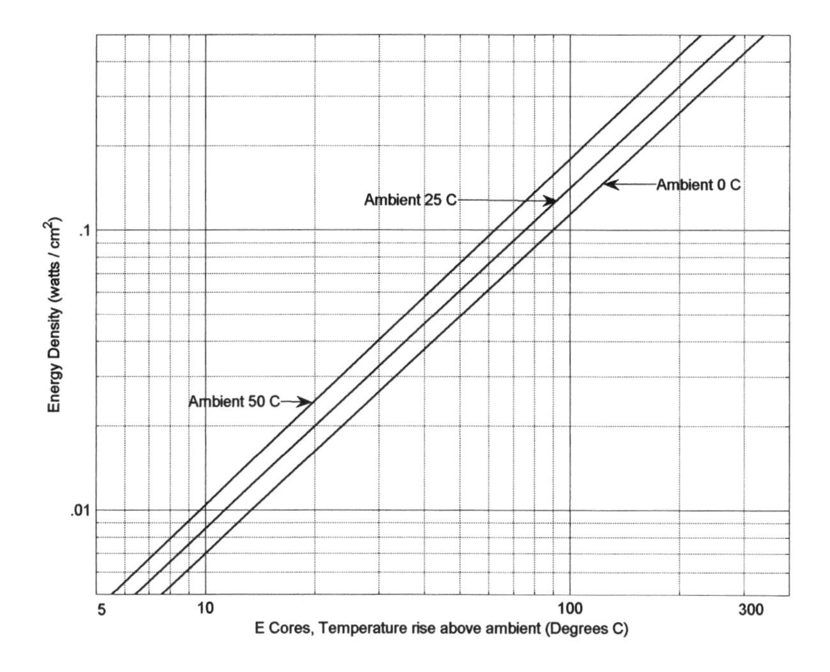

FIGURE 7.33 A chart showing the link between energy density (watts/cm^2 of surface area) and the predicted temperature rise, with ambient temperature as a parameter for E cores.

7.11.3.9 Step 9: Predict Temperature Rise by Area Product Method

The area product also allows the temperature rise to be predicted. Figure 7.34 shows the predicted temperature rise as a function of the power loss and area product. The surface area is also shown as a parameter.

The area product for the 77383 core is 31.8 cm^4 and the copper loss is 8 watts. Entering the chart with these values shows a predicted temperature rise of 30°C.

7.11.3.10 Step 10: Establish Core Loss

Up to this point, we have assumed that the core loss will be negligible. To complete the exercise, we will now calculate the actual core losses and check that this is a fair assumption.

To calculate the core loss we will use the buck regulator example shown in Figure 7.21. We have shown in Section 7.7.5 that, in general,

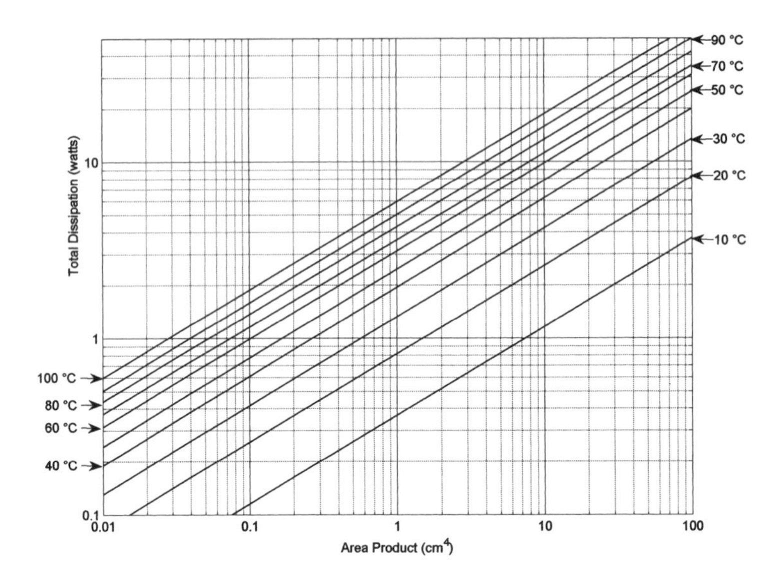

FIGURE 7.34 A chart for toroidal cores, showing the link between total dissipation (copper and core loss) and area product, with temperature rise above ambient as a parameter.

in a buck regulator, the peak AC stress conditions (B_{ac}) will be as follows:

$$B_{ac} = \frac{e \times t_{off}}{N \times A_e}$$

where e = the voltage across the choke
t_{off} = the off period of $Q1$ (in μs)
A_e = the area of the core (mm^2)
B_{ac} = the peak flux density (tesla)

In this example, V is 5.6 volts, t_{off} is 32 μs, N is 70, and A_e (the cross sectional area of the core) is 177 mm^2.

Hence

$$B_{ac} = \frac{5.6 \times 32}{70 \times 177} = 0.0146 \text{ tesla (146 gauss)}$$

To use the manufacturer's core loss shown in Figure 7.35, for this single-ended application, we divide this peak value by 2, giving an effective peak value of 73 gauss (146 gauss p-p). The chart shows the core loss for Kool Mμ with this value of the AC fluxing at 50 kHz, less than 10mW/cm^3, and the core loss can safely be neglected, as

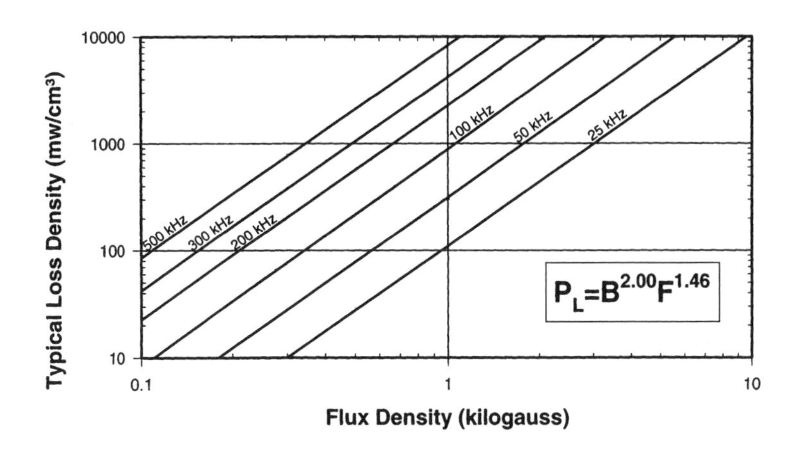

FIGURE 7.35 A chart for Kool Mµ material, showing the link between core loss and peak AC flux density with frequency as a parameter. (*Courtesy of Magnetics Inc.*)
Note: For chokes in single-ended applications, (such as buck regulators) divide the indicated loss by 2 when entering the chart using the AC flux density swing ΔB.

we expected. In fact, we can see that one of the higher loss, lower cost iron powder materials could probably be used in this particular design.

7.12 Choke Design Examples Using Various Powder E Cores

7.12.1 Introduction

In this section, to demonstrate the difference that may be expected using different core materials, we consider three design examples on the same size of E core for a choke meeting the same electrical specification, but using three different powder materials.

In the first example, we look at a core loss–limited design, where the core loss is high as a result of using a low-cost but high-loss iron powder material. In the second example, we show that changing to a lower permeability, lower loss, iron powder material can reduce the core losses to acceptable levels. In the third example, we show the same design using a Kool Mµ material.

7.12.2 First Example: Choke Using a #40 Iron Powder E Core

Here, we consider an example of a core loss–limited choke design using an iron powder E core. The same methods are suitable for the design of chokes using other low permeability materials such as Kool Mμ and molybdenum Permalloy (MPP) in both E core and toroidal core forms.

Core loss–limited designs have the majority of the total power losses generated within the core. However, unlike copper loss–limited designs, we must consider both core and copper losses, since the copper loss will always be significant. Hence, the temperature rise will be a result of the total loss in this type of design. Core loss can be large as a result of any combination of the following: core material, core size, inductance, and AC fluxing.

Once again, the first choice facing the designer is the selection of a suitable core size. The selection must satisfy the inductance requirement (that is the ripple current requirement), the mean load current, and the temperature rise limitations. We continue to use the area product method because it is universal and suitable for E cores and toroids.

Consider a choke design for a boost type switching regular application, as shown in Figure 7.36. Notice, power factor correction chokes also fit into this category. In this example, we design for an input voltage of 100 volts and an output of 200 volts at 50 kHz. This will result in high AC ripple stress on the choke $L1$, making the core work very hard, and producing significant core loss. We have seen previously that core loss is a function of both the type of core material and the AC ripple stress.

The following design specification will be assumed for this example:

1) Mean load current (I_{dc}) = 10 amps
2) Ripple current (ΔIL) = 15% (1.5 amps peak-peak)
3) Switching frequency = 50 kHz
4) The temperature rise is limited to 40°C above ambient

7.12.2.1 Step 1: Calculate Inductance for 1.5 Amps Ripple Current

We need to calculate the value of $L1$ to give a peak-to-peak ripple current of 1.5 amps at a mean load current of 10 amps.

Consider Figure 7.36 again. Under steady state conditions with an input voltage of 100 V and an output of 200 V, the duty cycle will be 50%, so the "on" period for the power device $Q1$ will be 10 μs. During

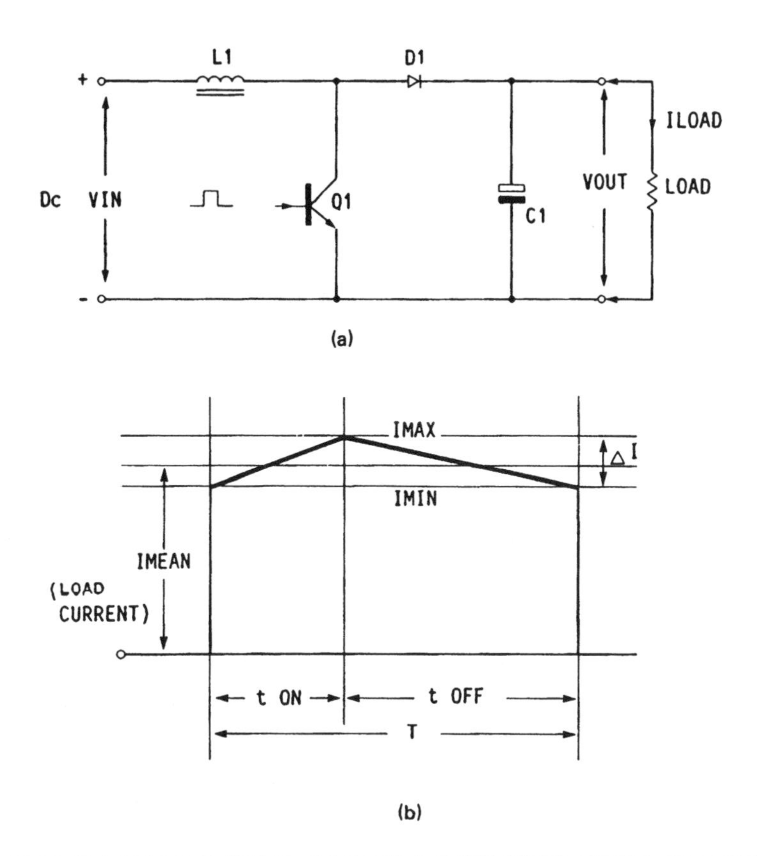

FIGURE 7.36 (*a*) A typical boost regulator power section. (*b*) The current waveform in the choke *L*1 for continuous-mode operation at a mean load current of 10 amps.

this 10-μs period, the inductor voltage is 100 V, and the ripple current will ramp up from 9.25 amps to 10.75 amps, an increase of 1.5 amps (as shown in the current waveform of Figure 7.36*b*). From this, we can calculate the inductance as follows:

In general, $V_L = L di/dt$, but in this example, the current ramp is essentially linear, so the relationship approximates to

$$V = \frac{L \Delta I_L}{\Delta t}$$

so

$$L = \frac{V \Delta t}{\Delta I_L} = \frac{100 \times 10 \times 10^{-6}}{1.5} = 0.666 \text{ mH}$$

7.12.2.2 Step 2: Calculate Energy Storage Number

With the above value of inductance and the mean DC current of 10 amps, we can calculate the energy storage number as follows:

$$W = \tfrac{1}{2} L I^2$$

where W = stored energy (millijoules)
L = inductance (millihenrys)
I = mean load current (amps)

In this example, the inductance is 0.666 mH and the mean load current is 10 amps, hence

$$W = \tfrac{1}{2} (0.666 \times 10^{-3})(10^2) = 33.3 \text{ millijoules}$$

With this energy storage number, from Figure 7.37, we can establish the area product and hence the core size.

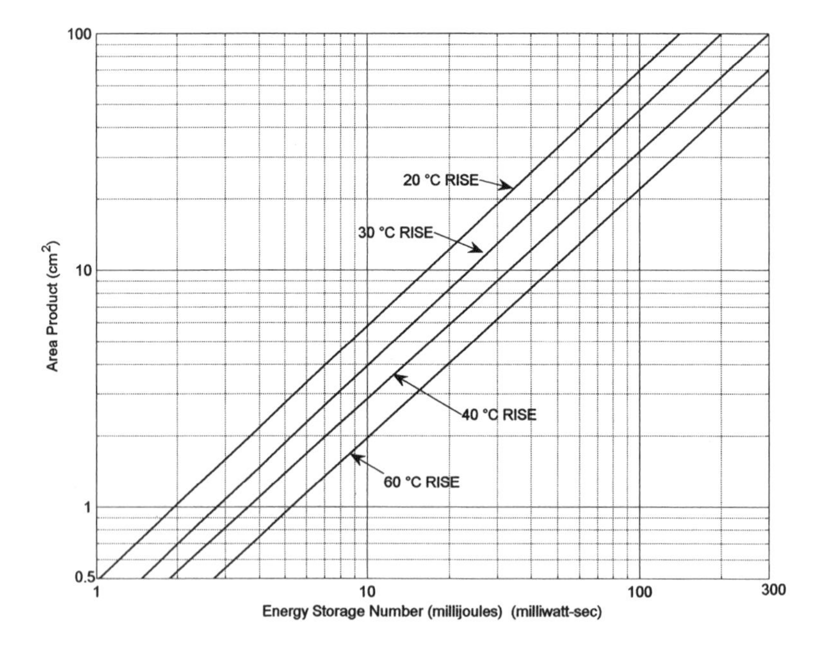

FIGURE 7.37 Link between energy storage $\tfrac{1}{2}LI^2$ and area product for standard E cores with temperature rise as a parameter. This chart allows area product and thus core size to be established.

7.12.2.3 Step 3: Establish Area Product and Select Core Size

Figure 7.37 was developed for standard E cores and shows the link between the energy storage number and the area product. The surface area for E cores is slightly greater than for toroidal cores of the same area product, so for the same dissipation, an E core will have a slightly lower temperature rise.

The temperature rise predictions shown in Figure 7.37 assume negligible core loss. Where the core loss is significant, however, the temperature rise will be a function of the total losses. Since we expect that the core loss will be significant in this first design, we will provide a margin by choosing the 30°C area product line, rather than the 40°C line, which will result in a larger core and allow for some additional loss.

We enter Figure 7.37 on the lower scale with the calculated energy storage value of 33.3 mJ, and project up to the diagonal area product line for 30°C rise. From this intercept, the area product indicated on the left scale is 14 cm^4.

With this area product, we can select a core size in Table 7.12 from the area product values provided by the core manufacturer, or we can calculate the area product of a selected core from the window and pole areas as shown in previous sections ($AP = A_w A_e$ cm^4). From the manufacturer's data for iron powder E cores shown in the table, we see the nearest larger core is the E220 with an area product of 14.2, so we select this core.

7.12.2.4 Step 4: Calculate Initial Turns

To establish the initial turns, we use the published permeability for the chosen core and core material. In this first example, we consider the #40 mix. From Table 7.12, we see that the E220 core has a reference A_l value for the #26 mix material of 275 nH/N^2. The correction factor for the #40 material is 87%, giving an initial permeability of $275 \times 0.87 = 240 \times 10^{-9}$. The initial turns are calculated as follows:

$$L = N^2 A_l$$

so

$$N = \sqrt{\frac{L}{A_l}} = \sqrt{\frac{0.666 \times 10^{-3}}{240 \times 10^{-9}}} = 53 \text{ turns}$$

At this point, we would normally consider Figure 7.38. This chart is for iron powder material and shows the relative permeability of the #40 core material with respect to DC magnetizing force H_{dc}. It is an indication of the curvature of the B/H loop. Due to this curvature, the turns calculation becomes an iterative process—until the turns are

Core #	Core Size cm	A_e cm^2	A_W cm^2	AP cm^4	MPL cm	MLT cm	Volume cm^3	A_l #40	A_l #2
E 450	114/35	12.2	12.7	155	22.9	22.8	280	480	132
E 305	77/31	7.5	8.1	60	18.5	16.3	139	339	
E 305	77/23	5.6	8.1	45	18.5	15.5	104	255	75
E 220	56/21	3.6	4.1	14	13.2	11.5	47.7	240	69
E 225	57/19	3.58	2.87	10	11.5	11.4	40.8	290	76
E 168	43/20	2.41	2.87	6.9	10.4	8.85	24.6	196	55
E 187	47/16	2.48	1.93	4.8	9.5	9.50	23.3	240	
E 162	41/13	1.61	1.7	2.7	8.4	8.26	13.6	175	105
E 137	35/10	0.91	1.55	1.4	7.4	6.99	6.72	113	32
E 118	30/7	0.49	1.27	0.63	7.14	5.38	4.60	80	
E 100	25/6	0.43	0.806	0.32	5.08	5.08	2.05	81	21

A_e = cross sectional area of core (cm^2); A_w = area of winding window (cm^2); AP = area product (cm^4); MPL = mean length of magnetic path (cm); MLT = mean length of turn (40% fill factor) (cm); Volume = volume of core for loss calculations (cm^3); A_l#40 = inductance factor for #40 material (nH/N^2); and A_l#2 = inductance factor for #2 material (nH/N^2)

TABLE 7.12 Basic Parameters for a Limited Selection of Iron Powder E Cores Suitable for Choke Designs (*Courtesy of Micrometals Inc.*)

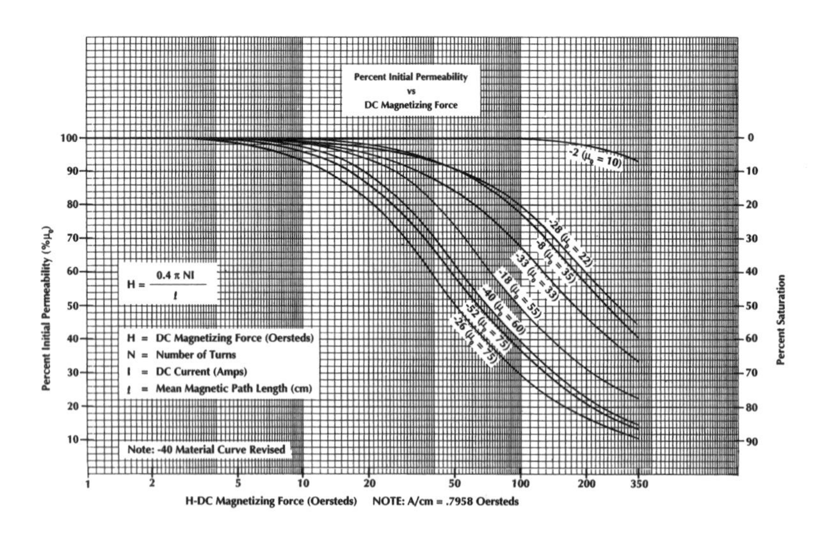

FIGURE 7.38 Magnetizing characteristics for iron powder materials. (*Courtesy of Micrometals Inc.*)

calculated, we do not know the value of H, so we do not know the relative permeability, and until the final permeability is known, we cannot finalize the turns calculation, and so on. For this example, however, since we only require the approximate turns to get an indication of core loss, we will move on to the core loss calculation and forgo the iterative process at this stage.

7.12.2.5 Step 5: Calculate Core Loss

We have seen in Section 7.7 that core loss is a function of frequency and the minor B/H loop swept out as a result of the flux density swing B_{ac}, which is proportional to the applied volt seconds.

In the boost regulator shown in Figure 7.36, when $Q1$ turns "on," the voltage across the inductor is the 100-V input voltage in this example, and the left side is positive. When $Q1$ turns "off," diode $D1$ conducts, taking the left side of $L1$ to ground, and the voltage across the inductor is the difference between the input voltage and the output voltage, which is also 100 V in this example, but now the polarity on $L1$ is reversed.

Assuming steady state conditions, during the 10 μs that $Q1$ is "on," and applying the input voltage across $L1$, the current will ramp up from its minimum value of 9.25 amps to the maximum 10.75 amps. We can calculate the peak flux density related to this stress B_{ac} as follows:

$$B_{ac} = \frac{V t_{on}}{N A_e}$$

where V = the voltage across $L1$ when $Q1$ is "on"
t_{on} = the period the voltage is applied (μs)
N = turns
A_e = area of core mm^2

In this example, V is 100 volts, $t_{on} = t_{off} = 10$ μs, N is 53, and Ae is 360 mm^2.

$$\text{so } B_{ac} = \frac{100(10)}{53(360)} = 0.0524 \text{ tesla (524 gauss)}$$

TIP *Remember that this flux change due to the AC conditions is centered around the mean flux level developed from the mean DC output current of 10 amps in this example. The flux density generated by the DC conditions is a function of the core permeability, but the flux change required as a result of the AC stress is independent of the properties of the core. ~K.B.*

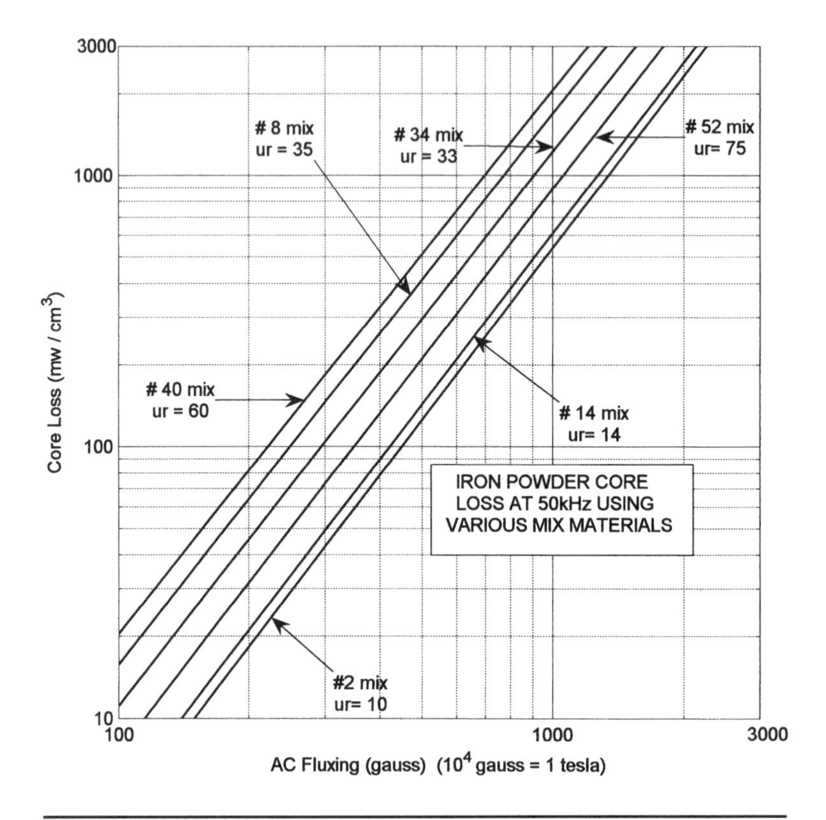

FIGURE 7.39 Material loss for iron powder material against peak AC flux density with frequency as a parameter. (*Courtesy of Micrometals Inc.*)

Figure 7.39 shows core loss for iron powder. We enter this chart with 524 gauss. The loss for the #40 mix at 524 gauss and 50 kHz is 600 mW/cm^3.

This chart shows the loss for push-pull operation, where the flux density swing is twice the peak value. For single-ended operation, we take half the indicated loss, 300 mw/cm^3. The E220 core has a volume of 47.7 cm^2, giving a total core loss of

$$47.7 \times 300 \times 10^{-3} = 13.4 \text{ watts}$$

The core loss looks large, but before we can compare the core loss with the copper loss, we must establish the resistance of the wound core, and calculate the copper losses. For this example at this stage, an approximate value is acceptable, and we can continue to use the initial 53 turns.

7.12.2.6 Step 6: Establish Wire Size

To minimize copper loss, we design for minimum winding resistance by using the maximum size of wire that will conveniently fit in the available window area.

For E cores using bobbins, the fill factor using round magnet wire ranges from a perfect but unrealistic fill factor of 87%, down to as low as 40%, depending on construction and insulation. For this example, we will use a realistic mean value of 60%.

From Table 7.12, the window area A_w for the E220 core is 4.09 cm^2. Hence 60% of the window area results in an effective usable copper area A_{cu} of 2.45 cm^2. With 53 turns, the area of a single copper wire will be

$$A_{cu}/N = 2.45/53 = 0.0462\ \text{cm}^2$$

From the winding table, Table 7.9, we see that the nearest wire is #11 AWG with area of 0.0464 cm^2 and resistance of 4.13 mΩ/meter.

With high AC ripple stress, skin effects should be considered, and we normally use several strands of a thinner wire to provide the same overall copper area, rather than a single large diameter wire. Also, #11 AWG wire would be very difficult to wind on this size of core.

7.12.2.7 Step 7: Establish Copper Loss

To calculate copper loss, we need to know the resistance of the winding. From Table 7.12, we see that the mean winding length per turn (MLT) for the E220 core is 11.5 cm. Hence the total length of the winding is

$$N \times \text{MLT} = 53 \times 11.5 = 610\ \text{cm}$$

The #11 AWG wire, or multiple strands of the same total cross sectional area, has a nominal resistance of 4.13 mΩ/meter to give a total winding resistance R_{cu} of $6.095 \times 0.00413 = 0.025$ ohms.

Hence, the copper loss $I^2 R_{cu}$ is $10^2 \times 0.025 = 2.5$ watts.

We see that the 13.4-watt core loss considerably exceeds the copper loss of 2.5 watts, as we expected. It is clearly not an optimum design. At this point, the designer has several options, as follows:

- Optimum efficiency results when copper and core losses are approximately equal. Using the existing core and increasing the turns will increase the copper loss, and decrease the core loss to the point of optimum efficiency. It will also increase the inductance and decrease the ripple current. This approach may yield an optimum design with this core and material. It may be satisfactory, but it will have a larger inductance than necessary and will probably exceed the temperature limitations.

- A better option, and one that will retain the original inductance and reduce the core loss, is to choose a core mix with a lower permeability, and hence a lower core loss. We will now look at this second option, substituting the lower loss #8 mix iron powder material.

7.12.3 Second Example: Choke Using a #8 Iron Powder E Core

We will now consider a design for the same choke using a #8 iron powder core material. The #8 mix has lower core loss, but also lower permeability. Hence we must recalculate the turns for the required inductance as follows.

7.12.3.1 Step 1: Calculate New Turns

The reference A_l value for the E220 core in #26 material is 275 nH/N^2. The correction factor for #8 mix is 51%, giving an initial permeability of $275 \times 10^{-9} \times 0.51 = 140 \times 10^{-9}$. The new initial turns are

$$N_1 = \sqrt{\frac{L}{A_l}} = \sqrt{\frac{0.666 \times 10^{-3}}{140 \times 10^{-9}}} = 69 \text{ turns}$$

We can now calculate the new core loss as follows in the next section.

7.12.3.2 Step 2: Calculate Core Loss with #8 Mix

$$B_{ac} = \frac{V t_{off}}{N A_e}$$

Where $V = 100$ volts
$t_{off} = t_{on} = 10\ \mu s$
$A_e = 360\ mm^2$

In this example, N is now 69 so

$$B_{ac} = \frac{100(10)}{69(360)} = 40 \text{ mT}(400 \text{ gauss})$$

Figure 7.39 shows that the core loss for the #8 mix at 50 kHz and 400 gauss is 190 mW/cm^3. For single-ended operation, we take half this value, 95 mw/cm^3. The E220 core has a volume of 47.7 cm^2, giving a total core loss of

$$47.7 \times 95 \times 10^{-3} = 4.53 \text{ watts}$$

TIP *Notice that there are two factors reducing the core loss. Core loss is lower because the turns have increased and the intrinsic material loss has also decreased. ~K.B.*

7.12.3.3 Step 3: Establish Copper Loss

At this stage, we can easily estimate the new copper loss since the usable winding window A_{cu} is being completely filled with wire. If we double the turns, we must halve the cross sectional area; this will double the resistance, and with twice as many turns at twice the resistance, the resistance will go up by a factor of four.

In general, for a fully wound bobbin, resistance changes as the ratio $(N_2/N_1)^2$, so the approximate resistance of the new winding will be

$$(69/52)^2 \times 0.025 = 1.76 \times 0.025 = 0.044 \text{ ohms}$$

Hence, the copper loss I^2R_{cu} is $10^2 \times 0.044 = 4.4$ watts.

Since the copper and core losses are now approximately equal, this would be considered an optimum efficiency design. To complete the design, it would be necessary to adjust the turns to allow for the loss in permeability at the working current. However, the curvature of the B/H loop for the number #8 mix is quite small and the adjustment is probably not necessary.

7.12.3.4 Step 4: Calculate Efficiency and Temperature Rise

We can now estimate the temperature rise as follows:

The total dissipation for core and copper is $4.5 + 4.4 = 8.9$ watts.

Figure 7.34 links the area product to the temperature rise for optimally wound toroidal cores. The surface area for E cores of the same area product is about 15% greater than for toroids, so the temperature rise will be approximately 15% lower when using E cores.

Entering Figure 7.34 with an area product of 14.2 cm^4 and a total loss of 8.9 watts, we see that the temperature rise for the toroidal core is predicted to be 47°C, so the E core will be 15% less, near 40°C.

We have satisfied our design requirements, and since the copper and core losses are now approximately equal, this would be considered an optimum efficiency design and is quite satisfactory.

7.12.4 Third Example: Choke Using #60 Kool Mμ E Cores

Once again, the area product method can be used to select a core size, using the Kool Mμ material. With this material, core loss is close throughout the range of permeabilities.

Core # 00K	Core Size cm	A_e cm^2	A_{WB} cm^2	AP cm^4	MPL cm	MLT cm	Volume cm^3	A_l #60	A_l #40
8020E	80/20	3.89	11.2	43.3	18.5	15.8	72.1	190	
6527E	65/27	5.40	5.4	29.0	14.7	14.18	79.4		
7228E	72/19	3.68	6.0	22.2	13.7	14.38	50.3		
5530E	55/25	4.17	3.8	15.9	12.3	12.4	51.4	261	
5528E	55/20	3.50	3.8	13.3	12.3	11.6	43.1	219	
4022E	43/20	2.37	2.8	6.60	9.84	10.1	23.3	194	281
4020E	43/15	1.83	2.8	5.10	9.84	9.2	18.0	150	217
4017E	43/11	1.28	2.8	3.56	9.84	8.26	12.6	105	151
4317E	41/12	1.52	1.64	2.49	7.75	8.16	11.8	163	234
3515E	35/9	0.84	1.52	1.28	6.94	6.86	5.83	102	146
3007E	30/7	0.60	1.25	0.75	6.56	5.36	3.94	71	92
2510E	25/6	0.38	0.78	0.30	4.85	5.00	1.87	70	100
1808E	19/5	0.23	0.52	0.117	4.10	3.78	0.914	48	69
1207E	13/4	0.13	0.23	0.030	2.96	2.48	0.385		

A_e = cross sectional area of core (cm^2); A_w = area of winding window (cm^2); AP = area product (cm^4); MPL = mean length of magnetic path (cm); MLT = mean length of turn (40% fill factor) (cm); Volume = volume for core loss calculations (cm^3); A_l#60 = inductance factor for #60 material (mH per 1000 turns); and A_l#40 = inductance factor for # 40 material (mH per 1000 turns)

TABLE 7.13 Basic Parameters for a Limited Selection of Iron Powder E Cores Suitable for Choke Designs (*Courtesy of Micrometals Inc.*)

7.12.4.1 Step 1: Select Core Size

We have already shown that the energy storage number for this application is 33.3 mJ, and the optimum area product is 14.2 cm^4. For the same ripple current, the inductance will be 0.666 mH.

From Table 7.13, we see that the nearest Kool Mμ core is the 5528E with an area product of 13.3 cm^4. It is a bit smaller than the previous E220 core, but we will try this one, because the next step up in core size is quite large.

7.12.4.2 Step 2: Calculate Turns

Since the core loss is relatively constant with permeability, we will choose the highest permeability #60 mix to minimize the number of turns. The initial permeability for this core in the #60 mix is 219, and we can calculate the turns for the required inductance as follows:

$$N = \sqrt{\frac{0.666 \times 10^{-2}}{219 \times 10^{-9}}} = 55 \text{ turns}$$

The #60 mix has a relatively large B/H characteristic curvature, and the permeability drops rapidly with increasing magnetizing force H. As a result, the turns will probably need to be adjusted later to compensate for the reduction in permeability.

7.12.4.3 Step 3: Calculate DC Magnetizing Force

We can calculate the initial value of H_{dc} as follows:

$$H_{dc} = \frac{0.4\pi NI}{MPL}$$

where H_{dc} = magnetizing force (oersteds)
N = initial turns
I = DC current (amps)
MPL = magnetic path length (cm)

From Table 7.13, the MPL for the 5528E core is 12.5 cm so the initial H_{dc} is

$$H_{dc} = \frac{0.4\pi 55(10)}{12.5} = 55 \text{ oersteds}$$

7.12.4.4 Step 4: Establish Relative Permeability and Adjust Turns

From Figure 7.25, the relative permeability for the #60 material at 55 oersteds is 70% of 219, so the new $\mu_r = 153 \text{ nH/N}^2$.

Calculate new turns N□:

$$N_2 = \sqrt{\frac{L}{\mu_r}} = \sqrt{\frac{0.666 \times 10^{-2}}{153 \times 10^{-9}}} = 66 \text{ turns}$$

7.12.4.5 Step 5: Calculate Core Loss with #60 Kool Mμ Mix

The core area A_e for this core is 350 mm^2.

Hence, the AC flux swing is

$$B_{ac} = \frac{Vt}{NA_e} = \frac{100(10)}{66(350)} = 0.0433 \text{ tesla (433 gauss)}$$

From Figure 7.35, the core loss for #60 material at 50 kHz and 433 gauss is 60 mw/cm^3 for push-pull operation, or 30 mw/cm^3 for single-ended operation. The volume of this core is 43.1 mm^3 so the loss is

$$43.1(30) = 1.3 \text{ watts}$$

7.12.4.6 Step 6: Establish Wire Size

From Table 7.13, the window area A_w for the 5528 Kool Mμ core is 3.81 cm^2. Hence 60% of the window area leaves an effective usable copper area A_{cu} of 2.28 cm^2. With 66 turns, the area of a single copper wire is

$$A_{cu}/N = 2.28/66 = 0.0345 \text{ cm}^2$$

From the wire table in Figure 7.9, we see that the nearest wire is #12 AWG, with an area of 0.037 cm^2 and a resistance of 0.00522 ohms per meter. With high AC ripple stress, skin effects should be considered, and we normally use several strands of a thinner wire to provide the same overall copper area rather than a single large diameter wire. Also the #12 AWG wire would be difficult to wind on this core.

7.12.4.7 Step 7: Establish Copper Loss

To calculate the copper loss, we need to know the resistance of the winding. From Table 7.13, the mean winding length per turn (MLT) for the 5528 core is 10.73 cm. Hence the total length of the winding is

$$66 \times 10.73 = 708 \text{ cm}$$

The #12 AWG wire, or several strands of wire of the same cross sectional area, has a nominal resistance of 0.00522 Ω/meter to give a total winding resistance R_{cu} of 7.08 meters × 0.00522 Ω/meter = 0.037 ohms.

Hence, the copper loss ($I^2 R_{cu}$) is $10^2 \times 0.037 = 3.7$ watts.

The core loss at 1.3 watts is considerably less than the copper loss, so the design is copper loss limited. A higher permeability material would reduce the turns and hence the copper loss. But this core size is not available in higher permeability material, so this is the best that we can do, without reducing the turns that would also reduce the inductance and increase the ripple current.

7.12.4.8 Step 8: Establish Temperature Rise

We can now estimate the temperature rise as follows:

The total dissipation for core and copper is $1.3 + 3.7 = 5$ watts.

Figure 7.34 links area product to temperature rise for optimally wound toroidal cores. The surface area for E cores is about 15% greater than for toroidal cores of the same area product, so the temperature rise will be approximately 15% lower when using E cores.

Entering this chart with an area product of 13.3 and a total loss of 5 watts, we see that the temperature rise for the toroidal core is predicted to be 35°C, so the E core at 15% less will be near 30°C.

So this third design on the Kool Mμ core is not only smaller, but also more efficient with less total power loss and a slightly reduced temperature rise.

7.13 Swinging Choke Design Example: Copper Loss Limited Using Kool Mμ Powder E Core

7.13.1 Swinging Chokes

Swinging chokes are used in continuous conduction applications. They have the property of increasing inductance as the DC bias current (load current) decreases. This has the advantage of reducing the ripple current at lower load currents and extending the range of current over which continuous conduction can be maintained. The "swing" is a function of the nonlinearity or curvature of the B/H characteristic.

In Figure 7.40, we see that the magnetization characteristics for Kool Mμ E core materials ranges in permeability from 26 μ to 90 μ. The curvatures in the characteristics of the various materials show that the permeability changes progressively with magnetizing force. We call this change "permeability swing" and take advantage of this characteristic in the design of "swinging chokes."

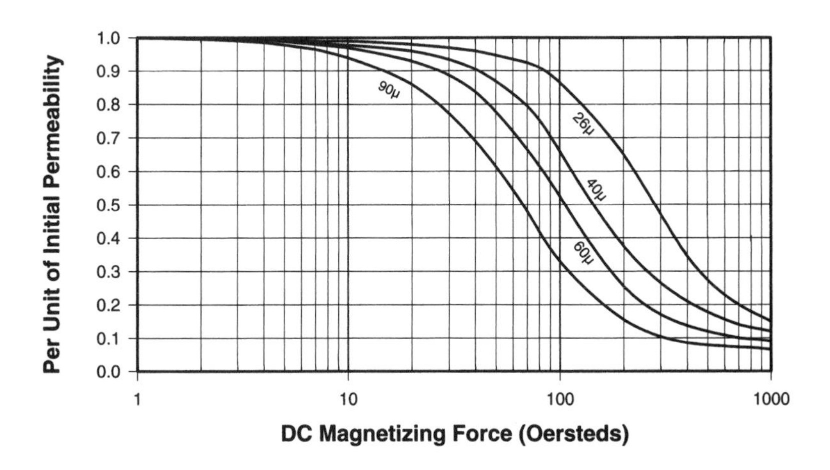

FIGURE 7.40 Magnetization characteristics for Kool Mμ E core powder materials, indicating how the core permeability decreases as the magnetization force (the DC bias current) increases. (*Courtesy of Magnetics Inc.*)

Further, we see that the higher permeability materials tend to saturate at a lower magnetizing force, as we have seen before. For example, at 100 oersteds, the 90 μ material shows only 32% of its initial permeability, whereas the lower permeability 26 μ material still shows over 87%. If we were to design a choke using the 90 μ material, such that H was at 100 oersteds with the maximum nominal load current, then when the load current decreased to say 5% of the maximum value (5 oersteds), the permeability and hence the effective inductance would increase by about 55%. So we see that the working permeability and the effective inductance swing with the change in mean load current.

A choke design of this nature is called a "swinging choke." It will maintain the choke in continuous conduction for a much larger range of load current. Clearly, this can be a great advantage in some regulator applications. The penalty paid is that the choke needs more turns and is larger to provide sufficient inductance at the maximum current, since the permeability of the 90 μ material is much lower at the higher currents. As a result, the turns and copper loss tend be greater and the saturation safety margin smaller in a swinging choke design.

7.13.2 Swinging Choke Design Example

For the following swinging choke design example, we will use a Kool Mμ powder E core because we need more turns to bring the working point on to the curved part of the B/H loop characteristic. The E core bobbin provides for easier winding when many turns are required.

We will design a swinging choke for position $L1$ in the buck regulator example shown in Figure 7.21, assuming the following design parameters:

1) Mean load current 10 amps
2) Required inductance 1 mH
3) Temperature rise to be limited to 40°C

7.13.2.1 Step 1: Calculate Energy Storage Number

Energy Storage $W = \frac{1}{2} LI^2 = \frac{1}{2} (1 \times 10^{-3})(10^2) = 50$ millijoules

7.13.2.2 Step 2: Establish Area Product and Select Core Size

With this energy storage number and Figure 7.32, we can establish the area product and, hence, the core size.

We enter this chart on the lower scale with the calculated energy storage number (50 mJ), and project up to the diagonal area product line, meeting the temperature rise requirements of 40°C. From this intercept, the area product value is indicated on the left scale, AP = 16. With this area product, we can select a core size from the area product values provided by the core manufacturer, or from Table 7.13. Alternatively, we can calculate the area product of a selected core from the window area and pole area as shown in previous sections.

For this example, from Table 7.13, we see that the nearest core is the 5530E with an area product of 15.9 cm^4, and we select this core, which has the following parameters:

Area product AP = 15.9 cm^4
Magnetic path length MPL = 12.3 cm
Mean length per turn MLT = 12.4 cm
Area of winding window bobbin $A_{wb} = 3.8\ cm^2$
Area of center pole $A_e = 4.17\ cm^2$
Inductance factor/turn $A_{lo} = 261\ nH/N^2$ for # 60 μ material

7.13.2.3 Step 3: Calculate Turns for 100 Oersteds

Normally we would choose the highest permeability material to get the largest "swing." However, in this core size, the highest permeability available is 60 μ, so we choose this material.

At nominal current, we need to be on the most curved part of the B/H characteristic shown in Figure 7.40, so we choose to work at 100 oersteds and 10 amps, which provides a permeability of 52% of the maximum value. The turns for 100 oersteds at 10 amps can be calculated as follows:

$$N = \frac{HMPL}{0.4\pi 1} = \frac{100(12.3)}{0.4\pi 10} = 98 \text{ turns}$$

7.13.2.4 Step 4: Calculate Inductance

From Table 7.13, we see that the 60 μ material has an A_{lo} value of 261 nH/N^2 (261×10^{-6}). At 100 oersteds, this relative permeability μ_r is reduced to 52% of its initial value, reducing the effective A_l value by the same ratio, giving 136 nH/N^2 at 10 amps. The inductance may be calculated as follows:

In general

$$L = N^2 A_l$$

Therefore

$$L = 98^2(136 \times 10^{-9}) = 1.33 \text{ mH(at 10 amps)}$$

At this point, we have the option to adjust turns, core size, or core material, if necessary, but this inductance is near enough to the design requirement and we will accept it.

From Figure 7.40, we see that at 20 amps (200 oersteds) the permeability has dropped to 25% and at 2 amps (2 oersteds) we have 100%, so the inductance swings from 0.65 mH at 20 amps to 2.5 mH at 2 amps, a 5:1 ratio.

7.13.2.5 Step 5: Calculate Wire Size

Since we believe this will be a copper loss–limited design, meaning that the copper loss will greatly exceed the core loss, we will design for minimum winding resistance by using the maximum gauge of wire that will conveniently fit in the available bobbin window area. Skin and proximity effects are likely to be quite small, because the AC ripple current is small compared with the mean DC current. Hence, we would normally use the largest wire gauge that will fit. For ease of winding, however, multiple strands making up the same area may be used.

For the E core bobbin, the normal fill factor using round wire is near 70%, and we will use this value to establish the wire size. From Table 7.13, the bobbin window area A_{wb} for the 5530E core is 3.8 cm^2. Hence, 70% of the window area provides an effective usable copper area A_{wcu} of 2.66 cm^2. With 98 turns, the area of a single copper wire will be

$$\frac{A_{wcu}}{N} = \frac{2.66}{98} = 0.026 \text{ cm}^2$$

From the winding table, Table 7.9, we see that the nearest wire size is #13 AWG with an area of 0.026 cm^2 so we choose this.

7.13.2.6 Step 6: Establish Copper Loss

To calculate the copper loss, we need to know the resistance of the winding. From Table 7.10, we see that the mean winding length *MLT* for a fully wound bobbin is 12.4 cm. Hence, the total length of the winding will be

$$NMLT = 98(12.4) = 1215 \text{ cm}$$

The #13 AWG wire, or multiple strands of the same total area, has a nominal resistance of 0.007 Ω/meter to give a total winding resistance R_{cu} of $12.15 \times 0.007 = 0.085$ ohms.

Hence, the copper loss $I^2 R_{cu}$ is $10^2 \times 0.085 = 8.5$ watts.

7.13.2.7 Step 7: Check Temperature Rise by Thermal Resistance Method

From Figure 7.37, we originally chose an area product and hence core size to give a temperature rise of not more than 40°C, and we can now check this selection as follows:

The temperature rise of the finished choke depends upon the total losses and the effective surface area of the wound component. Figure 7.14 shows the thermal resistance of the 5530E core with an area product of 16 is near 4.6°C/watt, the copper loss is 8.5 watts giving a predicted temperature rise of 39°C, and our design requirements are satisfied.

We have assumed negligible core loss, and we will now check this as follows.

7.13.2.8 Step 8: Establish Core Loss

Up to this point, we have assumed that the core loss is negligible; to complete the exercise, we will now calculate the actual core losses and check that this is a fair assumption.

To calculate core loss, we will use the buck regulator example shown in Figure 7.21. We have shown in Section 7.7.5 that, in general, in a buck regulator the peak AC stress conditions B_{ac} are

$$B_{ac} = \frac{V t_{off}}{N A_e}$$

where V = the voltage across the choke
t_{off} = the off period of Q1 (μs)
A_e = the area of the core (mm^2)
B_{ac} = the peak flux density (tesla)

In this example, $V = 5.6$ volts, $t_{off} = 32$ μs, N is 98, and A_e is 177 mm^2

$$\text{so } B_{ac} = \frac{5.6(32)}{98(417)} = 0.00438 \text{ tesla (43.8 gauss)}$$

The chart shows the core loss for Kool Mμ material with this value of the AC fluxing at 50 kHz is less than 1 mW/cm^3, and the core loss can safely be neglected.

References

1. Ferroxcube-Philips Catalog, *Ferrite Materials and Components*, Saugerties, NY.
2. Magnetics Inc. Catalog, *Ferrite Cores*, Butler, PA.
3. TDK Corp. Catalog, *Ferrite Cores*, M. H. & W. International, Mahwah, NJ.
4. Siemens Corp. Catalog, *Ferrites*, Siemens Corp., Iselin, NJ.
5. MMPA Publication PC100, *Standard Specifications for Ferrite Pot Cores*, MMPA, 800 Custer St., Evanston, IL.
6. MMPA Publication UE 1300 *Standard Specifications for Ferrite U, E And I Cores*, MMPA, 800 Custer St., Evanston, IL.
7. IEC Publications 133, 133A, 431, 431A, 647, American National Standards Institute, 1430 Broadway, New York.

8. A. I. Pressman, *Switching and Linear Power Supply, Power Converter Design*, pp. 116–120, Switchtronix Press, Waban, MA, 1977.
9. A. Kennelly, F. Laws, and P. Pierre, "Experimental Researches on Skin Effect in Conductors," *Transactions of AIEE*, 34: 1953, 1915.
10. F. Terman, *Radio Engineer's Handbook*, p. 30, McGraw-Hill, New York, 1943.
11. L. Dixon, *Eddy Current Losses in Transformer Windings and Circuit Wiring*, Unitrode Corp. Power Supply Design Seminar Handbook, Unitrode Corp., Watertown MA, 1988.
12. E. Snelling, *Soft Ferrites*, pp. 319–358, Iliffe, London, 1969.
13. P. Dowell, "Effects of Eddy Currents in Transformer Windings," *Proceedings IEE* (U.K.), 113(8): 1387–1394, 1966.
14. P. Venkatramen, "Winding Eddy Currents in Switchmode Power Transformers Due to Rectangular Wave Currents," *Proceedings Powercon 11*, 1984.
15. B. Carsten, "High Frequency Conductor Losses in Switchmode Magnetics," *High Frequency Power Converter Conference*, pp. 155–176, 1986.
16. A. Richter, "Litz Wire Use in High Frequency Power Conversion Magnetics," *Powertechnics Magazine*, April 1987.
17. Keith Billings, *Switchmode Power Supply Handbook*, Chapter 3.64, McGraw-Hill, New York, 1989.
18. Colonel Wm. T. McLyman, *Transformer and Inductor Design Handbook*, Marcel Dekker, New York, 1978.
19. Colonel Wm. T. McLyman, *Magnetic Core Selection for Transformers and Inductors*, Marcel Dekker, New York, 1982.
20. Jim Cox, "Power Conversion & Line Filter Applications," *Micrometals Catalog*, Issue 1, Feb. 1998.

CHAPTER 8

Bipolar Power Transistor Base Drive Circuits

8.1 Introduction

Since the 1980s, improvements in the technology of Metal Oxide Field Effect Transistors (MOSFETs or just FETs) have resulted in bipolar power transistors being progressively displaced by FET transistors in switching power supply applications. New designs in the coming years will probably use even more and better FETs.

However, there remain some niche areas (such as linear regulators and perhaps low-power applications) where, because of their lower cost and some advantages in linear applications, we will continue using bipolar transistors. Since there will remain areas where bipolar transistors still offer some advantages, and because the vast majority of the switching supplies still operating in the field were originally done with bipolars, in the event of field failures with these older but still operating designs, it is important for designers to remain familiar with their key characteristics.

The first consideration for the designer of a switching supply based on bipolar transistors is the selection of a device with the proper voltage and current ratings. Maximum voltage and current stresses are dependent on the topology being used as well as the input voltage, its tolerances, and output power. Equations giving these stresses have been derived and are presented in the discussions of each topology.

However, the means and detailed design of the bipolar base drive circuit are as important for overall reliability as choice of a transistor with adequate voltage and current ratings. The general principles of what constitutes a good base drive circuit and some widely used techniques are discussed herein.

After Pressman *FETs are essentially voltage-driven devices with current flowing in and out of the gate (essentially a modulated capacitor) on the leading and trailing edges of the drive pulse. The transient drive currents can be quite large, particularly when driving large devices at high frequencies, and the "on" state gate voltage can also be quite large—on the order of 8 V or so—with some devices. Hence, driving large FETs can be more difficult than might be expected. Further, FETs are more reliable when turned fully "on" or fully "off" and are not very good at dissipating the power developed in the substate in linear mode applications in which the device is only partly "on." Hence they are more suitable for switching applications.*

In contrast bipolar transistors are current-driven devices. The base drive current is approximately proportional to the collector current, as defined by the gain of the device. The base drive voltage tends to be quite small (0.6 V) with diode-like characteristics. Bipolar transistors are generally more robust in linear applications. There are still many applications where bipolar devices are a better choice. Hence, the need for bipolar transistors is likely to continue into the foreseeable future. ~K.B.

8.2 The Key Objectives of Good Base Drive Circuits for Bipolar Transistors

A good base drive circuit should have all the parameters described in the following six sections, 8.2.1 through 8.2.6.

8.2.1 Sufficiently High Current Throughout the "On" Time

The base current should be adequate to keep the lowest gain transistor fully saturated at the highest current it is required to conduct. With this drive the collector-to-emitter potential will be at its saturated value, typically 0.5 to 3.0 V at maximum current and minimum input voltage using the lowest beta transistor.

A good design should allow for a four-to-one production spread in transistor beta. The usual current-voltage curves (I_{ce} vs. V_{ce}) in the manufacturer's data sheets (Figure 8.1) are for a transistor of typical beta. It should be assumed that the minimum beta is one-half and the maximum beta is twice the typical shown in the curves.

After Pressman *It is not a good idea to overdrive the transistor as this will increase the storage time and can be a problem in switching applications. (This is not a problem in linear regulator applications.) The Baker clamp circuit shown in Figure 8.6 will prevent overdrive and is recommended for switching applications. ~K.B.*

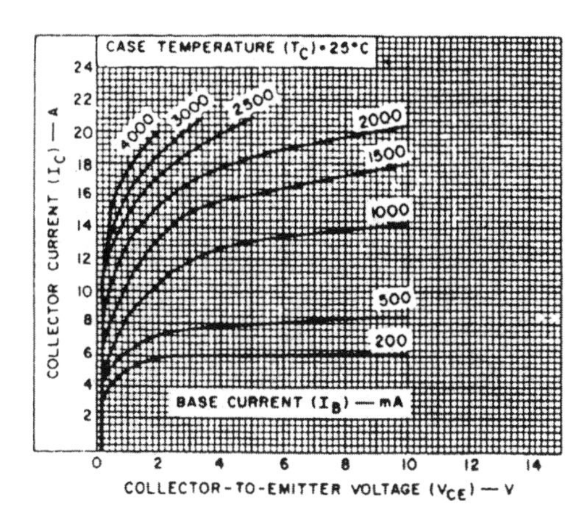

FIGURE 8.1 Curves of I_c versus V_{ce} for a typical high-voltage, high-current transistor: 2N6676, 15 A, 450 V (*Courtesy of RCA*). Curves such as this are usually for a typical device. Depending on the manufacturer's production spread, the lowest-beta device may have half the indicated beta, and the highest-beta device twice.

Power transistor currents in most topologies have the shape of a ramp starting from zero or a ramp on a step. Thus the input base current should be adequate to keep V_{ce} "bottomed" to about 0.5 to 1.0 V at the peak of the I_c ramp at maximum output power at minimum input voltage for the minimum beta transistor. This is especially true for discontinuous-mode flybacks where the ratio of peak to average current is high.

8.2.2 A Spike of High Base Input Current I_{b1} at Instant of Turn "On"

To ensure fast collector current turn "on," there should be a short spike of base current about two to three times "on" time average. This spike need last only about 2 to 3% of the minimum "on" time (Figure 8.2*a*).

The effect of this turn "on" overdrive can be seen in Figure 8.2*b*. If turn "on" speed is not a factor, base current (I_{b1}) for a desired collector current (I_{c1}) need be only that required to bottom V_{ce} to the saturation voltage $V_{ce(sat)}$ at the intersection of the collector load line and the I_c/V_{ce} curve.

At that I_{b1}, collector current will rise exponentially with some time constant τ_a and get to within 5% of I_{c1} in three time constants $3\tau_a$.

If base input current is $2I_{b1}$ (overdrive factor of 2), however, collector current will rise as if it were heading for $2I_{c1}$. It would reach $2I_{c1}$ in the same $3\tau_a$, but collector current is limited to I_{c1} by the supply voltage

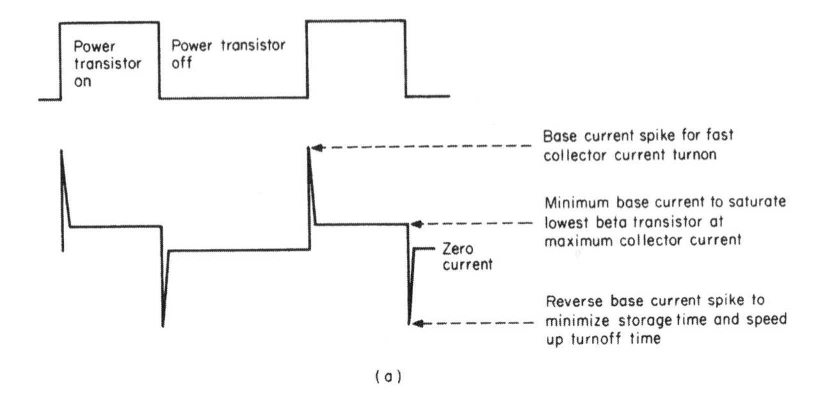

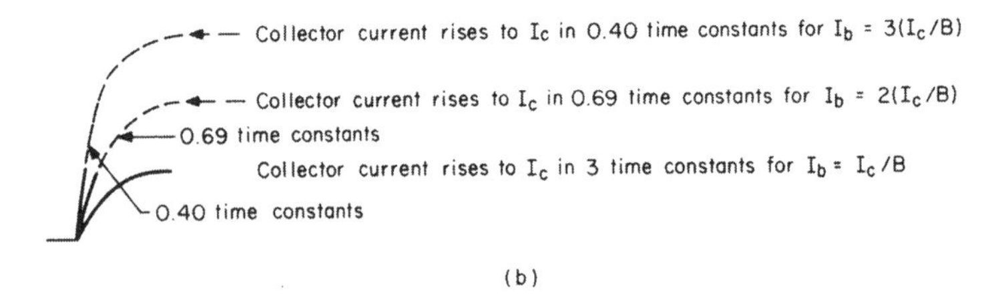

FIGURE 8.2 (*a*) Optimum base current waveforms. (*b*) Accelerating collector current rise time with base current overdrive.

and load impedance. Hence the current reaches I_{c1} (the desired value) in $0.69\tau_a$ instead of $3\tau_a$.

Similarly, if an overdrive factor of 3 is used ($I_b = 3I_{b1}$), collector current will rise as if it were headed to $3I_{c1}$ and, if not limited by the load resistance and supply voltage, would reach $3I_{c1}$ in the same $3\tau_a$. But since it is limited, it reaches the desired I_{c1} in $0.4\tau_a$ instead of $3\tau_a$.

Overdrive factors of 2 to 3 are usually used to speed up turn "on" time. The required base current overdrive can be calculated for a nominal-beta transistor. Low-beta transistors are faster and do not require as high an overdrive factor. High-beta transistors are slower, and an overdrive of 2 for a nominal beta device corresponds to an overdrive of 4 for a high-beta transistor as its beta value is generally twice that of the nominal-beta transistor.

TIP *The Baker clamp circuit in Figure 8.6 with overdrive also solves this problem without increasing the drive during the "on" period. Hence it does not increase the storage time. ~K.B.*

8.2.3 A Spike of High Reverse Base Current I_{b2} at the Instant of Turn "Off" (Figure 8.2*a*)

If base current is simply dropped to zero when it is desired to turn "off," collector current will remain unchanged for a certain time (storage time t_s). Collector voltage will remain at its low $V_{ce(sat)}$ value of about 0.5 V, and when it finally rises, it will have a relatively slow rise time.

This comes about because the base-emitter circuit acts like a charged capacitor. Collector current keeps flowing until the stored base charges drain away through the external base-to-emitter resistor. There is generally a large excess of stored base charges because the base current must be chosen sufficiently large to bottom the collector-to-emitter voltage to about 0.5 V for the lowest-beta transistor. Thus the highest or even the nominal-beta transistor has an excess of base current and long storage time.

A momentary spike of reverse base current I_{b2} is required to pull out the stored base charges. This reduces storage time and permits higher switching frequencies. It also significantly reduces power dissipation during the turn "off" interval.

This can be seen (Figure 8.3*a*) in the instantaneous I_c/V_{ce} curves during the turn "off" interval. There it is seen (t_1 to t_2) that before V_{ce} starts rising rapidly, it moves up out of saturation slowly while collector current hangs on at its peak value. During this interval, current is at its peak and collector-to-emitter voltage is considerably higher than its saturation level of 0.5 to 1.0 V.

The resulting high spike of power dissipation can be a large fraction of the total dissipation in the transistor. The spike of reverse base current (Figure 8.2*a*) reduces this dissipation by shortening the interval t_1 to t_2 and permits higher-frequency operation by reducing the storage time.

Manufacturers usually offer curves showing storage, rise, and fall times for their power transistors for values of I_c/I_{b1} and I_c/I_{b2} ranging from 5 to 10 at various values of collector current (Figure 8.3*b*, *c*).

8.2.4 A Base-to-Emitter Reverse Voltage Spike −1 to −5 V in Amplitude at the Instant of Turn "Off"

Bipolar transistors have three significant collector-to-emitter voltage ratings: V_{ceo}, V_{cer}, and V_{cev}. The V_{ceo} rating is the maximum collector-to-emitter voltage at turn "off" when the base to emitter is open-circuited at the instant of turn "off." It is the lowest voltage rating for the device.

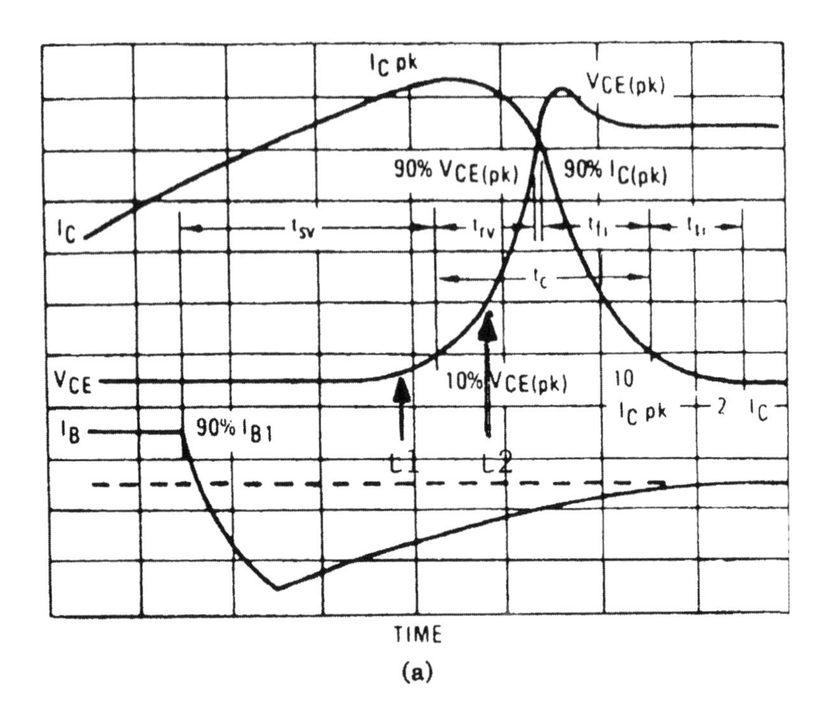

(a)

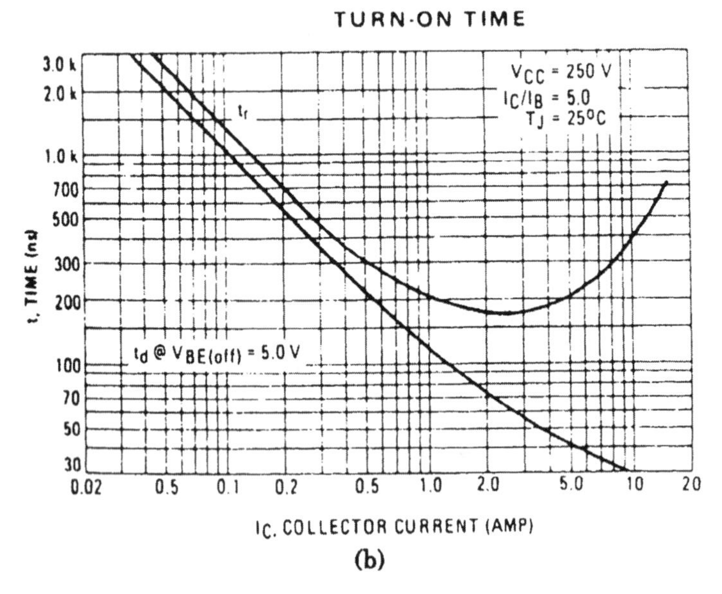

(b)

FIGURE 8.3 (*a*) Typical turn "off" transition falling current and rising collector-to-emitter voltage for a power transistor; no snubber at collector. (*b* and *c*) Typical switching times for a typical high-current high-voltage transistor: 2N6836, 15 A, 850 V (V_{cev}). (*Courtesy of Motorola Inc.*)

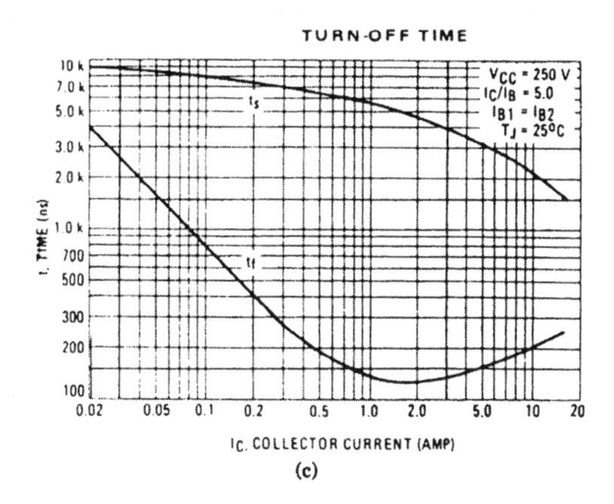

FIGURE 8.3 (*Continued*).

The transistor can tolerate a higher voltage (V_{cer} rating) during the "off" state if it has a "low" (usually 50 to 100 Ω) resistance from base to emitter.

The highest voltage the transistor can safely tolerate is its V_{cev} rating. This is the maximum voltage the transistor can tolerate at the instant of turn "off" during the leakage inductance spike (Figures 2.1 and 2.10). It can tolerate this voltage only if there is a –1 to –5 V reverse voltage spike at the base during the instant of turn "off" (Figure 8.4). This reverse-bias voltage or voltage spike must be supplied by the base drive circuit and must last at least as long as the leakage inductance spike.

8.2.5 The Baker Clamp (A Circuit That Works Equally Well with High-or Low-Beta Transistors)

Since production spread in beta may be four to one, base current that is sufficient to safely turn "on" a low-beta transistor will greatly overdrive a high-beta transistor and result in excessive storage time. Reducing this storage time adequately may require unacceptably large reverse base current. The Baker clamp drive circuit shown in Figure 8.6 solves this problem.

8.2.6 Improving Drive Efficiency

Since high-collector-current transistors generally have low beta, the base current driver must deliver high current. If this current comes directly from a high-voltage source without the benefit of current gain through a voltage step-down transformer, efficiency will be low.

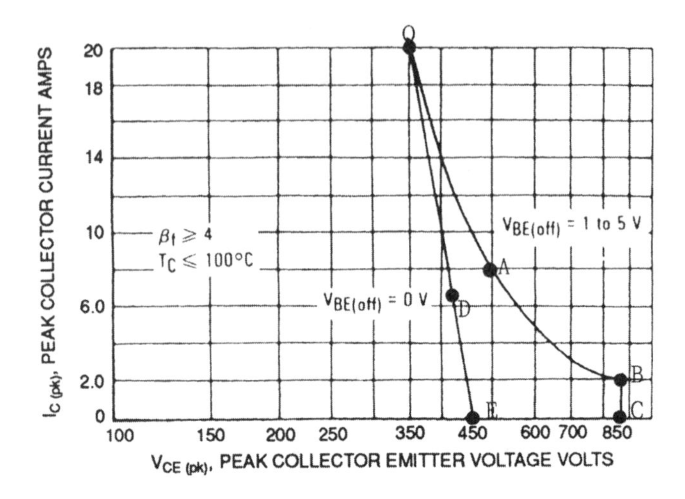

FIGURE 8.4 Reverse-bias safe operating area curves (RBSOA) for 2N6836. During turn "off," the $I_c - V_{ce}$ locus must not cross the boundaries shown. Even a single crossing may destroy the transistor because of the current crowding into a small part of the chip area and causing local hot spots. With a –1 to –5 V reverse bias at the instant of turn "off," the V_{cev} boundary *OABC* applies. For $V_{be} = 0$ at turn "off," the boundary *ODE* applies. (*Courtesy of Motorola Inc.*)

A widely used fast power transistor—the 2N6836—is rated at I_c of 15 A, and V_{cev} of 850 V. It is widely used in off-line switching supplies. At 15 A, it has a minimum current gain of 5, requiring at least 3 A of base current.

If this base drive current came from a 6-V source at, say, 80% duty cycle for a push-pull circuit, base drive dissipation at the source would be an unacceptably high 14.4 W. A good base drive scheme should couple the drive pulse from a DC housekeeping voltage source through a voltage step-down, current step-up transformer.

8.3 Transformer Coupled Baker Clamp Circuits

The transformer-drive Baker clamp[1–4] in Figure 8.7 is a widely used base drive scheme. It is inexpensive, low in component count, and provides all six features described in Sections 8.2.1 to 8.2.6.

Since it is transformer drive, it also nicely solves the problem of coupling a width-modulated pulse across the input-output boundary. In off-line supplies, the PWM chip and housekeeping supply are typically located on output common and the power transistor is on input common (see Figure 6.19 and Section 6.6.1).

Since a transformer is involved, it is relatively simple to get a voltage step-down, current step-up ratio of 10 or more. The secondary delivers about 1 to 1.8 V to the base, and the primary takes its current from the housekeeping power supply, which is usually 12 to 18 V.

Thus in the preceding example with a 10/1 turns ratio, the 3 A of base current at the transformer secondary is obtained at a cost of only 300 mA from the housekeeping supply referred to output common.

In Figure 8.7, a Baker clamp is located after the secondary of *T*1, between the collector and base of the power transistor. The operation of a Baker clamp is discussed below.

When not in saturation, the collector-to-base junction of an NPN transistor is reverse biased and the collector is positive with respect to the base. When it is hard "on" and in saturation, however, the collector is negative with respect to base, and the base-to-collector junction is forward biased, acting like a conducting diode.

This can be seen in Figure 8.5*a* and 8.5*b* for the 2N6386, which is a fast 15-A, 450-V transistor. Figure 8.5*a* shows the "on" collector-to-emitter voltage at various collector currents for two values of forced beta $B_f = I_c/I_b$ and temperature. It is seen that V_{ce} depends strongly on I_c, B_f, and temperature. At operating conditions of $I_c = 10$ A, $B_f = 5$, and 100°C junction temperature, the "on" V_{ce} potential is about 0.2 V. Figure 8.5*b* shows that at 10 A, the "on" base-to-emitter potential is about 0.9 V at 100°C.

With the resulting 0.7-V forward bias on the base-to-collector junction, there is an excess of stored base charge. Further, when the base current is simply reduced to zero, storage time for this very fast bipolar transistor is still a very long 3 μs (Figure 8.5*c*).

Baker clamping corrects this problem by not permitting the base-to-collector junction to take on a forward bias, or at the worst, by allowing only a 0.2- to 0.4-V forward bias—low enough to prevent significant storage time.

The Baker clamp can reduce storage time by a factor of 5 to 10.[4] It works nicely over a large temperature and collector current range, and very importantly, circuit operation is equally good with transistors whose production spread in beta is as large as 4 to 1. Its operation is described in the following section (see also Figure 8.6).

8.3.1 Baker Clamp Operation

In Figure 8.6, a large current I_1 of the desired pulse width is provided at the anode of *D*2. The current is large enough to overdrive *Q*1 "on" at the maximum current with the desired speed, when *Q*1 is a minimum beta transistor. As *Q*1 commences turning "on," *D*3 is reverse-biased,

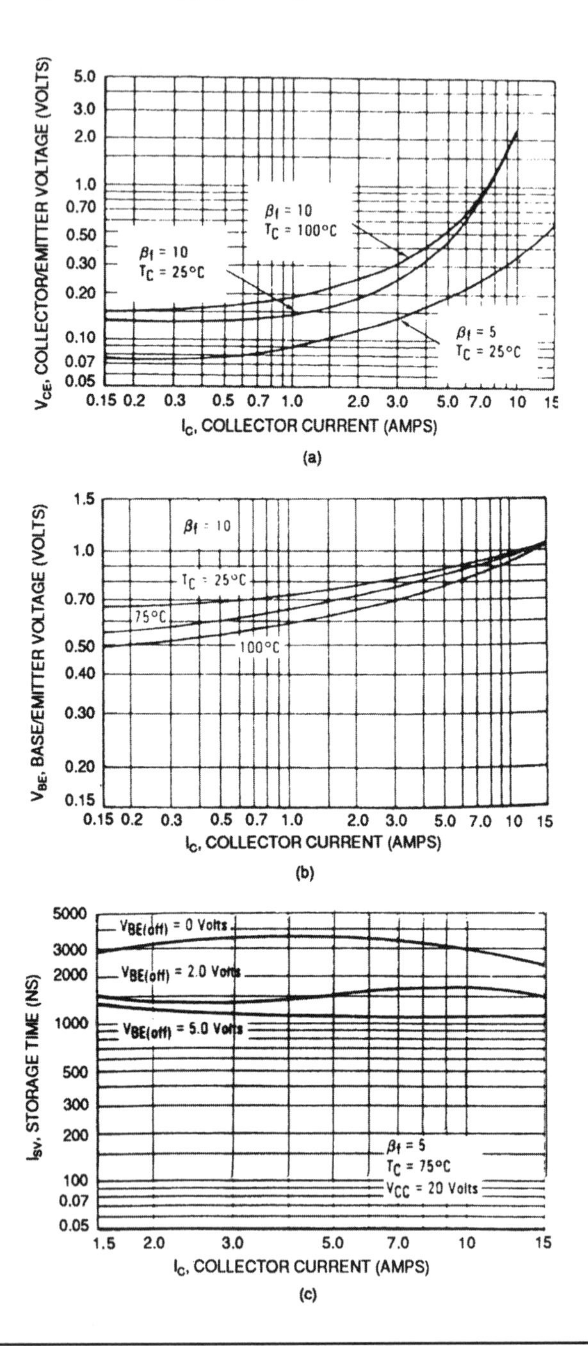

FIGURE 8.5 (*a* and *b*) Junction voltages for 2N6836 transistor at two values of forced beta. (*c*) Storage time for 2N6836 without Baker clamping. (*Courtesy of Motorola Inc.*)

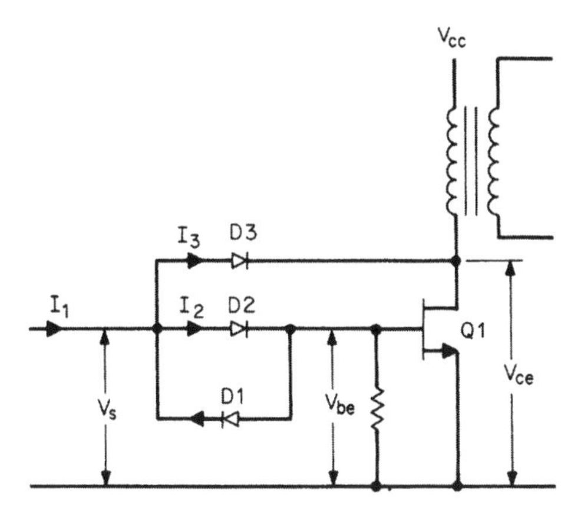

FIGURE 8.6 Baker clamping. The objective of the Baker clamp is to prevent a forward bias on the $Q1$ collector-to-base junction during the "on" time, and hence minimize storage time. If the voltage rise across $D2$ equals the drop across $D3$, V_{ce} is equal to V_{be}. Differences in the $D2$, $D3$ forward drops permit a small forward bias on the collection-to-base junction, but not enough to result in significant storage time. Currents I_2 and I_3 redistribute themselves by negative feedback so that I_2 keeps V_{ce} just low enough to $D3$ forward biased. The rest of $I_1(= I_3)$ flows through $D3$ and the $Q1$ collector to common.

draws no current, and is effectively out of the circuit. All the current I_1 current flows through $D2$ into the base, yielding very short collector current rise time.

However, when the collector voltage has fallen low enough to forward-bias $D3$, the current I_1 is redistributed. Only a fraction of I_1 sufficient to keep $D3$ forward-biased flows through $D2$ into the $Q1$ base. The balance, I_3, flows through $D3$ into the $Q1$ collector, and then through the emitter to common.

The circuit operates with negative feedback. As load current changes, or as transistors with different beta are used, the $Q1$ base demands from I_1 only enough current through $D2$ to keep $D3$ forward-biased. Since the forward drops in $D2$ and $D3$ change by only a few tens of millivolts with large forward current changes, the potential at the $Q1$ collector does not change significantly.

Now consider the $Q1$ junction potentials in a typical forward converter. Diode $D3$ must be a high-voltage, fast-recovery diode, as it is subjected to twice the supply voltage plus a leakage inductance spike. Diode $D2$ is also a fast-recovery type, but is never subjected to a reverse voltage of greater than about 0.8 V (forward voltage of $D1$). Thus, assume that $D3$ is an MUR450 (450 V, 3 A, 75 ns recovery time) and $D2$ is an MUR405 (50 V, 4 A, 35 ns recovery time).

	MUR450 V_f, V		MUR405 V_f, V	
I_f, A	25°C	100°C	25°C	100°C
0.5	0.89	0.75	0.71	0.61
1.0	0.93	0.80	0.74	0.65
2.0	1.01	0.90	0.78	0.70
3.0	1.10	0.95	0.80	0.73

TABLE 8.1 Forward Voltages of Ultra-Fast-Recovery Diodes *D*2, *D*3 in Figure 8.6

Assume for the moment that the *D*2, *D*3 forward voltages are independent of forward current and temperature and are equal to 0.75 V. This approximation is good enough as seen in Table 8.1. The small variations are not sufficient to change the reasoning described below.

In Figure 8.6, when *D*3 conducts, the voltage rise in *D*2 is close to the drop in *D*3, which is assumed to be 0.75 V. For V_{be} of 1.0 V in *Q*1, the *D*2 anode (V_s) is at +1.75, and the *Q*1 collector voltage is also 1.0 V. Thus, there is no forward bias on the *Q*1 base-to-collector junction and negligible storage time. If temperature rises, the forward drop in *D*2 decreases, but so does the drop in *D*3 and the collector-to-base junction still has no forward bias.

Now assume that I_1 is 3.5 A, and that *Q*1 is a maximum beta transistor and requires only 0.5 A of base current, leaving 3.0 A for *D*3. Table 8.1 shows the rise in *D*2 is 0.61 V at 100°C, and the drop in *D*3 is 0.95 V at 3 A, 100°C. This leaves a forward bias of only 0.34 V across the *Q*1 base-to-collector junction—not enough to cause diode-type conduction in it. Storage time is still negligible at that forward bias.

Note that Baker clamping holds the collector-to-emitter potential at about 1 V, as compared to the 0.2 to 0.5 V without the Baker clamp. This increases transistor loss during the "on" time, but the decreased AC loss during the transition from "on" to "off" more than makes up for it (Section 8.2 and Figure 8.3*a*).

Thus the Baker clamp has satisfactorily solved two significant problems. It prevents sufficient forward bias on the base-to-collector junction to cause appreciable storage time. It also permits the circuit to work equally well with large changes in load current and over a large production spread in transistor beta, because of the redistribution of currents between *D*2 and *D*3 as base current demands change.

However, it is still desired to provide –1 to –5 V reverse bias current to *Q*1 to speed up turn "off" time. This is prevented by the blocking action of *D*2, but by adding the "reach-around" diode *D*1 across *D*2, it becomes possible. It now only remains to find a low-parts-count scheme to provide the large forward current I_{b1} for turn "on," an equally large reverse current I_{b2}, and a reverse voltage bias at the base for turn "off." This is easily achieved with the transformer-coupled scheme of the following section.

8.3.2 Transformer Coupling into a Baker Clamp

8.3.2.1 Transformer Supply Voltage, Turns Ratio Selection, and Primary and Secondary Current Limiting

The Baker clamp circuit of Figure 8.7 provides all the required drive characteristics. It provides high forward and reverse base drive for $Q2$ with relatively low primary current drawn from the housekeeping supply V_h. It also provides the $Q2$ reverse base voltage which permits it to realize its V_{cev} rating. It works as follows.

First, the $T1$ turns ratio N_p/N_s is chosen as large as conveniently possible so as to provide the desired secondary current with a

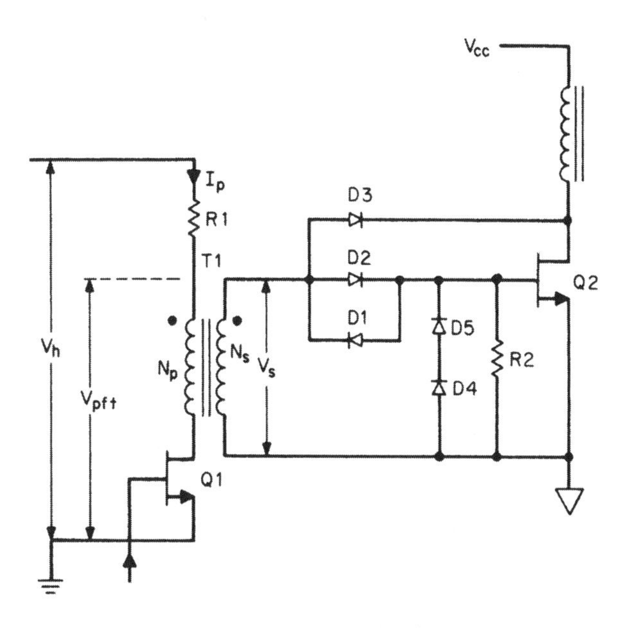

FIGURE 8.7 A transformer-driven Baker clamp. Transformer $T1$ provides a large current gain from primary to secondary so that secondary currents of 2 to 3 A can be obtained from primary currents of about 600 mA, permitting use of small, inexpensive transistors for $Q1$. By storing magnetizing current in N_p, large reverse currents are available in N_s at the instant of $Q2$ turn "off." Resistor $R1$ serves as a primary current limiter.

reasonably low primary current. Since the primary current will be taken from the housekeeping supply V_h, which also feeds the PWM chip, V_h should be kept low to keep dissipation low.

Choosing a high N_p/N_s ratio to get a large current gain in $T1$ may also force too high a value of V_h. A reasonable choice for V_h is the 15 to 18 V often needed for a PWM chip. This largely fixes N_p/N_s. It will soon be seen that the $T1$ primary voltage V_p should be considerably less than V_h because of $R1$, which plays a significant part in the circuit.

The $T1$ secondary forward voltage V_s is $V_{\text{be}(Q2)}$ plus the $D2$ forward drop V_{D2}. In other words, $V_s = V_{be(Q2)} + V_{D2} = 1.0 + 0.75 = 1.75$ V. The voltage at the top of the $T1$ primary is

$$\begin{aligned} V_{\text{pt}} &= \frac{N_p}{N_s} V_s + V_{\text{ce}(Q1)} \\ &= \frac{N_p}{N_s} 1.75 + V_{\text{ce}(Q1)} \qquad \text{V} \\ &\approx \frac{N_p}{N_s} 1.75 + 1.0 V \end{aligned} \tag{8.1}$$

Constant V_h can be provided by an inexpensive linear regulator fed from the secondary of a small 60-Hz transformer returned to output common as in Figure 6.19.

However, V_{pt} should be kept considerably lower than V_h so as to provide a relatively constant voltage across $R1$ (V_{R1}) when V_s varies as a result of temperature-induced variations in V_{be} and V_{D2}. The reason for keeping V_{R1} constant is that $R1$ serves to limit primary current

$$\begin{aligned} I_{p(Q1)} &= \frac{V_h - V_{\text{pt}}}{R_1} \\ &= \frac{V_h - (N_p/N_s)V_s - 1.0}{R_1} \end{aligned} \tag{8.2}$$

By limiting $T1$ primary current, $R1$ also limits secondary current and current into $D2$. Although negative feedback through the Baker clamp diode $D3$ allots current to the $Q2$ base only sufficient to supply the maximum collector current and to keep $D3$ in conduction, excess $T1$ secondary current is simply wasted by being diverted via $D3$ into the $Q2$ collector.

By choosing V_h large compared to $(N_p/N_s)V_s$, $R1$ approximates a constant-current source, relatively independent of temperature variations in V_s.

For an initial guess, assume $N_p/N_s = 5$. For a nominal $V_s (= V_{\text{sn}})$ of 1.75 V, the nominal I_p is

$$I_{\text{pn}} = \frac{V_h - 5V_{\text{sn}} - 1.0}{R_1}$$

and the change in $I_{pn}(= dI_{pn})$ is $dI_{pn} = 5\,dV_s/R_1$, where dV_s is the anticipated change in V_s due to temperature change. The fractional change in I_{pn} is

$$\begin{aligned}\frac{dI_{pn}}{I_{pn}} &= \frac{5dV_s}{V_h - 5 \times 1.75 - 1.0} \\ &= \frac{5dV_s}{V_h - 9.75}\end{aligned} \tag{8.3}$$

Then, for an anticipated temperature variation dV_s of 0.1 V, and a permissible fractional change of 0.1 in I_{pn}, from Eq. 8.3 we obtain

$$0.1 = \frac{5 \times 0.1}{V_h - 9.75}$$

or

$$\begin{aligned}V_h &\cong 14.75\ V \\ &= 15.0\ V\end{aligned}$$

Thus, if it were desired to limit the $Q1$ primary current to I_{pn}, $R1$ would be chosen as

$$R_1 = \frac{15 - (5 \times 1.75) - 1.0}{I_{pn}} \tag{8.4a}$$

$$R_1 = \frac{5.25}{I_{pn}} \tag{8.4b}$$

8.3.2.2 Power Transistor Reverse Base Current Derived from Flyback Action in Drive Transformer

A large reverse current to the base of power transistor $Q2$ (Figure 8.7) can be obtained by choosing a low magnetizing inductance in the $T1$ primary. During the $Q1$ "on" time, $T1$ primary current is limited by $R1$. Part of that current is multiplied by the N_p/N_s turns ratio and delivered to the secondary to turn the $Q2$ base "on."

But part of that current flows to the primary magnetizing inductance L_m and does not contribute to the secondary current. It simply ramps up linearly at a rate $dI_m/dt = (V_{pt} - 1)/L_m$. At the end of the "on" time t_{on} it has reached a peak $I_{pm} = (V_{pt} - 1)t_{on}/L_m$ and is sustained by energy stored in the magnetizing inductance.

Now when $Q1$ is turned "off," the magnetizing current I_{pm} multiplied by the turns ratio N_p/N_s is delivered by flyback action as a negative-going pulse to the secondary (note the $T1$ primary and secondary dots). At the secondary, it pulls reverse current from the base of $Q2$ through reach-around diode $D1$.

After the base current charge has been fully swept out, the base impedance is very high. As there is usually significant energy left in $T1$, the $Q2$ base voltage can be pulled sufficiently far negative to damage or destroy the transistor. This is prevented by the two series diodes $D4$, $D5$, which clamp the $Q2$ base to a negative bias of about 1.6 V—far enough negative to permit $Q2$ to sustain its V_{cev} rating.

8.3.2.3 Drive Transformer Primary Current Limiting to Achieve Equal Forward and Reverse Base Currents in Power Transistor at End of the "On" Time

Significant $T1$ current waveshapes are shown in Figure 8.8. Current through $R1$ is shown in Figure 8.8. For $V_h = 15$ V and $N_p/N_s = 5$, the nominal peak current is given by Eq. 8.4*b* as $I_{pn} = 5.25/R1$.

If the $Q2$ base reverse current at the instant of turn "off" is to be equal to its forward base current just before that, the $T1$ primary

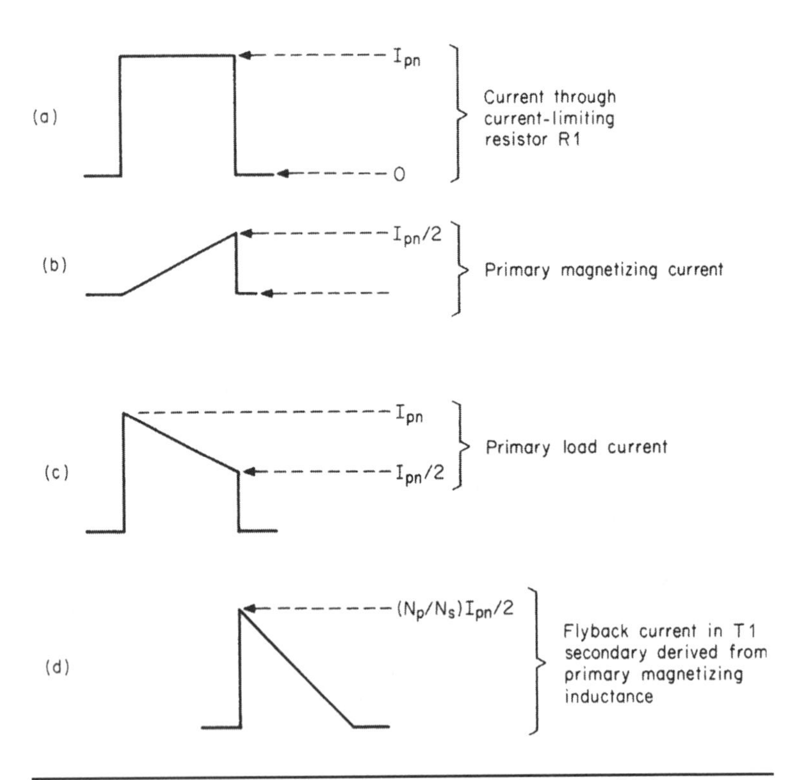

FIGURE 8.8 By choosing primary magnetizing inductance of $T1$ appropriately, the reverse base current to $Q2$ is made equal to its forward current by flyback action in $T1$.

magnetizing current at the end of "on" time should be permitted to ramp up linearly to $I_{pn}/2$ by proper choice of its magnetizing inductance.

The $T1$ magnetizing current is shown in Figure 8.8*b*; $T1$ primary load current is the difference between Figure 8.8*a* and 8.8*b*, and is as shown in Figure 8.8*c*. The $T1$ secondary current during $Q2$ "on" time is the $T1$ primary current of Figure 8.8*c* multiplied by N_p/N_s. Its amplitude at the end of the "on" time, $(N_p/N_s)I_{pn}/2$, is chosen to keep the minimum-beta $Q2$ saturated at the maximum collector current. Its amplitude at the start of the "on" time is twice that at its end, so turn "on" time is very fast.

At the end of the $Q1$ "on" time, $T1$ magnetizing current in L_m is $I_{pn}/2$. At the instant of $Q1$ turn "off," this reflects into the $T1$ secondary as $(N_p/N_s)I_{pn}/2$ and provides $Q2$ reverse base current (Figure 8.8*d*) equal to its current just prior to turn "off."

8.3.2.4 Design Example—Transformer-Driven Baker Clamp

Assume a 500-W forward converter operating from the rectified 115-V AC line that produces nominal rectified DC input of 160 V at minimum load, and minimum rectified DC input at maximum load of $0.85 \times 160 = 136$ V. From Eq. 2.28, peak primary current is

$$I_{pft} = \frac{3.13P_o}{V_{dc(min)}} = \frac{3.13 \times 500}{136} = 11.5\text{ A}$$

Let $Q2$ be a 2N6836—a 15-A, 450-V (V_{ceo}) device with a V_{cev} rating of 850 V. Its minimum beta at 10 A is 8—assume 7 at 11.5 A. Worst case minimum base current is $11.5/7 = 1.64$ A. For an N_p/N_s ratio of 5 in $T1$, primary load current is $1.64/5 = 0.328$ A. This is the base current that must flow at the end of the "on" time.

Also from Figure 8.8*a*, current in $R1$ (Figure 8.7) must be twice that so as to store 0.328 A of $T1$ magnetizing current for the $Q2$ reverse base drive. From Eq. 8.4b, for $V_h = 15$ V, $R1 = 5.25/2 \times 0.328 = 8.0\ \Omega$.

The $T1$ magnetizing inductance must permit a peak magnetizing current of 0.328 A at the end of the "on" time. Calculate the magnetizing inductance from the minimum "on" time of $Q1$.

Assume a switching frequency of 50 kHz, and a maximum "on" time of $0.8T/2$ or 8 μs at minimum AC input. For ± 10% variation in AC input, the minimum "on" time of $Q1$ will be $0.8 \times 8 = 6.4$ μs. Then

$$L_{m(T1)} = \frac{(N_p/N_s)V_s t_{on(min)}}{0.328} = \frac{5 \times 1.75 \times 6.4 \times 10^{-6}}{0.328} = 171\ \mu\text{H}$$

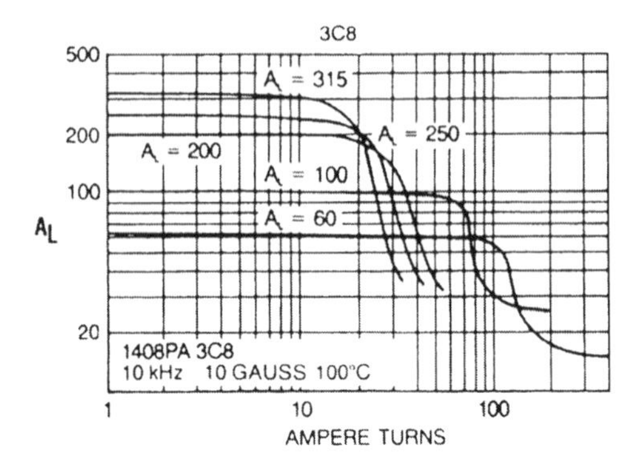

FIGURE 8.9 Inductance per 1000 turns A_l for Ferroxcube 1408PA3C8 pot core. This is a small core suitable for the transformer $T1$ of Figure 8.7.

Use a Ferroxcube 1408PA3153C8 core. It is small, pill-box in shape, with diameter 0.551 inch and height 0.328 inch. It has an A_l (inductance in milliHenries per 1000 turns) of 315 (Figure 8.9). For 0.171 mH, the required number of turns is $N_p = 1000\sqrt{0.171/315} = 23$ turns—say 25, and for a turns ratio of 5, $N_s = 5$ turns. The magnetizing ampere-turns product is $0.328 \times 25 = 8.2$ ampere turns. Figure 8.9 shows that the knee at which this specific core ($A_l = 315$) starts saturating is at about 12 ampere turns, which is safe.

Transistor $Q1$ conducts a peak current of only $2 \times 328 = 656$ mA. It can be a 2N2222A—an 800-mA, 40-V device whose rise and fall times are under 60 μs. It comes in an inexpensive TO-18 package.

8.3.3 Baker Clamp with Integral Transformer

By changing the circuit of Figure 8.7 to the simpler one of Figure 8.10, greatly improved performance with all the advantages of Baker clamping is achieved. Current gain in $T1$ can be doubled without increasing V_h, and better performance over a wider temperature range results. Also, the problem inherent in Figure 8.7—that the forward rise in $D2$ does not track the drop in $D3$ over large current changes—no longer exists. The circuit of Figure 8.10 works as follows. The secondary of $T1$ in Figure 8.10 is center-tapped ($N_{s1} = N_{s2}$). Thus $V_{Ns1} = V_{Ns2} = V_{\text{be}(Q2)}$, and when $D3$ is conducting, we obtain

$$\begin{aligned} V_{\text{ce}(Q)} &= V_{Ns1} + V_{Ns2} - V_{D3} \\ &= 2V_{\text{be}(Q2)} - V_{D3} \\ &= 2 \times 1.0 - 1.0 = 1.0 \qquad \text{(Table 8.1, Figure 8.5}b\text{)} \end{aligned}$$

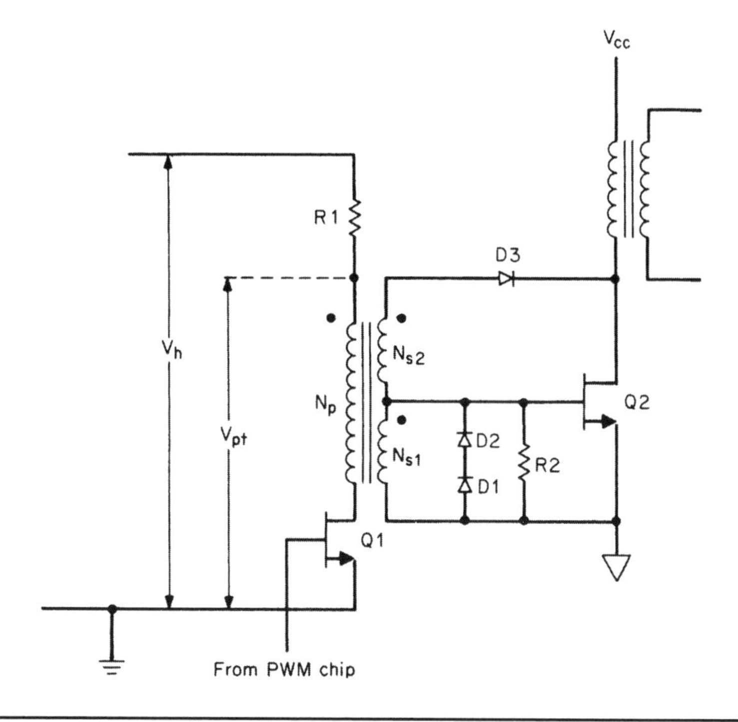

FIGURE 8.10 Transformer Baker clamp.

Since $V_{be} = 1.0$ also, there is no forward bias across the base-to-collector junction and storage time is minimized.

There are changes in V_{be} and V_{D3} with current and temperature, but since there is only one diode involved as compared to two for Figure 8.7, the maximum forward bias on the *Q*2 collector-to-base junction is considerably lower than that for Figure 8.7.

*D*2 was needed in Figure 8.7 only, to provide a forward-voltage rise equal to the *D*3 forward drop, so the reach-around diode *D*1 is not needed either.

The greatest advantage of Figure 8.10 is that since V_s is approximately 1.0 rather than the 1.75 V of Figure 8.7, the turns ratio N_p/N_{s1} can be roughly doubled without increasing V_h. Thus the same *Q*2 base current can be delivered at about half the current in *Q*1 and *T*1 primary. The advantage of Figure 8.10 can be appreciated by repeating the design example of Section 8.3.2.4 with the Figure 8.10 circuit as below.

8.3.3.1 Design Example—Transformer Baker Clamp

In Figure 8.10, $V_{pt} = (N_p/N_{s1})V_{Ns1} + V_{ce(Q1)}$. Now choose $N_p/N_{s1} = 10$, rather than 5 as in the Figure 8.7 circuit. With a larger $T1$ turns ratio, I_{Q1} will be smaller. Hence assume that $V_{ce(Q1)}$ will be 0.5 V, rather than the 1.0 V assumed for the Figure 8.7 circuit. Then $V_{pt} = 10 \times 1.0 + 0.5 = 10.5$ V, and keeping the same nominal 5.25 V across $R1$ (Eq. 8.4b), $V_h = 15.75$ V. For the same 1.64 A to the $Q2$ base as in the previous design example, with $N_p/N_{s1} = 10$, $Q1$ primary load current is now only 164 mA.

As in Section 8.3.2.4, choose the $T1$ magnetizing current at the end of the "on" time to equal the primary load current (Figure 8.8). Thus $R1$ is chosen to limit current to 328 mA and $R_1 = 5.25/0.328 = 16\Omega$. With 10 V across the $T1$ primary for the same 6.4 μs, the magnetizing inductance is $L_m = (10 \times 6.4 \times 10^{-6})/0.164 = 390$ μH.

For the same Ferroxcube 1408PA3153C8 core with an A_l value of 315 mH per 1000 turns, the required number of primary turns is $1000\sqrt{0.390/315} = 35$. For a turns ratio of 10 N_{s1}, N_{s2} is 3.5 turns.

Half turns are possible with pot cores, but they introduce other odd, undesirable effects. Hence choose $N_{s1} = N_{s2} = 4$ turns and $N_p = 40$ turns. This makes the magnetizing inductance $(40/35)^2 \times 390$ or 509 μH and the peak magnetizing current equal to $(390/509)164 = 126$ mA. For a 10/1 turns ratio, the reverse base current to $Q2$ is now 1.26 instead of 1.64 A. This still yields a sufficiently low storage time.

The number of magnetizing ampere turns in $T1$ is now $0.126 \times 40 = 5.04$ ampere turns, which is still safely below the saturation knee for the $A_l = 315$ core of Figure 8.9.

8.3.4 Inherent Baker Clamping with a Darlington Transistor

Using a Darlington-connected transistor pair, the output transistor $Q2$ is automatically Baker clamped by the base-emitter diode of the drive transistor $Q1$ acting as $D2$ of Figure 8.7, and the base-collector diode of $Q1$ acting as $D3$ in Figure 8.7. This can be seen in Figure 8.11*a* to 8.11*c*.

The output transistor in a Darlington has negligible forward bias on its base-collector junction and has a low storage time, but data sheets on integrated-circuit Darlingtons show storage times up to 3 or 4 μs. This is due mainly to the storage time in the Darlington drive transistor, which does saturate and has a forward-biased base-collector junction.

If a Darlington configuration with lesser storage time is needed, the 3- to 4-μs storage time can be lowered by using discrete transistors for the drive and output transistors.

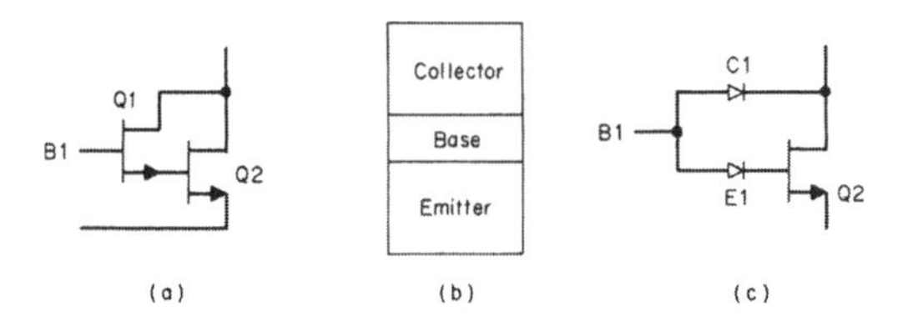

FIGURE 8.11 In a Darlington configuration, output transistor $Q2$ is inherently Baker clamped by the base-collector diode of $Q1$ acting as $D3$, and the base-emitter diode of $Q1$ acting as $D2$ in Figure 8.7. Storage time in a Darlington configuration is due to saturation of the Darlington driver which is not Baker clamped. Figure 8.11*b* shows the junctions in a junction transistor such as $Q1$.

By implementing a Darlington with discrete drive and output transistors, the drive transistor can be chosen as an ultra-high-frequency device which can have minimized storage time despite its forward-biased base-collector junction.

Most integrated-circuit Darlingtons have a built-in reach-around diode like $D1$ in Figure 8.7, which permits pulling reverse current from the output transistor base to improve its switching time.

8.3.5 Proportional Base Drive[2-4]

The base drive scheme shown in Figure 8.12 is widely used for high output power, or power transistor currents over 5 to 8 A.

The circuit does not attempt to keep the power transistor from saturating by use of Baker clamping. Rather, it ensures a large reverse base current to minimize storage time; and it always generates a base drive current proportional to its output current.

Thus even with a high base current required for high output current, when output current is reduced from maximum to minimum, so is the base current. Consequently, the base is never overdriven at low output currents and storage time is kept reasonably low throughout the load current range.

A particularly valuable feature is that the base current is obtained by positive feedback from the collector. For large output current which requires large base current, this results in far less base source dissipation than if the base current was derived from a housekeeping supply.

Circuit details are described in the following paragraphs.

8.3.5.1 Detailed Circuit Operation—Proportional Base Drive

In Figure 8.12, there is positive feedback between windings N_c and N_b (note the dots) of drive transformer $T1$. These windings act as a

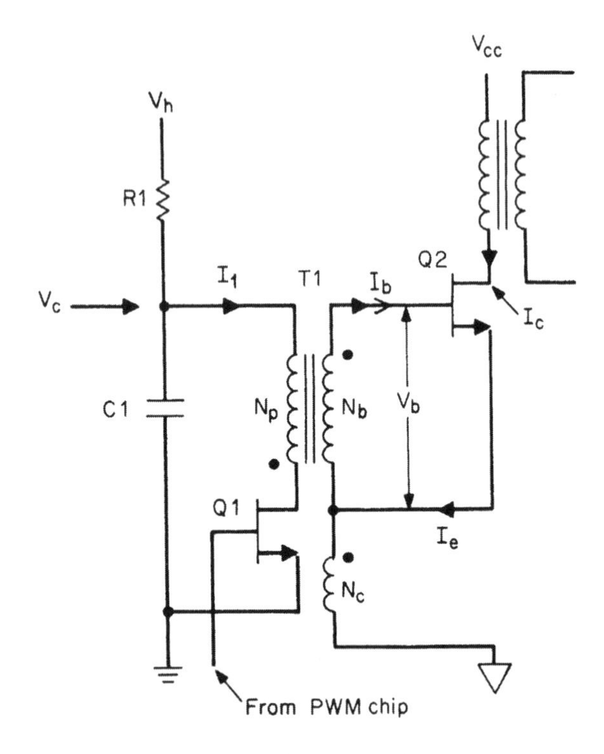

FIGURE 8.12 Proportional base drive. When $Q1$ turns "off," the magnetizing current stored in N_p provides a short impulse to turn $Q2$ "on" by flyback action in $T1$. Thereafter, positive feedback from N_c holds $Q2$ "on." The ratio N_b/N_c is chosen equal to $Q2$ minimum beta. When $Q1$ turns "on," it couples a negative impulse to N_b which starts a regenerative turn "off" sequence between N_c and N_b.

current transformer with a turns ratio N_c/N_b.If $Q2$ is turned "on," and a current $I_c(\approx I_e)$ flows in N_c, a base current $(N_c/N_b)I_c$ flows in N_b.

For $N_c/N_b = 0.1$ and $I_c = 10$ A, $Q2$ base current is 1 A. If I_c is reduced to 1 A, $Q2$ base current is only 0.1 A. When $Q2$ is to be turned "off," the stored base charge which must be removed corresponds to only 0.1 instead of 1.0 A, and turn "off" is still rapid.

The problem is how to initiate $Q2$ turn "on," and how to break the tightly coupled positive-feedback loop between N_c and N_b and supply a large reverse base current to minimize $Q2$ storage time at turn "off."

Transistor $Q2$ can be turned "on" and "off" in a number of ways. Some designers turn "on" an auxiliary transistor $Q1$ to turn $Q2$ "on," and turn $Q1$ "off" to turn $Q2$ "off." Here, however, $Q2$ is turned "on" by turning $Q1$ "off." Then the stored magnetizing current in the $T1$

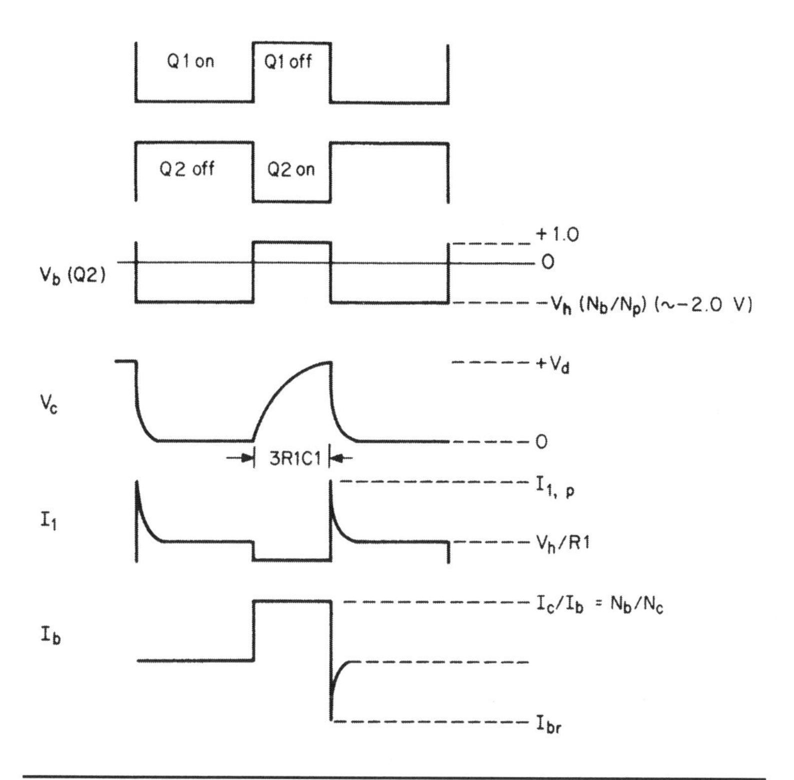

FIGURE 8.12 (*Continued*).

primary is multiplied by the N_p/N_b ratio, and delivered to N_b by flyback action (note the dots on N_p and N_b) to initiate the turn "on" operation.

This impulse need last only a short time—long enough for $Q2$ collector current to rise and establish a solid positive-feedback action between N_c and N_b. Then for the duration of the "on" time, base current is supplied by transformer coupling between N_c and N_b.

To turn $Q2$ "off," $Q1$ is turned "on." It is assumed that V_c, the voltage on capacitor C, is fully charged to the supply voltage V_h. When $Q1$ comes "on," the dot end of N_b goes negative with respect to its no-dot end (see dot orientation) and turns $Q2$ "off." The turns ratio N_p/N_b is chosen to yield a 2-V negative pulse at N_b so as to permit $Q2$ to sustain its V_{cev} rating as it turns "off."

The values of $R1$, $C1$, V_h, N_p magnetizing inductance, and the $T1$ turns ratios are critical to the design. The quantitative design of these components is covered in the following section.

8.3.5.2 Quantitative Design of Proportional Base Drive Scheme

The first decision is to select the turns ratio N_b/N_c (Figure 8.12). This is chosen equal to the minimum $Q2$ beta or

$$\frac{N_b}{N_c} = \beta_{\min(Q2)} \tag{8.5}$$

Proportional base drive is used mostly for $Q2$ collector currents over 5 to 8 A, at which minimum betas range between 5 and 10. Thus N_b/N_c is usually chosen in the range of 5 to 10.

The next decision is the choice of N_p/N_b. Below we will ensure that when $Q1$ is turned "on," the voltage V_c on capacitor $C1$ equals the supply voltage V_h. When $Q1$ is turned "on," a 2-V negative impulse to the $Q2$ base is needed. This permits $Q2$ to tolerate its V_{cev} voltage rating and prevents "second breakdown." Thus

$$\frac{N_p}{N_b} = \frac{V_h}{2} \tag{8.6}$$

With these ratios fixed, the transformer is almost designed. Then N_c is chosen at one turn, which fixes N_b and N_p. A core will be chosen and a gap will be specified as below.

When $Q2$ has been turned "on" and is carrying its maximum current $I_{c(\max)}$, the base current to support that collector current comes from positive feedback between N_c and N_b. To initiate that regenerative process, the current impulse delivered from N_p to N_b by flyback action when $Q1$ turns "off" must be large enough to ensure positive-feedback action. If the impulse from N_p is too small in amplitude or width, $Q2$ may not turn "on" fully, and fall back "off."

To ensure that this does not happen, the short current impulse delivered to N_b from N_p is made equal to that delivered to N_b from N_c after solid positive feedback has been established. Thus I_1 (Figure 8.12) is chosen, by proper selection of $R1$, to yield

$$\begin{aligned} I_1 &= \frac{N_b}{N_p}\frac{I_{c(\max)}}{\beta_{\min}} \\ &= \frac{N_b}{N_p}\left[I_{c(\max)}\left(\frac{N_c}{N_b}\right)\right] \end{aligned} \tag{8.7a}$$

$$I_1 = I_c\frac{N_c}{N_p} \tag{8.7b}$$

This value of I_1 is obtained by choosing the N_p inductance low enough (as will be calculated below) so that at the end of the minimum $Q1$ "on" time, the voltage across N_p collapses to zero, and I_1 is fixed by V_h and R_1 as

$$I_1 = \frac{V_h}{R_1} \tag{8.8a}$$

and from Eq. 8.7*b*

$$R_2 = \frac{V_h}{I_c}\frac{N_p}{N_c} \tag{8.8b}$$

8.3.5.3 Selection of Holdup Capacitor (*C*1, Figure 8.12) to Guarantee Power Transistor Turn "Off"

At the end of $Q1$ "on" time, when $Q1$ has been "on" long enough for the voltage across N_p to collapse to zero, N_p draws a current of only V_h/R_1. The corresponding current impulse delivered to N_b by flyback action is sufficient to safely close the regenerative feedback loop between N_c and N_b.

When $Q1$ has just turned "on," however, it must draw more current than I_1. The initial current in N_p has two tasks. When $Q1$ turns "on" initially, it must first break the positive-feedback loop between N_c and N_b by canceling I_b, the $Q2$ forward base current. Then to turn $Q2$ "off" quickly with minimum storage time, it should deliver an equal reverse current I_b to the $Q2$ base. Thus to deliver $-2I_b$ to N_b requires $2I_b(N_b/N_p)$ in N_p.

Further, in delivering $-2I_b$ to N_b, the voltage at the top of N_p must remain at V_h to deliver a momentary 2-V reverse-bias pulse to the $Q2$ base. Without capacitor $C1$, $R1$ would not be able to remain at V_h and supply $2I_b(N_b/N_p)$ to the primary. Thus, capacitor $C1$ is added at the junction of $R1$ and N_p to hold V_c up long enough to turn $Q2$ "off."

This requirement fixes the value of $C1$, which must supply the current $2I_b(N_b/N_p)$ at voltage V_h during $Q2$ turn "off" time t_{off}. This requires a stored energy in $C1$ of

$$\begin{aligned} {}^1\!/\!_2 C_1(V_h)^2 &= 2I_b\frac{N_b}{N_p}V_h t_{\text{off}} \\ &= 2I_c\frac{N_c}{N_b}\frac{N_b}{N_p}V_h t_{\text{off}} \end{aligned}$$

or

$${}^1\!/\!_2 C_1(V_h)^2 = 2I_c\frac{N_c}{N_p}V_h t_{\text{off}} \tag{8.9a}$$

or

$$C_1 = 4\left(\frac{I_c}{V_h}\right)\frac{N_c}{N_p}t_{\text{off}} \tag{8.9b}$$

For high-current bipolar transistors, the typical turn "off" time t_{off} is 0.30 μs, which fixes the value of $C1$.

Now that $R1$ and $C1$ have been selected, care must be taken that at the instant $Q1$ is turned "on" to turn $Q2$ "off," the voltage V_c (Figure 8.12) has risen to V_h. During the previous "on" time (Figure 8.12), the inductance of N_p has been chosen so low that V_c has fallen to zero to permit a buildup of I_1 in $R1$. Hence at the start of the $Q1$ "off" time, V_c is zero and has only the minimum "off" time to charge to V_h again.

To recharge $C1$ to within 5% of V_h, $3R_1C_1$ must equal the minimum $Q1$ "off" time.

If the preselected R_1C_1 time constant is too large, $C1$ can be recharged rapidly with an emitter-follower as in the scheme of Figure 8.13 suggested by Dixon.[6]

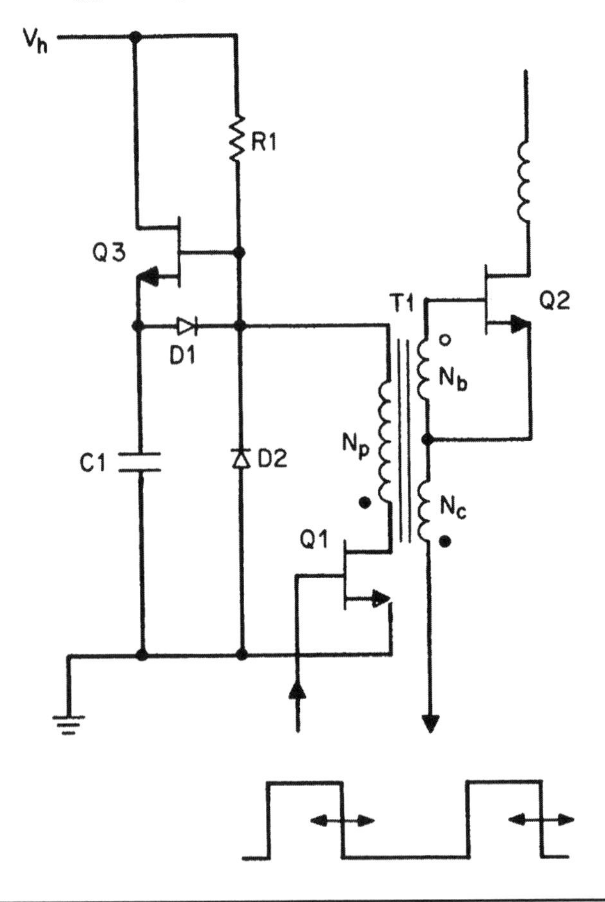

FIGURE 8.13 Fast $C1$ recharge circuit for proportional base drive. In Figure 8.12, if $C1$ cannot be recharged to V_h in the minimum $Q2$ "on" time, emitter-follower $Q3$ is interposed between $R1$ and $C1$.

8.3.5.4 Base Drive Transformer Primary Inductance and Core Selection

At the start of the $Q1$ "on" time V_c is at V_h, and at the end of the "on" time, V_c should collapse to zero to store a current I_1 in N_p. Now assume that V_c falls linearly from V_h to zero in the minimum $Q1$ "on" time $t_{\text{on(min)}}$. Then at the end of the "on" time, N_p must be carrying a current:

$$
\begin{aligned}
I(N_p) &= I_{R1} + I_{C1} \\
&= \frac{V_h}{R1} + \frac{C_1 V_h}{t_{\text{on(min)}}}
\end{aligned}
\tag{8.10}
$$

This voltage waveform is the volt-second equivalent of a voltage step $V_h/2$ applied to the inductance L_p for a time t_{on}. The current rise in the inductance is then $I_1 = V_h t_{\text{on}}/2L_p$. This current rise must equal the current $I(N_p)$ of Eq. 8.10.

$$\frac{V_h t_{\text{on}}}{2L_p} = \frac{V_h}{R_1} + \frac{C_1 V_h}{t_{\text{on(min)}}}$$

or

$$L_p = \frac{t_{\text{on}}}{2(1/R_1 + C_1/t_{\text{on(min)}})}$$

8.3.5.5 Design Example—Proportional Base Drive

Consider a forward converter with the base drive circuit of Figure 8.12. Assume a $Q2$ collector current of 12 A. Assume the circuit is a 115-V AC off-line converter with a minimum rectified DC supply voltage of 145 V. Then from Eq. 2.28, 12 A of collector current corresponds to an output power of $12 \times 145/3.13 = 556$ W.

Assume that $Q2$ is a 15-A device with a minimum beta of 6 at 12 A. Then from Eq. 8.5, $N_b/N_c = 6$ and $N_c = 6$ turns. Now assume a housekeeping supply voltage V_h of 12 V. From Eq. 8.6, $N_p/N_b = V_h/2 = 6$ and $N_p = 6N_b = 36$ turns.

From Eq. 8.7b, $I_1 = I_c(N_c/N_p) = 12/36 = 0.33$ A, and from Eq. 8.8b, $R_1 = (V_h/I_c)(N_p/N_c) = (12/12)(36/1) = 36\Omega$.

From Eq. 8.9b, for a $Q2$ turn "off" time t_{off} of 0.3 μs,

$$
\begin{aligned}
C_1 &= 4\frac{I_c}{V_h}\frac{N_c}{N_p}t_{\text{off}} \\
&= 4\left(\frac{12}{12}\right)\left(\frac{1}{36}\right)(0.3 \times 10^{-6}) \\
&= 0.033\ \mu\text{F}
\end{aligned}
$$

Suppose the switching frequency is 50 kHz. From Figure 8.11, minimum $Q1$ "on" time occurs at maximum $Q2$ "on" time, which will be assumed to be a half period or 10 μs. Then from Eq. 8.11

$$
\begin{aligned}
L_p &= \frac{t_{\text{on(min)}}}{2(1/R_1 + C_1/t_{\text{on}})} \\
&= \frac{10 \times 10^{-6}}{2[1/36 + (0.033 \times 10^{-6}/10 \times 10^{-6})]} \\
&= 162\ \mu H
\end{aligned}
\tag{8.11}
$$

Further, since N_p is calculated as 36 turns, A_l (inductance per 1000 turns) is $(1000/36)^2(0.162) = 125$ mH per 1000 turns.

Finally, the Ferroxcube 1408PA3C8-100 core from Figure 8.9 can be used. Its A_l is 100 mH per 1000 turns, which is close enough.

After Pressman *Proportional drive is very drive efficient but requires transistors with well-defined gain selections. Devices with high gain may be overdriven, increasing the storage time. Combining proportional drive with the antisaturation provided by Baker clamps will provide more flexibility in device selection and drive design. Such systems have the advantage of being dynamic so variations in gain are compensated dynamically, ensuring the drive is always optimum. ~K.B.*

8.3.6 Miscellaneous Base Drive Schemes

A wide variety of specialized bipolar base drive schemes have evolved through the years. They are often used at lower power levels, and by various circuit "tricks," seek to achieve two common goals: (1) a low-parts-count to obtain substantial reverse base voltage, reverse base current, or a base-emitter short circuit at turn "off" and at turn "on" and (2) forward base current adequate to drive lowest beta transistors at maximum current, without long storage times for high beta transistors at lowest current. Some examples are shown below.

P. Wood devised the circuit of Figure 8.14*a* for a 1000-W off-line power supply.[7] Its major features are current gain through transformer $T1$ and a 2-V reverse base bias for power transistor $Q2$ by turning $Q1$ "on" at the instant that $Q2$ is to be turned "off." It can be used for either the lower or upper transistors in a bridge. It has also been widely used at lower power levels. It works as follows. Assume (Figure 8.14*b*) the voltage at the dot end of N_s is positive for the part of a half period that $Q2$ is "on." For the balance of this half period that $Q2$ is to be "off" ($Q2$ dead time), the voltage across N_s is clamped to zero. Then, in the following half period, N_s reverses polarity to reset the core on its hysteresis loop.

Such a waveform can be obtained using the Unitrode UC3525A PWM chip by connecting the $T1$ primary across output pins 11 and 14.

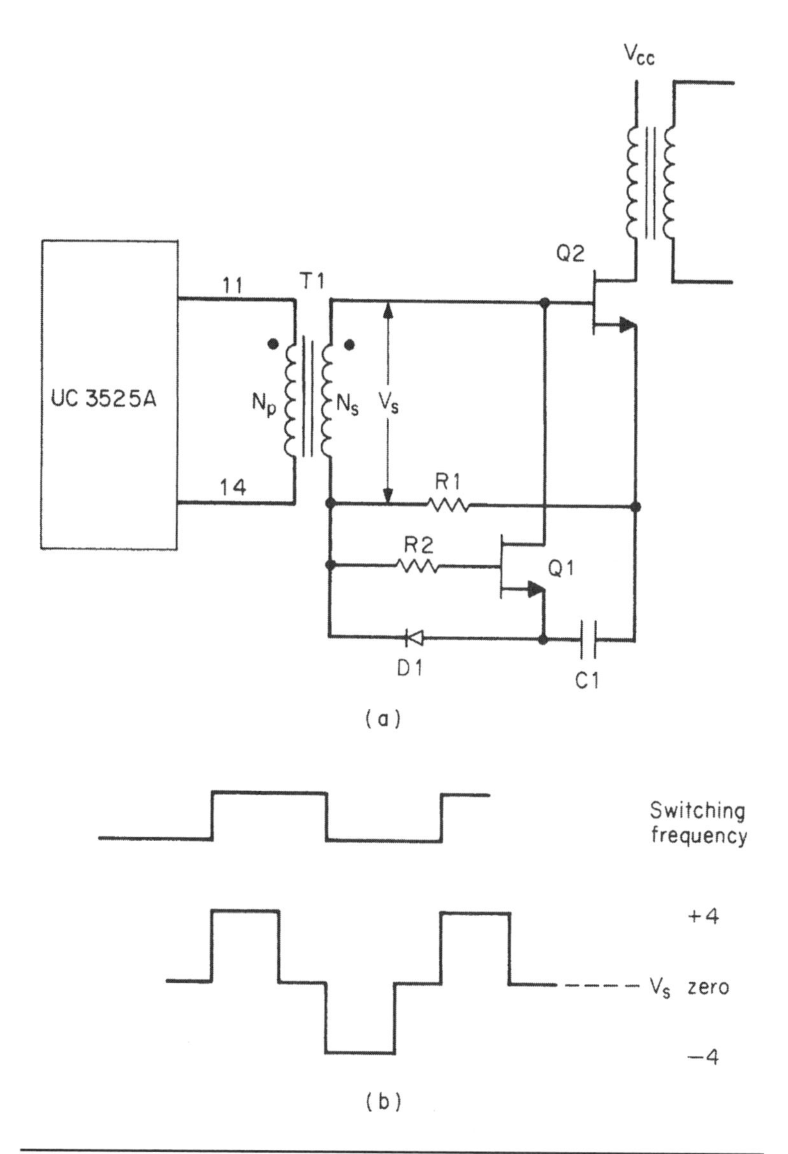

FIGURE 8.14 (*a*) Wood base drive circuit. When dot end of N_s goes positive, *Q*2 turns "on" with its base current limited by *R*1. Voltage V_s is chosen to provide about 4 V across *R*1 for the known base current. With a 4 V across *R*1, *C*1 charges to 3 V through *D*1. While voltage exists across N_s, *Q*1 is reverse-biased and is "off." When voltage across N_p drops to zero, so does voltage across N_s. Now 3-V charge on *C*1 turns "on" *Q*1 via *R*1 and *R*2. This brings *Q*2 base sharply down to a –3 V bias and turns it of rapidly.
(*c*) Adding *D*3, *Z*1, and *D*2 permits driving *T*1 from the "on" collector in the output transistor of a 3524 chip.

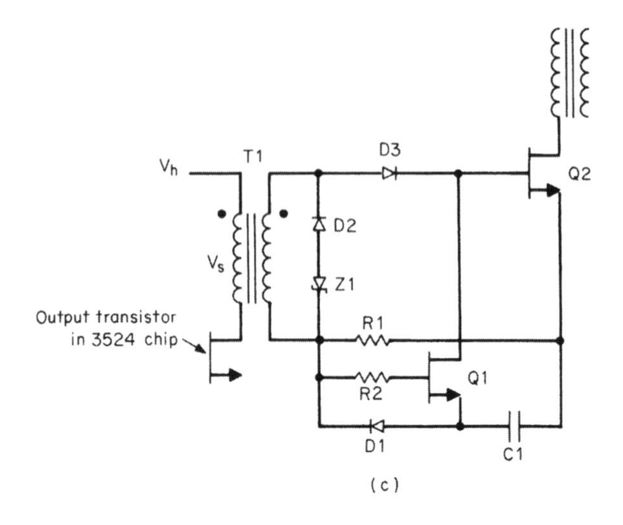

FIGURE 8.14 (*Continued*).

The $T1$ turns ratio is chosen to yield a forward secondary voltage V_s of about 4 V. Resistor $R1$ is chosen to limit $Q2$ base current to a value sufficient to turn "on" and saturate the lowest beta transistor at the maximum I_c with the required speed. Then

$$\begin{aligned} R_1 &= \frac{V_s - V_{be}}{I_{b(max)}} \\ &= \frac{(V_s - 1)\beta_{(min)}}{I_{c(max)}} \end{aligned} \tag{8.12}$$

For 4 V across N_s, the voltage across $R1$ is about 3 V and $C1$ charges to approximately –2 V with respect to the $Q2$ emitter. With 3 V across $R1$, $Q1$ has a –1 V bias and is kept "off" as long as current flows through $R1$.

At the start of the dead time when V_s falls to zero, $Q2$ turns "off" rapidly with small storage delay. $C1$ feeds the series combination of $R1$, $R2$, and the base-emitter of $Q1$, which turns "on" and pulls the $Q2$ base down to –2 V, turning it "off." The reverse base current to $Q2$ is not uniquely determined, but $Q1$ can be a small, fast transistor such as a 2N2222A with a minimum gain of 50 and collector current rating of 500 mA. This is sufficient to yield quite small storage and turn "off" time in $Q2$. As soon as the dot end of N_s goes positive again, the $Q1$ base is again reverse-biased and $Q1$ turns "off."

A circuit which permits the use of the less expensive SG3524 instead of the UC3525 chip is shown in Figure 8.14*c*. Here, when the V_s polarity reverses, zener diode *Z*1 (3.3 V) and diode *D*2 clamp its voltage and reset the core in a time approximately equal to its set time. Diode *D*3 blocks the reset voltage from being clamped by *Q*1.

Now the charge on *C*1 is a floating bias voltage which turns *Q*1 "on" via *R*1 and *R*2 in series. As *Q*1 turns "on," it pulls the *Q*2 base down to –2 V and turns it "off." The addition of diode *D*3 requires the *T*1 turns ratio to yield V_s of about 5 V.

If power transistor reverse base bias is not critical, *Q*1, *C*1, and *D*1 of the Wood circuit can be eliminated by the use of a UC3525 chip as shown in Figure 8.15. Although the scheme does not offer a reverse base bias, it does short-circuit the power transistor base at turn "off." This minimizes storage and turn "off" time, but does not provide the reverse base bias to sustain the V_{cev} rating.

The circuit in Figure 8.15 provides base drive to the upper and lower transistors in a half bridge, or alternatively to a pair of push-pull transistors.[8] Resistor *R*1 is used for current limiting, and to be a good constant-current source, the voltage across it should be relatively independent of voltage drops in the internal source and sink drivers at pins 11 and 14.

The source and sink drivers in the UC3525 have about a 2 V drop at 200 mA. For $V_s(= V_{be}) = 1$ V and a current gain of 10, the primary voltage at pins 11 and 14 is 10 V. The primary current flows out of the source driver and returns through the sink driver to common, so for 10 V across the primary, the top of the source driver on pin 13 is at 14 V. For *R*1 to be a fairly good constant-current source, the voltage across it is set at 6 V, making the housekeeping supply $V_h = 20$ V.

For primary current limiting at 200 mA, $R_1 = 6/0.2 = 30\ \Omega$. For a 10/1 turns ratio, this yields 2 A of power transistor base current. At maximum power transistor current, turn "off" and storage delay

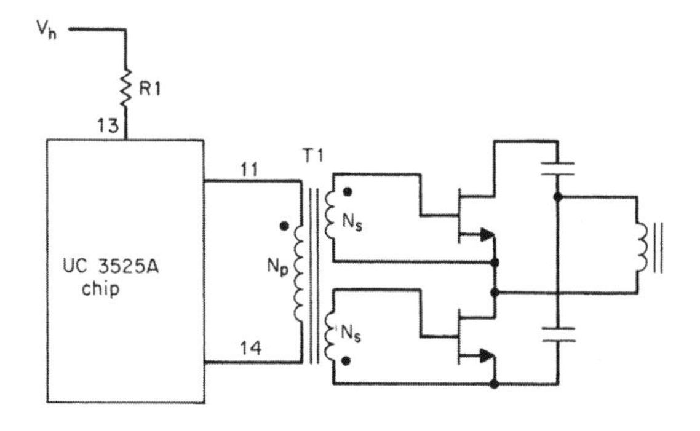

FIGURE 8.15 Base drive to upper and lower transistors in a bridge through a transformer whose primary is fed directly from UC3525A output.

are minimized because with the UC3525, both output terminals are short-circuited together immediately after the end of an "on" time.

Note that unlike Baker clamping or proportional base drive, the same power transistor base current is delivered at maximum and minimum collector currents. Thus, if the base current is adequate at maximum collector current, the base will be heavily overdriven at minimum current, and turn "off" time will be long.

Figure 8.16 shows another alternative with some desirable features. Turn "on" drive comes from the PNP emitter-follower *Q*2 (2N2907A—a small, fast 800-mA device). Resistor *R*3 is chosen to provide the desired maximum *Q*4 base current [$I_{R3} = (V_h - 2)/R3$]. When *Q*1, the PWM output transistor, turns "on," it turns "on" *Q*2, which then turns *Q*4 "on."

Normally *Q*3 is "off" and does not rob any base drive from *Q*4. When it is desired to turn *Q*4 "off," *Q*1 is turned "off," removing base drive from *Q*2. Its collector current ceases and forward drive is removed from *Q*4, which then starts turning "off."

As *Q*1 starts turning "off," its collector voltage rises steeply. A differentiated positive pulse is coupled via *C*1 into the *Q*3 base and turns it "on" momentarily. This short-circuits the *Q*4 base to common, minimizing its storage and current fall time.

As *Q*1 starts turning "off," *Q*2 does not turn completely "off" until the bottom end of *R*2 has risen almost to V_h. To speed up the *Q*4 turn "off" time, in addition to the differentiated positive pulse coupled to

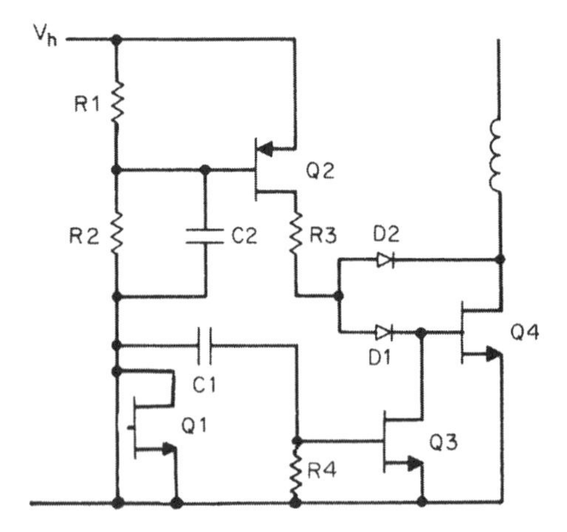

FIGURE 8.16 DC-coupled power transistor base drive. When *Q*1 in the PWM chip turns "on," it turns *Q*2 "on," which turns *Q*4 "on" via *R*3. When *Q*1 turns "off," a positive-going differentiated spike is coupled via *C*1 into *Q*3 base, which pulls *Q*4 base to common to turn it "off" rapidly. *Q*1 turn "off" also couples a positive spike into *Q*2 base to turn it "off" rapidly.

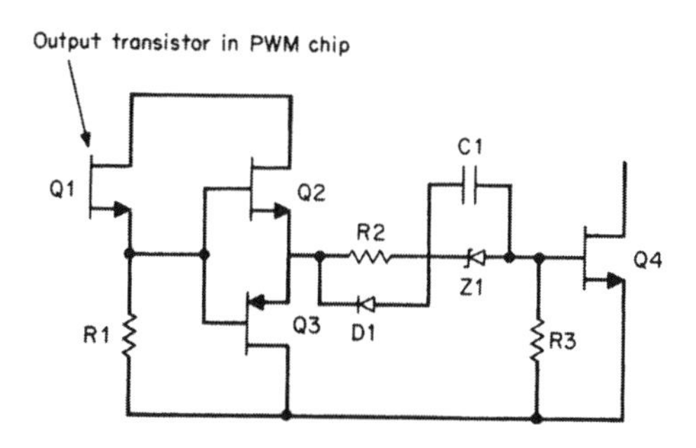

FIGURE 8.17 Direct coupling from emitter of output transistor in PWM chip. When *Q*1 is "on," totem-pole driver *Q*2 turns "on" power transistor *Q*4 with base current limiting determined by *R*2, *Z*1. Capacitor *C*1 takes on a charge equal to the zener voltage (–3.3 V). When *Q*1 turns "off," *Q*3 emitter falls to about +0.6 V and the right-hand side of *C*1 forces *Q*4 base down to about –3 V, turning it "off" rapidly.

the *Q*3 base via *C*1, a differentiated positive pulse is coupled to the *Q*2 base via *C*2 to accelerate its turn "off" time.

A final base drive scheme is shown in Figure 8.17, in which *Q*1 is the output transistor in the PWM chip, *Q*4 is the power transistor, and *Q*2 and *Q*3 constitute an NPN-PNP emitter-follower totem pole. With a 2N2222A for *Q*2, and a 2N2907A for*Q*3, it is capable of sourcing and sinking up to 800 mA. Both *Q*2 and *Q*3 are 300-mHz transistors.

When *Q*2 is turning "on," *Q*3 has a 0.6-V reverse base-emitter bias and is "off." When *Q*3 is turning "on," *Q*2 has a 0.6-V reverse base-emitter bias and is "off."

In Figure 8.17, *Z*1 is a 3.3-V zener diode. When *Q*1 and *Q*2 are "on," *C*1 charges up to 3.3 V and *R*2 limits *Q*4 base current $(= [V_h - V_{ce(Q2)} - V_{be(Q4)} - V_{Z1}]/R_2)$.

When *Q*1 turns "off," *R*1 brings the *Q*2, *Q*3 base quickly to common as the capacitance of the bases of the totem-pole is very low. The *Q*3 emitter comes to +0.6 and *Q*3 is hard "on." Now the 3.3-V negative charge at the right-hand side of *C*1 brings the *Q*3 base down to –3.3 V, and quickly turns *Q*4 "off."

References

1. A. I. Pressman, *Switching and Linear Power Supply, Power Converter Design*, pp. 322–323, Switchtronix Press, Waban, MA, 1977.
2. K. Billings, *Switchmode Power Supply Handbook*, pp. 1.132–1.133, McGraw-Hill, New York, 1989.
3. G. Chryssis, *High Frequency Switching Power Supplies*, 2d ed., pp. 68–71, McGraw-Hill, New York, 1989.

4. P. Wood, *High Efficiency, Cost Effective Off Line Inverters,* TRW Power Semiconductor Application Note 143–1978, Lawndale, CA, 1978.
5. R. Carpenter, "A New Universal Proportional Base Drive Technique for High Voltage Switching Transistors," *Proceedings Powerton 8,* 1981.
6. L. Dixon, "Improved Proportional Base Drive Circuit," Unitrode Power Supply Design Seminar, Unitrode Corp., 1985.
7. P. Wood, "Design of a 5 Volt, 1000 Watt Power Supply," TRW Power Semiconductor Application Note 122, Lawndale, CA, 1975.
8. Unitrode Corp. Application Note 89-1987, Unitrode Corp., Watertown, Mass.

CHAPTER 9

MOSFET and IGBT Power Transistors and Gate Drive Requirements

9.1 MOSFET Introduction

Since the early 1990s, the technology of power MOSFETs (Metal Oxide Silicon Field Effect Transistors) has advanced significantly and greatly changed the electronics industry in general. In particular, it has revolutionized the switching power supply industry. The faster switching speed of FETs has permitted increasing power supply switching frequencies from the typical 50 kHz of the bipolar transistor up to the low MHz range for FETs. It has thus made power supplies much smaller and made possible a host of new products which were previously feasible only for lower powers. The small-size power supplies in personal and laptop computers are prime examples of the advances in technology.

As a result the semiconductor industry changed dramatically. More research funds were spent on FETs, and as their voltage and current ratings were further improved and prices dropped, a large number of new applications became possible—even at lower frequencies.

9.1.1 IGBT Introduction

In the late 1980s, semiconductor designers developed the Insulated Gate Bipolar Transistor (IGBT), by combining a small, easily driven MOSFET and a bipolar-type power transistor. This marriage provided the advantages of both types of transistor in a single package. Although when first introduced, the early devices were not very suitable for switching power supplies due to excessive tail current, by continuous development the performance was slowly improved so that by the mid-1990s IGBTs began to rival both the MOSFET and the bipolar transistor in some applications.

Their improved drive performance and low conduction loss allows IGBTs to displace power bipolar transistors as the device of choice for high current and high voltage applications. The balance in tradeoffs between switching speed, conduction loss, and ruggedness is now being ever finely tuned, so that IGBTs are now encroaching upon the high frequency, high efficiency domain of power MOSFETs. In fact, the industry trend in the 2000s is for IGBTs to replace power MOSFETs except in very low current applications. To help understand the tradeoffs and to help circuit designers with IGBT device selection and application, Section 9.3 provides an overview of IGBT technology.

9.1.2 The Changing Industry

Magnetic materials with lower losses at high frequency and high magnetic flux density were developed. Pulse-width-modulation chips capable of operating at higher frequencies were introduced. Smaller transformers and smaller filter capacitors were developed and there was a greater emphasis on manufacturing processes such as surface-mount techniques.

A new industry and a new field for research were developed, and resonant mode power supplies emerged. Although resonant-mode power supplies using silicon controlled rectifiers at 20 to 30 kHz had been in use for many years, the higher frequencies possible with FETs encouraged the development of many new resonant circuit topologies operating at 0.3 to greater than 5 MHz.

9.1.3 The Impact on New Designs

Parasitic effects that could be ignored at lower frequencies now had to be scrutinized more closely by designers. Skin-effect and especially proximity-effect losses in transformer coils became a larger fraction of total transformer losses at higher frequencies. With faster rise-time current waveforms, $L\ di/dt$ spikes on ground buses and supply rails became more troublesome and more attention had to be paid to wiring layout, low-inductance tracking on common and supply rails, and capacitive decoupling at critical points.

With all the attractive new possibilities, the power supply designer familiar with bipolar design could quickly learn to design with MOSFETs and IGBTs by acquiring a surprisingly small amount of additional information about their characteristics. Details of the internal solid-state physics structure of a FET, which determines its behavior, is not of great importance to the circuit designer and is outside the scope of this book.

However, the functional properties are of great importance to the designer, and various DC volt-ampere characteristics, terminal capacitances, temperature characteristics, and turn "on" and turn "off" speed information required for good circuit design will be considered in some detail for both FETs and IGBTs. The converter topologies used with these devices are simpler than those used with bipolars.

The MOSFET and IGBT input terminal (the gate) has very high resistance, being a fully isolated oxide layer. For small devices, the drive circuitry is far simpler than the complex base drive schemes for bipolar transistors discussed in the previous chapter. For large power devices, however, the effective gate capacitance can be very large, typically several nF for FETs and greater than 1 nF for IGBTs. Hence, the design of drive circuits for large devices can be quite demanding. To their advantage, because FETs have no storage times, the complexities of Baker clamps and proportional base drive circuits are unnecessary, and the problems arising from the large production spread in the gain of bipolars do not exist with FETs.

During the short turn "off" transition in a MOSFET, the overlap of falling current and rising voltage occurs at a lower current. This decreases the V-I overlap area and reduces the AC switching losses (Section 1.3.4). This mitigates the need for aggressive snubbing and simplifies the design of load-line-shaping circuits (load line shaping and snubbers are discussed in Chapter 11).

Some basic MOSFET characteristics and design techniques will be discussed in the following section.

9.2 MOSFET Basics

The MOSFET[1,2] is a three-terminal voltage-controlled device—in contrast to the bipolar transistor, which is a three-terminal current-controlled device. In switching power supply circuits, it is used just as a bipolar transistor— as a switch, either fully "on" with an input drive that minimizes its "on"-voltage drop, or fully "off" at zero current and sustaining the supply voltage or some multiple of it.

The standard symbol for a MOSFET is shown in Figure 9.1*a*; herein, the simplified version shown in Figure 9.1*b* will be used.

The N-channel type (equivalent to a bipolar NPN type) is usually fed from a positive supply voltage. The load impedance is normally connected between the positive supply terminal and the drain. Current, controlled by a positive gate-to-source voltage, flows from the positive rail through the load impedance into the drain, and returns from the MOSFET source to the negative supply terminal (Figure 9.2*a*).

Most power MOSFETs are N-channel types. There are two further distinct types—either *enhancement* or *depletion* types. In the

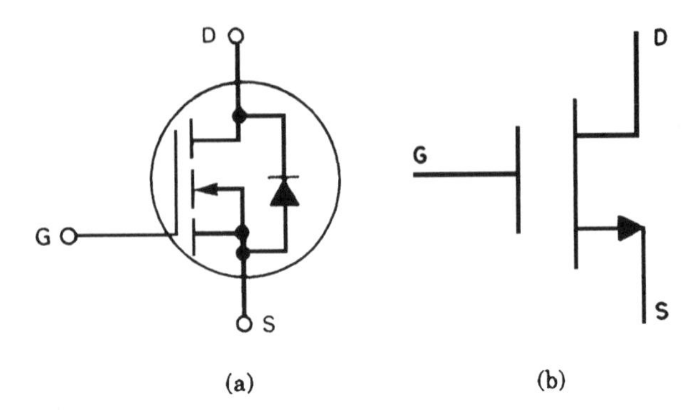

FIGURE 9.1 (*a*) Symbol for an N-type MOSFET. The diode is inherent in its structure and has forward current, and reverse voltage ratings close to those of the MOSFET. (*b*) The simplified N-type MOSFET symbol used herein. The three MOSFET terminals are called the *drain*, *gate*, and *source*, corresponding to the collector, base, and emitter of a bipolar transistor. Just like bipolars, MOSFETs are available for operation from either positive or negative power supply buses.

enhancement N-channel type, drain-to-source current is zero at zero gate-to-source voltage. It requires a positive gate-to-source voltage to turn "on" drain-to-source current.

In the depletion N-channel type, drain-to-source current is non-zero and often maximum at zero gate-to-source voltage. It requires a

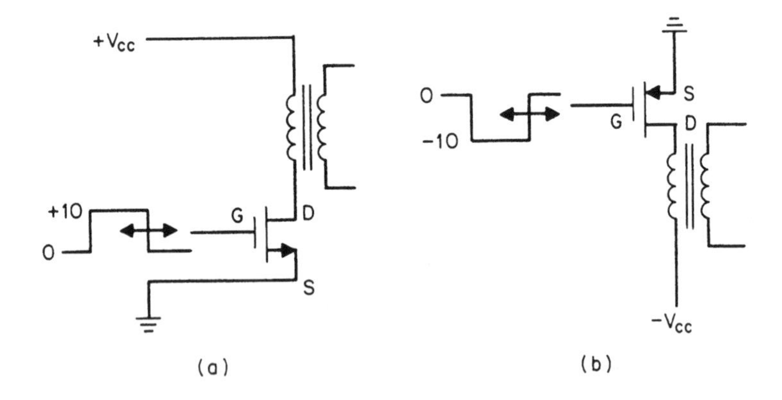

FIGURE 9.2 (*a*) A drain-loaded N-type MOSFET. (*b*) A drain-loaded P-type MOSFET. In a P-channel type (equivalent to a bipolar PNP type), current flow is controlled by a negative gate-to-source voltage and flows into the MOSFET source from the positive rail, out through the drain, and then through the load impedance to the negative rail.

negative gate-to-source voltage to turn "off" drain current. Depletion-type MOSFETs are not used for power transistors, and now are rarely used even in low-current single MOSFET applications. They still have applications for protecting (clamping to ground) the sensitive input ports of critical equipment when the power is "off."

9.2.1 Typical Drain Current vs. Drain-to-Source Voltage Characteristics ($I_d - V_{ds}$) for a FET Device

In Figure 9.3*a*, we see the $I_d - V_{ds}$ characteristics of a typical 7-A, 450-V device (Motorola MTM7N45). They correspond to the $I_c - V_{ce}$ curves of a bipolar transistor (see Figure 8.1). Note that drain current is turned "on" by a positive gate-to-source voltage. This can be seen more clearly in the "transfer" characteristics of Figure 9.3*b*, which shows drain current versus gate-to-source voltage.

The transfer characteristics show some of the advantages of the MOSFET over the bipolar transistor. Note in Figure 9.3*b* that drain current does not begin to turn "on" until the gate-to-source voltage reaches about 2.5 V. As a result, positive noise pickup spikes at the gate terminal cannot falsely turn drain current "on" until the 2.5-V threshold is reached. In bipolar transistors, the diode-like base-to-emitter input voltage characteristic permits false collector current to flow with noise voltage spikes as low as 0.6 V or even lower at temperatures above 25°C.

After Pressman *Although it is true that the noise voltage threshold for FET-like devices is much higher than for bipolar devices, the gate input impedance is much higher, which makes the FET more vulnerable to capacitively injected noise signals. Hence good attention to layout and noise rejection is essential. To reduce noise pickup and parasitic oscillation, FETs will normally have a series resistor in the gate feed near the device terminal. In high voltage, high power FET applications, a negative bias of a few volts is often applied to the gate during the "off" period. This is particularly important in IGBT applications to prevent "latch up." (See Section 9.3.) ~K.B.*

9.2.2 "On" State Resistance $r_{ds\,(on)}$

Note in Figure 9.3*a*, drain current characteristics have a "knee" somewhat like the bipolar transistor knee. Beyond the knee, drain current is constant over a large range of drain-to-source voltage and is determined only by the gate-to-source voltage. Below the knee, the $I_d - V_d$ curves converge asymptotically to a constant slope. The slope of that asymptote (dV/dI) is referred to as r_{ds}.

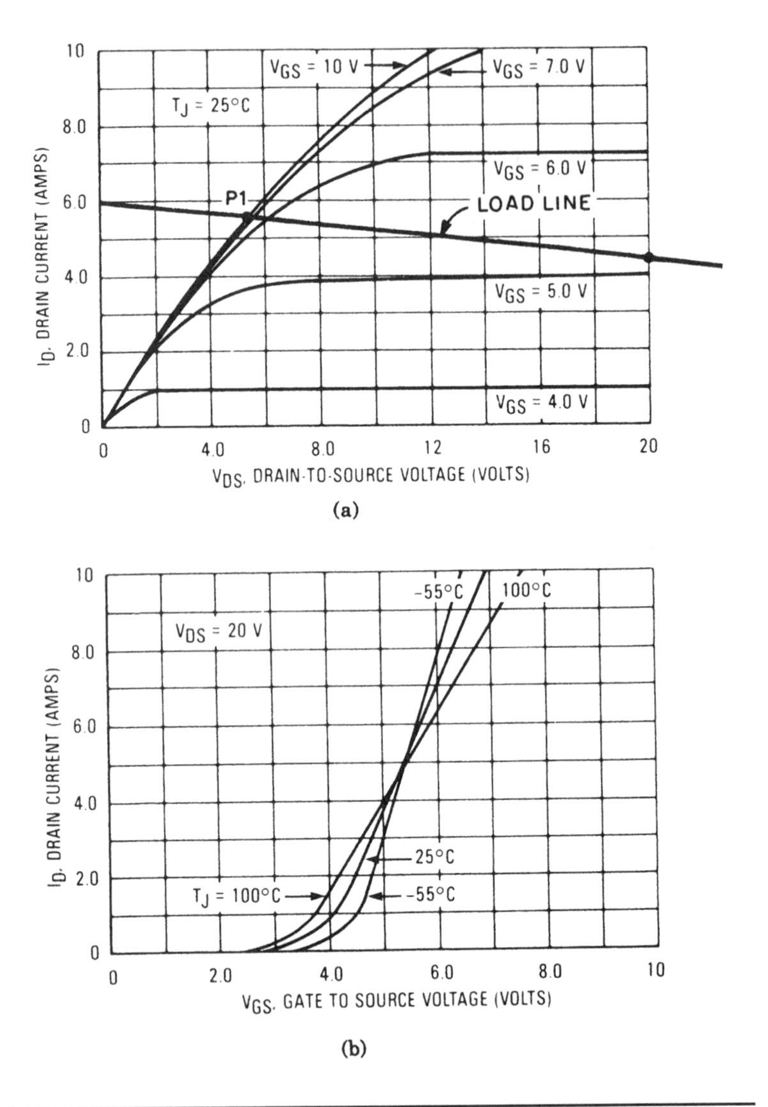

FIGURE 9.3 (*a*) The $I_d - V_{ds}$ characteristics of the MTM7N45—a typical 7-A, 450-V MOSFET. (*b*) The $I_d - V_{gs}$ or "transfer" characteristics of the MTM7N45. (*c*, *d*) $I_d - V_d$ and transfer characteristics of the MTM15N40. (*Courtesy of Motorola Inc.*)

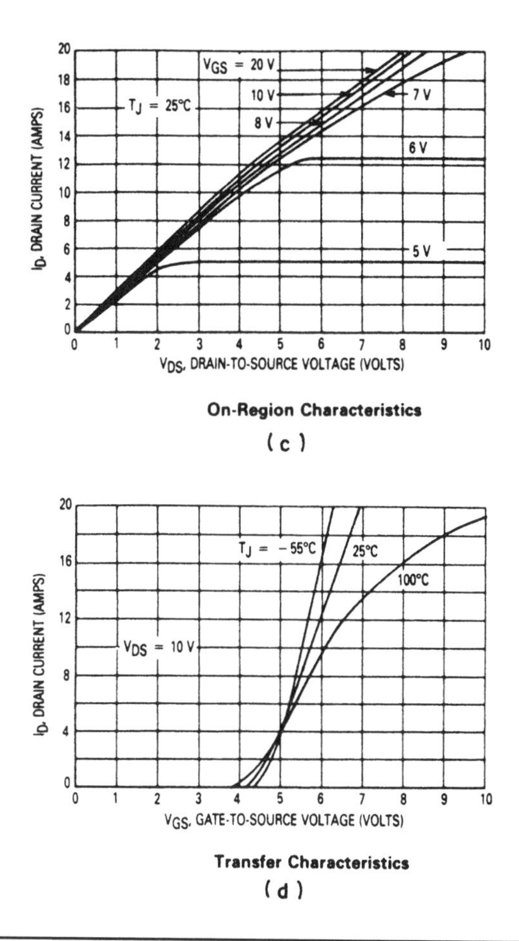

FIGURE 9.3 *Continued.*

It is seen in Figure 9.3*a* that a gate-to-source voltage of 10 V is sufficient to drive the drain-to-source voltage down to the intersection of a load line with the r_{ds} slope (point $P1$). Higher gate voltages will not decrease the "on" V_{ds} significantly unless the operating point is close to the maximum-rated current where the $I_d - V_d$ curve bends away from the r_{ds} slope.

Thus, in contrast to the bipolar transistor, in which the collector-to-emitter "on" voltage is about 0.3 to 0.5 V over a very large range of collector currents, the MOSFET drain "on" potential is equal to $I_{ds}\,R_{ds}$. Generally, to have an "on" V_{ds} voltage of about 1 V at a current I_d, a MOSFET with a maximum continuous current rating of about $3I_d$ to $5I_d$ should be selected because r_{ds} is inversely proportional to maximum current rating.

This can be seen in Figure 9.3*a* for the 7-A MTM7N45 ($r_{ds} = 0.8\ \Omega$). If it were to be used at 7 A, with a gate-to-source voltage V_{gs} of 10 V, V_{ds} is 7 V. This would yield an unacceptable 49 W of dissipation during the "on" time. The MTM15N40—a 15-A, 400-V Motorola device ($r_{ds} = 0.4\ \Omega$)—is shown in Figure 9.3*c* and 9.3*d*. There it is seen that at V_{gs} of 10 V and 7 A, its V_{ds} is still about 2.5 V.

It is customary to keep bipolar V_{ce} "on" voltages at 1 V or less to keep total dissipation low. With a bipolar, the "on" dissipation ($I_c V_{ce}$) may be only half to a third of the total—the balance being the overlap or AC switching dissipation. But MOSFETs generally can be operated with an "on" V_{ds} of up to 2 or 3 V. Drain current turn "off" time is so fast than the dissipation due to the overlap of falling current and rising voltage is generally negligible. Total dissipation is thus $I_{ds} V_{ds}(t_{on}/T)$ dissipation alone.

9.2.3 MOSFET Input Impedance Miller Effect and Required Gate Currents

The MOSFET DC input impedance is extremely high. At V_{gs} of 10 V, the gate draws only nanoamperes of current. Thus once the gate has been driven up to, say the 10 V "on" level, the current it draws is negligible.

However, there is considerable capacitance between the gate and source terminals. This requires relatively large short-lived transient currents to drive the gate voltage up and down by the 10 V required to switch drain current "on" and "off" with the required speed. The required gate drive currents are calculated as follows (Figure 9.4).

After Pressman *The Miller effect has a major influence on the performance of high voltage, high power FETs and IGBTs. The internal inter-electrode capacitances of the gate-to-source and gate-to-drain parasitic elements are similar. However, the gate-to-drain capacitance (shown as C2 in Figure 9.4) has a much more significant effect as follows: As the device turns "on," the drain voltage decreases in response to current flowing into the gate capacitance (C1). The reduction in voltage on the drain pulls current through C2 and robs gate drive current from the drive circuit that was intended to charge up C1. The faster the drain voltage falls, the more aggressive is the current robbing effect from C2. In effect, the input impedance of the gate becomes very low at the turn "on" threshold voltage (typically about 5 V). This is seen as a marked plateau on the gate drive voltage waveform at this transition voltage.*

The internal structure of the gate limits the maximum gate current, so the Miller effect is the major cause of turn "on" delay in power FETs, particularly in high voltage applications. High voltage IGBTs have much lower parasitic gate capacitances so the Miller problem is much less severe. ~K.B.

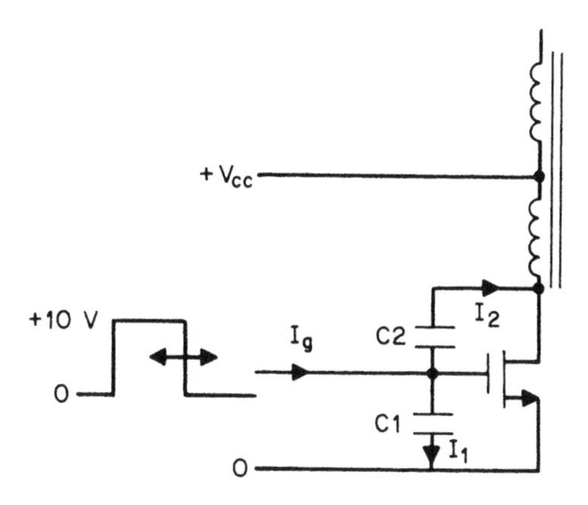

FIGURE 9.4 C1 and C2 simulate the internal parasitic capacitances of the power MOSFET. The gate driver must supply I_1 to C_1 and I_2 to C_2. Because of the Miller effect, I_2 is typically much greater than I_1 (It may be as much as 10 times larger than I_1).

In Figure 9.4, the current I_g required to drive the gate to 10 V above ground consists of the two currents I_1 and I_2. Two internal parasitic capacitors, C1 and C2, must be charged. C_1 is the capacitance from gate to source and is referred to in the data sheets as "C_{iss}," the input capacitance. Capacitance C_2 is the capacitance from gate to drain and is referred to in the data sheets as "C_{rss}," the reverse-transfer capacitance.

To drive the gate up to the 10-V "on" voltage in a time t_r, the required current I_1 is

$$I_1 = \frac{C_1 dV}{dt} = \frac{C_1 \times 10}{t_r} \tag{9.1}$$

However, in driving the gate up 10 V, the drain turns "on" and drops from the supply voltage V_{dc} (Figure 2.10) to the V_{ds} "on" voltage, which for simplicity will be taken as common return. Thus, as the top end of C_2 moves down V_{dc} volts, its bottom end moves up 10 V. The current required to achieve this is

$$\begin{aligned} I_2 &= C_2 \frac{dV}{dt} \\ &= \frac{C_2 (V_{dc} + 10)}{dt} \end{aligned} \tag{9.2}$$

It is revealing to calculate these currents in a typical case. Assume an off-line forward converter operating from a nominal 115-V-AC line with ± 10% tolerance. The maximum rectified DC voltage is 1.1 × 115 × 1.14 = 178 V. Assume an MTM7N45 whose C_1 is 1800 pF and whose C_2 is 150 pF. From Eq. 9.12, for a gate rise time of 50 ns, we obtain

$$I_1 = \frac{1800 \times 10^{-12} \times 10}{50 \times 10^{-9}}$$
$$= 360 \text{ mA}$$

and from Eq. 9.2

$$I_2 = \frac{(150 \times 10^{-12})(178 + 10)}{50 \times 10^{-9}}$$
$$= 564 \text{ mA}$$

The total current required from the gate driving source is $I_g = I_1 + I_2 = (0.36 + 0.564) = 0.924$ A. Thus, the smaller capacitor $C2$ requires about 50% more current than does $C1$. This is the well-known Miller effect, which multiplies the input-to-output capacitance by the voltage gain from input to output. A similar calculation shows the current required to move the gate voltage down 10 V is the same 0.924 A.

Generally, the effective input capacitance of low voltage MOSFETs is lower than that of high voltage devices. The input capacitance C_{iss} is higher and the reverse-transfer capacitance C_{rss} is lower than those for higher voltage devices. More importantly, there is less Miller effect multiplication of C_{rss} because the supply voltage and resulting voltage change across C_{rss} are lower.

Now consider an MTH15N20—a 15-A, 200-V device. Assume that it operates from a nominal 48-V telephone industry supply in which the minimum and maximum voltages are usually assumed to be 38 and 60 V, respectively.

The device capacitances are $C_{iss} = 2000$ pF and $C_{rss} = 200$ pF. Total effective input capacitance, assuming turn "on" from a supply voltage of 60 V, is

$$C_{in} = C_{iss} + \left(\frac{70}{10}\right) C_{rss} = 2000 + 7 \times 200 = 3400 \text{ pF}$$

and to charge the effective input capacitance by 10 V in the same 50 ns requires a gate input current of

$$I_g = C_{in}\, dv/dt = 3400 \times 10^{-12}(10/50 \times 10^{-9}) = 0.68 \text{ A}$$

Although the current is quite large it does not represent much power because the time is short (50 ns, or whatever gate rise time t_r is desired). However, the pulse current is a problem for the drive circuit and may exceed the specified maximum gate drive current of some devices.

9.2.4 Calculating the Gate Voltage Rise and Fall Times for a Desired Drain Current Rise and Fall Time

Although fast switching times are an advantage in reducing switching loss, excessively short drain current rise and fall times are undesirable as they cause large *L di/dt* spikes on ground buses and supply rails, and the accompanying short drain voltage rise and fall times cause large *C dV/dt* current spikes into adjacent wires or nodes. Some large devices have specified limits on drain current *di/dt* and drain voltage *dV/dt*. Exceeding these values with IGBT devices may "lock up" the device so it cannot be turned "off" by reducing the gate voltage.

The question thus arises as to what gate voltage rise time is required to yield a desired rate of change of drain current within the specified maximum value. This can be seen from the transfer characteristics shown in Figure 9.3*b* and 9.3*d*. Drain current switching time in a MOSFET between zero and I_d is just the time required for the gate voltage to move from the threshold to V_{gI} in Figure 9.3*b*. The time for the gate voltage to move from ground to the threshold of about 2.5 V is simply a delay time. Drain current turn "on" time is not accelerated by overdriving the input terminal in the same way as it is with bipolar transistors (Section 8.2 text and Figure 8.2*b*).

After Pressman *Driving the gate from a relatively high voltage, low resistance source will reduce the turn "on" delay time, and the larger gate current will speed up lags due to the Miller effect in both the drain current and voltage. However, most devices have maximum limits on drain* di/dt *and* dV/dt, *and care is required so as not to exceed these limits. If higher drive voltages are used then clamp diodes are required to prevent gate voltages exceeding the maximum gate voltage. To prevent high frequency parasitic oscillation, manufacturers recommend that the clamp diodes or zener diodes be fitted on the input end of the gate feed resistor away from the device gate terminal, and that the gate feed resistor be fitted near the gate terminal. The manufacturer normally specifies the minimum gate feed resistance, typically in the range 5 to 50 ohms. ~K.B.*

MOSFETs, being minority (electron) carrier devices, have negligible storage time. There is a turn "off" delay time corresponding to the time required for the gate voltage to fall from its uppermost level of about 10 V to the pinch-off level V_{gI} in Figure 9.3*b*, which corresponds to the current I_d. Once the gate voltage has fallen to V_{gI}, the drain current fall time is the time required for the gate voltage to fall from V_{gI} to the threshold. An IGBT conducts a combination of minority carriers and holes, and has significant storage time leading to a current tailing

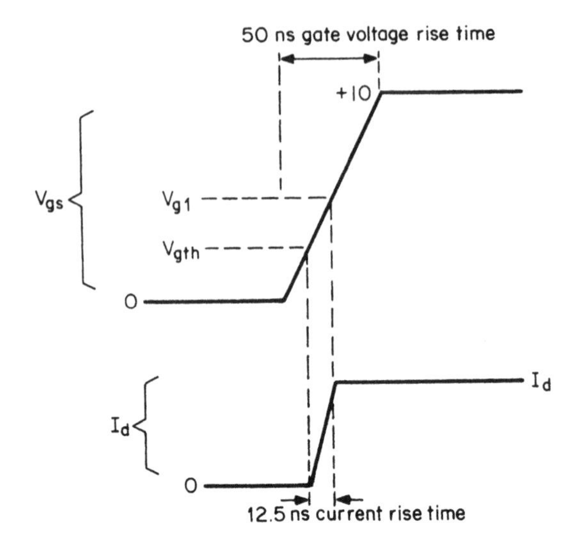

FIGURE 9.5 A 0- to 50-ns gate voltage rise time causing a 12.5-ns drain current rise time. The first 2.5 V on the gate voltage rise time to the gate threshold voltage is simply a delay time. At a gate voltage of about 5 to 7 V, most of the drain current is already flowing.

effect. In modern devices, however, the tailing has been much reduced and IGBTs now compare well with FETs.

We will now consider turning "on" an MTM7N45 to 2.5 A. If a 50-ns gate rise time were used for the first 2.5 V, no drain current at all would result (see Figure 9.3*b*). There would only be a delay equal to the time required for the gate to rise the first 2.5 V, and by the time the gate rose to about 5 V, most of the 2.5 A would have been turned "on."

Then, for a 0- to 10-V gate rise time in 50 ns, the drain current would rise from 0 to 2.5 A in about (2.5/10) 50 or 12.5 ns (Figure 9.5). Thus, 10-V gate voltage transition times can be two to three times the desired drain current transition times because of the narrow range of gate voltage across which drain current changes. The gate currents calculated above thus turn out to be only about one-half to one-third the actual.

9.2.5 MOSFET Gate Drive Circuits

As described above, the gate drive circuit must be able to deliver a high transient current in a positive direction to the gate of the FET to turn drain current "on." It must also be able to "sink" current to pull the gate voltage in a negative direction to discharge the gate capacitance and turn the drain current "off."

Most of the earlier PWM chips (e.g., SG1524 family) were unidirectional; they could source but not sink fairly large currents, as can be seen from the output stage circuit in Figure 9.6*a*.

The output stage of an SG1524-type PWM chip consists simply of a transistor with "uncommitted" emitter and collector. When the transistor is turned "on," it can sink 200 mA into its collector or source 200 mA out of its emitter. Most often, the power transistor must be "on" when the chip's output transistor is "on." Thus for an N-type MOSFET, its gate is driven from the emitter of the output stage as shown in Figure 9.6*b*. When the emitter is used to source gate current, it requires an external emitter resistor which can sink current from the gate to pull it low and turn the MOSFET "off" when the driver transistor turns "off." Such an emitter driving stage typically can source 200 mA of current to turn a MOSFET gate "on" relatively rapidly.

As mentioned in Section 9.2.2, it requires 0.924 A to drive the MTM7N45 gate up 10 V in 50 ns. The 200 mA limit of SG1524 source current can drive the gate up 10 V in (0.924/0.2) 50 = 231 ns. Also, as mentioned in Section 9.2.3, because drain current rise time takes place within the 2.5- to 5.0-V level of the gate voltage rise time, the drain current rise time will be only (2.5/10) = 58 ns. The first 2.5 V of gate rise time (until the threshold is reached) yields a further delay of 58 ns. Although this additional delay limits the maximum switching frequency, it does not contribute to the switching losses.

In the circuit shown in Figure 9.6*b*, the gate voltage rise time is 231 ns from 0 to 10 V, and this is fast enough. However, the gate voltage fall time is determined by the emitter resistor R_e and the gate input capacitance, as there is no active discharge. When the base of the chip output transistor goes low (internal to the chip), the emitter voltage is held up by the large MOSFET input capacitance. This biases the chip output transistor fully "off" and leaves only the external emitter resistor to discharge the MOSFET input capacitance. That capacitance (Section 9.2.2) is the input plus the Miller capacitance or $1800 + (180/10)$ $150 = 4500$ pf, so even with a 200-Ω emitter resistor, the MOSFET gate fall time is $3R_1C = 3 \times 200 \times 45000 = 270\,\mu\text{s}$. This is impossibly long if switching frequencies of over 100 kHz are to be achieved. Clearly, driving MOSFET gates requires the use of bidirectional drive circuits which can source and sink currents of 200 mA or more.

As the SG1524 PWM chip can only source or sink, but not both, additional external buffer circuits are required. A simple, but in most cases adequate, MOSFET gate driver can be formed from the NPN-PNP emitter-follower totem pole circuit shown in Figure 8.17. Here *Q*2 and *Q*3 are low cost 2N2222A, 2N2907A transistors that are capable of sourcing and sinking 800 mA with current rise times of about 60 ns. The upper transistor of the totem pole can be eliminated by using the PWM chip's output transistor as a source current driver as shown

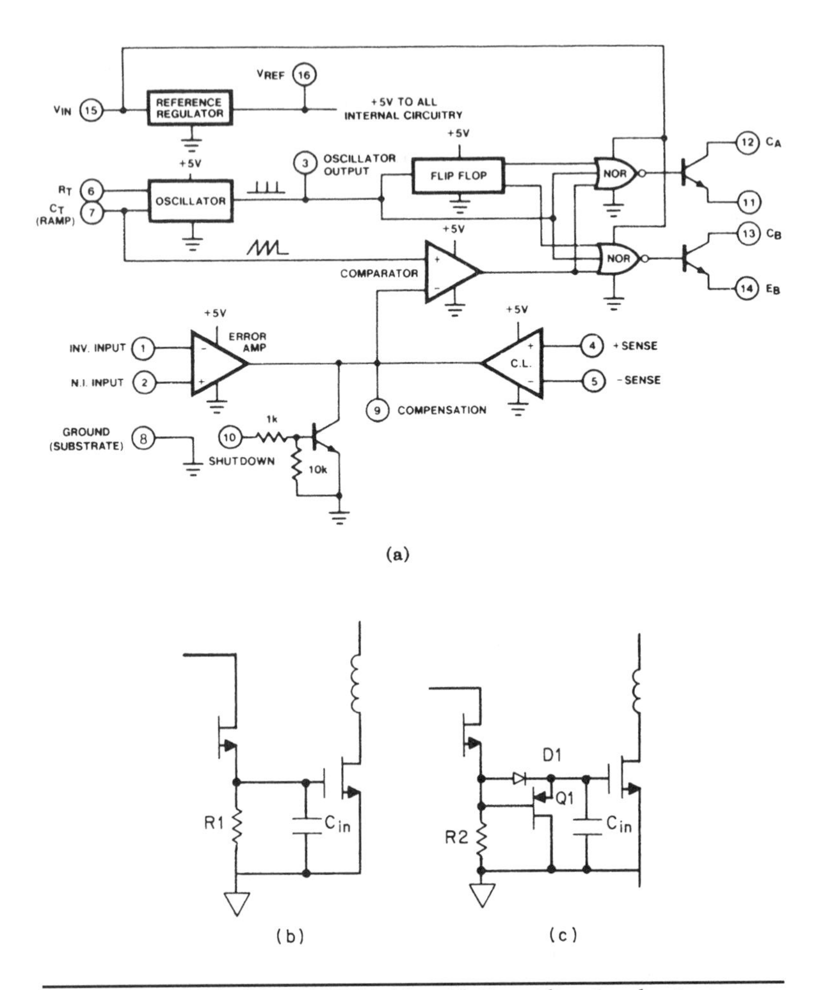

FIGURE 9.6 (*a*) The SG1524 PWM chip with its unidirectional output transistors (pins 11, 12 and 13, 14). The output transistors can either source or sink 200 mA, but not both. (*b*) The output transistor emitter driving the input capacitance of a MOSFET gate. The 200-mA source current can drive the MOSFET gate high in a positive direction rapidly, but only R_1 pulls the gate negative in a time $3R_1 C_{in}$. Part of the current must be used by R_2, and is thus not available to charge C_{in}. (*c*) An enhancement with a PNP emitter-follower, used during turn "off" to provide a larger current sink without additional loading to the drive IC. The full 200-mA chip output can be used to charge C_{in} because resistor R_2 can have a much higher value.

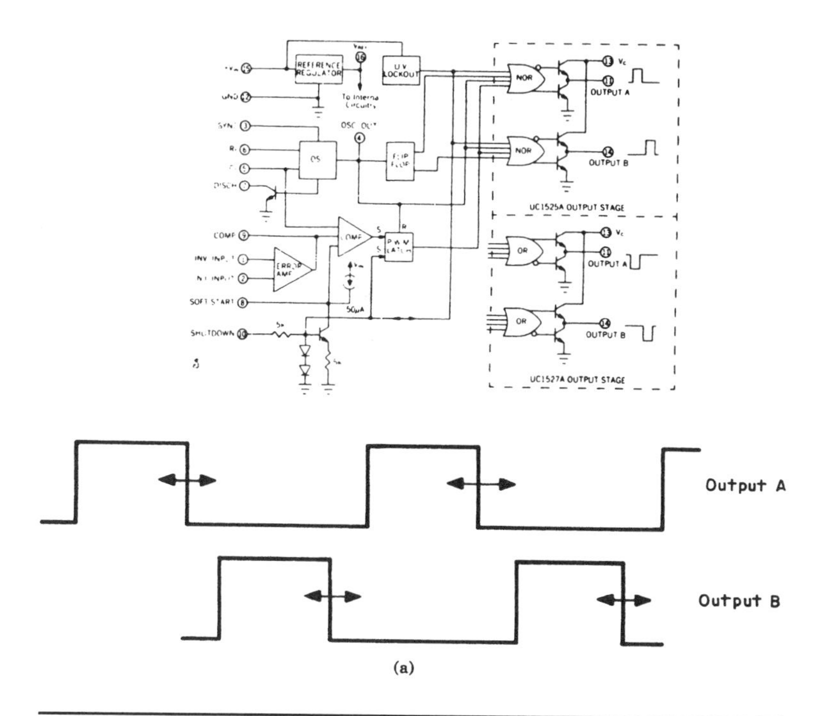

FIGURE 9.7 (*a*) Second-generation UC1525A PWM chip with totem-pole outputs (outputs *A* and *B*) capable of sourcing and sinking higher gate drive currents. (*Courtesy of Unitrode Corp.*) (*b*) Totem-pole outputs at *A*, *B* can drive high input capacitance of MOSFETs either directly when gates are at same DC voltage level as in a push-pull, or via a transformer when gates are at different DC voltage levels as in a half or full bridge.

in Figure 9.6*c* at the cost of somewhat less source capability (200 mA). Then the external 2N2222A can be used as the current sink. However, in many instances this is adequate.

Second-generation PWM chips such as the Unitrode Corp. UC1524[3] have a built-in totem pole consisting of an NPN emitter-follower resisting on top of an NPN inverter. These transistors can source and sink over 200 mA. The emitter-follower and the inverter are driven by 180° out-to-phase signals so that when either is "on," the other is "off" as shown in Figure 9.7*a*. By using a transformer with two isolated secondaries as in Figure 9.7*b*, the top and bottom transistors of a half or full bridge, which are at different DC voltage levels, can be driven simultaneously. Similarly, the MOSFET gates which are at the same DC voltage level in a push-pull circuit can also be driven in the same way.

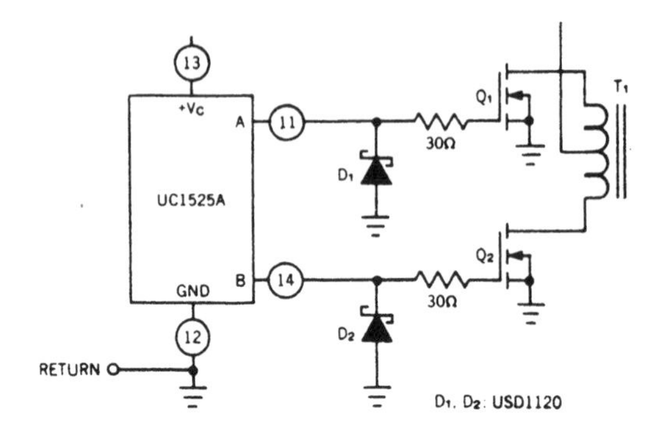

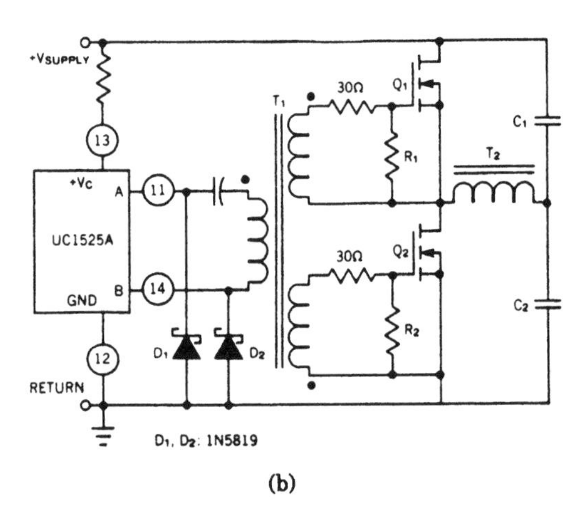

(b)

FIGURE 9.7 *Continued.*

In Figure 9.7*b*, the transformer primary is connected between pins 11 and 14 of the PWM chip. During the "on" time in one half period, when pin 11 is, say, positive with respect to pin 14 and is sourcing 200 mA, it is at about $+(V_h - 2)$ V with respect to ground and pin 14 is at about +2 V with respect to ground. In the next half period, the polarity across pins 11, 14 reverses for the "on" time in that half period. During the dead time within either half period, pins 11, 14 are both short-circuited to ground. Thus for ±10 V across the primary, the supply voltage V_h should be about 14 V.

9.2.6 MOSFET R_{ds} Temperature Characteristics and Safe Operating Area Limits[4,5]

The most common failure mode in bipolar transistors—secondary breakdown—comes about because their "on" state voltage $V_{ce(sat)}$ decreases with temperature. This imposes limits (RBSOA curve of Figure 8.4) that the $I_c - V_{ce}$ trajectory may not cross during the turn "off" transition. Manufacturers state that only a single crossing of this limit curve may cause a bipolar to fail in the secondary breakdown mode.

Because their "on"-voltage drop and r_{ds} increase with temperature however, MOSFETs do not suffer from secondary breakdown and consequently have a much larger switching SOA as shown in Figure 9.8. This is the boundary that, if crossed for over 1 μs during either the turn "on" or turn "off" trajectory, may damage or destroy the transistor. The boundary limits are the maximum pulsed current ($I_{dm,pulsed}$) and maximum drain-to-source V_{dss} voltage ratings for device.

An explanation of why the negative temperature coefficient of V_{ce} causes secondary breakdown in the bipolar, and the V_{ds} positive

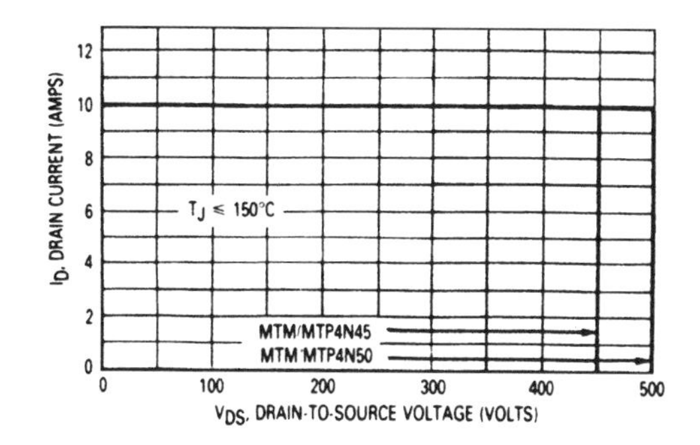

FIGURE 9.8 Reverse-biased safe operating area of 4-A, 400-V MOSFET. This area is far larger than that for a 4-A, 400-V bipolar transistor. For the MOSFET, it is a rectangle bounded on one side by I_{dm} (maximum pulsed current rating) which is two to three times I_d (maximum continuous current rating), and on the other by V_{dss}, the maximum drain-to-source voltage rating. The switching safe operating area is the area that the load line may traverse without incurring damage to the device. The fundamental limits are the maximum rated peak drain current I_{dm}, the minimum drain-to-source breakdown voltage $V_{(br)dss}$, and the maximum rated junction temperature. The boundaries are applicable to both turn "on" and turn "off" of the devices for rise and fall times of less than 1μs. (*Courtesy of Motorola Inc.*)

temperature coefficient prevents it in the MOSFET, is as follows. Secondary breakdown in bipolars comes about because of local hot spots on the chip. These hot spots are considerably hotter than the average junction temperature as calculated from the chip's junction-to-case thermal resistance and total transistor dissipation, because the calculation assumes a uniform distribution of current carriers throughout the collector area. However, the process by which a bipolar transistor turns "off" results in "current crowding" into an ever decreasing area of the chip. Since the current is not uniformly distributed but is crowded into a small fraction of the collector area, that area runs much hotter than the rest of the chip. Further, since the collector-to-emitter resistance decreases with increasing temperature, any incipient local hot spot has slightly less resistance than its surrounding areas and robs current carriers from adjacent areas. This results in a runaway situation, as the hot spot gets even hotter, causing a further decrease in resistance and robbing even more current from adjacent areas. This process builds up rapidly until the local hot spot reaches a high current density, and temperature sufficient (>200°C) to cause failure.

MOSFETs do not have the same current crowding mechanism. Their positive r_{ds} temperature coefficient tends to disperse and cool off incipient local hot spots. If a point on the chip started operating at a slightly higher current density than its neighbors, the temperature would rise slightly. Because r_{ds} has a positive temperature, its resistance would increase and it would shift some of its current carriers to neighboring areas and cool down. The result is the much larger SOA of Figure 9.8 for the MOSFET as compared to Figure 8.2 for the bipolar.

Curves showing the variation of r_{ds} with temperature and drain current are seen in Figure 9.9 for a typical 15-A, 450-V MOSFET (MTM15N45). The variation of r_{ds} with temperature is also dependent

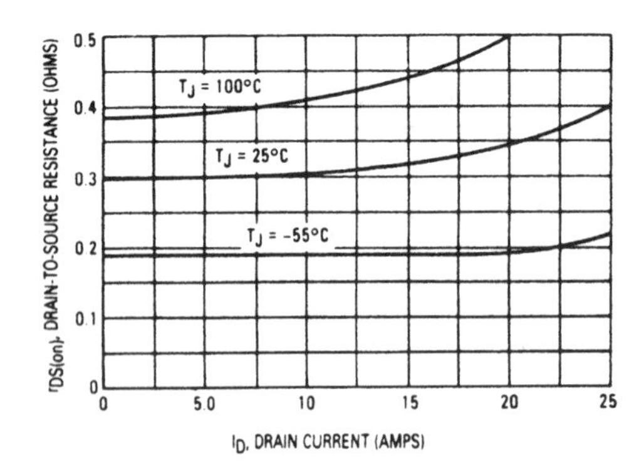

FIGURE 9.9 Variation in $r_{ds(on)}$ with drain current and temperature for the MTM15N45. (*Courtesy of Motorola Inc.*)

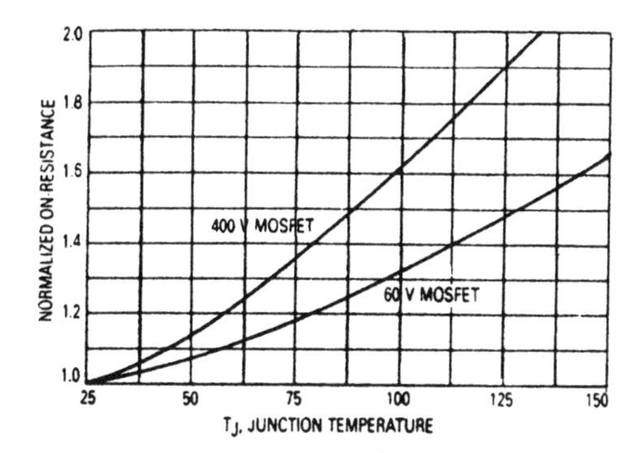

FIGURE 9.10 The influence of junction temperature on resistance varies with breakdown voltage. (*Courtesy of Motorola Inc.*)

on the voltage rating of the MOSFET as shown in Figure 9.10. It is seen there that higher voltage MOSFETs have larger r_{ds} temperature coefficients than do lower voltage devices.

9.2.7 MOSFET Gate Threshold Voltage and Temperature Characteristics[4,5]

The many MOSFET manufacturers specify the gate threshold voltage V_{gsth} in different ways. Some specify it as the gate-to-source voltage for which I_{ds} equals 1 mA at $V_{ds} = V_{gs}$. Others define it as the gate-to-source voltage for which $I_{ds} = 0.25$ mA at $V_{ds} = V_{gs}$. There appears to be a two-to-one production spread in V_{gsth}.

The gate threshold voltage V_{gsth} has a negative temperature coefficient; it falls about 5% for each 25°C rise in temperature (Figure 9.11)

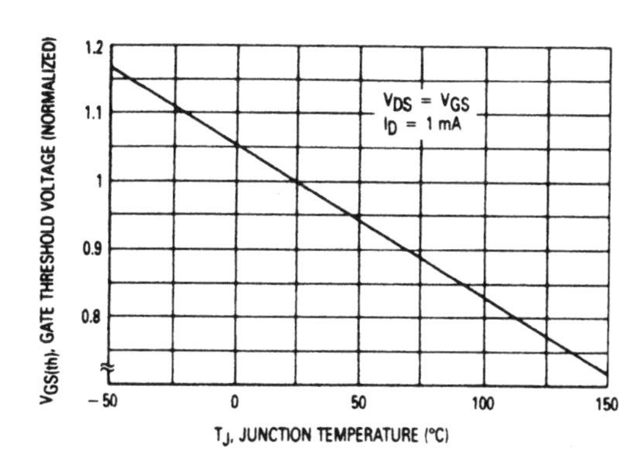

FIGURE 9.11 Gate threshold voltage variation with temperature. (*Courtesy of Motorola Inc.*)

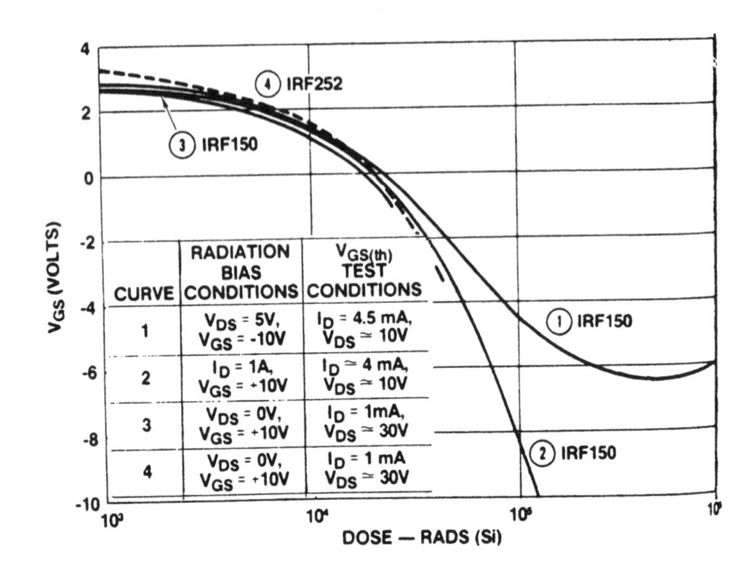

FIGURE 9.12 Typical gate threshold voltage versus total gamma dose. Under radiation, a negative gate bias is required to keep a MOSFET turned "off." Gate threshold voltage depends strongly on the type, intensity, and duration or radiation. (*Courtesy of Unitrode Corp.*)

and quickly falls to zero in a radiation environment.[6] A negative gate-to-source voltage is required to keep the MOSFET "off" under radiation. The transfer characteristic is strongly dependent on type, duration, and intensity of radiation (Figure 9.12). Discussion of MOSFETs in a radiation environment is beyond the scope of this text.

9.2.8 MOSFET Switching Speed and Temperature Characteristics

MOSFET switching speed is essentially independent of temperature. Drain current rise and fall times depend only on the time required for the gate voltage to cross the narrow band between the gate threshold voltage (V_{gsth}) and V_{gI} in Figure 9.3*b*.This depends on the total resistance of the gate drive circuit and the effective gate input capacitance. In many cases, the gate discharges via a discrete external resistance which has low temperature coefficient.

Further, gate input capacitance is also independent of temperature. Turn "on" and turn "off" delays are somewhat temperature-dependent. Turn "on" delay is the time for the gate voltage to rise from 0 to the threshold voltage V_{gsth}. Since V_{gsth} falls 5% for each 25°C rise in temperature, turn "on" delay will decrease with temperature.

Also since there is a two-to-one production spread in V_{gsth}, the turn "on" delay will vary from device to device even when they are at the same temperature. Note that devices with a large variation in V_{gsth} may not have a large variation in turn "on" delay, which is the delay to turn "on" a relatively large specific current. The lower tails of the transfer characteristic can vary significantly without changing V_{gI}—the gate voltage for a given current I.

Turn "off" delay is the time required for the gate to fall from its usual "on" voltage of 10 V to V_{gI} (Figure 9.3*b*). Since gate threshold voltage and transconductance vary with temperature, so will the turn "off" delays. Turn "on" and turn "off" delays must be considered when paralleling MOSFETs, to ensure current sharing for the full conduction period.

9.2.9 MOSFET Current Ratings

For bipolar transistors, maximum output current is limited by the fact that current gain falls drastically as output current rises. Very often, unacceptably high base input currents are required as collector current increases. This is shown in Figure 9.13 for the 2N6542—a typical 5-A, 400-V bipolar transistor. With MOSFETs, however, output-input gain (transconductance or dI_{ds}/dV_{gs}) does not decrease with output current, as can be seen in Figure 9.14. Thus the only limitation on drain

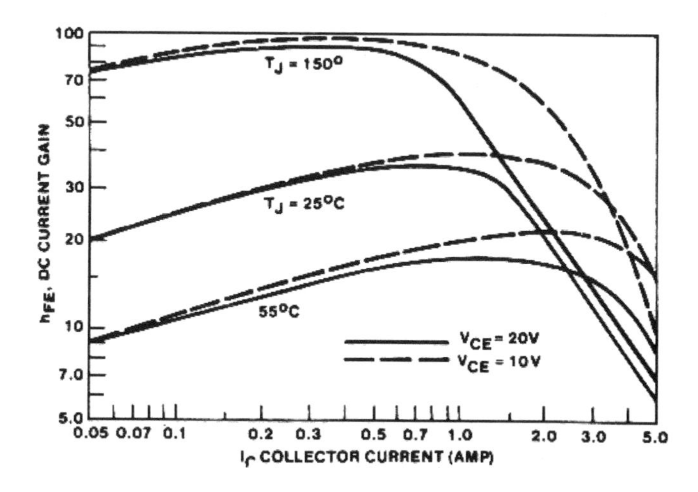

FIGURE 9.13 Typical DC current gain for a bipolar transistor: that of the 2N6542/3. Gain of a bipolar transistor falls off with increasing output current, but that of a MOSFET does not. Maximum current in a MOSFET is limited only by junction temperature rise. (*Courtesy of Motorola Inc.*)

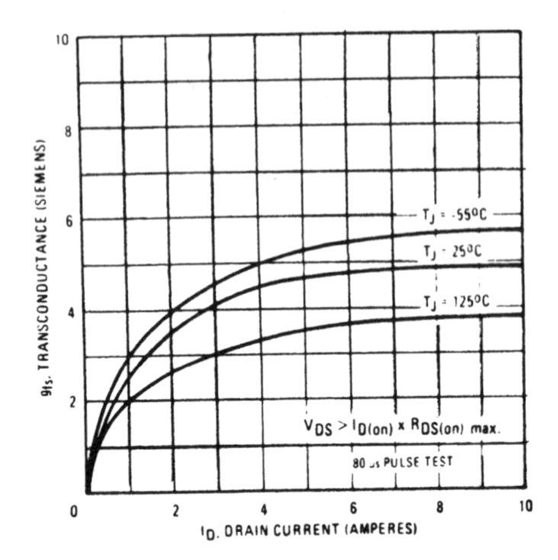

FIGURE 9.14 Transconductance versus drain current of the IRF330. Gain of a bipolar transistor falls off with increasing output current, but that of a MOSFET does not. Maximum current in a MOSFET is limited only by junction temperature rise and the manufacturer's rating. Maximum MOSFET junction temperature is 150°C. Good standard design practice is to de-rate this to 105°C or at most to 125°C. (*Courtesy of International Rectifier.*)

current is power dissipation, or maximum MOSFET junction temperature and the manufacturer's current rating based on construction details. Manufacturers rate the current carrying capability of their devices in terms of I_d, the maximum continuous drain current.

Many MOSFET manufacturers specify maximum I_d as that current, which at the maximum $V_{ds(on)}$ voltage for that current, yields a power dissipation at 100% duty cycle such that when multiplied by the thermal resistance brings the MOSFET junction temperature to the maximum of 150°C when the transistor case is at 100°C. Thus

$$dT = 50 = PDR_{th} = V_{ds(on)} I_d R_{th} \quad \text{or} \quad I_d = \frac{50}{V_{ds(on)} R_{th}}$$

in which $V_{ds(on)}$ is the maximum drain-to-source "on" voltage at 150°C and R_{th} is the thermal resistance from junction to case in degrees Celsius per watt.

This I_d rating is not a useful guide for selecting a MOSFET for a given peak current in a switching supply application. In such usage, duty cycle is never 100%. For reliability, it is desired to operate at junction temperatures de-rated to 125°C, or the usual military-specified 105°C. But it is a useful measure as it does show the relative current-carrying capability of various MOSFETs when operated at 100% duty cycle.

Device	I_d A	V_{dss}, V	$r_{ds}\Omega$ at 25°C
MTH7N45	7	450	0.8
MTH13N45	13	450	0.4

TABLE 9.1 Motorola MOSFET Current, Voltage and "On" State Resistance Ratings

As a general guide, there are two ways of selecting a MOSFET for a specified output power in a switching supply. First, the equivalent flat-topped primary current pulse I_{pft} is calculated for specified output power and minimum DC input voltage. (This current is given in Eqs. 2.9, 2.28, 3.1 and 3.7 for the push-pull, forward converter, half and full bridge, respectively.) Then for these currents, a MOSFET is chosen for r_{ds} so that the "on" drain-to-source voltage $I_{pft}r_{ds}$ is a small percentage (usually no more than 2%) of the minimum supply voltage so as to rob no more than 2% of the transformer's minimum primary voltage.

In selecting a device with a desired r_{ds}, it should be recalled that the data sheets typically give it at a case temperature of 25°C. Also noteworthy is the variation of r_{ds} with temperature and device voltage rating as shown in Figures 9.9 and 9.10. Figure 9.10 shows that the r_{ds} of a 400-V MOSFET at 100°C is 1.6 times its value at 25°C.

As a design example, consider a 150-W forward converter operating from a nominal 115-V AC line. Assume that maximum and minimum rectified DC voltages are 184 and 136 V, respectively. Peak flat-topped pulse current from Eq. 2.28 is $I_{pft} = 3.13(150/136) = 3.45$ A. Then for the MOSFET "on" voltage to be 2% of the minimum supply voltage, $V_{on} = 0.02 \times 136 = 2.72 = I_{pft}r_{ds} = 3.45r_{ds}$ or $r_{ds} = 0.79\,\Omega$ at, say, 100°C or $0.79/1.6 = 0.49\,\Omega$ at 25°C.

Possible plastic-cased 450-V MOSFET choices from the Motorola catalog are shown in Table 9.1.

The choice would be made on an engineering judgment of the relative importance of cost and performance. The 7-A MTH7N45 is not quite good enough. It would have more than the sought "on" drop of 2.72 V at 3.45 A. That in itself is not prohibitive; it would run at a junction temperature somewhat higher than the MTH13N45. That would have to be weighed against the higher cost of the MTH13N45.

An alternative way to select the MOSFET is to define a maximum junction temperature for the required reliability, keeping in mind the drop in reliability of typically 50% for each 10 degree rise. Say 100°C is chosen for the reliability we want. Then select a heat sink for a reasonably low MOSFET junction-to-case temperature rise.

In this example, we will assume a reasonably low 5°C junction-to-case temperature rise. The temperature difference between the junction and the case $dT = 5 =$ *power dissipation* × *thermal resistance* (junction to case) and assuming that there are negligible AC switching losses

$$dT = (I_{\text{rms}})^2 (r_{\text{ds}}) R_{\text{th}} \text{ or } r_{\text{ds}} = \frac{5}{(I_{\text{rms}})^2 R_{\text{th}}}$$

in which I_{rms} is the rms current in the MOSFET.

For a forward converter with maximum "on" time per period of $0.8T/2$, the rms current is $I_p(\sqrt{t_{\text{on}}/T}) = 0.632 I_p$ For the preceding design example of a 150-W forward converter with I_p of 3.45 A, we obtain

$$r_{\text{ds}} = \frac{5}{(0.632 I_p)^2 R_{\text{th}}}$$

and for the typical thermal resistance of 0.83°C/W for MOSFETs of this current and package size

$$r_{\text{ds}} = \frac{5}{(0.632 \times 3.45)^2 \times 0.83}$$
$$= 1.26\,\Omega \text{ at } 100^\circ\text{C}$$

This is the r_{ds} at 100°C junction temperature that causes a 5°C junction-to-case temperature differential with a 3.45-A peak current pulse at 0.4% duty cycle. The r_{ds} at 25°C junction temperature is then 1.26/1.6 or .78 Ω. Thus, on the basis of a 5°C junction-to-case temperature differential, Table 9.1 shows that the MTH7N45 would be an adequate choice. The MTH13N45 would be a better choice if its somewhat higher cost were acceptable.

9.2.10 Paralleling MOSFETs[7]

In paralleling MOSFETs, two situations must be considered: (1) whether the paralleled devices share current equally—in the static case when they are fully "on" and (2) whether they share current equally during the dynamic turn "on"–to–turn "off" transitions. Unequal static current sharing comes about because of unequal r_{ds} of the paralleled devices. The lowest r_{ds} device draws the largest share of the total current—just as with a group of paralleled discrete resistors, the smallest resistor draws the most current. With paralleled MOSFETs, in either the static or dynamic case, the concern is that if one MOSFET hogs a disproportionate part of the current, it will run hotter. In the long term, reliability will decrease, and in severe examples it may fail after a short time.

Earlier it was pointed out that the absence of MOSFET secondary breakdown results because of the positive temperature coefficient

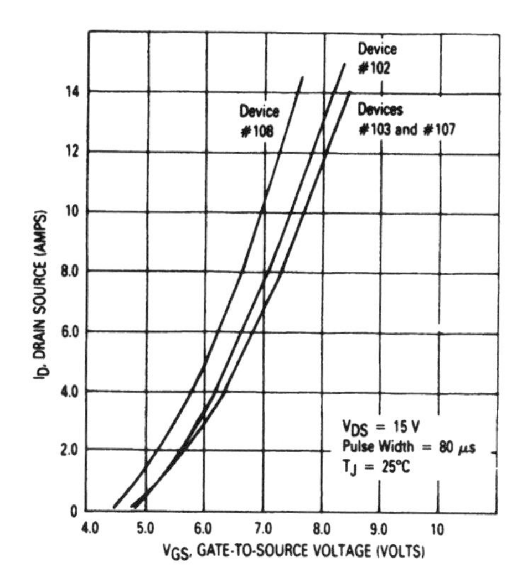

FIGURE 9.15 Variation in transconductance curves for 250 MTPBN2O devices. For equal dynamic current sharing, MOSFET transconductance curves must coincide. (*Courtesy of Motorola Inc.*)

of r_{ds}. Thus, if a small portion of the chip tends to hog a disproportionate part of the total current, it runs hotter, its r_{ds} increases, and it shifts off some of its current to some neighboring areas to equalize current density. This mechanism also works to some extent with paralleled discrete MOSFETs. By itself, however, it is not sufficient to minimize the temperature of the hottest device. This is because the temperature coefficient of r_{ds} is not very large and a large temperature differential between devices is required to shift off excess current. With a large temperature differential, however, the hottest device is at a high temperature, and this is exactly what reduces reliability and is to be avoided. The mechanism does work well within a chip because all elementary areas of the chip are thermally coupled. However, it does not work so well if thermal coupling between the parallel MOSFETs is poor, as with discrete MOSFETs that are physically separated on a common heat sink, or worse, on separate heat sinks.

To improve static current sharing, discrete MOSFETs should be located as close as possible on the same heat sink. Packages containing multiple paralleled MOSFETs on a common substrate are currently available from a number of manufacturers. As a last resort, if discrete MOSFETs must be used, and close location on a common heat sink does not suffice, matching the r_{ds} of paralleled devices will much improve current sharing.

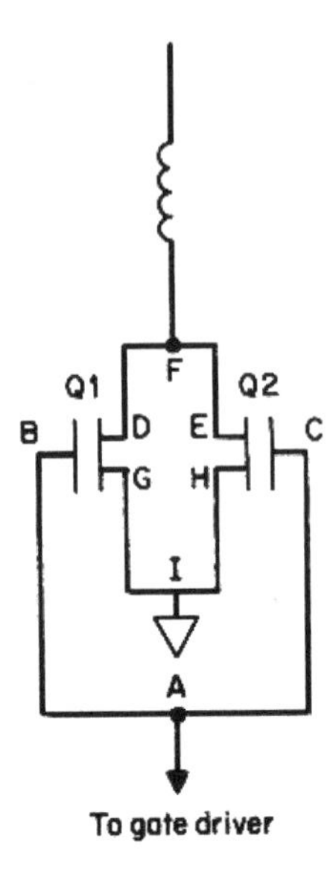

FIGURE 9.16 To ensure equal dynamic current sharing in parallel MOSFETs, circuit layout should be symmetrical. Thus $AB = AC$, $GI = HI$, $DF = EF$.

For equal dynamic current sharing, the transconductance curves of paralleled devices must lie exactly on top of one another, that is, they must be identical. This is shown in Figure 9.15. If all gates have identical voltage at the same time and the transconductance curves are superimposed, the drains will carry the same currents at that time, on either turn "on" or turn "off." It is not essential that the gate thresholds match exactly. If n devices are to be paralleled for a total current of I_t, they should be matched as closely as possible for the same I_t/n at the same gate voltage—even if there is a large mismatch in gate threshold voltages.

Symmetrical circuit layout is also important for equal dynamic current sharing (Figure 9.16). Lead lengths from the common output point of the gate driver to the gate terminals should be equal. Lead lengths from the source terminals of the MOSFETs to a common tie point should be equal, and that common tie point should be brought as directly as possible to a common tie point on the ground bus. That ground bus tie point should be common (with as short a lead as possible) with the negative rail of the housekeeping supply. Finally, to prevent oscillations with paralleled MOSFETs, resistors of 10 to 20 Ω or ferrite beads should be placed in series with gate leads.

9.2.11 MOSFETs in Push-Pull Topology

MOSFETs in the push-pull topology improve the transformer flux-imbalance problem significantly. It will be recalled (Section 2.2.5) that if the volt-second product applied to the transformer during "on" half period is not equal to that in the next half period, the transformer core moves off the center of its hysteresis loop. After a number of such periods, the core saturates, cannot support the supply voltage, and the transistors are destroyed. (This is referred to as "staircase saturation.")

MOSFETs reduce the staircase saturation problem in two ways. First, there is no storage time with MOSFETs and for equal gate "on" times, drain voltage "on" times are always equal in alternate half periods. There is hence no inequality in volt-second product applied to the transformer due to unequal transistor "on" times. In contrast, with bipolars, the main reason for unequal volt-seconds on alternate half periods is the inequality in storage times.

Also, the staircase saturation problem is reduced by the positive temperature coefficient since r_{ds} acts in a negative-feedback way to reduce flux imbalance. If there is a certain amount of flux imbalance, the core walks partly up its hysteresis loop. This causes the magnetizing current and hence total current on one half period to be larger than that on the alternate half period (Figure 2.4*b* and 2.4*c*). Now the MOSFET with the larger peak current runs hotter and its r_{ds} increases, increasing its "on" state voltage drop. This robs voltage from its half primary, its volt-second product decreases, and the core moves back down toward the center of its hysteresis loop.

Qualitatively, both these effects help to prevent a catastrophic flux imbalance. But it is not easy or even feasible to demonstrate quantitatively that it works at all power levels, at all temperatures, and for all core materials. One solution to the flux-imbalance problems is to use a current-mode topology (Section 2.2.8.5), but if for some reason this is not desired, many designers have successfully used MOSFETs in conventional push-pull circuits with acceptable flux imbalance up to 150 W.

One final interesting drive scheme for MOSFETs in a push-pull circuit is shown in Figure 9.17. This simple arrangement virtually eliminates any residual flux imbalance due possibly to differences in r_{ds} as follows. If there is an incipient flux imbalance, it manifests itself by the current in one transistor being somewhat larger than the current in the other (Figure 2.4*b* and 2.4*c*). The MOSFET with the higher peak current has, for some reason, a larger volt-second product than the other device. This is corrected as illustrated in Figure 9.17 by adding a small, empirically selected resistor r_b in series with the gate of the MOSFET with the larger peak current. This integrates away some of the front edge of its gate turn "on" pulse, narrows it, equalizes the volt-second products on each side, and equalizes peak currents on

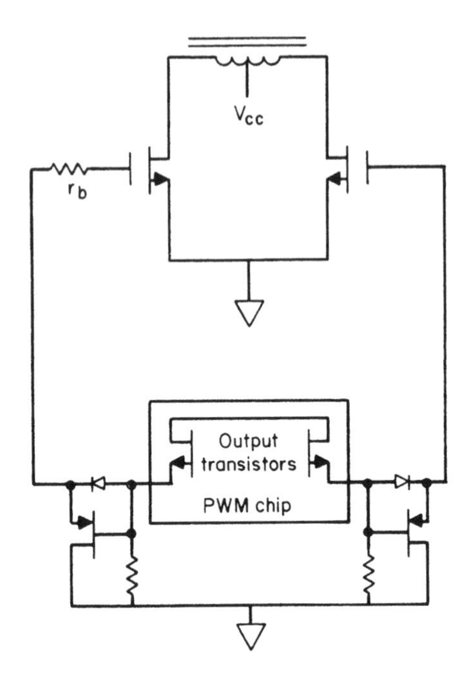

FIGURE 9.17 A flux-imbalance correction method for a MOSFET push-pull circuit. An empirically selected resistor, inserted in series with the gate of the MOSFET that originally had the higher peak current, forces currents in Q_1, Q_2 to be exactly equal. It does this by integrating away part of the front end of the gate drive pulse and forces the volt-second product on each transformer half primary to be equal.

alternate periods. The main objection to the scheme is that if either transistor is changed, a new "Select At Test" resistor must be chosen. This may be unacceptable in the field where the proper test equipment is not available. Also variations as parts age and with temperature changes may result in drift away from ideal balance. Some (military) programs do not permit empirical "Select At Test."

After Pressman *The designer should not forget that this problem is completely eliminated by using current-mode control. See Chapter 5. ~K.B.*

9.2.12 MOSFET Maximum Gate Voltage Specifications

Most MOSFETs have a maximum gate-to-source voltage specification of ±20 V. The gate is easily destroyed if that limit is exceeded. A problem can arise with this limit when the device is turned "off." When a MOSFET has a gate input resistor and is turned "off" rapidly in a circuit with a large working voltage, the internal Miller capacitance couples a voltage spike back to the gate. When added to the existing gate voltage, this spike may overstress the gate.

Consider a forward converter operating from a nominal supply voltage of 160 DC V (maximum of 186 V). When the MOSFET turns "off" at maximum supply voltage, its drain goes to twice the supply voltage or 372 V. A fraction of this positive-going front edge is coupled back and voltage divided by C_{rss} and C_{iss}. For the MTH7N45, $C_{rss} = 150$ pF, $C_{iss} = 1800$ pF, so the voltage coupled back down to the gate will be

$$372 \times 150/(150 + 1800) = 29 \text{ V}$$

This may damage the gate, as it exceeds the 20-V limit. The gate resistor will decrease the amplitude but may still be close to the point of causing a possible failure as line transients and the leakage inductance spike have not yet been considered. Thus, good design practice is to shunt the gate to source with a zener diode (18-V is a good choice). Some manufacturers recommend that the clamp zener be fitted at the drive input end of the series gate feed resistor, providing the recommended resistor value is used—typically in the range 5 to 50 ohms. Note that a high frequency oscillation may result from the capacitive drain to gate feedback if higher value series resistors are used.

9.2.13 MOSFET Drain-to-Source "Body" Diode

In the solid-state structure of a MOSFET, a parasitic "body" diode is located inherently across the drain-source terminals as shown in Figure 9.18.

The diode polarity is such as to prevent reverse voltage across the MOSFET. The forward current handling capability and reverse voltage rating of the diode are similar to those of the MOSFET. Its reverse recovery time is shorter than that of a conventional power rectifier diode, but not as fast as discrete fast-recovery types. Manufacturers' data sheets show the diode reverse recovery times for specific MOSFETs.

The diode is of no importance for most switching supply topologies as the drain to source is never subjected to a reverse voltage (drain negative with respect to source for an N-type MOSFET, positive with respect to source for a P-type MOSFET). There are some exceptions—specifically, the half or full bridge topologies of Figures 3.1 and 3.3. In these circuits, there is normally a dead time between the time the diode

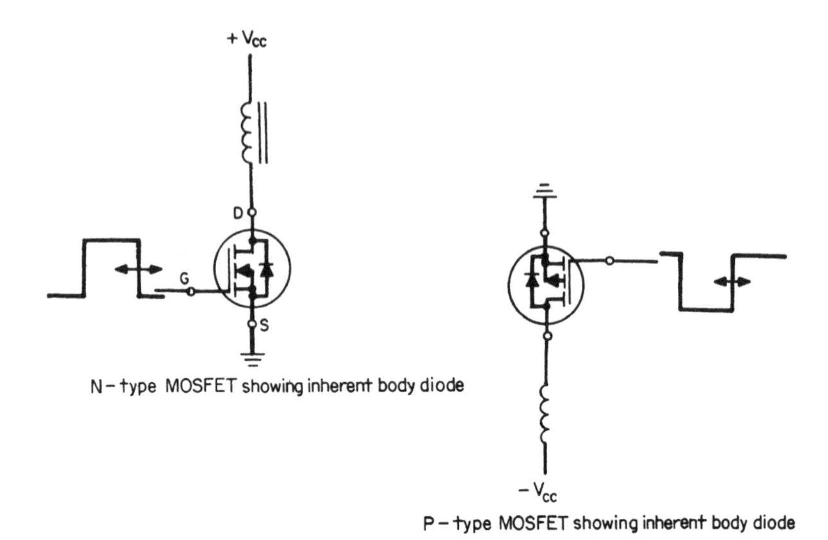

FIGURE 9.18 Inherent body diodes in N- and P-type MOSFETs. In the N-channel MOSFET, the diode prevents a negative drain-to-source voltage. In the P-channel MOSFET, the diode prevents a positive drain-to-source voltage.

conducts (when it is returning the energy stored in the transformer leakage inductance to the supply line) and the time it is subjected to reverse voltage. Because of this delay between forward current and reverse voltage, the relatively poor reverse recovery time of the MOSFET body diode is not harmful.

However, if a specific new circuit configuration requires reverse voltage across the MOSFET, a blocking diode must be placed in series with the drain. Various motor drive circuits or circuits with highly inductive loads may have problems because of the body diode.[8] High-frequency resonant circuit topologies (Chapter 13) frequently must be able to support reverse voltage immediately after carrying forward current. When this is necessary, the circuit of Figure 9.19 is used. Diode D_1 prevents forward current from flowing through the MOSFET body diode, and fast-reverse-recovery-time diode D_2 carries the required forward current.

After Pressman *Modern devices manufactured from the mid-1990s and onward have much better substrate diodes with fully specified forward and reverse recovery characteristics. When such devices are used, the preceding methods may no longer be necessary. ~K.B.*

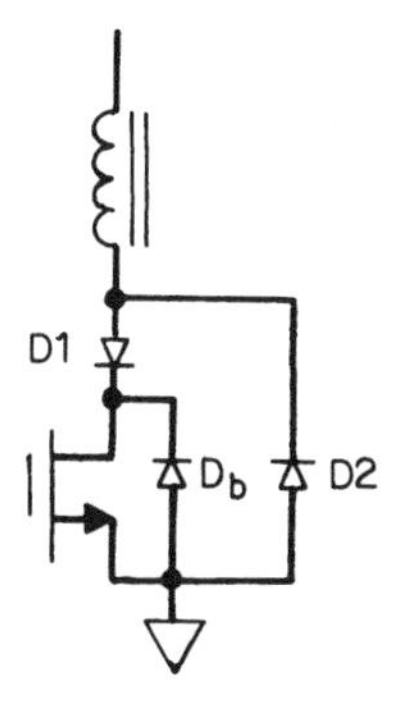

FIGURE 9.19 A scheme that will avoid energizing the MOSFET body diode, when reverse current must be allowed to flow around a MOSFET. Db is the internal substrate diode. A blocking diode D_1 is added in series with the drain to block reverse current. D_2, a diode with better reverse recovery characteristics, is shunted around D_1 and the FET to provide for reverse current flow.

9.3 Introduction to Insulated Gate Bipolar Transistors (IGBTs)

The following section on IGBTs owes much to a paper by Jonathan Dodge and John Hess of APT.[10] However, any errors are entirely mine. The data and graphs presented here are intended as a general guide to IGBT performance and do not represent a particular device. For design purposes always refer to the manufacturer's data for the chosen device.

In the mid-1980s, the Insulated Gate Bipolar Transistor (IGBT), a combination of an easily driven MOSFET gate and low conduction loss power bipolar transistor, started to become the device of choice for high current and high voltage switching power supply applications.

The balance in tradeoffs among switching speed, conduction loss, and ruggedness continues to be finely tuned so that IGBTs are encroaching upon the high frequency, high efficiency domain of power MOSFETs. In fact, the future industry trend is for IGBTs to replace power MOSFETs except in very low current applications.

To help circuit designers understand the tradeoffs in device selection, this section provides a relatively painless overview of IGBT technology, a walkthrough of IGBT datasheet information, and methods of how to select an IGBT. This section is intentionally placed before the technical discourse. Answers to the following set of critical questions will help determine which IGBT is appropriate for a particular application.

9.3.1 Selecting Suitable IGBTs for Your Application

The designer should consider the following questions before selecting a device:

1. What is the difference between non-punch-through (NPT) and punch-through (PT) devices?
   For a given $V_{CE(on)}$, PT IGBTs have higher speed switching capability with lower total switching energy. This is due to higher gain and minority carrier lifetime reduction, which quenches the tail current.
   NPT IGBTs are slower but are more rugged, and typically, they are short circuit rated, whereas PT devices often are not. NPT IGBTs can absorb more avalanche energy than PT IGBTs. NPT technology is more rugged due to the wider base and lower gain of the PNP bipolar transistor.
2. What is the recommended maximum operating voltage?
   The highest voltage the IGBT has to block should be no more than 80% of the V_{CES} rating.
3. Should I use a PT device or an NPT device in my application?
   This depends on questions 4 and 5 (below) and if your application is hard or soft switched. A PT device is better suited for fast switching due to its reduced tail current and reduced switching loss. An NPT device may also work quite well in your application; it is more rugged but will have higher switching loss.
4. What is the desired switching speed for your application?
   If your answer is "the higher, the better," then a PT device is the best choice. Again, the usable frequency versus collector current graph in Figure 9.30 can help answer this question for hard switching applications.
5. Is short circuit withstand capability required?
   For applications such as motor drives, the answer is probably yes, but the switching frequency tends to be relatively low. For this application, the more rugged NPT device would be a better choice. However, switch mode power supplies often do not require short circuit capability, so the faster PT device would be a better choice there.
6. How do I select the current rating for the IGBT?
   This depends on how much current will flow through the device in your application. For soft switching applications, the I_{C2} rating could be used as a starting point. For hard switching applications, the usable frequency versus collector current graph (Figure 9.30) is helpful in determining whether a device

will fit the application. Any difference between datasheet test conditions and the application conditions should be taken into account.

9.3.2 IGBT Construction Overview

An N-channel IGBT is basically an N-channel power MOSFET constructed on a p-type substrate, as illustrated by the generic IGBT cross section in Figure 9.20. Consequently, operation of an IGBT is very similar to that of a power MOSFET. A positive voltage applied between the emitter and gate terminals will cause electrons to be drawn toward the gate terminal in the body region. If the gate-emitter voltage is at or above what is called the threshold voltage, enough electrons are drawn toward the gate to form a conductive channel across the body region, allowing current to flow from the collector to the emitter. (To be precise, it allows electrons to flow from the emitter to the collector.) This flow of electrons draws positive ions, or holes, from the p-type substrate into the drift region toward the emitter. This leads to a couple of simplified equivalent circuits for an IGBT as shown below.

The left circuit of Figure 9.21 shows an N-channel power MOSFET directly driving a wide-base PNP bipolar transistor in a Darlington configuration. The advantages are very clear. The FET provides the normal high resistance gate input characteristic, and since the FET part is quite small, the input capacitance will also be quite small. This FET then drives the PNP power transistor with its low saturation characteristic so the PNP part does the power handling work. The result is a combination of the best properties of each type.

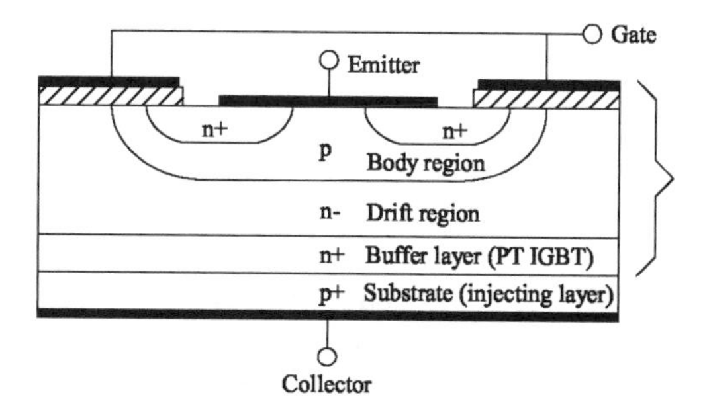

FIGURE 9.20 N-channel IGBT cross section (PT IGBTs have an additional N+ layer as well, as will be explained later).

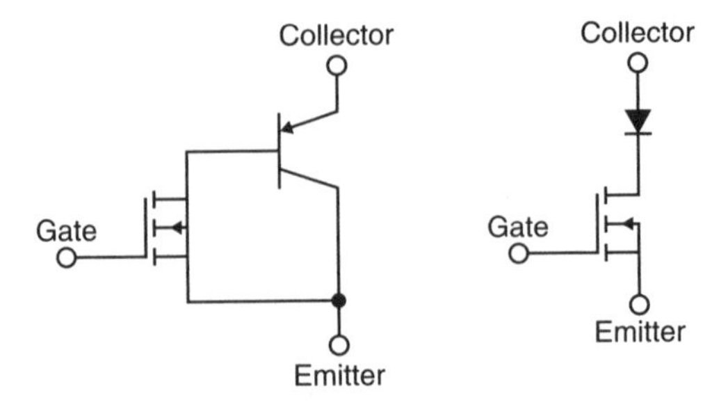

FIGURE 9.21 IGBT simplified equivalent circuits.

9.3.2.1 Equivalent Circuits

The right side circuit shows a diode in series with the drain of an N-channel power MOSFET. In this arrangement It would appear that the "on" state voltage across the combination would be one diode drop higher than it would be for the N-channel power MOSFET without the diode. However, although it is true that the "on" state voltage across the combined IGBT is always at least one diode drop, when compared to a power MOSFET of the same die size, operating at the same temperature and current, the combination IGBT can have significantly lower "on" state voltage drop.

9.3.3 Performance Characteristics of IGBTs

A normal N-channel MOSFET is a majority carrier device so that only electrons contribute to the current flow. In the IGBT combination, the p-type substrate injects holes (minority carriers) into the drift region so that current flow in the IGBT is composed of both electrons and holes. This injection of holes significantly reduces the effective resistance in the drift region, significantly increasing the conductivity. The reduction in "on" state voltage drop and the high gate resistance are the main advantages of IGBTs over discrete power MOSFETs and power bipolar transistors. However, nothing comes for free and the price for lower "on" state voltage is slower switching speed due to higher switching loss, especially at turn "off," due to current tailing.

9.3.3.1 Turn "Off" Characteristics of IGBTs

During turn "off," the electron flow can be stopped quite rapidly by reducing the gate-emitter voltage below the threshold voltage, just as it would be in a power MOSFET. However, the holes are left in the drift region, and the only way to remove them is by recombination and voltage gradient drift. As a result the IGBT exhibits a tail current during turn "off" until all the holes are swept out or recombined.

For many years this current tail current was the parameter that limited the IGBT to low frequency applications. However, the rate of recombination can be controlled by the addition of an N+ buffer layer as shown in Figure 9.20. This buffer layer quickly absorbs trapped holes during turn "off" and modern PT-type IGBTs now provide excellent high frequency performance.

9.3.3.2 The Difference Between PT- and NPT-Type IGBTs

Not all IGBTs incorporate an N+ buffer layer; those that do are called *punch-through (PT),* or *asymmetrical IGBTs.* Those without an N+ buffer layer are called *non-punch-through (NPT),* or *symmetrical IGBTs.*

9.3.3.3 The Conduction of PT- and NPT-Type IGBTs

For a given switching speed, the NPT technology generally has higher $V_{CE(on)}$ ratings than PT technology. This difference is magnified further by the fact that $V_{CE(on)}$ increases with temperature for NPT, giving a positive temperature coefficient, whereas it decreases with temperature for PT devices, giving a negative temperature coefficient. However, for any IGBT, whether PT or NPT, there is a trade off between switching loss and $V_{CE(on)}$. Higher speed IGBTs have a higher $V_{CE(on)}$; lower speed IGBTs have a lower $V_{CE(on)}$. In fact, a very fast PT device can have a higher $V_{CE(on)}$ than an NPT device of slower switching speed.

9.3.3.4 The Link Between Ruggedness and Switching Loss in PT- and NPT-Type IGBTs

For a given $V_{CE(on)}$, PT IGBTs have higher speed switching capability, with lower total switching loss. This is due to higher gain and reduction in minority carrier lifetime, as a result of the N+ buffer layer which quenches the tail current.

While NPT IGBTs are generally slower than PT devices, the NPT type is typically short circuit rated, whereas PT devices often are not. NPT IGBTs can absorb more avalanche energy and are more rugged due to the wider base and lower gain of the PNP bipolar transistor. This is the main advantage gained by trading off the higher switching speed of PT technology with the slower but more rugged NPT technology.

It is difficult to make a PT IGBT with a V_{CES} rating greater than 600 V, whereas it is easily done with NPT technology. However, several manufacturers now offer very fast 1200-V PT type IGBTs.

9.3.3.5 IGBT Latch-Up Possibilities

Latch-up is a failure mode in which the IGBT can no longer be turned "off" by the gate. Latch-up can be induced in any IGBT through misuse. Thus, the latch-up failure mechanism warrants some further explanation.

In IGBTs, a price is paid for the lower "on" state voltage. In a poor device design, there is a possibility of latch-up if the IGBT is operated outside the datasheet ratings. The basic structure of an IGBT resembles a thyristor, namely a PNPN series of junctions. A parasitic NPN bipolar transistor exists within all N-channel power MOSFETS. The base of this transistor is the body region, which is shorted to the emitter to prevent the parasitic NPN transistor from turning "on." However, the body region has some resistance, called "body region spreading resistance." The P-type substrate, drift and body regions form the PNP portion of the IGBT. This PNPN structure forms a parasitic thyristor. As a result, if the parasitic NPN transistor ever turns "on" and the sum of the gains of the NPN and PNP transistors are greater than one, the parasitic thyristor turns "on" and latch-up occurs.

Normally, latch-up is avoided through good design of the IGBT. By optimizing the doping levels and geometries of the various regions (shown in Figure 9.20), the gains of the PNP and NPN transistors are set so that their sum is less than one. However, as the temperature increases, the PNP and NPN gains will increase, as will the body region spreading resistance. Excessive localized heating of the die increases the parasitic transistor gains so their sum exceeds one. Then high collector current can cause sufficient voltage drop across the body region to turn "on" the parasitic NPN transistor. If this happens, the parasitic thyristor latches "on," and the IGBT cannot be turned "off" by the gate. This is static latch-up, and can result in destruction of the IGBT due to over-current heating.

High dv/dt during turn "off," combined with excessive collector current, can also effectively increase gains and turn "on" the parasitic NPN transistor. This is referred to as dynamic latch-up, and it is this effect which actually limits the safe operating area, since it can happen at a much lower collector current than static latch-up and depends mainly on the turn "off" dv/dt.

The bottom line is that by staying within the maximum current and safe operating area ratings, static and dynamic latch-up will be avoided regardless of turn "off" dv/dt. Note that turn "on" and turn "off" dv/dt, overshoot, and excessive ringing can be caused by external stray inductances in the circuit and gate resistor, as well as by poor circuit layout.

9.3.3.6 Temperature Effects

In both PT and NPT IGBTs, the turn "on" switching speed and loss are essentially unaffected by temperature. However, the reverse recovery current in external diodes normally increases with temperature, so the temperature effects of any external diodes in the power circuit can affect IGBT circuit turn "on" loss. For NPT IGBTs, turn "off" speed and switching loss remain relatively constant over the operating temperature range. For PT IGBTs, turn "off" speed degrades and switching loss consequently increases with temperature. However, in the PT device the switching loss is low to begin with, due to good tail current quenching, so this effect is minimal.

9.3.4 Parallel Operation of IGBTs

As mentioned previously, NPT IGBTs typically have a positive temperature coefficient, which makes them well suited for paralleling. A positive temperature coefficient is desirable for paralleling devices because a hot device will conduct less current than a cooler device, so the parallel devices naturally tend to share current equally.

It is a common misconception that PT IGBTs cannot be paralleled because of their negative temperature coefficient. In fact, as shown by the following points, with some care they can be paralleled:

1. Their temperature coefficients tend to be almost zero, and in fact they are sometimes even positive at higher currents.
2. Heat sharing through a common heat sink tends to force devices to share current, because a hot device will heat its neighbors, thus lowering their "on" state voltages.
3. Parameters that affect the temperature coefficient tend to be well matched between devices.

For the power circuit designer, the selection of the appropriate device, as well as predicting its performance in a particular application, is essential for reliable design. Graphs are provided to enable the designer to extrapolate from one set of operating conditions to another. It should be noted that test results are very strongly circuit dependent, especially with respect to stray collector inductance and stray emitter inductance and also on gate drive circuit design and layout. Different test circuits will yield different results.

9.3.5 Specification Parameters and Maximum Ratings

V_{CES} – Collector-Emitter Sustaining Voltage This is the maximum rating of voltage between the collector and emitter terminals with the gate shorted to the emitter at nominal temperature. It is temperature dependent and could actually be less than the V_{CES} rating at high temperature. See also the description of BV_{CES} in Section 9.3.6.

V_{GE} – Gate-Emitter Voltage V_{GE} is the maximum continuous voltage rating between the gate and emitter terminals. The purpose of this rating is to prevent breakdown of the gate oxide and to limit short circuit current.

Normally, the actual gate oxide breakdown voltage is significantly higher than the rating, but staying within this rating at all times ensures application reliability.

V_{GEM} – Gate-Emitter Voltage Transient V_{GEM} is the maximum pulsed voltage between the gate and emitter terminals. The purpose of this rating is to prevent breakdown of the gate oxide. Transients on the gate can be induced not only by the applied gate drive signal, but also more significantly, by stray inductance in the gate drive circuit as well as Miller feedback through the gate-collector capacitance.

If you find there is more ringing in test on the gate than V_{GEM}, stray circuit inductances probably need to be reduced, and/or the gate resistance should be increased to slow down the switching speed. In addition to the power circuit layout, gate drive circuit layout is critical in minimizing the effective gate drive loop area to reduce stray inductances.

If a clamping zener is used, it is recommended to connect it between the gate driver and the gate resistor rather than directly to the gate terminal. Negative gate drive is not essential, but may be used to achieve the utmost in switching speed while avoiding dv/dt induced turn "on."

Continuous Collector Current Ratings I_{C1} and I_{C2} I_{C1} and I_{C2} are ratings of the maximum continuous DC collector current, with the die at 25°C (I_{C1}) and at maximum die temperature (I_{C2}). They are based on the case temperature, the continuous DC collector current, and the junction to case thermal resistance. The limit depends on the internal dissipation that will just cause the die to heat up to its maximum rated junction temperature. These ratings do not include any dissipation caused by switching loss.

Thermal De-Rating I_{C1} and I_{C2} must be de-rated for heat sink temperatures above ambient. To assist designers in the selection of devices

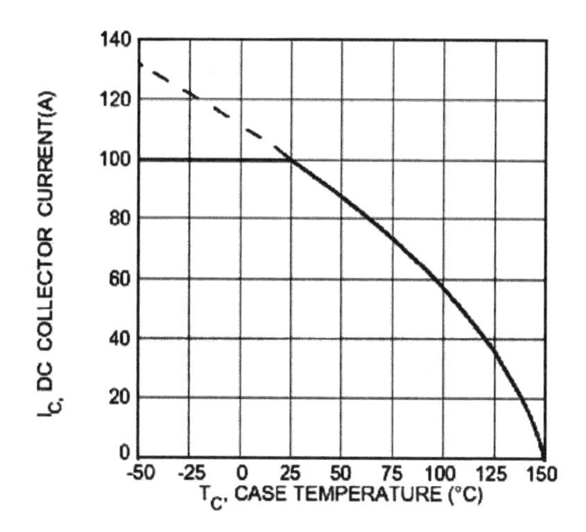

FIGURE 9.22 Typical thermal de-rating of maximum collector current. The horizontal 100-amp limit is an internal connection-related limitation.

for a particular application, most manufacturers provide a graph of maximum collector current versus case temperature.

As an example, Figure 9.22 shows a typical de-rating curve for a 100-Amp APT Power MOS 7 IGBT device. It indicates the maximum theoretical continuous DC current that the device can carry, based on the maximum junction to case thermal resistance and the heat sink (or case) working temperature. Note that in this figure, the package leads limit the current to 100 amps at low temperature, not the die temperature. Since Figure 9.22 does not include switching loss, it serves mainly to provide figures of merit for comparing devices, but it does provide a good starting point for selecting a device. In a hard or soft switching application, the device might safely carry more or less current depending upon the following switching-related losses:

- Switching losses
- Duty cycle
- Switching frequency
- Switching speed
- Heat sinking capacity
- Thermal impedances and transients

One must not assume that the device can safely carry the same current in a switch-mode power converter application as that indicated in the simple I_{C1} or I_{C2} DC ratings, or as shown in Figure 9.22, because the dissipation due to switching loss must be included.

I_{CM} – Pulsed Collector Current This rating indicates how much pulsed current the device can handle. The pulse rating is significantly higher than the continuous DC rating. The purposes for the I_{CM} rating are as follows:

a. To keep the IGBT operating in the "linear" region of its transfer characteristic. See Figure 9.23. There is a maximum collector current that an IGBT will conduct for a corresponding gate-emitter voltage.

   Note *If the operating point at a given gate-emitter voltage goes above the linear region "knee" as shown in Figure 9.23, any further increase in collector current results in significant rise in collector-emitter voltage and consequent rise in conduction loss with possible device destruction. Hence, for typical gate drive voltages, the I_{CM} rating is set below the "knee."*

b. To prevent burnout or latch-up. Even if the pulse width is theoretically too short to overheat the die, significantly exceeding the I_{CM} rating can cause enough localized die feature heating to result in a burnout site or latch-up.

c. To prevent overheating the die. The footnote "Repetitive rating: Pulse width limited by maximum junction temperature" implies that I_{CM} is based on a thermal limitation depending on pulse width. This is always true for two reasons:

   1. There is some margin in the I_{CM} rating to take into account for potential damage factors other than exceeding maximum junction temperature.

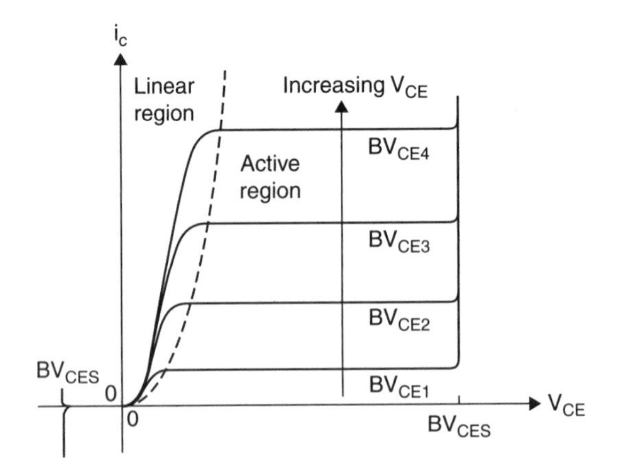

FIGURE 9.23 IGBT transfer characteristic.

2. No matter what the failure mechanism really is, overheating is almost always the observed end result anyway.

Regarding the thermal limitation on I_{CM}, the temperature rise depends upon several factors as follows:

- The pulse width
- The time between pulses
- The heat dissipation
- The voltage drop, $V_{CE(on)}$
- The shape and magnitude of the current pulse

Simply staying within the I_{CM} limit does not ensure that the maximum junction temperature will not be exceeded.

I_{LM} RBSOA, FBSOA, and Switching Safe Operating Area (SSOA) These ratings are all related. I_{LM} is the clamped inductive load current the device can safely switch in a snubberless hard switching application. The circuit conditions for this rating are as specified by the manufacturer. They include case temperature, gate resistance, and clamp voltage. The I_{LM} rating is limited by the turn "off" transient, given that the gate was positive-biased and switches to zero or negative bias. Hence the I_{LM} rating and the Reverse Bias Safe Operating Area (RBSOA) are similar. The I_{LM} rating is a maximum current, while the RBSOA boundary is a set of maximum currents specified voltages.

Switching safe operating area (SSOA) is simply RBSOA at the full V_{CES} voltage rating. Forward bias safe operating area (FBSOA), which covers the turn "on" transient, is typically much higher than the RBSOA, so it is not normally listed in IGBT datasheets. In terms of IGBT reliability, the circuit designer does not need to worry about snubbers, minimum gate resistance, or limits on dv/dt as long as the above ratings are not exceeded.

E_{AS} – Single Pulse Avalanche Energy Any device that is avalanche energy rated should have an E_{AS} rating. Avalanche energy rated is synonymous with unclamped inductive switching (UIS) rated. E_{AS} is both thermally limited and defect limited and indicates how much reverse avalanche energy the device can safely absorb with the case at 25°C and with the die at or below the maximum rated junction temperature. In modern devices the cell structure mitigates the defect limitation on E_{AS}. On the other hand, a defect in a closed cell structure can cause the cell to latch-up under avalanche conditions. So do not operate an IGBT intentionally in the avalanche region without thorough testing.

The E_{AS} rating is equal to $\frac{1}{2}L I_C^2$, where L is the value of any external inductor carrying a peak current I_C.

For testing the E_{AS} rating, inductor current is suddenly diverted into the collector of the device under test. At that time, the inductor's voltage will exceed the breakdown voltage of the IGBT, and drive it into the avalanche condition. The avalanche condition allows the inductor current to flow through the IGBT, even though the IGBT is in the fully "off" state.

Energy stored in the external test inductor is analogous to energy stored in any circuit-related leakage and/or stray inductances and is dissipated in the device under test. In an application, if ringing due to leakage and stray inductances does not exceed the breakdown voltage, then the device will not avalanche and hence does not need to dissipate avalanche energy. Avalanche energy rated devices offer a safety margin between the voltage rating of the device and system voltages, including transients.

P_D – Total Power Dissipation This is a rating of the maximum power that the IGBT device can dissipate and is based on the maximum junction temperature and the thermal resistance from junction to case, with the case maintained at a temperature of 25°C (an infinite heat sink).

$$P_D = (T_{J(max)} - 25°C)\, R_{\theta JC}$$

T_J, T_{STG} – Operating and Storage Junction Temperature Range This is the range of permissible storage and operating junction temperatures. The limits of this range are set to ensure a minimum acceptable device service life. Operating within the limits can significantly enhance service life. As a "rule of thumb," for thermally induced effects every 10°C reduction in the junction temperature below the upper limit doubles the device life.

9.3.6 Static Electrical Characteristics

BV_{CES} – Collector-Emitter Breakdown Voltage BV_{CES} has a positive temperature coefficient, rising about 10% from 25°C to 150°C. At a fixed leakage current, an IGBT can block more voltage when hot than when cold.

RBV_{CES} – Reverse Collector-Emitter Breakdown Voltage This is the reverse collector-emitter breakdown voltage specification, i.e. when the emitter voltage is positive with respect to the collector. In IGBTs, RBV_{CES} is not normally specified, since an IGBT is not designed for reverse voltage blocking. A PT-type IGBT cannot block very much reverse voltage due to the N+ buffer layer.

$V_{GE(th)}$ – Gate Threshold Voltage This is the gate-source voltage at which collector current begins to flow. Test conditions (collector current, collector-emitter voltage, junction temperature) are also specified. All MOS gated devices exhibit variation in $V_{GE(th)}$ between devices, which is normal. Therefore, a range of $V_{GE(th)}$ is specified, with the minimum and maximum representing the edges of the $V_{GE(th)}$ distribution. $V_{GE(th)}$ has a negative temperature coefficient, meaning that as the die heats up, the IGBT will turn "on" at a lower gate-emitter voltage. This temperature coefficient is typically about minus 12 mV/°C, the same as for a power MOSFET.

$V_{CE(on)}$ – Collector-Emitter "On" Voltage This is the collector-emitter voltage across the IGBT at a specified collector current, gate-emitter voltage, and junction temperature. Since $V_{CE(on)}$ is temperature dependent, it is specified both at room temperature and hot. Most manufacturers provide graphs that show the relationship between typical collector-emitter voltage and collector current, temperature, and gate-emitter voltage. From these graphs, a circuit designer can estimate conduction loss and the temperature coefficient of $V_{CE(on)}$. Conduction power loss is $V_{CE(on)}$ times collector current I_C. The temperature coefficient is the slope of $V_{CE(on)}$ versus temperature. NPT IGBTs have a positive temperature coefficient, meaning that as the junction temperature increases, $V_{CE(on)}$ increases. PT IGBTs on the other hand tend to have a slightly negative temperature coefficient. For both types, the temperature coefficient tends to become more positive with increasing collector current. As current increases, the temperature coefficient of a PT IGBT can transition from negative to positive.

I_{CES} – Collector Cutoff Current This is the leakage current that flows from collector to emitter when the device is "off," at a specified collector-emitter and gate-emitter voltage. Since leakage current increases with temperature, I_{CES} is specified both at room temperature and hot. Leakage power loss is I_{CES} times the collector-emitter voltage.

I_{GES} – Gate-Emitter Leakage Current This is the leakage current that flows through the gate terminal at a specified gate-emitter voltage.

9.3.7 Dynamic Characteristics

Input, output, and reverse transfer capacitances are combinations of the capacitances shown in Figure 9.24.

C_{ies} – Input Capacitance C_{ies} is the input capacitance measured between the gate and emitter terminals with the collector shorted to the

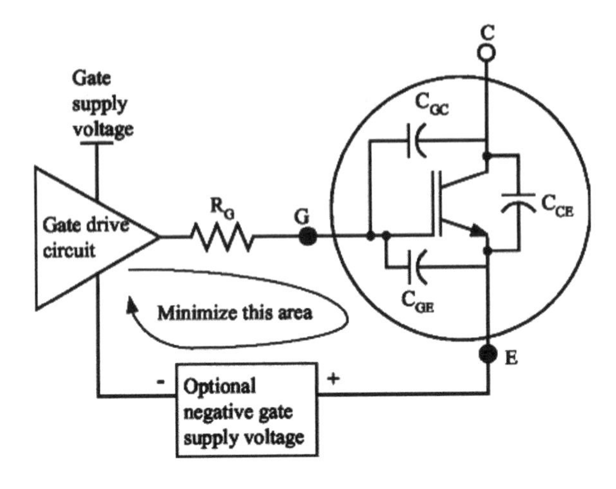

FIGURE 9.24 Internal IGBT capacitances. This shows an equivalent IGBT model that includes the capacitances between the terminals.

emitter for AC signals. C_{ies} is made up of the gate to collector capacitance C_{GC} in parallel with the gate to emitter capacitance C_{GE}, so $C_{ies} = C_{GE} + C_{GC}$.

The input capacitance must be charged to the threshold voltage before the device begins to turn "on," and discharged to the plateau voltage before the device begins to turn "off." Therefore, the impedance of the drive circuitry and the value of C_{ies} have a direct relationship to turn "on" and turn "off" delays.

C_{oes} – Output Capacitance C_{oes} is the output capacitance measured between the collector and emitter terminals with the gate shorted to the emitter for AC voltages. C_{oes} is made up of the collector to emitter capacitance (C_{CE}) in parallel with the gate to collector capacitance (C_{GC}), so $C_{oes} = C_{CE} + C_{GC}$.

For soft switching applications, C_{oes} is important because it can affect the resonance of the circuit.

C_{res} – Reverse Transfer Capacitance C_{res} is the reverse transfer capacitance measured between the collector and gate terminals with the emitter connected to ground. The reverse transfer capacitance is equal to the gate to collector capacitance, so $C_{res} = C_{GC}$. The reverse transfer capacitance, often referred to as the Miller capacitance, is one of the major parameters affecting voltage rise and fall times during switching.

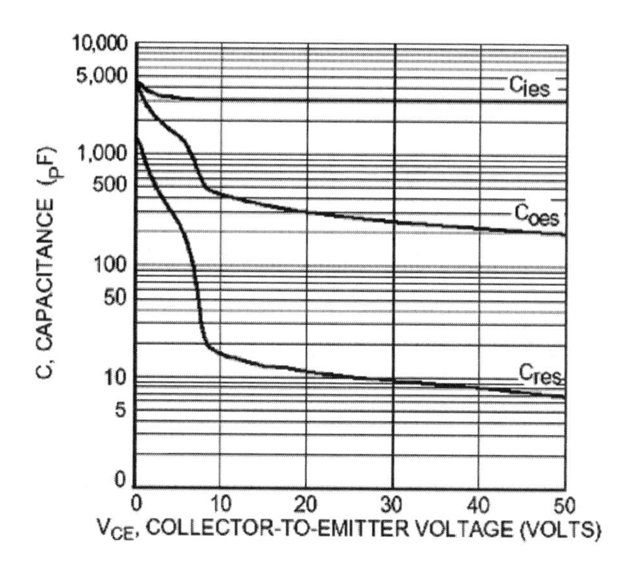

FIGURE 9.25 Reverse transfer capacitance (C_{res}) versus collector-emitter voltage.

Figure 9.25 shows a typical example of C_{res} versus collector-emitter voltage. The capacitances decrease over a range of increasing collector-emitter voltages, especially the output and reverse transfer capacitances. This variation is the basis for gate charge data.

V_{GEP} – Plateau Voltage Figure 9.26 shows the gate-emitter voltage as a function of gate charge. The method for measuring gate charge is described in JEDEC standard 24-2. The gate plateau voltage V_{GEP} is defined as the gate-emitter voltage when the slope of the gate-emitter voltage first reaches a minimum during the turn "on" switching transition for a constant gate current drive condition. In other words, it is the gate-emitter voltage where the gate charge curve first straightens out after the first inflection in the curve, as shown in Figure 9.26. Alternatively, V_{GEP} is the gate-emitter voltage at the last minimum slope during the turn "off." The plateau voltage increases with current but not with temperature. Beware when replacing power MOSFETs with IGBTs. A 10- or 12-V gate drive may work fine for a high voltage power MOSFET, but depending upon its plateau voltage, an IGBT at high current might switch surprisingly slowly or not completely during turn "on," unless the gate drive voltage is increased.

Q_{GE}, Q_{GC}, and Q_G – Gate Charge Referring to Figure 9.26, Q_{GE} is the charge from the origin to the first inflection in the curve, Q_{GC}

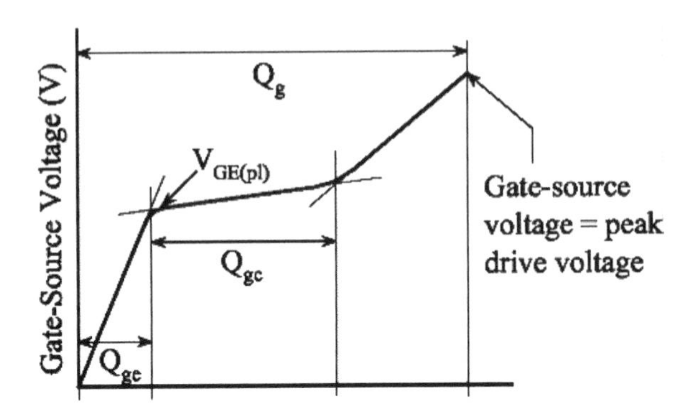

FIGURE 9.26 V_{GE} as a function of gate charges Q_{GE}, Q_{GC}, and Q_G – gate charge.

(also known as the "Miller" charge) is the charge from the first to second inflections in the curve, and Q_G is the charge from the origin to the point on the curve at which V_{GE} equals the peak drive voltage. Gate charge values vary with collector current and collector-emitter voltage but not with temperature. Test conditions are specified, and a graph of gate charge is typically included in the datasheet showing gate charge curves for a fixed collector current and different collector-emitter voltages. The gate charge values reflect charges stored on the inter-terminal capacitances described earlier. Gate charge is often used for designing gate drive circuitry, since it takes into account the changes in capacitance with changes in voltage during a switching transient.

Switching Times and Energies In general, turn "on" speed and energy are relatively independent of temperature; they increase in speed (decrease in energy) very slightly with increasing temperature. External diode reverse recovery current increases with temperature, resulting in the increase in E_{on2} with temperature. E_{on1} and E_{on2} are defined below. The turn "off" speed decreases with increasing temperature, corresponding to an increase in turn "off" energy. Both turn "on" and turn "off" switching speeds decrease with increasing gate resistance, corresponding to an increase in switching energies. Switching energy can be scaled directly for variation between application voltage and the datasheet switching energy test voltage. If the datasheet tests were done at 400 V, for example, and the application is at 300 V, simply multiply the datasheet switching energy values by the ratio 300/400 to interpolate.

Switching times and energies also vary strongly with stray inductances in the circuit, including the gate drive circuit. In particular, stray inductance in series with the emitter significantly affects switching times and energies. Therefore, switching time and energy values given in a datasheet are representative only and may differ from observed results in an actual power supply.

$t_{d(on)}$ – Turn "on" Delay Time Turn "on" delay time is the time from when the gate-emitter voltage rises past 10% of the drive voltage to when the collector current rises past 10% of the specified current.

$t_{d(off)}$ – Turn "off" Delay Time Turn "off" delay time is the time from when the gate-emitter voltage drops below 90% of the drive voltage to when the collector current drops below 90% of the specified current.

t_r – Current Rise Time Current rise time is the time it takes for the collector current to rise from 10% to 90% of that specified.

t_f – Current Fall Time Current fall time is the time it takes for the collector current to drop from 90% to 10% of that specified.

g_{fe} – Forward Transconductance Forward transconductance is the ratio of collector current to gate-emitter voltage. Forward transconductance varies with collector current, collector-emitter voltage, and temperature. High transconductance leads to low plateau voltage and fast current rise and fall times.

Both MOSFETS and IGBTs exhibit relatively high gain at high gate-emitter voltages. However, unlike the MOSFET, the IGBT retains control of current even at high gate voltages and high currents as shown in Figure 9.27.

In IGBTs, increasing the gate-emitter voltage increases the flow of both electrons and holes, modulating the effective resistance of the junction. This parameter provides a simple means for detecting and protecting against a transient overcurrent condition well above the intended working range.

Effective transient overload current protection can be implemented by turning the drive to the IGBT "off" if a current stress drives the device well above its intended working range. This is done by limiting the maximum gate voltage and detecting when the collector to emitter voltage starts to rise, due to the transient increase in collector current moving the IGBT into the intrinsic current limited region shown in Figure 9.9. However, this method is not so effective with power MOSFETs because the drain current is very insensitive to gate voltage once the device is fully "on."

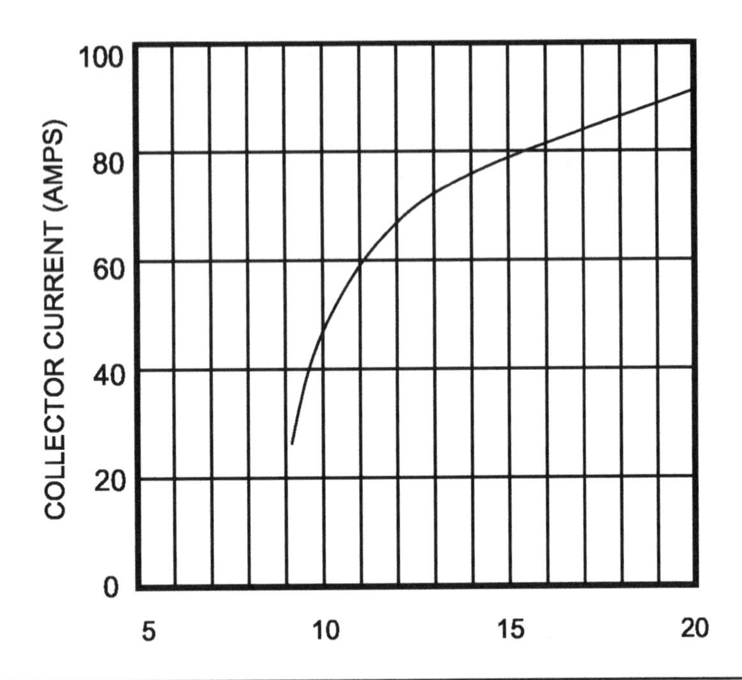

FIGURE 9.27 Typical collector current versus gate voltage overcurrent protection.

9.3.8 Thermal and Mechanical Characteristics

$R_{\theta JC}$ – Junction to Case Thermal Resistance This is the thermal resistance from the junction of the die to the outside of the device case. Heat is the result of the total power lost in the device substrate, and thermal resistance relates how hot the die gets based on this power

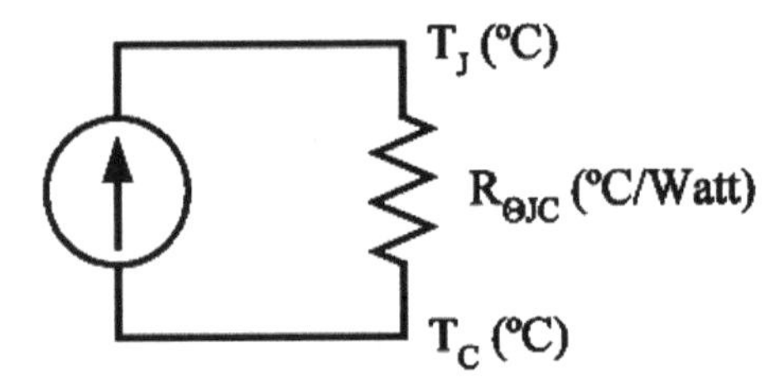

FIGURE 9.28 Thermal resistance model.

loss. It is called thermal resistance because an electrical model is used to predict temperature rise based on steady state power loss as shown in Figure 9.28.

Device Power Loss (W) In Figure 9.28, power loss is modeled as current flowing through a thermal resistance, resulting in a voltage rise. This voltage rise is the analogue of temperature rise, and additional resistors could be added in series to model case-to-sink and sink-to-ambient thermal resistances. The temperatures at various physical locations are analogous to the voltages at the same nodes in the thermal resistance circuit model. Thus, on a steady-state basis, junction temperature can be calculated as

$$T_J = T_C + P_{Loss}(R_{\theta JC})$$

Device power loss is the sum of averaged switching, conduction, and leakage losses. Typically, leakage losses can be ignored. Since case-to-sink and sink-to-ambient thermal resistances depend entirely upon the application (thermal compounds, heat sink type, etc.), only $R_{\theta JC}$ is specified in the datasheet. The thermal resistance from case-to-free air typically includes the mounting hardware, and heat sink properties must be included to get the total effective resistance $R_{\theta JA}$ for temperature rise predictions. Ratings such as maximum continuous DC current, total power dissipation, and frequency versus current are based on a maximum $R_{\theta JC}$. The maximum $R_{\theta JC}$ is used because it incorporates margin to account for normal manufacturing variations and to provide some application margin as well.

$Z_{\theta JC}$ – Junction to Case Thermal Impedance Thermal impedance is the dynamic analogue of thermal resistance. Thermal impedance takes into account the heat capacity (or specific heat) and mass of the material used to form the substrate or die. It relates the temperature rise of the die material itself to the instantaneous dissipation within the die material. For longer periods more of the heat is conducted away from the die, reducing the effective thermal impedance. This can be seen in Figure 9.29. Thermal impedance can be used to estimate instantaneous junction temperatures resulting from power loss, caused by pulse loading conditions of various duty periods, or on a transient basis.

Transient thermal impedance is determined by applying power pulses to the device of various magnitudes and durations. The result is the transient impedance family of curves, an example of which is shown in Figure 9.29. Note that the family of curves is based on the maximum $R_{\theta JC}$, which incorporates safety margin as discussed

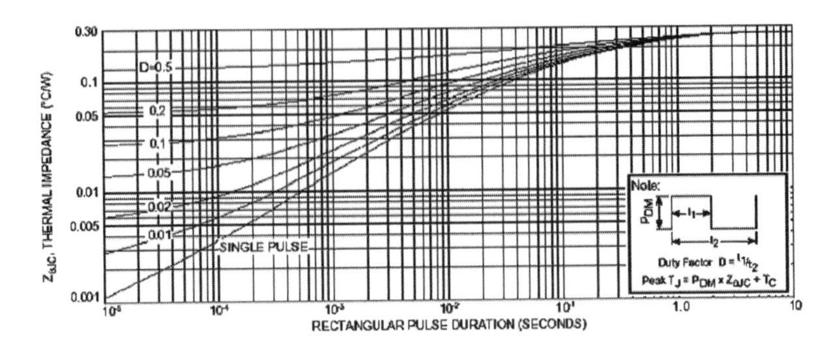

FIGURE 9.29 Typical junction to case thermal impedance.

previously. The method of calculating peak junction temperature is shown in Figure 9.29. For nonrectangular power pulses, a pulse by pulse linear approximation must be used.

In an inductive hard switching application, switching frequency is limited by minimum and maximum pulse widths as well as conduction and switching losses. The pulse width limitation is due to transient thermal response in the die. Back-to-back switching transients do not allow the die time to cool between the large hard-switching power loss spikes. Also, not allowing the switching transient to complete before switching the other way repetitively can overheat the die. Depending upon operating temperatures and transient thermal impedance, the die junction may become overheated even if the duty cycle is very small. The minimum duty cycle limitation is a challenge for motor drives such as in an electric vehicle, in which an exceptionally small duty cycle is required at very low power unless the switching frequency is dropped into the audible range or some type of pulse skipping scheme is implemented.

A parameter, developed from the above, that is more useful to the designer is the relationship between the collector current and the maximum working frequency as shown in Figure 9.30.

The usable frequency versus current curve, shown in Figure 9.30, is one of the more useful items in the datasheet. Even though it is limited to certain conditions specified in the datasheet, in general it provides a realistic indication of how the device will perform in a particular application. The trend in the industry is towards using this parameter as a figure of merit for comparing devices rather than relying so much on I_{C1} and I_{C2} ratings.

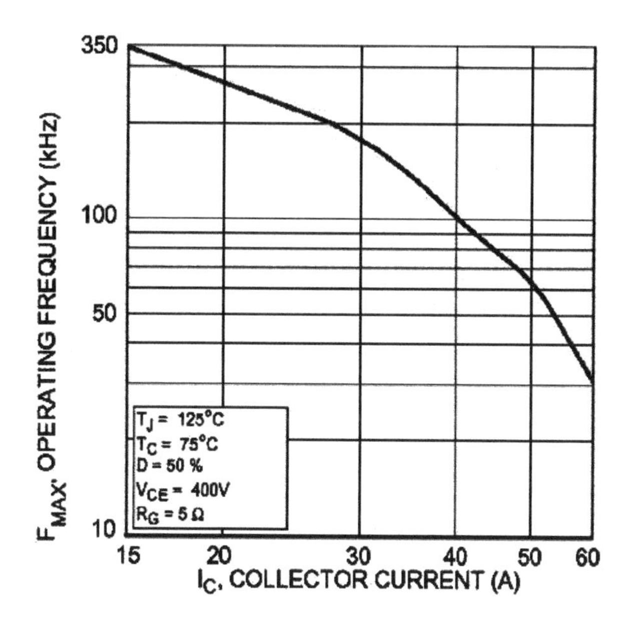

FIGURE 9.30 Typical maximum working frequency versus collector current.

To arrive at a frequency limit based on minimum pulse width, one major manufacturer (APT)[10] defines the minimum limit on pulse width such that the total switching time (the sum of turn "on" and turn "off" switching times) must be no more than 5% of the switching period. This is a reasonable limitation in most cases that can be verified by transient thermal analysis. The question is: what is the total switching time? It can be estimated by adding the turn "on" and turn "off" current delay times and current rise and fall times, which gives a good approximation of total switching time. The voltage fall time during turn "on" is not accounted for, but this is relatively short. The limitation on total switching time of 5% of the switching period provides plenty of margin for this approximation. The frequency is typically limited thermally except at very low current.

The Effect of Gate Source Resistance Switching loss is also a function of gate source resistance, as shown in Figure 9.31. Increasing resistance slows down switching speed, as it takes longer to charge and discharge the gate input capacitance, increasing the switching loss.

"On" state voltage temperature effects are shown in Figure 9.32.

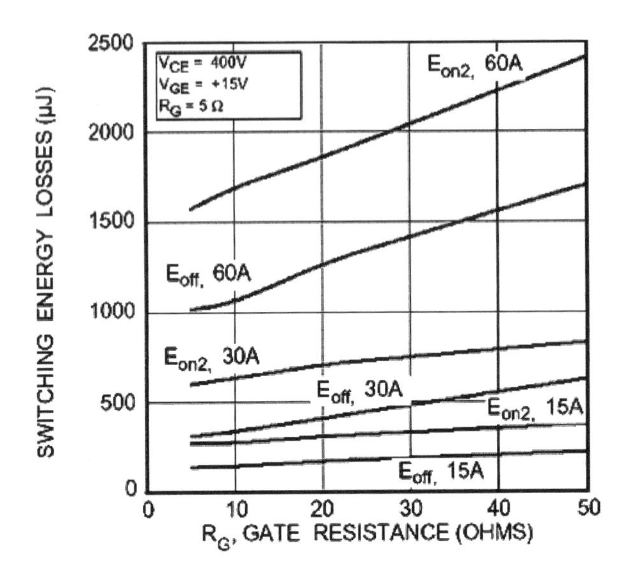

FIGURE 9.31 Typical switching loss versus gate source resistance.

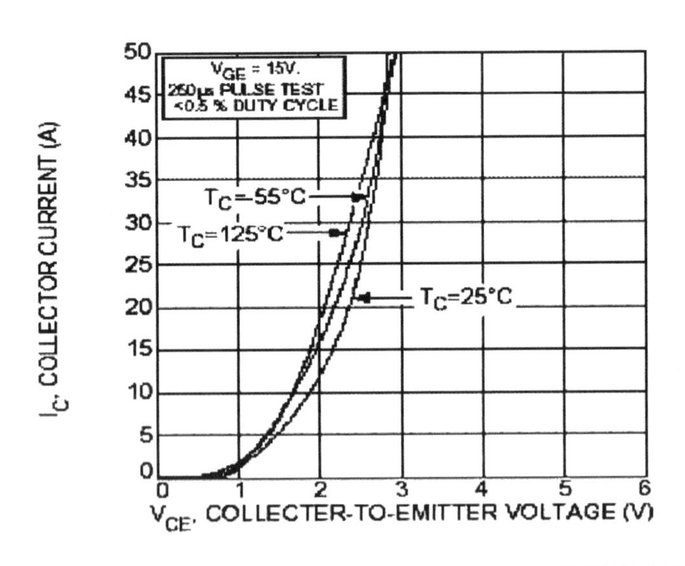

FIGURE 9.32 Typical collector emitter voltage versus collector current.

References

1. *International Rectifier Power MOSFET Data Book*, HDB-3, International Rectifier, El Segundo, CA.
2. *Power MOSFET Transistor Data*, Motorola Inc., Motorola Literature Distribution, Phoenix, AZ.
3. *Linear Integrated Circuits Data Book—PWM chip UC3525A*, Unitrode Corp., Merrimack, NH.
4. Reference 1, Chapters 1–4.
5. Reference 2, Chapter 2.
6. "Radiation Resistance of Hexfets," Reference 1, p. B10.
7. Reference 1, Chapter 1; Reference 2, Chapter 7; International Rectifier Application Note AN-941.
8. "Hexfet's Integral Body Diode," Reference 1, p. A65.
9. Reference 1, p. A65.
10. Jonathan P.E. Dodge and John Hess, Advanced Power Technology Application Note, APT0201, July 2002.

CHAPTER 10

Magnetic-Amplifier Postregulators

10.1 Introduction

TIP *Although "magnetic amplifier" has become the term generally used for this application, the action is more correctly that of a saturable reactor (or magnetic switch). True magnetic amplifiers control large AC currents on a main power winding with smaller currents on one or more control windings. A very interesting range of magnetic amplifier devices were in common use well before semiconductors were available. Unfortunately, a study of these is outside the scope of this book. ~K.B.*

In Chapter 2, Sections 2.2.1 and 2.3.3, multiple output voltage push-pull and forward converter topologies were discussed. As was described, in either circuit, a feedback loop is usually closed around a main or "master" (usually the highest current or 5-V) output. The feedback loop keeps the master output constant against line or load changes.

Additional secondaries on the power transformer yield "slave" output voltages that are proportional to their respective numbers of turns. These slaves operate in a semiregulated mode. The "on" time, or duration of conduction in such secondaries, is being defined by the master feedback loop that is acting to keep the master output voltage constant. Hence, the "on" time is largely independent of the slave output currents and is inversely proportional to the DC supply voltage. As a result the slave voltages are as well regulated against input line changes as the master.

However, the slaves are not well regulated against load current changes—either in the master or in themselves. Slave output voltage change due to current changes in the master is referred to as *cross regulation*, and may be as high as ±8% for the maximum specified current change in the master.

Slave output changes due to current change in themselves are considerably less so long as the master or slave output inductor does not enter discontinuous conduction (Section 2.2.4). If either the master or an individual slave output inductor enters discontinuous mode by decreasing its output current, that slave DC output voltage may change by up to 50%.

TIP *Much better cross regulation and a wider current range can be obtained by using a coupled output choke (a single common output choke with all secondaries wound on a single core).*[17] *~K.B.*

If the inductances are chosen to be very large, they can be kept in continuous mode for a wider current range, but this causes larger and longer lasting output voltage transients in response to step output current changes.

A final drawback to open-loop slave outputs is that their voltages are not precisely controlled, and can be set only to within a few percent of a specific value.

The preciseness of setting depends in part on the volts per turn of the transformer core. Further, since both the primary and secondary number of turns can be changed only by an integer, output voltage can be changed only in coarse steps. From Faraday's law, moreover, since the volts per turn is directly proportional to switching frequency, the coarseness of these steps increases with increasing frequency.

This type of multi-output supply with the master output well regulated against line and load changes, but with slaves poorly regulated against master or slave load current changes, is nevertheless widely used.

Usually it is only the master—typically a 5-V output that feeds crucial logic circuits—which must be well regulated against line and load changes. Slaves usually feed motors on disk or tape drives, or error amplifiers. Such loads often can tolerate a DC voltage which is 1 or 2 V off a specified nominal value. For motor drives, this changes the motor acceleration times only slightly. For various linear circuits, it changes internal dissipation only somewhat.

Yet there are numerous applications where the slave outputs must be precisely set to a specified value, and must be well regulated (better than 1%) against line and load changes. Typically, when slaves well regulated against line and load changes are required, the solution is to postregulate a semiregulated slave with either a linear regulator for output currents under 1.5 A, or a buck regulator for higher output currents.

These approaches have their merits and drawbacks, which are discussed below. A better solution for poorly regulated slaves is the magnetic-amplifier postregulator.[1,2] It uses an old basic technique, but with a simpler circuit and better magnetic material, it made a dramatic reappearance in the mid-eighties and has rapidly been adopted throughout the industry.

10.2 Linear and Buck Postregulators

A linear postregulator is the best approach for output currents up to 1.5 A, because of its low cost and acceptably low internal dissipation.

Inexpensive linear regulators with up to 1.5 A of output current are available as integrated circuits in plastic TO220 packages. They require no additional external components other than a small filter capacitor.

They are usually specified for a minimum input-output differential, or headroom, of 2 or 3 V as discussed in Section 1.2.3. Thus, for 1.0 A of output current with 3 V of headroom, the internal dissipation is 3.0 W.

They are also available as integrated circuits at much higher currents in metal TO66 or TO3 cases. However, they are not widely used at currents above 1.0 A—not so much because of excessive junction temperature, for that can be handled by heat-sinking—but rather because of the excessive dissipation and consequent inefficiency due to the 3-V headroom requirement.

Integrated-circuit linear regulators with only 0.5- to 1.0-V headroom (Section 1.2.5) are available, but these are considerably more expensive.

Postregulators for output currents above 1.5 A or so are most often implemented as buck regulators. The slave output voltage is usually set to a minimum of about 4 V above the desired output, which is then bucked down (Section 1.3.1) to the desired output voltage.

This yields higher efficiency than a linear step-down regulator, but it is more expensive, more complicated, and bigger. Further, the buck transistor introduces an additional source of RFI, and may produce beats against the main switching frequency if it is not synchronized, causing problems in other parts of the frequency spectrum.

The magnetic-amplifier postregulator, discussed next, is a better approach than a buck regulator at currents over 1.5 A, and is a credible alternative even at lower currents.

10.3 Magnetic Amplifiers—Introduction

Referring to Figures 2.1 and 2.10, it is seen that the main output (V_{om}) is controlled by adjusting the effective conduction period of *D*4 (or more correctly the duty cycle) by adjusting the period that forward voltage is applied to the secondary winding. The rectangular pulse chain from *D*4 is averaged by the inductor and output capacitor and provides a stabilized output voltage by closed loop control of the duty cycle. However, the slave output voltages (V_{os}) are not well regulated. This is because they are operated open-loop, and their duty cycle is that of the master, which is determined by the master feedback loop. If each slave duty cycle were independently controlled through separate feedback loops, they too could have constant output voltage.

The duty ratio of the slave secondary of a forward converter can be controlled by a generic switch $S1$ as in Figure 10.1. Switch $S1$ is shown in series between the slave secondary winding and the output rectifying diode $D1$. By independently controlling the duty cycle of the pulse chain applied to the slave LC filter, the slave DC output voltage can be controlled. Notice $S1$ can only reduce the applied duty period and hence can only reduce the output voltage. Therefore, the secondary voltage for the slave output must be greater than otherwise required under all conditions, and is reduced to the required output by reducing the conduction period.

Assume that the conduction time at the slave secondary output applied to $S1$ is t_h out of a period T as in Figure 10.1. The time t_h is set by the main feedback loop to keep the main DC output voltage V_{om} constant. Thus

$$V_{om} = \left[(V_{dc} - V_{ce})\frac{N_{sm}}{N_p} - V_{D4}\right]\frac{t_h}{T} \tag{10.1a}$$

$$V_{om} \approx \left[(V_{dc} - 1)\frac{N_{sm}}{N_p} - 1\right]\frac{t_h}{T} \tag{10.1b}$$

Of this time t_h, assume that $S1$ is open and blocking the slave secondary voltage V_{sp} from getting through to $D1$ anode for a time t_b (Figure 10.1*c*), and that $S1$ is closed with zero resistance for a firing time t_f (Figure 10.1*d*). The slave *DC* output voltage is then

$$V_{os} = (V_{sp} - V_{D1})\frac{t_f}{T} \tag{10.2a}$$

$$V_{os} \approx (V_{sp} - 1)\frac{t_f}{T} \tag{10.2b}$$

Now

$$t_f + t_b = t_h, \text{ or}$$

$$t_f = t_h - t_b \tag{10.3}$$

The slave output voltage will be kept constant, as indicated by Eq. 10.2, by controlling t_f. But the physical nature of the switch $S1$ is such that it is not t_f directly that is controlled, but rather t_b, the blocking time.

Thus in Eq. 10.2*a*, if the peak secondary voltage $V_{sp} = (V_{dc} - V_{Q1})(N_s/N_p)$ increases, because the supply V_{dc} increases, t_f will be decreased by increasing t_b or by cutting away a larger piece of the front end of the t_h pulse.

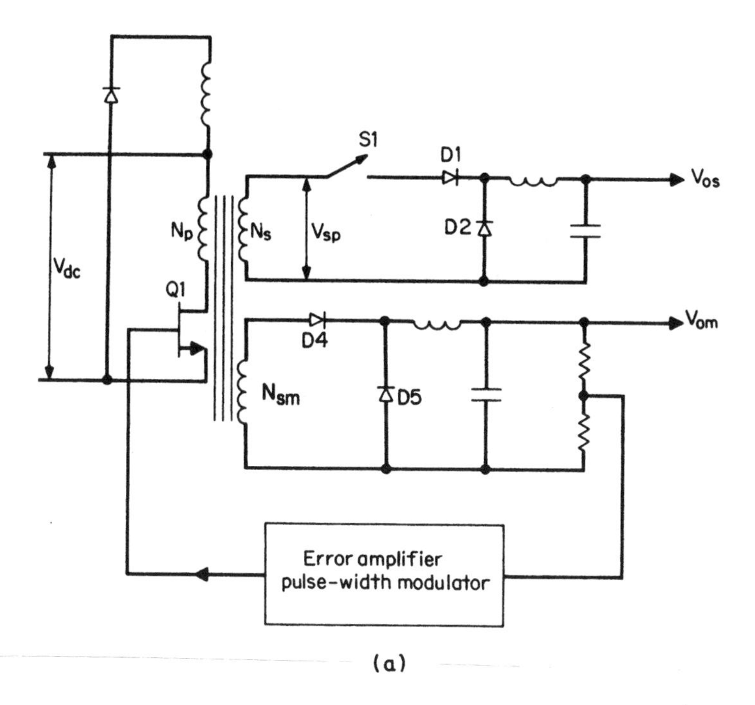

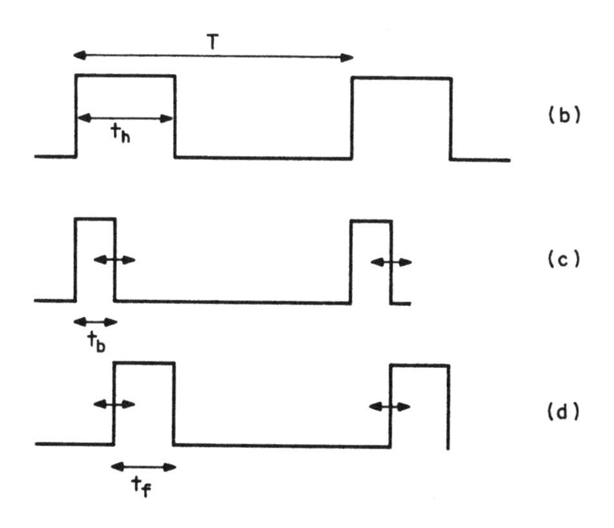

FIGURE 10.1 Width modulation of a slave secondary pulse with a generic switch *S*1. When the open and closed times of *S*1 are controlled by a separate independent feedback loop, the slave output voltage is regulated independent of the master.

In this example, the switch $S1$ is to be a *magnetic amplifier* (or more correctly a saturable reactor; see Ref. 1). It consists simply of a toroidal magnetic core of square hysteresis loop material with a few turns of wire. It works as follows.

10.3.1 Square Hysteresis Loop Magnetic Core as a Fast Acting On/Off Switch with Electrically Adjustable "On" and "Off" Times

Figure 10.2 shows the *BH* loop of a typical square hysteresis loop material (Toshiba MB amorphous core material[3]). Other square-loop materials usable in magnetic amplifiers will be discussed below.

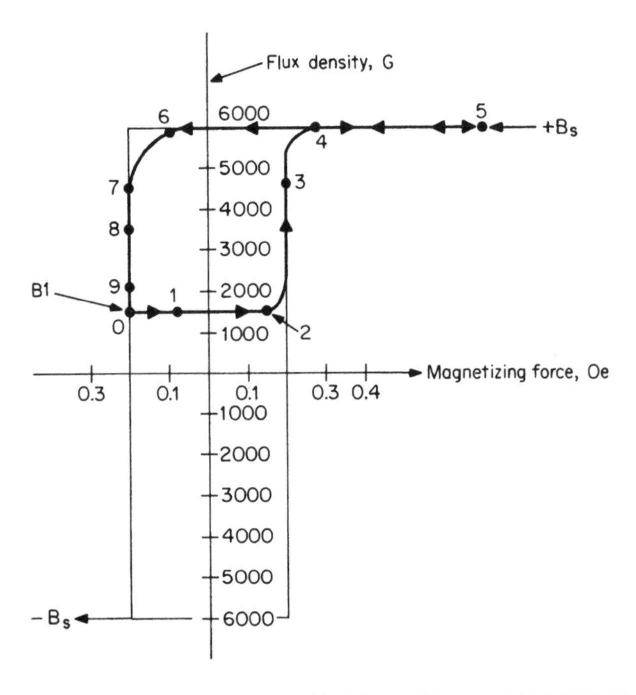

FIGURE 10.2 Toshiba MB amorphous core, *BH* loop at 100 kHz. In magnetic-amplifier operation, the core moves along a minor loop 01234567890. In going from 1 to 4, the core is on the steep part of the hysteresis loop and the magnetic amplifier MA (see Figure 10.3) has high impedance. At point 4, the core saturates and the MA has essentially zero impedance. At the end of the $Q1$ "on" time (Figure 10.1), the core is reset to B_1. The time to move from B_1 to $+B_s$ is the switch-open time. The further down B_1 is pushed, the longer the blocking or switch-open time. The level to which B_1 is reset is determined by the current forced into the no-dot end of MA by $Q2$ (Figure 10.3). That current is controlled by the error amplifier.

TIP *In Figure 10.2 the vertical scale is flux density* B, *which is a function of the applied volt seconds per turn. The time it takes to saturate the core (point 0 to 6) is a function of the voltage applied by the secondary winding, the number of turns, and the area of the core. In SI units the time is*

$$t_d = \frac{N \Delta B A_e}{V_s}$$

where t_d = *delay time (microseconds)*
N = *turns*
ΔB = *change in flux density* (*tesla*)
A_e = *effective core area* (mm^2)
V_s = *secondary voltage*

With N, A_e, *and* V_s *fixed*, t_d *is a function of* ΔB *only, and the delay to saturation depends on where the core is on the* B/H *loop at the start of a pulse.*

The horizontal scale is the magnetizing force H, *which is proportional to current. As the core moves from point 0 to point 7 on the vertical scale, the change on the horizontal scale is very small, and the current change is small. The switch is effectively "off." But the change on the vertical scale is proportional to delay time as shown by the equation above. When the core gets to point 7, there is a rapid increase in the movement along the horizontal scale from point 7 to point 5, say, which translates to a rapid increase in current as the core saturates. Hence, the name* saturating reactor. *The switch is now effectively "on." This is explained further in Reference 16. ~K.B.*

The slope of the *BH* loop is its permeability $\mu = dB/dH$. A coil wound around the core has an impedance proportional to the core permeability μ. So long as the core is on the vertical part of its hysteresis loop, its permeability and hence the coil impedance is very high. It is effectively a single-pole switch in the open position.

When the core is in saturation on the horizontal part of its hysteresis loop (beyond point 4 in Figure 10.2), the *BH* loop is so square that the slope dB/dH or permeability is unity. The coil impedance is thus the very low impedance of an air core coil of an equal number of turns. The coil is thus effectively a single-pole switch in the closed position.

Such a core with a few turns of wire constitutes the switch *S*1 of Figure 10.1. It is shown in Figure 10.3 with an error amplifier and the scheme for controlling the "on" and "off" times of the switch.

Throughout one switching cycle (t_0 to t_3 in Figure 10.4) the core moves around a so-called minor hysteresis loop—the path 01234567890 shown in Figure 10.2. The blocking (t_b) and firing (t_f) times of the magnetic core switch are controlled as follows. Assume that at the start of a cycle (t_0), the core has been pushed down to B_1 on Figure 10.2. When *Q*1 turns "on," a voltage $V_{sp} \approx (V_{dc} - 1)N_{ss}/N_p$ appears across the slave secondary.

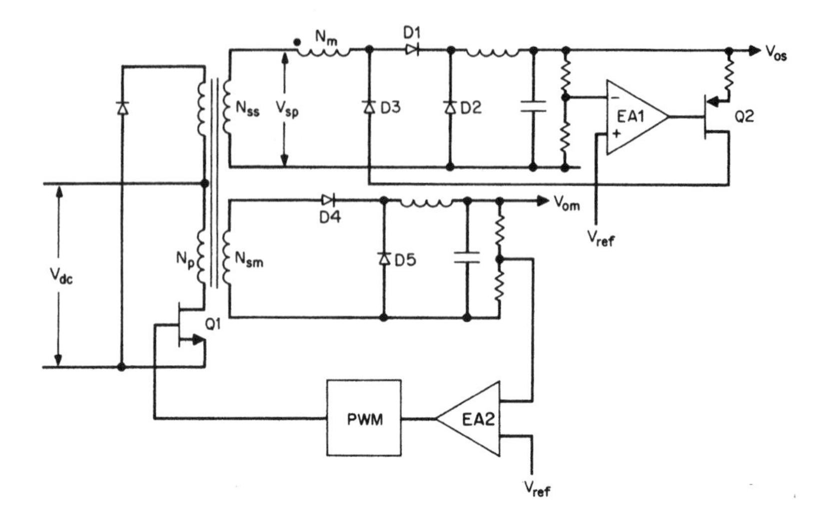

FIGURE 10.3 Magnetic-amplifier regulation of a slave output. The switch *S*1 of Figure 10.1 is implemented with magnetic amplifier MA, shown as N_m in the figure. The MA has high impedance as long as it is below saturation on the steep part of its hysteresis loop. That high-impedance time is determined by the applied forward volt seconds, and where the core is on the B/H loop at the start of the forward pulse. This reset is determined by the amount of reset current pushed into the no-dot end of MA during the Q1 "on" time. The MA is simply a square hysteresis loop core with a few turns of wire.

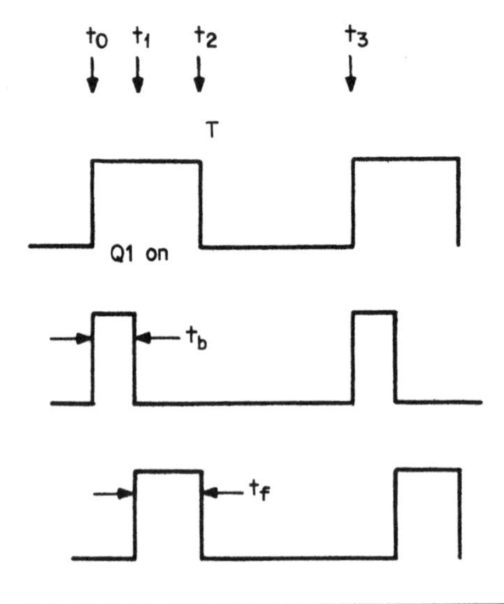

FIGURE 10.4 Critical timing intervals for magnetic amplifier (Figure 10.3).

Just prior to t_0, $D2$ was conducting and its cathode was one diode drop below ground. At t_0, a voltage ($V_{sp} + 1$) appears across the magnetic amplifier MA and $D1$ in series, in the direction to drive the core up toward saturation at point 4 (Figure 10.2).

Until the core gets up to point 4, only a very small "coercive" current flows through MA and $D1$ into $D2$. This setting current is considerably smaller than the free-wheeling current $D2$ is carrying. Thus the $D2$ cathode voltage does not change substantially; it remains at one diode drop below ground and the voltage V_{sp} is blocked from getting through and raising the voltage at the front end of L_s.

Diode $D1$ carries the MA coercive current, and although it is much smaller than the $D2$ current, assume that forward drops in $D1$ and $D2$ are 1 V. The full V_{sp} voltage thus appears across MA, whose right-hand end remains at ground. The MA is now in its blocking phase and remains there for a time t_b equal to the time required to move the core up from B_1 to saturation at B_s (point 4).

Once the core has reached saturation at point 4, the MA impedance becomes negligible within a few nanoseconds and it cannot support the voltage V_s. A large current flows from the top end of N_{ss} into $D2$ and unclamps it, and its cathode voltage rises to one diode drop ($D1$) below V_{sp} and remains there for the duration of the $Q1$ "on" time.

The time from the onset of saturation to the instant $Q1$ turns "off" is the firing time t_f. The LC output filter is thus presented with a square pulse of amplitude ($V_{sp} - V_{D1}$), duration t_f, and duty cycle t_f/T. It averages this pulse to produce the slave DC output voltage of $(V_{sp} - V_{D1})t_f/T$.

Regulation against line and load variations is achieved by controlling the firing time t_f. This is done indirectly by controlling the blocking time t_b.

10.3.2 Blocking and Firing Times in Magnetic-Amplifier Postregulators

Blocking time t_b is calculated from Faraday's law:

$$V_{sp} = N_m A_e \frac{dB}{dt} 10^{-8}$$

$$= N_m A_e \frac{(B_s - B_1)}{t_b} 10^{-8}$$

or

$$t_b = N_m A_e \frac{(B_s - B_1)}{V_{sp}} 10^{-8} \tag{10.4}$$

where N_m = number of turns on magnetic amplifier MA
A_e = MA core area, cm^2
B_s = saturation flux density, G
B_1 = starting point in flux density, G
V_{sp} = peak slave secondary voltage, V
t_b = blocking time, s (Figure 10.4)

Note (from Eq. 10.2b) $V_{os} = (V_{sp} - 1)t_f/T$ and (from Eq. 10.3) $t_f = t_h - t_b$.

If V_{sp} increases for any reason, V_{os} can be kept constant by decreasing the firing time t_f (Eq. 10.3), and from Eq. 10.2, t_f can be decreased by increasing t_b. From Eq. 10.4, t_b can be increased by decreasing B_1, that is, by pushing B_1 further down on the hysteresis loop. Similarly, if for some reason V_{os} decreases, t_f must be increased by decreasing the blocking time t_b. This is accomplished by increasing the flux level B_1 so that less time is required for the voltage V_{sp} to drive the core up into saturation.

10.3.3 Magnetic-Amplifier Core Resetting and Voltage Regulation

Thus far only the transition of the core from its starting point B_1 up to saturation has been discussed. During t_f, the MA impedance is essentially zero and it delivers the characteristic ramp-on-a-step current waveform that is applied to the output LC filter.

While $Q1$ is "off," the core must be restored to the B_1 level on the hysteresis loop, which yields the correct blocking time on the next switching cycle. If no current were delivered into MA in the reverse direction (into its no-dot end) immediately after the end of the ramp-on-a-step pulse, the MA core would return to $+B_s$ at 0 O_e.

The core is reset to B_1 by a current-reset technique in this example. In magnetic-amplifier practice, cores are reset to the desired flux level by either a voltage-reset or current-reset scheme. In voltage reset, the core is reset to any desired flux level by applying the correct volt-second product (Faraday's law: $dB = E\ dt/NA$). In this particular circuit, the core is reset more simply by current reset.

Line and load regulation are achieved, and current is reset to the appropriate B_1 by the voltage error amplifier, blocking diode $D3$, and voltage-controlled current source $Q2$ of Figure 10.3. The core is reset to that B_1 level which yields a blocking t_b, and hence firing time t_f that gives the desired DC output voltage.

Reset is accomplished by sourcing the appropriate DC current into the no-dot end of MA, starting at the end of the $Q1$ "on" time.

When $Q1$ is "on," $D3$ is reverse-biased and blocks $Q2$ from conducting and loading down the secondary. When $Q1$ turns "off," the top end of N_{ss} goes negative and current from the $Q2$ collector is driven into the no-dot end of MA. The amount of this current is controlled by the voltage error amplifier EA1 such as to make the fraction of the DC output voltage sampled by the resistive divider equal to the reference voltage.

If, for any reason, the DC output voltage goes up, t_f must go down, which means that t_b must go up. Thus, the error-amplifier output goes down and the $Q2$ collector current goes up, pushing more reset current into the no-dot end of MA. This pushes the initial flux level B_1 down and, by Eq. 10.4, increases t_b, thus decreasing t_f and bringing the DC output voltage back down.

Similarly, of course, a decrease in V_{os} causes an increase in error-amplifier output voltage and a decrease in $Q2$ reset current. As B_1 flux level rises, t_b decreases, t_f increases, and V_{os} is brought back up.

This action occurs over a number of switching cycles, and is accomplished in a time that is dependent on the error-amplifier bandwidth.

TIP *It is noteworthy that this simplified circuit does not take into account for a practical concern: When D3 blocks the Q2 collector current, Q2 emitter current and base currents immediately become equal. Since the error amplifier supplies the base current, but cannot support the normal emitter current, the internal circuits of the error amplifier will saturate. While most opamps are not damaged in this situation, it is nonetheless an undesirable operation, and the circuit would need to be refined to prevent it. ~T.M.*

Stabilization of this negative-feedback loop will not be considered here. It is treated by C. Jamerson[4] and C. Mullett.[2]

10.3.4 Slave Output Voltage Shutdown with Magnetic Amplifiers

In the preceding sections, the magnetic amplifier was presented only as a means of voltage regulating the slave output voltage. That was done by controlling the flux level B_1 to which the core is reset at the end of the power transistor "on" time. The further down B_1 was pushed, the longer the blocking time t_b, the shorter the firing time t_f, and the lower the DC output voltage were.

The magnetic amplifier can also be used to shut down the DC output voltage completely. This is done by pushing the initial flux level $+B_1$ down to $-B_s$. Blocking time, from Eq. 10.4, is $t_b = N_m A_e (2B_s) 10^{-8}/V_{sp}$. The core area A_e and N_m are chosen so that this blocking time is greater than the maximum $Q1$ "on" time.

Flux level B_1 can be brought down to $-B_s$ in a number of different ways, by forcing the $Q2$ current to be sufficiently large. The error-amplifier output can be overridden by forcing it down with a sufficiently low-impedance, low-voltage source, or if an isolating resistor is connected between V_{ref} and the error-amplifier input terminal, the terminal can be short-circuited to ground.

It should be noted that if the magnetic-amplifier core is used only for voltage regulation (as it generally is), it moves around a minor hysteresis loop as in Figure 10.2. The area of this loop may be a small fraction of the total hysteresis loop area—depending on the minimum B_1 flux level, which is dependent on the maximum and minimum supply voltage and load current specifications.

Generally, for a voltage-regulation-only design, the minor loop area will be about one-fourth the total hysteresis loop area. Since core losses are proportional to the area of the hysteresis loop traversed, ordinarily this will result in relatively low core dissipation and temperature rise.

However, if the design includes shutting the slave voltage down completely to zero, and the core flux excursion covers the full area of the major hysteresis loop from $+B_s$ to $-B_s$, higher core losses and temperature rise will result. Losses and core temperature rise should thus be calculated from the manufacturer's curves of core loss versus total flux excursion. This will be demonstrated below.

Note also that in Figure 10.3, supply voltage for the error amplifier and Q2 is the slave output voltage itself. If it is desired to shut the slave output down completely to zero, supply voltage for the error amplifier and Q2 will have to be taken from a source which is always present—possibly another slave output.

TIP *The magnetic amplifier can control the duty cycle (or "on" period) by delaying the time to conduction on the leading edge as described above, or by turning "off" before the end of a conducting period thus acting on the trailing edge. These are referred to as set or reset types. The set type, as described above, requires one of various types of high permeability metallic core. These cores have high core loss at high frequency. The second, reset type can use low loss ferrite cores that are more suitable in very high frequency applications.*[16] *~K.B.*

10.3.5 Square Hysteresis Loop Core Characteristics and Sources

When interest in magnetic amplifiers resumed in the 1970s, only a few materials with the required characteristics for an efficient high-frequency set type magnetic amplifier were available.

For many years cold rolled, grain orientated, high permeability materials were made for pulse transformers and similar applications. Typically these materials have high nickel content, and a good example is an alloy of 79% nickel, 17% iron, and 4% molybdenum. This material is available from various manufacturers under their particular brand names. Magnetics Inc. of Butler, Pennsylvania[3] is the foremost American supplier and has the largest range of available core sizes and iron areas. It calls its material Square Permalloy 80. Other manufacturers' brand names for similar materials are 4-79 Moly-permalloy, Square Mu 79, Square Permalloy, and Hy Ra 80.

Square Permalloy 80 is available from Magnetics Inc.[5] in tapes of various thicknesses wound on toroidal bobbins. Available tape thicknesses are 0.5, 1.0, 2.0, 4.0, 6.0, and 14.0 mils. Since the material is electrically conductive, eddy currents contribute a large fraction of the total losses, and at high frequencies, the thinner tapes must be used to keep losses down.

Generally, 1-mil tape thickness is used up to a switching frequency of 50 kHz, and ½-mil for frequencies of 50 to 100 kHz. Beyond 100 kHz, the newer "amorphous" core materials, to be discussed below, are used because of their lower losses.

The two characteristics required for an efficient, high-frequency magnetic amplifier are a very square hysteresis loop, and low losses at high peak flux density. The ½-mil tape is more expensive than 1-mil tape and should be used only if low loss is more important than low cost. If the design is only to provide voltage regulation by traversing minor loops rather than the full major loop, the higher loss 1-mil tape can be used even above 50 kHz.

The square hysteresis loop requirement is necessary for a very low impedance in the saturated state. If the hysteresis loop is not sufficiently square, its permeability dB/dH, as shown in Figure 10.2, is appreciable in saturation and its impedance at the top of the loop is significantly greater than that of an air core coil of an equal number of turns.

Thus, in Figure 10.3, the voltage at the output of the MA is less than V_{sp} during the firing time and the drop across MA depends on the secondary current. Further, if the loop is not square, the transition from high to low impedance may take a considerable time, prohibiting its use at high frequencies.

Core losses in watts per pound as a function of peak flux density and operating frequency are shown in Figure 10.5*a* and 10.5*b* for 1- and ½-mil Square Permalloy tapes. Flux density on those curves is half the peak-to-peak excursion. The Magnetics Inc. catalog does not give the tape weight for each core, but this can be calculated from the core area, mean path length, and material density of 8.75 g/cm^3 that are given in the catalogs.

Temperature rise of the outer surface of the case enclosing the magnetic toroid can be estimated from the core losses, and thermal resistance as read from Figure 7.4*a* or 7.4*c*. A reasonable estimate of the temperature differential between the magnetic core itself and the outer surface of the case is about 15°C. A relatively large core temperature

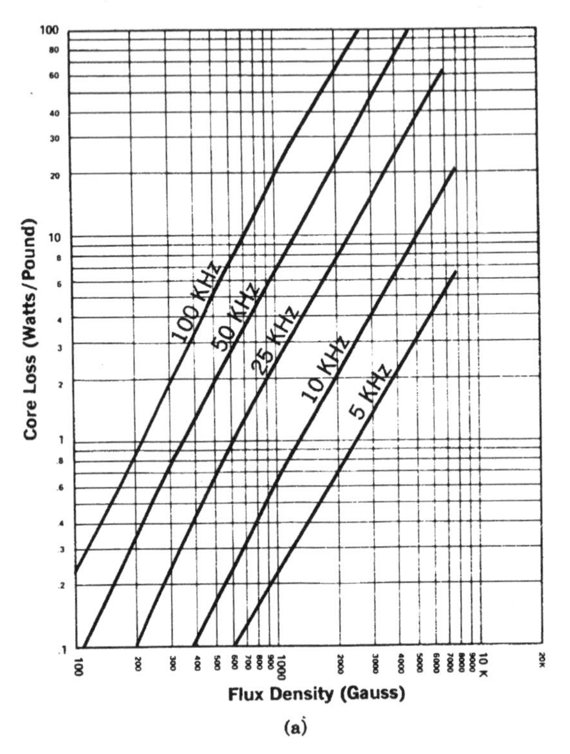

(a)

FIGURE 10.5 (*a*) Core losses, 1-mil Square Permalloy. (*Courtesy of Magnetics Inc.*) (*b*) Core losses, $^1/_2$-mil Square Permalloy. (*Courtesy of Magnetics Inc.*) (*c*) Core losses, amorphous core material, Metglas 27144. (*Courtesy of Magnetics Inc.*) (*d*) Core losses, Toshiba MA (watts/lb = 56.8 × watts/cm^3). (*Courtesy of Toshiba Corp.*)

rise can be tolerated, since the Curie temperature of Square Permalloy is 460°C. Thus the limiting factor for core losses is either the temperature rating of the wire, or the specified magnetic-amplifier efficiency.

TIP *Amorphous magnetic material is a good example of serendipity in which the development of material for one application can benefit other applications. This material came about in the 1970s as a result of the research into a better method for producing thin metal strips for reinforcing vehicle tires. In this process, a jet of molten metal was directed on to a rapidly rotating super-cooled metal drum to produce a continuous thin strip of metal. The rapid cooling did not provide time for the development of the crystal structure normally found in metal production. The noncrystalline structure was glass-like or amorphous, and was found to have extraordinary magnetic*

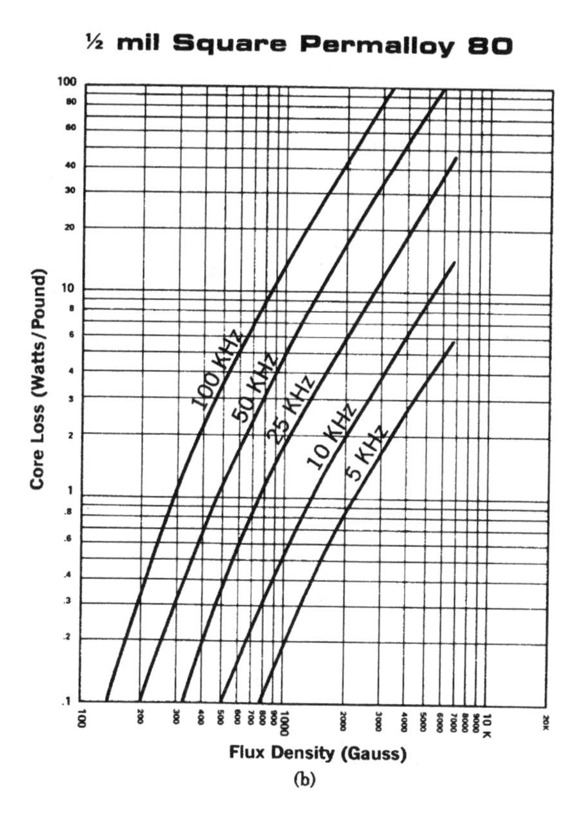

FIGURE 10.5 *Continued.*

properties. Vacuumschmelze in Germany were quick to exploit these properties with their Vitrovac 6025, with permeability up to 2 × 10⁶, and at the same time Allied Signal USA introduced Sq Metglas. ~K.B.

Shortly after the renewed interest in magnetic amplifiers, a new type of magnetic material was introduced. It is not crystalline in structure, but amorphous, and has lower core losses and a squarer hysteresis loop at higher frequencies than Square Permalloy. Although 1/2-mil Permalloy can be used up to 100 kHz or so, beyond that frequency this new amorphous material is preferable, and is available from many sources including Allied Signal in Parsippany, New Jersey,[6] Toshiba Corporation[7] (the American sales agent is Mitsui in New York), and Vacuumschmelze in Germany.

Allied calls its product Metglas 2714A. Toshiba has two amorphous core materials: MA and MB. Toshiba MB material is closely identical to Metglas 2714A in core losses, coercive force, and squareness of its hysteresis loop. Toshiba MA material is midway between 1/2-mil

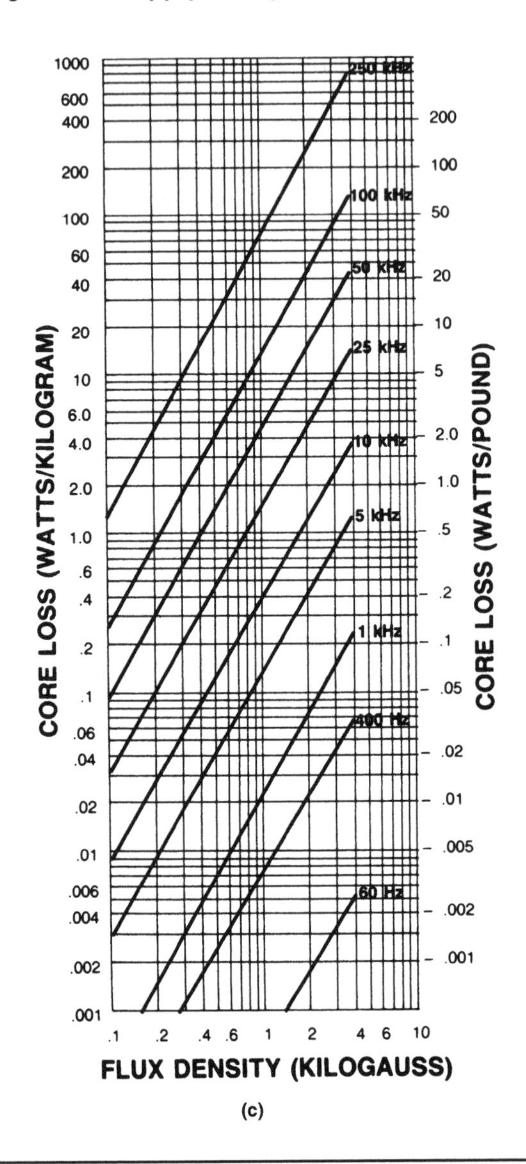

FIGURE 10.5 *Continued.*

Permalloy and MB material. Metglas 2714A material is also used by Magnetics Inc. for magnetic-amplifier cores in its own line of standard-sized cores.

Curves of core loss versus peak flux density at various frequencies for Metglas 2714A are shown in Figure 10.5*c*. Data comparing core loss versus frequency at a peak flux density of 2000 G for Toshiba MA,

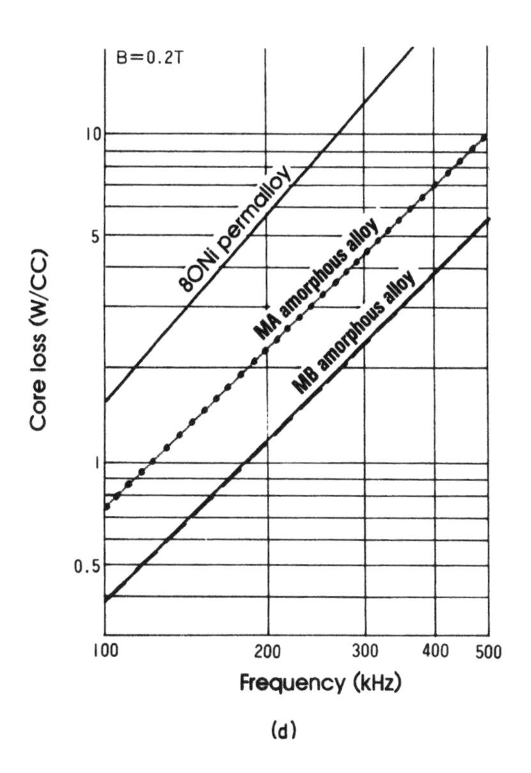

FIGURE 10.5 *Continued.*

MB material, and 1-mil Permalloy are given in Figure 10.5*d*. Note that figure gives loss in watts per cubic centimeter. For MA, MB density of 8.0 g/cm^3, loss in watts per pound is 56.8 × loss in watts per cubic centimeter.

The *BH* loops for increasing frequency have a characteristic appearance. As frequency increases, coercive force increases but saturation flux density remains fixed. This, of course, explains the increase in core loss with frequency, as the loss is proportional to the area of the hysteresis loop. The increase in coercive force with frequency is shown in Figure 10.6*b* for Toshiba MA, MB, and Permalloy materials.

Standard-sized MA and MB cores available from Toshiba are shown in Figure 10.7*a* and 10.7*b*. Metglas 2714A cores available from Magnetics Inc.[8] are shown in Figure 10.8*a*, and from Allied Signal in Figure 10.8*b*.

It is interesting to note that the temperature rise given by Toshiba in Figure 10.12 coincides to within a few degrees with the rise as calculated from Figure 7.4*c* for any power dissipation and core area as calculated from Figure 10.7*b*.

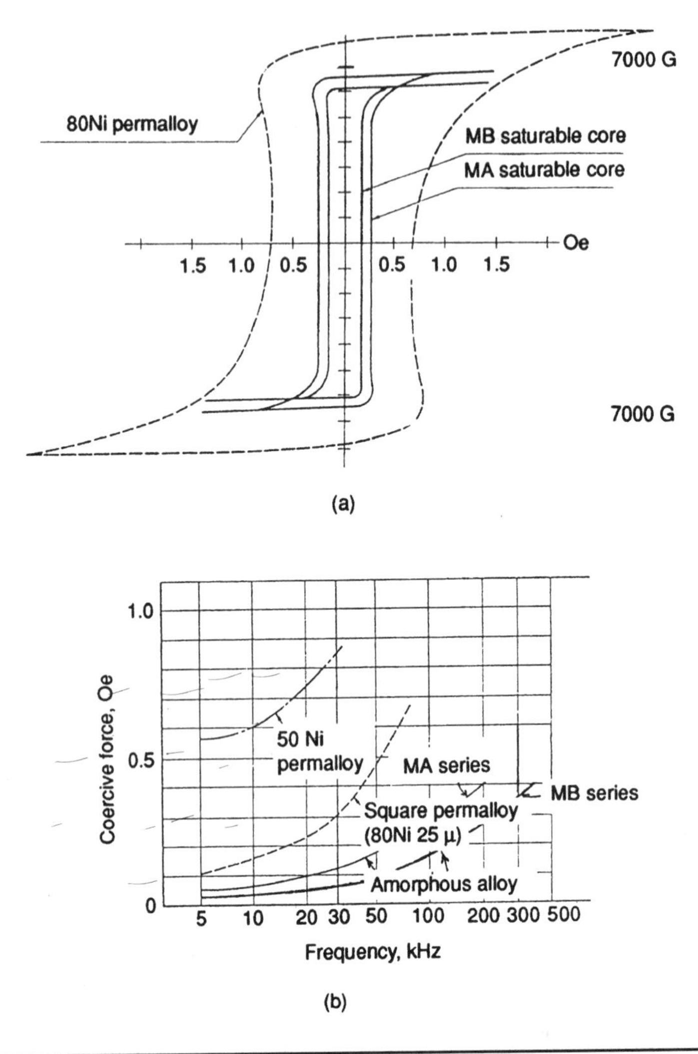

FIGURE 10.6 (*a*) *BH* loops at 100 kHz. (*Courtesy of Toshiba Corp.*) (*b*) Coercive force versus frequency. (*Courtesy of Toshiba Corp.*) (*c*) *BH* loops at 100 kHz, 1-mil and ½-mil Permalloy, Metglas 2714A. (*Courtesy of Magnetics Inc.*)

Recall that flux change in maxwells equals flux density change in gauss multiplied by core area in square centimeters. Thus, dividing the maxwells shown in the curves by the core area gives the flux density change in gauss. The maximum maxwells shown on the curves then correspond to the total maximum flux density change from negative

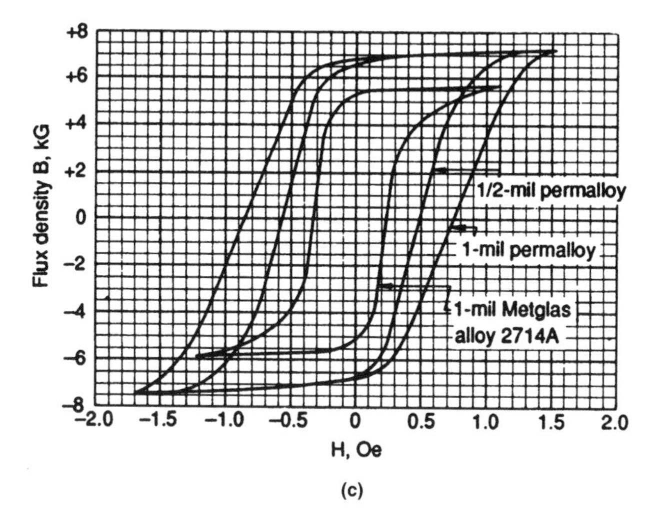

FIGURE 10.6 *Continued.*

to positive saturation flux density ($-B_S$ to $+B_S$). Toshiba's MA, MB materials have B_S of 6500 and 6000 G, respectively.

Losses at the maximum maxwells, shown later in Figures 10.9 to 10.11, correspond to operation around the total major loop, as when the magnetic amplifier is used to shut down the slave output voltage fully. Losses at the lower flux level correspond to operation around a minor loop as in Figure 10.2, where the magnetic amplifier is used only for voltage regulation.

The *BH* loops at 100 kHz for Toshiba MA, MB material and ½-mil Permalloy are shown in Figure 10.6*a*. Figure 10.6*c* compares 100-kHz *BH* loops for Metglas 2417A, and ½- and 1-mil Permalloy.

10.3.6 Core Loss and Temperature Rise Calculations

Toshiba provides curves for each of its cores that are useful in calculating core temperature rise. Figures 10.9, 10.10, and 10.11 show core loss versus total flux change in maxwells for its three largest MB cores. Thus, whatever the total flux excursion is, Figures 10.9 to 10.11 give core loss for each core at that flux excursion. From this core loss, Figure 10.12 gives the core temperature rise. Actually, Figure 10.12 is a measure of the thermal resistance of each core and is related to the radiating surface area of the core, which can be calculated from its outer dimensions as given in Figure 10.7*b*.

Type No.	Standard core dimensions (mm)			Finished dimensions (mm)			Effective core cross section (mm²)	Mean flux path length (mm)	Total flux $2\Phi_1$ ($\times 10^{-8}$WB)	Insulating covers
	Outer dia.	Inner dia.	Height	Outer dia.	Inner dia.	Height				
MA 26×16×4.5W	26	16	4.5	29.5 max.	13.0 min.	8.0 max.	16.9	66.0	1800	Resin casing
〃 22×14×4.5W	22	14	4.5	25.5 〃	11.0 〃	8.0 〃	13.5	56.5	1440	〃
〃 18×12×4.5X	18	12	4.5	20.5 〃	10.2 〃	8.0 〃	10.1	47.1	1080	Epoxy resin coating
〃 14× 8×4.5X	14	8	4.5	16.3 〃	6.3 〃	7.5 〃	10.1	34.6	1080	〃
〃 10× 6×4.5X	10	6	4.5	12.3 〃	4.4 〃	7.5 〃	6.75	25.1	720	〃
〃 8× 6×4.5X	8	6	4.5	10.0 〃	4.4 〃	7.5 〃	3.38	22.0	360	〃
〃 7× 6×4.5X	7	6	4.5	9.0 〃	4.4 〃	7.5 〃	1.69	20.4	180	〃

(a)

Type No.	Standard core dimensions (mm)			Finished dimensions (mm)			Effective core cross section (mm²)	Mean flux path length (mm)	Total flux $2\Phi_1$ ($\times 10^{-8}$WB)	Insulating covers
	Outer dia.	Inner dia.	Height	Outer dia.	Inner dia.	Height				
MB 21×14×4.5	21	14	4.5	24.0 max.	12.0 min.	8.5 max.	11.8	55.0	1105	Epoxy resin coating
〃 18×12×4.5	18	12	4.5	20.5 〃	10.2 〃	8.0 〃	10.1	47.1	935	〃
〃 15×10×4.5	15	10	4.5	17.5 〃	8.3 〃	7.5 〃	8.44	39.3	790	〃
〃 12× 8×4.5	12	8	4.5	14.3 〃	6.3 〃	7.5 〃	6.75	31.4	629	〃
〃 10× 7×4.5	10	7	4.5	12.3 〃	5.4 〃	7.5 〃	5.06	26.7	476	〃
〃 9× 7×4.5	9	7	4.5	11.0 〃	5.4 〃	7.5 〃	3.38	25.1	315	〃
〃 8× 7×4.5	8	7	4.5	10.0 〃	5.4 〃	7.5 〃	1.69	23.6	158	〃
〃 15×10×3W	15	10	3.0	17.0 〃	8.0 〃	5.0 〃	5.63	39.3	526	Resin casing
〃 12× 8×3W	12	8	3.0	14.0 〃	6.0 〃	5.0 〃	4.5	31.4	420	〃

(b)

FIGURE 10.7 (*a*) Available MA amorphous cores for magnetic amplifiers. (*Courtesy of Toshiba Corp.*) (*b*) Available MB amorphous cores for magnetic amplifiers. (*Courtesy of Toshiba Corp.*)

Part Number*	DIMENSIONS						Core loss (w) @ 50KHz, 2000 gauss (Max.)	ml cm	Ac cm^2	Wa cr m	Core wt. grams	Wa Ac cr m cm^2 ($\times 10^{-6}$)
	I.D. (in.)		O.D. (in.)		Ht. (in.)							
	core	case (Min.)	core	case (Max.)	core	case (Max.)						
50B10-5D	.650	.580	.900	.970	.125	.200	.118	6.18	.051	348,000	2.7	0177
50B10-1D	.650	.580	.900	.970	.125	.200	.22	6.18	.076	348,000	4.0	.0264
50B10-1E	.650	.580	.900	.970	.125	.200	.092	6.18	.076	348,000	3.5	.0264
50B11-5D	.500	.430	.625	.695	.125	.200	.044	4.49	.025	194,000	1.0	.0048
50B11-1D	.500	.430	.625	.695	.125	.200	.083	4.49	.038	194,000	1.5	.0074
50B11-1E	.500	.430	.625	.695	.125	.200	.034	4.49	.038	194,000	1.3	.0074
50B12-5D	.375	.305	.500	.570	.125	.200	.035	3.49	.025	99,000	8	.0025
50B12-1D	.375	.305	.500	.570	.125	.200	.066	3.49	.038	99,000	1.2	.0038
50B12-1E	.375	.305	.500	.570	.125	.200	.027	3.49	.038	99,000	1.04	.0038
50B45-5D	.500	.430	.750	.820	.250	.325	.194	4.99	.101	194,000	4.4	.0143
50B45-1D	.500	.430	.750	.820	.250	.325	.363	4.99	.151	194,000	6.6	.0214
50B45-1E	.500	.430	.750	.820	.250	.325	.149	4.99	.151	194,000	5.7	.0214
50B66-5D	.500	.430	.750	.820	.125	.200	.097	4.99	.050	194,000	2.2	.0071
50B66-1D	.500	.430	.750	.820	.125	.200	.182	4.99	.076	194,000	3.3	.0108
50B66-1E	.500	.430	.750	.820	.125	.200	.075	4.99	.076	194,000	2.9	.0108

*For other sizes, refer to factory.

FIGURE 10.8 Available Square Permalloy 80 and Metglas 2714A cores for magnetic amplifiers. (*Courtesy of Magnetics Inc.*) (*b*) Standard Metglas 2714A amorphous magnetic-amplifier cores. (*Courtesy of Allied Signal*)

PART NUMBER CODE

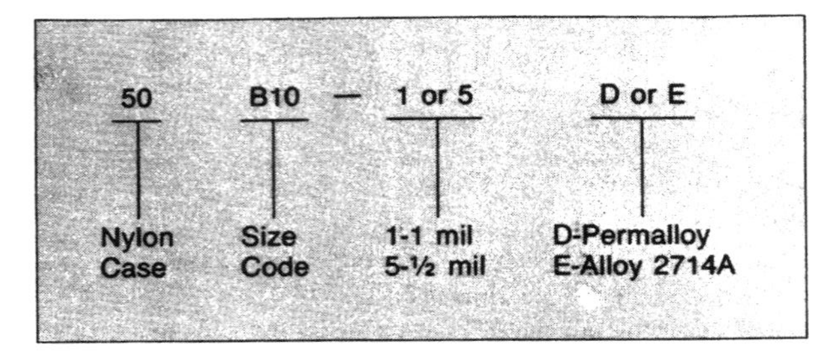

MATERIAL CHARACTERISTICS

	Alloy 2714A	½ mil Permalloy	1 mil Permalloy
Bm (gauss min.)	5000	7000	7000
Br/Bm (min.)**	.9	.83	.80
H_1 (oersted max.)**	.025	.045	.040
Core loss (w/lb. max. @ 50 kHz, 2000 gauss)	12	20	25

**Measured @ 400 Hz, CCFR Test

FIGURE 10.8 *Continued.*

CORE NUMBER	DIMENSIONS (mm)*				(cm)	A (cm²)	Mass (g)	Flux Capacity 2φ (μWb)	W (cm²)	W A (cm⁴)
		OD	ID	HT						
MP1303	CORE CASE	12.8 14.6	9.5 7.9	3.2 5.1	3.50	0.041	1.1	4.7	0.49	0.021
MP1603	CORE CASE	15.9 17.8	12.7 11.1	3.2 5.1	4.50	0.041	1.4	4.7	0.96	0.039
MP1903	CORE CASE	19.2 21.0	12.7 11.1	3.2 5.1	5.00	0.082	3.1	9.3	0.96	0.079
MP2303	CORE CASE	22.9 25.0	16.5 14.6	3.2 5.1	6.19	0.081	3.8	9.2	1.68	0.14
MP1305	CORE	12.5	9.5	4.8	3.46	0.057	1.5	6.5	0.49	0.028
MP1505	CORE	15.1	9.5	4.8	3.87	0.11	3.1	12	0.49	0.049
MP1805	CORE CASE	18.4 20.8	12.7 10.8	4.8 6.7	4.88	0.11	4.0	12	0.92	0.10
MP1906	CORE CASE	19.1 21.3	12.7 10.7	6.4 8.4	4.99	0.16	6.1	18	0.90	0.14
MP3506	CORE CASE	35.2 37.3	25.4 23.4	6.4 8.4	9.52	0.241	17.4	27.5	0.49	1.04
MP2510	CORE CASE	25.6 27.8	19.1 17.0	9.5 11.8	7.01	0.241	12.8	27.5	2.28	0.552

NOTE: 1 circular mil = 5.067×10^{-6} cm²

PART NUMBER CODE:

MP	**13 03**	**P or E**	**4A**	**S**
Metglas Products	OD HT	P - Plastic Box E - Encapsulated	Alloy 2714A	Square Loop

(b)

FIGURE 10.8 *Continued.*

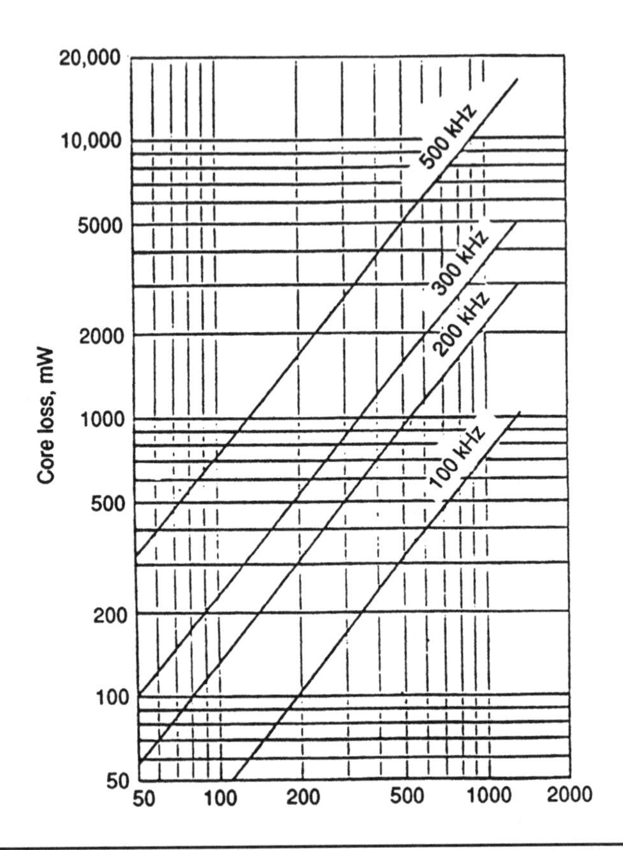

FIGURE 10.9 Core loss versus total flux change. Toshiba MB 21 × 14 × 4.5 core. Core area = 0.118 cm^2; $\Delta B = \Delta\phi$ (maxwells)/0.118. (*Courtesy of Toshiba Corp.*)

10.3.7 Design Example—Magnetic-Amplifier Postregulator

Design a magnetic-amplifier postregulator for the output of the forward converter shown in Figure 10.13*a*. Specifications are

Forward converter switching frequency	100 kHz
Slave output voltage	15 V
Slave output current	10 A

The main output voltage is $V_{\text{om}} = V_{\text{dc}}(N_{\text{sm}}/N_p)(t_{\text{on}}/T)$. The main feedback loop, in keeping V_{om} constant, must keep the product $V_{\text{dc}}t_{\text{on}}$ constant, so t_{on} is a maximum when V_{dc} is a minimum.

In the usual case, the number of turns on the $T1$ reset winding N_r is equal to the turns on the power winding N_p. This forces the voltage

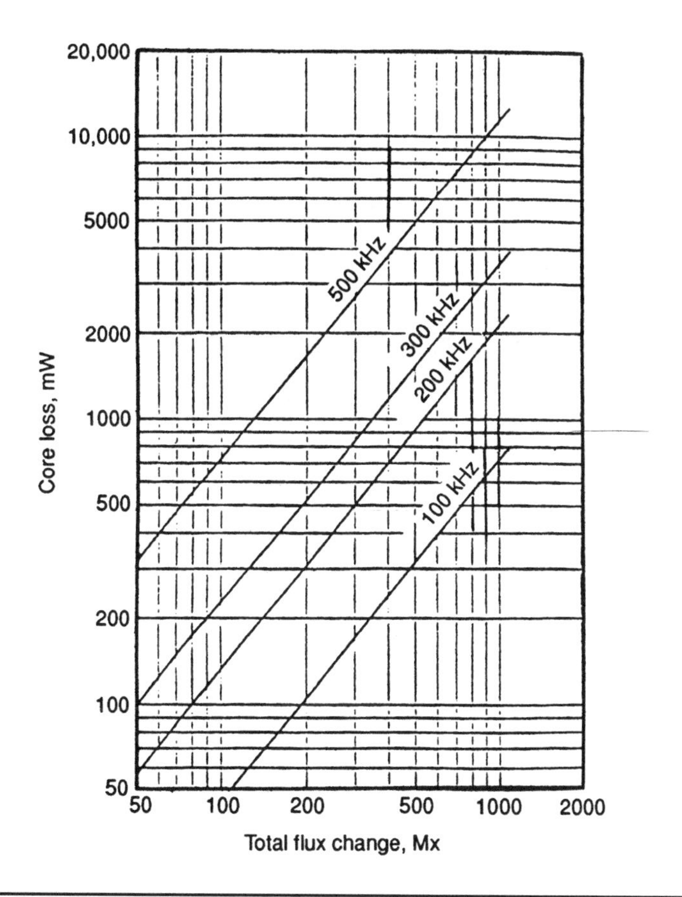

FIGURE 10.10 Core loss versus total flux change. Toshiba MB 18 × 12 × 4.5 core. Core area = 0.101 cm^2; $\Delta B = \Delta\phi$(maxwells)/0.101. (*Courtesy of Toshiba Corp.*)

across N_p when $Q1$ is "off" to be equal and opposite to its voltage when $Q1$ is "on."

Over a complete cycle, the volt-second product across N_p when $Q1$ is "on" must be equal and opposite to the volt-second product across it when $Q2$ is "off." Otherwise, from Faraday's law, the core flux density would increase in one direction on the hysteresis loop, and at the start of the next cycle would not have been returned to its starting point. After a number of such cycles, the core would drift up the hysteresis loop, saturate, and—being unable to support voltage—destroy the power transistor the next time it is turned "on." Thus the absolute maximum $Q1$ "on" time at minimum DC input voltage is $0.5T$ or 5 μs so that $+V_{dc}t_{on}$ can equal $-V_{dc}t_{off}$.

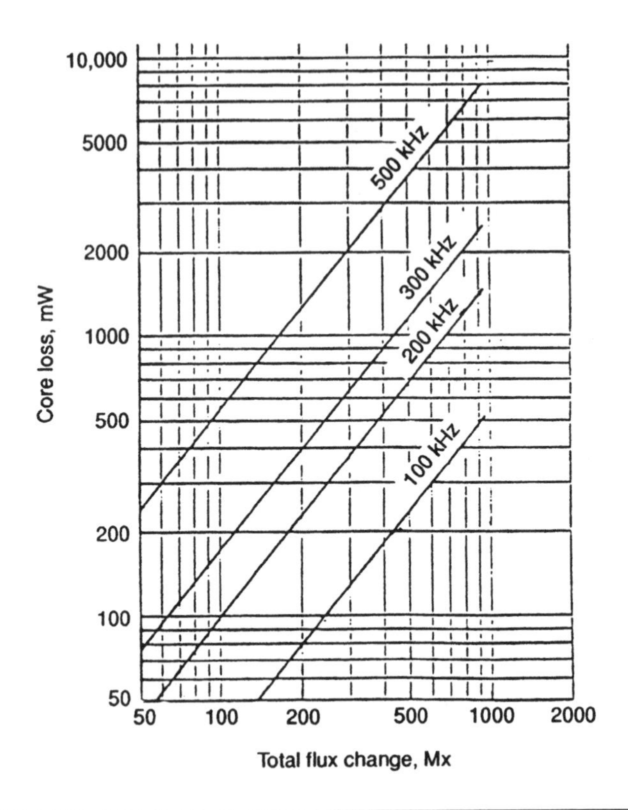

FIGURE 10.11 Core loss versus total flux change. Toshiba MB 15 × 10 × 4.5 core. Core area = 0.0843 cm^2; $\Delta B = \Delta\phi$(maxwells)/0.0843. (*Courtesy of Toshiba Corp.*)

To ensure that the core can always be reset during transient power-line dips below its specified minimum, in the above expression for V_{om}, N_{sm} is chosen so that V_{om} is obtained for a maximum "on" time of $0.4T$ or 4 μs at the specified minimum V_{dc}. This yields a guard band of $0.1T$ or 1 μs to allow for line input dips.

The slave peak voltage V_{sp} at the input end of MA is then high for 4 μs at minimum DC input voltage. The MA will block this voltage for a time t_b (Figure 10.13*c*), leaving a high voltage V_{sp1} for a time t_f at the no-dot end of the MA after it has saturated.

If the MA has zero impedance when saturated, then $V_{sp1} = V_{sp}$. Assuming a 1-V drop across rectifier diode $D1$, the peak voltage at the front end of L_s when MA has fired is ($V_{sp} - 1$). Of the 4-μs duration of V_{sp} at low V_{dc} input, we arbitrarily set $t_f = 3$ μs, and $t_b = 1$ μs. This permits t_1 (Figure 10.13b) to move either left or right, increasing or decreasing t_f to regulate against load changes.

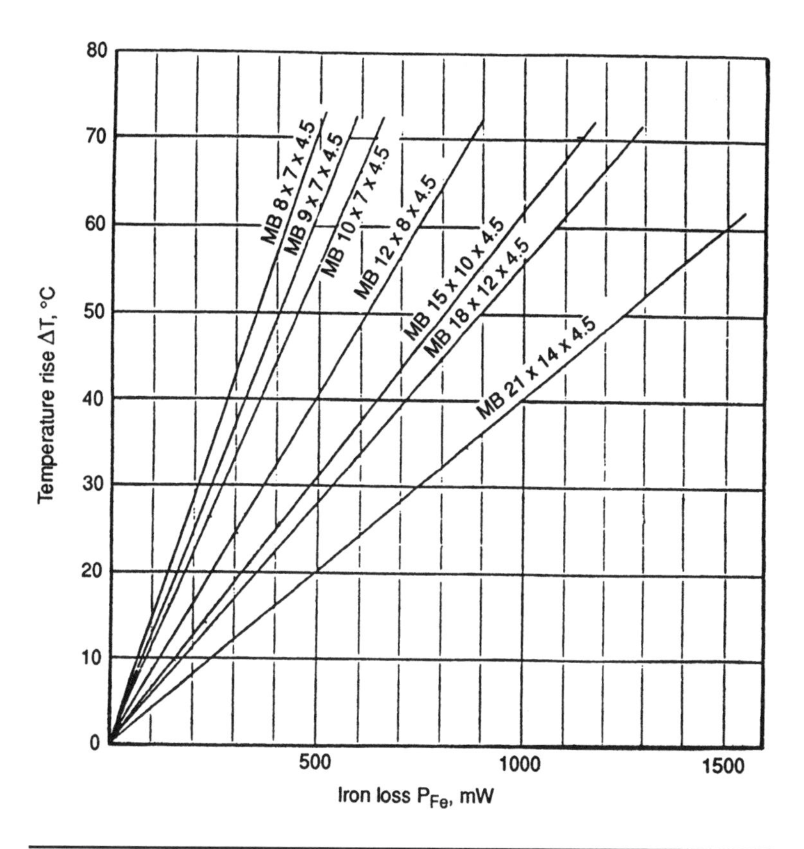

FIGURE 10.12 Toshiba amorphous MB cores. Temperature rise versus core losses. (*Courtesy of Toshiba Corp.*)

For a slave output voltage of the specified 15 V

$$15\text{ V} = (V_{\text{SP}} - 1)\left(\frac{t_f}{T}\right) = (V_{\text{SP}} - 1)\left(\frac{3}{10}\right) \quad \text{or} \quad V_{\text{SP}} = 51\text{ V}$$

The number of turns on MA and its iron area must be chosen to block 51 V. The turns will be chosen to block not for just the 1 μs minimum blocking time, but for the full 4 μs maximum duration of V_{sp}, on the assumption that the MA may be used to force the slave output voltage completely down to zero.

For a 100-kHz magnetic amplifier, let us use an amorphous core such as the Toshiba MB (Figure 10.7*b*). Arbitrarily select the one with the largest iron area, as it will require the least turns to block the 51 V for 4 μs for a given flux change. This minimizes the residual air core inductance of the MA in its saturated state. Thus the MB 21 × 14 × 4.5, of iron area 0.118 cm^2, is chosen.

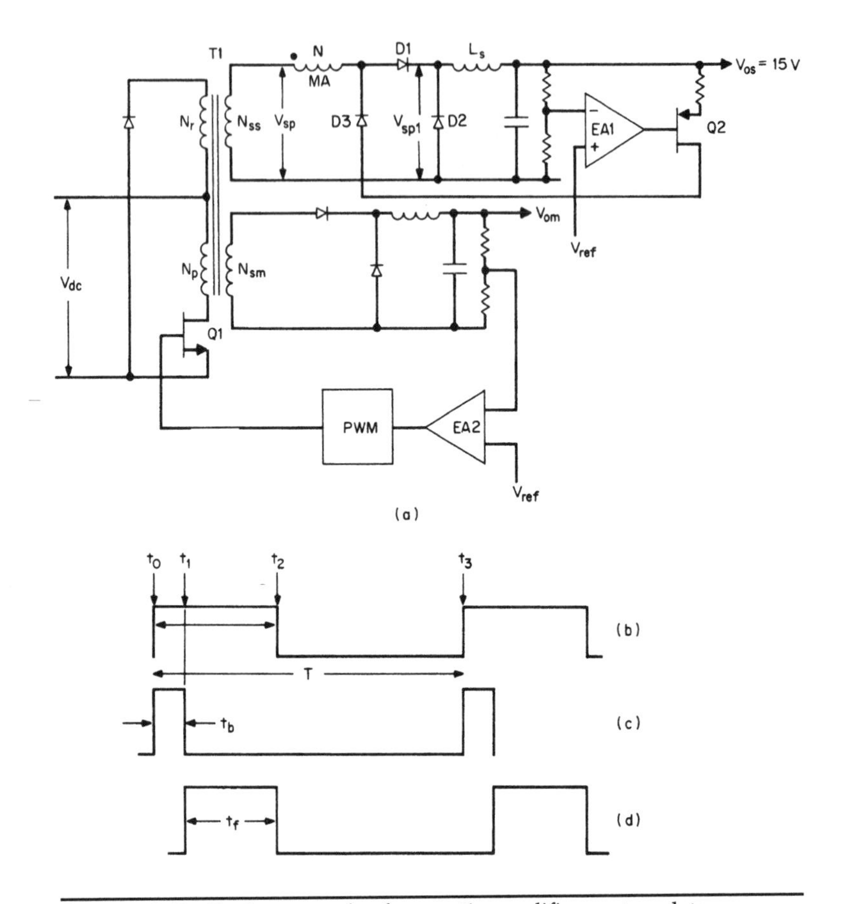

FIGURE 10.13 Design example of magnetic-amplifier postregulator. Magnetic amplifier blocks V_{so} for a time t_b and is a short circuit for a time t_r. Then $V_{os} = V_{so}\,(t_f/T)$. Time t_f is controlled in a negative-feedback loop by the current that Q2 forces into MA via D_m. That current is controlled by error amplifier EA1.

From Faraday's law, to minimize the number of turns, the flux change should be maximized or the core should traverse the full *BH* loop from $-B_s$ to $+B_s$. From Figure 10.6*a*, B_s for the Toshiba MB material is 6000 G. Then from Faraday's law, to block 51 V for 4 μs with a core area of 0.118 cm^2:

$$\begin{aligned} E &= 51\text{V} = NA_e\frac{dB}{dt}10^{-8} \\ &= N(0.118)\frac{2B_s}{4\times10^{-6}}10^{-8} \\ &= 14 \text{ turns} \end{aligned}$$

For DC output current of 10 A, MA will be carrying the 10 A for a maximum of only 3 μs (maximum t_f) at minimum DC input. Maximum RMS current is then $10\sqrt{3/10} = 5.48$ A. At 500 circular mils per RMS ampere, a wire area of 2739 circular mils is required.

Two No. 19 wires in parallel provide 2 × 1290 or 2580 circular mils, which is close enough. The inner periphery of the Toshiba MB21 × 14 × 4.5 core is $\pi \times 0.55 = 1.73$ in. For the 0.0391-in diameter of No. 19 wire, the inner periphery can hold 1.73/0.0391 or 44 turns on a single layer. The 14 turns of two paralleled No. 19 wires can thus easily be accommodated in a single layer on the inner periphery.

If the core is operated in the shutdown mode, the full major loop is traversed each cycle. Figure 10.9 shows that for the MB 21 × 14 × 4.5, this corresponds to a total flux change of 1400 mW or 12000 G ($-B_s$ to $+B_s$) and core loss of 1 W, and Figure 10.12 shows that the core temperature rise is only 40°C.

10.3.8 Magnetic-Amplifier Gain

When the MA has saturated, it has close to zero impedance, and the DC current through it is determined only by the DC output impedance and the slave output voltage. That is simply the specified DC output current. But to bring the MA to its saturated state, a current equal to twice the coercive current I_c is required to force the core from the left to the right side of the hysteresis loop (Figure 10.2). That current comes from the transformer secondary V_{sp}. Similarly, when the core is reset to the left side of the hysteresis loop (Figure 10.2), a current equal to the coercive current must be supplied from $Q2$ via $D3$.

Magnetic-amplifier gain from $Q2$ to the output is then I_o/I_c. From Ampere's law, the coercive force H is

$$H_c = 0.4\pi N_m I_c/l_p$$

where N_m is the number of turns
I_c is the coercive current
l_p is the mean path length in centimeters

For the MB core at 100 kHz, the coercive force is 0.18 O_e (Figure 10.6*b*). Mean path length l_p of the MB 21 × 14 × 4.5 core is 5.5 cm, from Figure 10.7*b*. Then for N_m of 14 turns, the coercive current is

$$\begin{aligned} I_c &= \frac{H_c l_p}{0.4\pi N_m} \\ &= \frac{0.18 \times 5.5}{0.4\pi \times 14} \\ &= 56.3\,\text{mA} \end{aligned}$$

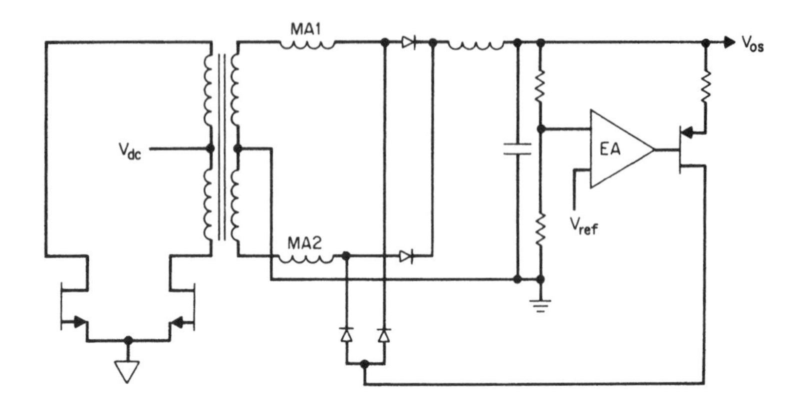

FIGURE 10.14 Two magnetic-amplifier cores driven from the same EA and PNP current source are required for a push-pull output.

Thus a current of 56.3 mA from *Q*2 can control the 10-A output—or stated another way, the magnetic-amplifier gain is $10/0.056 = 178$.

It is of interest to realize the physical significance of this and to appreciate that a magnetic amplifier is very different from the usual saturable reactor. In the magnetic amplifier, when the control current (from *Q*2) flows, no load current flows as *D*1 is reverse-biased. Thus the control current does not have to "buck out" the load current.

In a saturable reactor (a variable-inductance reactor controlled by current in a control winding) in series with a load, however, current flows to the load at the same time the control current flows. Thus the control winding ampere turns must buck out the load ampere turns and gain is low.

10.3.9 Magnetic Amplifiers for a Push-Pull Output

For a full-wave output (push-pull or half bridge topology), the circuit of Figure 10.14 is often presented. However, it has serious problems in that during the dead time between transitions, the magamps carry primary magnetizing current (Figure 2.6). This is fully discussed in Reference 15.

10.4 Magnetic Amplifier Pulse-Width Modulator and Error Amplifier

Thus far in this chapter, magnetic amplifiers have been considered only as postregulators. In this section, an interesting example of a magnetic amplifier used simultaneously as a pulse-width modulator and error amplifier is described.[9]

It may be puzzling why, with the current proliferation of inexpensive semiconductor pulse-width modulating chips with their built-in error amplifiers, there is interest in magnetic elements to perform these functions.

The circuit to be described below does have advantages in specialized applications where a semiconductor integrated circuit cannot be used for some reason. The circuit consists of only square hysteresis loop core material and wire, which is far more reliable than an integrated circuit.

There are some environmental conditions—such as excessive temperature—where a magnetic amplifier can survive more easily than an integrated circuit. Since discrete transistors that can tolerate high temperatures are found easily, a circuit consisting of discrete transistors and a magnetic amplifier PWM–error amplifier is a more robust circuit than one with discrete transistors and an integrated-circuit PWM chip.

Finally, the circuit described below offers a simple solution to an omnipresent problem in switching power supplies. That problem is how to sense a voltage on output ground, and deliver the appropriate width-modulated pulse to the power transistor on input ground, without requiring a housekeeping power supply to power the error amplifier on output ground.

The circuit works as follows.

10.4.1 Circuit Details, Magnetic Amplifier Pulse-Width Modulator–Error Amplifier[9]

The circuit shown in Figure 10.15 was originally devised by Dulskis and Estey.[9] It is a conventional push-pull topology using Darlington power transistors. A 40-kHz, ±8-V, 50/50 duty cycle square wave is applied at points *AB* via transformer *T*1.

The heart of the design is the magnetic amplifier *M*1. It consists of two square hysteresis loop cores sitting one above another. There are two equal-turn gate windings N_{g1}, N_{g2} and a control winding N_c. Gate winding N_{g1} links core *A* only, N_{g2} links core *B* only, and N_c links both cores.

On alternate half cycles, a +8-V, 12.5-μs pulse appears at points *A*, *B* relative to *C*. Consider the half period when *A* is positive. That 8 V is divided across *D*1, the base-emitters of *Q*3, *Q*1, and *R*5. Resistor *R*5 is a current limiter.

At the start of the half period, magnetic-amplifier core *M*1*A* is on the steep part of its hysteresis loop (B_0 of Figure 10.16*a*) and has sufficiently high impedance so that it does not short out the voltage from the *Q*3 base to ground. The Darlington, *Q*3, *Q*1, is energized and a voltage of roughly $V_{dc} - 1$ is applied to its half primary of *T*2.

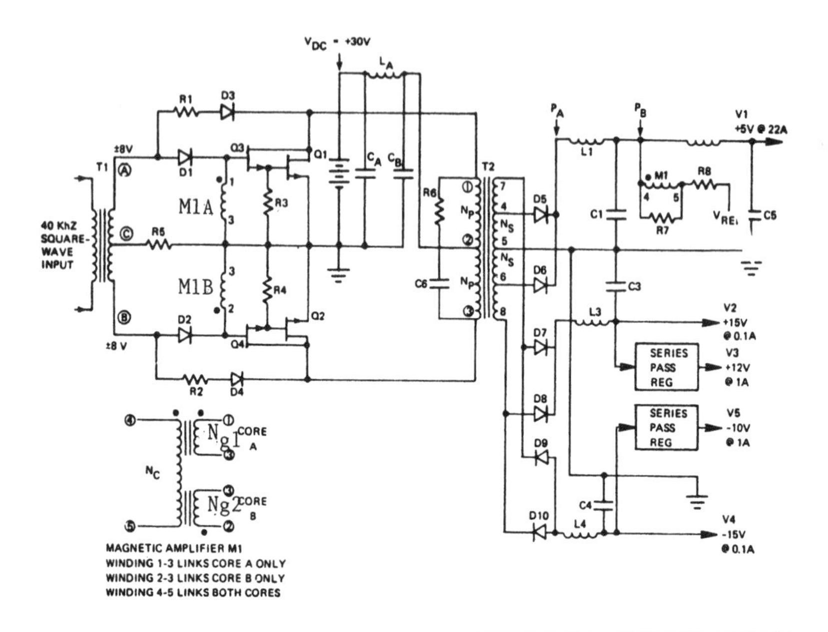

FIGURE 10.15 Magnetic-amplifier-controlled switching regulator. The magnetic amplifier serves both as a pulse-width modulator and an error amplifier. DC current through control winding N_c determines the initial flux bias in gate windings and hence their time to saturate, which fixes the Darlington "on" times. *(From R. Dulskis, J. Estey, and A. Pressman, "A Magnetic Amplifier Controlled 40 kHz Switching Regulator," Wescon Proceedings, San Francisco, 1977.)*

The voltage across $M1A$ is the sum of the base-emitter drops of $Q1$ and $Q3$ (~1.6 V). This voltage drives $M1A$ upward toward saturation along the minor hysteresis loop $B0$–$B1$–$B2$–$B3$–B_S. At B_S, $M1A$ saturates, $Q3$ loses its base drive, and Darlington $Q3$, $Q1$ turn "off." The $Q3$, $Q1$ "on" time is given by Faraday's law as

$$t_{\mathrm{on}(Q1,Q3)} = \frac{N_g A_e (B_S - B_0)}{1.6} 10^{-8}$$

where N_g = number of turns of $M1A$
A_e = core iron area
B_0 = starting point on the hysteresis loop
1.6 = voltage across $M1A$ driving it toward saturation

The further down B_0 is from B_S, the longer the "on" time. "On" time control is accomplished by controlling the level B_0 to which the core is reset at the start of the positive half cycle.

As $M1A$ is being pushed up toward saturation with a voltage of 1.6 V across its N_g turns, it induces a voltage of $(N_c/N_g)1.6$ across N_e.

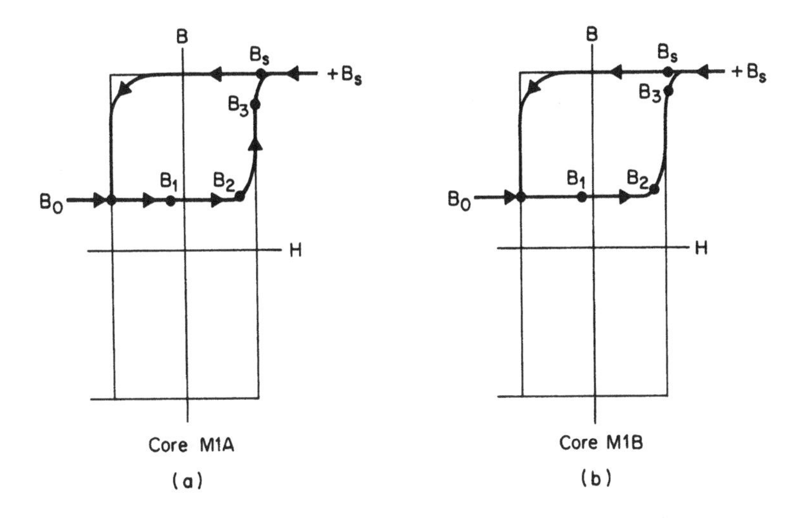

FIGURE 10.16 The *BH* loop operating locus of magnetic-amplifier cores (Figure 10.15). As core $M1A$ moves from B_0 to $+B_s$, core $M1B$ is pushed down from $+B_s$ to B_0. Time to move from B_0 to B_s is the "on" time of the power transistors. "On" time is determined by how far down from B_s the starting point B_0 has been pushed. This is determined by the DC current in control winding N_c, which is proportional to the output voltage (Figure 10.15).

That couples a voltage back down to 1.6 V across $M1B$, but because of the winding polarities, that voltage is in the direction to push $M1B$ down to B_0 if in the previous half period it had been pushed up to $+B_4$ by point B having been positive.

Thus in one half period, as—say—$M1A$ goes from B_0 to B_s, its Darlington is "on" and $M1B$ is driven down from B_s to B_0. But the $M1B$ Darlington is "off," as its input diode $D2$ is reverse-biased. During the next half period the sequence reverses; $M1B$ is driven up from B_0 to $+B_s$, its Darlington is "on," and $M1A$ is forced down from B_s to B_0. Power Darlington "on" time is fixed by B_0, the starting point on the *BH* loop, and that is determined by the DC current in the control winding N_c.

Because of the winding polarities, there is no net AC voltage across N_c, which is bridged between the DC output voltage and a reference voltage as in Figure 10.15. The polarity of N_c is such that if the DC output voltage goes up, the gate windings are biased further up toward B_s, time to saturate the core decreases, and the Darlington "on" time decreases, bringing the DC output voltage back down.

Thus the control winding serves as the error voltage sensor. It is referenced to output ground, and its DC current serves to control the Darlington "on" time, on input ground, by controlling the gate winding's initial flux density B_0. It is this last feature which is perhaps the most useful characteristic of the circuit. Resistors $R7$, $R8$ serve to set the control winding error-amplifier gain.

In the circuit of Figure 10.15, the MA cores are 1/4-mil Square Permalloy, wound on bobbins of 0.290 OD, 0.16 ID, and 0.175 height. The gate windings—N_{g1} that linked core A only, and N_{g2} that linked core B only—consist of 40 turns of No. 35 wire. The control winding N_c, which links both cores, consists of 250 turns of No. 37 wire.

References

1. C. Mullett, "Design and Analysis of High Frequency Saturable Core Magnetic Regulators," *Proceedings Powercon 10*, 1983.
2. C. Mullett, "Design of High Frequency Saturable Reactor Output Regulators," *High Frequency Power Conversion Conference Proceedings*, 1986.
3. Toshiba Corp. Bulletin, *Toshiba Amorphous Magnetics Parts*, Mitsui & Co., New York.
4. C. Jamerson, "Calculation of Magnetic Amplifier Post Regulator Voltage Control Loop Parameters," *High Frequency Power Conversion Conference Proceedings*, 1987.
5. Magnetics, Inc. Bulletin TWC 300S, Butler, PA.
6. Allied Signal Technical Bulletin, *Metglas Amorphous Alloy Cores*, Parsippany, NJ.
7. Toshiba Corp. Bulletin, *Toshiba Amorphous Saturable Cores*, Mitsui & Co., New York.
8. Magnetics Inc. Bulletin, *New Magnetic Amplifier Cores and Materials*, Butler, PA.
9. R. Dulskis, J. Estey, and A. Pressman, "A Magnetic Amplifier Controlled 40 kHz Switching Regulator," *Wescon Proceedings*, San Francisco, CA, 1977.
10. R. Taylor, "Optimizing High Frequency Control Magamp Design," *Proceedings Powercon 10*, 1983.
11. R. Hiramatsu, K. Harada, and T. Ninomiya, "Switchmode Converter Using High Frequency Magnetic Amplifier," *International Telecommunications Energy Conference*, 1979.
12. S. Takeda and K. Hasegawa, "Designing Improved Saturable Reactor Regulators with Amorphous Magnetic Materials," *Proceedings Powercon 9*, 1982.
13. R. Hiramatsu and C. Mullett, "Using Saturable Reactor Control in 500 kHz Converter Design," *Proceedings Powercon 10*, 1983.
14. A. Pressman, "Amorphous Magnetic Parts for High Frequency Power Supplies," *APEC Conference 1990* (Toshiba Seminar), Mitsui & Co., New York.
15. C. Jamerson and D. Chen, "Magamp Post Regulators for Symmetrical Topologies with Emphasis on Half Bridge Configuration," *Applied Power Electronics Conference (APEC)*, 1991.
16. Keith Billings, *Switchmode Power Supply Handbook*, Chapter 2.1.4, McGraw-Hill, New York, 1998.
17. H. Matsuo and K. Harada, "New Energy Storage DC-DC Converter with Multiple Outputs," *Solid State Power Conversion*, November 1978, pp. 54–56.

CHAPTER 11

Analysis of Turn "On" and Turn "Off" Switching Losses and the Design of Load-Line Shaping Snubber Circuits

11.1 Introduction

Topologies that have a transformer winding or inductor in series with the power transistor have switching losses due to the overlap of current and voltage in the power transistor during the "on/off" transitions. Transistor dissipation during the turn "off" edge accounts for most of the switching losses.

The turn "off" power loss is given by the integral $\int I(t)V(t)dt$ over the turn "off" and turn "on" interval, which may last anywhere from 0.2 to 2 μs (for bipolar transistors). Circuits to minimize these overlap losses at turn "off" are called turn "off" snubbing or load-line shaping circuits and are the main subject of discussion in this chapter.

The switching loss generally has a very high peak value. Even when averaged, it can be larger than the average transistor conduction time dissipation. This switching loss, occurring once per period, is a larger fraction of the total transistor dissipation at higher frequencies, and is one of the prime limitations to the use of bipolar transistors at frequencies above 50 kHz.

Switching losses during the turn "on" transition are often considerably smaller for topologies with a power transformer because the transformer leakage inductance reduces the rate of rise of current. At the instant of turn "on," the large instantaneous impedance of the leakage inductance forces the voltage across the transistor to drop rapidly to zero while the leakage inductance slows up current rise time. Thus, throughout most of the current rise time, the voltage across the transistor is close to zero, and switching losses due to voltage-current overlap are small.

In the buck regulator (Figure 1.4), however, there are large voltage-current overlap losses in the transistor at both turn "on" and turn "off." In the buck, the power transistor turns "on" into the negligibly low impedance of the conducting free-wheeling diode *D*1, and the overlap of rising current and falling voltage generates a large spike of instantaneous power dissipation. Switching losses in the buck transistor at turn "off" are minimized by the same turn "off" snubber circuits used in transformer-type topologies. (Turn "on" snubbers are discussed in Sections 1.3.4 and 5.6.6.3 to 5.6.6.8.)

With MOSFETs, the turn "off" switching losses are considerably lower than those with bipolar transistors. Current fall time with a MOSFET is so rapid that with a small snubber, or even the disrupted capacitance seen at the drain, the current will have fallen considerably by the time the voltage across it has risen significantly.

Although turn "off" snubbers are used with MOSFETs, their prime function is not to reduce overlap dissipation, which is already low. Rather, the function of the MOSFET turn "off" snubber is to reduce the amplitude of the leakage inductance voltage spike. Since leakage inductance voltage spikes are proportional to dI/dt in the transistor, a MOSFET with much faster current turn "off" time than a bipolar will have a larger voltage leakage spike. Thus, although a MOSFET also requires a turn "off" snubber, it is less dissipative than that for a bipolar transistor.

With MOSFETs, there is considerable dissipation at turn "on"; this is due not to overlap of simultaneous high voltage and current, but to the relatively high MOSFET drain to source capacitance C_o. This capacitance is charged, often to twice the supply voltage, and thus stores energy $\frac{1}{2}C_o(2V_{dc})^2$ at turn "off." At the subsequent turn "on," this energy is dissipated in the MOSFET, which averaged over a period T amounts to $\frac{1}{2}C_o(2V_{dc})^2/T$.

Unfortunately, this dissipation is further increased by the small, benign snubber the MOSFET requires for minimizing the leakage inductance voltage spike, as it adds capacitance at the transistor output. All this will become more obvious after considering the design of the usual *RCD* (resistor, capacitor, diode) turn "off" snubber covered later in this section.

11.2 Transistor Turn "Off" Losses Without a Snubber

Consider the primary elements of a forward converter shown in Figure 11.1*a*, which include a typical RCD turn "off" snubber *R*1, *C*1, *D*1. Assume that the output power is 150 W and the DC supply

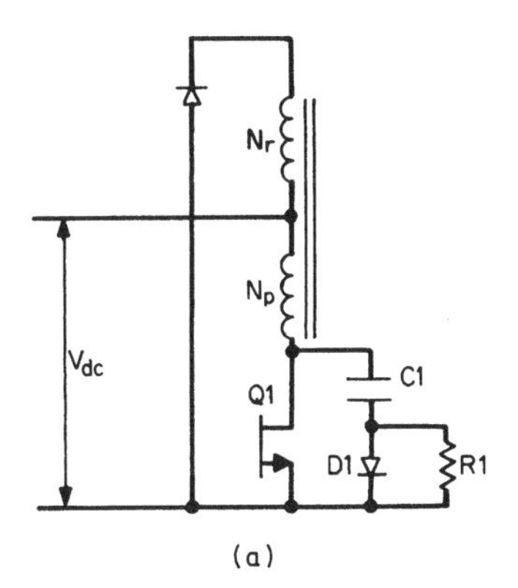

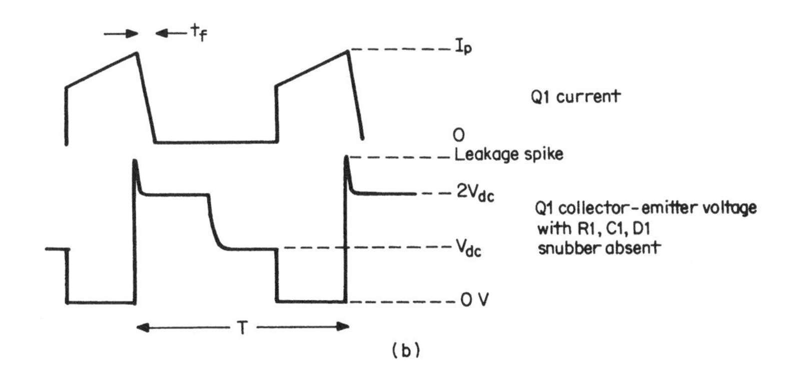

FIGURE 11.1 (*a*) A forward converter with snubber *R*1, *C*1, *D*1. When *Q*1 turns "off," the collector voltage starts to rise, *D*1 turns "on" immediately, and *C*1 slows the voltage rise time to minimize the overlap of rising voltage and falling current, thus reducing the *Q*1 switching loss. The next time *Q*1 turns "off," *C*1 must lose the $2V_{dc}$ charge it picked up the previous turn "off." It loses that charge in the previous *Q*1 "on" time. Capacitor *C*1 discharges through *Q*1 and *R*1. (*b*) With snubber absent, the collector voltage rises almost instantaneously; *Q*1 dissipation is $(2V_{dc}I_p/2)(t_f/T)$.

voltage is developed from a nominal 115-VAC line. Typically, the rectified DC supply voltage for an off-line power supply will range from 136 V to 184 V.

From Eq. 2.28, the peak current I_p in $Q1$ is

$$I_p = 3.13P_o/\underline{V_{dc}} = 3.13 \times 150/136 = 3.45 \text{ A}$$

Assume that $Q1$ is a very fast bipolar transistor such as the third-generation Motorola 2N6836. It has a V_{ceo} rating of 450 V (850 V V_{cev}), and a 15-A collector current rating. Its data sheet shows that nominal collector current fall time from 3.5 A with a 5-V reverse bias is 0.15 μs. We will assume that the worst case is twice that, or 0.3 μs.

Consider operation with the snubber network $R1, C1, D1$ absent. At the instant of turn "off," the collector rises, as in any forward converter whose power and reset windings have equal turns, to twice V_{dc}. The force that drives the collector to $2V_{dc}$ is the current that is stored in the magnetizing inductance and leakage inductances in series. As with any inductor which is carrying a current, when a series switch opens to interrupt that current, the polarity across the inductors reverses and drives the collector voltage toward $2V_{dc}$.

Since the collector output capacitance is low, this voltage rise time is essentially instantaneous. Assuming a maximum V_{dc} of 184 V, the voltage rises instantaneously to 368 V, and the current falls linearly from 3.45 A in 0.3 μs, as seen in Figure 11.1*b*.

This amounts to a dissipation, when averaged over the 0.3 μs switching period, of $368 \times 3.45/2 = 635$ W. Further, assuming a 100-kHz switching frequency, the average power dissipation is $63 \times 0.3/10 = 19$ W. This is excessive and would require, in most cases, an unacceptably large heat sink to keep the transistor junction temperature to a reasonable value.

Note also that this is an optimistic calculation. As discussed in Section 1.3.4, the estimated dissipation depends on the scenario assumed in the timing of current turn "off." Current generally hangs on at its peak for a short time before starting to fall. Thus the 19-W dissipation calculated above may well be about 50% greater in actuality.

The $R1, C1, D1$ snubber of Figure 11.1*a* reduces transistor dissipation by slowing up the collector voltage rise time so that the current fall-time waveform intersects the voltage rise-time waveform as low as feasible on the rising voltage waveform. How it works and how the component magnitudes are calculated can be seen as follows.

11.3 *RCD* Turn "Off" Snubber Operation

In Figure 11.1*a*, when the $Q1$ base receives its turn "off" command, the transformer leakage inductance maintains the peak current which had been flowing just before turn "off." That peak current divides in

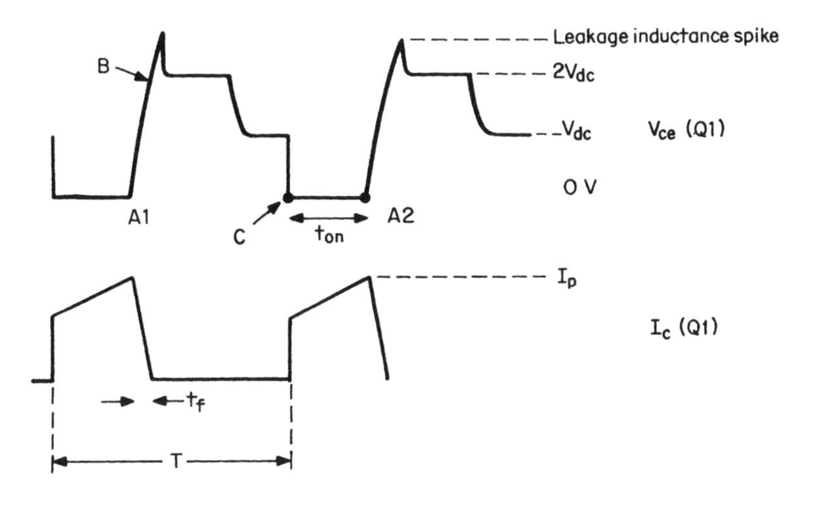

FIGURE 11.2 Snubber capacitor $C1$ slows up collector voltage rise time from $A1$ to B at turn "off." At the next turn "off" at $A2$, $C1$ must lose the $2V_{dc}$ charge it picked up at time B. It loses that charge in the "on" time t_{on} just before turn "off." With the snubber in, $Q1$ dissipation is $(2V_{dc}I_p/12)(t_f/T)$.

some way between the "off"-turning collector and $C1$, which is now conducting through diode $D1$.

The amount of current I_{C1} flowing into $C1$ slows up the collector voltage rise time, and by making $C1$ large enough, the rising collector voltage and falling collector current intersect sufficiently low on the rising collector voltage waveform that transistor dissipation is decreased significantly.

There is a limitation on how large $C1$ can be made, as can be seen in Figure 11.2, where it is shown that at the start of any turn "off"—say at $A1$—$C1$ must have no charge. At the end of the turn "off" at B, $C1$ has slowed up the voltage rise time but has accumulated a voltage $2V_{dc}$ (neglecting for the moment the leakage inductance spike).

At the start of the next turn "off" at $A2$, $C1$ must again have no voltage across it. Thus, at some time between B and $A2$, $C1$ must be discharged. It is discharged in the interval C to $A2$ by resistor $R1$. When $Q1$ turns "on" at C, the top end of $C1$ goes immediately to ground and $C1$ discharges through $Q1$ and $R1$.

Thus, once $C1$ has been selected large enough to yield a sufficiently long collector voltage rise time, $R1$ is chosen to discharge $C1$ to within 5% of its full charge in the minimum t_{on}. Or

$$3R1C1 = t_{on(min)} \tag{11.1}$$

If $C1$ accumulates a voltage $2V_{dc}$ at each turn "off," it stores an energy of $0.5C1(2V_{dc})^2$ joules. If it dissipates that energy in $R1$ during each "on" time, power dissipation in $R1$ in watts (for T in seconds) is

$$PD_{R1} = \frac{0.5C1(V_{dc})^2}{T} \tag{11.2}$$

It will be seen in the following section that the power dissipation in $R1$ given by Eq. 11.2 seriously limits how much collector voltage rise time can be increased by increasing $C1$.

11.4 Selection of Capacitor Size in *RCD* Snubber

Equation 11.2 shows that the power dissipation in $R1$ is proportional to $C1$. Hence, there is need to select $C1$ large enough to adequately lengthen collector voltage rise time, but not cause excessive dissipation in $R1$.

There is no best way to select $C1$. It is sized in different ways by different designers who make different assumptions as to how much of the peak current in $Q1$ is available to charge $C1$, how much the $Q1$ collector voltage rise time is to be increased, and how fast the collector current falls to zero. The latter depends strongly on the turn "off" speed of the transistor used, and the reverse base drive voltage. However, the following method has been found to be a satisfactory way to select $C1$ for bipolar transistors.

TIP *Some modern oscilloscopes have real time or digital VI multiplying capability, and can measure the actual switching edge dissipation, displaying the power trace over a switching period. This permits the optimization of the snubber components. ~K.B.*

When the $Q1$ base receives its turn "off" command, the peak current in the collector is partly diverted into $C1$ as the voltage across it starts to increase. It is assumed that half the initial peak current I_p is diverted into $C1$ and half remains flowing into the gradually "off"-turning collector (since the transformer leakage inductance maintains the total current I_p for some time).

Capacitor $C1$ is selected so that the collector voltage is permitted to rise to $2V_{dc}$ in the same time t_f that the collector current falls to zero. This last is read or estimated from data sheets for the specific transistors. Thus

$$C1 = \frac{I_p}{2}\frac{t_f}{2V_{dc}} \tag{11.3}$$

Now capacitance $C1$ can be calculated from Eq. 11.3, resistance $R1$ can be calculated from Eq. 11.1 from the known minimum "on" time, and the dissipation in resistor $R1$ can be calculated from Eq. 11.2.

Overlap dissipation in $Q1$ can be estimated as in Section 1.3.4 on the assumption that during the turn "off" interval t_f, $Q1$ current $I_p/2$ starts falling toward zero at the same time its voltage starts rising toward $2V_{dc}$, and that the collector current reaches zero at the same time the collector voltage reaches $2V_{dc}$.

In Section 1.3.4, it was noted that the collector dissipation during the interval t_f is

$$\frac{I_{max}V_{max}}{6} = \frac{I_p}{2}\frac{2V_{dc}}{6}$$

and averaged over one period T, transistor dissipation is

$$\frac{(I_p/2)(2V_{dc})t_f}{6T} \tag{11.4}$$

11.5 Design Example—*RCD* Snubber

Design the *RCD* snubber for the forward converter of Section 11.2. Recall that the peak current I_p just before turn "off" was 3.45 A, transistor fall time was 0.3 μs, and transistor dissipation without a snubber was 19 W. From Eq. 11.3

$$\begin{aligned} C1 &= \frac{(I_p/2)t_f}{2V_{dc}} \\ &= \frac{(3.45/2)(0.3\times 10^{-6})}{2\times 184} \\ &= 0.0014\ \mu F \end{aligned}$$

and from Eq. 11.1

$$R1 = \frac{t_{on(min)}}{3C1}$$

Recall that a forward converter transformer is designed so that maximum transistor "on" time occurs at minimum DC voltage, and is forced to be $0.8T/2$. For a switching frequency of 100 kHz, this is 4 μs. In Section 11.2, maximum and minimum input voltages were 15% above and below the nominal value, so $t_{on(min)}$ is $4/1.3$ or 3 μs. Then from Eq. 11.1

$$\begin{aligned} R1 &= \frac{3\times 10^{-6}}{3\times 0.0014\times 10^{-6}} \\ &= 714\ \Omega \end{aligned}$$

and from Eq. 11.2

$$PD_{R1} = \frac{0.5C1(2V_{dc})^2}{T}$$
$$= \frac{0.5(0.0014 \times 10^{-6})(2 \times 184)^2}{10 \times 10^{-6}}$$
$$= 9.5\,W$$

and from Eq. 11.4, $Q1$ overlap dissipation is

$$PD_{Q1} = \frac{(I_p/2)(2V_{dc})t_f}{6T}$$
$$= \frac{(3.45/2)(2 \times 184)(0.3 \times 10^{-6})}{6 \times 10 \times 10^{-6}}$$
$$= 3.2\,W$$

Thus, although 9.5 W has been added in the resistor $R1$, dissipation in the transistor, which is far more failure-prone, has been reduced from 19 to 3.2 W. Actually overlap dissipation in the transistor may be more than the calculated 3.2 W, as a best-case scenario has been assumed for the relative timing of the current fall time and voltage rise time (Section 1.3.4).

If temperature measurement on the $Q1$ case indicates that it is running too warm, $C1$ can be increased at the expense of more dissipation in $R1$, but this is far more acceptable than dissipation in $Q1$.

It is sometimes thought that decreasing $R1$ decreases its dissipation. This is not so, as Eq. 11.2 indicates. Making $R1$ smaller only causes $C1$ to be totally discharged earlier than needed (earlier than $A2$ in Figure 11.2). Energy dissipated in $R1$ is equal to the energy stored in $C1$ per cycle, and energy stored is proportional only to the magnitude of $C1$ and the square of the voltage to which it is charged at turn "off." Dissipation in $R1$ is thus unrelated to its resistance.

11.5.1 *RCD* Snubber Returned to Positive Supply Rail

The *RCD* snubber is often (and preferably) returned to the positive supply rail as shown in Figure 11.3. It works exactly in the same way as when it is returned to ground as in Figure 11.1. At turn "off," $D1$ conducts, and $C1$ increases collector voltage rise time with its charging current flowing into V_{dc}. At the following turn "on," $C1$ is discharged through $Q1$ and the supply source V_{dc}.

The advantage of returning $R1$, $D1$ to V_{dc} instead of ground is that the maximum voltage stress on $C1$ is now only V_{dc} instead of $2V_{dc}$.

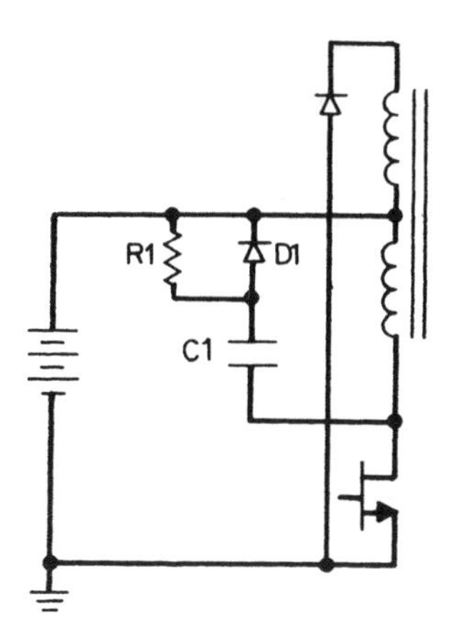

FIGURE 11.3 Returning the snubber to the positive rail reduces the voltage stress on *C*1 to half that of returning it to ground.

11.6 Non-Dissipative Snubbers[1-8]

The conventional *RCD* snubber for off-line switching supplies operating at over 50 kHz most often ends up dissipating 10 W or more. This is troublesome not only for the added dissipation, but also because of the size and required location of the snubber resistor. General practice is to de-rate power resistors by a factor of 2, so 10 W of dissipation usually requires the use of a 20-W resistor.

A 20-W resistor is quite large, and finding a location for it is often difficult. Also, in dissipating 10 W, it heats any surrounding nearby components, which further complicates the selection of a satisfactory location for it.

"Dissipationless snubbers," as covered in Figure 11.4, are more complex but provide a good solution to this problem. Just as in the conventional *RCD* dissipative snubber, a capacitor is used to slow down the collector voltage rise time, but the capacitor is not discharged through a resistor as before, as this would waste power.

Instead, the stored electrostatic energy in the capacitor *C*1 is transferred to the inductor *L*1 in the form of electromagnetic energy as current in the inductor *L*1. Then, before the next cycle when the capacitor must again be discharged, the inductor by resonant discharge action will discharge its stored energy back into the DC input bus. Thus no energy is wasted—it is first stored on a slow-up capacitor, and then returned with negligible loss to the input bus.

The details can be seen in Figure 11.4. When *Q*1 turns "off," its collector voltage starts rising, *D*1 conducts, and *C*1 slows down the

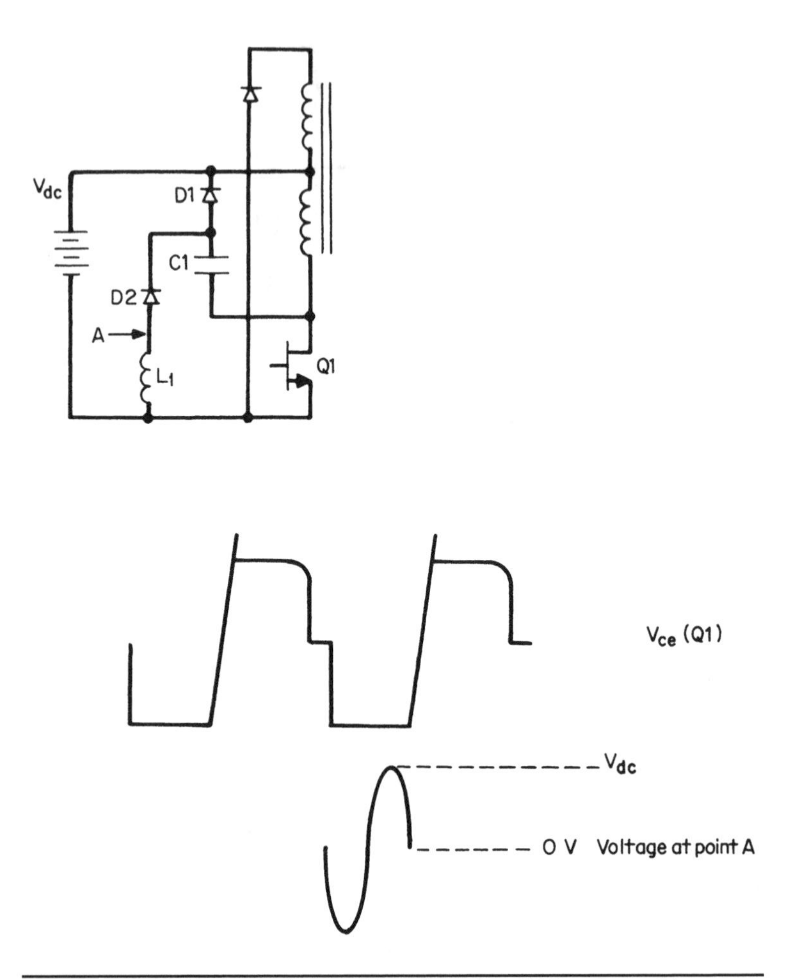

FIGURE 11.4 A dissipationless snubber. Capacitor $C1$ lengthens $Q1$ collector voltage rise time as in an *RCD* snubber, but at $Q1$ turn "on," the stored energy on $C1$ is converted to stored magnetic energy in $L1$ during the first half cycle of a "ring." During the second half cycle of the ring, point A goes positive and returns this energy to V_{dc} without loss.

voltage rise time just as in Figure 11.3. The bottom end of $C1$ is driven up to $2V_{dc}$, and its top end is clamped to V_{dc} through $D1$, so current flows from the bottom end of the transformer up through $C1$ and $D1$ to slow down the change in voltage across $Q1$. Energy $0.5C1(V_{dc})^2$ joules is stored in the capacitor.

When $Q1$ turns "on" again, the bottom end of capacitor $C1$ is pulled from $2V_{dc}$ to ground and the top end goes negative by V_{cc} and this negative voltage is applied across $L1$ and $D2$ in series so that current builds up in the inductor as the capacitor discharges via $D2$, the current flowing from bottom to top in $L1$. An resonant oscillatory current "ring" commences with a frequency $f_r = 1/2\pi\sqrt{L1C1}$.

At the end of a half period of this ring, the electrostatic energy in the form of voltage on $C1$ has been changed into electromagnetic energy in the form of stored current in $L1$ and the current in $L1$ is at its maximum. During the next half period of this ring, the voltage at the top end of $L1$ rings towards twice the voltage, in the normal resonant manner, so that it goes positive enough for $D1$ to conduct and the current in $L1$ now flows via D1 back into the supply bus. If $L1$ is a high-Q inductor, all the energy stored in it in the first half cycle of the ring is returned back to V_{dc} in the second half cycle.

Capacitor $C1$ is first chosen large enough to lengthen the $Q1$ voltage rise time as required. Then $L1$ is chosen so that the full ring period is somewhat less than the minimum $Q1$ "on" time.

11.7 Load-Line Shaping (The Snubber's Ability to Reduce Spike Voltages so as to Avoid Secondary Breakdown)

The snubber offers a second very important advantage in addition to increasing voltage rise time and thus decreasing average transistor dissipation. It prevents secondary breakdown, which occurs if the instantaneous voltage and current cross the reverse-bias safe operating area (RBSOA) boundary given in the manufacturer's data sheets (Figure 11.5).

This boundary can be crossed by the leakage inductance spike (Figure 2.10) that occurs at the instant of turn "off." Transistor manufacturers state that if the boundary is crossed even once, secondary-breakdown failure may occur.

An exact analysis of the sequence of collector voltage changes and their magnitudes at the instant of turn "off" is not possible without computer analysis. However, the following inexact discussion illustrates the magnitude of the problem and how the snubber capacitor reduces the leakage inductance spike.

When $Q1$ turns "off," the transformer leakage inductance keeps the current which had been flowing in it from falling for some short time. It was assumed above that approximately half that current continues to flow into the slowly "off"-turning transistor, and half flows into the snubber capacitor $C1$.

In Figure 11.6, when $Q1$ turns "off," the voltage across the magnetizing inductance reverses and this reverses the voltage across the reset winding N_r. The top end of N_r immediately goes negative, and

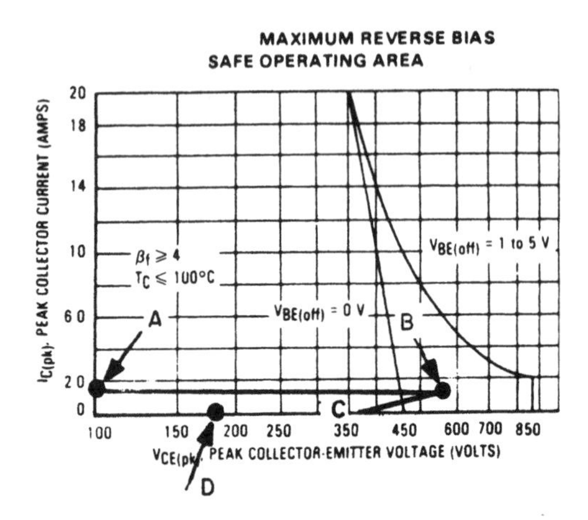

FIGURE 11.5 Reverse-bias safe operating area (RBSOA) for 15-A, 450-V fast transistor type 2N6836. At turn "off," because the transformer leakage inductance maintains current, operation is from *A* to *B* at the tip of the leakage inductance spike. If not for reverse base bias, operation would have crossed the RBSOA boundary and the transistor would have failed in secondary-breakdown mode. The *RCD* snubber, in addition to decreasing overlap dissipation, decreases the amplitude of the leakage inductance spike. *(Courtesy of Motorola Inc.)*

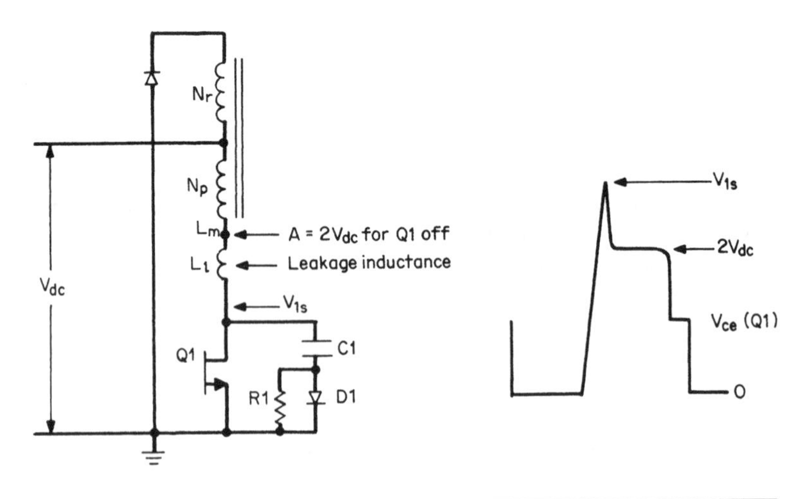

FIGURE 11.6 The leakage spike is roughly $(I_p/2)(L_l/C1)^{1/2}$ above $2V_{dc}$.

is clamped to ground by $D4$. Since $N_p = N_r$, this clamps the voltage across L_m to V_{dc}, and the voltage at point A rises to $2V_{dc}$.

This first half-cycle ring, which is the leakage inductance spike, sits on top of the voltage $2V_{dc}$ at point A. The quantity $\sqrt{L_l/C1}$ is often referred to as the *characteristic impedance* of the LC circuit. Thus increasing $C1$ to increase collector voltage rise time also decreases the leakage inductance spike.

Half the original peak current $I_p/2$ flowed into the series combination of the leakage inductance L_l, the snubber capacitor $C1$, and diode $D1$. This causes a sinusoidal ring whose half period is $\pi\sqrt{L_l C_1}$. To a close approximation, the amplitude of the first half cycle is $1/2\pi(\sqrt{L_l/C1})$.

It is of interest to calculate the magnitude of the leakage inductance spike on the basis of these observations. The leakage inductance of a 100-kHz transformer for the design example at the end of Section 11.4 is expected to be about 15 μH, so in the preceding example the characteristic impedance of the LC circuit is

$$\sqrt{L_l/C1} = \sqrt{15 \times 10^{-6}/0.0014 \times 10^{-6}} = 103\,\Omega$$

Hence, for the 3.45-A peak current in $Q1$ just prior to turn "off," the leakage inductance spike amplitude is $(3.45/2)103 = 178$ V. The peak voltage at the top of the leakage inductance spike is then $2V_{dc} + 178 =$ 547 V. Although not too precise an analysis, this yields the amplitude of the leakage inductance spike sufficiently accurately to explain the possibility of secondary breakdown.

In Figure 11.5 just prior to turn "off," the operating point of $Q1$ on its volt-ampere curve is at point A—3.45 A at close to 0 V. As the transistor is turned "off," its load-line locus is along the path $ABCD$. The leakage inductance maintains total current, but half starts flowing into $C1$, leaving 1.73 A flowing into the transistor. The transistor operation then moves horizontally along the line AB. At B, the current is still 1.73 A, and the voltage is 547 V. After the short duration of the leakage spike, the locus drops to $2V_{dc}$, until the transformer core resets, and then back to V_{dc}, where it stays until the next turn "on" (Figure 2.10). Figure 11.5 shows that if the transistor is reverse biased by 5 V at turn "off," the point at 547 V, 1.73 A is still within the manufacturer's RBSOA curve and secondary breakdown should not occur. With no reverse bias, the operating point goes outside the RBSOA boundary and secondary breakdown failure will occur.

Thus, although the initial function of the snubber capacitor $C1$ (Figure 11.1) was to lengthen collector voltage rise time, it may have to be increased above the value calculated from Eq. 11.3 to reduce the leakage inductance spike. At higher output powers, even with a 5-V reverse base bias at turn "off," the leakage inductance spike may cross the RBSOA curve, as the turn "off" locus is at a higher current along its horizontal portion. This would require a larger snubber capacitor.

11.8 Transformer Lossless Snubber Circuit

Figure 11.7 shows a scheme which reduces or even totally eliminates the leakage inductance spike for an *RCD* snubber. It does this without increasing the snubber capacitance, which would increase snubber resistor dissipation.

The tradeoff for this is the addition of the small transformer *T*1. It still requires the conventional *RCD* snubber, but *C*1 can be considerably smaller. It works as follows. As shown in Figure 11.7, *T*1 is a small 1/1 transformer. Its core area and number of turns must be chosen to sustain the maximum volt-second product across the *T*2 primary when *Q*1 is "off." Since the core size should be as small as possible, this means a small core area and hence a relatively large number of turns. Since the power that *T*1 carries is small, primary and secondary wire sizes can be small, thus permitting a small core.

Note the dot polarities on *T*1 primary and secondary. Diode *D*1 does not conduct until the dot end of N_s reaches V_{dc}, which does not happen until the voltage at the dot end of *T*1 primary with respect to its no-dot end rises to V_{dc}. That occurs only when the *Q*1 collector reaches $2V_{dc}$.

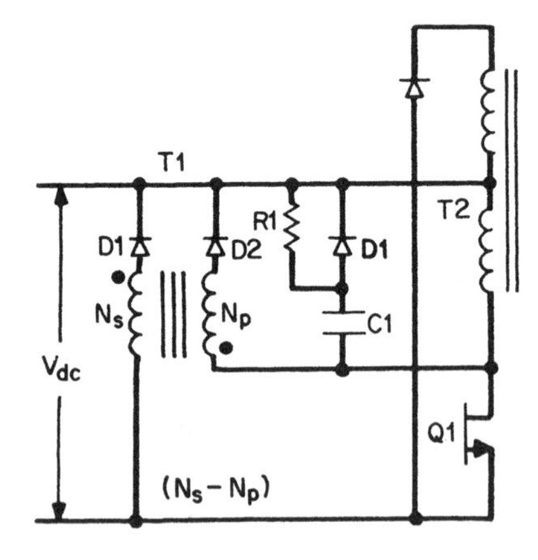

FIGURE 11.7 A leakage spike clipping aid to an *RCD* snubber. *T*1 is a small 1/1 transformer. When *Q*1 collector voltage reaches $2V_{dc}$, *D*2 conducts, forcing *D*1 to conduct and clamp $V_{ce(Q1)}$ to $2V_{dc}$. Thus, there is no leakage spike at the *Q*1 collector above $2V_{dc}$ if *T*1 has low leakage inductance. This minimizes the size of *C*1.

Thus, when the $Q1$ collector reaches $2V_{dc}$, $D1$ conducts and clamps the voltage across N_s to V_{dc}. Since the transformer turns ratio is 1/1, the collector voltage is clamped at $2V_{dc}$—i.e., there is no leakage spike at the collector above $2V_{dc}$. For this to be effective, the leakage inductance of $T1$ must be very small.

When $D1$ conducts, the energy stored in the $T2$ leakage inductance is returned with no loss to the V_{dc} supply bus.

References

1. E. Whitcomb, "Designing Non-Dissipative Current Snubbers for Switched Mode Converters," *Proceedings Powercon 6*, 1979.
2. W. Shaunessy, "Modeling and Design of Non-Dissipative *LC* Snubber Networks," *Proceedings Powercon 7*, 1980.
3. L. Mears, "Improved Non-Dissipative Snubbers for Buck Regulators and Current Fed Inverters," *Proceedings Powercon 9*, 1982.
4. M. Domb, R. Redl, and N. Sokal, "Non-Dissipative Turn Off Snubber Alleviates Switching Power Dissipation, Second Breakdown Stress," *PESC Record*, 1982.
5. T. Ninomiya, T. Tanaka, and K. Harada, "Optimum Design of Non-Dissipative Snubber by Evaluation of Transistor's Switching Loss," *PESC Record*, 1985.
6. T. Tanaka, T. Ninomiya, and K. Harada, "Design of a Non-Dissipative Snubber in a Forward Converter," *PESC Record*, 1988.
7. Keith Billings, *Switchmode Power Supply Handbook*, Chapter 2.32, McGraw-Hill, New York, 1999.

CHAPTER 12

Feedback Loop Stabilization

12.1 Introduction

Before going into the details of stabilizing a feedback loop, we will first consider in a semi-quantitative way why a feedback loop may oscillate.

Consider the negative-feedback loop for a typical forward converter as shown in Figure 12.1. Although the essential error-amplifier and PWM functions are contained in pulse-width-modulating chips, the chips also provide many other functions. However, for our initial understanding of the stability problem, we will consider only the error amplifier and pulse-width modulator at this stage.

A small, slow variation of V_o due to either line input or load changes is sensed by the inverting input of error amplifier EA via the sampling network $R1$, $R2$, and compared to a reference voltage at the non-inverting EA input. This will cause a small change in the relatively slow-changing voltage V_{on} at the EA output, and at the A input to the pulse-width-modulator PWM. The PWM, as described heretofore, typically compares that DC voltage to a 0- to 3-V triangle V_t at its B input. This generates a rectangular pulse, whose width t_{on} is the time from the start of the triangle t_o until t_1, when the triangle crosses the voltage at the B input of the PWM. That pulse fixes the "on" time of the chip output transistor and that of the power transistor. Thus a slow increase (e.g.) in V_{dc} causes a slow increase in V_{y} and hence a slow increase in V_o, since $V_o = V_{\text{y}} t_{\text{on}}/T$. The increase in V_o causes an increase in V_s and hence a decrease in V_{ea}. Since t_{on} is the time from the start of the triangle to t_1, this causes a decrease in t_{on} and restores V_o to its original value. Similarly, a decrease in V_{dc} causes an increase in t_{on} to maintain V_o constant and the output voltage is stabilized as required.

Drive to the power transistor often may be taken from either the emitter or the collector of the chip's output transistor via a

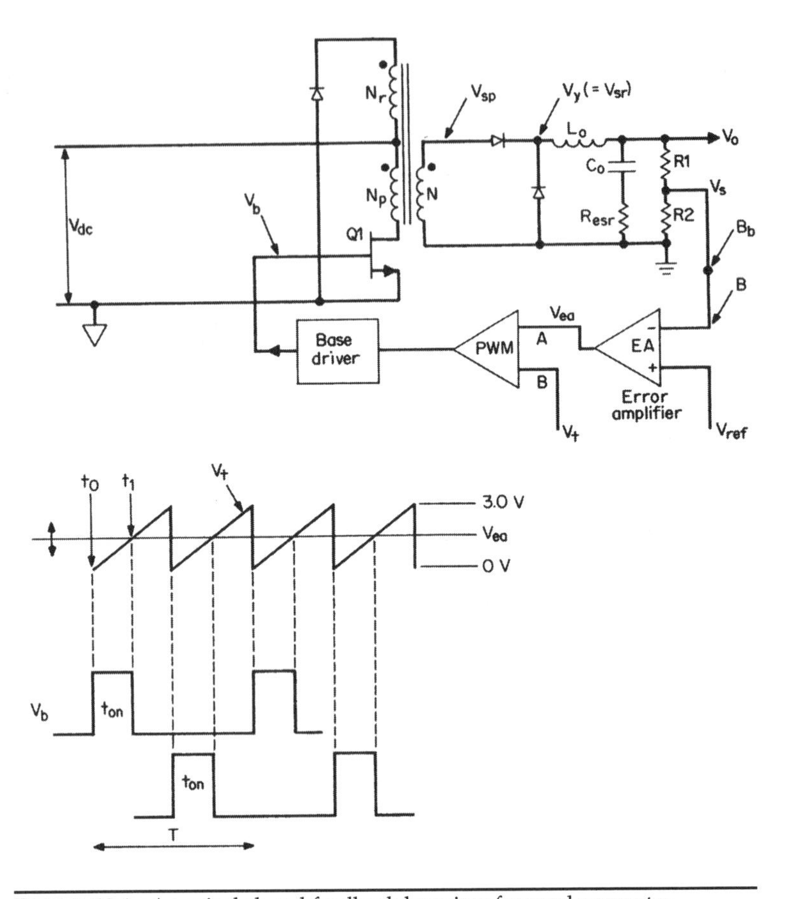

FIGURE 12.1 A typical closed feedback loop in a forward converter.

current-amplifying base driver. Whichever of the emitter or collector is chosen for the output, it must be ensured that polarities are such that an increase in V_o causes a decrease in t_{on}. Note that most PWM chip's output transistors are "on" for the time from t_0 to t_1. With such chips, V_s is fed to the inverting EA input, and for an NPN power transistor, its base (or gate if it is an N-channel MOSFET) is driven from the emitter of the chip's output transistor.

The circuit in Figure 12.1 thus provides negative feedback and a stable output voltage at low frequencies. Within the loop, however, there are low-level noise voltages and voltage transients possible, which have a wide spectrum of sinusoidal Fourier components. The gain changes and phase shifts for all these Fourier components by different amounts in the L_o, C_o output filter, the error amplifier, and the

PWM from V_{ea} to V_{sr}. If one of these Fourier components has a loop gain of 1 and a phase shift of a further 180° (the first 180 being provided by the negative feedback connection), the total phase shift will be 360°. Then the feedback signal will be in phase with the original input, resulting in positive rather than negative feedback, which will result in oscillation as described below.

12.2 Mechanism of Loop Oscillation

Consider the forward converter feedback loop of Figure 12.1. Assume for a moment that the loop is broken open at point B, the inverting input to the error amplifier. At any frequency, there is gain and phase shift from B to V_{ea}, from V_{ea} to the average voltage at V_{sr}, and from the average voltage at V_{sr} through the L_o, C_o filter around back to B_b, just before the loop break.

Now assume that a signal of some frequency f_1 is injected into the loop at B and comes back around as an echo at B_b. The echo is modified in phase and amplitude by all the previously mentioned elements in the loop. If the modified echo has returned exactly in phase with, and equal in amplitude to, the signal which started the echo, and the loop is now closed (B_b closed to B) with the injected signal removed, the circuit will continue to oscillate at the frequency f_1. The initial signal that starts the echo and maintains the oscillation is the f_1 Fourier component in the noise spectrum.

12.2.1 The Gain Criterion for a Stable Circuit

The first criterion for a stable loop is that at the *crossover frequency*, where the total open-loop gain is unity, the total open-loop phase shift must be less than 360°. This includes the necessary 180° negative feedback connection. The amount by which the total phase shift is less than 360° at the crossover frequency is called the *phase margin*.

To ensure a stable loop under worst-case variation of the associated components, the usual practice is to design for 35° to 45° phase margin.

12.2.2 Gain Slope Criteria for a Stable Circuit

At this point, we will describe some universally-used stability criteria involving the gain slope. Consider the networks shown in Figure 12.2. The gain V_o/V_{in} versus frequency is usually plotted in decibels (dB) on semilog paper. The scales are such that a linear distance of 20 dB (a numerical gain of 10) is equal to the linear distance of a factor of 10 in frequency, so lines representing gain variations of ±20 dB/decade

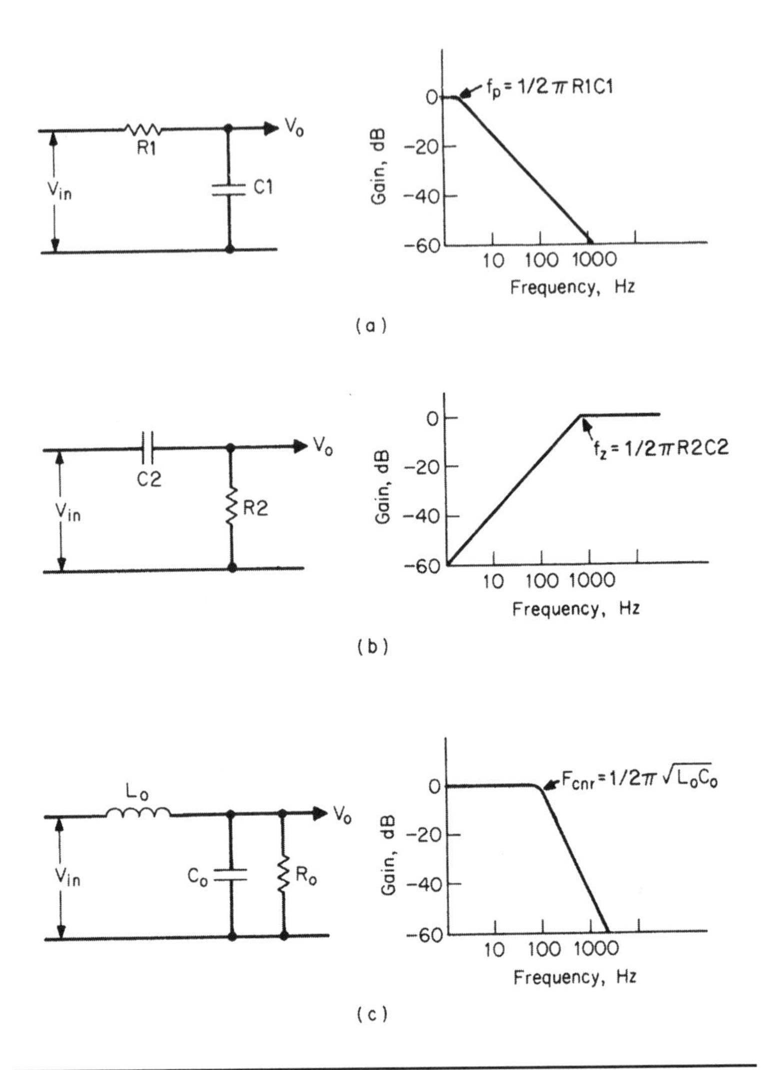

FIGURE 12.2 Some networks with their corresponding transfer functions: (*a*) an R/C network, (*b*) a C/R network, (*c*) an L/C/R network.

have slopes of ±1. Circuit configurations having a gain variation of ±20 dB per decade are thus described as having "±1 gain slopes."

Plot (*a*) shows an *RC* integrator, which has a gain dV_o/dV_{in} of −20 dB/decade above $f_p = 1/(2\pi R1\,C1)$. With the scales such that 20 dB of gain is the same linear distance as 1 decade in frequency, −20 dB/decade is a −1 gain slope. Such a circuit is referred to as a −1 *slope circuit.*

Plot (*b*) shows an *RC* differentiator, which has a gain of +20 dB/decade below $f_z = 1/(2\pi R2C2)$, where $X_{c2} = R_2$. Above f_z, gain asymptotically approaches 0 dB. With scales such that 20 dB is the same linear distance as 1 decade in frequency, +20 dB/decade has a +1 gain slope. Such a circuit is referred to as a +1 slope circuit.

Plot (*c*) shows an *L*/*C*/*R* filter with gain dV_o/dV_{in} of unity (0 dB) below its corner frequency of $F_{cnr} = 1/(2\pi\sqrt{L_oC_o})$, when critically damped ($R_o = \sqrt{L_oC_o}$). Above F_{cnr} the gain falls at a rate of −40 dB/decade. This happens because for every decade increase in frequency, X_L increases and X_C decreases, both by a factor of 10. With scales such that 40 dB is the same linear distance as 1 decade in frequency, −40 dB/decade is a −2 gain slope. Such a circuit is referred to as a −2 slope circuit.

An elementary circuit having a gain slope of −1 above the crossover frequency is the *RC* integrator of Figure 12.2*a*. The *RC* differentiator of Figure 12.2*b* has a +1 gain slope below the crossover frequency or a gain variation of +20 dB/decade. Such circuits have only 20 dB/decade gain variations because as frequency increases or decreases by a factor of 10, the capacitor impedance decreases or increases by a factor of 10 but the resistor impedance remains constant.

A circuit which has a −2 or −40 dB/decade gain slope above the corner frequency is the output *LC* filter (Figure 12.2*c*), which has no resistance (ESR) in its output capacitor. This is because as frequency increases by a factor of 10, the inductor impedance increases and the capacitor impedance decreases, both by a factor of 10.

Gain and phase shift versus frequency for an L_oC_o filter are plotted in Figure 12.3*a* and 12.3*b* for various values of output resistance R_o. The gain curves are normalized for various ratios of $k_1 = f/F_o$ where $F_o = 1/(2\pi\sqrt{L_oC_o})$ and for various ratios $k_2 = R_o/\sqrt{L_o/C_o}$.

Figure 12.3*a* shows that whatever the value of k_2, all gain curves beyond the so-called corner frequency of $F_o = 1/(2\pi\sqrt{L_oC_o})$ asymptotically approach a −2 slope (−40 dB/decade). The circuit with $k_2 =$ 1.0 is referred to as the *critically damped* circuit. The critically damped circuit has a very small resonant "bump" in gain, which starts falling at −2 slope immediately above the corner frequency F_o.

For k_2 greater than 1, the circuit is described as *underdamped*. Underdamped *LC* filters can have a very high resonant bump in gain at F_o.

Circuits with k_2 less than 1.0 are *overdamped*. Figure 12.3*a* shows that overdamped *LC* filters also asymptotically approach a gain slope of −2, but for a heavily overdamped (k_2 = 0.1) filter, the frequency at which the gain slope has come close to −2 is about 20 times the corner frequency F_o.

Figure 12.3*b* shows phase shift versus normalized frequency (f/F_o) for various ratios of $k_2 = R_o/\sqrt{L_o/C_o}$. For any value of k_2, the phase

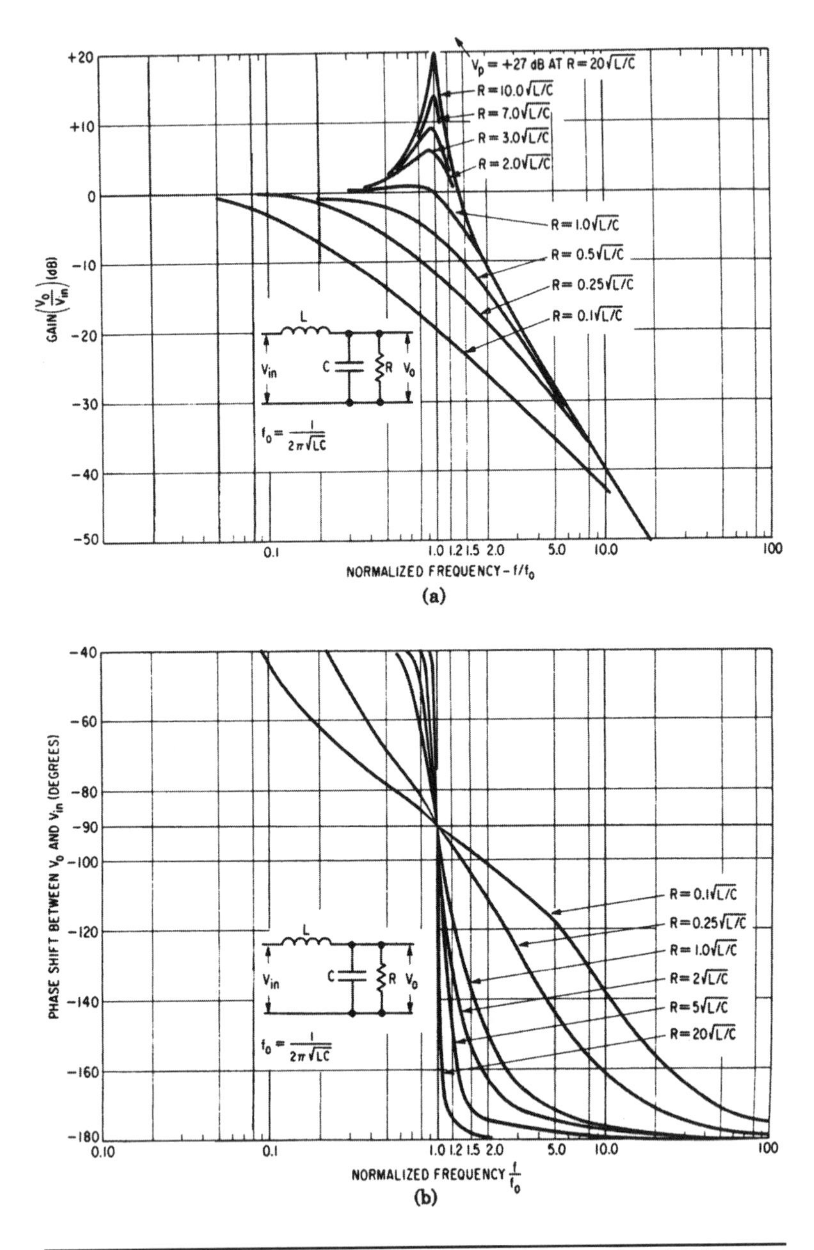

FIGURE 12.3 (*a*) Gain versus frequency for switching regulator *LC* filter. (*b*) Phase shift versus frequency for switching regulator *LC* filter. (*Courtesy of Switchtronix Press*)

shift of the output relative to the input is 90° at the corner frequency $F_o = 1/(2\pi\sqrt{L_o C_o})$, and for highly underdamped filters (R_o greater than $\approx 5\sqrt{L_o C_o}$) phase shift varies very rapidly with frequency. The shift is already 170° at a frequency of 1.5 F_o for $R_o = 5\sqrt{L_o/C_o}$.

In contrast, a circuit with a −1 gain slope can never yield more than a 90° phase shift, and its rate of change of phase shift with frequency is far lower than that of a −2 gain slope circuit, as exemplified in Figure 12.3*b*.

This leads to the second criterion for a stable circuit. The first criterion was that the total phase shift at the crossover frequency (frequency where total open-loop gain is unity or 0 dB) should be less than 360° by the "phase margin," which is usually taken as at least 45°.

This second criterion for a stable circuit is that the slope of the open-loop gain-frequency curve of the entire circuit as it passes through the crossover frequency should be −1. This gain curve is obtained from the arithmetic sum in decibels of all involved elements' gain curves. The criterion prevents rapid changes of phase shift with frequency, which are characteristic of a circuit with a −2 gain slope. An example is shown in Figure 12.4.

It is not an absolute requirement that the total open-loop gain curve have a −1 slope at crossover, but it does provide insurance that if any phase-shift elements have been overlooked, the phase margin will still be adequate.

The third criterion for a stable loop is to provide the desired phase margin, which will be set at 45° herein (Figure 12.4).

To satisfy all three criteria, it is necessary to calculate gains and phase shifts of all elements shown in Figure 12.1. This is shown below.

12.2.3 Gain Characteristic of Output *LC* Filter with and without Equivalent Series Resistance (ESR) in Output Capacitor

Aside from the flyback, which has an output filter capacitor only, all topologies discussed here have an output *LC* filter.[1] The gain versus frequency characteristic of this output *LC* filter is of fundamental importance. It must be calculated first as it determines how the frequency characteristics of the error amplifier must be shaped to satisfy the three criteria for a stable loop.

The gain characteristic of an output *LC* filter with various output load resistances is shown in Figure 12.3*a*. This curve is for an output *LC* filter whose capacitor has zero ESR. For the purpose of this discussion, it is sufficiently accurate to assume that the filter is critically damped, that is, $R_o = 1.0\sqrt{L_o/C_o}$. If the circuit is made stable for the gain curve corresponding to $R_o = 1.0\sqrt{L_o/C_o}$, it will be stable at other loads.

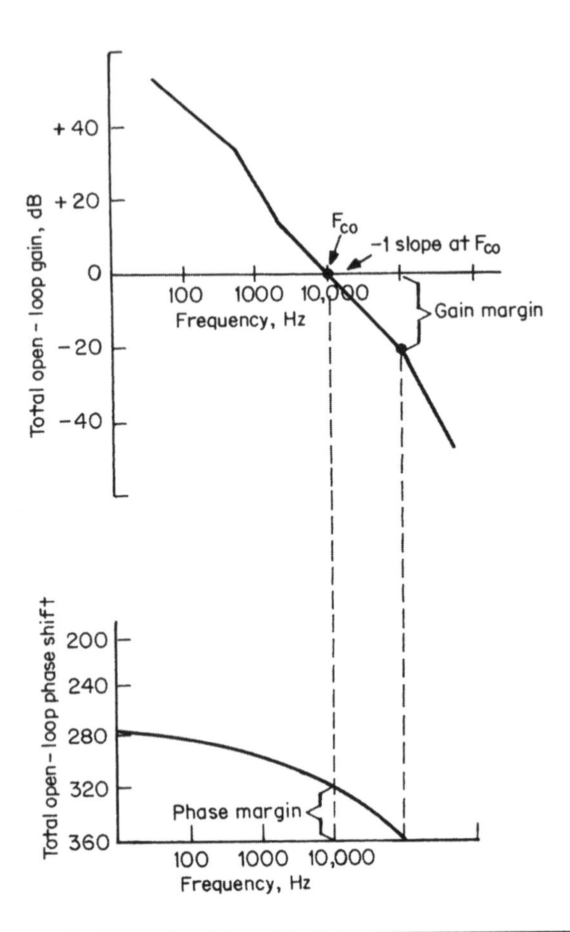

FIGURE 12.4 Total open-loop gain and phase shift. The frequency at which the total open-loop gain is forced to cross 0 dB is usually chosen one-fourth or one-fifth the switching frequency. For the loop to be stable, the *phase margin* should be maximized and at least 45°. Also, the total open-loop gain should pass through F_{co} at a −1 slope.

Nevertheless, the circuit merits examination for light loads ($R_o \gg 1.0\sqrt{L_o/C_o}$) because of the resonant bump in gain at the *LC* corner frequency $F_o = 1/(2\pi\sqrt{L_oC_o})$. This will be considered below.

The gain characteristic of the output *LC* filter with zero ESR will be drawn as curve 12345 in Figure 12.5*a*. There it is seen that the gain is 0 dB (numerical gain of 1) at any frequency below 2, up to the corner frequency $F_o = 1/(2\pi\sqrt{L_oC_o})$. At DC and frequencies less than F_o, the impedance of C_o is much greater than that of L_o and the output/input gain is unity.

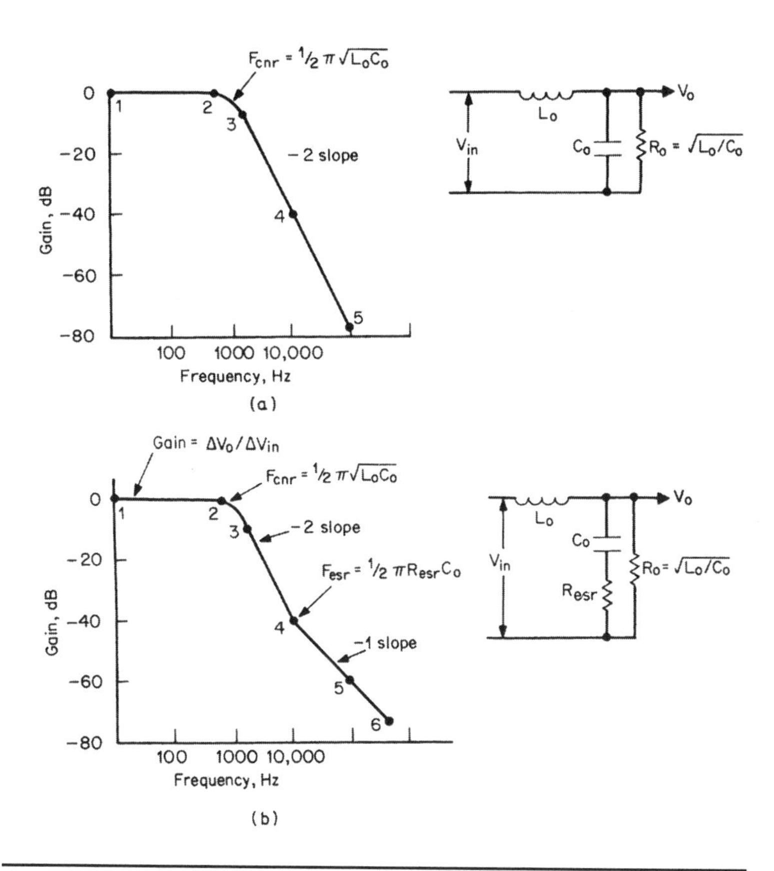

FIGURE 12.5 (*a*) Gain versus frequency for a critically damped *LC* filter in which the output capacitor has zero ESR. (*b*) Gain versus frequency for a critically damped *LC* filter in which the output capacitor has an equivalent series resistance (R_{esr}).

Beyond F_o, the impedance of C_o decreases and that of L_o increases at the rate of 20 dB/decade each, making the gain fall at the rate of −40 dB/decade, or at a −2 slope. The transition to a −2 slope at F_o is not abrupt. The actual gain curve leaves 0 dB smoothly just below F_o and asymptotically approaches the −2 slope shortly above F_o.

Most filter capacitor types have an internal resistance R_{esr} in series with their output leads as shown in Figure 12.5*b*. This modifies the gain characteristic between the output and input terminals in a characteristic way.

Immediately above F_o, the impedance of C_o is much greater than R_{esr}, and looking down to ground from V_o, the only effective impedance is that of C_o. In this frequency range, the gain still falls at a −2 slope. At higher frequencies, where the impedance of C_o is less than R_{esr}, the effective impedance looking down from V_o to ground is that of R_{esr} alone. Hence in that frequency range, the circuit is an *LR* rather than an *LC*, and the impedance of L_o increases at the rate of 20 dB/decade while that of R_{esr} remains constant. Thus in that frequency range, gain falls at a −1 slope.

The break from a −2 to a −1 gain slope occurs at the frequency $F_{esr} = 1/(2\pi R_{esr} C_o)$, where the impedance of C_o (without ESR) is equal to R_{esr}. This is shown as F_{esr} in curve 123456 in Figure 12.5*b*. This break in slope from −2 to −1 is smooth, but it is sufficiently accurate to assume it is abrupt as shown.

12.2.4 Pulse-Width-Modulator Gain

In Figure 12.1, the gain from the error-amplifier output to the average voltage at V_{sr} (input end of the output inductor) is the PWM gain and is designed as G_{pwm}.

It may be puzzling how this can be referred to as a *voltage gain*. At V_{ea}, there are slow voltage level variations proportional to the error-amplifier input at point *B*, and at V_{sr}, there are fixed-amplitude pulses of adjustable width.

The significance and magnitude of this gain can be seen as follows. In Figure 12.1, the PWM compares the DC voltage level from V_{ea} to a 3-V triangle at V_t. In all PWM chips which produce two 180° out-of-phase adjustable-width pulses (for driving push-pulls, half or full bridges), these pulses occur one per triangle wave cycle, and have a maximum "on" or high time of a half period. After the PWM, the pulses are alternately routed to two separate output terminals (see Figure 5.2*a*). In a forward converter, only one of these outputs is used.

When C_o has significant ESR, the gain slope still breaks from horizontal to a −2 slope at F_{cnr}. But at a frequency $F_{esr} = 1/(2\pi R_{esr} C_o)$, it breaks into a −1 slope. This is because at F_{esr}, $X_{co} = R_{esr}$, and the impedance of C_o becomes increasingly small with frequency compared to R_{esr}. The circuit above F_{esr} is an *LR* rather than an *LC* circuit. The gain of an *LR* circuit falls at a −1 slope because as frequency increases, the impedance of the series *L* increases, but that of the shunt *R* remains constant.

In Figure 12.1*b*, when V_{ea} is at the bottom of the 3-V triangle, the "on" time or pulse width at V_{sr} is zero. The average voltage V_{av} at V_{sr} is then zero, since $V_{av} = (V_{sp} - 1)(t_{on}/T)$, where V_{sp} is the secondary peak voltage. When V_{ea} has moved up to the top of the 3-V triangle, $t_{on}/T = 0.5$ and $V_{av} = 0.5\,(V_{sp} - 1)$. The modulator gain G_m between V_{av} and V_{ea} is

$$G_m = \frac{0.5(V_{sp} - 1)}{3} \qquad (12.1)$$

This gain is independent of frequency.

There is also a gain loss G_s due to the sampling network R_1, R_2 in Figure 12.1. Most of the frequently used PWM chips use 2.5 V at the reference input to the error amplifier (point A). Thus, when sampling a +5-V output, $R_1 = R_2$ and gain G_s between V_s and V_o in Figure 12.1 is −6 dB.

12.2.5 Gain of Output *LC* Filter Plus Modulator and Sampling Network

From the above, the total gain G_t in decibels of the output LC filter G_f, plus modulator gain G_m, plus sampling network gain G_s is plotted as in Figure 12.6. It is equal to $G_m + G_s$ from DC up to $F_o = 1/(2\pi\sqrt{L_o C_o})$. At F_o, it breaks into a −2 slope and remains at that slope up to the frequency F_{esr} where the impedance of C_o equals R_{esr}. At that frequency, it breaks into a −1 slope.

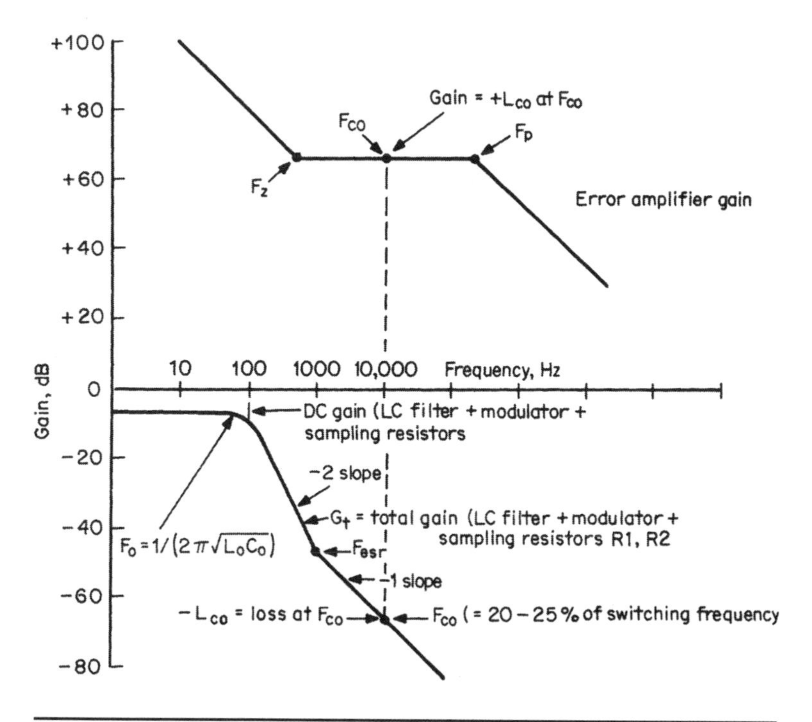

FIGURE 12.6 Gain G_t = gains of LC filter + modulator + output voltage sampling resistors is used to determine error-amplifier gain. Error-amplifier gain at F_{co} is made equal and opposite to G_t. Error-amplifier gain slope at F_{co} is made horizontal, with upward and downward breaks at F_z and F_p. Locations of F_z and F_p determine total circuit phase margin.

From this curve, the error-amplifier gain and phase-shift versus frequency characteristic is established to meet the three criteria for a stable loop as described below.

12.3 Shaping Error-Amplifier Gain Versus Frequency Characteristic

Recall that the first criterion for a stable loop is that at F_{co}, where the total open-loop gain is unity (0 dB), total open-loop phase shift must be less than 360° by the desired *phase margin*, which will herein be taken as 45°.

The sequence of steps is first to establish the crossover frequency F_{co} where the total open-loop gain should be 0 dB. Then choose the error-amplifier gain so that the total open-loop gain is 0 dB at that frequency. Next, design the error-amplifier gain slope so that the total open-loop gain passes through F_{co} at a -1 slope (Figure 12.4). Finally, tailor the error-amplifier gain versus frequency so that the desired phase margin is achieved.

Sampling theory shows that F_{co} must be less than half the switching frequency for the loop to be stable. Actually, it must be considerably less than that, or there will be large-amplitude switching frequency ripple at the output. Thus, the usual practice is to fix F_{on} at one-fourth to one-fifth the switching frequency.

Refer to Figure 12.6, which is the sum of the open-loop gains of the *LC* filter, the PWM modulator, and the sampling network. The capacitor in the output filter of Figure 12.6 is assumed to have an ESR which causes a break in the slope from -2 to -1 at $F_{esr} = 1/(2\pi R_{esr} C_o)$. Assume that F_{co} is one-fifth the switching frequency, and read the loss in decibels at that point.

In most cases, the output capacitor has a significant ESR, and F_{esr} will occur at a frequency lower than F_{co}. Thus at F_{co}, the $G_t = (G_{lc} + G_{pwm} + G_s)$ curve will already have a -1 slope.

When gains are plotted in decibels, both gains and gain slopes of elements in cascade are additive. Hence, to force crossover frequency to be at the desired one-fifth the switching frequency, choose the error-amplifier gain at F_{co} to be equal and opposite in decibels to $G_t = (G_{lc} + G_{pwm} + G_s)$, which is a loss at that frequency.

That forces F_{co} to occur at the desired point. Then, if the error-amplifier gain slope at F_{co} is horizontal, since the G_t curve at F_{co} already has a -1 slope, the sum of the error amplifier plus the G_t curve passes through the crossover frequency at the desired -1 slope and the second criterion for a stable loop has been met.

The error-amplifier gain has been fixed equal and opposite to G_t at F_{co}, where it has a horizontal slope (Figure 12.6). At F_{co}, this gain

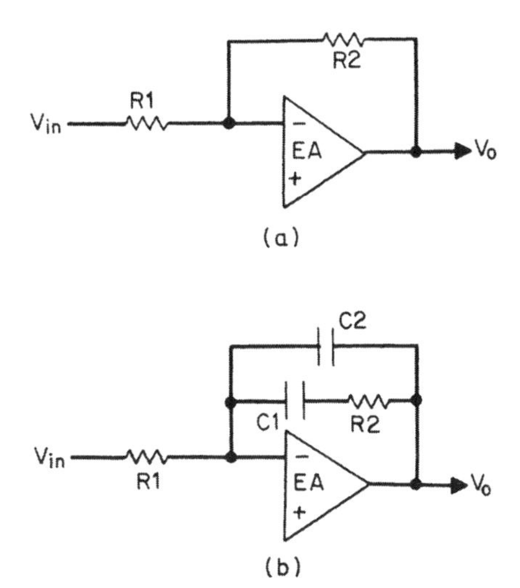

FIGURE 12.7 (*a*) Error amplifier with feedback resistor *R*2 and input resistor *R*1 has gain equal to *R*2/*R*1, which is independent of frequency up to the frequency where the open-loop gain of the op-amp commences falling off. (*b*) Using a complex feedback network permits shaping the gain-versus-frequency and phase shift–versus-frequency curves. The configuration above has the gain-versus-frequency characteristic of Figure 12.6.

characteristic can be achieved using an operational amplifier with input and feedback resistors as in Figure 12.7*a*. Recall that the gain of such an amplifier is $G_{ea} = Z2/Z1 = R2/R1$. But how far in frequency to the left and right of F_{co} should it continue to have this constant gain?

Recall that the total open-loop gain is the sum of the error-amplifier gain plus G_t gain. If the error-amplifier gain were constant down to DC, the total open-loop gain would not be very large at 120 Hz—the frequency of the AC power line ripple. Power line ripple should be highly attenuated at the output. To degenerate the 120-Hz ripple sufficiently, the open-loop gain at that frequency should be as high as possible. Thus at some frequency to the left of F_{co}, the error-amplifier gain should be permitted to increase rapidly.

This can be done by placing a capacitor *C*1 in series with *R*2 (Figure 12.7*b*). Ignoring the effect of *C*2 for the moment, this yields the low-frequency gain characteristic shown in Figure 12.6. In the frequency range where the impedance of *C*1 is small compared to *R*2, the gain is horizontal and equal to *R*2/*R*1. At lower frequencies, where the impedance of *C*2 is much higher than *R*2, *R*2 is effectively out of the circuit, and the gain is X_{c1}/R_1. This gain has a slope of −20 dB/decade

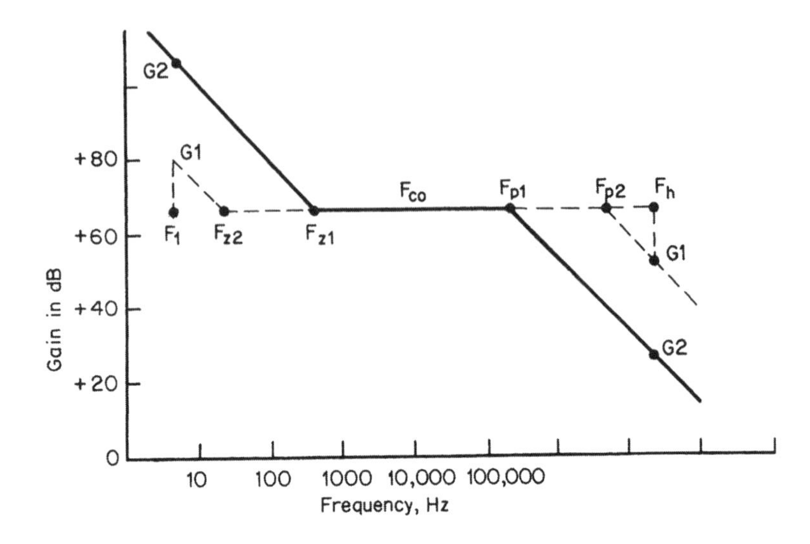

FIGURE 12.8 Locations of break frequencies F_z and F_p. The farther apart F_z and F_p are spread, the greater the phase margin, but this reduces low-frequency gain and the degeneration of low-frequency line ripple. It also increases high-frequency gain, which amplifies noise spikes.

and yields the higher gain at 120 Hz. At $F_z = 1/(2\pi R2C1)$, the −1 gain slope breaks and becomes horizontal.

If the error-amplifier gain curve were permitted to remain horizontal to the right of F_{co} (Figure 12.6), total open-loop gain would remain relatively high at high frequencies. But high gain at high frequencies is undesirable, because high-frequency noise spikes would be picked up and transmitted at large amplitudes to the output. Thus gain should fall off at high frequencies.

This is done by placing a capacitor $C2$ across the series combination of $R2$ and $C1$ (Figure 12.7*b*). At F_{co}, X_{c1} is small compared to $R2$ and X_{c1}, and has no effect. At higher frequencies where X_{c2} is small compared to $R2$, however, $R2$ is effectively out of the circuit and gain is X_{c2}/R_1. The gain characteristic above F_{co} is horizontal up to a frequency $F_p = 1/(2\pi R2C2)$, where it breaks and falls at a −1 slope, as seen in Figure 12.6. This diminishing gain at high frequencies keeps high-frequency noise spikes from coming through to the output.

The break frequencies F_z and F_p are chosen so that $F_{co}/F_z = F_p/F_{co}$. The farther apart F_z and F_p are, the greater is the phase margin at F_{co}. A large phase margin is desirable, but if F_z is too low, the gain at 120 Hz will be insufficient (Figure 12.8), and 120-Hz attenuation will be poor. If F_p is too high (Figure 12.8), high-frequency gain will be excessive and noise spikes will be amplified.

Thus a compromise is sought. This compromise, and a more exact analysis of the problem, is made easy by introducing the concept of transfer functions, poles, and zeros as shown below.

12.4 Error-Amplifier Transfer Function, Poles, and Zeros

An operational amplifier circuit, with input arm complex impedance Z_1 and feedback arm complex impedance Z_2, is shown in Figure 12.9. Its gain is $-Z_2/Z_1$. If Z_1 is a pure resistor $R1$ and Z_2 is a pure resistor R_2 as in Figure 12.7*a*, gain is $-R_2/R_1$ and is independent of frequency. Phase shift between V_{in} and V_o is 180°, since the input is applied to the inverting terminal.

Impedances Z_1, Z_2 are expressed in terms of the complex variable $s = j2\pi f = j\omega$. Thus the impedance of capacitor $C1$ is $1/sC1$, and that of resistor $R1$ and capacitor $C1$ in series is $R1 + 1/sC1$.

The impedance of an arm consisting of capacitor $C2$ in parallel with series combination $R1$, $C1$ is

$$Z = \frac{(R1 + 1/sC1)(1/sC2)}{R1 + 1/sC1 + 1/sC2} \tag{12.2}$$

The transfer function of the error amplifier is written in terms of its Z_1, Z_2 impedances, which are expressed in terms of the complex variable s. Thus $G(s) = -Z_2(s)/Z_1(s)$, and by algebraic manipulation, $G(s)$ is broken down into a simplified numerator and denominator which are functions of s: $G(s) = N(s)/D(s)$. The numerator and denominator, again by algebraic manipulation, are factored and $N(s)$,

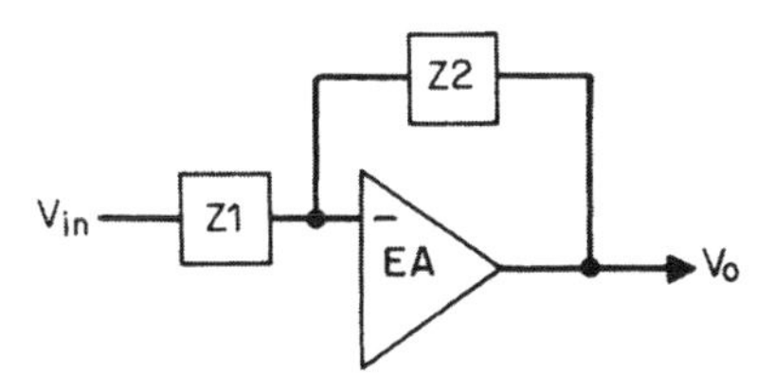

FIGURE 12.9 Various gain-versus-frequency and phase shift–versus-frequency curves are possible by connecting different *RC* combinations in the input and feedback arms. By expressing impedances Z_1 and Z_2 in terms of the $s = j\omega$ operator, and performing a number of algebraic manipulations, a simplified expression for the gain arises. From this simplified gain expression (transfer function) the gain-versus-frequency and phase shift–versus-frequency curves can be drawn easily.

$D(s)$ are expressed in terms of these factors. Thus

$$G(s) = \frac{N(s)}{D(s)} = \frac{(1+sz_1)(1+sz_2)(1+sz_3)}{sp_0(1+sp_1)(1+sp_2)(1+sp_3)} \tag{12.3}$$

These z and p values are RC products, and the corresponding frequencies are obtained by setting the factors equal to zero. Thus

$$1 + sz_1 = 1 + j2\pi f z_1 = 1 + j2\pi f\, R1C1 = 0 \quad \text{or} \quad f_1 = 1/(2\pi\, R1C1)$$

The frequencies corresponding to the z values are called *zero* frequencies, and those corresponding to the p values are called *pole* frequencies. There is always a factor in the denominator which has the "1" missing (note sp_0 above). This represents an important pole frequency, $F_{po} = 1/(2\pi\, R_o/C_o)$, which is called *the pole at the origin*.

From the location of the pole at the origin, and the zero and pole frequencies, the gain-versus-frequency characteristic of the error amplifier can be drawn as discussed below.

12.5 Rules for Gain Slope Changes Due to Zeros and Poles

The zero and pole frequencies are points where the error-amplifier gain slope changes. A zero represents a +1 change in gain slope. Thus (Figure 12.10*a*), if a zero appears at a point in frequency where the gain slope is zero, it turns the gain into a +1 slope. If it appears where the original gain slope is −1 (Figure 12.10*b*), it turns the gain slope to zero. Or if there are two zeros at the same frequency (two factors in the numerator of Eq. 12.3 having the same RC product) where the original gain slope is −1, the first zero turns the gain slope horizontal, and the second zero at the same frequency turns the gain into a +1 slope (Figure 12.10*c*).

A pole represents a −1 change in gain slope. If it appears at a frequency where the original gain slope is zero, it turns the slope to −1 (Figure 12.10*d*). If there are two poles at the same frequency at a point where the original gain slope is +1, the first turns the slope horizontal and the second turns the slope to −1 (Figure 12.10*e*).

A pole at the origin, like any pole, represents a gain slope of −1. It also indicates the frequency at which the gain is 1 or 0 dB. Thus, drawing the total gain curve for an error amplifier starts as follows. At 0 dB and the frequency of the pole at the origin $F_{po} = 1/(2\pi\, R_o C_o)$, draw a line backward in frequency with a slope of −1 (Figure 12.11). Now if somewhere on this line the transfer function has a zero at a frequency $F_z = 1/2\pi\, R1C1$, turn the gain slope horizontal above F_z. Extend the horizontal gain indefinitely, or until some higher frequency

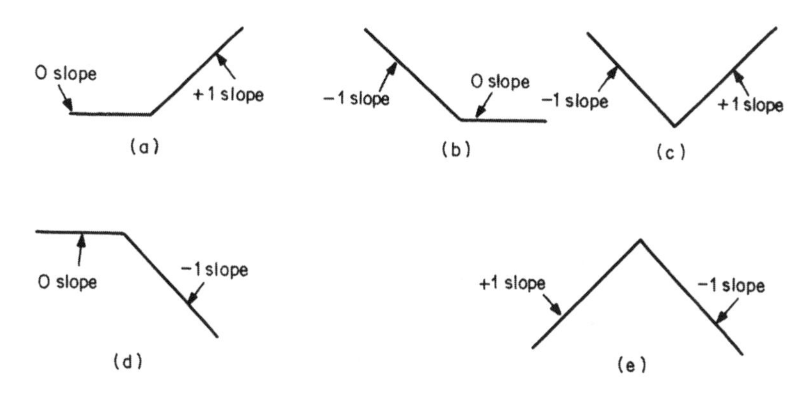

FIGURE 12.10 (*a*) A zero appearing on a gain curve where the original slope was horizontal turns that gain slope to +1 or +20 dB/decade. (*b*) A zero appearing on a gain curve where the original slope was −1 turns that gain slope horizontal. (*c*) Two zeros at the same frequency appearing on a gain curve where the original slope was −1 turns that slope to +1. (*d*) A pole appearing on a gain curve where the original slope is horizontal turns that slope to −1 or −20 dB/decade. (*e*) Two poles at the same frequency appearing on a gain curve where the original slope is +1 turns that slope to −1.

where there is a pole in the transfer function at $F_p = 1/(2\pi R2C2)$, turn the horizontal slope into a −1 slope (Figure 12.11).

The gain along the horizontal part of the transfer function is $R2/R1$, which is made equal and opposite in decibels to that of the G_t curve (Figure 12.6) at F_{co}.

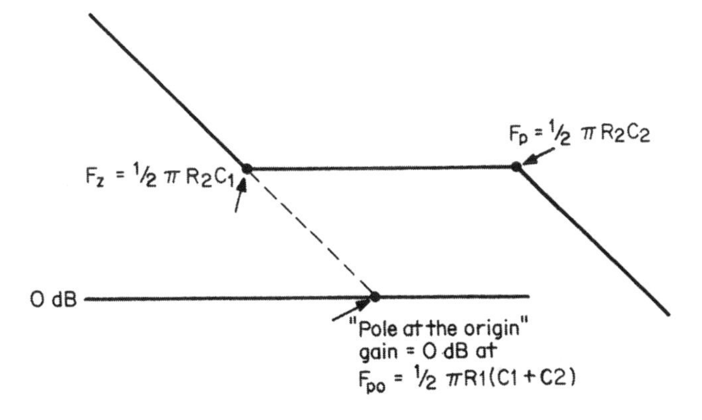

FIGURE 12.11 Drawing the gain curve for the error amplifier of Figure 12.7*b* directly from its transfer function of Eq. 12.3.

Thus an error-amplifier gain curve having a single pole at the origin, a single zero below this, and another single pole above has the desired shape shown in Figure 12.11. It is implemented with the circuit of Figure 12.7*b*. It remains only to select the zero and pole frequencies which yield the desired phase margin. This will be discussed below.

12.6 Derivation of Transfer Function of an Error Amplifier with Single Zero and Single Pole from Its Schematic

It has been shown above that an error amplifier with a single zero, a pole at the origin, and another single pole in that order, has a gain-versus-frequency curve as in Figure 12.11.

Now it will be demonstrated how the transfer function of an error amplifier is derived, and that the circuit of Figure 12.7*b* does have a single zero, a pole at the origin, and another single pole. Gain of the circuit in Figure 12.7*b*, ignoring polarity, is

$$G = \frac{dV_o}{dV_i}$$

$$= \frac{Z_2}{Z_1}$$

$$= \frac{(R2 + 1/j\omega C1)(1/j\omega C2)}{R1(R2 + 1/j\omega C1 + 1/j\omega C2)}$$

Now introduce the complex variable $s = j\omega$. Then

$$G = \frac{(R2 + 1/sC1)(1/sC2)}{R1(R2 + 1/sC1 + 1/sC2)}$$

And by algebraic manipulation

$$G = \frac{1 + sR2C1}{sR1(C1 + C2)(1 + sR2C1C2/(C1 + C2))}$$

And since generally $C2 \ll C1$

$$G = \frac{1 + sR2C1}{sR1(C1 + C2)(1 + R2C2)} \tag{12.4}$$

The error amplifier of Figure 12.7*b*, having the transfer function of Eq. 12.4, is commonly referred to as a *Type 2 amplifier* in conformance with the designation introduced by Venable in his classic paper.[1] A Type 2 error amplifier is used when the output filter capacitor has an ESR such that F_{co} lies on a −1 slope of the G_1 curve (Figure 12.6).

Examination of this transfer function for the circuit of Figure 12.7*b* permits immediate drawing of its gain characteristic as follows (Figure 12.11). Equation 12.4 shows that this circuit has a pole at the origin at $F_{po} = 1/(2\pi R1(C1 + C2))$. Start at 0 dB at this frequency, and draw a line toward lower frequency with a slope of -1.

From Eq. 12.4, the circuit has a zero at $F_z = 1/(2\pi R2C1)$. The sloped line just drawn should turn horizontal above F_z. Again from Eq. 12.4, the circuit has a pole at $F_p = 1/(2\pi R2C2)$. The horizontal line just drawn should turn into a -1 slope above F_p.

Now that the transfer function of the Type 2 error amplifier can be drawn from its pole and zero frequencies, it remains to locate them, by choosing $R1, R2, C1, C2$ to achieve the desired phase margin. This is demonstrated below.

12.7 Calculation of Type 2 Error-Amplifier Phase Shift from Its Zero and Pole Locations

Adopting Venable's scheme[1], the ratios $F_{co}/F_z = F_p/F_{co} = K$ will be chosen.

A zero, like an *RC* differentiator (Figure 12.2*b*), causes a phase lead. A pole, like an *RC* integrator (Figure 12.2*a*), causes a phase lag.

The phase lead at any frequency F due to a zero F_z is

$$\theta_{ld} = \tan^{-1}\frac{F}{F_z}$$

We are interested in the phase lead at F_{co} due to a zero at F_z. This is

$$\theta_{ld}(\text{at } F_{co}) = \tan^{-1} K \tag{12.5}$$

The phase lag at a frequency F due to a pole at F_p is

$$\theta_{lag} = \tan^{-1}\frac{F}{F_p}$$

and we are interested in the lag at F_{co} due to the pole at F_p. This is

$$\theta_{lag}(\text{at } F_{co}) = \tan^{-1}\frac{1}{K} \tag{12.6}$$

The total phase shift at F_{co} due to the lead of the zero at F_z and the lag due to the pole at F_p is the sum of Eqs. 12.5 and 12.6.

These shifts are in addition to the inherent low-frequency phase shift of the error amplifier with its pole at the origin. Also, the error amplifier is an inverter, which causes a 180° phase shift.

K	Lag (from Eq. 12.7)
2	233°
3	216°
4	208°
5	202°
6	198°
10	191°

TABLE 12.1 Phase Lag Through a Type 2 Error Amplifier for Various Values of $K(= F_{co}/F_z = F_p/F_{co})$

The pole at the origin causes a 90° phase shift. This is another way of saying that at low frequencies, the circuit is just an integrator with resistor input and capacitor feedback. This is seen from Figure 12.7*b*. At low frequencies, the impedance of $C1$ is much greater than that of $R2$. The feedback arm is thus only $C1$ and $C2$ in parallel.

Thus the phase lag is 180° because of the phase inversion, plus 90° inherent low-frequency lag due to the pole at the origin, for a total lag of 270°. Total phase lag, including the lead due to the zero and lag due to the pole, is then

$$\theta_{(\text{total lag})} = 270° - \tan^{-1} K + \tan^{-1} \frac{1}{K} \tag{12.7}$$

Note that this is always a net phase lag, because when K is large (zero and pole frequencies far apart), the lead due to the zero is a maximum of 90°, and the lag due to the pole tends to 0°.

Total phase lag through the error amplifier, calculated from Eq. 12.7, is shown in Table 12.1.

12.8 Phase Shift Through *LC* Filter with Significant ESR

The total open-loop phase shift consists of that of the error amplifier plus that of the output LC filter, since the contribution of the modulator is small and generally neglected. Figure 12.3*b* showed for $R_o = 20\sqrt{L_o/C_o}$ and no ESR in the filter capacitor, the lag through the filter only is already 175° at $1.2F_o$.

This lag is modified significantly if the output capacitor has an ESR as in Figure 12.5*b*. In that figure, the gain slope breaks from a −2 to

F_{co}/F_{esro}	Phase lag	F_{co}/F_{esro}	Phase lag
0.25	166°	2.5	112°
0.50	153°	3	108°
0.75	143°	4	104°
1.0	135°	5	101°
1.2	130°	6	99.5°
1.4	126°	7	98.1°
1.6	122°	8	97.1°
1.8	119°	9	96.3°
2.0	116°	10	95.7°

TABLE 12.2 Phase Lag Through an *LC* Filter at F_{co} Due to a Zero at F_{esro}

a −1 slope at the so-called ESR zero frequency $F_{esr} = 1/(2\pi R_{esr} C_o)$. Recall that at F_{esr}, the impedance of C_o equals that of R_{esr}. Beyond F_{esr}, the impedance of C_o becomes smaller than R_{esr} and the circuit becomes increasingly like an *LR* rather than an *LC* circuit. Moreover, an *LR* circuit can cause only a 90° phase lag as compared to the possible maximum of 180° for an *LC* circuit.

Thus the ESR zero contributes a phase lead to the possible maximum 180° of the LC filter. Phase lag at a frequency F due to an ESR zero at F_{esro} is

$$\theta_{ic} = 180° - \tan^{-1} \frac{F}{F_{esro}}$$

and since we are interested in the phase lag at F_{co} due to the zero at F_{esro}

$$\theta_{lc} = 180° - \tan^{-1} \frac{F_{co}}{F_{esro}} \tag{12.8}$$

Phase lags through an *LC* filter having an ESR zero are shown in Table 12.2 for various values of F_{co}/F_{esro} (from Eq. 12.8).

By setting the error-amplifier gain in the horizontal part of its gain curve (Figure 12.6) equal and opposite to the G_t (Figure 12.6) at F_{co}, the location of F_{co} is fixed where it is desired. Since F_{co} is located on the −1 slope portion of the G_t curve, the total open-loop gain curve will pass through F_{co} at a −1 slope. From Tables 12.1 and 12.2, the proper value of K (locations of the zero and pole) is established to yield the desired phase margin.

12.9 Design Example—Stabilizing a Forward Converter Feedback Loop with a Type 2 Error Amplifier

The design example presented below demonstrates how much of the material discussed in previous chapters is interrelated.

Stabilize the feedback loop for a forward converter with the following specifications:

V_o	5.0 V
$I_{o(\text{nom})}$	10 A
Minimum I_o	1 A
Switching frequency	100 kHz
Minimum output ripple (peak to peak)	50 mV

It is assumed that the filter output capacitor has significant ESR, and that F_{co} will occur on the −1 slope of the LC filter. This permits the use of a Type 2 error amplifier with the gain characteristics of Figure 12.6. The circuit is shown in Figure 12.12.

First, L_o, C_o will be calculated and the gain characteristic of the output filter will be drawn. From Eq. 2.47

$$L_o = \frac{3V_oT}{I_{\text{on}}} = \frac{3 \times 5 \times 10^{-5}}{10} = 15 \times 10^{-6}\text{H}$$

and from Eq. 2.48

$$C_o = 65 \times 10^{-6}\frac{dI}{V_{\text{or}}}$$

where dI is twice the minimum output current $= 2 \times 1 = 2$ A and V_{or} is the output ripple voltage $= 0.05$ V. Then $C_o = 65 \times 10^{-6} \times 2/0.05 =$ 2600 microfarads.

Corner frequency of the output LC filter, from Section 12.2.3, is

$$\begin{aligned} F_o &= 1/(2\pi\sqrt{L_oC_o}) \\ &= 1/(2\pi\sqrt{15 \times 10^{-6} \times 2600 \times 10^{-6}}) \\ &= 806 \text{ Hz} \end{aligned}$$

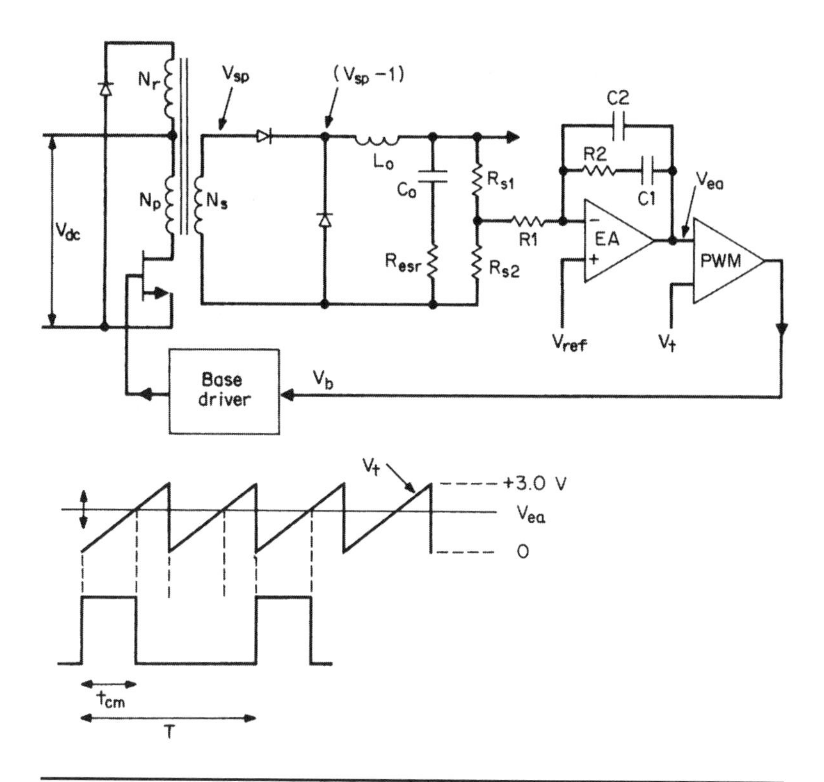

FIGURE 12.12 Forward converter design example schematic for stabilizing the feedback loop.

Again from Section 12.2.3, the frequency of the ESR zero is

$$F_{\text{esr}} = 1/(2\pi R_{\text{esr}} C_o)$$
$$= 1/(2\pi(65 \times 10^{-6}))$$
$$= 2500 \text{ Hz}$$

This assumes, as in Section 2.3.11.2, that over a large range of aluminum electrolytic capacitor magnitudes and voltage ratings, $R_{\text{esr}}C_o$ is constant and equal to 65×10^{-6}.

From Eq. 12.1, the modulator gain is $G_m = 0.5(V_{\text{sp}} - 1)/3$, and when the duty cycle is 0.5, for $V_o = 5\text{V}$, $V_{\text{sp}} = 11$ V since $V_o = (V_{\text{sp}} - 1)T_{\text{on}}/T$. Then $G_m = 0.5(11 - 1)/3 = 1.67 = +4.5$ dB.

For the typical SG1524-type PWM chip, which needs 2.5 V at the reference input to the error amplifier, $R_{s1} = R_{s2}$ for $V_o = 5$ V. Sampling network gain is $G_s = -6$ dB. Then $G_m + G_s = +4.5 - 6.0 = -1.5$ dB.

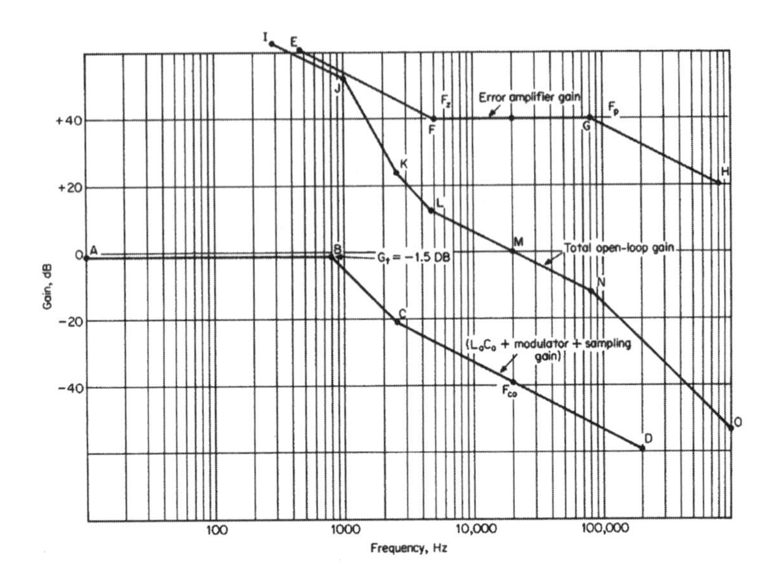

FIGURE 12.13 Design example—stabilizing the feedback loop for Figure 12.12.

The open-loop gain curve of everything but the error amplifier is $G_t = G_{1c} + G_m + G_s$, and is drawn in Figure 12.13 as curve *ABCD*. From *A* to the corner frequency at 806 Hz (*B*) it has a gain of $G_m + G_s = -1.5$ dB. At *B*, it breaks into a −2 slope and continues at that slope up to the ESR zero at 2500 Hz (*C*). At point *C*, it breaks into a −1 slope.

Crossover frequency is set at one-fifth the switching frequency or 20 kHz. From the G_t curve, gain at 20 kHz is −40 dB (numerical gain of 1/100). To make 20 kHz the crossover frequency, the error-amplifier gain at that frequency is set at +40 dB. Since the total open-loop gain of the error amplifier plus curve *ABCD* must pass through the crossover point M at a −1 slope, the error-amplifier gain curve must have zero slope between points *F* and *G* in curve *EFGH*, since *ABCD* already has a −1 slope at $F_{co} = 20$ kHz.

This horizontal gain slope between points *F* and *G* is obtained as described above with a Type 2 error amplifier. The gain of the Type 2 error amplifier in the horizontal part of its slope is *R*2/*R*1. If *R*1 is arbitrarily set at 1 kΩ, *R*2 is 100 kΩ.

A zero is located at *F* to increase low-frequency gain and degenerate 120-Hz line ripple, and a pole is located at *G* to decrease high-frequency gain and minimize thin noise spikes at the output. The zero and the pole are located to give the desired phase margin.

We will design for 45° phase margin. Then total phase shift around the loop at 20 kHz is 360 − 45 = 315°. The *LC* filter by itself causes a phase lag given by Eq. 12.7. From that equation, the lag with $F_{co} =$ 20 kHz and $F_{esro} = 2500$ Hz is 97° (Table 12.2). Thus the error amplifier is permitted only 315 − 97 or 218° of lag. Table 12.1 shows that for an error-amplifier lag of 218°, a K factor of slightly less than 3 would suffice.

To provide somewhat more insurance, assume a K factor of 4, which yields a phase lag of 208°. This, plus the 97° lag of the LC filter, yields a total lag of 305° and a phase margin of 360 − 305° or 55° at F_{on}.

For a K factor of 4, the zero is at $F_z = 20/4 = 5$ kHz. From Eq. 12.3, $F_z = 1/(2\pi R2C1)$. For $R2$ determined above as 100 k, $C1 = 1/(2\pi$ (100 k)(5 k)) $= 318 \times 10^{-12}$.

Again for the K factor of 4, the pole is at $F_{po} = 20 \times 4 = 80$ kHz. From Eq. 12.3, $F_{po} = 1/(2\pi R2C2)$. For $R2 = 100$ k, $F_{po} = 80$ kHz, $C2 = 1/(2\pi(100\text{ k})(80\text{ k})) = 20 \times 10^{-12}$. This completes the design; the final gain curves are shown in Figure 12.13. Curve *IJKLMO* is the total open-loop gain. It is the sum of curves *ABCD* and *EFGH*.

12.10 Type 3 Error Amplifier—Application and Transfer Function

In Section 2.3.11.2, it was pointed out that the output ripple $V_{or} = R_o dI$ where R_o is the ESR of the filter output capacitor C_o and dI is twice the minimum DC current. Most aluminum electrolytic capacitors do have significant ESR. Study of many capacitor manufacturers' catalogs indicates that for such capacitors, $R_o C_o$ is constant and has an average value of 65×10^{-6}.

Thus, using conventional aluminum electrolytic capacitors, the only way to reduce output ripple is to decrease R_o, which can be done only by increasing C_o. This increases the size of the capacitor, which may be unacceptable.

Within the past few years, capacitor manufacturers have been able (at considerably greater cost) to produce aluminum electrolytic capacitors with essentially zero ESR for those applications where output ripple must be reduced to an absolute minimum.

When such zero ESR capacitors are used, it affects the design of the error amplifier in the feedback loop significantly. When the output capacitor had significant ESR, F_{co} usually was located on the −1 slope of the output filter. This required a Type 2 error amplifier with a horizontal slope at F_{co} in its gain-versus-frequency characteristic (Figure 12.6).

With a zero ESR capacitor, the LC gain-versus-frequency curve continues falling at a −2 slope above the corner frequency (curve *ABCD* in Figure 12.14). An error amplifier can be designed to have gain equal

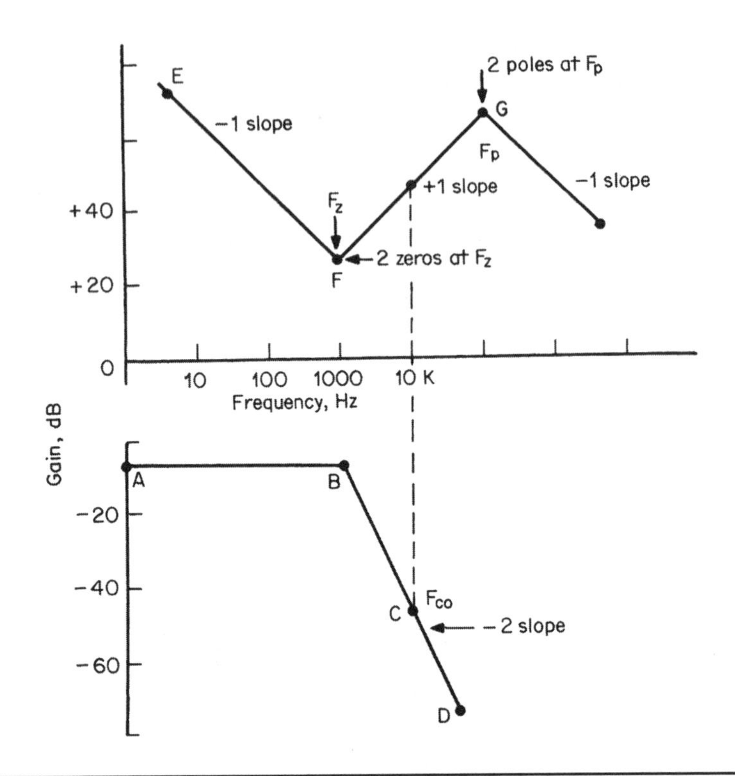

FIGURE 12.14 When the output capacitor has no ESR, its gain continues to fall at a −2 slope. This requires an error amplifier with a +1 slope at F_{co} for the total open-loop gain to pass through F_{co} at a −1 slope. To achieve the above error-amplifier gain curve, two zeros are located at F_z and two poles at F_p.

and opposite to the *LC* loss at the desired F_{co}, but for the total gain to pass through F_{co} at a −1 slope, the error-amplifier gain curve must be designed to have a +1 slope in its central region at F_{co} (curve *EFGHI* in Figure 12.14).

The error-amplifier EA must have sufficient gain at lower frequencies to reject 120-Hz line frequency input ripple. Also the total open-loop gain must be zero at F_{co}, where the EA gain has a +1 slope. Thus below the frequency F_z (Figure 12.14), the EA gain curve slope must be −1. As described in Section 12.5, this is done by providing two zeros at the same frequency (F_z) in the EA transfer function. Below F_z, the gain falls at a −1 slope because of a pole at the origin which will be provided. At F_z, the first zero turns the gain slope horizontal; the second one turns it to a +1 slope.

The gain cannot be permitted to continue upward at a +1 slope much beyond F_{co}. If it did, gain would be high at high frequencies, and noise spikes would get through to the output. Thus, as described in Section 12.5, two poles are provided at point H at frequency F_p. The first pole turns the +1 gain slope horizontal; the second pole turns it to a −1 slope.

An EA with the gain-versus-frequency *EFGHI* in Figure 12.14 is referred to as a *Type 3 error amplifier*, again following the widely used Venable designation.[1]

As for the Type 2 error amplifier, location of the two zeros at F_z and the two poles at F_p determines the phase lag at F_{co}. The wider the separation between F_p and F_z, the greater the phase margin.

Like the Type 2 error amplifier, locating F_z at too low a frequency reduces low-frequency gain and prevents sufficient degeneration of 120-Hz line ripple. Placing F_p at too high a frequency increases gain at high frequencies and permits high-frequency noise spikes to come through at greater amplitude.

Again a K factor is introduced to define the locations of F_z and F_p. This factor is a ratio set to $K = F_{co}/F_z = F_p/F_{co}$. In the following section, phase boost at F_{co} due to the double zero at F_z and phase lag at F_{co} due to the double pole at F_p will be calculated.

12.11 Phase Lag Through a Type 3 Error Amplifier as Function of Zero and Pole Locations

In Section 12.7, it was pointed out that the phase boost at a frequency F_{co} due to a zero at a frequency F_z is $\theta_{zb} = \tan^{-1}(F_{co}/F_z) = \tan^{-1} K$ (Eq. 12.4). If there are two zeros at the frequency F_z, the boosts are additive. Thus boost at F_{co} due to two zeros at the same frequency F_z is $\theta_{2zb} = 2\tan^{-1} K$.

Similarly, the lag at F_{co} due to a pole at F_p is $\theta_{1p} = \tan^{-1}(1/K)$ (Eq. 12.5). The lags due to two poles at F_p are also additive. Thus lag at F_{co} due to two poles at F_p is $\theta_{2p} = 2\tan^{-1}(1/K)$. The lag and boost are in addition to the inherent low-frequency 270° lag, which is the 180° phase inversion plus the 90° due to the pole at the origin.

Thus total phase lag through a Type 3 error amplifier is

$$\theta = 270^\circ - 2\tan^{-1} K + 2\tan^{-1}(1/K) \tag{12.9}$$

Total phase lag through the Type 3 error amplifier is calculated from Eq. 12.9 for various values of K (see Table 12.3).

Comparing Tables 12.3 and 12.1, it is seen that a Type 3 error amplifier with two zeros and two poles has considerably less phase lag than does the Type 2 error amplifier, which has only a single zero and a single pole.

K	Lag (from Eq. 12.9)
2	196°
3	164°
4	146°
5	136°
6	128°

TABLE 12.3 Phase Lag Through Type 3 Error Amplifier for Various Values of $K = F_{co}/F_z = F_p/F_{co}$

However, the Type 3 error amplifier is used with an *LC* filter which has no ESR zero to decrease the lag. Thus the lower lag of the Type 3 error amplifier is essential because of the higher lag of an *LC* filter with no ESR.

12.12 Type 3 Error Amplifier Schematic, Transfer Function, and Zero and Pole Locations

The schematic of a circuit which has the gain-versus-frequency characteristic of Figure 12.14 is shown in Figure 12.15. Its transfer function can be derived in the manner described in Section 12.6 for the Type 2 error amplifier. Impedances of the feedback and input arms Z_2 and Z_1, respectively, are expressed in terms of the *s* operator, and the transfer function is $G(s) = -Z_2(s)/Z_1(s)$. Algebraic manipulation yields the following expression for the transfer function:

$$G(s) = \frac{dV_o}{dV_{in}} = \frac{-(1 + sR2C1)[1 + s(R1 + R3)C3]}{sR1(C1 + C2)(1 + sR3C3)[1 + sR2C1C2/(C1 + C2)]} \tag{12.10}$$

This transfer function is seen to have

(a) A pole at the origin at a frequency of

$$F_{po} = 1/[2\pi R1(C1 + C2)] \tag{12.11}$$

This is the frequency where the impedance of $R1$ is equal to that of $C1$ and $C2$ in parallel.

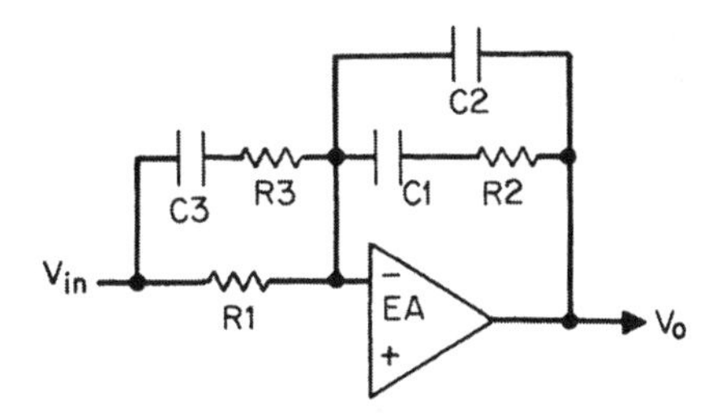

FIGURE 12.15 Type 3 error amplifier. It has a pole at the origin, two zeros, and two poles. Its transfer function is
$G = dV_o/dV_i = \frac{-(1+sR2C1)[1+s(R1+R3)C3]}{sR1(C1+C2)(1+sR3C3)[1+sR2C1C2/(C1+C2)]}$

(b) A first zero at a frequency of

$$F_{z1} = 1/(2\pi R2C1) \tag{12.12}$$

This is the frequency where the impedance of $R2$ equals that of $C1$.

(c) A second zero at a frequency of

$$F_{z2} = 1/[2\pi(R1 + R3)C3] \approx 1/(2\pi R1C3) \tag{12.13}$$

This is the frequency where the impedance of $R1 + R3$ equals that of $C3$. $R1$ is generally much greater than $R3$.

(d) A first pole at a frequency of

$$F_{p1} = \frac{1}{2\pi R2C1C2/(C1 + C2)} \approx 1/(2\pi R2C2) \tag{12.14}$$

This is the frequency where the impedance of $R2$ equals that of the series combination of $C1$ and $C2$. $C1$ is generally much greater than $C2$.

(e) A second pole at a frequency of

$$F_{p2} = 1/(2\pi R3C3) \tag{12.15}$$

This is the frequency where the impedance of $R3$ equals that of $C3$.

To yield the gain-versus-frequency curves of Figure 12.14, the RC products will be chosen so that $F_{z1} = F_{z2}$ and $F_{p1} = F_{p2}$. The location of the double-zero and double-pole frequencies will be fixed by the K factor, which yields the desired phase margin. Gain of the error amplifier on the +1 slope of Figure 12.14 will be set equal to the loss of the *LC* filter (Figure 12.14) at the desired F_{co}.

From Table 12.3 and the transfer function of Eq. 12.10, the *RC* products which set the zero and pole frequencies at the desired points are determined as in the design example below.

12.13 Design Example—Stabilizing a Forward Converter Feedback Loop with a Type 3 Error Amplifier

Design the feedback loop for a forward converter having the following specifications:

V_o	5.0 V
I_o(nom)	10 A
I_o(min)	1.0 A
Switching frequency	50 kHz
Output ripple (peak to peak)	< 20 mV

Assume that the output capacitor is of the type advertised as having zero ESR.

First the output *LC* filter and its corner frequency are calculated. Refer to Figure 12.15. From Eq. 2.47

$$\begin{aligned} L_o &= \frac{3V_oT}{I_o} \\ &= \frac{3 \times 5 \times 20 \times 10^{-6}}{10} \\ &= 30 \times 10^{-6}\ \text{H} \end{aligned}$$

It was assumed that the output capacitor had zero ESR, so ripple due to ESR should be zero. But there is a small capacitive ripple component (Section 1.3.2). This is usually very small, so a filter capacitor much smaller than the 2600-μF capacitor used in the Type 2 error-amplifier design example can be used. To be conservative, for this design assume the same 2600-μF capacitor is used, and that it has zero ESR. Then

$$\begin{aligned} F_o &= 1/(2\pi\sqrt{L_oC_o}) \\ &= 1/(2\pi\sqrt{30 \times 10^{-6} \times 2600 \times 10^{-6}}) \\ &= 570\ \text{Hz} \end{aligned}$$

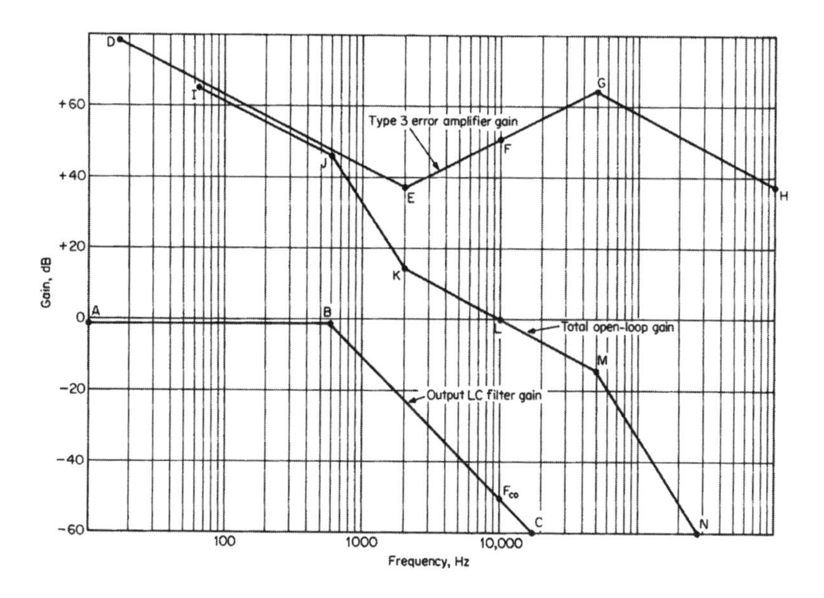

FIGURE 12.16 Gain curves—design example of Section 12.13. Output capacitor with zero ESR and Type 3 error amplifier.

Assume, as for Type 2 error-amplifier design example, that the modulator plus sampling divider gain is −1.5 dB. The gain of the *LC* filter plus modulator plus sampling divider is plotted in Figure 12.16 as curve *ABC*. It is horizontal at a level of −1.5 dB up to the corner frequency of 570 Hz at point *B*. There it changes abruptly to a −2 slope and remains at that slope since the capacitor has no ESR.

Frequency F_{co} is chosen as one-fifth the switching frequency or $50/5 = 10$ kHz. On curve *ABC* of Figure 12.16, loss at 10 kHz is −50 dB. To force 10 kHz to be F_{co}, the error-amplifier gain at 10 kHz is set at +50 dB (point *F* in Figure 12.16). However, the error amplifier must have a +1 slope at F_{co} to yield a net −1 slope when added to the −2 slope of the *LC* filter. Thus, at point *F* draw a line of +1 slope. Extend this in the direction of lower frequency to F_z—the frequency of the double zero. Extend it in the direction of higher frequencies to F_p, the frequency of the double pole. Then determine F_z and F_p from the *K* factor (Table 12.3) required to yield the desired phase margin.

Assume a phase margin of 45°. Then at F_{co} the total phase lag of the error amplifier plus the *LC* filter is $360 - 45 = 315°$. But the *LC* filter, not having an ESR zero, has a lag of 180°. This leaves a permissible lag of $315 - 180 = 135°$ for the error amplifier. From Table 12.3, a *K* factor of 5 yields a lag of 136°, which is close enough. For $F_{co} = 10$ kHz,

$K = 5$, F_z is 2 kHz, and F_p is 50 kHz. Thus in Figure 12.16, the +1 sloped line is extended down to 2 kHz at E, where it breaks upward to a −1 slope due to the pole at the origin. It is extended on a +1 slope from F_z to the double-pole frequency at 50 kHz. There it turns down to a −1 slope because of the two poles.

The curve *IJKLMN* is the total open-loop gain and is the sum of curves *ABC* and *DEFGH*. It is seen to have a gain of 0 dB at 10 kHz (the crossover frequency F_{co}) and to pass through F_{co} at a −1 slope. The K factor of 5 yields the required 45° phase margin. Components must now be selected to yield the error-amplifier gain curve *DEFGH* in Figure 12.16.

12.14 Component Selection to Yield Desired Type 3 Error-Amplifier Gain Curve

There are six components to be selected ($R1$, $R2$, $R3$, $C1$, $C2$, $C3$), and four equations for zero and pole frequencies (Eqs. 12.12 to 12.15).

Arbitrarily choose $R1 = 1\ \text{k}\Omega$. The first zero (at 2000 Hz) occurs when $R2 = X_{C1}$ and the impedance of the feedback arm above that frequency is mainly that of $R2$ itself. Using the asymptote approximations of Figure 12.16, gain at 2000 Hz is $R2/R1$. From Figure 12.16, gain of the error amplifier at 2000 Hz is +37 dB, or a numerical gain of 70.8. Then for $R1 = 1K$, $R2 = 70.8K$, and from Eq. 12.12 we obtain

$$\begin{aligned} C1 &= 1/(2\pi R2 F_z) \\ &= 1/(2\pi(70,800)2000) \\ &= 0.011\ \mu\text{F} \end{aligned}$$

from Eq. 12.14

$$\begin{aligned} C2 &= 1/(2\pi R2 F_p) \\ &= 1/(2\pi(70,800)50,000) \\ &= 45\ \text{pF} \end{aligned}$$

from Eq. 12.13

$$\begin{aligned} C3 &= 1/(2\pi R1 F_z) \\ &= 1/(2\pi(1000)2000) \\ &= 0.08\ \mu\text{F} \end{aligned}$$

and finally from Eq. 12.15

$$R3 = 1/(2\pi C3F_p)$$
$$= 1/(2\pi(0.08 \times 10^{-6})50,000)$$
$$= 40\ \Omega$$

12.15 Conditional Stability in Feedback Loops

A feedback loop may be stable under normal operating conditions when it is up and running, but can be shocked into continuous oscillation at turn on, or by a line input transient. This odd situation, called *conditional stability*, can be understood from Figure 12.17*a* and 12.17*b*.

Figures 12.17*a* and 12.17*b* contain plots of total open-loop phase shift and total open-loop gain versus frequency, respectively. Conditional stability may arise if there are two frequencies (points *A* and *C*) at which the total open-loop phase shift reaches 360° as in Figure 12.17*a*.

Recall that the criterion for oscillation is that at the frequency where the total open-loop gain is unity or 0 dB, the total open-loop phase shift is 360°. The loop is still stable if the total open-loop phase shift is 360° at a given frequency but the total open-loop gain at that frequency is greater than 1.

This may not be obvious, as it might appear that if, at some frequency, the echo of a signal coming around the loop is exactly in phase with the original signal but larger in amplitude, it would grow larger in amplitude each time around the loop. It would thus build up to a level where the losses would be such to limit the oscillation to some high level and remain in oscillation. This does not occur, as can be demonstrated mathematically. However, this demonstration is outside the scope of the book, so we will simply accept that oscillations do not occur if the total open-loop gain is greater than unity at a frequency where the total open-loop phase shift is 360°.

Thus the loop is unconditionally stable at *B* in Figure 12.17*a*, because the open-loop gain there is unity but the open-loop phase shift is less than 360° by about 40°—i.e., there is a phase margin of 40° at point *B*. The loop is also stable at point *C*, since the open-loop phase shift there is 360°, but the gain is less than unity—i.e., there is gain margin at point *C*. At point *A* the loop is conditionally stable. Although the total open-loop phase shift is 360°, the gain is greater than unity (about +16 dB), and as stated, the loop is stable for those conditions.

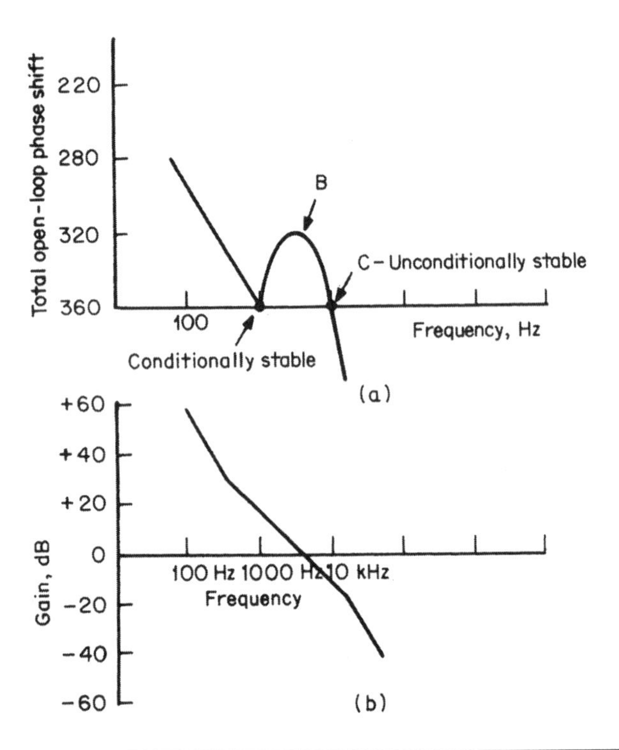

FIGURE 12.17 A loop may be conditionally stable if there are two frequencies where the total open-loop phase shift is 360°. The loop is conditionally stable at point *A* if there is a momentary drop in gain to 0 dB, such as may occur at initial turn "on," and this may result in the conditions for oscillation: 360° total open-loop phase shift and 0 dB gain. Once oscillation breaks out, it will continue. Circuit is unconditionally stable at *C*, as a momentary increase in gain is unlikely.

Under certain conditions, however—for example, at initial turn "on" when the circuit has not yet come to equilibrium and open-loop gain momentarily drops 16 dB at the frequency of point *A*—the condition for oscillation exists: gain is unity and phase shift is 360°. The circuit will break into oscillation and remain oscillatory. Point *C* is not a likely location for such conditional oscillation, as it is not possible for gain to increase momentarily.

If conditional stability exists (most likely at initial turn "on"), it is likely to occur near the corner frequency of the output *LC* filter under conditions of light load. It is seen in Figure 12.3*a* and 12.3*b* that a lightly loaded *LC* filter has a large resonant bump in gain and high slope phase shifts near its corner frequency. The large phase shifts can result in a total of 360° near the *LC* corner frequency. Total open-loop gain is not easily predictable during a turn "on" transient, and momentarily may be unity—and then the loop can break into oscillation.

It is rather difficult to calculate whether this may occur. The safest way to avoid the possibility is to provide a phase boost at the *LC* corner frequency, by introducing a zero there to cancel some of the phase lag in the loop. This can be done easily by adding a capacitor in shunt with the upper resistor in the output voltage sampling network (Figure 12.12).

12.16 Stabilizing a Discontinuous-Mode Flyback Converter

12.16.1 DC Gain from Error-Amplifier Output to Output Voltage Node

Before proceeding, the reader should note some differences in symbols used in the following sections from those in previous sections. Here, R_o refers to the load resistance connected to the converter, whereas in the forward converter it was the internal ESR of the output capacitor. In the flyback discussed here, R_c is the output capacitor ESR. Refer to Figure 12.18.

The essential elements of the loop are shown in Figure 12.18*a*. The first step in designing the feedback loop is to calculate its DC and low-frequency gain from the error-amplifier output V_{ea} to the output voltage node V_o. Assume an efficiency of 80%. Then from Eq. 4.2*a*

$$P_o = \frac{0.8(1/2L)(I_p)^2}{T} = \frac{(V_o)^2}{R_o} \tag{12.16}$$

However, $I_p = \underline{V_{dc}}\overline{T_{on}}L_p$; so

$$P_o = \frac{0.8L_p(\underline{V_{dc}}\overline{T_{on}}/L_p)^2}{2T} = \frac{(V_o)^2}{R_o} \tag{12.17}$$

Referring to Figure 12.18*b*, the PWM compares the output of the error-amplifier V_{ea} to a 0- to 3-V triangle. It generates a rectangular pulse (T_{on}, Figure 12.18*c*) whose width is equal to the time from the start of the triangle to its intersection with the largely unchanging voltage level V_{ea}. This T_{on} will be the "on" time of power transistor Q1. It is seen in Figure 12.18*b* that $V_{ea}/3 = T_{on}/T$ or $T_{on} = V_{ea}T/3$.

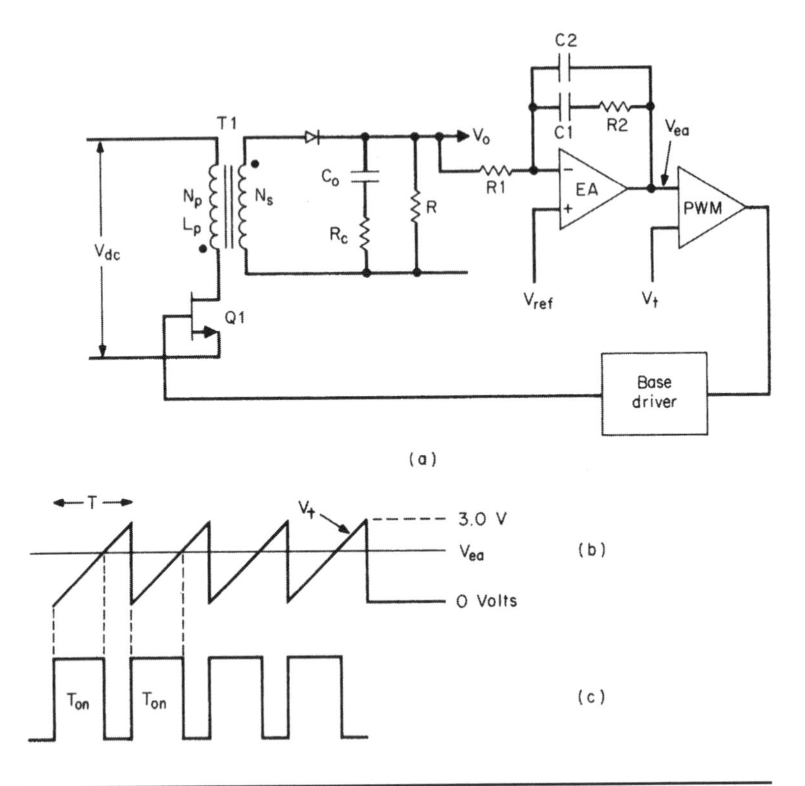

FIGURE 12.18 Discontinuous-mode flyback feedback loop. Note that R_o in the text refers to R in the figure.

Putting this into Eq. 12.16, we have

$$P_o = \frac{0.8L_p(V_{\mathrm{dc}}/L_p)^2(V_{\mathrm{ea}}T/3)^2}{2T} = \frac{(V_o)^2}{R_o}$$

or

$$V_o = \frac{V_{\mathrm{dc}}V_{\mathrm{ea}}}{3}\sqrt{\frac{0.4R_oT}{L_p}} \tag{12.18}$$

and the low-frequency gain from the error-amplifier output to the output node is

$$\frac{\Delta V_o}{\Delta V_{\mathrm{ea}}} = \frac{V_{\mathrm{dc}}}{3}\sqrt{\frac{0.4R_oT}{L_p}} \tag{12.19}$$

12.16.2 Discontinuous-Mode Flyback Transfer Function from Error-Amplifier Output to Output Voltage Node

Assume a small sinusoidal signal of frequency f_n inserted in series at the error-amplifier output. This will cause sinusoidal pulse-width modulation of T_{on}, and amplitude modulation of the triangular current pulses of peak amplitude I_p in the $T1$ primary. Consequently, there is sinusoidal amplitude modulation of the triangular secondary current pulses, whose instantaneous amplitude is $I_p N_p/N_s$.

These triangular secondary current pulses are modulated at the same sinusoidal frequency f_n. There is thus a sinusoidal current of frequency f_n flowing into the top of the parallel combination R_o and C_o. Thus the output voltage across C_o falls off in amplitude at the rate of -20 dB/decade, or at a -1 slope above $F_p = 1/(2\pi R_o C_o)$.

This is simply another way of saying that the transfer function from the error-amplifier output to the output voltage node has a pole at

$$F_p = 1/(2\pi R_o C_o) \tag{12.20}$$

and gain below the pole frequency is given by Eq. 12.19.

This is in contrast to topologies with an output LC filter, in which a sinusoidal voltage inserted at the error-amplifier output node results in a sinusoidal voltage at the input to the LC filter. That voltage, coming through the LC filter, falls off in amplitude at -40 dB/decade rate or -2 slope above the corner frequency. To use the common expression, the LC filter has a *two-pole rolloff* at the output node.

This -1 slope or *single-pole rolloff* of the flyback topology output circuit makes the error-amplifier transfer function required to stabilize the feedback loop different from that for a forward converter. The flyback converter output filter capacitor, in most cases, also has an ESR zero at a frequency of

$$F_p = 1/(2\pi R_c C_o)$$

A complete analysis of the stabilization problem should consider maximum and minimum values of both DC input voltage and of R_o. Equation 12.19 shows low-frequency gain is proportional to V_{dc} and to the square root of R_o. Further, the output circuit pole frequency is inversely proportional to R_o. In the following graphical analysis, all four combinations of V_{dc} and R_o should be considered individually, as the output circuit transfer function may vary significantly with them.

For one output circuit transfer function (one set of line and load conditions), the error-amplifier transfer function is designed to establish F_{co} at a desired frequency and to have the total gain curve pass

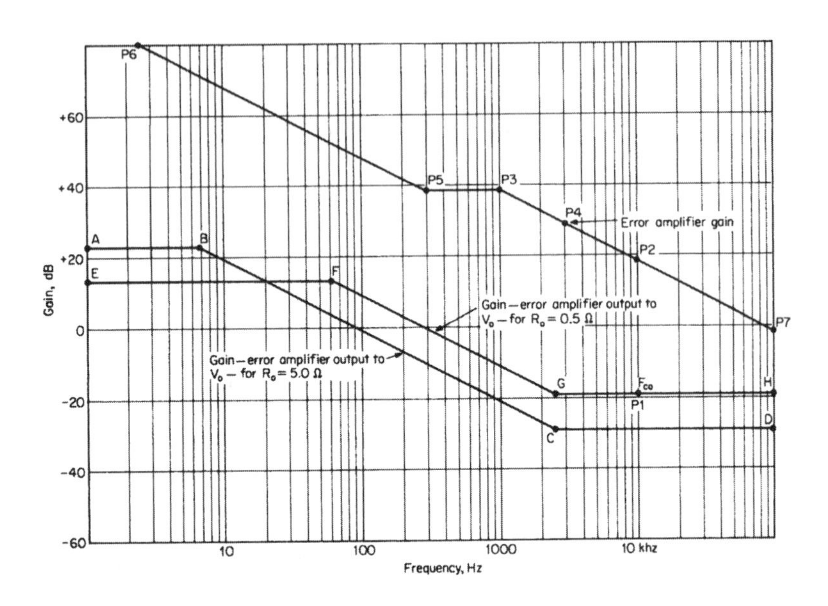

FIGURE 12.19 Gain curves for stabilizing the feedback loop for the discontinuous-mode flyback of the design example in Section 12.18.

through F_{co} with a -1 slope. Care must be taken, then, that under different load and line conditions, the total gain curve does not pass through F_{co} with a -2 slope and possibly cause oscillation.

For this example, we assume that V_{dc} variations are small enough to be neglected. We then calculate low frequency gain from Eq. 12.19, and output circuit pole frequency from Eq. 12.20. Assume $R_{o(max)} = 10R_{o(min)}$.

In Figure 12.19, curve *ABCD* is the output circuit transfer function for $R_{o(max)}$. It has a gain given by Eq. 12.19 from *A* to *B*. At *B*, it breaks into a -1 slope because of the output pole given by Eq. 12.20. At *C*, its slope turns horizontal because of the ESR zero of the output capacitor. Frequency at point *C* is given by Eq. 12.21, whereas in Section 12.7, $R_c C_o$ was taken as 65×10^{-6}, which is typical for an aluminum electrolytic capacitor over a large range of voltage and capacitance ratings.

Also in Figure 12.19, curve *EFGH* is the output circuit transfer function for $R_{o(min)} = R_{o(max)}/10$. Its pole frequency is 10 times that for R_o, because F_p is inversely proportional to R_o. Low-frequency gain below point *F* is 10 dB below that for $R_{o(max)}$ below point *B*, as this gain is proportional to the square root of R_o ($20 \log \sqrt{10} = 10$ dB).

The output circuit transfer function for $R_{o(min)}$ is drawn as follows. Go to point *F*, which is at a frequency 10 times that of point *B* and 10 dB below point *B*. Draw a horizontal line back toward DC for the low-frequency gain (line *FE*). Again at *F*, draw a line of -1 slope

(−20 dB/decade) down to the right, and continue it to the ESR zero frequency at *G*. At *G*, draw a line of horizontal slope toward higher frequency.

Using the output circuit transfer functions *ABCD* and *EFGH* of Figure 12.19, the error-amplifier gain or transfer function curve is drawn as described below (Section 12.17).

12.17 Error-Amplifier Transfer Function for Discontinuous-Mode Flyback

In Figure 12.19, for $R_{o(\min)}$ on curve *EFGH*, F_{co} is established at one-fifth the switching frequency (point *P*1, 10 kHz in this example) as stated in Section 12.3. Most often, F_{co} will occur on this horizontal slope section of the output circuit transfer function.

The error amplifier is designed to have a gain at F_{co} (point *P*2) equal and opposite to the output circuit loss at point *P*1. Since the slope of *EFGH* at F_{co} is horizontal, the error-amplifier gain slope must be −1 at point *P*2.

Thus, go to point *P*2 and draw a line with a slope of −1 in the direction of lower frequencies. Extend it to a frequency (point *P*3) somewhat lower than the frequency at point *C*. At $R_{o(\max)}$, the output circuit transfer function is *ABCD*. Since this new total gain curve also must come through the new F_{co} at a −1 slope, this new F_{co} will occur where the loss along the horizontal line *CD* is equal and opposite (at point *P*4) to the gain of the error amplifier on its −1 slope.

The exact frequency for point *P*3 is not critical. It must be lower than the frequency at point *C* to ensure that for the maximum R_o, this maximum loss can be matched by the equal and opposite gain of the error amplifier somewhere along its −1 slope.

Thus a pole is located at F_p, corresponding to the pole at point *P*3. A Type 2 error amplifier is used. The input resistor *R*1 (Figure 12.18*a*) is arbitrarily selected sufficiently high so as not to load down the sampling resistor network (not shown).

Gain along the horizontal section (points *P*3–*P*5) is read from the graph and made equal to *R*2/*R*1 (Figure 12.18*a*). This fixes resistor *R*2. From the pole frequency F_p and *R*2, the value of $C2 = 1/(2\pi F_p R2)$ in Figure 12.18*a* is fixed.

Now the gain is extended along the horizontal line *P*3–*P*5 and a zero is introduced at point *P*5 to increase low-frequency gain and provide a phase boost. Frequency of the zero F_z at point *P*5 is not critical; it should be about a decade below F_p. To locate F_z, calculate $C1 = 1/(2\pi R2F_z)$.

The design example of the following section will clarify all the above.

12.18 Design Example—Stabilizing a Discontinuous-Mode Flyback Converter

Stabilize the feedback loop of the design example in Section 4.6.3. It is assumed that the output capacitor has ESR, so a Type 2 error amplifier will be used. The circuit is shown in Figure 12.18*a*. The specifications are repeated here:

V_o	5.0 V
$I_{o(\text{nom})}$	10 A
$I_{o(\text{min})}$	1.0 A
$V_{\text{dc(max)}}$	60 V
$V_{\text{dc(min)}}$	38 V
$V_{\text{dc(av)}}$	49 V
Switching frequency	50 kHz
L_p (calculated in Section 4.3.2.7)	56.6 μH

In Section 4.3.2.7, C_o was calculated as 2000 μF, but it was pointed out there that at the instant of turn "off," the peak secondary current of 66 A would cause a thin spike of 66 × 0.03 = 2 V across the anticipated ESR of 0.03 Ω for a 2000-μF capacitor. It was noted that either this thin spike could be integrated away with a small *LC* circuit, or C_o could be increased to lower its ESR.

Here both will be done. Capacitance C_o will be increased to 5000 μF to decrease R_c to (2/5)0.03 or 0.012 Ωs since R_c is inversely proportional to C_o. The initial spike at *Q*1 turn "off" is then 66 × 0.012 or 0.79 V peak. This can easily be filtered down to an acceptable level with a small *LC* which will not affect the feedback loop.

Now the output circuit gain curve can be drawn, first for $R_{o(\text{min})}$ of 5 V/10 A = 0.5 Ω. The low-frequency gain from Eq. 12.19 is

$$
\begin{aligned}
G &= \frac{V_{\text{dc}}}{3}\sqrt{\frac{0.4R_oT}{L_p}} \\
&= \frac{49}{3}\sqrt{\frac{0.4 \times 0.5 \times 20 \times 10^{-6}}{56.6 \times 10^{-6}}} \\
&= 4.3 \\
&= +12.8\,\text{dB}
\end{aligned}
$$

Pole frequency, from Eq. 12.20, is

$$F_p = 1/(2\pi R_o C_o)$$
$$= 1/(2\pi 0.5 \times 5000 \times 10^{-6})$$
$$= 63.7\,\text{Hz}$$

and ESR zero frequency, from Eq. 12.20, is

$$F_{\text{esro}} = 1/(2\pi R_o C_o)$$
$$= 1/(2\pi 65 \times 10^{-6})$$
$$= 2500\,\text{Hz}$$

The output circuit gain curve for $R_o = 0.5\,\Omega$ is then drawn as *EFGH* in Figure 12.19. It is horizontal at a level of +12.8 dB up to $F_p = 63.7$ Hz. There it breaks to a −1 slope down to the ESR zero at 2500 Hz. The error-amplifier gain curve can now be drawn.

Choose F_{co} as one-fifth the switching frequency or 50/5 = 10 kHz. On *EFGH*, the loss is −19 dB at 10 kHz. Hence make the error-amplifier gain +19 dB at 10 kHz. Go to 10 kHz and +19 dB (point *P*2) and draw a line with a slope of −1 (−20 dB/decade) in the direction of lower frequency. Now extend that line to a frequency somewhat lower than F_{esro}—say, to point *P*3 at 1 kHz, +39 dB. At point *P*3, draw a horizontal line back to—say—300 Hz at point *P*5 where a zero will be located.

The location of the zero is not critical. In Section 12.17, it was suggested the zero at point *P*5 should be one decade below point *P*3. Some designers actually neglect the zero at point *P*5[5], but here it is added to gain some phase boost. Thus, for a zero at point *P*5, draw the gain slope toward lower frequency at a −1 slope.

Now verify that for $R_{o(\text{max})}$ of 5 Ω, the total gain curve (output circuit plus error-amplifier transfer function) comes through F_{co} at a −1 slope.

For $R_o = 5\,\Omega$, Eq. 12.19 gives a DC gain of 13.8 or +23 dB. Eq. 12.20 gives the pole frequency as 6.4 Hz. The frequency of the ESR zero remains at 2500 Hz. The output circuit transfer function for $R_o = 5\,\Omega$ is *ABCD*.

The new F_{co} is the frequency where the gain of the error amplifier equals the loss on *ABCD*. This is seen to be at point *P*4 (3200 Hz), where the output filter loss is −29 dB and the error-amplifier gain is +29 dB. The sum of the error-amplifier gain and total gain *ABCD* has a −1 slope as it passes through F_{co}.

It should be noted, however, that if R_o were somewhat larger, the curve *ABCD* would be depressed to a lower gain along its entire length. Then the point at which the previously fixed error-amplifier gain is equal and opposite to the output filter loss would occur on the -1 slope of each curve.

The total gain curve would then come through the new F_{co} at a -2 slope and oscillations could occur. Thus, as a rule, discontinuous-mode flybacks should be tested carefully for stability at minimum load current (maximum R_o).

The error-amplifier transfer function $P6–P5–P3–P7$ is implemented as follows. In Figure 12.18*a*, arbitrarily choose $R_1 = 1000\ \Omega$. Gain at point $P3$ is seen in Figure 12.19 to be +38 dB, or a numerical gain of 79. Thus $R_2/R_1 = 79$ or $R_2 = 79$ kΩ. For the pole at point $P3$ at 1 kHz, $C_2 = 1/(2\pi F_p R_2) = 2000$ pF. For the error-amplifier zero at 300 Hz, $C1 = 1/(2\pi F_z R_2) = 6700$ pF.

Because of the single-pole rolloff characteristic of the output circuit, its absolute maximum phase lag is 90°. Because of the ESR zero, usually it is much less and there rarely is a phase-margin problem in the discontinuous-mode flyback.

Consider the situation for $R_o = 0.5\ \Omega$. Lag at F_{co} (10 kHz) due to the pole at 64 Hz and the ESR zero at 2500 Hz is

$$\begin{aligned}\text{Output circuit lag} &= \tan^1\left(\frac{10,000}{64}\right) - \tan^1\left(\frac{10,000}{2500}\right)\\ &= 89.6 - 76.0\\ &= 13.6^\circ\end{aligned}$$

and the error-amplifier lag at 10,000 Hz due to the zero at 300 Hz and the pole at 1000 Hz (see Figure 12.20, curve $P6–P5–P3–P7$) is

$$270 - \tan^{-1}\left(\frac{10,000}{300}\right) + \tan^{-1}\left(\frac{10,000}{1000}\right) = 270 - 88 + 84 = 266^\circ$$

Total phase lag at 10 kHz is then $13.6 + 266 = 280^\circ$. This yields a phase margin at F_{co} of 80°.

12.19 Transconductance Error Amplifiers

Many of the commonly used PWM chips (1524, 1525, 1526 family) have *transconductance* error amplifiers. Transconductance g_m is the change in output current per unit change in input voltage. Thus

$$g_m = \frac{dI_o}{dV_{\text{in}}}$$

For shunt impedance Z_o at the output node of the error amp to ground

$$dV_o = dI_o Z_o = g_m d V_{\text{in}} Z_o$$

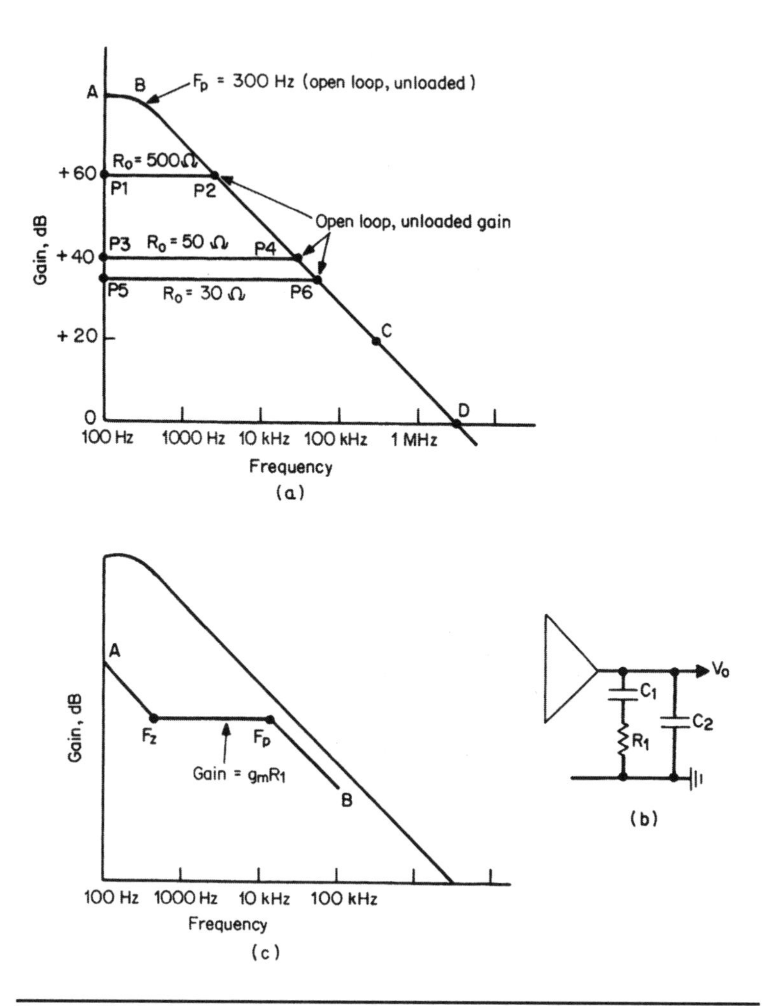

FIGURE 12.20 (*a*) Open-loop unloaded gain curve for PWM chip 1524, 1525 error amplifiers shown as *ABCD*. When loaded with indicated shunt resistance to ground, gain is constant at $G = g_m R_o$. (*Courtesy of Silicon General*) (*b*) A Type 2 error-amplifier gain curve with shunt network to ground. (*c*) Gain with circuit of Figure 12.20*b* is $A - F_z - F_p - B$; $F_z = 1/(2\pi R1C1)$; $F_p = 1/(2\pi R1C2)$.

or gain G is

$$G = \frac{dV_o}{dV_{\text{in}}} = g_m Z_o$$

The unloaded, open-loop 1524, 1525-family amplifiers have a nominal low-frequency gain of +80 dB, a pole at 300 Hz, and gain slope above the pole of −1, or −20 dB/decade slope. This is seen as curve *ABCD* in Figure 12.20*a*.

A pure resistance R_o in shunt from output node of the error amp to ground yields a gain curve which is constant and equal to $g_m R_o$ from DC up to the frequency where it intersects the curve *ABCD* in Figure 12.20*a*. For the 1524, 1525 family, g_m is nominally 2 mA/V. Thus gains for R_o = 500 *k*, 50 *k*, and 30 *k* are 1000, 100, and 60 respectively and are shown as curves *P*1–*P*2, *P*3–*P*4, and *P*5–*P*6 in Figure 12.20*a*.

In most cases, Type 2 error-amplifier gain characteristics are required. This is easily obtained with the network shown in Figure 12.20*b*.

At low frequencies, X_{C1} is much greater than *R*1, and *C*1 and *C*2 are effectively in parallel with the internal 100 pF to ground which causes the open-loop 300-Hz pole. This shifts the 300-Hz pole to a lower frequency, and above that lower frequency, gain resumes falling at a −1 slope. At a frequency $F_z = 1/(2\pi R1C1)$, where $X_{C1} = R_1$, there is a zero and gain slope turns horizontal at a magnitude $g_m R1$. Higher in frequency at $F_p = 1/(2\pi R1C2)$ where $X_{C2} = R1$, the pole turns the gain slope to −1.

The gain curve with the circuit configuration of Figure 12.20*b* is shown in Figure 12.20*c*.

Most frequently, in the 1524, 1525 family of PWM chips the error-amplifier gain curves are shaped in the above-mentioned fashion with the network of Figure 12.20*b* shunted to ground, rather than being used in the conventional operational-amplifier mode.

Whether a network is shunted to ground as in Figure 12.20*b*, or returned to the inverting input terminal as in a conventional operational amplifier, there is a restriction on the magnitude of *R*1 arising from the following. The internal error amplifiers in the above-mentioned chips cannot source or sink more than 100 μA. With a 3-V triangle at the PWM comparator, the error-amplifier output may have to move the entire 3 V for sudden line or load changes. For *R*1 less than 30 kΩ, this fast 3-V swing would demand more than the available 100 μA. Response time to fast load or line changes would be sluggish.

Because of this 100-μA limit on output current, many designers prefer not to use the error amplifier internal to the PWM chip. Since the chip's EA output node is brought out to one of the pins, some prefer to use a better, external error amplifier and connect it to the chip's error–amplifier output node.

However, it may be essential from a cost viewpoint to use the chip's internal error amplifier. Calculation of the converter output filter may show that its loss at F_{co} is so great that to match the error-amplifier gain to it, R_1 must be less than 30 kΩ. If this happens, *R*1 can be set to 30 kΩ to match an arbitrarily increased output filter loss at F_{co}. This increased output filter loss at F_{co} can easily be achieved by shifting its pole frequency to a lower value by increasing the output filter inductance or capacitance.

References

1. D. Venable, "The *K* Factor: A New Mathematical Tool for Stability Analysis and Synthesis," *Proceedings Powercon 10*, 1983.
2. A. Pressman, *Switching and Linear Power Supply, Power Converter Design*, pp. 331–332, Switchtronix Press, Waban, Mass., 1977.
3. K. Billings, *Switchmode Power Supply Handbook*, Chapter 8, McGraw-Hill, New York, 1989.
4. G. Chryssis, *High Frequency Switching Power Supplies*, 2nd ed., Chapter 9, McGraw-Hill, New York, 1989.
5. Unitrode Corp., *Power Supply Design Seminar Handbook*, Apps. B, C, Watertown, Mass., 1988.

CHAPTER 13
Resonant Converters

13.1 Introduction

As new integrated circuits lead to more electronic functions in smaller packages, it becomes essential for power supplies to become smaller. Power supplies get smaller mainly by increasing their operating frequency to decrease the size of the power transformer and output *LC* or capacitive filter. Supplies also get smaller by increasing their efficiency so as to require smaller heat sinks.

Thus a major objective in present-day power supply technology is to operate at switching frequencies higher than the currently commonplace 100 to 200 kHz.

However, going to higher switching frequencies with the conventional square current waveform topologies discussed up to this point increases transistor switching losses at both turn "off" and turn "on." Turn "on" losses, due to charging and discharging MOSFET output capacitances (Section 11.1), become very important at frequencies over 1 MHz.

As discussed in Chapters 11 and 1, the overlap of falling collector current and rising collector voltage during turn "off" yields a high spike of dissipation. As switching frequency increases, there are more high-dissipation spikes per second, which results in higher average transistor dissipation. The higher losses require larger heat sinks, so that there may not be any size decrease, despite the smaller power transformer and output filter. Further, the junction-to-case temperature rise with the usual 1°C/W junction-to-case thermal resistance may still result in dangerous transistor junction temperature.

Adding snubbers (Chapter 11) at collector or drain outputs reduces transistor switching losses. If a dissipative *RCD* snubber is used (Section 11.3), it does not decrease total dissipation—it simply shifts losses from the transistor to the snubber resistor. Nondissipative snubbers (Section 11.6) do reduce transistor switching losses but are troublesome at frequencies over 200 kHz.

13.2 Resonant Converters

To operate at higher frequencies, which will permit smaller power supplies, transistor switching losses at turn "off" and turn "on" must be fundamentally decreased. This is achieved in *resonant converters* by associating a resonating *LC* circuit with the switching transistor, to render its current sinusoidal rather than square wave in shape. It is then arranged to turn the transistor "on" and "off" at the zero crossings of the current sine wave. As a result there is little overlap of falling current and rising voltage at turn "off" or at turn "on" and hence very much reduced switching losses. Circuits which turn "on" and "off" at zero current are referred to as *zero current switching* (ZCS) types.[1]

We have seen previously (Section 11.1) that switching losses can occur at turn "on" even though there is no overlap of rising voltage and falling current at the zero crossing of the current sine wave. In Section 11.1 it was pointed out that considerable energy $[0.5C_o(2V_{dc})^2]$ is stored on the relatively large output capacitance of a MOSFET. When the MOSFET is turned "on" once per period T, it dissipates $0.5C_o(2V_{dc})^2/T$ watts in the MOSFET. Circuits designed to cope with this problem are called *zero-voltage-switching* (ZVS) types.[2]

ZVS types work by designing the transistor output capacitance to be part of that in a resonant *LC* circuit. Then the energy stored in the capacitance is returned without loss to the power supply bus. Operation is similar to that of the nondissipative snubber of Section 11.6.

The industry found intense interest in resonant converters in the mid 1980's. Since then many researchers have produced a large number of articles on the subject. Dozens of new resonant converter topologies have been proposed and mathematically analyzed. Most of these have been built and have achieved high efficiencies (80 to 97%) with very high power density. Some claims exceeding 50 W/in^3 have been made for DC/DC converters, which do not have the large input filter capacitor required of off-line converters.

However, these high density converters normally require cooling by an external "cold plate" whose size and means of cooling are seldom reckoned in calculating the published power densities.

Covering the very large number of resonant converter topologies available and their large range of operating modes is beyond the scope of this book, so I will present here only an overview of some of the more well-proven topologies and their operating modes as examples of what the resonant mode can provide.

It is worth noting here that in any new field of research, articles on a new approach very frequently end with comments on its restricted usage, with limits on line and load variations and excessive component stress. As with any emerging technology at the next conference the following year, new solutions to last year's problems are offered by numerous other investigators, and so on. As yet, resonant converters do not have the flexibility of PWM converters. They do not cope well with large line and load changes. Further, component tolerances are more critical. A major limitation is that they operate at higher peak transistor currents for the same output power than do conventional PWM square-wave inverters, and in some circuit configurations they subject devices to larger voltage stresses.

13.3 The Resonant Forward Converter

First we will consider the simplest resonant circuit, the resonant forward converter. We are interested primarily to see how transistor turn "off" is arranged to occur at zero current, and to see how critical the exact turn "off" time may be. Figure 13.1 shows a simple resonant forward converter[3] operating in the discontinuous mode.

In the *discontinuous mode* the current in the resonant *LC* circuit is not a continuous sine wave, but a sequence individual half or full cycles of sine-wave current, separated by a large time interval T_s as in Figure 13.1.

The resonant frequency of the circuit is $F_r = 0.5/(t_1 - t_0)$ and is fixed by the passive resonant elements L_r and the resonant C reflected into the primary, $(C_r)(N_s/N_p)^2$. The resonant frequency of the circuit is

$$F_r = 1/(2\pi\sqrt{L_r C_r (N_s/N_p)^2}) \tag{13.1}$$

where L_r is the transformer leakage inductance, or that plus some externally added small inductance to make the total L_r relatively independent of production variations in the leakage inductance itself.

Transistor $Q1$ is turned "on" at a switching frequency $F_s = 1/T_s$. The circuit works as follows. First, L_r and $C_r(N_s/N_p)^2$ form a series resonant circuit, with DC secondary current reflected by the transformer turns ratio in shunt across the reflected capacitance in the primary. In resonant converter parlance, this is a *parallel resonant converter* (PRC), as the load is placed in parallel or shunt with the resonating capacitor. Other circuits, to be discussed later, place the load in series with the series resonating *LC* elements and are called *series resonant converters* (SRCs).

Just prior to $Q1$ turn "on," no current is flowing in the resonant circuit, as it is discontinuous—i.e., there is a long time between half sine waves (Figure 13.1). Thus when $Q1$ does turn "on" at t_0, a half sine wave of current starts through it. The current amplitude is zero at time t_0 since there has been no current flow in the resonant circuit just prior to turn "on."

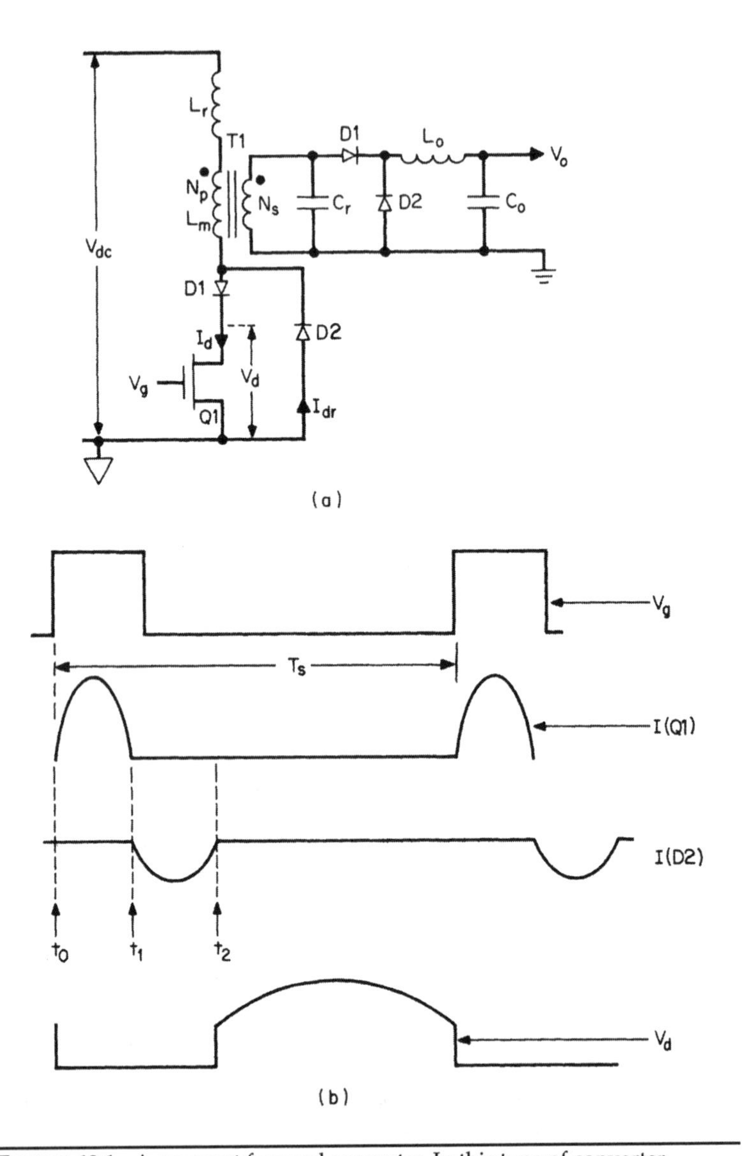

FIGURE 13.1 A resonant forward converter. In this type of converter, capacitance C_r is reflected into the primary and resonates with L_r, the transformer leakage inductance. The MOSFET gate is turned "off" shortly after the zero crossing of the positive half cycle of drain current. When the drain voltage rises at the end of the negative half cycle of current in $D2$, a half cycle of resonant ring of the magnetizing inductance L_m with the capacitance C_r reflected into the primary resets the $T1$ core.

The current goes through its first positive half sine wave. At t_1, it passes through zero and reverses. This negative current is forced to flow by the voltage stored on the resonating capacitor in the primary. It flows up through the anode of $D2$ and completes its loop through the supply source V_{dc}. For the half cycle that current flows through $D2$, the $Q1$ drain to source voltage remains clamped to about −1 V (the $D2$ forward drop). Between t_1 and t_2, there is no current in $Q1$ and its gate may be turned "off." The $Q1$ "on"-time duration is not at all critical. It must be greater than a half period of the resonant current sine wave and less than a full period. At t_2, the negative half sine wave of current in $D2$ has returned to zero and now current in the $T1$ magnetizing inductance drives the drain up toward $2V_{dc}$ to reset the core. The magnetizing inductance and the capacitance reflected across it from the secondary form another resonant circuit. When the drain finally rises at t_2, a negative half sine wave of voltage across that resonant circuit resets the core exactly to its starting point on the BH loop.

The objective of zero current turn "off" has thus been achieved, and there is no turn "off" dissipation. The reverse recovery time of diode $D2$ is short enough so that the dissipation is not significant—especially since the current in it is already zero, or close to zero, when the drain rises toward $2V_{dc}$.

For the duration of the positive resonant half cycle ($t_1 - t_0$), the primary is delivering a pulse of power to the secondary and the load. The DC output voltage is regulated by varying the spacing between these pulses—i.e., varying the switching frequency $F_s = 1/T_s$. If V_{dc} goes up or the DC output load current goes down, the spacing between pulses must increase (F_s must decrease)—and vice versa; as V_{dc} goes down or load current increases, F_s must increase.

This method of voltage regulation—varying the switching frequency F_s—can be considered a drawback of this resonant converter. In many resonant topologies, pulse width (the half period of the resonant LC circuit) is maintained constant and its repetition frequency is varied to get regulation.

Variable-frequency regulation is objectionable in some circumstances. Where a computer is involved, the designers often require the power supply switching frequency to be synchronized to a submultiple of the computer clock. This reduces the probability of power supply noise spike pickup generating false ones or zeros in the computer logic circuits. In conventional pulse-width-modulated converters, the switching frequency remains constant and the "on"-time pulse width is varied.

Further, where a cathode-ray tube (CRT) screen is involved, it is desirable to have the power supply switching frequency phase locked with the CRT horizontal line frequency. When it not synchronized, it is possible that power supply switching frequency noise pickup can appear as "herringbone interference" running in a random fashion across the screen. When power supply switching frequency is constant and phase locked to the CRT horizontal line rate, any noise pickup remains at a fixed location on the screen and is far less disconcerting to an operator.

13.3.1 Measured Waveforms in a Resonant Forward Converter

The circuit in Figure 13.2*a* is a forward converter with maximum output power of 32 W (5.2 V, 6.2 A). The circuit is designed for input voltage V_{dc} of 150 V, and is shown at a switching frequency of 856 kHz. The transformer turns ratio is 10/1. It is of interest to see actual measured waveforms in a resonant forward converter of the type discussed above. The waveforms in Figure 13.2*b* are from Reference 3, and are reproduced with the courtesy of the authors, F. Lee and K. Liu.

In the third waveform of Figure 13.2*b*, primary current is a half sine wave whose resonant half period is 0.2 μs ($F_r = 2.5$ Hz). The secondary capacitor is 0.15 μF, which reflects into the primary as $0.15(0.1)^2 = 0.0015$ μF. For $F_r = 2.5$ MHz, the transformer leakage inductance plus any added discrete inductance must be

$$L_r = 1/\left(4\pi^2 F_r^2 C\right) = 1/(4\pi(2.5)^2(0.0015)10^{-6})$$
$$= 2.7\ \mu\text{H}$$

In Figure 13.2*b* Waveform 1, the MOSFET gate is turned "off" shortly after the first positive half cycle of the current sine wave has passed through zero, thus meeting the objective of zero current turn "off." It is also seen in Waveform 4 that the drain voltage has started rising shortly after the negative primary current through *D*2 has returned to zero as discussed above.

The voltage across the secondary capacitor in Waveform 2, which is a replica of the voltage across the primary magnetizing inductance, has reversed polarity and resets the core as discussed above. That waveform also indicates the maximum switching frequency F_s. The minimum spacing between *Q*1 current pulses must be sufficient to permit the negative half cycle in Waveform 2 to return to zero, after the core has been reset fully.

The foregoing shows some of the difficulties in making resonant converters work over large line and load variations. Consider the 2.7 μH calculated for the resonant inductor above. This is so small that changes in wire lengths and routing to the transformer can result in large percentage changes in the total inductance. For a value as

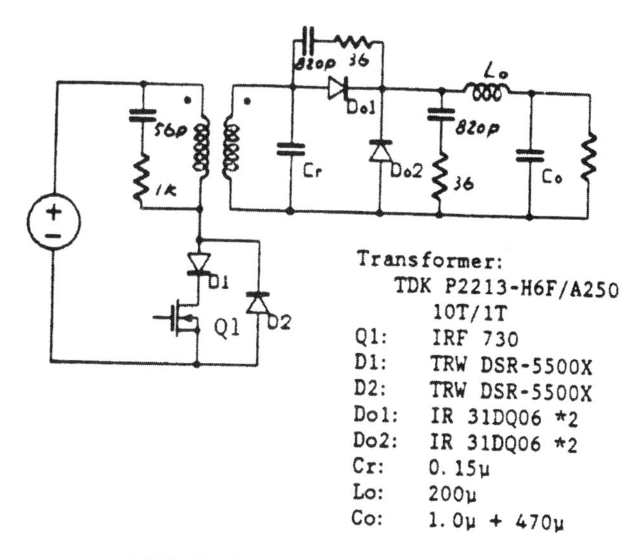

A 800kHz, 30W forward quasi-resonant converter

(a)

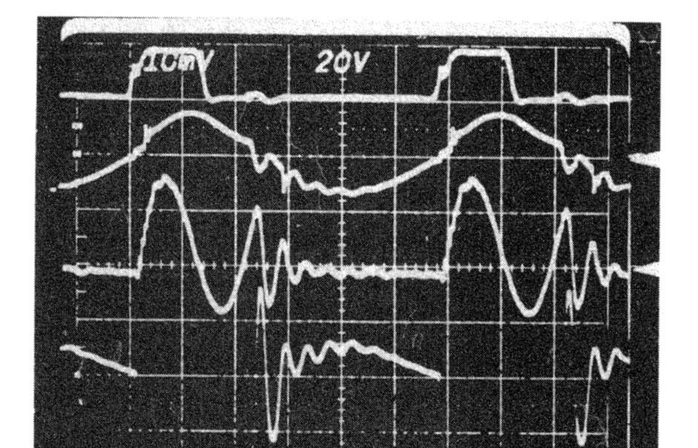

Vin = 120V, Vo = 3.56V
fs = 856kHz, Io = 4.2A
1st waveform: Vgs(20V/Div)
2nd waveform: Vcr(20V/Div)
3rd waveform: Ip (1A/Div)
4th waveform: Vds(100V/Div)

(b)

FIGURE 13.2 Measured waveforms on a circuit such as in Figure 13.1. (*Courtesy of F. Lee and K. Liu.*)

small as 2.7 μH, production spread in the transformer can cause large percentage variation in its leakage inductance. If the inductance increases, the resonant half period increases. Then the MOSFET "on" time, which should be greater than a resonant half period, may be too short. The MOSFET may turn "off" before the end of a resonant half period—i.e., before the current zero crossing.

L_r may be increased to make it less susceptible to variations from transformer production spread and wire length and routing. This increases the resonant half period and decreases the maximum switching frequency, which must be sufficiently low to permit complete core reset.

If L_r is increased and C_r is decreased to maintain the same resonant half period, the ratio L_r/C_r is increased. Since the peak of the Q1 sine-wave current (Figure 13.2*b*, Waveform 3) is roughly inversely proportional to $\sqrt{L_r/C_r}$, required maximum peak current and maximum DC output current may not be obtainable if L_r/C_r is increased.

Of course, many such problems may be solved in specific cases. This is mentioned only as an indication that even such a simple resonant converter has less flexibility than a square-wave circuit to cope with varying specifications, line and load conditions, and manufacturing tolerances.

It is of interest to note that a conventional forward converter operating at a relatively low frequency could yield equal output power at a lower primary current than shown in Figure 13.2*b*. That figure shows (for full-wave mode) peak current is 1.5 A for $V_o = 3.56$ V, $I_o = 4.2$ A, or 15.0 W at V_{in} of 120 V.

Equation 2.28 gives the peak primary current for a conventional PWM square-wave forward converter as $I_p = 3.13P_o/\underline{V_{dc}} = 3.13 \times 15/120 = 0.39$ A. Table 7.2*a* shows that the next core smaller than the 2213 pot core used for Figure 13.2*b* could easily be used, and at a lower frequency. The table shows that the 1811 core at 150 kHz could deliver 19.4 W in a forward converter operating at only 150 kHz.

This is not to put down resonant converters, but only to point out the need to weigh them against conventional, proven topologies.

13.4 Resonant Converter Operating Modes

13.4.1 Discontinuous and Continuous: Operating Modes Above and Below Resonance

Operating modes can be continuous or discontinuous as shown in Figure 13.1. In the discontinuous mode (DCM), as noted, output voltage regulation is accomplished by varying the switching frequency.

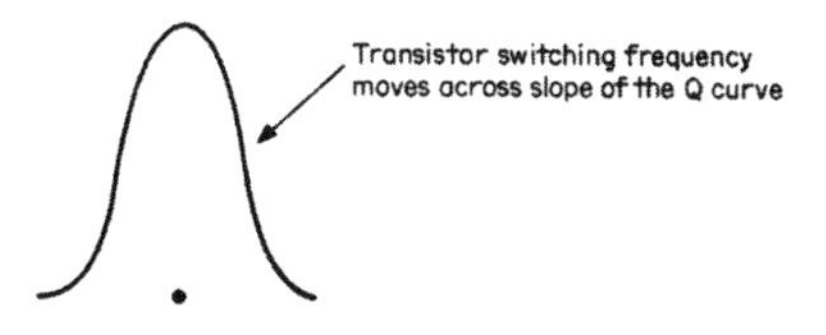

FIGURE 13.3 Output voltage regulation by shifting switching frequency along the slope of the resonant curve.

Power is delivered to the load as a sequence of discrete pulses whose duration is short compared to the period.

If the output voltage must be raised because V_{dc} has gone down or DC load current has been increased, the switching frequency or repetition rate of the discrete pulses is increased. Conversely, if output voltage has gone up because input voltage or output load resistance has gone up, the repetition rate of the discrete pulses is decreased.

SCR resonant mode converters have been operating successfully in the discontinuous mode for many years at frequencies in the range of 20 to 30 kHz. (These are discussed in Chapter 6.) However, in discontinuous-mode operation, large load and line changes result in large changes in switching frequency. An improvement in this respect is obtained by operating in the continuous-conduction mode (CCM). In CCM, there is negligible gap between successive square-wave voltage pulses from the switching transistors or between successive current sinusoids.

Fourier has shown that the fundamental component of any square wave is a sine wave whose frequency is that of the square wave. A resonant *LC* circuit has the characteristic impedance versus frequency curve shown in Figure 13.3. To obtain amplitude control in CCM, the average switching frequency is set either above or below resonance on one side of the curve.

Direct-current output voltage in the CCM mode is proportional to either the peak AC voltage on the resonating capacitor or the peak AC current in the series resonating *LC* circuit. Output voltage regulation is accomplished by moving the switching frequency along one side of the resonant curve to change the output amplitude.

When the switching frequency is set above the resonant peak, it is referred to as *above-resonance mode* (ARM). Operation below the resonant peak is referred to as *below-resonance mode* (BRM).

In Figure 13.3, with a relatively steep or high *Q* curve, small changes in frequency cause large changes in output amplitude.

Note that if the average switching frequency is above the resonant peak (ARM), switching frequency must decrease to increase the output, and that in BRM, switching frequency must increase to increase the output voltage or current.

We can now consider one of the problems with resonant converters. Much of the current literature indicates that CCM is becoming the preferred mode of operation, because it results in a smaller frequency range to achieve the usually desired load and line regulation. However, if the feedback system has been designed for ARM, a major problem can arise if operation shifts to BRM, because the control will now be in the wrong direction.

In ARM, a decrease in the DC output voltage would be corrected by the variable-frequency oscillator in the control loop decreasing switching frequency to move higher up on the resonant curve. However, since the curve is relatively steep, a small change in the magnitudes of the resonant L and C, due perhaps to production tolerances, can shift the resonant frequency. If the resonant peak is shifted sufficiently, operation could fall to the other side of the curve, and now the feedback loop, sensing a decrease in output voltage, would still try to correct it by decreasing switching frequency. This, of course, would result in a further decrease in output voltage—i.e., positive rather than negative feedback.

13.5 Resonant Half Bridge in Continuous-Conduction Mode[4]

Much resonant converter development effort seems to be for half bridge topologies, which will now be considered. The following discussion is based on a classic article by R. Steigerwald.[4]

13.5.1 Parallel Resonant Converter (PRC) and Series Resonant Converter (SRC)

We have seen that the output power can be taken from the resonant LC circuit in either of two ways. When the output load (reflected into the power transformer primary) is reflected in parallel with the resonating capacitor, the circuit is referred to as a *parallel resonant converter* (PRC). When the load is reflected in series with the resonating LC circuit, it is referred to as a *series resonant converter* (SRC).

A parallel loaded resonant half bridge is shown in Figure 13.4*b*; C_{f1}, C_{f2} are the input filter capacitors used in the scheme for generating a rectified 320 V whether operation is from 120 or 220 VAC (Figure 5.1). They are large capacitors used only to split the rectified DC and have nothing to do with the resonant LC circuit.

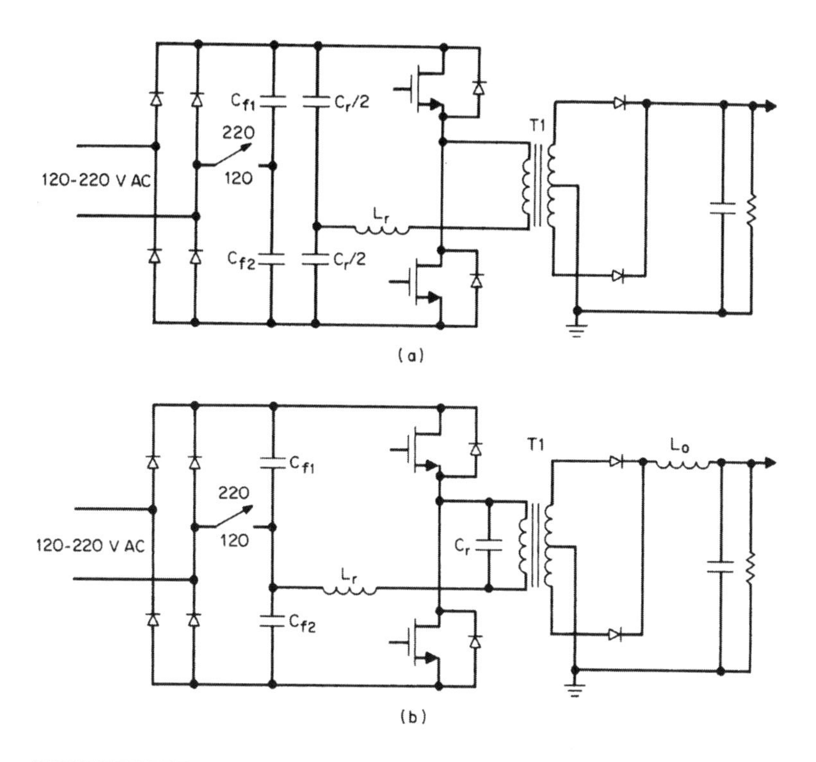

FIGURE 13.4 (*a*) A series-loaded resonant half bridge. Inductance L_r resonates with capacitance C_r. The load is reflected by $T1$ in series with the resonant circuit. Transistors are turned "off" directly after the end of the first half cycle of resonant current to achieve zero current switching. In series loading, the output filter is capacitive. (*b*) A parallel-loaded resonant half bridge. Inductance L_r resonates with capacitance C_r. The load is reflected by $T1$ in shunt with the resonating capacitor. In parallel loading, the output filter has a high-impedance inductor input to avoid lowering the Q of the resonant circuit.

The capacitor C_r across the power transformer primary resonates with an external inductor L_r at a frequency $F_r = 1/(2\pi\sqrt{L_r C_r})$. The output inductor L_o is large and has a high impedance at F_r so that it does not load down C_r and kill the Q of the resonant LC circuit. L_o is sufficiently large that it runs in the continuous-conduction mode (Section 1.3.6). The impedance seen across C_r is the output load resistance multiplied by the turns ratio squared. The $T1$ magnetizing inductance is much larger than this and does not affect circuit operation.

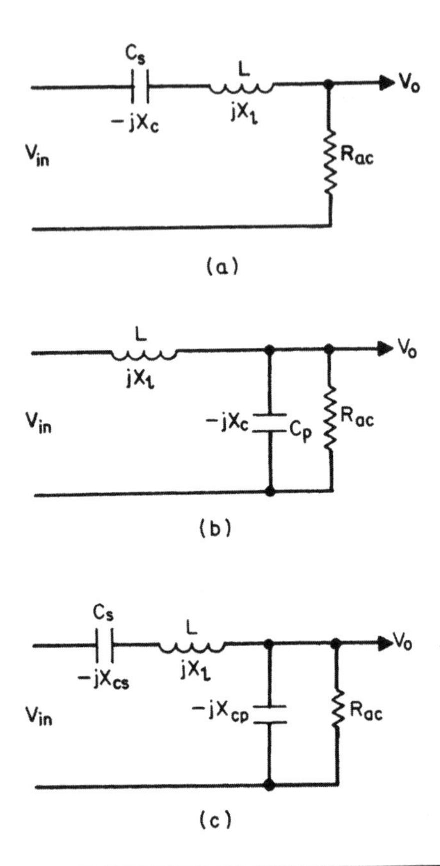

FIGURE 13.5 (*a*) The AC equivalent circuit for the series-loaded resonant half bridge shown in Figure 13.4*a*. (*b*) The AC equivalent circuit for the parallel-loaded resonant half bridge shown in Figure 13.4*b*. (*c*) The AC equivalent circuit for a series-parallel-loaded resonant half bridge, often called an *LCC circuit*. Steigerwald has shown that the AC gain between outputs and inputs of the above equivalent circuits fixes the ratio between DC output voltage, and half the DC input voltage. (*Courtesy of R. Steigerwald.*)

A series-loaded resonant half-bridge converter (SRC) is shown in Figure 13.4*a*. Here an external inductor L_r resonates with the equivalent capacitance C_r at the junction of the two $C_r/2$ capacitors. Again, the C_f capacitors are large line frequency filter capacitors that have nothing to do with the resonant circuit operation. The load in an SRC is the load resistor reflected by the turns ratio squared in series with the resonating *LC* elements. In the series resonant circuit, the secondary side inductor is omitted.

The SRC is often used for high-voltage supplies as it requires no output inductor. An output inductor for high output voltage is bulky. The PRC is usually used for low-voltage, high-current supplies as the output inductor limits ripple current in the output capacitor.

Both the series and parallel half bridges discussed below will operate in the continuous-conduction mode. Since DC voltage is regulated by varying the switching frequency, it is necessary to know how AC voltage across the reflected load varies with frequency as operation moves along the side of the resonant curve. The rectified DC output voltage is proportional to the AC voltage across the reflected resistance.

13.5.2 AC Equivalent Circuits and Gain Curves for Series-Loaded and Parallel-Loaded Half Bridges Operating in the Continuous-Conduction Mode[4]

Figures 13.5*a* and 13.5*b* show the equivalent AC circuits for the series- and parallel-loaded half bridges of Figures 13.4*a* and 13.4*b*, respectively. Inputs to these circuits are square waves of amplitude $\pm V_{dc}/2$ generated by the switching transistors. Following the analysis presented by Steigerwald[4], we will consider only the fundamental of the square-wave frequency and thus calculate gain (the ratio of output to input voltage) as a function of frequency.

From the equivalent circuits of Figure 13.5, the ratios for series- and parallel-loaded circuits are

$$\text{Series-loaded:} \quad \frac{V_o}{V_{in}} = \frac{1}{1 + j[(X_l/R_{ac})} - (X_c/R_{ac})] \tag{13.2}$$

$$\text{Parallel-loaded:} \quad \frac{V_o}{V_{in}} = \frac{1}{1 - (X_l/X_c) + j(X_l/R_{ac})} \tag{13.3}$$

where R_L is the secondary load reflected into the primary, and $R_{ac} = 8R_L/\pi^2$ for series loading, $R_{ac} = \pi^2 R_L/8$ for parallel loading.

From these relations, Steigerwald plots the ratio $NV_{odc}/0.5V_{in}$, where N is the power transformer turns ratio, V_{in} is the input supply voltage, and V_{odc} is the DC output voltage.

For the series-loaded case in Figure 13.6, $Q = w_o L/R_L$ where $w_o = 1/\sqrt{LC_s}$. For the parallel-loaded case in Figure 13.7, $Q = R_L/w_o L$, and $w_o = 1/\sqrt{LC_p}$.

From Figures 13.6 and 13.7, some of the problems with resonant converters can be seen.

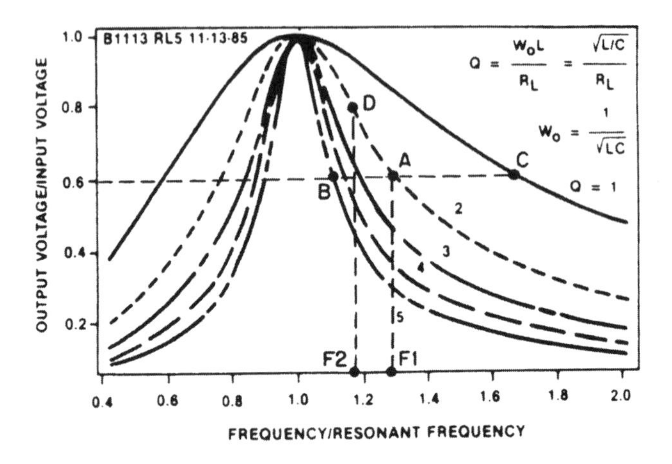

FIGURE 13.6 Series resonant converter gain curve from AC equivalent circuit of Figure 13.5*a*. (*Courtesy of R. Steigerwald.*)

13.5.3 Regulation with Series-Loaded Half Bridge in Continuous-Conduction Mode (CCM)

For a number of reasons, Steigerwald states that operating above the resonant peak (ARM) is preferable to operating below resonance.

Figure 13.6 shows how the continuous conduction mode SRC half bridge regulates. If initial operation were at A with $Q = 2$ at normalized frequency 1.3, the output/input voltage ratio would be

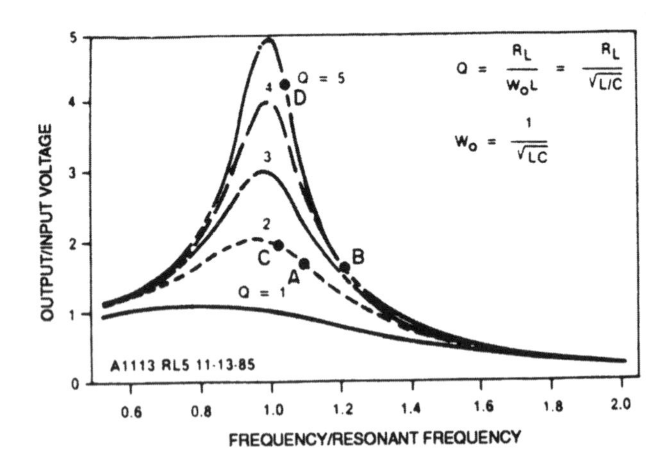

FIGURE 13.7 Parallel resonant converter gain curve from AC equivalent circuit of Figure 13.5*b*. (*Courtesy of R. Steigerwald.*)

0.6. Now if the load resistance R_L decreased so as to increase Q to 5, normalized switching frequency would have to be decreased to about 1.15 at point B to yield the same output voltage. If R_L increased to a value which yielded $Q = 1$, for the same output voltage, operation would have to shift to point C, where normalized frequency is about 1.62.

It is obvious that as load R_L increases, Q decreases and the gain curve approaches a horizontal line, making the required frequency change so large as to be impractical. Obviously, regulation at open circuit is impossible as the Q curve has no resonant peak or *selectivity*.

The actual operating point on the Figure 13.6 gain curve depends on the DC input supply and output voltages. For a given output voltage at F1, if input voltage dropped from A to D, for example, operation would shift out on the same curve to a lower normalized frequency at F2. Operation is thus very nonlinear over the slopes of the resonant curve. Some combinations of line and load may not even be possible. Choosing the exact location to operate on the resonant curve is very tricky.

The feedback loop automatically selects the location on the resonant curve which yields the correct output voltage. However, that location may be dangerously close to the resonant peak on a low-Q curve where selectivity is minimal, or on the bottom tail of a high-Q curve where large frequency changes are required.

The scheme is sensitive to tolerances in the resonant components. Operation close to the resonant peak may cause switchover to the opposite side of the peak during load or line transients, and result in positive rather than negative feedback.

13.5.4 Regulation with a Parallel-Loaded Half Bridge in the Continuous-Conduction Mode

From Eq. 13.3, Steigerwald plots the gain curve of the parallel-loaded half bridge (Figure 13.7). Here it is seen that operation at light load and even no load is possible. If operation were initially at point A ($Q = 2$, normalized frequency $= 1.1$), and R_L increased so that $Q(= R_L/w_o L)$ was 5, operation would shift to B at a normalized frequency of about 1.23. Obviously from the shape of the curve, even larger values of Q or open-circuit operation could be tolerated.

One problem with the PRC circuit is that if operation is close to the resonant peak at low Q (say, point C) and the load momentarily opens or gets much larger to, say, $Q = 5$ before the feedback loop can correct frequency, output voltage could rise dangerously high at point D.

It is sometimes stated that the PRC is naturally short-circuit proof, since even when there is an output short circuit, there is also a short circuit across the transformer primarily (Figure 13.4*b*) and the resonating inductor limits transistor current.

Further examination indicates that this would not occur. If operation is above resonance, a short circuit at the output would force the feedback loop to move higher up on the resonant curve to increase output (Figure 13.3).

This would move switching frequency lower and eventually over the peak of the curve into the positive-feedback region. But the loop would not "know" that; it would continue trying to drive the switching frequency lower, "thinking" that lower frequency would move operation higher up to the top or the resonant peak.

Further, at lower frequency, the resonant inductor would have lower impedance, which would result in increased current drawn from the transistors. A clamp to limit the minimum frequency would not be practical, as the resonant peak frequency is subject to large variations because of production spreads in the resonant L and C.

13.5.5 Series-Parallel Resonant Converter in Continuous-Conduction Mode

One disadvantage of the PRC is that at light load (large shunt resistor across C_r in Figure 13.4*b*), the circulating currents (and currents in the transistor) are no less than those at heavy load. In either case, the effective resistance reflected across C_r must be high so as not to kill the Q of the circuit. If this is true at heavy loads, it is even more true at light loads. This is simply another way of saying that the current in the reflected load resistor across C_r at light load is a small fraction of the current in C_r. Thus at light load, power losses in the transistors do not decrease and efficiency is poor.

This is not so in the SRC. For constant output voltage across the load resistor in series in the resonating *LC* circuit, current through the load resistor (which is also reflected as current in the transistors) decreases as load current decreases. Thus efficiency remains high at light load (low output power) in the SRC.

A circuit which takes advantage of the good light-load efficiency of the SRC and the ability to regulate at light or open load of the PRC is the series-parallel continuous-conduction mode converter of Figure 13.5*c*. It is also referred to as the *LCC circuit*.[5] As seen in Figure 13.5*c*, it has both a series capacitor C_s and a shunt capacitor C_p. With proper selection of C_s and C_p, the advantages of both SRC and PRC can be obtained to an acceptable degree. Notice that if C_p is zero, the circuit is that of an SRC. As C_p becomes larger than C_s, it takes on more of the characteristics of a PRC. If it becomes higher than C_s, it develops the poor efficiency at light load characteristic of a PRC.

Again from Steigerwald's article[4], the AC gain characteristics of the *LCC* is calculated from its AC equivalent circuit in Figure 13.5*c* as

$$\frac{V_o}{V_{\text{in}}} = \frac{1}{1 + (X_{\text{cs}}/X_{\text{cp}}) - (X_l/X_{\text{cp}}) + j[(X_l/R_{\text{ac}}) - (X_{\text{cs}}/R_{\text{ac}})]}$$

With $Q_s = X_l/R_l$, $R_{\text{ac}} = 8R_l/\pi^2$, $w_o = 1/\sqrt{LC_s}$, and R_l equal to the DC output resistance reflected by the square of the turns ratio into the primary, the gain from half the input supply voltage to the DC output voltage reflected into the primary is

$$\frac{NV_{\text{odc}}}{0.5V_{\text{in}}} = \frac{8/\pi^2}{1 + (C_p/C_s) - (w^2LC_p) + jQ_s[(w/w_s) - (w_s/w)]}$$

This is plotted in Figures 13.8*a* and 13.8*b* for $C_s = C_p$ and $C_s = 2C_p$.

It is seen in both Figures 13.8*a* and 13.8*b*, for $Q = 1$ and less, that a resonant bump and some selectivity remains on the gain curves, so that no load operation is possible. This removes one of the drawbacks of the pure SRC circuit: no regulation for no-load or light-load conditions (compare to Figure 13.6 for the pure SRC circuit).

Figures 13.8*a* and 13.8*b* illustrate some of the subtleties and complexities of continuous-mode operation.

Consider operation at point *A* in Figure 13.8*a*, with DC load resistance and *Q* constant, and DC input voltage decreases causing a slight decrease in output voltage. The feedback loop will attempt to correct this by moving higher up on the same curve to get more output. To do this, the switching frequency will be decreased. If it decreases too much, it will fall over the top of the resonant peak at *B*, and thereafter further frequency decrease will decrease the output.

To avoid this, it must be ensured that at lowest DC input and minimum *Q*, operation never requires a frequency lower than that corresponding to point *B* in Figure 13.8*a*. A better appreciation for the practical problems in CCM is gained when one attempts to make the circuit work over all tolerances in the *LC* product which shift the resonant peak in frequency and amplitude.

Steigerwald concludes that $C_p = C_s$ is a best-compromise design. Additional analyses of the *LCC* resonant converter, discussed above, can be found in References 6 to 9.

13.5.6 Zero-Voltage-Switching Quasi-Resonant (CCM) Converters[2]

The zero-current-switching (ZCS) converters, discussed above, force the transistor current to be sinusoidal by having the square-wave drive from the transistor switches drive a resonant *LC* circuit. Further, by turning the transistor "off" as the sine wave of current passes through

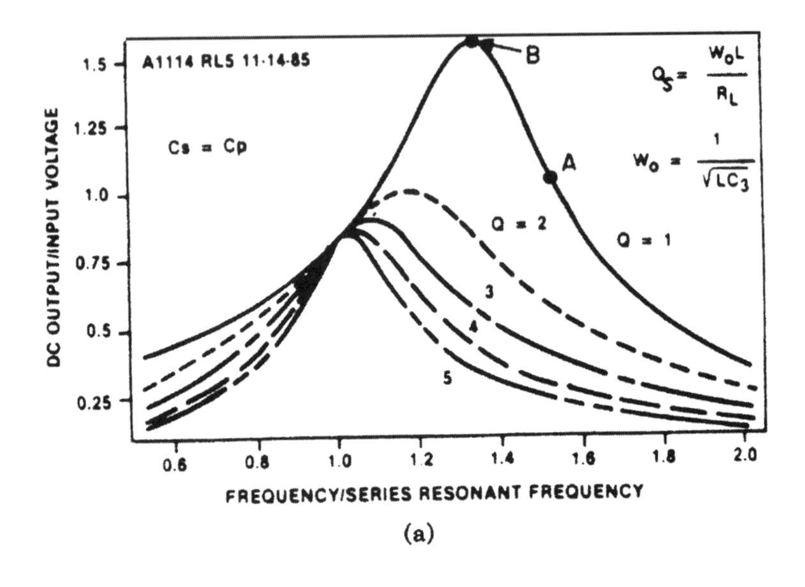

(a)

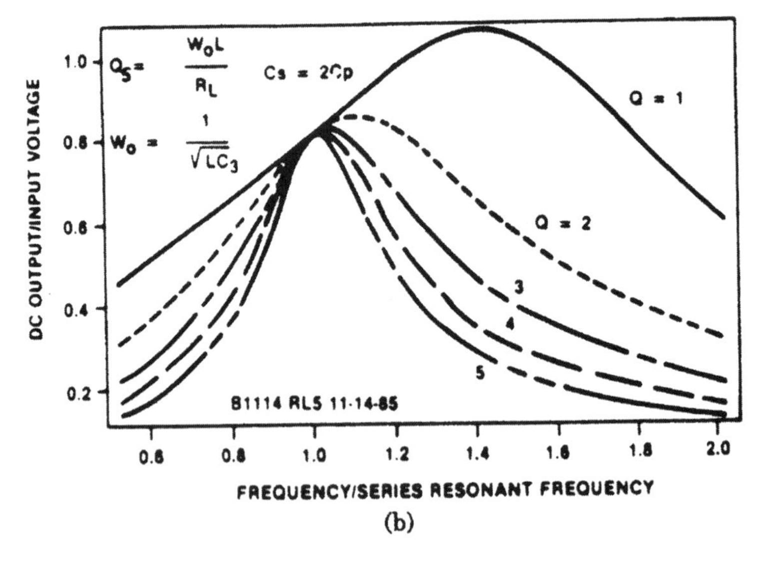

(b)

FIGURE 13.8 (*a*) Gain curves for series-parallel half-bridge resonant converter (*LCC*) of Figure 13.5*c* for $C_s = C_p$. (*b*) Gain curves for series-parallel half-bridge resonant converter (*LCC*) of Figure 13.5*c* for $C_s = 2C_p$. (*Courtesy of R. Steigerwald.*)

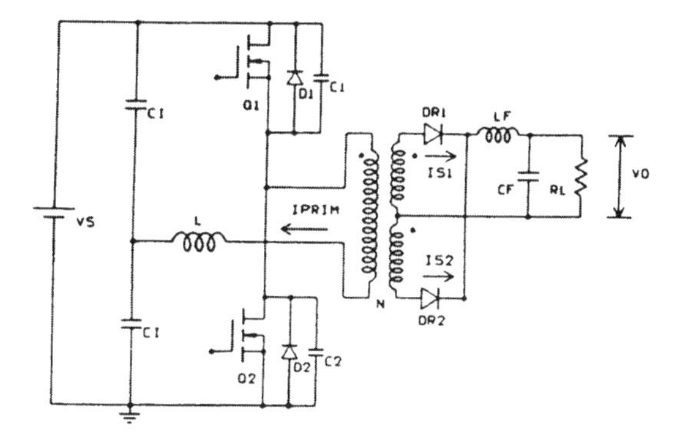

FIGURE 13.9 A zero-voltage-switching half-bridge resonant converter. *(Courtesy of Jovanovic, Tabisz, Lee.)*

zero at the end of the first half period (or shortly after when current has reversed and flows into the antiparallel diode as in Figure 13.1), there is no overlap of high voltage and current and turn "off" losses are eliminated.

However, there are turn "on" losses even though there is no overlap of high voltage and current because the leakage inductance speeds up drain (or collector) voltage fall time and slows up drain current rise time. These losses occur because the drain capacitance stores an energy at turn "off" of $0.5C(V_{max})^2$, and dissipates it in the transistor at the next turn "on." This happens once per cycle and results in average dissipation of $0.5C(V_{max})^2/T$, which becomes significant over 500 kHz to 1 MHz. As designers move to these higher frequencies to reduce size, this becomes a problem.

A new technique—ZVS—has been proposed to circumvent this problem. It has been proposed for single-ended (flyback or buck) circuits[10], but its value is mainly for half bridges.

A ZVS half bridge is shown in Figure 13.9.[2] Its basic principle is that the MOSFET output capacitor is used as part of the capacitor of the resonant *LC* circuit. It stores a voltage and hence energy in one part of the switching cycle. In a following part of the switching cycle, the energy in the capacitor is discharged through the resonant inductor back into the supply bus without dissipation. The following discussion is based on Reference 2.

In Figure 13.9, when *Q*1 is "on" and *Q*2 is "off," *C*2 is charged up to V_S. Transistor *Q*1 is first turned "off." Transformer *T*1 magnetizing current continues to flow through *C*1, pulling its voltage down. Halfway down, there is no voltage across the primary and in the *T*1 secondary, the output inductor tries to maintain a constant current.

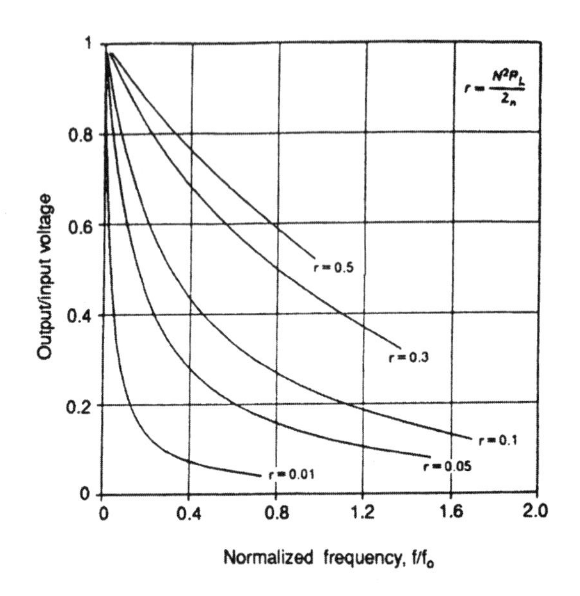

FIGURE 13.10 Gain curve for the zero-voltage-switching circuit of Figure 13.9. (*Courtesy of Jovanovic, Tabisz, Lee.*)

Both rectifier diodes turn "on," which provides a short circuit across the secondary and hence across the primary.

At this point, the energy stored on the top end of $C2$ discharges through the short-circuited primary, through the resonant L, through the bottom filter capacitor C_{r2}, and back into the bottom end of $C2$. Since there are no resistors in this path, the discharge is lossless. The negative resonant voltage impulse on the right hand of L pulls the junction of $C1$, $C2$ down to ground, and now $Q2$ is turned "on" at zero voltage.

Capacitor $C1$ has slowed up $Q1$ voltage fall time sufficiently, so that there is no simultaneous high voltage and current during its turn "off."

The circuit, a CCM type, operates on the slope of the resonant curve of the circuit consisting of L and $C1$, $C2$ in parallel. Jovanovic et al. give the DC voltage conversion curve for the circuit (the equivalent of Figures 13.5 to 13.8) in Figure 13.10.[2]

Despite the many articles written on CCM, which is regulated by changing frequency along the steep slope of a resonant curve, this author feels it is not a reliable scheme. It will not yield similar results over a large production run of supplies with all the various production spreads in the resonant components and with the limited range of V_{dc} and R_l in the conversion curves such as those in Figures 13.5 to 13.8 and 13.10.

13.6 Resonant Power Supplies—Conclusion

To the engineer who must decide whether to consider a resonant supply for the next design, the following should be considered:

1. Can I get the same (or acceptable) power density in watts per cubic inch with a conventional PWM supply operating at, say, 200 to 300 kHz? Is the added complexity, limited line and load capability, and difficultly to design for worst-case conditions worth the extra 3 to 6% better efficiency possible with resonant supplies?
2. Will all units coming off a production line have identical characteristics? Or because of tolerances and production spread in component values, will fine-tuning of each unit or a large fraction of the production run be required?
3. Do resonant supplies actually generate less RFI (because their currents are sinusoidal rather than square wave) after you factor in their higher di/dt? Consider that most resonant supplies have sine wave currents three to four times the amplitude of PWM square-wave supplies of equal power. It is possible that the RFI problem with resonant supplies may be as severe since di/dt at the zero crossing of a sine wave is proportional to its peak value.
4. How serious is the problem that most resonant supplies regulate by varying the frequency? Will users accept this or insist that the switching supply be synchronized to a clock signal which they will supply?
5. If the decision is to attempt a resonant design, which of the bewildering number of topologies advocated would be the safest approach? Are continuous-mode designs too unpredictable because they require operation on the slippery slope of a narrow resonant curve for regulation? There seems to be general agreement that discontinuous-mode operation is more predictable and reproducible. Is the larger frequency range required in discontinuous mode an important drawback?
6. There is no question that high-frequency resonant converters will continue to be studied and improved. Until configurations are found which lend themselves to simple, worst-case designing and which are as insensitive as PWM circuits to component tolerances, wiring layout, and parasitic inductances and capacitances, high-frequency resonant converters will not be widely adopted—certainly not in programs with a large production run. They will occupy only a narrow niche in the power supply field where higher cost and finely tuned component selection and wiring layout are acceptable.

In summary, in any design including switching power supplies, complexity equals higher cost and possible unreliability.

References

1. F. Lee, "High Frequency Quasi-Resonant and Multi Resonant Converter Topologies," *Proceedings of the International Conference on Industrial Electronics*, 1988.
2. M. Jovanovic, W. Tabisz, and F. Lee, "Zero Voltage Switching Technique in High Frequency Inverters," Applied Power Electronics Conference, 1988.
3. K. Liu and F. Lee, "Secondary Side Resonance for High Frequency Power Conversion," Applied Power Electronics Conference, 1986.
4. R. Steigerwald, "A Comparison of Half Bridge Resonant Topologies," *IEEE Transactions on Power Electronics*, 1988.
5. R. Seven, "Topologies for Three Element Resonant Converters," Applied Power Electronics Conference, 1990.
6. F. Lee, X. Batarseh, and K. Liu, "Design of the Capacitive Coupled LCC Parallel Resonant Converter," *IEEE IECON Record*, 1988.
7. X. Bhat and X. Dewan, "Analysis and Design of a High Frequency Converter Using LCC Type Commutation," *IEEE IAS Record*, 1986.
8. X. Batarseh, K. Liu, F. Lee, and X. Upadhyay, "150 Watt, 140 kHz Multi Output LCC Type Parallel Resonant Converter," IEEE APEC Conference, 1989.
9. B. Carsten, "A Hybrid Series-Parallel Resonant Converter for High Frequencies And Power Levels," *HFPC Conference Record*, 1987.
10. W. Tabisz, P. Gradski, and F. Lee, "Zero Voltage Switched Quasi-Resonant Buck and Flyback Converters," PESC Conference, 1987.
11. F. Lee, ed., "High Frequency Resonant Quasi-Resonant and Multi-Resonant Converters," Virginia Power Electronics Center, 1989.
12. Y. Kang, A. Upadhyay, and D. Stephens, "Off Line Resonant Power Supplies," *Powertechnics Magazine*, May 1990.
13. Y. Kang, A. Upadhyay, and D. Stephens, "Designing Parallel Resonant Converters," *Powertechnics Magazine*, June 1990.
14. P. Todd, "Practical Resonant Power Converters—Theory and Application," *Powertechnics Magazine*, April–June 1986.
15. P. Todd, "Resonant Converters: To Use or Not to Use? That Is the Question," *Powertechnics Magazine*, October 1988.

PART 3

Waveforms

CHAPTER 14

Typical Waveforms for Switching Power Supplies

14.1 Introduction

In previous chapters, we have shown the voltage and current waveforms at critical points throughout the various circuits. Some of the major topologies are shown in Figures 2.1, 2.10, 3.1, and 4.1. In many cases these are idealized waveforms, and newcomers to switching power supply design may wonder how closely the actual waveforms, on which much of the circuit design is based, resemble the theoretical ones shown.

Question arise as to how these waveshapes vary with line voltage and load current and whether there are noise spikes on outputs with respect to ground. As a result of observed behavior, designers will question if there should be decaying, oscillatory, ringing waveforms at sharp transitions in voltage and current, and whether there is time jitter at the leading or trailing edges of the waveforms. Also, how closely should the "on" volt-second product equal the reset volt-second product in a transformer or an inductor? What does a leakage inductance spike really look like? Since output inductors and flyback transformers are designed to yield certain current waveshapes, how close are these actual waveshapes to the theoretical ones? Since much of the power transistor dissipation at high frequencies comes from simultaneous high voltage and current at turn "off" and turn "on," can this be observed on a fast time base? What sort of waveform oddities may be expected? It may be very instructive for designers who see strange high-frequency switching waveforms for the first time, to provide some actual oscilloscope waveforms at critical points in the circuits for some of the major topologies. The waveforms are taken mostly at points in the so-called power train—the circuitry from the input to the power transistors to the output of the output filters, as that is where most of the energy is handled and where most of the potential failures occur.

Critical waveforms for topologies selected below are for some of the more often used, such as the forward, the push-pull, and the flyback converters as well as the buck regulator. These show the impact of operating output inductors in the continuous mode and were shown in the oscilloscope photographs of Figures 1.6 and 1.7.

Offline converters operating from AC voltages of 120 or 220 V will not be considered here. Supplies operating from those rectified voltages have similar waveshapes at corresponding points, but since the rectified DC supply voltages are higher, current amplitudes are lower and voltage waveshape amplitudes are higher than in telecommunications power supplies.

The selected circuits operate at switching frequencies above 100 kHz, and those shown are powered from telecommunications industry DC supply voltages—nominal 48 V, minimum 38 V, and maximum 60 V. Output powers in all cases are under 100 W, as voltage and current waveforms at higher powers differ only in amplitude but not significantly in shape. Since the waveshapes shown here are from DC-powered circuits, they have somewhat less time and amplitude jitter than do circuits powered from rectified alternating supplies, since the rectified DC in an off-line converter has line frequency ripple which will cause additional amplitude ripple. The feedback loop, in reducing the input line ripple, necessarily injects some jitter in the pulse widths.

Since all the waveshapes shown are taken on circuits using MOSFETs, they do not display any power transistor storage delay effects. In all cases, the feedback loop was open and the power transistor(s) was (were) driven with a pulse of the desired frequency. The pulse width was manually adjusted at each supply voltage input (38, 48, 60 V) so as to maintain the 5-V output within a few millivolts of 5 V, just as if the feedback loop were closed. The slave output voltage was also recorded at that pulse width.

The power transistor driver for all photos is the UC3525 PWM control chip. Pulse widths were set by feeding the error-amplifier output node (which is available at one of the chip pins) from a well-regulated DC voltage source adjusted to yield the pulse width required to set the output to 5.00 V at each DC input voltage.

14.2 Forward Converter Waveshapes

The circuit schematic for the forward converter waveshapes is shown in Figure 14.1. It is a 125-kHz forward converter designed for 100 W, and waveshapes are shown at 80 and 40% of full load. Full-load

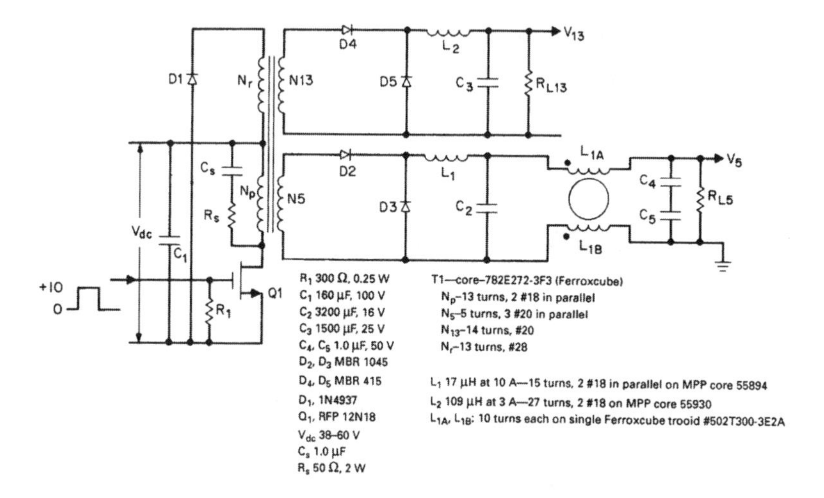

FIGURE 14.1 A 125-kHz forward converter providing + 5.0 V at 10 A, + 13 V at 3.8 A.

outputs are 5 V at 10 A, and 13 V at 3.8 A. Waveshapes are shown for the nominal input voltage of 48 V, the minimum of 38 V, and the maximum of 60 V. The transformer core was selected from Table 7.2*a* and the numbers of turns and wire sizes from Sections 2.3.2 to 2.3.10. The output filters ($L1$, $C2$ and $L2$, $C3$) were chosen from the relations given in Section 2.3.11.

14.2.1 V_{ds}, I_d Photos at 80% of Full Load

Photos 1 to 3 in Figure 14.2 show drain-to-source voltages and drain currents at low, nominal, and maximum DC supply voltages.

Drain currents have the ramp-on-a-step waveshape characteristic of secondaries with output *LC* filters. Drain current is the sum of the secondary currents reflected by their turns ratios into the primary. Also, since both secondaries have output inductors which yield ramp-on-a-step currents (Section 1.3.2), these reflect into the primary as ramp-on-a-step current.

Drain current amplitude at the center of the ramp should be (from Eq. 2.28) equal to $I_{pft} = 3.12P_o / V_{dc}$. For 80-W output and a minimum DC voltage of 38 V, this current should be 6.57 A. Photos 1, 2, and 3 (Figure 14.2) show that to be the current at the center of the ramp as accurately as can be read.

As DC supply voltage is increased, the photos show that pulse width (transistor "on" time) decreases but peak current and current at the ramp center remain unchanged—as theoretically they should.

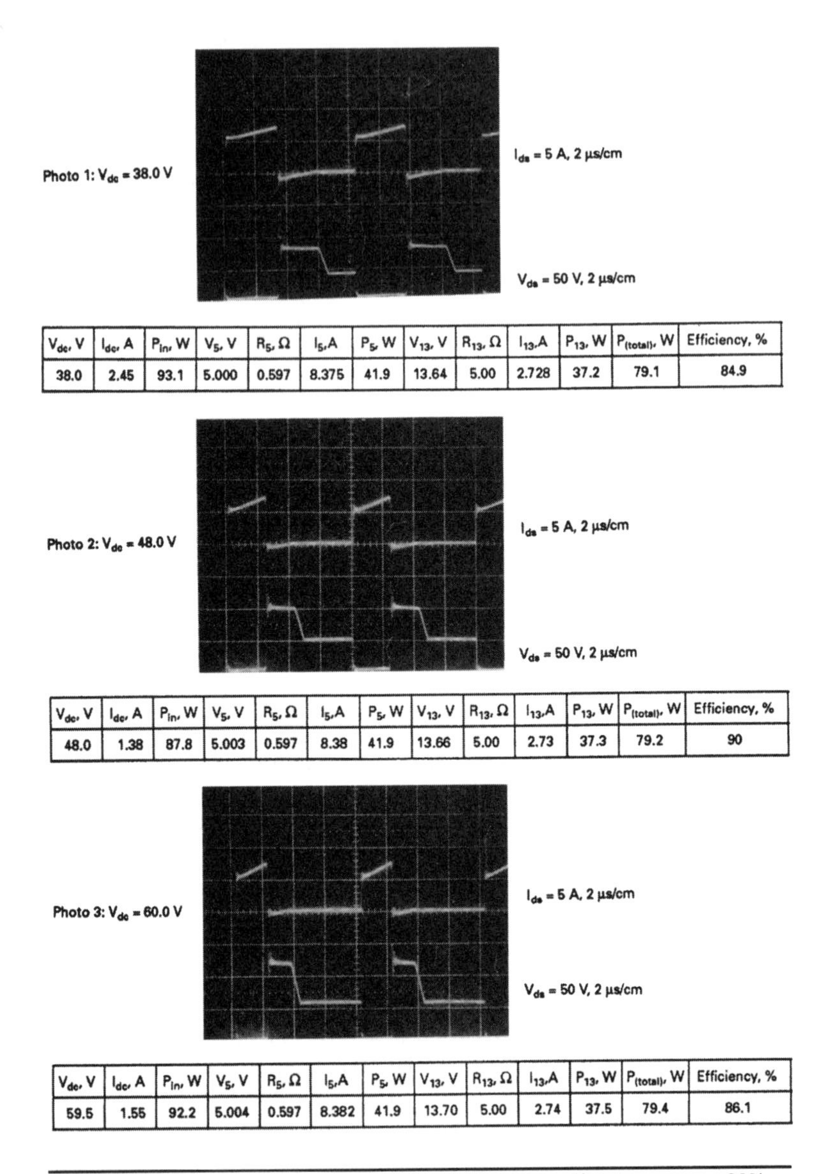

V_{dc}, V	I_{dc}, A	P_{in}, W	V_5, V	R_5, Ω	I_5, A	P_5, W	V_{13}, V	R_{13}, Ω	I_{13}, A	P_{13}, W	$P_{(total)}$, W	Efficiency, %
38.0	2.45	93.1	5.000	0.597	8.375	41.9	13.64	5.00	2.728	37.2	79.1	84.9

V_{dc}, V	I_{dc}, A	P_{in}, W	V_5, V	R_5, Ω	I_5, A	P_5, W	V_{13}, V	R_{13}, Ω	I_{13}, A	P_{13}, W	$P_{(total)}$, W	Efficiency, %
48.0	1.38	87.8	5.003	0.597	8.38	41.9	13.66	5.00	2.73	37.3	79.2	90

V_{dc}, V	I_{dc}, A	P_{in}, W	V_5, V	R_5, Ω	I_5, A	P_5, W	V_{13}, V	R_{13}, Ω	I_{13}, A	P_{13}, W	$P_{(total)}$, W	Efficiency, %
59.5	1.55	92.2	5.004	0.597	8.382	41.9	13.70	5.00	2.74	37.5	79.4	86.1

FIGURE 14.2 The 125-kHz 100-W forward converter of Figure 14.1 at 80% full load.

Drain-to-source voltages also correspond to their theoretical values. Transistor "on" time at low line (V_{dc} = 38 V) is seen to be very close to 80% of a half period as discussed in Section 2.3.2. It is not always exactly that because of the inevitable rounding up or down of fractional secondary turns to the nearest integral number.

In this transformer, the calculated 4.5 turns on the secondary were rounded up to 5 turns. This yielded a larger peak secondary voltage and a somewhat shorter "on" time than called for by Eq. 2.25.

Narrow and barely discernible leakage spikes at the instant of turn-off are seen in Photos 1, 2 and 3. At V_{dc} of 60 V, the leakage spike is only about 21% above $2V_{dc}$. But at $V_{dc} = 38$ V, it is about 64% above $2V_{dc}$.

The voltage waveshapes also show that V_{ds} at turn "off" falls back down to $2V_{ds}$ immediately after the leakage inductance spike and remains there until the "on" volt-second product equals the reset volt-second product $(V_{dc}t_{on}) = (2V_{dc} - V_{dc})t_{reset}$. When those volt-second areas are equal, the drain voltage drifts back down to V_{dc}.

Figure 14.2 shows the average efficiency from $V_{dc} = 38$ to 60 V to be 87%. This is achieved at a peak flux density of 1600 G with Ferroxcube 3F3 core material. Core temperature rise was under 25°C.

Such high efficiency at 125-kHz and 1600-G peak flux density could not have been achieved with the very widely used, higher-loss core material—3C8. Even with the lower-loss 3F3 material, as discussed in Section 7.3.5.1, larger-sized cores could not operate at 1600 G at 125 kHz. Peak flux density would probably have to be reduced to 1400 or possibly 1200 G.

14.2.2 V_{ds}, I_d Photos at 40% of Full Load

Photos 4 to 6 in Figure 14.3 give much the same information as Photos 1 to 3. At the lower output currents and power, efficiencies average 90% over the low, nominal, and maximum DC input voltages.

Leakage inductance spikes are considerably smaller and transistor "on" times are slightly shorter, as the forward drop in the output rectifier diodes is somewhat less as a result of the lower output current. This increases the peak square-wave voltage at the cathode of the 5-V rectifier diode and permits a shorter "on" time to generate the 5-V output.

14.2.3 Overlap of Drain Voltage and Drain Current at Turn "On"/Turn "Off" Transitions

The simultaneous high voltage and current at the turn "on"/turn "off" transitions result in spikes of high instantaneous power dissipation. Even though the high-dissipation spikes are very narrow, especially with MOSFETs, when they come at a high repetition rate their average dissipation can be high and can exceed the "conduction" dissipation of $V_{ds}I_{ds}t_{on}/T$.

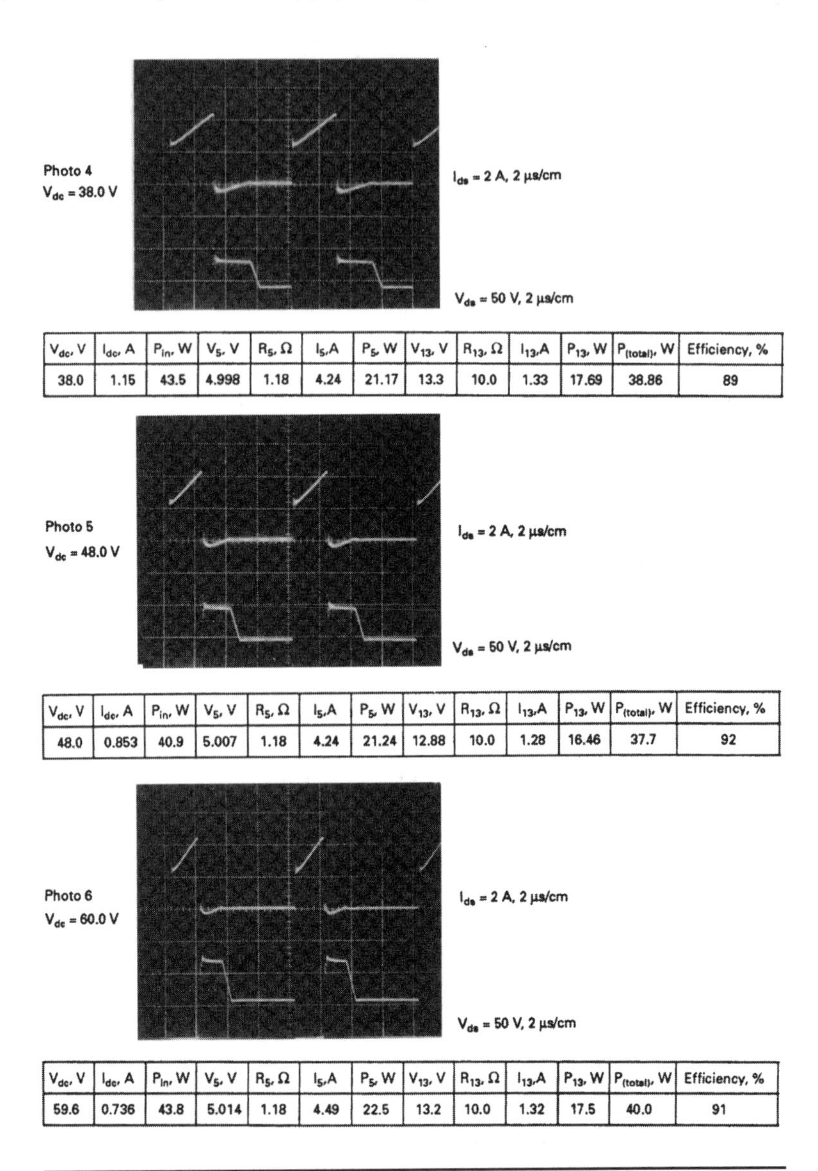

V_{dc}, V	I_{dc}, A	P_{in}, W	V_5, V	R_5, Ω	I_5,A	P_5, W	V_{13}, V	R_{13}, Ω	I_{13},A	P_{13}, W	$P_{(total)}$, W	Efficiency, %
38.0	1.15	43.5	4.998	1.18	4.24	21.17	13.3	10.0	1.33	17.69	38.86	89

V_{dc}, V	I_{dc}, A	P_{in}, W	V_5, V	R_5, Ω	I_5,A	P_5, W	V_{13}, V	R_{13}, Ω	I_{13},A	P_{13}, W	$P_{(total)}$, W	Efficiency, %
48.0	0.853	40.9	5.007	1.18	4.24	21.24	12.88	10.0	1.28	16.46	37.7	92

V_{dc}, V	I_{dc}, A	P_{in}, W	V_5, V	R_5, Ω	I_5,A	P_5, W	V_{13}, V	R_{13}, Ω	I_{13},A	P_{13}, W	$P_{(total)}$, W	Efficiency, %
59.6	0.736	43.8	5.014	1.18	4.49	22.5	13.2	10.0	1.32	17.5	40.0	91

FIGURE 14.3 The 125-kHz 100-W forward converter of Figure 14.1 at 40% of full load.

The overlap dissipation at turn "on" is not as serious as at turn "off." At turn "on" the power transformer leakage inductance presents very high impedance for a short time and causes a very short drain-to-source voltage fall time. The same leakage inductance does not permit a very fast current rise time. Thus the falling V_{ds} intersects the rising

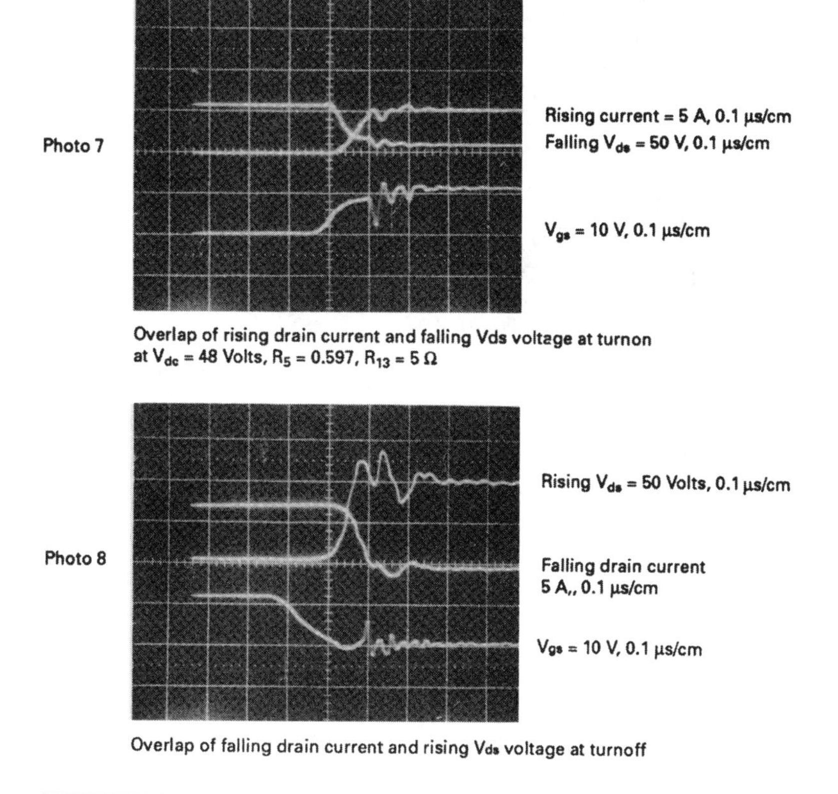

FIGURE 14.4 The overlap of rising drain current and falling V_{ds} at turn "on," and that of falling drain current and rising V_{ds} at turn "off," causes the AC switching losses. Losses are greater at the turn "off" transition because current falls more slowly and V_{ds} rises more rapidly than at turn "on."

I_{ds} when both are quite low, the integral $\int V_{ds} I_{ds}\, dt$ taken over the turn "on" time is small, and dissipation averaged over a full cycle is small. This can be seen on the fast time base of 0.1 μs/cm in Photo 7 of Figure 14.4.

At turn "off" (Photo 8 in Figure 14.4), however, the drain current remains constant at its peak for a while (because leakage inductance tends to maintain constant current) as V_{ds} rises to about V_{dc}. Then V_{ds} continues to rise at a fast rate and reaches $2V_{dc}$ before I_d has fallen significantly below its peak. Thus, as seen in Photo 8, the integral $\int V_{ds} I_d\, dt$ over the turn "off" time is much greater than the same integral over the turn "on" time.

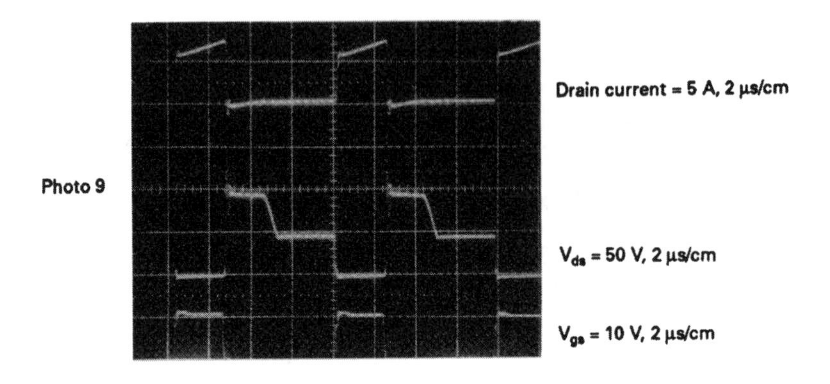

FIGURE 14.5 Relative timing of drain current, drain-to-source voltage, and gate-to-source voltage for 125-kHz forward converter of Figure 14.1 at $V_{dc} = 48$ V, $R_5 = 0.597\ \Omega$, $R_{13} = 5\ \Omega$.

Performing an approximate "eyeball" integration of the two integrals above, it is seen that the average dissipation due to the overlap of falling current and rising voltage at turn "off" is 2.18 W. At turn "on," the average dissipation due to rising current and falling voltage is 1.4 W.

14.2.4 Relative Timing of Drain Current, Drain-to-Source Voltage, and Gate-to-Source Voltage

Photo 9 in Figure 14.5 shows the negligible delay between the gate input voltage transitions and the drain voltage–drain current transitions at turn "on" and turn "off."

14.2.5 Relationship of Input Voltage to Output Inductor, Output Inductor Current Rise and Fall Times, and Power Transistor Drain-Source Voltage

Photo 10 in Figure 14.6 shows the output inductor upramp of current during the transistor "on" time and the downramp of current during the transistor "off" time. For L_1 of 17 μH, input voltage of 16 V, and "on" time of 2.4 μs, the ramp amplitude should be $dI = (16 - 5)(2.4/17) = 1.55$ A. As accurately as Photo 10 (Figure 14.6) can be read, the ramp amplitude is 1.4 A. Because of the scales on the photo, this is the best correlation which can be obtained between the calculated and measured values.

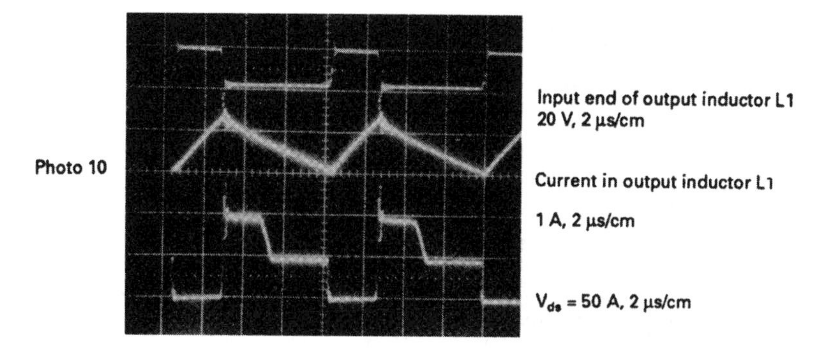

FIGURE 14.6 Relative timing of output inductor voltage, its ripple current, and drain-source voltage.

14.2.6 Relative Timing of Critical Waveforms in PWM Driver Chip (UC3525A) for Forward Converter of Figure 14.1

This demonstrates how the UC3525A PWM chip generates output pulses whose widths are inversely proportional to the DC supply voltage (see Figure 14.7).

Internal to the chip, a 3-V peak-to-peak triangle (~0.5 to 3.5 V) is generated and occurs once per half period of the switching frequency. This triangle is compared in a voltage comparator to the voltage at the output of the internal error amplifier. The error amplifier compares a fraction of the DC voltage to be regulated (at its inverting input) to a reference voltage.

The voltage comparator generates two 180° out-of-phase rectangular pulses. These pulses commence at the start of the triangle, and terminate when the triangle crosses the voltage at the error-amplifier output.

Thus, when the sampled fraction of the regulated voltage goes slightly positive, the error-amplifier output goes slightly negative, the triangle crosses the lowered voltage earlier in time, and the PWM comparator output pulse widths decrease. Similarly, a decrease in sampled input to the error amplifier raises the error-amplifier output voltage slightly, the triangle crosses that higher voltage later in time, and the pulse widths increase.

These adjustable-width pulses are alternately routed by a binary counter to two chip output pins for driving the transistors in a push-pull topology. For the forward converter of Figure 14.1, only one of these pulse outputs is used.

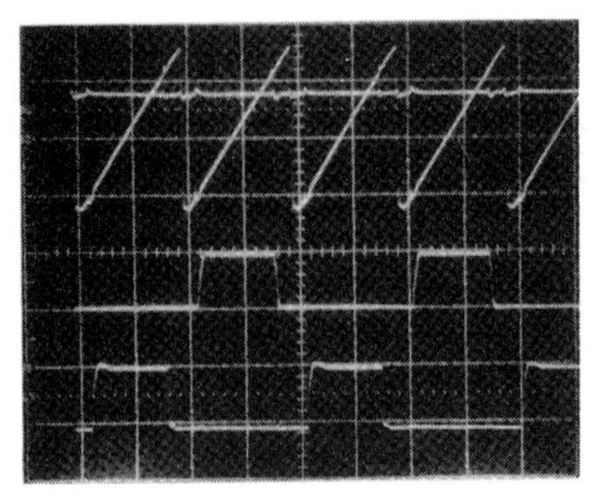

Error amplifier output voltage
1 V/cm

PWM comparator input triangle
1 V, 2 μs/cm

Zero DC volts

Chip "A" output—10 V, 2 μs/cm

Chip "B" output—10 V, 2 μs/cm

Photo 11

FIGURE 14.7 Significant waveforms in UC3525 pulse-width-modulating control chip. The PWM comparator internal to the chip compares an internally generated 3-V triangle to the voltage at its internal error-amplifier output node. The PWM comparator generates two 180° out-of-phase pulses at the chip *A* and *B* output pins. The width of these pulses is the time duration between the start of the triangle and the instant the triangle reaches the voltage at the error-amplifier output node. As the error voltage output moves across the 3-V height of the triangle in response to the difference between an internal reference voltage and a fraction of the output voltage being regulated, the output pulse widths at outputs *A* and *B* are varied.

14.3 Push-Pull Topology Waveshapes—Introduction

The next topology for which significant waveforms are presented is a push-pull whose circuit schematic is shown in Figure 14.8.

The circuit is a 200-kHz DC/DC converter of 85-W maximum output power, and minimum of one-fifth of that. It was designed to operate from standard telecommunications industry supply voltages of 48 V nominal, 60 V maximum, and 38 V minimum. Outputs are +5 V at a maximum of 8 A and +23 V at a maximum of 1.9 A.

Significant waveforms are shown at maximum, nominal, and minimum supply voltages, at maximum and 20% maximum output currents. Some waveforms are also shown at 100-W output power (15 W above maximum) to demonstrate that at this power level, efficiency is still high, transformer temperature rise is acceptable, and waveshapes remain as expected.

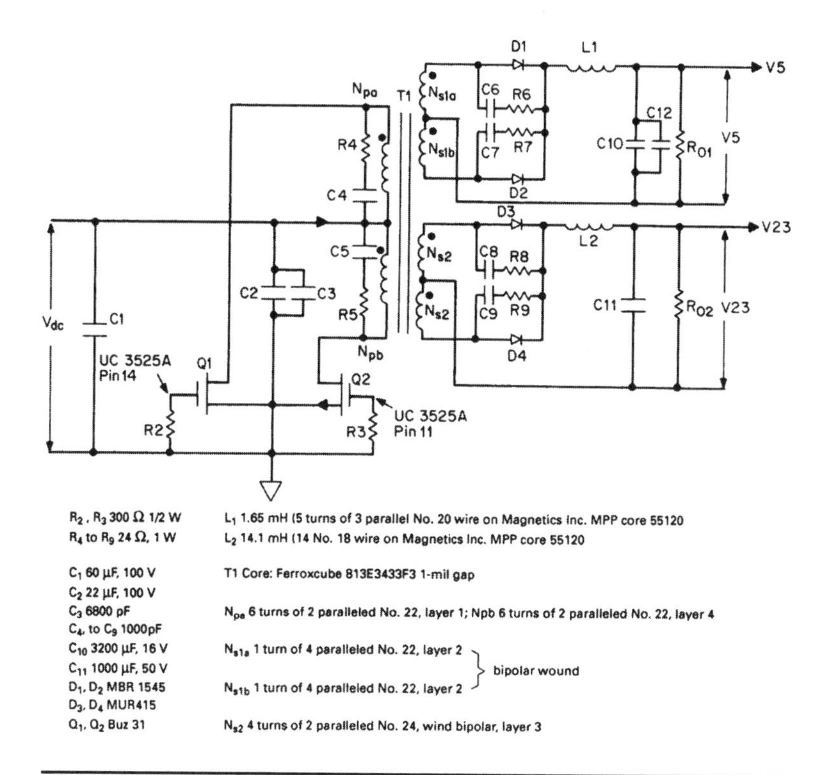

FIGURE 14.8 A 200-kHz 85-W DC/DC converter: +5 V at 8.0 A, +23 V at 1.9 A.

For the design, the transformer core was selected from Table 7.2*a* and the transformer was designed (peak flux density selection, numbers of turns, wire sizes) from Sections 2.2.9 and 2.2.10. The output filters were designed from the equations in Section 2.2.14.

A 200-kHz operating frequency was chosen arbitrarily. Above 200 kHz, the transformer and output filter sizes and efficiencies all decrease rapidly, and transformer core and copper losses increase sharply. Above 200 kHz, it is questionable whether the advantage of a small decrease in size is worth the penalty of increased dissipation and transformer temperature rise.

As for the forward converter of Figure 14.1, the feedback loop was not closed. At each DC input voltage, the pulse width was manually adjusted as described in Section 14.1 to yield 5.00 V at the 5-V output. The resulting slave secondary voltage was also recorded at the pulse width which yielded 5.00 V at the 5-V output.

14.3.1 Transformer Center Tap Currents and Drain-to-Source Voltages at Maximum Load Currents for Maximum, Nominal, and Minimum Supply Voltages

Alternate current pulses, as monitored in the transformer center tap, correspond to alternate transistors (*Q*1, *Q*2) turning "on."

These waveshapes are shown to demonstrate that with equal-width pulses at the two inputs, and with MOSFETs, which have no storage times, there is no sign of flux imbalance. As discussed in Section 2.2.5, flux imbalance would show up as an inequality in amplitude of current pulses monitored in the transformer center tap (as in Figures 2.4*b* and 2.4*c*).

It is seen in Photos PP1 to PP3 in Figure 14.9 that alternate current pulses at any supply voltage are of equal amplitude, as closely as can be read in the photos. No attempt was made to match the r_{ds} of MOSFETs *Q*1 *Q*2.

The V_{ds} waveshapes show negligible leakage inductance spikes at the end of the "on" times. The largest leakage inductance spike (at $V_{dc} = 41$ V) is only about 5 V above $2V_{dc}$ (Photo PP5).

Negligible leakage inductance spikes result from the fact that at high frequencies, leakage inductance is minimal because the number of turns is small, and coupling between primary and secondaries is better than would be possible at lower frequencies. Also, sandwiching secondaries between the two half primaries (Figure 14.8) has helped reduce leakage inductance.

Photos PP1 to PP3 show that as supply voltage increases, "on" times must decrease to maintain a constant 5.00 V output. Peak currents remain constant for constant DC output currents. It is only pulse widths that change as supply voltage changes.

The primary currents are seen to have the characteristic shape of a ramp on a step. They have this shape because they are the sum of the ramp-on-a step secondary currents, reflected into the primaries by the respective turn ratios, plus the primary magnetizing current which is a triangle (Figure 2.4*e*). The secondary currents have ramp-on-a step waveshape because there are inductors in all outputs.

It should be noted, however, that although the equal current peaks in Photos PP1 to PP3 indicate equal peak currents in *Q*1, *Q*2 during their "on" times, those photos give a spurious picture of the absolute value of the transistor currents, especially during the dead time between turn "ons." This results from the magnetically coupled current probe used to monitor currents in the transformer center tap.

Apparently because of the short dead time between alternate transistor turn "ons," the current probe does not have time to recover to

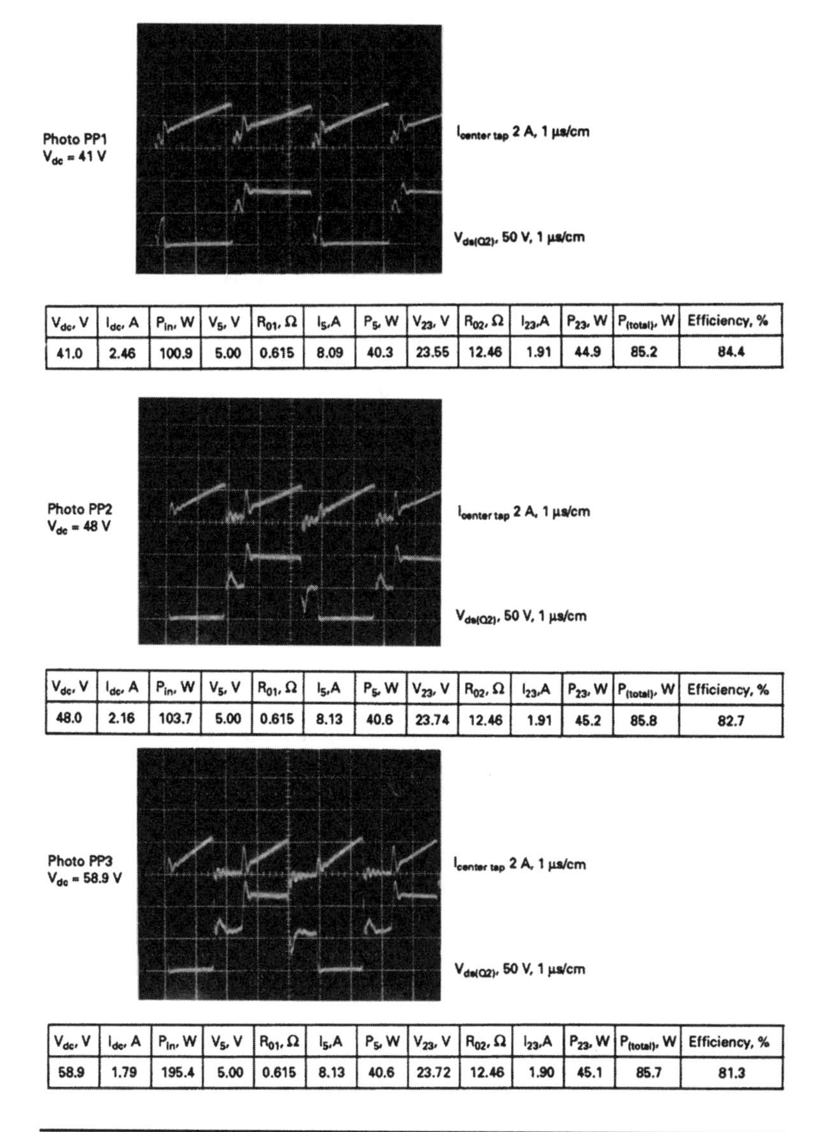

V_{dc}, V	I_{dc}, A	P_{in}, W	V_5, V	R_{01}, Ω	I_5, A	P_5, W	V_{23}, V	R_{02}, Ω	I_{23}, A	P_{23}, W	$P_{(total)}$, W	Efficiency, %
41.0	2.46	100.9	5.00	0.615	8.09	40.3	23.55	12.46	1.91	44.9	85.2	84.4

V_{dc}, V	I_{dc}, A	P_{in}, W	V_5, V	R_{01}, Ω	I_5, A	P_5, W	V_{23}, V	R_{02}, Ω	I_{23}, A	P_{23}, W	$P_{(total)}$, W	Efficiency, %
48.0	2.16	103.7	5.00	0.615	8.13	40.6	23.74	12.46	1.91	45.2	85.8	82.7

V_{dc}, V	I_{dc}, A	P_{in}, W	V_5, V	R_{01}, Ω	I_5, A	P_5, W	V_{23}, V	R_{02}, Ω	I_{23}, A	P_{23}, W	$P_{(total)}$, W	Efficiency, %
58.9	1.79	195.4	5.00	0.615	8.13	40.6	23.72	12.46	1.90	45.1	85.7	81.3

FIGURE 14.9 Transformer center tap current and drain-to-source voltage (Q2) with maximum output currents, at minimum (Photo PP1), nominal (Photo PP2), and maximum (Photo PP3) input voltage.

its original starting point on its hysteresis loop and hence gives a false picture of absolute current.

A true measure of absolute current amplitude in each transistor is obtained by monitoring its drain current with the same probe as in

Photo PP6 of Figure 14.10. In that photo, it is seen with the longer time between current pulses (more than a half period), the peak current is 4.4 A. This compares to 2.4 A in Photo PP2 in Figure 14.9, which is taken at the same supply voltage and output load currents.

The assumption that the amplitude measured with a current probe in the drain as in Photo PP6 is more valid than the measurement with the same probe in the transformer center tap (as in Photos PP1 to PP3) is verified by measuring voltage drop across a small current monitoring resistor in series with the transistor source. A current probe in series in the drain measures exactly the same absolute currents as measuring the voltages drop across a known resistor in the source.

Figure 14.10 shows efficiency exceeds 81.9% at any supply voltage for maximum current in the two output nodes. If it were seriously attempted, efficiency could be raised by 3 to 4% by going to larger wire size and perhaps decreasing peak flux density. However, the efficiency achieved is quite good for an operating frequency of 200 kHz and a peak flux density of 1600 G.

14.3.2 Opposing V_{ds} Waveshapes, Relative Timing, and Flux Locus During Dead Time

Photo PP4 in Figure 14.10 shows the relative timing of the *Q*1, *Q*2 drain voltages. This is classic and just what would be expected. As one transistor turns "on," its drain voltage falls to near zero and the other drain rises to $2V_{dc}$.

Delays between turn "on" of one drain and turn "off" of the other are seen to be negligible. The highest leakage inductance spike rises to about 20 V above $2V_{dc}$ (Figure 14.17, Photo PP24). Note that for this photo, total output power was increased to 112 W (5 V, 4.85 A and 21 V, 4.05 A).

Note that at the start of dead time between turn "ons," after one transistor turns "off," there is a leakage inductance spike and immediately thereafter, drain voltage falls back to V_{dc}. It remains at V_{dc} until the opposite transistor turns "on," driving it to $2V_{dc}$.

One may ask: What is the transformer core's flux density during the dead time between transistor turn "ons"? At the end of a turn "on," the flux density has been driven through a change of $2B_{max}$— say, up from $-B_{max}$ to $+B_{max}$. During the dead time, neither transistor is "on." Hence, does the flux density remain at $+B_{max}$ or fall back to remanence (Figure 2.3) at 0 Oe?

If flux density fell back to remanence (~100 G from Figure 2.3), at the end of next turn "on," an application of the same $2B_{max}$ change would drive the flux density to a peak of $(100\text{ G} + 2B_{max})$. This would saturate the core and destroy the transistors.

Photo PP4
V_{dc} = 48.0 V

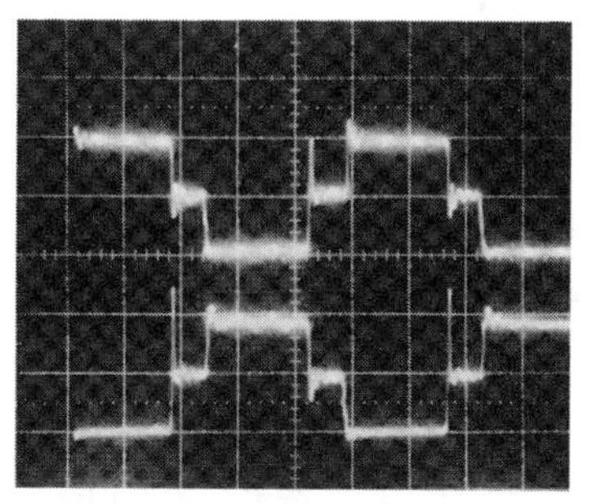

$V_{ds(Q1)}$, 50 V, 1 μs/cm

$V_{ds(Q2)}$, 50 V, 1 μs/cm

V_{dc}, V	I_{dc}, A	P_{in}, W	V_5, V	I_5, A	P_5, W	V_{23}, V	I_{23}, A	P_{23}, W	$P_{o\,(total)}$, W	Efficiency, %
48.0	2.17	104.1	5.00	8.13	40.6	23.73	1.90	45.1	85.7	82.3

Photo PP5
V_{dc} = 41.0 V

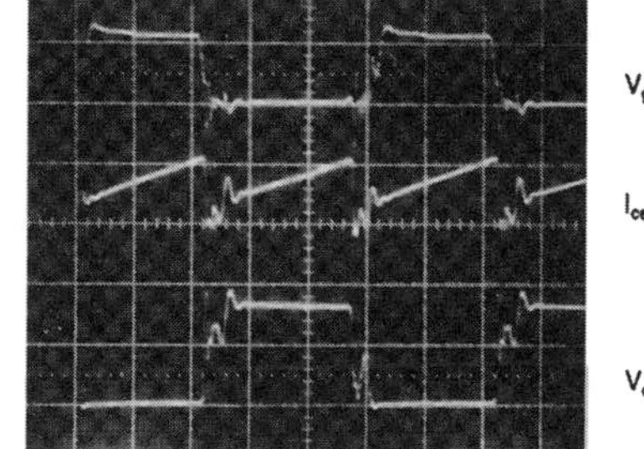

$V_{gs(Q1)}$, 10 V, 1 μs/cm

$I_{center\,tap}$ 2 A, 1 μs/cm

$V_{ds(Q1)}$, 50 V, 1 μs/cm

V_{dc}, V	I_{dc}, A	P_{in}, W	V_5, V	I_5, A	P_5, W	V_{23}, V	I_{23}, A	P_{23}, W	$P_{o\,(total)}$, W	Efficiency, %
41.0	2.46	100.9	5.00	8.13	40.6	23.55	1.19	44.9	84.9	84.1

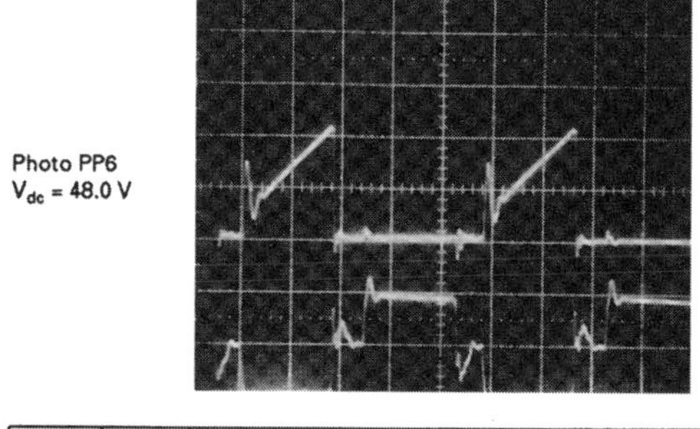

$I_{ds(Q2)}$, 2 A, 1 μs/cm

$V_{ds(Q)}$, 50 V, 1 μs/cm

V_{dc}, V	I_{dc}, A	P_{in}, W	V_5, V	I_5, A	P_5, W	V_{23}, V	I_{23}, A	P_{23}, W	$P_{o\,(total)}$, W	Efficiency, %
48.0	2.19	105.1	5.00	8.13	40.7	23.79	1.91	45.4	86.1	81.9

FIGURE 14.10 Significant waveforms in 200-kHz 85-W converter of Figure 14.8.

This is not the case. At the end of a transistor turn "on," the core flux density, having been driven up to—say—$+B_{max}$, remains locked there throughout the dead time. At the end of the dead time, the opposite transistor turns "on" and drives the flux density down from $+B_{max}$ to $-B_{max}$ and the cycle repeats.

One may then ask: Since neither transistor is "on" during the dead time, where is the current coming from to keep the flux density locked at $+B_{max}$ or $-B_{max}$? To keep the core there, there must be some magnetizing force holding it up—and since this is proportional to ampere turns, there must be some current flowing during the dead time.

Since both transistors are "off" during the dead time, the current holding the core up at $+B_{max}$ or $-B_{max}$ must flow in the secondaries. This current is the core primary magnetizing current reflected into the secondaries.

During the transistor "on" time (see Figure 2.4*e*), the DC input voltage supplies primary load current (all the secondary currents reflected by their turns ratios into the primary) plus magnetizing current which flows in the primary magnetizing inductance.

However, since current in an inductor cannot change instantaneously, when the "on" transistor turns "off," the current that supports that stored energy must continue elsewhere. The magnetizing current continues to flow where it finds a closed path.

That closed path is in the secondaries. As a transistor turns "off" the current in each secondary output inductor cannot change. Thus all output inductors reverse voltage polarity at the end of an "on" time. If there were free-wheeling diodes at the output as for forward converters (see Figure 2.10), the output inductor current would continue to flow through those diodes.

In a push-pull, the output rectifiers serve a similar function (Section 2.2.10.3). As the output inductors reverse polarity, when they reach a point one diode drop below ground (or above for a negative output voltage), the rectifier diodes conduct and serve as free-wheeling diodes, carrying the output inductor current.

However, the diodes carry more than the output inductor current. By flyback action, the magnetizing current built up in the primary during the "on" time is reflected by the turns ratio into the half secondary that just previously had not been carrying current. It is this current flowing in one of the half secondaries that supports B_{max} in the core during the dead time.

Thus, during the dead time, the output inductor current continues to flow through the half-secondary windings, with the rectifier diodes acting as free-wheeling diodes. That "ledge" current divides roughly equally (Figure 2.6*d* and 2.6*e*) between the two secondaries as described in Section 2.2.10.3. One of those ledge currents is always larger than the other, as can be seen dramatically in Figure 14.15.

The primary magnetizing current does not result in significant dissipation. It increases in one direction during one transistor "on" time, switches to the secondary and decreases slightly during the dead time, then switches back and reverses direction in the primary during the next transistor "on" time, repeating the previous half cycle.

As discussed, during the dead time, output inductor current divides between the two half secondaries. Since those half secondaries have low impedance, current flowing through them produces no voltage drop. Hence there is no voltage drop across either of the two half primaries, and voltage at the two "off" drains during the dead time must equal V_{dc} as seen in Photos PP1 to PP3 and PP4.

14.3.3 Relative Timing of Gate Input Voltage, Drain-to-Source Voltage, and Drain Currents

This timing is shown in Photo PP5 (Figure 14.3). It shows negligible delays between gate voltage rise and fall times, and the corresponding drain current and voltage transitions.

14.3.4 Drain Current Measured with a Current Probe in the Drain Compared to that Measured with a Current Probe in the Transformer Center Tap

As discussed in Section 14.3.1, a current probe in the drain measures absolute drain current correctly, as in Photo PP6.

As in Photo PP1, however, when drain current is measured with the current probe in the transformer center tap where both transistor currents are seen, the short transistor dead time between alternate "on" times does not permit the flux in the current probe transformer to reset. Consequently in such a measurement, the current indicated during the dead time is not zero and absolute values of current cannot be obtained from it.

14.3.5 Output Ripple Voltage and Rectifier Cathode Voltage

Photo PP7 in Figure 14.11 shows the 5-V output ripple voltages and noise of about 80 mV peak to peak. Such measurements are difficult to make and often spurious, because they can be masked by common-mode noise.

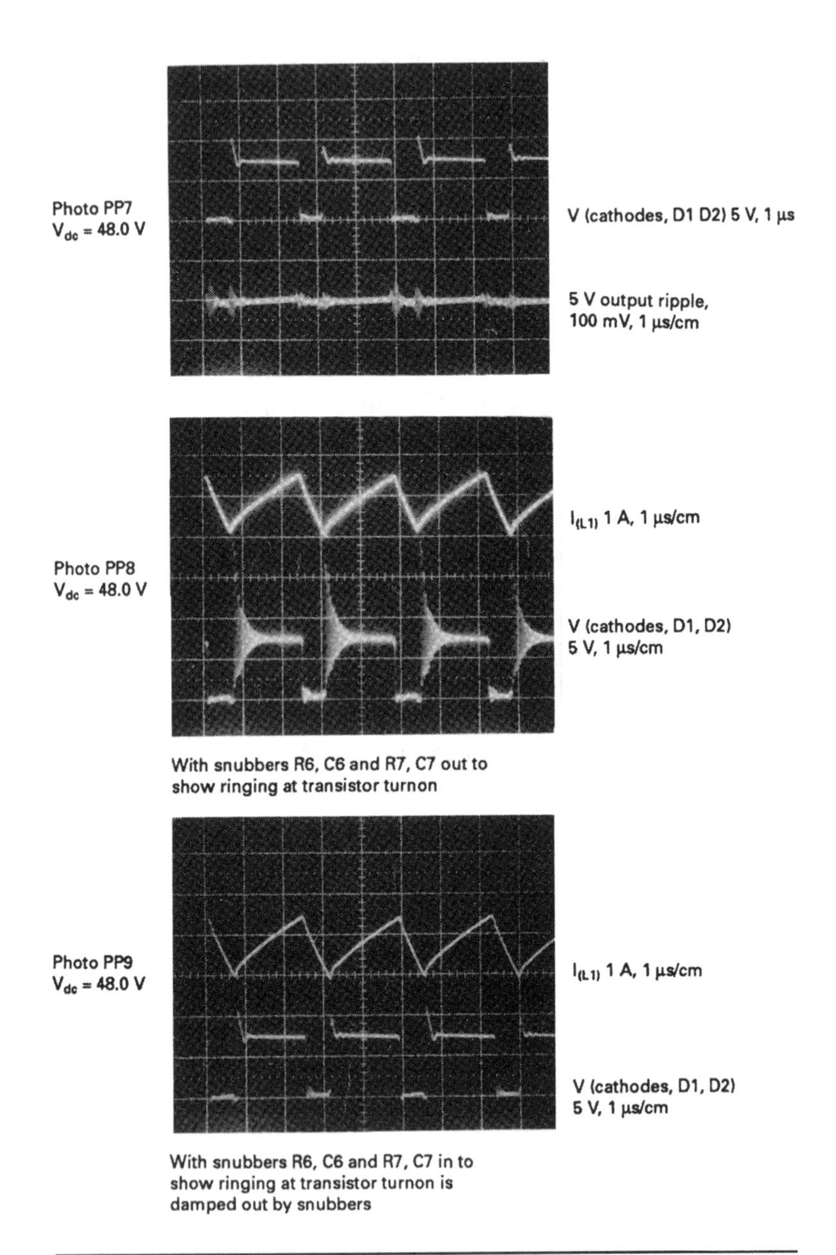

FIGURE 14.11 Significant waveforms in 200-kHz converter of Figure 14.8.

TIP *Common-mode noise is any noise voltage that appears simultaneously on both outputs with respect to the common line (typically the ground line or the oscilloscope common line). Very often the noise is developed by currents flowing down the oscilloscope probe ground wire. This type of noise can be identified by removing the probe ground wire and connecting the ground ring of the probe directly to the measurement point. Common-mode noise is a problem in that it results in RFI radiation. It can be difficult to eliminate in high-frequency square wave type converters and requires careful attention at places where the noise couples to the chassis. Typical injection points are where switching devices are mounted to the chassis for cooling requirements. (See Reference 1, Chapter 4.)*

Such common-mode noise can cause problems at the loads and can be minimized by a common-mode filter or *balun* as shown in Figure 14.1.

To determine whether a noise voltage is truly differential or common-mode, the "hot" end of the voltage probe is short-circuited to its shortest ground lead and those two points are taken to the return rail of the output voltage.

If the oscilloscope still indicates almost the same large noise voltage (as it most often will), that noise is common-mode noise. It will change in amplitude as oscilloscope ground connections are changed to various points on the power supply ground rail and as grounding lead lengths are changed.

Output ripple voltage measurements should be made with a differential probe which has a good common-mode rejection ratio at high frequencies. With the fast rise and fall times of MOSFETs, the common-mode "ringing" or noise on ground buses can occur at frequencies over 50 MHz.

Photo PP7 also shows the voltage at the 5-V output rectifier cathodes. It is this voltage which is averaged by the output *LC* filter to yield the desired DC output voltage.

If this waveform has notches, bumps, or odd ledges along its supposedly vertical sides, those contribute area to the voltage being averaged. The feedback loop will then alter the transistor "on" time so that the averaged volt-second area at the rectifier cathodes yields the desired DC voltage.

Thus the master output voltage will always be correct regardless of whether there are odd bumps, notches, or ledges during the dead time or along the vertical sides of the rectifier cathode waveshapes. However, any slave rectifier cathode voltage may not have the same proportion of extraneous bumps or notches as the master, and the slave DC output voltage would differ from what would be expected from the relative turns ratio.

For the slave DC output voltages to be what is expected from their turns ratios, the voltages at the rectifier cathodes of the master and all slaves should have steep vertical sides (Photo PP9, Figure 14.11) and no bumps, or notches during the dead time which alter their volt-second area. Examples of such aberrations of the rectifier cathode voltages are shown in Photo PP11 of Figure 14.12 and PP18 of Figure 14.14. These waveform aberrations will be explained below.

14.3.6 Oscillatory Ringing at Rectifier Cathodes after Transistor Turn "On"

This is shown in Photos PP8 and PP10 (Figures 14.11 and 14.12) and its elimination in Photos PP9 and PP11.

At the instant of transistor turn "on," the "on"-turning rectifier diode, say diode *D*1 in Figure 14.8, cancels the free-wheeling current in the opposite diode—say, *D*2. As the *D*2 forward current is canceled and its cathode voltage starts rising, there is an exponentially decaying oscillation or "ring" at the common cathodes as seen in Photos PP8 and PP10.

The oscillation is at a frequency determined by the inherent capacitance of the "off"-turning diode *D*2 and the value of the output inductor. The amplitude and duration of the ring are determined by the rectifier diode reverse recovery times and DC output current.

The ring can cause RFI problems, drive the rectifier diodes too close to their maximum reverse voltage rating, and increase their dissipation. The oscillation can easily be eliminated by *RC* snubbers (*R*6, *C*6; *R*7; *R*8; and *R*9, *C*9) across the diodes as shown in Figure 14.8. Cathode waveforms before the snubbers were added are shown in Photos PP8 and PP10 and after the snubbers were added, in Photos PP9 and PP11.

14.3.7 AC Switching Loss Due to Overlap of Falling Drain Current and Rising Drain Voltage at Turn "Off"

This is shown in Photo PP12 (Figure 14.12). Because MOSFETs have negligible storage and very fast current turn "off" time, this is close to the best-case scenario as described in Section 1.3.4.

In that scenario, current starts falling at the same instant the drain voltage starts rising, and current has fallen to zero at the same instant voltage has risen to its maximum. As mentioned in Section 1.3.4, for this case, the AC switching loss averaged over the current fall time is $\int_0^{tf} IV\,dt = I_{max}V_{max}/6 = 4.2 \times 85/6 = 59.5$ W. For a current fall time of about 40 ns and a switching period of 5 μs, this AC switching loss averaged over a full cycle is only $59.6 \times 0.045/5$ or 0.48 W.

Photo PP10
V_{dc} = 48.0 V

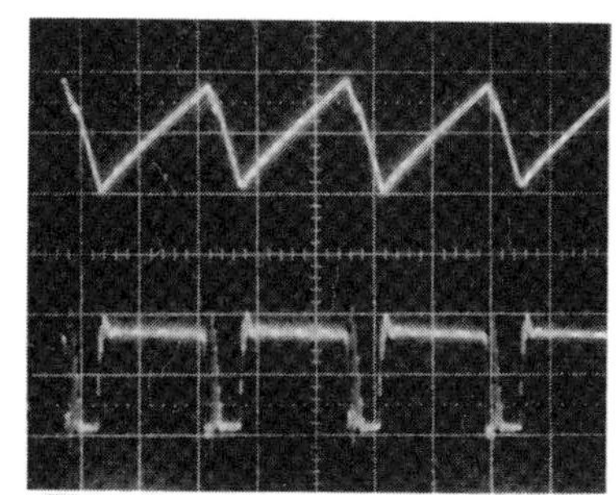

With snubbers R8, C8 and R9, C9 out to show ringing at transistor turnon

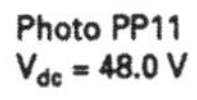
Photo PP11
V_{dc} = 48.0 V

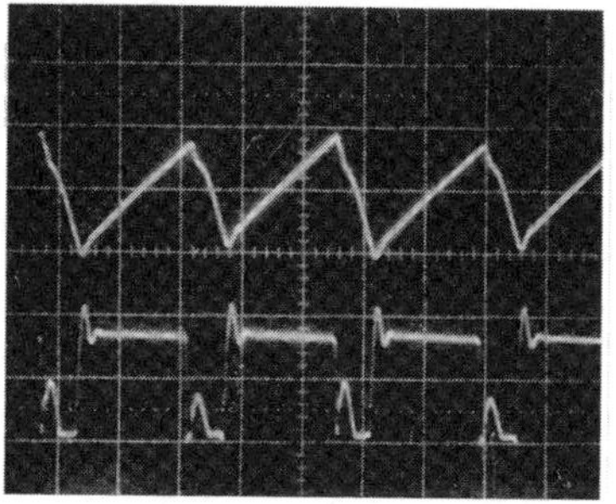

With snubbers R8, C8 and R9, C9 in to show ringing at transistor turn on is damped out by snubbers

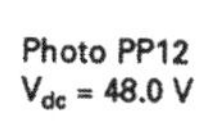
Photo PP12
V_{dc} = 48.0 V

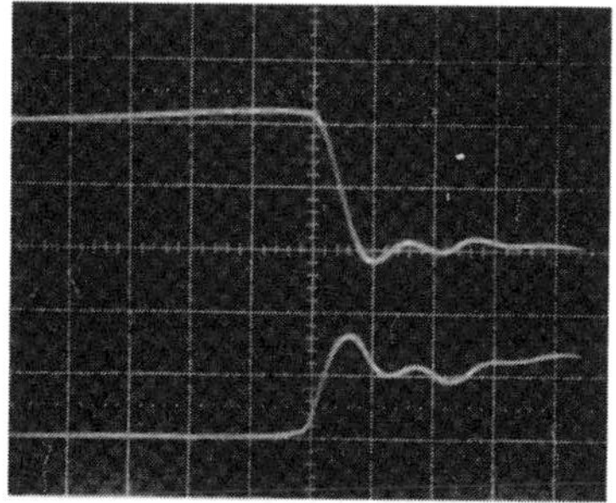

To show overlap of falling current and rising V_{ds} voltage at turnoff; even with the fast I_{ds} fall time of MOSFETs, this overlap yields AC switching losses

FIGURE 14.12 Significant waveforms in 200-kHz converter of Figure 14.8.

14.3.8 Drain Currents as Measured in the Transformer Center Tap and Drain-to-Source Voltage at One-Fifth of Maximum Output Power

Waveshapes for output currents of one-fifth the maximum (a typical power supply specification for minimum output currents) are shown in Photos PP13 to PP15 in Figure 14.13.

Efficiencies are still close to 80% as shown in Figure 14.13. For the worst efficiency of 78.7% at V_{dc} of 59.8, total internal losses are only 43 W—which is quite good.

The photos shown here and those following show an interesting and subtle problem. The problem is not a catastrophic failure mode, but one which may cause slave voltages to depart significantly from specified values. This problem arises from too large a transformer primary magnetizing current, or perhaps, too low a DC output current. The magnetizing current can become larger than originally specified if the two transformer halves inadvertently separate slightly. This will decrease the magnetizing inductance and increase the magnetizing current. It will also happen if too large a transformer gap was used to achieve a desired magnetizing inductance, or if the minimum load current is made lower than the value originally specified.

The problem can be seen in Photos PP13 to PP15. In those photos, the drain voltage during the dead time (which should be V_{dc} as in Photo PP14) starts at V_{dc} but gradually increases before the end of the dead time. The result is a manifestation of the problem but is not the actual cause. The basic cause is seen in Photo PP18 of Figure 14.14, which shows the voltage at the cathodes of the 5-V output rectifiers. There it is seen that during the dead time, the rectifier cathodes do not remain clamped to ground as they should be, but gradually pull away from ground. This is the bump ledge discussed in Section 14.3.5. It is this voltage at the rectifier cathodes that is averaged by the output *LC* filter to yield the 5-V DC output voltage. If that voltage has a bump during the dead time, before the normal transistor turn "on" time, its volt-second area increased. Thus the feedback loop, which is forcing the controlled output to be precisely 5.00 V, will decrease the normal "on" time. Then, since the slave outputs do not have this increased volt-second area due to a bump at their rectifier cathodes, their DC output voltages will decrease. Unlike Photo PP18 (Figure 14.14), it can be seen in Photo PP9 (Figure 14.11) that the rectifier cathodes voltage has no bump during the dead time, and the voltage rises vertically from ground at the start of the normal "on" time. For this case, which is at maximum load current in both outputs, the 23-V output is 23.74 V at V_{dc} of 48.0 V (Photo PP2, Figure 14.1) In Photo PP18 at the same V_{dc}

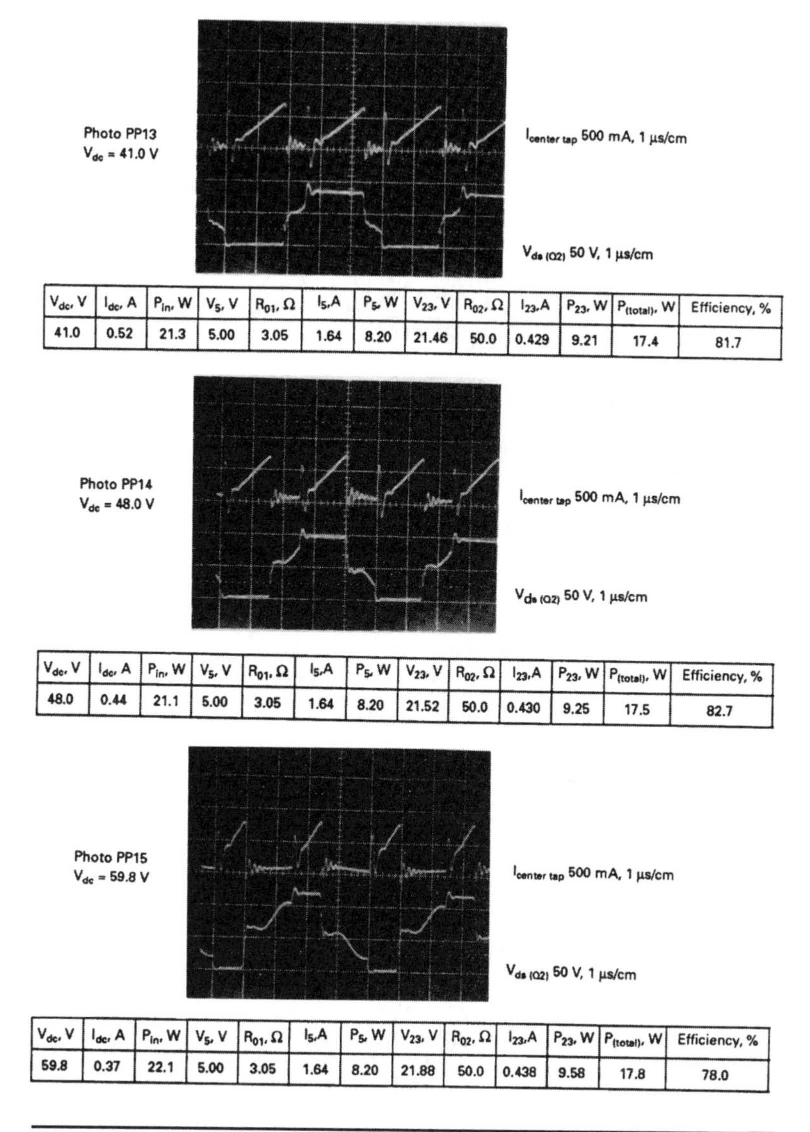

V_{dc}, V	I_{dc}, A	P_{in}, W	V_5, V	R_{01}, Ω	I_5, A	P_5, W	V_{23}, V	R_{02}, Ω	I_{23}, A	P_{23}, W	$P_{(total)}$, W	Efficiency, %
41.0	0.52	21.3	5.00	3.05	1.64	8.20	21.46	50.0	0.429	9.21	17.4	81.7

V_{dc}, V	I_{dc}, A	P_{in}, W	V_5, V	R_{01}, Ω	I_5, A	P_5, W	V_{23}, V	R_{02}, Ω	I_{23}, A	P_{23}, W	$P_{(total)}$, W	Efficiency, %
48.0	0.44	21.1	5.00	3.05	1.64	8.20	21.52	50.0	0.430	9.25	17.5	82.7

V_{dc}, V	I_{dc}, A	P_{in}, W	V_5, V	R_{01}, Ω	I_5, A	P_5, W	V_{23}, V	R_{02}, Ω	I_{23}, A	P_{23}, W	$P_{(total)}$, W	Efficiency, %
59.8	0.37	22.1	5.00	3.05	1.64	8.20	21.88	50.0	0.438	9.58	17.8	78.0

FIGURE 14.13 Transformer center tap current and drain-source voltage (*Q*2) at minimum (Photo PP13), nominal (Photo PP14), and maximum (Photo PP15) input voltage for one-fifth of maximum output currents.

and at minimum load, the 5-V output is 5.00 V, and the 23-V output is 21.52 V (tabular data for Photo PP14 in Figure 14.13).

The final question to be answered is why this bump during the dead time occurs for too high a magnetizing current or too low a

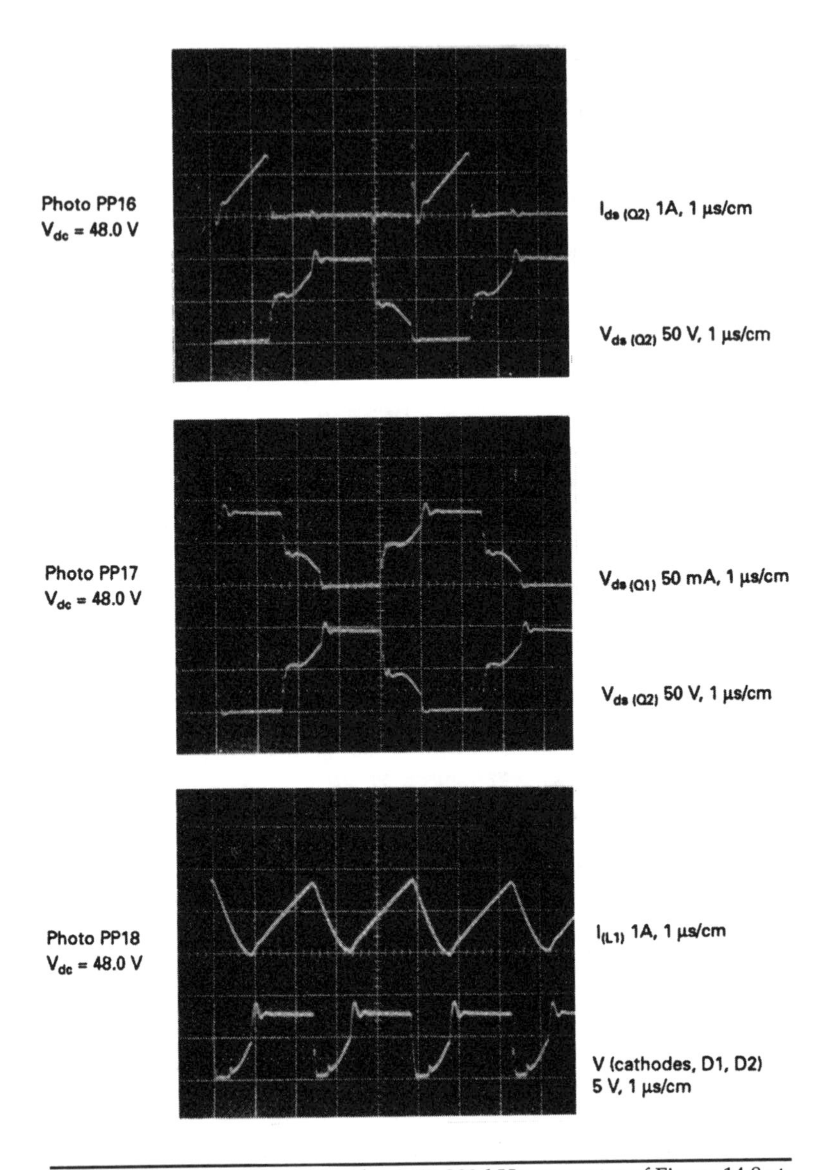

FIGURE 14.14 Significant waveforms in 200-kHz converter of Figure 14.8 at minimum (Photo PP16), nominal (Photo PP17), and maximum (Photo PP18) input voltages for one-fifth of maximum output current.

DC output current. This can be understood from the discussion in Section 14.3.2 as follows. As the "on" transistor turns "off," a fraction of the total primary magnetizing current, multiplied by the turns ratio, continues flowing in one of the half secondaries. Rectifier diodes of all secondaries are now free-wheeling and carrying about half their associated output inductor currents. In each secondary, one rectifier diode now carries a portion of the primary magnetizing current which has been reflected into that secondary as well.

In Figure 14.8, consider the 5-V output rectifiers. Assume that *Q*1 is "on." This makes the *D*2 anode positive and *D*2 delivers current to the load via *L*1. When *Q*1 turns "off," the dead time starts and the *L*1 current is divided between *D*1 and *D*2 acting as freewheeling diodes. As long as the sum of the *D*1, *D*2 currents is equal to the *L*1 current, their cathodes remain clamped one diode drop below ground.

However, as *T*1 turns "off," a fraction of the total primary magnetizing current is transferred by flyback action into N_{s1a} and *D*1 into *L*1. As long as this current is less than the current in *L*1, the *D*1, *D*2 cathodes remain clamped one diode drop below ground and the balance of the *L*1 current is supplied via *D*1, *D*2.

When the current reflected by flyback action from the primary into N_{s1a} exceeds the current in *L*1, the impedance seen at the input end of *L*1 increases and the common cathodes of *D*1, *D*2 pull up from ground before the end of the dead time as seen in Photo PP18.

With a closed feedback loop, the increased volt-second area due to the bump at the rectifier cathodes during the dead time causes the PWM chip to decrease the "on" time, so that the voltage averaged by the *LC* filter yields the desired 5.00-V output. This decreased "on" time then results in lower DC voltages at the slave outputs.

14.3.9 Drain Current and Voltage at One-Fifth Maximum Output Power

This is shown in Photo PP16 (Figure 14.14). Note that with the current probe in the transistor drain lead, a true measure of drain current is obtained (15 A peak). When drain current is measured in the transformer center tap as in Photo PP14 (Figure 14.13), a false value of 700 mA is observed. This has been discussed in Sections 14.3.1 and 14.3.4.

14.3.10 Relative Timing of Opposing Drain Voltages at One-Fifth Maximum Output Currents

This is shown in Photo PP17 (Figure 14.14), which is presented to show relative timings.

14.3.11 Controlled Output Inductor Current and Rectifier Cathode Voltage

The inductor current waveform in Photo PP18 shows the upslope of inductor current during the transistor "on" time [$di/dt = (V_{cathode} - V_o)/L_1$], and the downslope of its current during the dead time ($di/dt = V_o/L1$).

It is of interest to calculate $L1$ and verify that the inductor is as designed. As closely as can be read from the "on" time cathode voltage waveform, $V_{cathode} = 7.5$ V, and from the inductor current waveform during the upslope, di is 1.8 A and dt is 1.45 μs. Then $L_1 = (7.5 - 5)(1.45 \times 10^{-6})/1.8 = 2.0$ μH.

The inductor has 5 turns on an MPP 55120 core (see Magnetics Inc. MPP core catalog) which has an A_l of 72 mH/1000 turns. A 5-turn winding should have an inductance of $(0.005)^2 \times 72000 = 1.8$ μH. This is as close as can be expected from reading di/dt on Photo PP18.

14.3.12 Controlled Rectifier Cathode Voltage Above Minimum Output Current

This is shown in Photo PP19 (Figure 14.15), and is presented to show that when output current is increased, the voltage ledge during the dead time shown in Photo PP18 vanishes. The voltage remains clamped at ground throughout the dead time and rises steeply to its peak at transistor turn "on." As discussed in Section 14.3.8, the 23-V output voltage rises from 21.50 V in Photo PP18 to 22.97 V in Photo PP19.

14.3.13 Gate Voltage and Drain Current Timing

Photo PP20 in Figure 14.15 is presented to show the relative timing of both gate voltages and both drain currents. Delays between gate voltage transitions and the corresponding drain current rise and fall times are seen to be negligible.

14.3.14 Rectifier Diode and Transformer Secondary Currents

This has been discussed in Section 14.3.2 (see also Photo PP21 in Figure 14.15). It shows the ledge currents during the transistor dead time. Note that the ledge current immediately after an "on" time is greater than that before the other "on" time. This is because the previously "on" diode carries the primary magnetizing current plus half

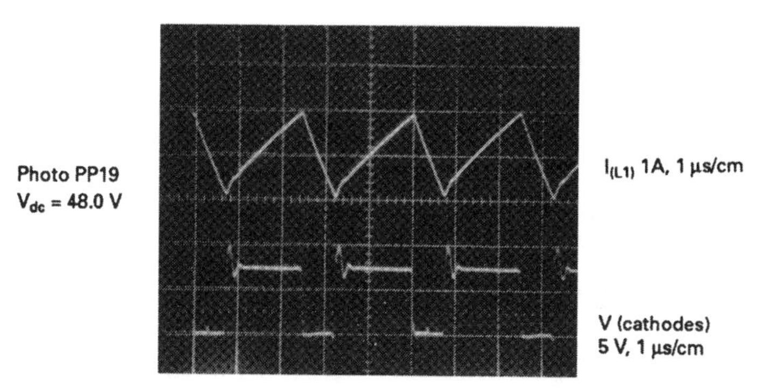

For I_5 = 5 A, I_{23} = 0.425 A
Showing the D1, D2 cathode voltage remaining clamped down to ground during the transistor off time for larger output currents because the larger diode free-wheeling current is not unclamped by the transformer magnetizing current (compare to photo PP18)

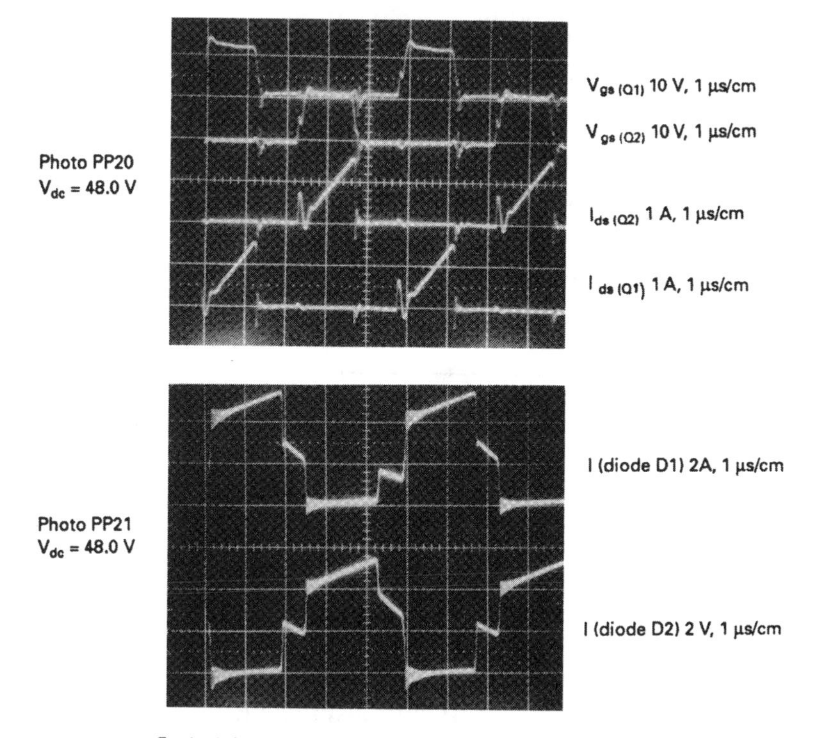

For I_{dc} (5 V output) of 5 A
Showing "ledge" currents during dead time between transistor on times (refer to Section 2.2.10.3 and Fig. 2.6)

FIGURE 14.15 Significant waveforms in 200-kHz converter of Figure 14.8.

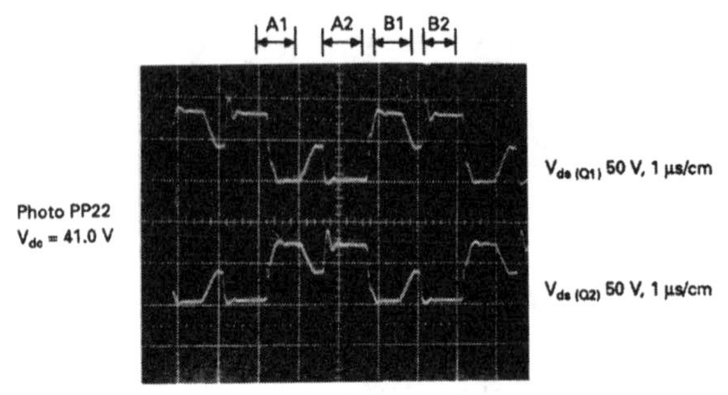

Showing an apparent failure mode at minimum output current or too high a magnetizing current: each transistor apparently turns on twice per half period (at A1, A2 for Q1 and at B1, B2 for Q2); the drains dropping to zero at A1, B1 are not transistor turnons—they occur for too large a transformer magnetizing current or too low a DC load current, as explained in the accompanying text

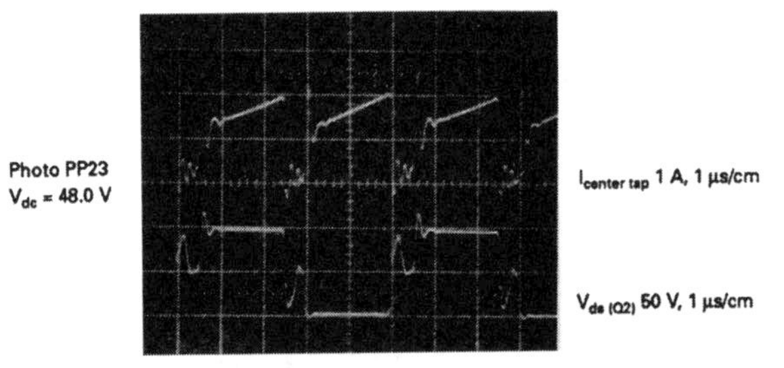

Showing critical waveforms on Fig. 14.8 circuit at output power 15% above specified maximum; converter efficiency is still a relatively high 83.8% and transformer temperature rise is only 54°C

V_{dc}, V	I_{dc}, A	P_{in}, W	V_5, V	R_{01}, Ω	I_5, A	P_5, W	V_{23}, V	R_{02}, Ω	I_{23}, A	P_{23}, W	$P_{(total)}$, W	Efficiency, %
48.0	2.51	120.5	5.00	0.339	14.7	73.7	26.12	25.0	1.04	27.3	101.0	83.8

FIGURE 14.16 Significant waveforms in 200-kHz converter of Figure 14.8.

the output inductor current during the dead time immediately after turn "off."

14.3.15 Apparent Double Turn "On" per Half Period Arising from Excessive Magnetizing Current or Insufficient Output Currents

This is shown in Photo PP22 of Figure 14.16. It is an extreme example of the case of Photo PP13 (Figure 14.13). It is seen in Photo PP22 that there are apparently two turn "ons" per half period—at *A*1 and *A*2 for *Q*1 and *B*1 and *B*2 for *Q*2.

The transistors turn "on" only at $A2$ and $B2$, and this brings the drain voltages to ground. The apparent turn "on" events at $A1$ and $B1$ are spurious. At those times, the corresponding drain voltage is driven to ground by the positive drain bump after turn "off" of the opposite transistor. This phenomenon is an extreme example of the situation seen in Photo PP13, which was obtained by increasing the primary magnetizing current by increasing the transformer gap with low DC output currents.

As discussed in Section 14.3.8, this phenomenon occurs with a large magnetizing current reflected into the secondary. When this exceeds the DC load current in the controlled output inductor, the rectifier diodes are unclamped from ground and produce a large positive bump above V_{dc} at the drain.

This, through the transformer coupling, produces a negative dip at the opposite drain. When that negative dip at the opposite drain falls as low as ground, the inherent body diode of that MOSFET conducts and holds the drain at ground. It appears thus that the MOSFET has turned "on."

The circuit may continue to work in this odd mode, but oscillations may occur with the feedback loop closed. "On" time will jump erratically from a true $B2$ aided by the spurious $B1$ to a true "on" time of $B1 + B2$ at a slightly higher load current or slightly lower primary magnetizing current.

The problem can be avoided by ensuring that there is no inadvertent increase in transformer gap, which would increase magnetizing current, and that the lowest DC load current is always larger than the magnetizing current reflected into the master secondary.

14.3.16 Drain Currents and Voltages at 15% Above Specified Maximum Output Power

Photo PP23, Figure 14.16 is presented to show that efficiency is still above 83%, transformer temperature rise is still only 54°C, and critical waveforms are still clean at this higher power level. The circuit of Figure 14.8 has also been run at 112 W of output power (Photo PP24 in Figure 14.17) with an efficiency of over 86%, clean waveforms, and transformer temperature rise of only 65°C.

14.3.17 Ringing at Drain During Transistor Dead Time

To avoid ringing, RC snubbers ($R1, C4$ and $R5, C5$) are required across each half primary. Without them, a high-frequency oscillatory ring occurs throughout the transistor dead time (Photo PP25 in Figure 14.17).

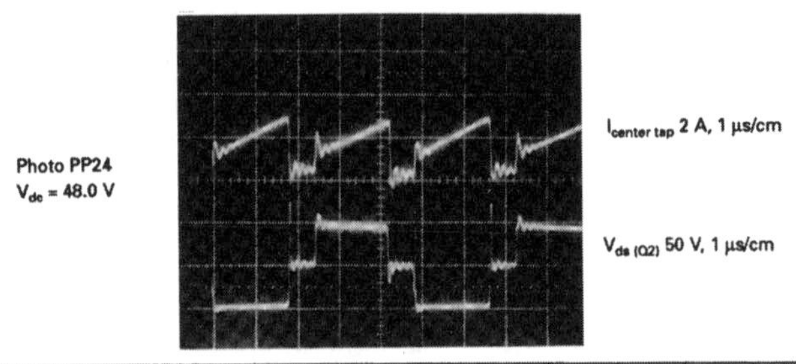

V_{dc}, V	I_{dc}, A	P_{in}, W	V_5, V	R_{01}, Ω	I_5, A	P_5, W	V_{23}, V	R_{02}, Ω	I_{23}, A	P_{23}, W	$P_{(total)}$, W	Efficiency, %
48.0	2.71	130	5.00	1.03	4.85	24.3	21.7	5.36	4.05	87.9	112.1	86.2

Drain currents and drain-to-source voltage at 30% above maximum specified output power

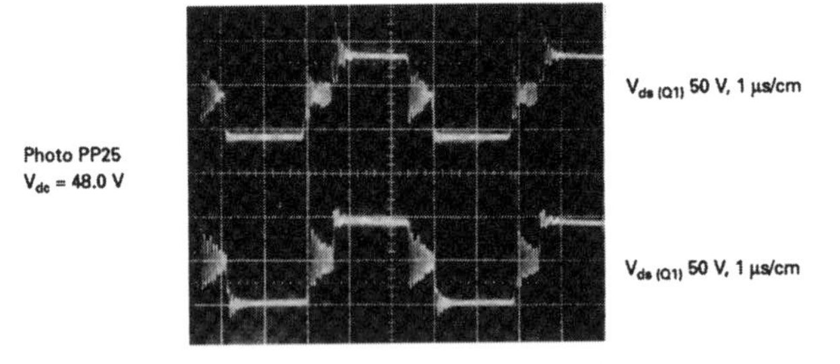

Showing ringing during dead time when RC snubbers (R4, C4 and C5) are removed from primary

FIGURE 14.17 Ringing drain voltage during transistor dead time.

This worsens the RFI problem, and affects slave DC output voltages in the same way as the dead time ledge of Photo PP18.

14.4 Flyback Topology Waveshapes

14.4.1 Introduction

Typical waveforms are presented for a relatively low-power, discontinuous mode, single-ended flyback.

A discontinuous mode, single-ended flyback was selected as it is the simplest topology for output powers up to about 60 W—the area of greatest usage for flybacks.

A serious drawback of the single-ended flyback is that energy stored in the transformer leakage inductance must be absorbed and subsequently dissipated in an *RCD* snubber (Chapter 11). If it is not, the leakage spike can destroy the transistor. Above 60 W, the high snubber dissipation is a significant drawback.

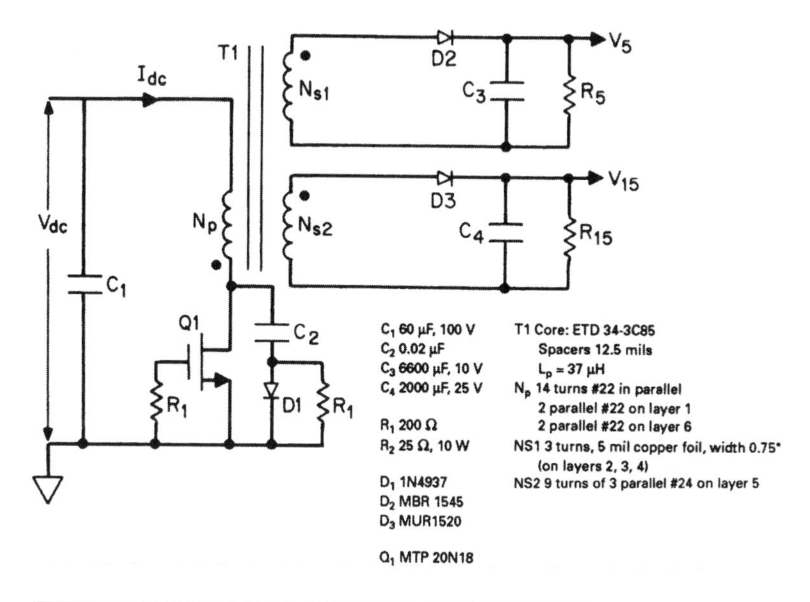

FIGURE 14.18 A 50-kHz 50-W flyback supply.

The double-ended flyback (Section 4.6) solves this problem not by storing leakage inductance energy in a snubber capacitor and then dissipating it in a resistor, but rather by returning the leakage inductance energy without dissipation to the input supply bus.

Thus, the double-ended flyback is a widely used approach for output powers above 60 to 75 W. Significant waveforms in the single-ended flyback are much like those of the superior double-ended flyback, so waveshapes of the single-ended circuit only will be shown.

Waveforms shown herein were taken from the circuit of Figure 14.18. The circuit is a 50-W, 50-kHz supply with one master output and one slave. Here again, the feedback loop was not closed. The transistor input was driven by a 50 kHz, manually adjustable pulse width generator using a 3525A PWM chip.

The pulse width at all input voltages and output load conditions was manually adjusted to set the master output at 5.00 V. The voltage of the slave was read and recorded.

The slave winding turns were chosen to yield 15 V when the master was 5.00 V. As will be noted, the slave output voltage is not determined entirely by the master/slave turns ratio. Due to secondary leakage inductances and interaction between master and slave windings, the slave output voltage is also dependent on the master DC output current.

14.4.2 Drain Current and Voltage Waveshapes at 90% of Full Load for Minimum, Nominal, and Maximum Input Voltages

These are shown in Figure 14.19. They show the characteristic linear ramp of primary current ($di_p/dt = V_{primary}/L_{primary}$) during the transistor "on" time. At the instant of turn "off," they show the leakage inductance voltage spike at the rising drain. The amplitude of this spike is controlled by the *RCD* snubber capacitor $C2$, which is chosen large enough to limit the spike to a safe amplitude without causing too much dissipation ($= 0.5C2(V_{peak})^2/T$) in snubber resistor $R1$.

The waveforms show that as V_{dc} increases, the pulse width required to maintain a constant master output voltage decreases. It is also seen that the peak ramp current is constant for all input voltages at constant output power.

It is also seen that after the leakage spike, the drain voltage falls (except for the small pedestal for a short time after the spike) to a level of $V_{dc}+(N_p/N_{s1})(V_5+V_{D2})$. It remains at that level until the reset volt-second product equals the set volt-second product $[V_{dc}t_{on} = (N_p/N_s)]$ $(V_5 + V_{D2})$ and then it falls back to V_{dc}.

Figure 14.20 shows the same waveforms at the three input voltages for a lower total output power (17 W). All "on" time pulse widths are narrower, and consequently all peak primary currents ($di = V_{primary}t_{on}/L_p$) are lower. This is because less power $[0.5L_p(I_p)^2/T]$ is required from the input bus, so the peak primary current at the end of the "on" time is less—as is the leakage inductance spike.

14.4.3 Voltage and Currents at Output Rectifier Inputs

These are shown to demonstrate why slave output voltages are not entirely dependent on the master/slave turns ratios, and why they are dependent on the master output current.

When the transistor turns "off," all the energy stored in the primary $0.5L_pI_p^2$ is delivered to the secondaries, except for some stored in the leakage inductance that is diverted to the snubber capacitor $C2$. That energy is delivered to the master and slave outputs in the form of currents shown in Figure 14.21. The currents to the master and slave outputs via their rectifiers are not of equal duration and certainly not of similar waveshapes.

Note that when the DC output current of the master is increased from 2.08 A (Photo FB8) to 6.58 A (Photo FB10), the small voltage pedestal on the slave output rectifier increases from about 20 to 28 V (Photos FB7 and FB9). This occurs because of secondary leakage

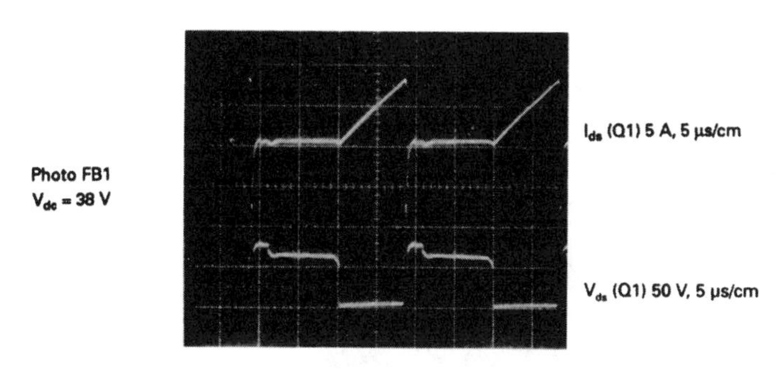

V_{dc}, V	I_{dc}, A	P_{in}, W	V_5, V	I_5, A	P_5, W	V_{15}, V	I_{15}, A	P_{15}, W	$P_{o\ (total)}$, W	Efficiency, %
38.0	1.67	63	5.002	6.58	32.9	25.3	0.504	12.8	45.7	72.1

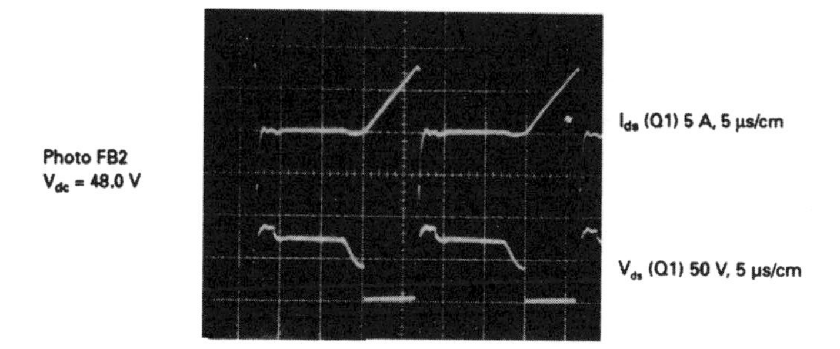

V_{dc}, V	I_{dc}, A	P_{in}, W	V_5, V	I_5, A	P_5, W	V_{15}, V	I_{15}, A	P_{15}, W	$P_{o\ (total)}$, W	Efficiency, %
48.0	1.31	62.6	5.001	6.57	32.9	25.42	0.505	12.8	45.7	72.0

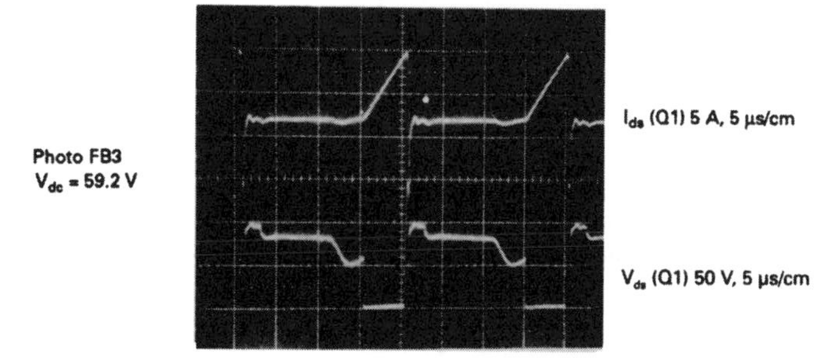

V_{dc}, V	I_{dc}, A	P_{in}, W	V_5, V	I_5, A	P_5, W	V_{15}, V	I_{15}, A	P_{15}, W	$P_{o\ (total)}$, W	Efficiency, %
59.2	1.07	63.1	5.003	6.58	32.9	25.36	0.504	12.8	45.7	72.4

FIGURE 14.19 Typical waveforms in the 50-kHz flyback supply of Figure 14.18.

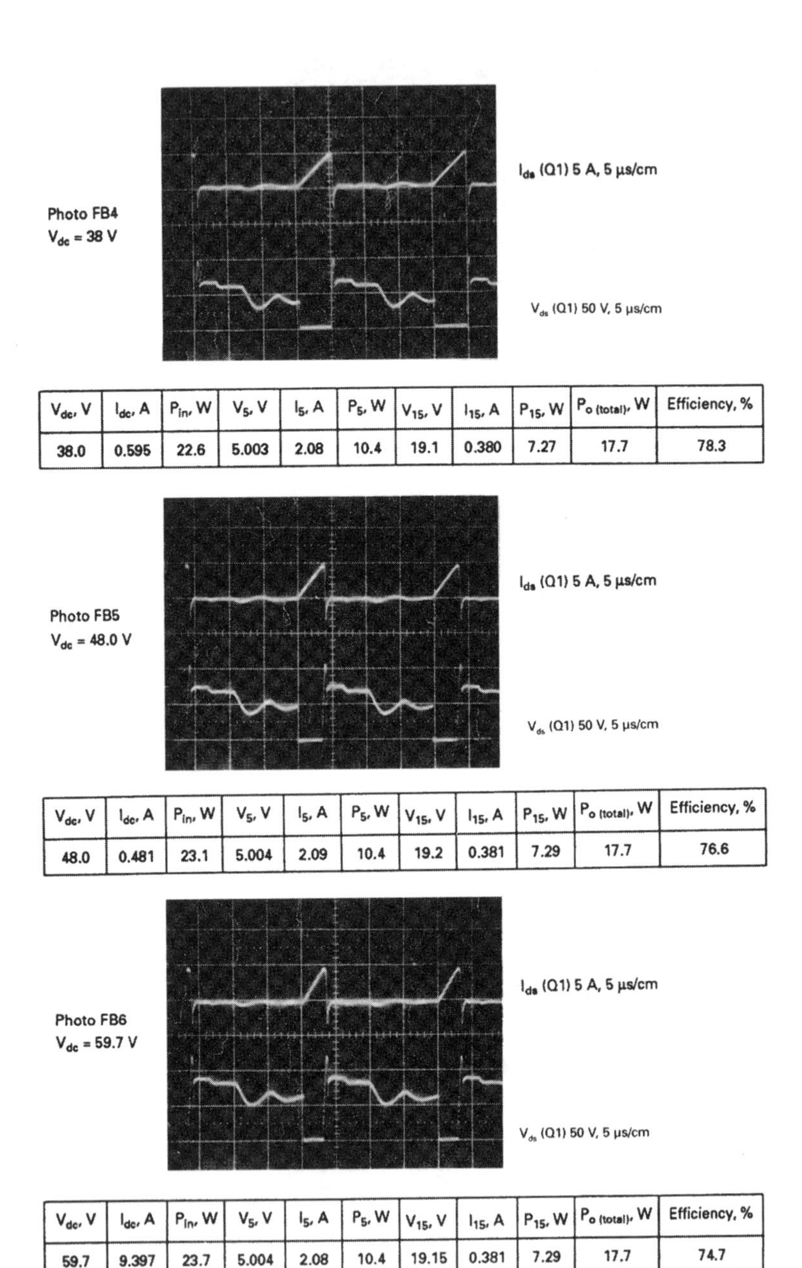

V_{dc}, V	I_{dc}, A	P_{in}, W	V_5, V	I_5, A	P_5, W	V_{15}, V	I_{15}, A	P_{15}, W	$P_{o\,(total)}$, W	Efficiency, %
38.0	0.595	22.6	5.003	2.08	10.4	19.1	0.380	7.27	17.7	78.3

V_{dc}, V	I_{dc}, A	P_{in}, W	V_5, V	I_5, A	P_5, W	V_{15}, V	I_{15}, A	P_{15}, W	$P_{o\,(total)}$, W	Efficiency, %
48.0	0.481	23.1	5.004	2.09	10.4	19.2	0.381	7.29	17.7	76.6

V_{dc}, V	I_{dc}, A	P_{in}, W	V_5, V	I_5, A	P_5, W	V_{15}, V	I_{15}, A	P_{15}, W	$P_{o\,(total)}$, W	Efficiency, %
59.7	9.397	23.7	5.004	2.08	10.4	19.15	0.381	7.29	17.7	74.7

FIGURE 14.20 Significant waveforms for the 50-kHz flyback supply shown in Figure 14.18.

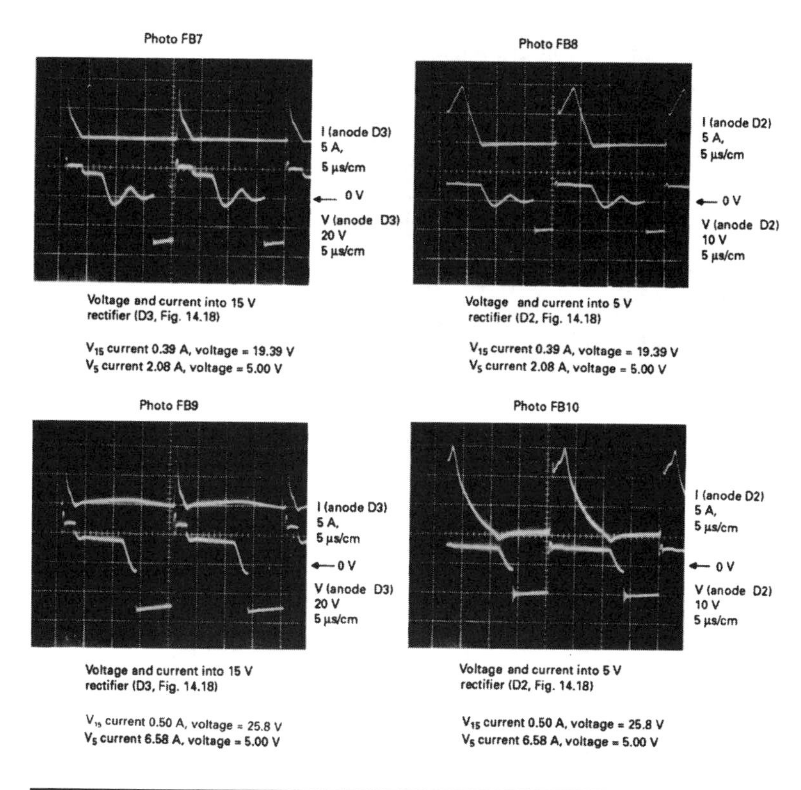

FIGURE 14.21 Why the slave DC output voltage is not directly related to the master DC output voltage and the corresponding turns ratio. Master secondary leakage inductance diverts some of its peak secondary current into the slave rectifier anode immediately after turn "off." As long as this diverted current persists, it couples a pedestal voltage into the slave rectifier anode. The slave output capacitor charges up to the pedestal peak which is higher than the induced secondary voltage and is dependent on the master current. The remedy is to minimize secondary leakage inductances. The effect is minimized by an inductor in series in the slave secondaries.

inductance and mutual coupling between master and slave secondaries, and with the large voltage at the pedestal at the slave rectifier anode (compare Photos FB7 and FB9), the slave DC output voltage is increased.

14.4.4 Snubber Capacitor Current at Transistor Turn "Off"

Figure 14.22 shows that at the instant of transistor turn "off," all the transformer primary current (5 A from Photo FB5) is immediately transferred to the snubber capacitor $C2$ and through snubber diode $D1$

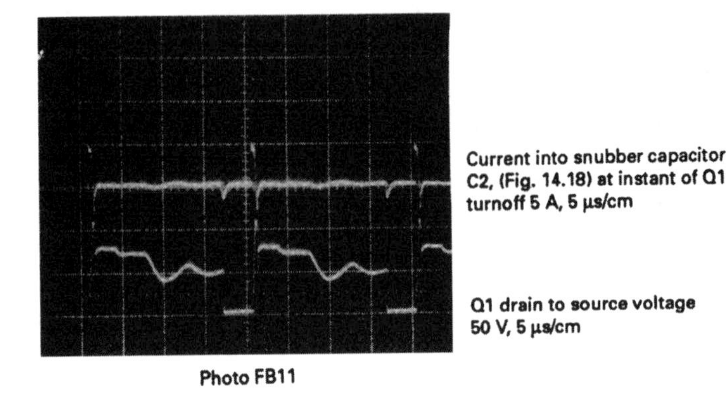

FIGURE 14.22 At the instant of turn "off," all the current which had been flowing in the transformer magnetizing and leakage series inductances is driven immediately into the snubber capacitor of Figure 14.18*c*.

to common. That current represents energy stored in the primary leakage inductance $[0.5L_{\text{leakage}}(I_p)^2]$. Until it is transferred to $C2$ as electrostatic energy $[0.5C2(V_p)^2]$, all the energy stored in the transformer magnetizing inductance cannot be delivered to the secondaries.

If $C2$ is made too small, the current reflected from the primary at the instant of turn "off" may charge it to a dangerously high voltage. Thus, $C2$ is selected large enough to limit the leakage spike to a safe value, but not so large as to result in excessive dissipation.

References

1. Keith Billings, *Switchmode Power Supply Handbook,* Chapter 4, McGraw-Hill, New York, 1989.

PART 4

More Recent Applications for Switching Power Supply Techniques

CHAPTER 15

Power Factor and Power Factor Correction

15.1 Power Factor—What Is It and Why Must It Be Corrected?

We are all familiar with the concept of power when dealing with DC currents and voltages, where the VI product gives power, or the rate of doing work in watts, directly. However, when dealing with AC conditions the calculation of power is not so straightforward.

For AC conditions we are also familiar with the term RMS (the square root of the mean of the squares), a value allocated to any voltage or current waveform that produces the same heating effect (power) into a resistive load as would a DC voltage or current of the same value.

However, for an AC waveform, the product $V_i I_i$ that is the product of RMS input voltage V_i and RMS input current I_i yields *apparent power*, which is the same as real power only for a purely resistive load.

A component of input current normal to the voltage across the load resistor (I_i sin x) does not contribute to the actual load power. In the case of sinusoidal voltage and current, this is a current that flows 90° out of phase with the voltage. This current represents energy drawn from the input source which is stored temporarily in a reactive component of the circuit. Later, this stored energy is returned to the input source. This current, which does not contribute to load power, wastes power in the winding resistances of the input power source and power lines.

The term *power factor* stems from elementary AC circuit theory. When a sinusoidal AC power source feeds either an inductive or a

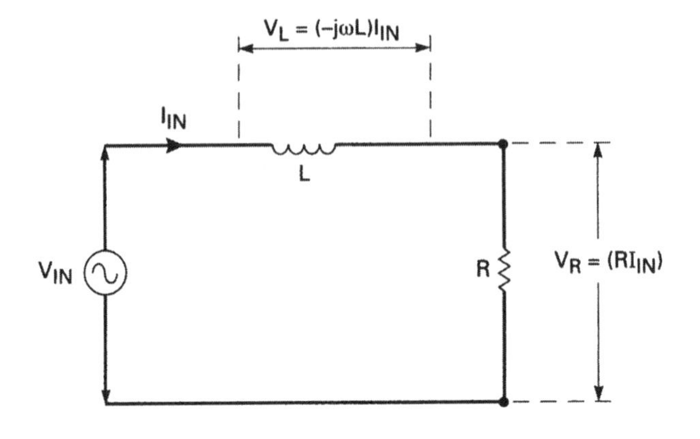

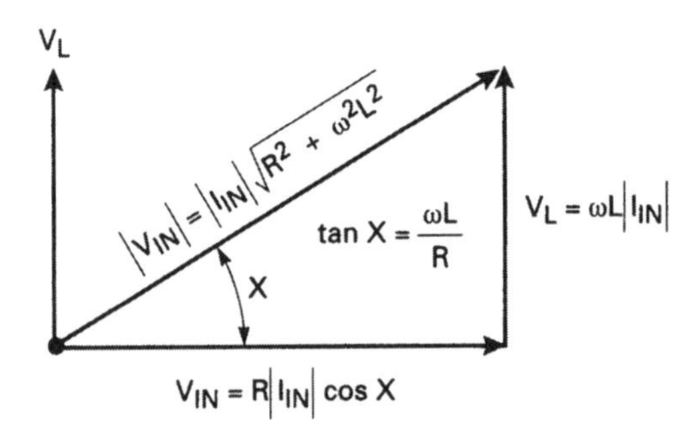

FIGURE 15.1 Defining Power Factor. In AC circuit terminology, *power factor* for a sinusoidal waveform is defined as the cosine of the phase angle between input voltage and input current. In circuits with pure resistive load, there is no phase difference between input voltage and current, and the power factor is unity. If current lags or leads input voltage, it is only the component of input current in phase with the voltage that contributes power to the load.

capacitive load, the load current is also sinusoidal, but lags or leads the input voltage by some phase angle x. The actual power delivered to the load is $V_i I_i$ cosine x. It is only the component of input current which is in phase with the voltage across the load impedance ($I_i \cos x$) that contributes to the true load power. The power factor is defined as $\cos x$, and the true power is obtained by multiplying the apparent power by the power factor. Reactive and power generating components of a sinusoidal input line current can be seen in Figure 15.1

In AC power circuit parlance, $\cos x$ is referred to as the *power factor*. To minimize power loss it is desirable to keep the power factor as close to unity as possible. To do this, we need to keep the input line current

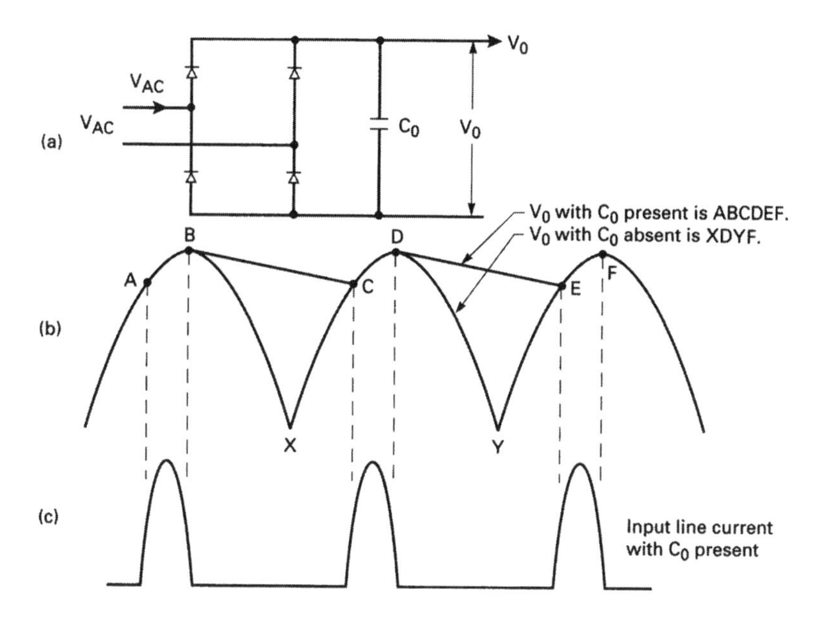

FIGURE 15.2 Full Wave Rectifier Waveforms. (*a*) Full-wave rectifier and filter circuit. (*b*) Output voltages with and without a capacitor filter after the input bridge rectifier. (*c*) Input line current with capacitor C_o present.

sinusoidal and in phase with the sinusoidal input line voltage. The means to achieve this is referred to as *power factor correction.*

The circuit techniques described in this chapter maximize or "correct" the power factor by forcing the input line current to be sinusoidal, and in phase with the input line voltage such that real power and apparent power are the same. This keeps the line input free from excessive line harmonics. These techniques are especially useful—indeed, mandatory in certain new designs for "off-the-AC power line" power supplies.

The reason for mandating such correction measures is that severely distorted input line current waveforms caused, for example, by the capacitor filter that follows the input bridge rectifier in a typical "off line" power supply (see Figure 15.2*c*.) provoke excessive loss in distribution systems and generating equipment.

15.2 Power Factor Correction in Switching Power Supplies

In the field of switching regulators, any circuit configuration that causes the input line current to be nonsinusoidal (or even sinusoidal but out of phase with the sinusoidal input voltage) has harmonics resulting in a lowered power factor and consequent waste of power. Any component of input current normal to the applied voltage does not contribute to delivered power, and wastes power in the resistance of the supply input network and source generator.

In power supplies with a capacitor filter across the input bridge rectifier, the input line current consists of very narrow spikes with fast rise and fall time. These current spikes have a high RMS value, waste power, and give rise to RFI/EMI problems. Borrowing a term from AC circuit theory, power supplies with such input line currents are said to have poor power factor. The object of *power factor correction* is to force the input current to track the applied voltage (typically this will be sinusoidal) as closely as possible, so that it will be in phase with the line voltage, and generate a regulated DC output voltage somewhat greater than the peak of the line voltage.

In Figure 15.2*a* and *b*, if the filter capacitor C_o were absent and the load were a pure resistor, the voltage at V_o would have the half sinusoidal waveshape *ABXCDYEF.* Rectified current coming out of the rectifier would have the same half sinusoidal waveshape (referred to as a haversine), and the line current drawn from the input source would be almost purely sinusoidal and in phase with the sinusoidal input voltage. The power factor would be unity; and if V_i and I_i were the input voltage and current measured with RMS meters, the input and output load power would be $V_i I_i$.

Half sinusoids of rectified output voltage such as *ABXCDYEF* (Figure 15.2*b*) are not useful. The sole purpose of the rectifier and filter is to convert the input AC voltage to a DC voltage with as low a ripple content as possible. The capacitor C_o is thus added to yield the waveform *ABCDEF.* This results in a higher DC voltage component (midway between amplitudes at B and C or D and E) and a lower peak-to-peak ripple of $B - C$ or $D - E$. Between instants B and C or D and E, all rectifiers are reverse-biased, no line current flows, and all load current is drawn from the filter capacitor C_o. At instants A, C, and E, the rising input voltage forward-biases the rectifiers, and line current now flows to the load, and into C_o to replenish the charge lost when it alone was supplying load current.

The line current with the filter capacitor in place is shown in Figure 15.2*c*. It is a sequence of narrow current pulses before the peak of each half sinusoid of input voltage. The larger the filter capacitor, the shorter the duration, rise, and fall times of the pulses, and the higher their peak and RMS values.

It is these narrow line current pulses that power factor correction aims to eliminate. Their fast rise time causes radio-frequency interference (RFI) problems, and more important, their RMS value is higher than what is needed to supply the required output load power and causes excessive temperature rise and decreased reliability in the filter capacitor.

15.3 Power Factor Correction—Basic Circuit Details

Power factor correction eliminates the large filter capacitor C_o after the bridge rectifier, and thereby permits the voltage after the rectifier to rise and fall in half-sinusoidal fashion, called a haversine waveform. It then converts this haversine to a constant regulated DC output voltage.

The essence of the technique is that by monitoring the rectified input haversine voltage and forcing the current waveform to track it, the instantaneous input line current is directly proportional to the voltage in the same way as if it were driving a resistive load.

A boost regulator (Section 1.4), can do this by using a special control circuit. During the half sinusoids of voltage, the "on" time of the boost regulator is modulated by a PWM control chip in such a way as to force the input line current to track the voltage and be half sinusoidal also. At the same time it maintains the output voltage constant at a value somewhat greater than the incoming sine wave peak value.

The basic scheme used to implement power factor correction (PFC) is shown in Figure 15.3. First, the large input filter capacitor of Figure 15.2*a* is replaced with a much smaller value, allowing the voltage immediately after the bridge rectifier to fall to zero each half cycle, as shown in (Figure 15.3*a*).

By removing the input capacitor C_o, the line current flows continuously and sinusoidally, avoiding the narrow current pulses of Figure 15.2*c*. The resulting half sinusoids of voltage drive a continuous-mode boost converter.

The first task of the power factor correction circuit is to use the boost converter to convert the varying input voltage (the half sinusoids) to constant, fairly well-regulated DC voltage somewhat higher than the input since wave peak. It does this by using a continuous-mode *boost converter* (Section 1.4) in the following way.

A boost converter "boosts" a low voltage to a higher voltage. It does this by turning "on" the transistor $Q1$ for a time T_{on} out of a period T, and storing energy in inductor $L1$. When $Q1$ turns "off," the voltage polarity across $L1$ reverses, and the dot end of $L1$ rises to a voltage V_o higher than the input voltage V_{in}. Energy stored in $L1$ during T_{on} is transferred via $D1$ to the load and $C1$ during the $Q1$ "off" time. It can be shown that the output-input voltage relation of such a boost converter is given by

$$V_o = \frac{V_{in}}{(1 - T_{on})/T} \qquad (15.1)$$

During the half sinusoids of V_{in}, the $Q1$ "on" time T_{on} is width-modulated in accordance with Eq. 15.1 to yield a constant DC voltage

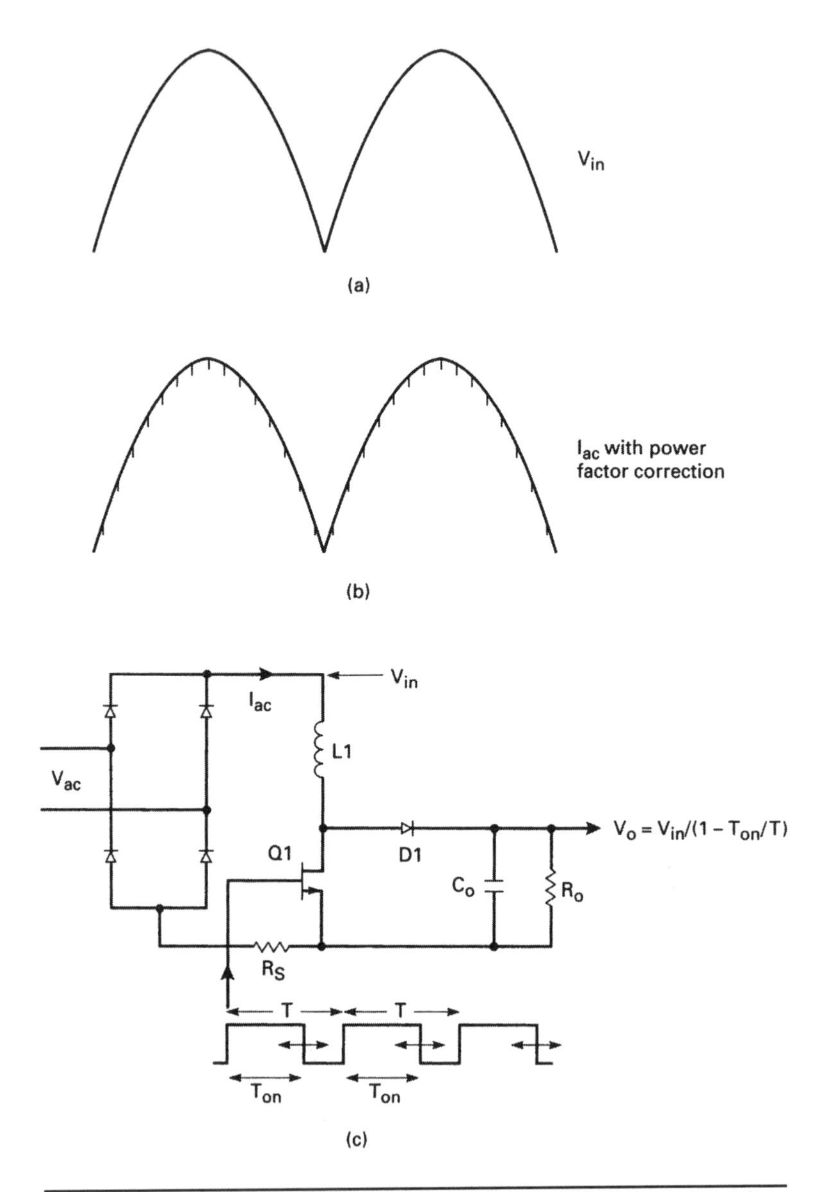

FIGURE 15.3 A typical boost-type power factor correction circuit. Providing a boost regulator after the input bridge rectifier will force a sinusoidal line current in phase with the voltage, and yield a regulated DC output voltage somewhat higher than the peak line voltage.

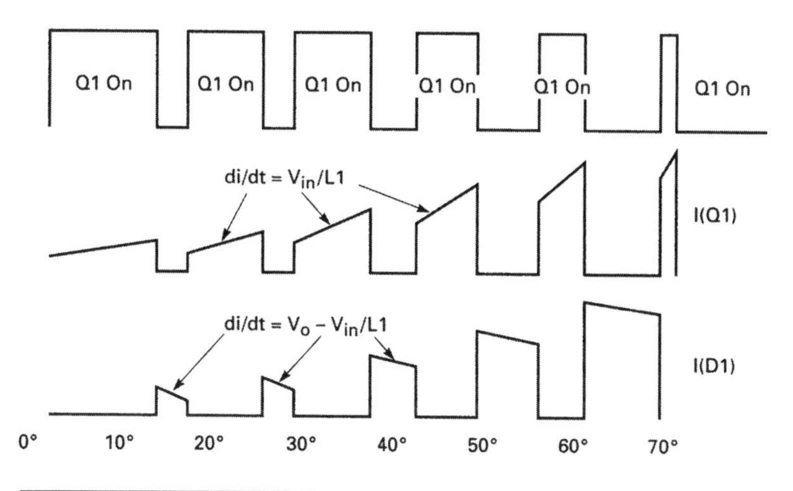

FIGURE 15.4 Boost Converter Waveforms. Showing the significant waveforms in the boost converter of Figure 15.3 as the input voltage increases toward the peak from zero.

V_o somewhat higher than the peak of the input voltage sine wave. The "on" time is controlled continuously by a PFC control chip whose DC voltage error amplifier senses V_o, compares it to an internal reference, and sets T_{on} to keep V_o constant at the selected value in a negative feedback loop.

From Eq. 15.1 it is seen that at the lower voltage portions of the half sinusoids of Figure 15.3*a*, the *Q*1 "on" time must be large to boost it to a value higher than the peak of the sinusoid. As V_{in} rises toward its peak, the PFC chip automatically decreases the *Q*1 "on" time so that the input voltage at each moment is boosted to the same output voltage. The progression of the "on" times throughout the half sinusoids is seen in Figure 15.4.

The second task of the power factor correction circuit is to sense input line current and force it to have a sinusoidal waveshape in phase with the input line voltage. This, too, is done by width modulation of the same boost regulator's "on" time. The "on" time is determined in a negative feedback loop which compares a sample of the actual input line current to the amplitude of a reference sine wave. The difference between these two sine waves is an error voltage which modulates the "on" time to force the two sine waves to be equal in amplitude.

The total error voltage that controls the boost regulator's "on" time is a mix of the output voltage and the input current error voltages. This mixing is done in a real-time multiplier such that the output is proportional to the product of the two error voltages.

15.3.1 Continuous- Versus Discontinuous-Mode Boost Topology for Power Factor Correction

Boost converters can be operated in either the discontinuous or continuous mode (Section 1.4). The continuous-mode boost topology is far better suited to yield relatively smooth, ripple-free half sinusoids of input line current is this application. This can be seen from Figure 15.5, which shows a continuous-mode boost converter fed from a constant DC input voltage. The continuous-mode boost topology differs significantly from the discontinuous mode (Figure 1.10).

In the discontinuous mode, the inductor $L1$ is made small to yield a steep ramp ($di/dt = V_{in}/L1$) of input current (Figure 1.10*c*) to $Q1$. When $Q1$ turns "off," all the current or energy stored in $L1$ is transferred via $D1$ to the load (Figure 1.10*d*) Since $L1$ is small, the downward ramp of current through $D1$ [$di/dt = (V_o - V_{in})/L1$] is also steep and $D1$ current falls to zero before the next $Q1$ turn "on." The input line current, which is the sum of the $Q1$ current when it is "on" and the $D1$ current when $Q1$ is "off," is not constant over one complete switching cycle. It consists of steep up and down ramps with zero current gaps between a turn "off" and the next turn "on."

In the continuous mode of Figure 15.5, however, the inductor $L1$ is made significantly larger. As a result, the $Q1$ current (Figure 15.5*c*) has the shape of a large step of current with a slow upward ramp on it, and the $D1$ current has the shape of a large step with a slow downward ramp. Importantly, there is no gap of zero current between a turn "off" and the next turn "on." The input line current (Figure 15.5*e*) is the sum of the I_{Q1} and I_d currents, and if the ramps are made small by using a large $L1$, the line input current averaged over one switching cycle is I_{av} with small peak-to-peak ripple ΔI. The input power is $V_{in}\ I_{av}$.

With an AC input, a continuous-mode boost converter is used after the input bridge rectifier (as shown in Figure 15.3). At any point on the half sinusoid input voltage, the $Q1$ "on" time will be forced by the PWM control chip to boost that instantaneous voltage to the desired DC output voltage. A voltage error amplifier, a DC reference voltage, and a pulse width modulator in the control chip modulate the $Q1$ "on" time in a negative feedback loop to yield a constant DC output voltage.

The instantaneous input line current is sensed by R_s and is controlled to be proportional to the instantaneous input voltage. During any one "on" time, current flows through $L1$, $Q1$, and R_s back to the negative end of the bridge, and during the following "off" time it flows through $L1$, $D1$, R_o and C_o in parallel, and R_s back to the negative end of the bridge.

By making $L1$ large, the peak-to-peak ripple current during each switching cycle is kept small. Depending on the switching speed of $Q1$,

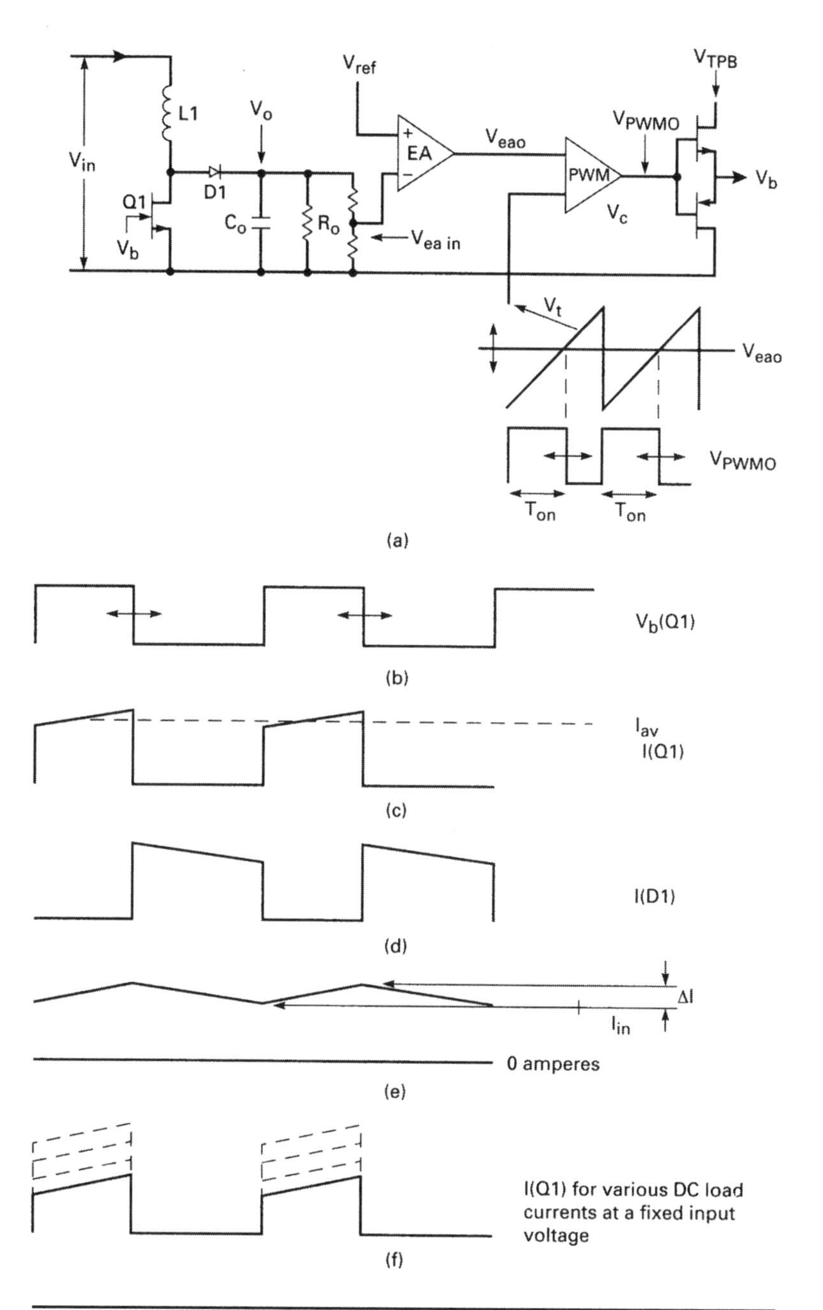

FIGURE 15.5 A continuous conduction mode boost converter. This shows a boost converter fed from a fixed DC input voltage. The circuit regulates against DC input voltage changes by varying *Q*1 "on" time. It regulates against load current changes by maintaining a fixed "on" time, while building up the average current delivered by *Q*1 over a number of cycles (Figure 15.5*f*).

there may be narrow spikes on the half sinusoids of current monitored in R_s (seen in Figure 15.3*b*). If present, these can cause an RFI problem, but a small capacitor across R_s can be used to minimize them.

15.3.2 Line Input Voltage Regulation in Continuous-Mode Boost Converters

Before we consider more details of the boost regulators shown here, it is of interest to see how a continuous-mode boost regulator corrects against line and load changes with a constant DC input voltage.

First, consider how the output/input voltage relation of Eq. 15.1 comes about. In Figure 15.5, the switch transistor $Q1$ is "on" for a time T_{on} and "off" for T_{off} out of the total period T. Neglect the "on" voltage drops of $Q1$ and $D1$. Since inductor $L1$ has negligible resistance, the voltage across it averaged over one switching cycle must be zero. Since the voltage at the top end of $L1$ is V_{in}, the voltage averaged over one cycle at the bottom end must also be V_{in}. This means that area $A1$ in Figure 15.6*a* must equal area $A2$. Since the top end of $L1$ is at V_o during T_{off}:

$$V_{\text{in}} T_{\text{on}} = (V_o - V_{\text{in}}) T_{\text{off}}$$
$$= (V_o - V_{\text{in}})(T - T_{\text{on}})$$

solving for V_o we have

$$V_o = \frac{V_{\text{in}}}{(1 - T_{\text{on}})/T}$$

which is the previously mentioned Eq. 15.1.

In Figure 15.5*a*, output voltage regulation against V_{in} changes is achieved by changing T_{on} with the pulse width modulator in accordance with Eq. 15.1. If V_{in} momentarily changes, so does V_o. A fraction of V_o is sensed and compared to a reference voltage V_{ref} by error amplifier EA to yield an error voltage V_{eao}. This error voltage is compared to a chip-generated triangle voltage V_t in voltage comparator V_c. The V_c output is a square wave which is high from the bottom of the triangle until it crosses error voltage output V_{eao}. While the V_c output is high, $Q1$ is turned "on" via a totem pole driver (TPD).

Thus, if V_{in} goes momentarily low, so do V_o and the EA inverting input. Then the V_{ea} output goes higher, the triangle V_t crosses the error amplifier output later, the "on" time increases, and V_o moves back up in accordance with Eq. 15.1. Conversely, if V_{in} goes high, V_o goes high, V_{ea} goes lower, T_{on} decreases, and V_o moves back down.

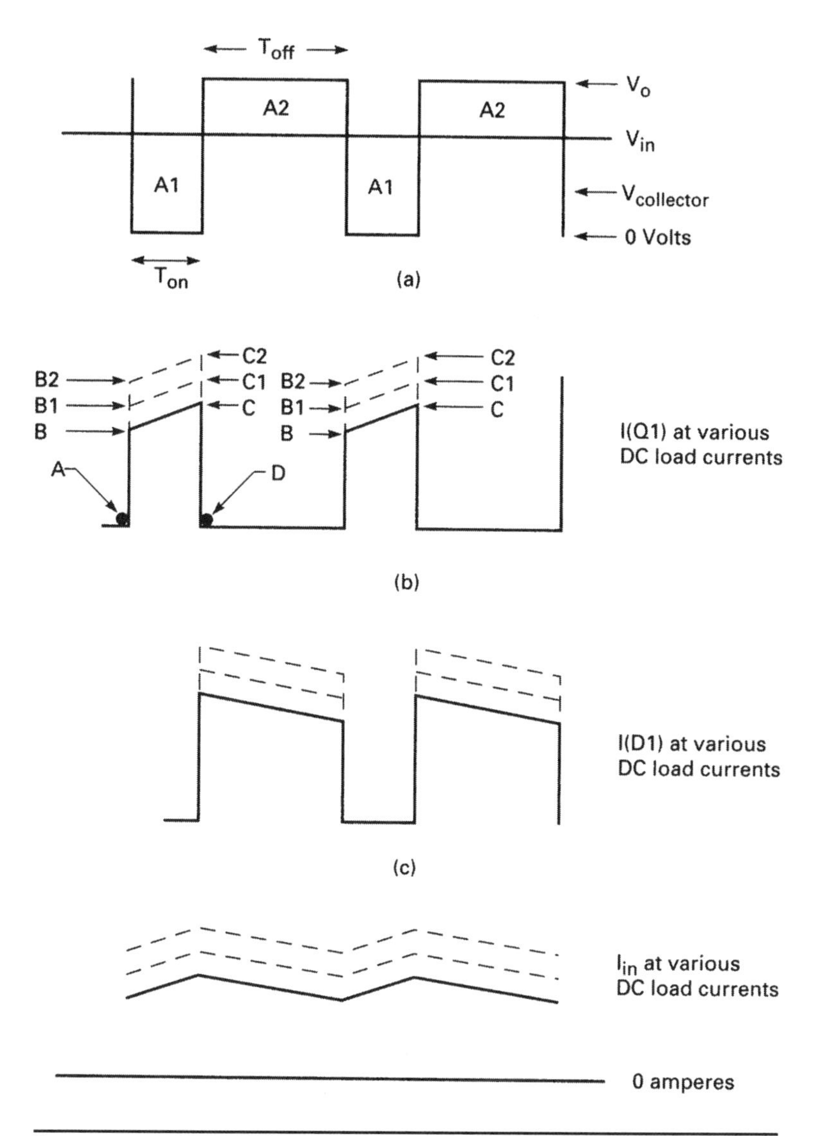

FIGURE 15.6 Showing regulation against load current changes in a continuous conduction mode boost converter.

15.3.3 Load Current Regulation in Continuous-Mode Boost Regulators

Continuous-mode boost converters also operate to correct for load current changes in a less obvious way. Referring to Eq. 15.1, note that V_o and T_{on} are independent of the load current. However, if the DC load current changes, the transistor and output diode currents must also change despite the constant "on" time. To achieve this the circuit responds to a change in load current in the following way.

Prior to, say, an increase in load current, assume the *Q*1 current is like *ABCD* in Fig 15.6*b*. For a small increase in steady state load current, *Q*1 current will move up to, say, *AB*1*C*1*D*. For a larger load current change, the *Q*1 current will move up to *AB*2*C*2*D*. To cause these changes, T_{on} changes over a few switching cycles but returns to its original value in the steady state. The steady-state current in diode *D*1 for these three different load currents is shown in Figure 15.6*c*. The output load current is the sum of I_{Q1} and I_{D1}, and its peak-to-peak ripple I_{or} can be made small as desired by increasing *L*1.

The "ramp on a step" waveforms (shown in Figure 15.6*b* and *c*) change over a number of switching cycles as follows (Figure 15.5*a*). If the DC load current increases, V_o goes down momentarily because of its source impedance. Then $V_{ea\ in}$ goes down, $V_{ea\ o}$ goes up, the V_t triangle crosses $V_{ea\ o}$ later in time, and T_{on} increases. Now the I_{Q1} current ramps up for a longer time to a higher value. Then I_{D1} starts later in time from a higher value and, with a shorter "off" time, has a higher value at the end of the "off" time. Hence the current I_t is larger at the start of next turn "on."

This progresses over a number of cycles with the average currents at the center of the I_{Q1}, I_{D1} ramps in Figure 15.5*c* and *d* increasing until they equal the increased DC load, at which time T_{on} and T_{off} slowly fall back to their initial values, as called for by Eq. 15.1. Thus for any changes in DC load, the T_{on} and T_{off} times temporarily change, but slowly relax back to their original values.

Thus, it can be seen that the bandwidth of the output voltage error amplifier must not be too large. If it were, it would respond too quickly and not permit the output voltage to shift for adequate time from its normal value at a fixed input voltage. This time must be sufficient for the above-described current buildup to occur over a number of switching cycles.

Various designs of PFC chips, available from a number of different manufacturers, will normally provide voltage and current sensing error amplifiers, error signal mixing, and width-modulated transistor turn "on" pulses so as to simplify the design of power factor correction circuits.

Adding power factor correction to a power supply entails removing the filter capacitor (C_o) of Figure 15.2*a*, and adding one of the available chips, together with a suitably designed boost inductor, a boost transistor, current-sensing resistor, and an output capacitor, as shown in Figure 15.3*c*, plus about half a dozen small resistors and capacitors.

After Pressman *There is a common misconception that the actual efficiency of the power factor–corrected supply is better than its uncorrected counterpart. This is not true. The designer and customer should be aware that due to the additional components, the actual power loss in the power factor–corrected power supply is normally greater than the uncorrected counterpart, so the temperature rise will be greater. The power savings are to be found in the external RFI filters, supply lines, and distribution equipment not in the actual power supply. If a true wattmeter is used to measure input power (such as a dynamometer wattmeter), the real input power will be greater for the power factor-corrected unit. ~K.B.*

15.4 Integrated-Circuit Chips for Power Factor Correction

A number of major manufacturers provide integrated-circuit (IC) chips to perform all the functions required for power factor correction. They more often are designed to support a continuous-mode boost regulator as described above, and use a scheme to sense and control the DC output voltage and input line current by width modulation of the boost "on" time.

The earliest and hence the most widely used of these chips, the Unitrode UC 3854 is typical of most of the others and is discussed here in detail. Other chips—the Motorola MC 34261 and MC 3426— are mentioned briefly. The Microlinear ML 4821 (now TI), the Linear Technology LT 1248, and the Toko 83854 Linear are similar in concept to the Unitrode UC 3854 but differ in important details. Hence the manufacturer's data sheet and application notes should be carefully studied in all new designs.

15.4.1 The Unitrode UC 3854 Power Factor Correction Chip

A simplified block diagram of the major elements of the chip is shown in Figure 15.7, based on Unitrode Application Note U-125 by Claudio de Silva. We will consider the functions of the various components as follows.

Transistor $Q1$, inductor $L1$, diode $D1$, and output capacitor C_o comprise the boost converter. A sawtooth voltage oscillator sets its switching frequency at $F_s = 1.25/(R_{14}C_t)$. Power switch $Q1$ is turned "on" and "off" by totem pole output drivers $Q2$ and $Q3$.

An "on" time commences when FF (flip-flop) is set by a narrow spike at the start of each sawtooth from the oscillator. The PWM resets the FF at the end of the "on" time, when the sawtooth at its

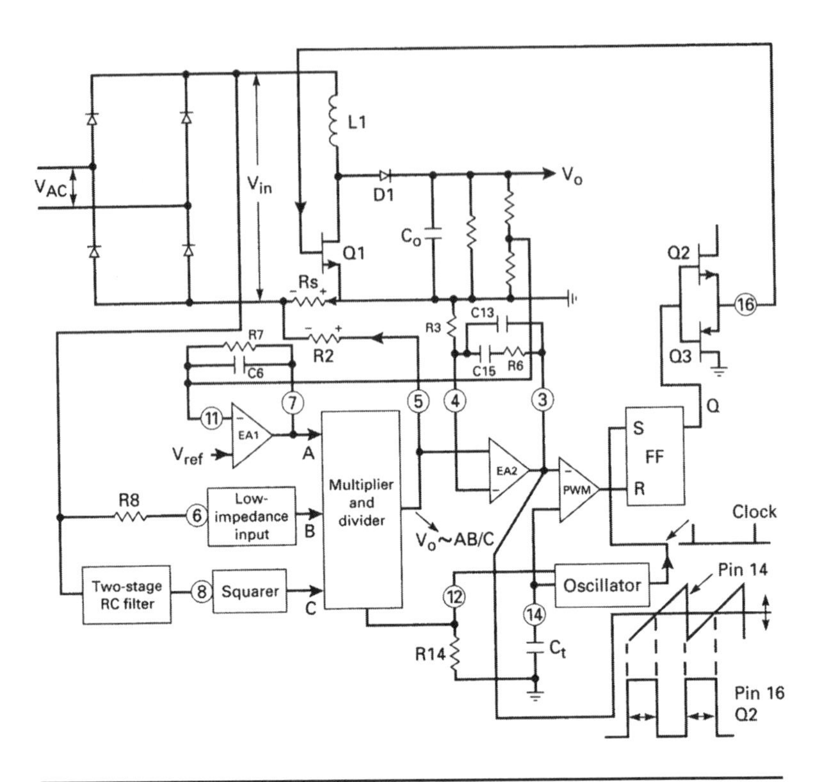

FIGURE 15.7 Showing a simplified block diagram of the Unitrode UC 3854 power factor correction chip.

noninverting input terminal crosses the voltage at the output (pin 3) of linear current amplifier EA2. Voltage at pin 3 is a noninverted, amplified version of the instantaneous difference between voltage drop across R_s and voltage rise across $R2$.

Width modulation of this "on" time by the PWM boosts the half sinusoids of input voltage from the bridge rectifier to a constant output voltage. It also forces the input line current to be accurately sinusoidal and in phase with the input line voltage.

15.4.2 Forcing Sinusoidal Line Current with the UC 3854

The current out of pin 5 is a continuous sequence of positive-going half sinusoids whose amplitude at any instant is proportional to the product of the DC voltage at point A and the current in pin 6. The input at pin 6 is a reference half sinusoid of current in phase with the half sinusoid line voltage after the bridge. Voltage at pin 5 is then a continuing sequence of half sinusoids in phase with the half sinusoids

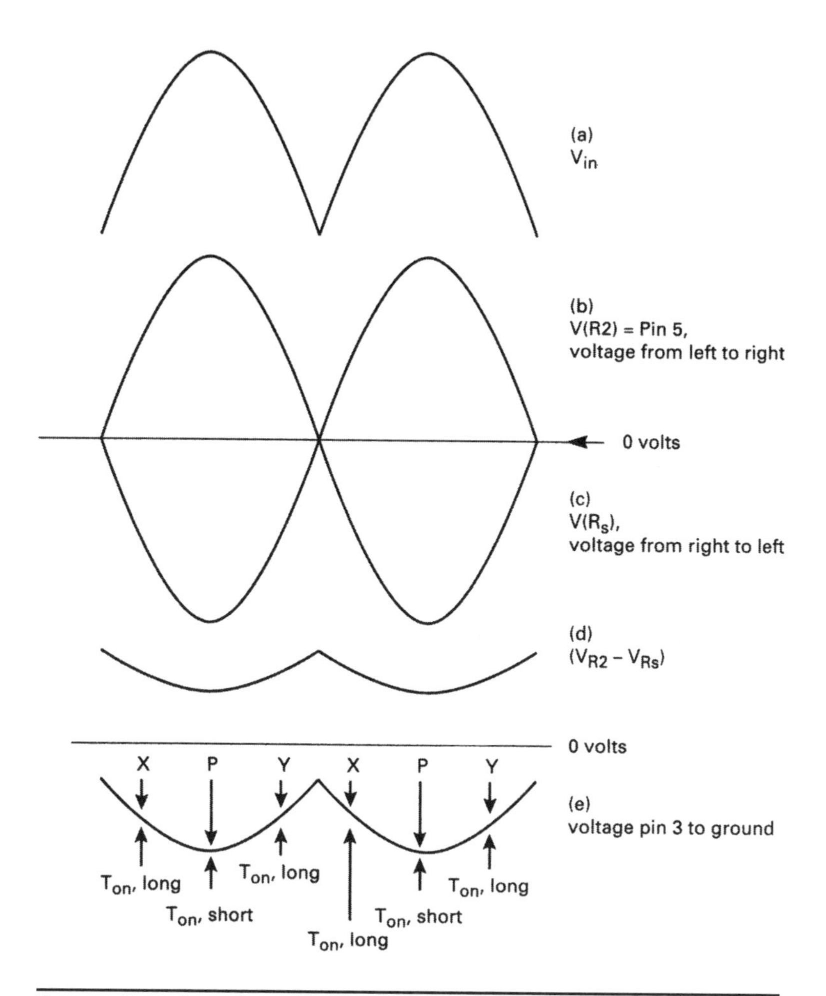

FIGURE 15.8 Critical waveforms in Unitrode 3854 power factor correction chip.

of line voltage the bridge output. The amplitude of the sinusoids is proportional to the voltage at the output of error amplifier EA1.

The line current is made sinusoidal by making it track the voltage throughout each half sinusoid. The voltage drop across R_S from right to left (Figure 15.8*c*) is very close to the voltage rise across *R*2 from left to right (Figure 15.8*b*).

The current in R_S is the rectified input line current. This rectified line current is equal to the sum of the *Q*1 current when it is "on" and the *D*1 current when *Q*1 is "off."

Hence when the voltage drop across R_S is forced to equal the voltage rise in R_2, the line current is also half sinusoidal and in phase with the voltage after the bridge rectifier. It can be seen from Figures 15.5*c*, *d* and *e* that since the boost regulator part operates in the continuous mode with a large inductor, the ripple current over one switching cycle is quite small. Thus, when the current is controlled such that the voltage drop across R_S is made equal to the voltage rise across $R2$ throughout the half period, and since the voltage across $R2$ is a smooth haversine, the line current in R_S will also become a smooth haversine with very little switching frequency ripple.

During the 60-Hz half cycle, the voltage rise across $R2$ is slightly higher than the drop across R_S, as the voltage drop across R_S is continually adjusting so as to keep up with the reference voltage rising across $R2$. This difference—the instantaneous error voltage—is shown in Figure 15.8*d*. It is a positive voltage with respect to ground throughout the half sinusoid, and is concave-upward. It is amplified by noninverting current amplifier EA2 and has the concave-upward waveshape shown in Figure 15.8*e*.

In the PWM comparator, the waveform at pin 3 is compared to the roughly 5-V peak triangle at pin 14. At points X and Y (Figure 15.8*e*), the triangle crosses the higher voltages there later in time, and the "on" time is long. At the sine wave peak (point P), the voltage level is lower and hence the triangle crosses it earlier in time and the "on" time is shorter. Thus during the half cycle, the cusplike waveform at pin 3 yields an "on" time that is maximum at the zero crossing, decreases as input voltage from the bridge rises toward its peak, and falls again as the input voltage falls toward the zero crossing. These varying "on" times boost the half sinusoid input voltage to the constant DC output voltage at C_o as called for by Eq. 15.1.

These "on" times (as dictated by the error voltage signal at pin 3) are the average taken over a few switching cycles. As the current demanded by the sinusoidal voltage across $R2$ changes, so must the ramp-on-a-step current pulses in R_S change. This occurs as discussed in Section 15.3.3 by momentarily altering the error voltage at pin 5 and hence at pin 3. The PWM comparator momentarily alters the "on" time so that the ramp-on-a-step current pulse flowing through R_S causes the voltage across it averaged over one switching cycle to equal the voltage across $R2$. After a few cycles when those voltages are equal, the "on" time relaxes back to the value required by Eq. 15.1 to boost the instantaneous input voltage to the desired constant DC output voltage.

15.4.3 Maintaining Constant Output Voltage with UC 3854

Without going into the details of the multiplier and divider, it is sufficient to say that the output voltage at pin 5 is simply the product of voltage at input point A and current at input point B.

Regulation against V_o changes takes place as follows (Figure 15.7.): Point A is the output of the V_o error amplifier, which compares a fraction of V_o to a fixed reference voltage. The voltage at point 5 is a sequence of distortion-free half sinusoids of voltage whose amplitude is proportional to the DC level at pin 7, the output of error amplifier EA1. Then if V_o goes up, say, the voltage at pin 7 goes down and the half sinusoid voltages at pin 5 get smaller in amplitude. The difference in error voltage between pin 5 and ground (Figure 15.8*d*) goes closer to ground, and so does the voltage at pin 3 in the PWM comparator. The sawtooth crosses the voltage at pin 3 earlier in time, the "on" time for each switching cycle does down throughout each half period, and in accordance with Eq. 15.1, V_o goes back up to the required value. So the output at pin 5 contains the information needed to keep both the output voltage V_o constant and the input line current sinusoidal.

Current into pin 6 is sinusoidal and in phase with the input voltage because the impedance at that point is low and the large resistor $R8$ is driven by the sinusoidal voltage after the bridge rectifier.

15.4.4 Controlling Power Output with the UC 3854

Figure 15.9 is the schematic for a 250-W power factor corrector using the UC 3854. The maximum obtainable output power is determined by setting the peak of the sinusoidal current I_{p1} that flows through sensing resistor R_S. This determines the maximum obtainable RMS line input current and hence the maximum obtainable output power at any RMS input voltage. Output power for the PFC circuit of Figure 15.9 is given in the following equation, in which lines above or below any terms indicate the maximum or minimum values for those terms.

$$P_o = EP_{\text{in}} = E\underline{V_{\text{RMS}}}\,\overline{I_{\text{RMS}}} = E\underline{V_{\text{RMS}}}(0.707 I_{p1}) \tag{15.2}$$

Where E is the efficiency, and I_{p1} is the current flowing in the current-sensing resistor R_s at its peak at V_{RMS}.

First I_{p1} is chosen from Eq. 15.2. Then R_s is selected for minimum dissipation at low line and maximum load with peak voltage drop at low line not less than 1 V. Then for, say, a 1-V peak drop across R_s,

$$R_s = \frac{1}{I_{p1}} \tag{15.3}$$

Now because of internal design details, the maximum current I_{pmd} available at MD output (pin 5) is fixed by

$$I_{\text{pmd}} = \frac{3.75}{R_{14}} \tag{15.4}$$

and I_{pmd} can be up to 0.5 mA but is usually set at 0.25 mA.

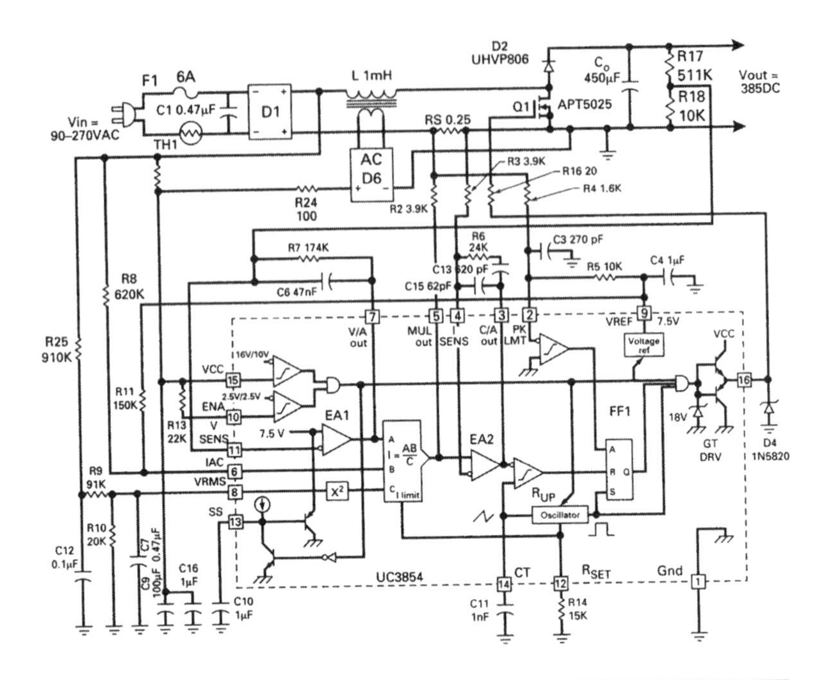

FIGURE 15.9 A detailed schematic for a 250-watt power factor controller using the Unitrode UC 3854 control chip. (*Courtesy of Unitrode Integrated Circuits Inc.*)

At every instant, the feedback loop keeps the voltage drop across R_s ($= R_s I_1$) equal to the rise across $R2$ ($= R2 I_{\text{md}}$). For maximum line and MD currents of I_{p1} and I_{pmd},

$$R2 = \frac{I_{p1} R_s}{I_{\text{pmd}}} \tag{15.5}$$

Thus for $P_o = 250$ W, with $\underline{V_{\text{RMS}}} = 90$ V, and $E = 0.85$: From Eq. 15.2, $\overline{I_{\text{RMS}}} = 250/0.85 \times 90 = 3.27$ A and $I_{p1} = 1.41 \times 3.27 = 4.61$ A. For a 1-V drop in R_s from Eq. 15.3:

$$R_s = \frac{1.0}{I_{p1}} = \frac{1}{4.61} = 0.22\,\Omega$$

We will select the closest standard value of .25 Ω. Then from Eq. 15.4 for an I_{pmd} of 0.25 mA,

$$R14 = \frac{3.75}{0.00025} = 15\,\text{k}\Omega$$

From Eq. 15.5, for I_{pmd} of 0.25 mA, I_{p1} of 4.61 A, and R_s of 0.25 Ω:

$$R2 = \frac{4.61 \times 0.25}{0.00025} = 4.61\ \text{k}\Omega$$

To minimize drift in EA2 (Figure 15.7), $R3$ is set equal to $R2$.

15.4.5 Boost Switching Frequency with the UC 3854

In addition to determining the current out of the MD output at pin 5, $R14$ also sets the boost switching frequency because of internal circuit details. Once $R14$ has been fixed, the boost switching frequency is fixed by

$$F_s = \frac{1.25}{R14C11} \tag{15.6}$$

where $C11$ is the capacitor to ground at pin 14. For $R14$ in ohms and $C11$ in farads, F_s is in hertz. The UC 3854 can be used up to somewhat above 200 kHz but is generally used at closer to 100 kHz.

15.4.6 Selection of Boost Output Inductor *L1*

Boost inductor $L1$ of Figure 15.5 or 15.7 is chosen for a desired minimum line ripple current at the peak of the sinusoidal input voltage. It is chosen at maximum input power and minimum input voltage, when the sine wave current peak $\overline{I_{p1}}$ is at its maximum. This ripple is ΔI; the ramp amplitude is either I_{Q1} or I_{D1} in Figure 15.5, and is the current change in $L1$ for minimum voltage $\underline{V_p}$ across $L1$ when the $Q1$ "on" time $\overline{T_{on}}$ is a maximum. Thus again for lines above and below the terms, signifying maximum and minimum values:

$$L1 = \frac{\underline{V_p}\,\overline{T_{on}}}{\Delta I} = \frac{1.41\,\underline{V_{rms}}\,\overline{T_{on}}}{\Delta I} \tag{15.7}$$

and

$$I_{p1} = \frac{1.41 P_o}{E\,\underline{V_{rms}}} \tag{15.8}$$

Arbitrarily choose ΔI as 20% of I_p.

$$\Delta I = \frac{0.2 \times 1.41 P_o}{E\,\underline{V_{RMS}}} = \frac{0.282 P_o}{E\,\underline{V_{RMS}}} \tag{15.9}$$

Then from Eqs. 15.7 and 15.9,

$$L1 = \frac{5.0\,(\underline{V_{RMS}})^2 E\,\overline{T_{on}}}{P_o} \tag{15.10}$$

and from Eq. 15.1,

$$\overline{T_{\text{on}}} = T\left(\frac{1 - \underline{V_P}}{V_o}\right) \qquad (15.11)$$

Now set V_o at 10% above $\overline{V_p}$—the sine wave peak at maximum RMS input voltage. Then

$$\overline{T_{\text{on}}} = T\left(1 - \frac{\underline{V_p}}{1.1\,\overline{V_p}}\right) \qquad (15.12)$$

Since $V_p/\overline{V_p} = \underline{V_{\text{RMS}}}/V_{\text{RMS}}$, taking $\underline{V_{\text{RMS}}} = 90$ V and $\overline{V_{\text{RMS}}} = 250$ V from Eq. 15.12,

$$\overline{T_{\text{on}}} = T\left(1 - \frac{90}{1.1 \times 250}\right) = 0.673\,T \qquad (15.13)$$

From Eqs. 15.10 and 15.13,

$$L1 = \frac{3.37(\underline{V_{\text{RMS}}})^2 TE}{P_o} \qquad (15.14)$$

Thus for $V_{\text{RMS}} = 90$ V, frequency $= 100$ kHz ($T = 10$ μs), $E = 85\%$, and $P_o = 250$ W, from Eq. 15.14

$$L1 = \frac{3.37(90)^2(10 \times 10^{-6})(0.85)}{250} = 928\,\mu\text{H}$$

15.4.7 Selection of Boost Output Capacitor

Refer to Figure 15.10. The boost capacitor C_o usually feeds a DC/DC converter—generally a half bridge for output powers under 600 W and a full bridge for higher powers. Recall that the nominal output voltage V_{on} is generally set at least 10% above the peak at maximum RMS input voltage V_{RMS}. Then for $\overline{V_{\text{rms}}} = 250$ V, $V_{\text{on}} = 1.1 \times 1.41 \times 250 = 388$ V. This voltage is not well regulated, as the voltage error amplifier gain bandwidth is kept low to improve response to changes in load current. Hence assume the minimum output voltage $\underline{V_o}$ is 370 V.

If AC input is lost at the instant V_o is at its minimum value, C_o should be large enough to hold up the output voltage to a value (V_{mhu}) still permitting all DC/DC converter outputs to remain within specifications for a time T_{mhu}. This time is often specified at 30 ms.

As V_o droops from $\underline{V_o}$ toward V_{mhu}, the DC/DC converter "on" time increases to maintain all its outputs within specification. A small droop of V_o would require a large capacitor, and excessive permitted drop would force an increased "on" time too close to the maximum permissible value of a half period for most converter topologies. A usual

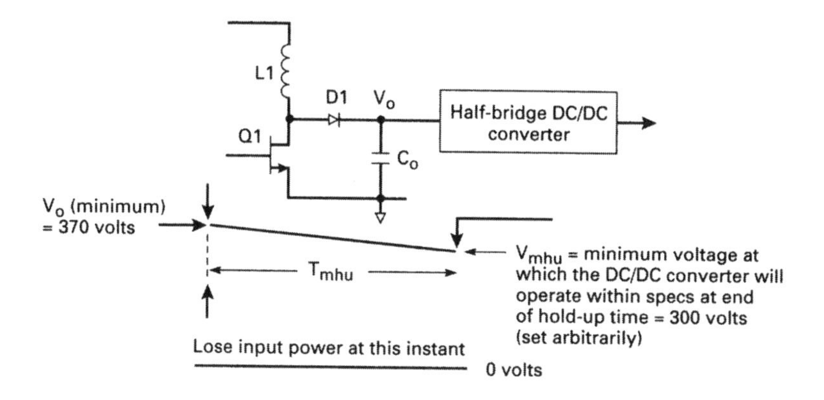

FIGURE 15.10 Showing how to select capacitor C_o to meet a specified hold-up time.

compromise is for V_{mhu} to be chosen 60V to 80V below $\underline{V_o}$ and to design the converter transformer to have sufficient secondary turns that the "on" time for an input of V_{mhu} is still only 80% of a half period (Section 3.2.2.1).

Thus C_o is selected using the following equation:

$$C_o = \frac{I_{\mathrm{av}} T_{\mathrm{hu}}}{\Delta V} = \frac{I_{\mathrm{av}} T_{\mathrm{hu}}}{\underline{V_o} - V_{\mathrm{mhu}}} \tag{15.15}$$

where I_{av} is the average output current during the droop from $\underline{V_o}$ to V_{mhu}. For converter output power of P_c and an efficiency of E_c,

$$I_{\mathrm{av}} = \frac{2P_c}{E_c(\underline{V_o} + V_{\mathrm{mhu}})} \tag{15.16}$$

Thus for $V_{\mathrm{mhu}} = \underline{V_o} - 70 = 370 - 70 = 300\ V$ and $T_{\mathrm{hu}} = 30$ ms, from Eq. 15.15,

$$C_o = \frac{I_{\mathrm{av}} \times 0.03}{370 - 300} = 429 \times 10^{-6} I_{\mathrm{av}}$$

For $P_c = 250$ W and $E_c = 0.85$, from Eq. 15.16,

$$I_{\mathrm{av}} = \frac{2 \times 250}{0.85(370 + 300)} = 0.88\ \mathrm{A}$$

$$C_o = 0.88(429 \times 10^{-6}) = 378\ \mu\mathrm{F}$$

We need only to select the closest standard value of 390 μF.

The transformer in the DC/DC converter must be designed so that the outputs remain within specification at the specified V_{mhu}. This will be safely achieved, as discussed in Section 3.2.2.1, if the number of secondary turns is sufficient to yield the required output voltages at an "on" time of 80% of a half period.

Capacitor C_o, in addition to being specified to yield the desired holdup time, must also have an adequate ripple current rating. It can be shown that in the boost diode $D2$ (Figure 15.7) the current consists of the DC load current component, plus a 120-Hz component whose peak amplitude is equal to the DC load current. The DC component flows to the load, but the 120-Hz component flows into capacitor C_o. Thus the RMS ripple current rating for C_o is $I_{RMS} = 0.707 I_{dc}$.

Thus for a DC/DC converter output of 250 W at $V_o = 388$ V and 85% efficiency, $I_{dc} = 250/[388(0.85)] = 0.76$ A, and the RMS ripple current rating for C_o is then $0.707 \times 0.76 = 0.54$ A.

15.4.8 Peak Current Limiting in the UC 3854

The peak limiting comparator at pin 2 and resistors $R4$ and $R5$ (Figure 15.9) provide peak current limiting. Note that the power transistor $Q1$ is turned "off" and no further current is taken from $Q1$ when the FF is reset. The flip-flop is reset when the output of the current limit comparator goes positive. This occurs when the DC voltage at the inverting input pin 2 of the comparator falls below the voltage at the non-inverting input. Resistors $R4$ and $R5$, fed from the +7.5-V reference at pin 9, provide an upward level shift so that as pin 2 falls to ground it limits at a peak current I_{1p} such that

$$I_{p1} R_s = I_{R4} R4$$

With pin 2 at ground

$$I_{R4} = I_{R5} = \frac{7.5}{R5}$$

Then for, say, $R5 = 10$ kΩ, $I_{R4} = 0.75$ mA

$$R4 = \frac{R_s I_{1p}}{I_{R_4}} \qquad (15.17)$$

For a P_o of 250 W, it was calculated that the peak current I_p was 4.61 A. Then for peak current limiting at 5.5 A, say, from Eq. 15.17, $R4 = 0.25 \times 5.5/0.00075 = 1.8$ kΩ.

15.4.9 Stabilizing the UC 3854 Feedback Loop

Stabilizing the feedback loops is beyond the scope of this discussion. Let us note that there are two feedback loops—a fast wide-bandwidth inner loop (EA2) which forces the input line current to be sinusoidal, and a slow low-bandwidth outer loop (EA1) that maintains constant output voltage.

The EA2 (Figure 15.7 or 15.9) is a type 2 linear amplifier (Section 12.6). It has a zero at $F_z = 1/(2\pi R6C15)$, a pole at $F_p = 1/(2\pi\ R6C13)$, and a pole at the origin at $F_p = 1/[(2\pi R3(C13 + C15)]$.

The voltage error amplifier EA1, in addition to maintaining constant DC output voltage, minimizes harmonic distortion of the 60-Hz line current by having low bandwidth, and low gain beyond the third harmonic of the line frequency.

Detailed design of the feedback loops is discussed in References 1 through 4.

15.5 The Motorola MC 34261 Power Factor Correction Chip

To demonstrate a different principle, we will consider the obsolete Motorola MC 34261 chip that was widely used in previous designs (see Reference 5); it is shown in Figure 15.11.

As with the Unitrode chip, it is designed to complement a boost converter that produces an output voltage somewhat higher than the

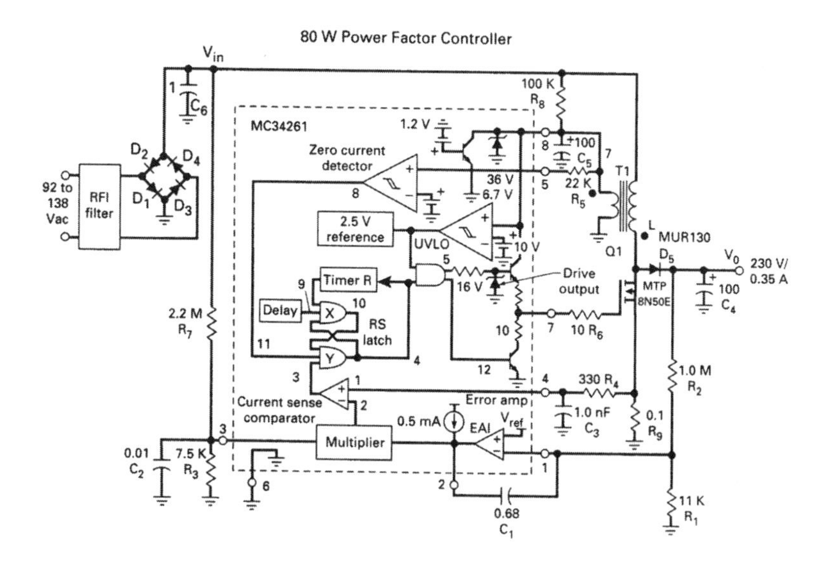

FIGURE 15.11 Motorola MC 34261 Power Factor Controller. It can be used for inputs of 85 to 265 V_{ac}. *(Courtesy of Motorola, Inc.)*

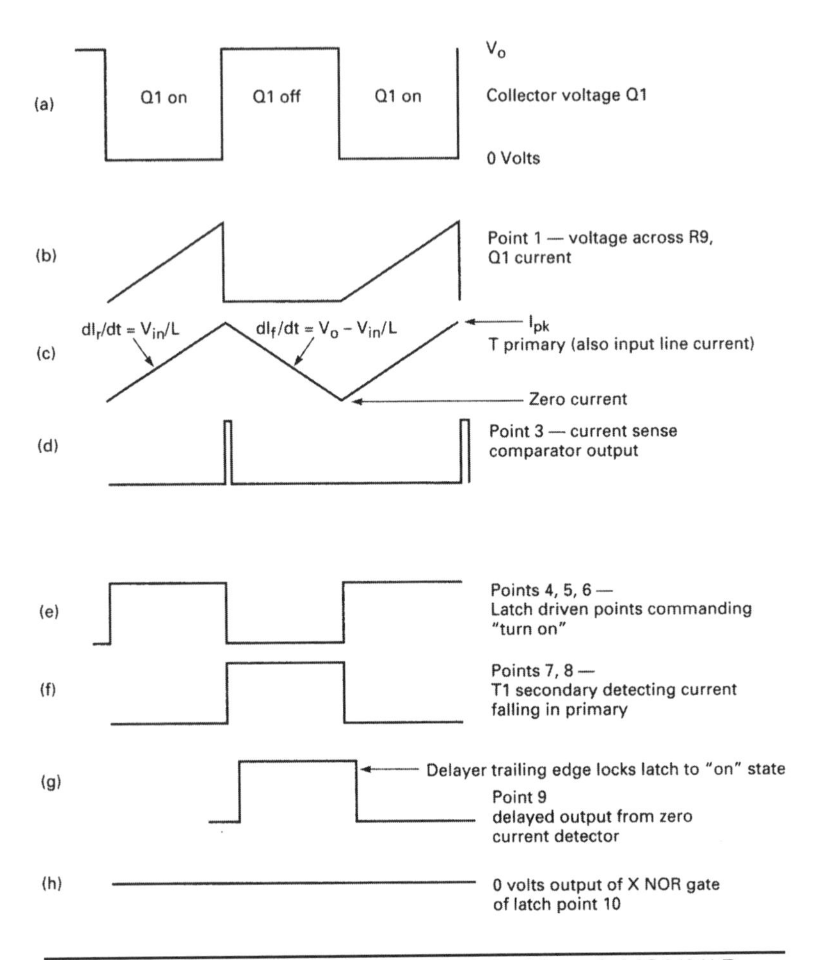

FIGURE 15.12 Showing the critical waveforms in Motorola MC 34261 Power Factor Controller, over one switching cycle within a 120-Hz haversine. Points refer to locations in the schematic shown in Figure 15.11.

peak of the incoming half sinusoids from the bridge rectifier. It also monitors input line current amplitude and forces it to track an internally generated reference haversine, at every instant. Unlike the Unitrode chip (which uses a fixed-frequency, continuous-mode boost converter as shown in Figure 15.4), it supports a boost converter operating on the edge of the discontinuous mode. This can be seen in Figure 15.12*c*, which shows the line current drawn through the *T*1 primary through one switching cycle. It ramps up to a peak when the switch transistor is "on" and falls back to zero when the transistor turns "off." There is no gap between the time the current has fallen to zero and the time of the start of the next upward ramp, hence the frequency is variable.

In this respect, it has the potential to generate more switching noise because of the very large up-and-down current ramps. To its advantage, the current is zero at the switching instant. In contrast, the continuous-mode boost of the Unitrode circuit has very flat current ramps, and over one switching cycle, the line current change is very small. Motorola and Unitrode claim similar third-harmonic distortion for their power factor controllers of equal output power from identical input power sources. Both suppliers quote power factors better than 0.99.

15.5.1 More Details of the Motorola MC 34261 (Figure 15.11)

The large filter capacitor, fitted after the input bridge rectifier in an uncorrected supply, is replaced by a very small capacitor ($C6$) which permits the voltage output from the bridge to follow the input line voltage in haversine fashion down to about 1 V above ground. This haversine must be clean and sinusoidal and must be in phase with the input line voltage. This waveshape, after multiplication by the EA1 output voltage sensor, is compared in the current sense comparator, to the line current haversines developed by $R7$. This forces the input line current to be sinusoidal, distortion-free, and in phase with the input line voltage.

The peak voltage at pin 3 must be kept less than 3 V to prevent the internal circuitry from distorting the haversines. The current into $R9$ is a sequence of discrete triangles (Figure 15.12*b*) whose peak is equal to the peak of the up and down triangles of Figure 15.12*c*. The average of the current triangle of Figure 15.12*c* during one switching cycle is the average of the input line current over that cycle. Thus by forcing the average of the current triangles, converted to voltage triangles by $R9$, to equal the instantaneous amplitude of the reference haversines at the multiplier output, the line current is forced to be sinusoidal and in phase with the input line voltage.

15.5.2 Logic Details for the MC 34261 (Figures 15.11 and 15.12)

Assume the RS latch has been set, and points 4 and 5 have gone high. Output pin 7 goes high, turning "on" main power switch $Q1$. The $Q1$ collector falls to ground (Figure 15.12*a*), and current in $T1$ primary starts rising at a rate $dI/dt = V_{in}/L$ (Figure 15.12*b* and *c*). When that current, flowing through $R9$, reaches a peak value (point 1) equal to the instantaneous voltage at the multiplier output (point 2), the current sense comparator output (point 3) goes positive and resets the latch and point 4 goes low. This pulls point 5 and chip output pin 7 low, turning "off" $Q1$.

Now the $Q1$ collector (Figure 15.12*a*) goes high to V_o, and current in the $T1$ primary falls at a rate of $dI/dt = (V_o - V_{in})/L$ (Figure 15.12*c*). As the dot end of the $T1$ primary went positive, so did the dot end of its secondary (point 7). Both inputs (9 and 11) to the RS latch are positive, so both its set and reset outputs (points 4 and 10) go low. Importantly, point 4 remains low as long as point 7, and hence point 8, remains high. The low at point 4 keeps point 12 high and point 5 low, so point 7 is low. This keeps $Q1$ turned "off" as long as point 7 is high, or as long as current still flows in the $T1$ primary.

When current in the $T1$ primary falls to zero (Figure 15.2*c*), the dot end of the secondary (point 7) falls to zero and so does the output of the zero current detector (point 8). With three "lows" at NOR gate Y, its output at point 4 goes high, driving point 5 high, and turning $Q1$ "on" again to repeat the cycle. The latches must be locked with point 4 high and point 10 low after point 7 falls low. This happens as follows.

There is a delay between points 8 and 9, so for a moment after point 8 falls to ground, point 9 still holds up, keeping point 10 low. After the delay, when point 9 falls low, point 4, being high, takes over at the input to NOR gate X and keeps its output low, locking the latch with point 4 high and point 10 low, until the current in $Q1$ has driven the voltage at point 1 above the instantaneous voltage from the multiplier at point 2.

15.5.3 Calculations for Frequency and Inductor *L1*

When $Q1$ in Figure 15.12*c* is "on," the voltage across $L1$ is V_{in}, and its current ramps up at a rate $dI_r/dt = V_{in}/L1$. When $Q1$ is "off," the voltage across $L1$ is $V_o - V_{in}$ and its current, which is the AC line input current, ramps downward at a rate $dI_f/dt = (V_o - V_{in})/L1$. If the current rises to a peak I_p in a time T_{on}, then before it is turned "on" again, it must fall the same I_p in a time T_{off}, or

$$\frac{V_{in}T_{on}}{L1} = \frac{(V_o - V_{in})T_{off}}{L1}$$

then

$$T_{off} = \frac{T_{on}V_{in}}{V_o - V_{in}} \tag{15.18}$$

Here V_{in} is the instantaneous haversine input voltage as time progresses. It simplifies matters to operate at a fixed "on" time and permit the "off" time to vary as in Eq. 15.18. Frequency $1/(T_{on} + T_{off})$ will then vary with the haversine voltage.

A reasonably small inductance L is chosen that can tolerate the peak current with the required output power. Large inductances (say, over 1 mH) that do not saturate at currents over 2 A are large and expensive. Then

$$P_{in} = \frac{P_o}{E} = \frac{\underline{V_{RMS}}\,\overline{I_{RMS}}}{E}$$

where current is a maximum for minimum voltage.

It follows

$$I_{rms} = \frac{P_o}{E\,\underline{V_{RMS}}} = 0.707 I_{pk} \text{ or } I_{pk} = \frac{1.41 P_o}{E\,\underline{V_{RMS}}}$$

where I_{pk} is the peak 60-Hz input line current at $\underline{V_{RMS}}$. As seen in Figure 15.12*c* that since the current averaged over one switching cycle is only one-half of I_{pkt}, then I_{pkt} must be $2I_{pk}$ or

$$I_{pkt} = \frac{2.82 P_o}{E\,\underline{V_{RMS}}} \tag{15.19}$$

This is the peak transistor ramp current at the peak of the input haversine at $\underline{V_{RMS}}$. When the transistor turns "on" at the peak of the haversine, its current ramps up in a time T_{on} to

$$I_{pkt} = \frac{\underline{V_p} T_{on}}{L}$$

where $\underline{V_p} = 1.41\,\underline{V_{RMS}}$. Then L is

$$L = \frac{1.41\,\underline{V_{RMS}}\,T_{on}}{I_{ptt}}$$

Then from Eq. 15.19, $I_{pkt} = 2.82 P_o / E\,\underline{V_{RMS}}$

$$L = \frac{(\underline{V_{RMS}})^2 T_{on} E}{2 P_o} \tag{15.20}$$

Now assume nominal, minimum, and maximum RMS input voltages of 120, 92, and 138 V. For $\underline{V_{RMS}} = 92$ V, $P_o = 80$ W, and $E = 0.95$,

$$L1 = \frac{(92)^2 \times 0.95 T_{on}}{2 \times 80} = 50 T_{on} \tag{15.21}$$

If a T_{on} of 10 μs is selected, $L1$ is 500 μH from Eq. 15.21, which is reasonably small for a peak ramp current of

$$I_{ptt} = 1.41 \times 92 \times 10 \times 10^{-6}/(500 \times 10^{-6} = 2.59\ A.$$

The boosted voltage must be above the sine wave peak at maximum line input. For $\overline{V_{RMS}} = 138$ V, the sine wave peak is $1.41 \times 138 = 195$ V. If it is not boosted far enough, Eq. 15.18 shows the "off" time will be many times the "on" time and frequency will be low. This large frequency change must be balanced against the higher voltage stress on the "off" transistor, and more recovery time losses in rectifier diode $D5$ for higher boost voltages.

Thus at low line of 92 V, the sine wave peak is $1.41 \times 92 = 129$ V. If this boosted to 50 V above the high-line peak of 195 V, Eq. 15.18 shows the time:

$$T_{off} = T_{on} \quad V_{in}/(V_o - V_{in}) = 10 \times 129/(245 - 129) = 11.1\ \mu s$$

This yields a period of 21.1 μs, or frequency of 48 kHz. At high line of $V_{AC} = 138$, where peak $V_{in} = 1.14 \times 138 = 195$ V, from Eq. 15.18

$$T_{off} = \frac{T_{on} V_{in}}{V_o - V_{in}} = \frac{10 \times 195}{245 - 195} = 39\ \mu s$$

The switching frequency F_s would then be $1/(T_{on} + T_{off}) = 20$ kHz. This change from 48 kHz at the peak of the 92-V AC line to 20 kHz at the peak of a 138-V line may not be desirable.

The frequency change during a haversine is much greater. Assume $V_{AC} = 92$ V and the output is boosted to 245 V—only 50 V above the peak at high line of 138 V. As shown above, this yields a switching frequency of 48 kHz. Now calculate the switching frequency close to the notch of the haversine, at the 10° point, say, V_{in} is 1.41×92 sin $10 = 23$ V. At that point, Eq. 15.18 shows $T_{off} = T_{on} \times 23/(245–23) = 0.1\ T_{on}$, and the switching frequency is $1/10.1 = 99$ kHz.

Such frequency variations with AC input voltages and haversine tracking at the 120-Hz rate could present an RFI/EMI problem as they represent a wide frequency spectrum. Even with these limitations, however, the MC 34261 chip has found widespread acceptance throughout the industry.

15.5.4 Selection of Sensing and Multiplier Resistors for the MC 34261

The Motorola data sheet recommends calculating current sense resistor $R9 = V_{cs}/I_{pk\ t}$, where $I_{pk\ t}$ is the peak transistor current (Eq. 15.19), and the current threshold V_{cs} is 0.5 V for $V_{AC} = 92$ to 138 V RMS.

From Eq. 15.19, for $P_o = 80$ W, $V_{RMS} = 92$ V, and efficiency of 95%

$$I_{pkt} = \frac{2.82 \times 80}{0.95 \times 92} = 2.58 \text{ A}$$

then

$$R9 = \frac{0.5}{2.58} = 0.19\ \Omega$$

The data sheet suggests the multiplier voltage (chip pin 3) be 3.0 V (V_M) at the haversine peak at high line. Thus,

$$3 = \frac{1.141\ V_{AC} R3}{R3 + R7}$$

And for $V_{AC} = 138$ V,

$$\frac{R3}{R7} = 0.016$$

This is suggested as a starting point. First set V_M at 3 V with a high value of $R7$ (say 1 MΩ) and vary $R3$ for the lowest distortion in the AC line current.

References

1. Claudio De Silva, "Power Factor Correction with the UC 3854," Application Note U-125, Unitrode Corporation.
2. B. Mammano and L. Dixon, "Choose the Optimum Topology for High Power Factor Supplies," *PCM*, March 1991.
3. Motorola Inc. data sheet, "MC 34261 Power Factor Controller."
4. Motorola Inc. data sheet, "MC 34262 Power Factor Controller."
5. N. Nalbant and W. Chom, "Theory and Application of the ML 4821 Average Current Mode Controller," Application Note 16, Micro Linear Inc.
6. Toko Inc. data sheet, "TK 83854 High Power Factor Pre-regulator."
7. Silicon General data sheet, "Power Factor Controller."
8. Keith Billings, *Switchmode Power Supply Handbook*, 2nd Ed., Part 4, McGraw-Hill, New York, 1999.

CHAPTER 16

Electronic Ballasts

High-Frequency Power Regulators for Fluorescent Lamps

16.1 Introduction: Magnetic Ballasts

In an age of "green energy" more efficient electronic ballasts (high-frequency efficient switchmode power regulators for fluorescent lamps) are becoming an increasingly high-volume market for switching power supplies. By as early as the mid-1980s, shipments of fluorescent lamps of all types exceeded 300 million annually. It is estimated the more than 30% of energy used in the U.S. goes toward lighting.

From the introduction of the fluorescent lamp in 1938 until the late 1970s, fluorescent lamps were driven directly from the 60-Hz power line via a series inductor or via a 60-Hz step-up transformer/inductor combination. The inductor or transformer/inductor element is referred to as *magnetic ballast* (Figure 16.1).

When the fluorescent lamp is fed from the 60-Hz power line, a "ballast" (a device or electronic circuit that provides current limiting) is required in series with the lamp. Current limiting is required because the lamp itself has a negative slope resistance in its working range. The magnetic ballast employs an inductor (as shown in Figure 16.1*a*) or the enhanced leakage inductance of a special autotransformer (as shown in Figure 16.1*b*).

In "preheat" lamps, the switch is momentarily closed to preheat the filament and then the switch is opened. The current established in the inductor results in a voltage spike that strikes the lamp.

The filaments of rapid-start lamps are powered up during and after starting. Instant-start lamps have no heated filaments and depend on field emission to supply the electron cloud required to start the lamp.

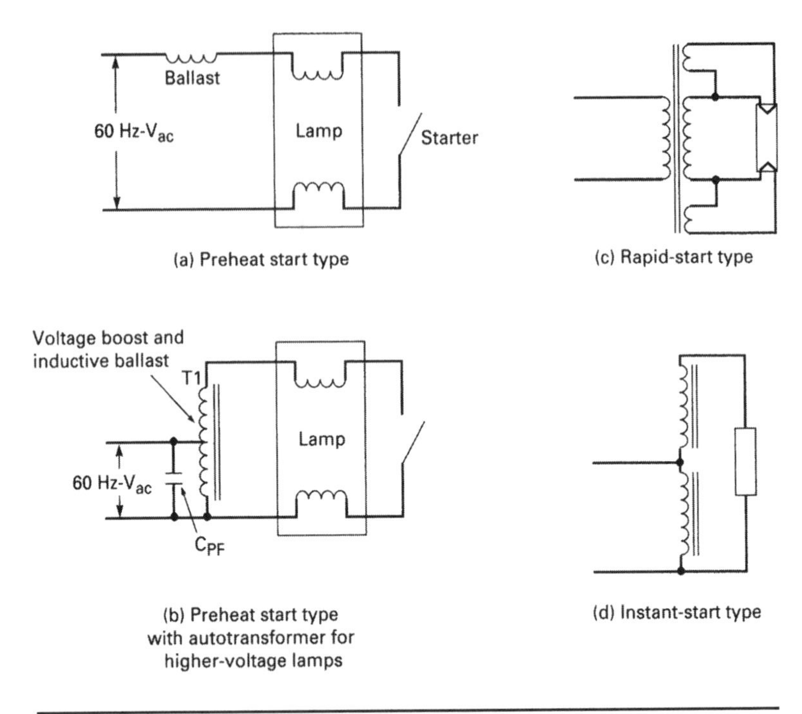

FIGURE 16.1 Examples of fluorescent lamps with magnetic ballasts.

This saves filament heating power, but the necessarily higher starting voltage for instant-start types results in stripping of the cathode material and shorter lifetimes.

The series inductor is essential for current-limiting, because of the lamp's negatively sloping resistive volt/ampere characteristic. This characteristic means that as the lamp input current increases, its voltage drop decreases. Hence it cannot be fed from a low impedance voltage source, because any momentary increase in lamp current would cause a decrease in lamp voltage, and the lower lamp voltage would cause a further increase in lamp current. This is a runaway condition that would continue until the lamp failed.

The fluorescent lamps shipped from 1938 to the mid-1980s required an equal number of magnetic ballasts that are inexpensive in large quantities. The major problem with low-cost magnetic ballasts, however, is low efficiency due to power loss in the choke windings and laminations. Also, the lamp efficiency is low because the lamp extinguishes at each zero crossing of the 60-Hz waveform, and this produces noticeable flicker. Flicker decreases the average light intensity, and is dangerous where rotating machinery is in use due to strobe effects. Also, the iron laminations tend to vibrate and cause an audible buzz, so they are potted in tar or varnish to suppress the noise, which is a fire and disposal problem. Finally, the typical weight of such a ballast is 3 to 4 lb, and although this has been tolerated in the industry, it is a significant shortcoming.

These drawbacks led to the development of electronic high-frequency alternating current ballasts to power the lamp. The advantages are increased lamp efficiency (measured as light output power in lumens per watt of input power), smaller size, and lower weight.

Early experiments with capacitor ballasts showed that lamp efficiency increased with increasing frequency up to about 20 kHz, and then leveled off at about 14% (Figure 16.2). Also the capacitor ballast was smaller and lighter, had no audible noise, and was less expensive. At high frequency the lamp showed no flicker, and conducted and radiated EMI were easier to suppress. The advantages of high-frequency operation, though significant, could not be fully realized until a practical, inexpensive method of generating a high-frequency source was available. Small SCR inverters for each lamp were considered, as well as larger ones or even high-frequency rotating AC generators for the large banks of lamps in factories or office buildings. With the rapid drop in transistor and ferrite core prices, DC/AC inverters powered from the rectified AC power line for each lamp or two-to-four lamp assembly became possible, instead of a large central high-frequency power source.

In recent years, most of these goals have been achieved with high-frequency electronic ballasts, but their cost is still greater than that of the equivalent magnetic ballasts. With the lower operating cost from improved efficiency, and longer lamp lifetime, the high-frequency electronic ballast is now increasingly displacing magnetic types in new installations. Further, since electronic ballasts can replace magnetic ones in old installations without replacing the fluorescent lamps, there is a large market in replacing the enormous number of magnetic ballasts in large office buildings and factories.

It is estimated that the cost of replacing an older magnetic ballast with an electronic ballast is recovered through reduced power cost in about 1 year. The total reduction in power cost with the new electronic ballasts can amount to 20 to 25% when all factors are considered. Contributions to lower power cost come from improved lamp efficiency due to flicker elimination and the ability to operate lamps at higher current, and the inherently higher efficiency of a capacitor ballast over a magnetic one.

Finally, in 2008 many governments began legislating the replacement of incandescent lighting with high-efficiency electronic ballasts and CFLs (compact fluorescent lamps) for both domestic and industrial lighting. There is an additional advantage where air conditioning is in use, as the improved efficiency of the lighting reduces the demand

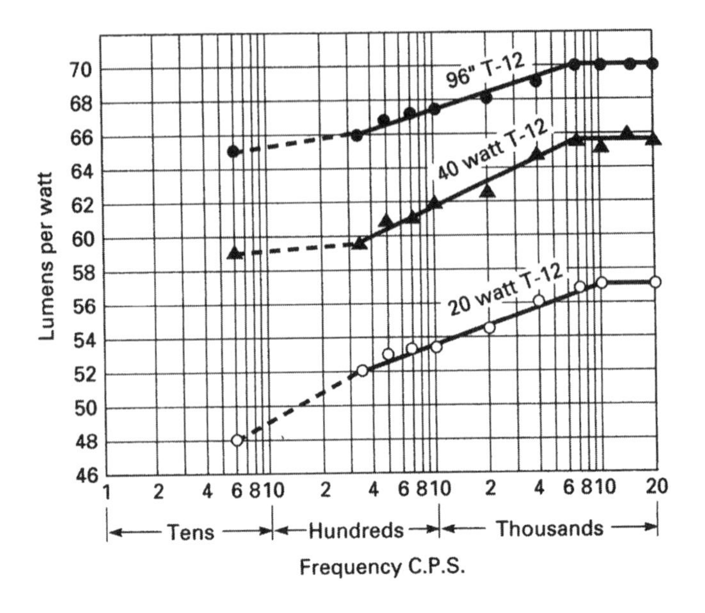

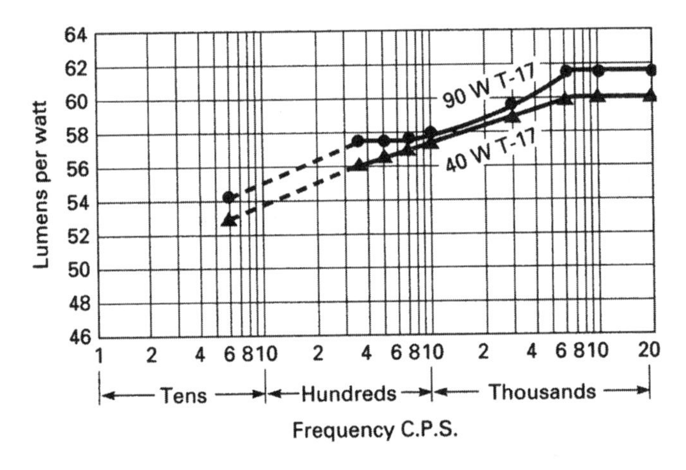

FIGURE 16.2 Fluorescent lamp light output in lumens per watt for T12(1.5-in lamp diameter) and T17(1.88-in lamp diameter) versus frequency. *(From Campbell, Schultz, Kershaw, "High-Frequency Fluorescent Lamps,"* Illumination Engineering, *February 1953.)*

on the air conditioning system. The next step will probably be towards solid state lighting.

One major manufacturer reports that after replacing magnetic ballasts with electronic ones in a building with 960 eight-foot fluorescent lamps, each fixture then drew only 87 W rather than 227 W for the same light output. At 6,000 hours of lamp operation per year and $0.106 per kilowatt hour, the annual savings in power cost for the 960 lamps amounted to $86,000. Even taking into account the higher costs of electronic ballasts and the changeover labor, the manufacturer recovered costs in 1 year. The manufacturer also estimates another $8,814 saving in decreased burden on the air-conditioning system.[2]

16.2 Fluorescent Lamp—Physics and Types

Most fluorescent lamps come in tubes of standard diameters, lengths, and wattage ratings. Diameters come in increments of 1/8 in, and standard lengths are 2, 3, 4, and 8 ft. The various manufacturers all use a four-character type code *FPTD*, in which *F* designates fluorescent, *P* designates input power in watts, *T* signifies tubular, and *D* is diameter in eighths of an inch. Thus an F32T8 is a 32-W, 1-in-diameter lamp, and an F40T12 is a 40-W, 1.5-in-diameter lamp. Fluorescent lamps are also available in circular and U shapes. Electronic ballasts available from the various manufacturers are designed to operate the various lamp types from the same line voltages as their original magnetic counterparts.

Fluorescent lamps consist of glass tubes in the above-described standard lengths, diameters, and shapes and are filled with argon or krypton gas at low pressure. A small amount of liquid mercury, which vaporizes when heated by the low-energy arc in the gas, is also enclosed (Figure 16.3).

Lamps are coated inside with various phosphors, which emit the desired visible light when irradiated with ultraviolet light generated by a high-current mercury arc flowing through the lamp. The current is drawn through the lamp by a high voltage applied across the electrodes at each end.

These electrodes are passive, unheated coils of wire in instant-start lamps, or oxide-coated filaments in rapid-start lamps.

Before the electrode voltage is applied, there are relatively few current carriers in the lamp gas, as none of the gas molecules are ionized. Current carriers are supplied in quantity when a few free electrons in the gas are accelerated to high speeds by the voltage applied across the end electrodes. An accelerated electron colliding with a neutral gas atom ionizes it, providing a free electron and a massive positively charged ion. The released electron that is accelerated toward the anode, and the positive ion toward the cathode, now produce more *ionization by collision*. Each collision produces more current carriers, and each current carrier causes more ionizing collisions. The result is an avalanche of current or arc.

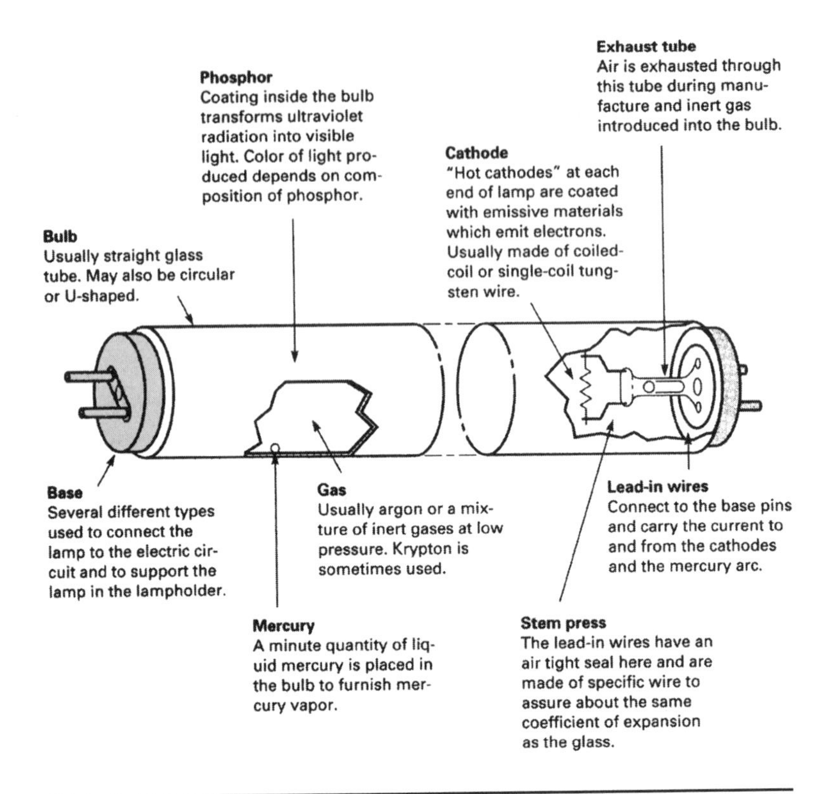

FIGURE 16.3 A typical rapid-start fluorescent lamp. Heated filaments at each end emit free electrons to permit starting the lamp at lower voltages. Instant-start lamps do not use a heated filament to supply the starting electrons. Instead, they use a higher voltage across the end electrodes to supply the electrons by field emission. This saves filament heating power but results in shorter lamp lifetime. (*Courtesy GTE/Sylvania, "Fluorescent Lamps."*)

Instant-start (Figure 16.1*d*) lamps do not have a large initial supply of free electrons to start the above process. They depend on the few free electrons produced by cosmic rays and a large voltage gradient between the end electrodes to start the ionization-by-collision process. They consequently require a high electrode voltage to start or "strike" the arc, but start instantly on application of the voltage.

Rapid-start (Figure 16.1*c*) lamps have current-carrying, oxide-coated filaments, which supply a large reservoir of electrons to initiate the arc. They require a lower accelerating voltage to light the lamp, and do not light as quickly as instant-start types, but are acceptably fast. The filaments of rapid-start lamps remain powered after the lamp starts.

Older "preheat" lamps had filaments that were heated at turn "on," but used a starter device to turn off filament power automatically after the lamp lighted, to conserve power.

Instant- and rapid-start lamps have their pluses and minuses. The rapid-start type is slower in starting, but requires lower starting voltage than the instant-start type. Its filament makes it more expensive to manufacture and it needs a source of filament power, which may require a separate filament transformer or winding on the power transformer. The instant-start lamp, with its higher striking voltage, strips material from the cathode at each start. After many starts, this darkens the ends of the lamp and reduces its lifetime. Considering the infrequent starts in its usual applications, its lower cost makes it competitive. Both types are commercially available, and require ballasts tailored to their characteristics.

The purpose of the low-pressure argon or krypton gas in the lamp is to achieve faster lamp starting. Initially, the ionization by collision is started in those gases. As the temperature rises due to the arc, the mercury vaporizes and ionization by collision of its atoms produces significantly more electrons and positively charged ions. More importantly, the mercury vapor atoms produce the ultraviolet light that stimulates the phosphors to emit visible light.

Electrons driven to the anode, and positively charged mercury ions driven to the cathode, collide with other mercury atoms and excite some of their orbital electrons to various higher energy levels. On falling back to their original energy levels, radiation is emitted at frequencies corresponding to the energies of the transitions. Transitions across lower-energy differences produce relatively long-wavelength, visible light; those across higher-energy differences produce short-wavelength ultraviolet light.

One particular transition produces high-energy ultraviolet light at a wavelength of 2537 angstroms ($1\ \Delta = 10^{-8}$ cm). This is below the visible spectrum of about 4000 to 7000 Δ, but the high energy of this short wavelength is very effective in producing visible light from the phosphors.

Figure 16.4 shows the spectral distribution of energy in watts per nanometer ($1\ \text{nm} = 10\Delta$) from a 40-W white fluorescent lamp. The smooth curve represents the continuous spectrum emitted by the phosphor powder, stimulated by the ultraviolet light. The 10-nm wide discrete bands represent radiation emitted by the mercury atoms in transitions between low-energy differences. These transitions produce visible light but are not as effective in generating light from the phosphors as the mercury atom energy transitions at 2537 Δ. Most of the visible light output comes from the phosphors stimulated by the ultraviolet light.

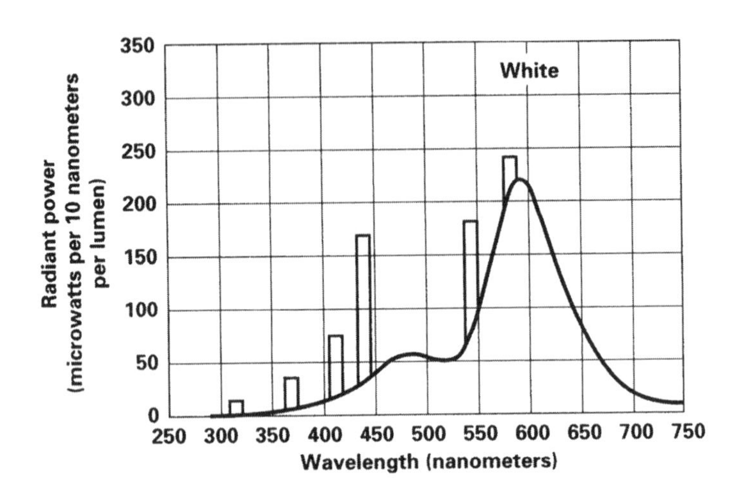

FIGURE 16.4 Spectral energy distribution from a 40-W white fluorescent lamp, in microwatts per nanometer (1 nm = 10Δ). The smooth curve is the continuous spectrum of energy generated by the white phosphorous. The 10-nm wide discrete bands represent energy generated by the mercury atoms in transition from a high to a low energy level. (*Courtesy "General Electric Bulletin Fluorescent Lamps."*)

16.3 Electric Arc Characteristics

Further details of an electric arc in a gas are presented here. Although not essential, they are of value to the electronic ballast circuit designer in making some design decisions. The nature of electrical conditions in gases began to be studied intensively at the end of the 19th century. This led to understanding the nature and properties of the electron and atomic structure, and to the use of X rays for medical diagnosis.

In 1989, Paschen studied the DC voltage required to initiate a spark between a pair or electrodes in air as a function of pressure. He used spherical electrodes of diameter larger than the electrode separation (Figure 16.5), to avoid high-voltage gradients in the vicinity of sharp points or edges. His result—the well-known Paschen's law—is shown in Figure 16.6. For an electrode spacing of 0.3 to 0.5 cm at atmospheric pressure, the voltage required to initiate a spark is close to 1000 V. As pressure is decreased, the sparking potential falls continuously toward a minimum of about 300 V and then rises again steeply. Gases other than air exhibit the same general characteristics—the minimum or critical pressure may be different for other gases.

Paschen's law offers an experimental explanation of the phenomena described above—ionization by collision. At high pressures, the mean spacing between neutral atoms is so small that an ion—an electron or a

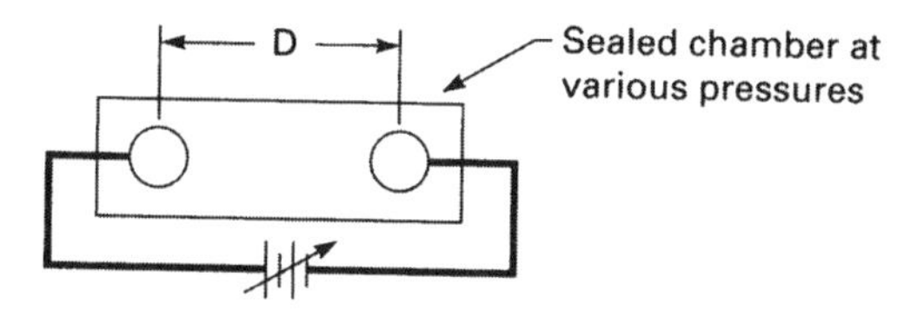

FIGURE 16.5 Paschen's classic experiment for measuring sparking potential between a pair of spherical electrodes at various spacings as a function of air pressure in the chamber.

positive ion—collides with neutral atoms before it can be accelerated to a velocity sufficient to ionize them. As pressure is reduced, the mean free path between atoms is increased, and accelerating electrons or positive ions can travel longer distances and gather speed before being slowed down by collisions. When they finally do collide, their energy is sufficient to ionize the neutral atoms, there is an avalanche of charge carriers, and an arc occurs.

16.3.1 Arc Characteristics with DC Supply Voltage

As early as the late 19th century, physicists studied the visible appearance of an arc discharge with DC voltage at the electrodes. Their original experiments used solid electrodes—not heated, electron-emitting cathodes—at each end of a glass tube. Several hundred volts were applied across the electrodes through a current-limiting resistor.

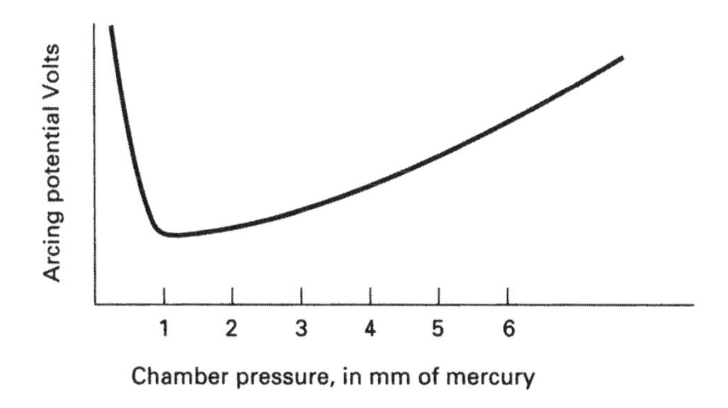

FIGURE 16.6 Classical Paschen curve, showing required potential between electrodes to start an arc discharge. Minimum arcing potential is about 300 V at 3 to 5 mm of spacing in air.

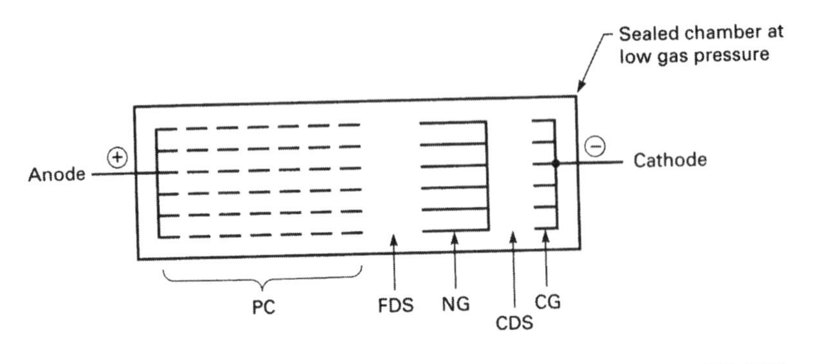

FIGURE 16.7 Pattern of light and dark regions in an arc discharge.

As the air pressure inside the tube was decreased sufficiently, physicists observed the following pattern of light and dark regions stretching out from the cathode to the anode (Figure 16.7). Starting close to the cathode, there was a short glowing region CG followed by a longer dark segment CDS. This was followed by a longer glowing region NG, then an equally long dark region FDS. Between the FDS and the anode, there were alternate bands (PC) of luminosity separated by dark spaces. These acronyms are from *cathode glow* (CG), *Crookes' dark space* (CDS), *negative glow* (NG), *Faraday dark space* (FDS), and *positive column* (PC).

This was explained in subsequent years as follows:

As the minimum pressure on the Paschen curve was approached, stray free electrons (from cosmic rays or a high-voltage gradient close to the cathode) were accelerated sufficiently to ionize neutral gas atoms. The resulting positive ions, being massive, do not move very rapidly or far from the cathode, and build up a positive space charge and hence a high-voltage gradient near the cathode. This voltage gradient—nowadays called the *cathode fall*—accelerates positive ions and drives them into the cathode, sputtering off some of its material.

When neutral or ionized mercury atoms close to the cathode are bombarded with electrons of sufficient energy, some of their orbital electrons absorb this energy and are driven up to higher energies within the atom. When these electrons fall back to their initial energy level, they emit the visible light seen in the CG region.

At the outer edge of the cathode glow, all the electrons on their way to the anode have given up their energy and slowed to the point where they can no longer excite atoms to higher energy levels. Throughout the Crookes' dark space, they are being accelerated again, and at the edge of the negative glow region they again have sufficient energy to excite atoms to higher energy levels. Throughout the negative glow, those atoms emit visible radiation as they fall back to their initial energy states. This happens again as the electrons speed up through the Faraday dark space, where they do not have sufficient energy to excite the atoms to luminosity.

And at the start of the positive column, there is again a luminous region. Thereafter, throughout the positive column, there are alternate dark and luminous regions. The dark spaces are the speed-gathering regions; the bright ones occur when the electrons have achieved enough energy to stimulate the atoms to emit visible light on their fallback to their initial energy level. Most of the voltage applied across the electrodes is dropped across the positive column, which occupies 80 to 90% of the distance between them.

16.3.2 AC-Driven Fluorescent Lamps

Electrodes in a fluorescent lamp are driven by an AC voltage—60 Hz with a magnetic ballast, and above 20 kHz for an electronic ballast. Thus throughout a cycle, each end of the lamp is alternately an anode and a cathode, and the above pattern of bright and dark regions flips and flops right and left so that the brightness striations are not visible in the phosphor coating.

The secondary of a flyback transformer can be used to power a fluorescent lamp. This appears attractive at first, as the output voltage of a flyback secondary after the rectifying diode is unlimited in amplitude and is easily high enough to strike an arc in most fluorescent lamps. After the arc is struck, the lamp voltage falls back to its operating voltage, which is generally in the range of 100 to 300 V.

This would shorten the lamp lifetime because the same end of the lamp would always be the cathode and that end would quickly darken. It has been pointed out that there is a large voltage gradient at the cathode because of the positive space charge accumulating close to it. This voltage gradient would drive heavy positive ions into the cathode, sputtering away material, and would soon darken the lamp close to the cathode. Thus alternating anode and cathode at opposite ends of a fluorescent lamp is advantageous.

A fluorescent lamp driven from a high-frequency source produces more light and is easier to start than one driven from a 60-Hz source at the same input power level. At 60 Hz, there is no voltage across the lamp at the zero crossings of the input sine wave. The lamp thus extinguishes, and the arc must be restruck twice per cycle after the zero crossings. This lowers the average light output, especially at low temperatures, and makes continuous restarting necessary. When driven at frequencies above 20 kHz, the ionized atoms do not have time to recombine at the zero crossings, and the lamp does not extinguish, but maintains its light output.

The greater dissipation in a lamp driven from a standard 60-Hz magnetic ballast than in a 25-kHz electronic ballast can be seen in

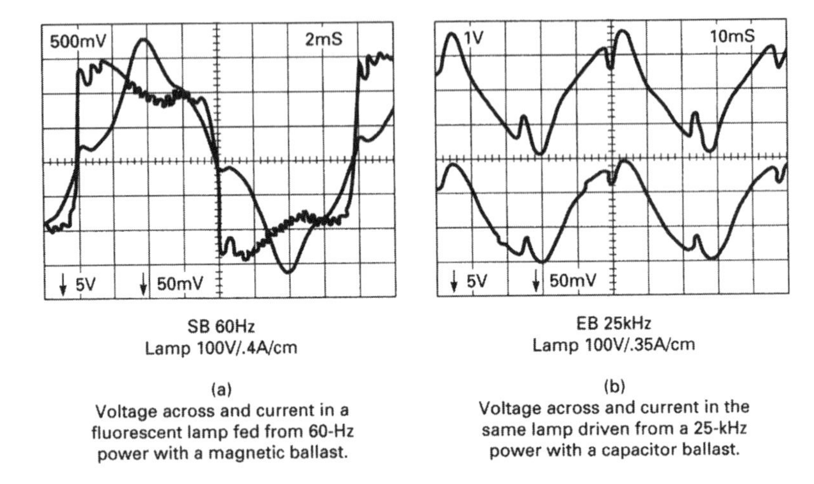

FIGURE 16.8 Lamp waveforms. At every zero crossing of the current waveform with a 60-Hz ballast, voltage across the lamp rises steeply. At the zero crossing, the lamp extinguishes, its impedance increases, and the voltage across the lamp must rise to a high value to reignite the lamp. With a high-frequency power source, the lamp never extinguishes at the zero crossings, and the lamp voltage is instantaneously proportional to the lamp current. With a 60-Hz power source, the lamp extinguishing at every zero crossing results in diminished light output power efficiency. (*From R. J. Haver,* Power Conversion/Intelligent Motion, *April 1987.*)

Figure 16.8*a* and *b*. In Figure 16.8*a* the lamp is driven by a magnetic ballast, where shortly after the zero crossing of the current, the lamp voltage rises steeply to ignite the lamp.

These high-voltage episodes shortly after the zero crossings waste power. In contrast, Figure 16.8*b* for a 25-Hz electronic ballast shows there are no such high-voltage intervals immediately after the current zero crossings. Also with the electronic ballast, voltage and current waveforms are fairly sinusoidal and in phase. Further, in the 60-Hz magnetic ballast, the *crest factor,* or ratio of peak to RMS current, is much higher than that for the 25-kHz electronic ballast. It has been widely reported in the literature[3] that high current crest factors yield poor lamp efficiencies. A perfect sine wave has a crest factor of 1.41.

It is estimated (Figure 16.9) that in a 40-W fluorescent lamp at 60 Hz with a magnetic ballast, 23% of the input energy or 9.3 W is converted to visible light via the conversion to ultraviolet light and the consequent stimulation of the lamp phosphors, while 41% or 16.3 W is converted to convected and conducted heat, leaving 36% or 14.4 W radiated as infrared energy. For comparison, a 300-W incandescent

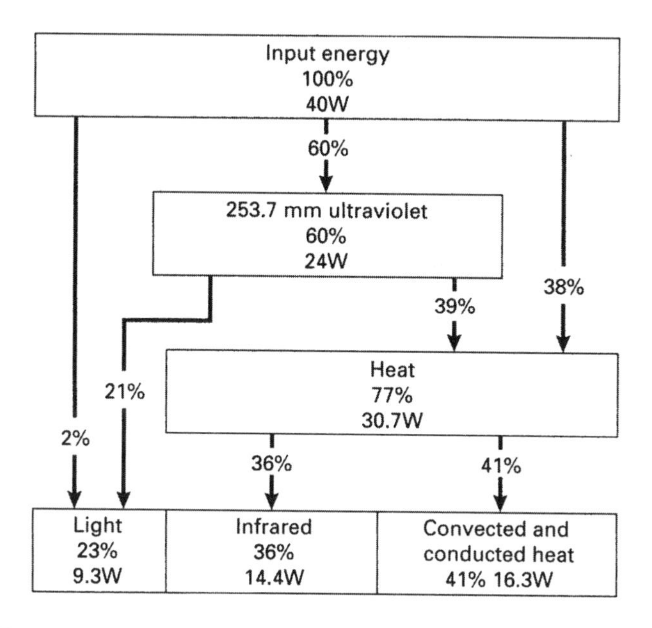

FIGURE 16.9 Distribution of energy in a 40-W fluorescent lamp. (*Courtesy GET/Sylvania Bulletin, "Fluorescent Lamps."*)

light bulb yields 11% of its input power as visible light and 89% as heat. In terms of light efficiency, a fluorescent lamp delivers 75 lm/W, or up to 90 to 100 lm/W for the newest lamps with electronic ballasts[4], compared to 18 lm/W for an incandescent lamp.

16.3.3 Fluorescent Lamp Volt/Ampere Characteristics with an Electronic Ballast

Before a fluorescent lamp lights, it has high impedance as there are few current carriers. It takes a high voltage—V_{ns}, the *nominal striking voltage*—to light the lamp. After lighting, the voltage across it with an electronic ballast falls to a lower *operating voltage* V_{op}. The operating current I_{op} drawn by the lamp is largely determined by X_b, the impedance of the ballast at operating frequency, and is given by

$$I_{op} = \frac{V_{ns} - V_{op}}{X_b} \tag{16.1}$$

The voltages and currents here are the RMS values, and hence the actual power drawn by the lamp with an electronic ballast is

$$P_{in} = V_{op} I_{op} \tag{16.2}$$

because the voltage and current are proportional and in phase.

The nominal striking voltage V_{ns} is given by the lamp manufacturer, usually at 50°F. To ensure lighting the hardest-to-start lamp, the striking voltage at its minimum should be about 10% higher to accommodate manufacturing tolerances. The American National Standards Institute (ANSI), in "Fluorescent Lamp ANSI Specifications,"[5] sets the values for V_{op} and I_{op} for each specific lamp type so that the product meets the maximum wattage rating for that lamp.

Thus with V_{op} and I_{op} from the ANSI specifications, and V_{ns} from the manufacturer's data sheet, Eq. 16.1 fixes the ballast impedance X_b. For electronic ballasts in which the ballast is a capacitor, its value is calculated from

$$X_b = \frac{1}{2\pi f C_{bT}} \tag{16.3}$$

where C_{bT} is the value of the effective capacitance in series with the lamp, as the lamp may be driven from the junction of two capacitors.

Note that by choice of the ballast impedance, a lamp can be operated at a higher or lower power level than the rated maximum for that type, at different combinations of V_{op} and I_{op}. Thus, for any desired power level, I_{op} can be selected arbitrarily, and V_{op} calculated from Eq. 16.2. Then from V_{ns} and these values of I_{op} and V_{op}, the ballast impedance X_b is calculated from Eq. 16.1. Then for a capacitor ballast, C_{bT} is calculated from Eq. 16.3.

Although any fluorescent lamp can be operated at input power levels greater than those specified in the ANSI specification to yield more lumens of light output power, its lifetime will be decreased. Lamp lifetime may also be shorter than the manufacturer's specification even if it is operated at the specified wattage, but at other than the specified current. The manufacturer's specified lifetime is based on life tests operating at specified currents and voltages.

Figures 16.10*a* and *b* show lamp operating voltages and currents for a number of T8 and T12 hot-cathode (rapid-start) lamps. The negative input impedance of a fluorescent lamp can be seen. With a constant arc length of about 90% of the lamp length, lamp voltage decreases and input power increases as the lamp current increases. Figures 16.11*a* and *b* show the important ANSI specifications (V_{op}, I_{op}, and V_{ns}) for a number of instant- and rapid-start lamps. Note again with the negative input impedance of instant-start lamps in Figure 16.11*a*, that as operating input current increases, operating voltage decreases.

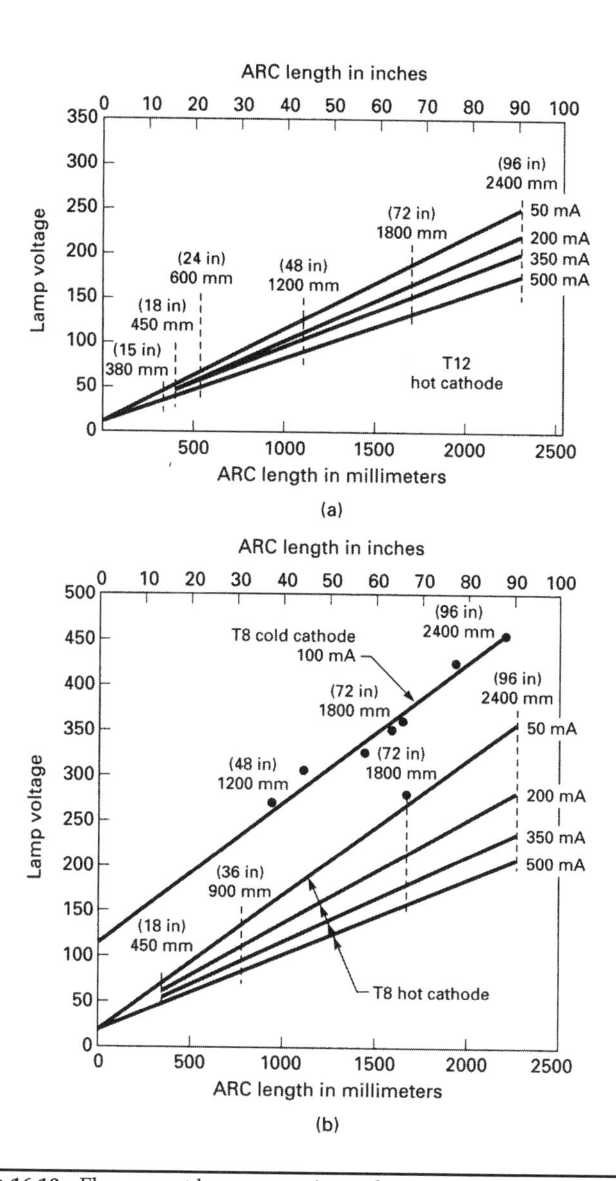

FIGURE 16.10 Fluorescent lamp operating voltages and currents. The source voltage and ballast impedance determine the ballast operating voltage at the manufacturer's specified operating current. Operating below the manufacturer's specified current results in lower input and light output powers. At higher than specified current, input and light output powers increase but lamp lifetime is decreased. (*From "Fluorescent Lamp Light Sources,"* Illumination Engineering Magazine.)

Nominal Power (Watts)	T Size	Length (Inches)	V op (Volts)	I op (Amperes)	Vop x I op (Watts)	V st (Volts)
Rapid Start:						
25	T8	36	100	0.265	26.5	230
30	T12	36	77	0.43	33.1	205
32	T8	48	137	0.265	36.3	204
34	T12	48	79	0.46	36.3	260
37	T12	24	41	0.80	32.8	325
95	T12	96	126	0.83	106	
100	T12	84	135	0.80	108	280
113	T12	96	153	0.79	121	295
113	T12	96	139	1.00	139	295
116	T12	48	84	1.50	126	160
168	T12	72	125	1.50	188	225
215	T12	96	163	1.50	245	300
Instant Start:						
40	T12	48	104	0.425	44	385
40	T12	60	107	0.425	39	385
60	T12	96	157	0.44	69	565
38	T8	72	195	0.300	59	540
38	T8	72	220	0.200	44	540
38	T8	72	245	0.120	29	540
51	T8	96	263	0.300	79	675
51	T8	96	295	0.200	59	675
51	T8	96	325	0.120	39	675

(a)

Lamp Type	MA – RMS Current	Volts-RMS Voltage	Watts Power
48" Hot Cathode – T12	500	75	37.5
"	350	95	33.3
"	250	110	22.0
"	50	120	6.0
72" Hot Cathode – T12	500	120	60
"	350	140	49
"	200	160	32
"	50	185	9.3
96" Hot Cathode – T12	500	160	80
"	350	196	67
"	200	210	42
"	50	246	12
72" Hot Cathode – T8	500	150	75
"	350	175	61
"	200	210	42
"	50	265	13
96" Hot Cathode – T8	500	200	100
"	350	240	84
"	200	280	56
"	50	350	18
48" Cold Cathode – T8	100	260	26
72" Cold Cathode – T8	100	380	38
96" Cold Cathode – T8	100	460	46

(b)

FIGURE 16.11 (*a*) American National Standards Institute (ANSI) specifications for various fluorescent lamps; (*b*) Volt/ampere characteristics at different operating currents for various hot and cold cathode lamps.

16.4 Electronic Ballast Circuits

The basic block diagram of a modern electronic ballast is shown in Figure 16.12. The DC/AC converter that drives the lamp is not powered from the AC power line directly, but through a power factor correction building block (Chapter 15).

Recall from Chapter 15 that without power factor correction, the input bridge rectifier requires a large filter capacitor. This capacitor results in high amplitude line current pulses with fast rise and fall times of high harmonic content. The current pulses cause EMI and RFI problems in adjacent electronic equipment. The RMS value of the non-sinusoidal line current is higher than that required to supply the actual DC load, and consequently causes unnecessary heating of the input power lines and generator windings. For the many fluorescent lamps in a large office building or smaller generators as on naval vessels, this could be a problem.

Power factor correction (PFC) solves this problem by eliminating the large input capacitor and forcing the input line current to be sinusoidal and in phase with the input line voltage. Lamp ballast manufacturers are currently required to have power factor correction meeting IEC555-2 specification, which limits the harmonic content of the input power line. Electronic ballasts are also required to meet EMI/RFI limits set by FCC (CFR 47, part 18).

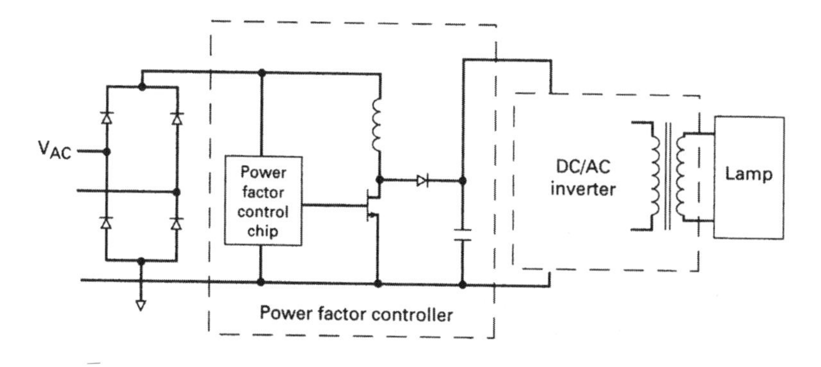

FIGURE 16.12 Block diagram of a modern fluorescent lamp ballast. Output frequency of the DC/AC inverter is set by a series or parallel self-resonant oscillator in the range of 20 to 50 kHz. The ballast is usually a capacitor or the controlled source impedance of a series *LC* resonant circuit.

16.5 DC/AC Inverter—General Characteristics

The DC/AC inverter topologies usually used in electronic ballasts are the push-pull for 120-V AC line input, and the half-bridge for 220-V line input. Unlike switching power supplies that use fixed-frequency, chip-driven, square wave circuits for their DC/AC inverters, ballast inverters are series or parallel, self-resonant *LC* oscillators. There are a number of reasons for this.

Probably most important is that the lamp power efficiency is highest with sinusoidal current drive. Thus there are cost and space savings in generating a sinusoidal current rather than generating a square wave and adding filter components to remove higher harmonics. Although using a chip is probably the simplest and most direct way of generating alternating current from the rectified AC line, the cost would be prohibitive. The *LC* oscillators actually used are not constant in frequency, but that is no problem as long as the frequency remains above 20 kHz, beyond which there is little further gain in lamp efficiency.

Also, with sinusoidal voltages and currents, there are significant advantages in reduced switching losses at turn "on" and turn "off," and higher permissible "off" voltage stresses with alternating base drive voltage. The negative half cycle of the base drive automatically provides reverse bias, which permits the transistor to sustain the V_{cer} voltage rating rather than the lower V_{ceo} rating.

Higher permissible "off" collector voltages with sinusoidal base drive can be appreciated from Figure 16.14*a* for one of the above topologies. There it can be seen that during the collector "off" time, the negative half sinusoid of the base drive, aided by the negative bias on the input capacitor, automatically forces a negative bias on the "off" base. This permits the usual high-voltage transistor to safety sustain its V_{cev} rather than its V_{ceo} rating. The V_{ceo} rating is the maximum "off" voltage that a bipolar transistor can withstand if it has a high impedance or open circuit at its base in the "off" condition. On the other hand, the V_{cev} rating for high-voltage bipolar transistors is generally 100 to 300 V higher than the V_{ceo} rating and pertains if the base has a 2- to 5-V negative bias in the "off" state. Such a negative bias is automatically obtained from the negative half cycle of the voltage from the feedback winding (Figure 16.14*a*) of the oscillation transformer. This negative bias on the "off" transistor is also automatically obtained in the alternative base drive scheme in Figure 16.14*a*.

It will be shown later (Figures 16.14*b* and 16.22*b*) that two of the most frequently used topologies achieve transistor switching at the zero crossing of the collector voltage waveform and thus minimize transistor switching loss.

16.6 DC/AC Inverter Topologies

Figure 16.13 (Reference 6) shows the four most commonly used topologies for electronic ballasts. They comprise two versions—current-fed and voltage-fed variations of the push-pull for 120-V AC line input, and half bridge for 220-V line input.

For both push-pull and half bridge, the current-fed version requires either one or two extra inductors and subjects the "off" transistors to higher "off" voltage stress than the voltage-fed types. Reliable designs are harder to achieve with voltage-fed types, however, because they suffer from larger start-up voltage and current transients whose magnitudes depend on the resonant circuit Q. Start-up current transients in voltage-fed circuits may be 5 to 10 times the operating currents.[7] Further, voltage-fed circuits have greater difficulty coping with open- or short-circuited lamp loads, while current-fed schemes can operate indefinitely with such loads.[6,7]

Further, current-fed circuits yield cleaner sinusoidal waveforms and hence offer higher lamp efficiency.[6] Importantly, the current-fed circuit can drive a number of lamps in parallel, compared to only a single lamp drive capability with the voltage-fed scheme.

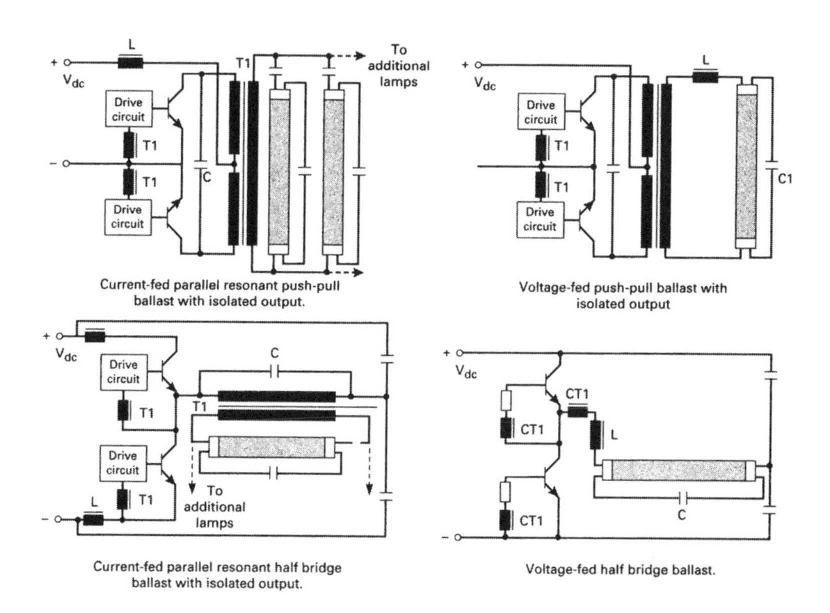

FIGURE 16.13 Usual DC/AC inverter topologies for fluorescent lamps. Half bridges are used for 220-V AC input, push-pulls for 120-V AC input. *(Courtesy Ferroxcube/Philips, Inc.)*

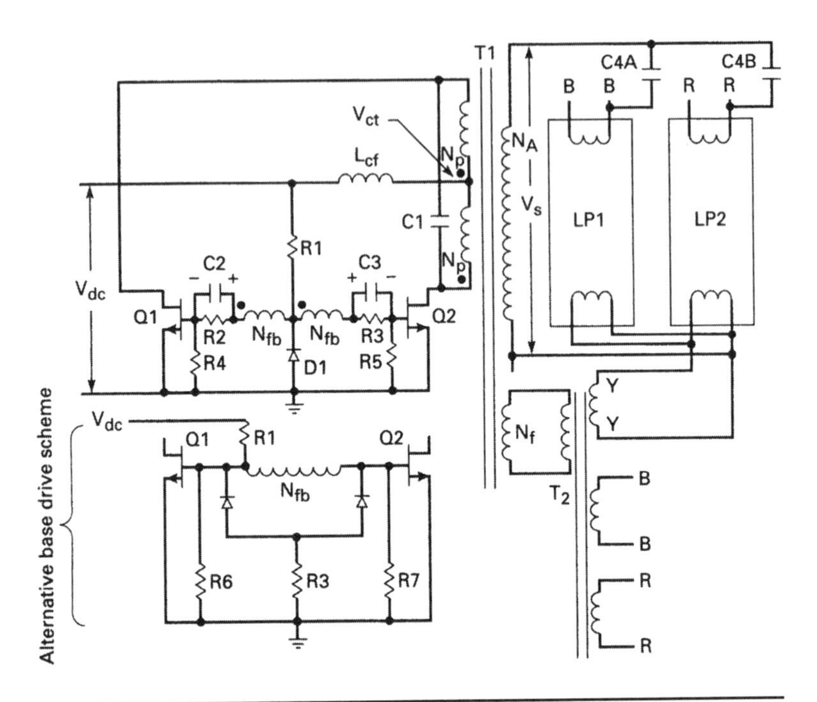

FIGURE 16.14a Current-fed parallel resonant DC/AC inverter. $C1$ plus the reflected ballast capacitors $C44$ and $C4B$ form the parallel resonant tank together with the $T1$ magnetizing inductance in shunt. L_{CF} is the feed-current inductor.

With the availability of ever higher-voltage and lower-cost bipolar transistors, the higher-voltage stress with current-fed topologies is not a drawback to their use. Their major drawback remains the few extra components and hence somewhat higher cost.

16.6.1 Current-Fed Push-Pull Topology

This is shown in Figure 16.14*a* and *b*. It was first described in detail in a classic article by R. J. Haver.[8] The Haver version did not use an isolation transformer to drive the lamps, but rather a tapped choke connected directly between collectors with the lamps across the choke.

Capacitor $C1$, in shunt with the magnetizing inductance L_m of the entire primary, comprises the parallel resonant tank circuit with initial resonant frequency of $f_r = 1/(2\pi\sqrt{L_m C1})$. This is the resonant frequency before the lamps light, as ballast capacitors $C4$ do not reflect back into the primary until the lamps are lighted. When the

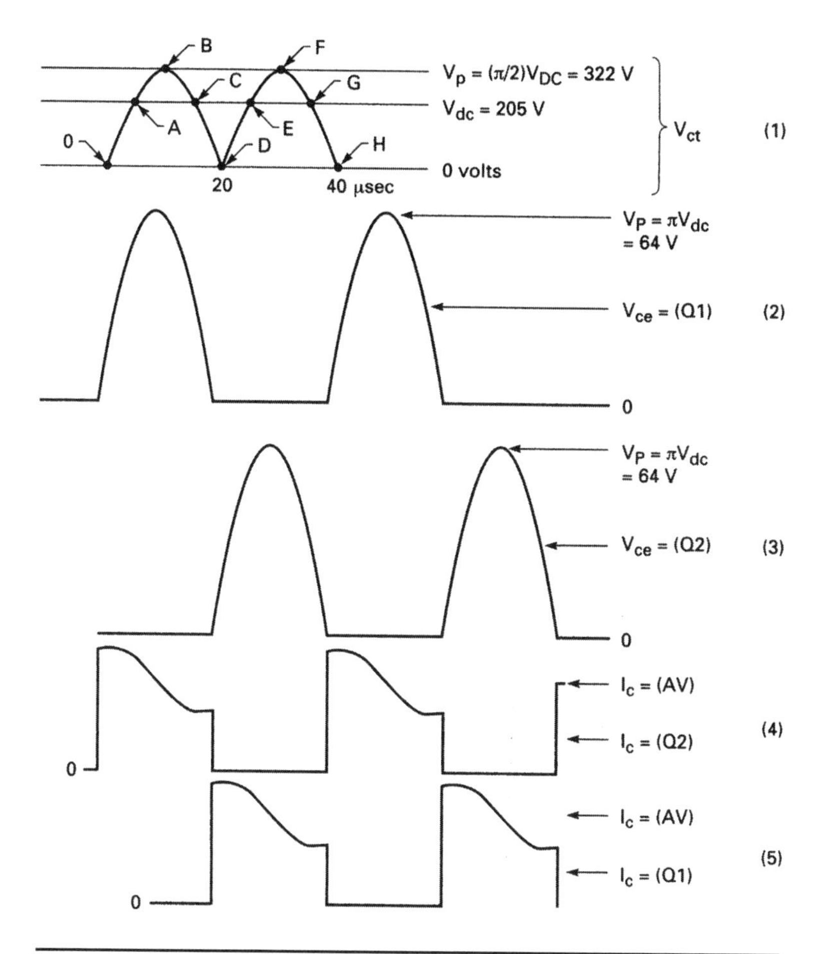

FIGURE 16.14b Critical waveforms for the current-fed parallel resonant DC/AC inverter of Figure 16.14*a*. In Figure 16.14*a*, the power transformer center tap is not fed directly from the low-impedance DC output of the boost regulator, but via a constant-current feed inductor L_{cf}. The circuit is an oscillator with positive feedback from a pair of windings (N_{fb}) on the power transformer to drive the bases.

lamps light, the ballast capacitors draw current, and an indeterminate capacitance reflects back to the primary in shunt with C_4 and lowers the operating frequency.

Two alternative base drive schemes are shown. The upper one uses an *RC* combination (*R2C2*, *R3C3*) at each base in series with the drive winding. The negative bias at the base side of the capacitors helps provide more rapid turn "off" and reduced storage time for the "off"-turning transistor. It also adds to the negative bias, which helps ensure it is the higher "off" voltage specification V_{cev}, rather than the V_{ceo} rating that pertains.

The secondary is shown driving two rapid-start lamps in parallel. Filament power for the lamps is derived from the secondaries of a second transformer $T2$ whose primary is connected to an additional winding on $T1$. Though $T2$ is an added-cost item, it may be preferable to the alternative of having the filament windings themselves as additional secondaries on $T1$. This would make for a bulkier, less efficient, and harder and more expensive transformer to wind.

Drawing a small current from V_{dc}, $R1$ supplies turn "on" current to start the oscillation. The ever-present inequality in beta between transistors Q_1 and Q_2 ensures that one device turns "on" first, and thereafter the base drive for the continuing oscillation comes from positive feedback winding N_{fb} via $D1$.

Collector voltage and current waveforms are shown in Figure 16.14b2–b5. Turn "on" and turn "off" are achieved at the zero voltage points on the collector voltage waveform, minimizing AC switching losses.

The peak secondary voltage across N_S is set to the specified nominal lamp striking voltage V_{ns} at 50°F for the specific lamp type. Capacitor $C4$ is the ballast capacitor that limits the lamp RMS operating current (I_{1RMS}) to its specified value at its specified RMS operating voltage (V_{1RMS}) to yield the specified lamp power $P = (I_{1RMS})(V_{1RMS})$. Thus $C4$ will be selected from

$$\begin{aligned} I_{1rms} = I_{C4rms} &= \frac{0.707V_{ns} - V_{1rms}}{X_{c4}} \\ &= (0.707V_{ns} - V_{1\,rms})(2\pi f_r C4) \end{aligned} \tag{16.4}$$

16.6.2 Voltage and Currents in Current-Fed Push-Pull Topology

Figure 16.14b1 shows the waveform at the center tap of power transformer $T1$ in Figure 16.14a. It comes about because of the current feeding inductor L_{cf}, and is a full wave rectified sine wave. Since L_{cf} has negligible DC resistance and hence cannot support a DC voltage, the average voltage at the output end of L_{cf} must equal that at its input end, or V_{dc}. The average of a full wave rectified sine wave of amplitude V_p is $V_{av} = V_{dc} = (2/\pi)V_p$, so the peak voltage at the center tap is $V_p = (\pi/2)V_{dc}$. Also, the peak center tap voltage is $(2/\pi)V_{dc}$, so the opposite "off" transistor is subjected to a peak V_{ce} voltage of $\pi\, V_{dc}$ (Figure 16.14b2, b3).

A nominal AC input of 120 V RMS with ±15% tolerance yields a maximum peak of $1.41 \times 1.15 \times 120 = 195$ V. Recall that a PFC circuit generates a DC output voltage about 20 V above the peak AC input

(Chapter 15). Then V_{dc} will be about 195 + 20 = 205 V, and from the above, the transistors must safely sustain a peak "off" voltage of π (205) or 644 V. There are currently a number of transistors with adequate voltage, current, and f_t ratings to meet this requirement.

Figure 16.14*b*2–*b*5 show collector currents rising and falling at the zero points of the collector voltage waveforms. This minimizes AC switching losses. Since the half sine waves of voltage across each half primary are equal in volt-second area (Figure 16.14*b*1) and with negligible storage time, there is no possibility of *flux imbalance* (Section 2.2.5) or momentary simultaneous condition.

Collector currents in each half cycle are shown in Figures 16.14*b*4 and *b*5. The sinusoidal "wiggle" at the top of the square pulse of current is characteristic and will be discussed below. The average value of the current at the center of the sinusoidal wiggle (I_{cav}) is calculated from the lamp power. Assuming two lamps of specified power P_1, a DC/AC converter efficiency of E, and input voltage V_{dc}, the collector current is

$$I_{cav} = \frac{2P_1}{EV_{dc}} \tag{16.5}$$

Thus, for two 40-W lamps, a converter efficiency of 90%, and an input voltage V_{dc} of 205 V from the power factor corrector circuit, $I_{cav} = 2 \times 40/(0.9 \times 205) = 434$ mA.

16.6.3 Magnitude of "Current Feed" Inductor in Current-Fed Topology

The feed inductor L_{CF} in Figure 16.14*a* is calculable from Figure 16.14*b*1. There it is seen that the output end of L_{CF} swings in half sinusoidal fashion around V_{dc}. Voltage across L_{CF} at any instant is $V_l = L_{CF}\, dI/dt$. Hence dI, the change in current in L_{CF} between any two instants t_1 and t_2, is

$$dI = \frac{1}{L_{CF}} \int_{t_1}^{t_2} V_1 dt \tag{16.6}$$

Note in Figure 16.14*b*1 that from times A to C the voltage at the output end of L_{CF} is above V_{dc}. During this interval the volt-second area across the inductor in Eq. 16.6 is positive, signifying that the inductor current increases from times A to C. Between times C and E, the volt-second area across the inductor is negative, so the current decreases.

The inductor L_{CF} is chosen so that the current change dI in Eq. 16.6 is an arbitrarily chosen small fraction of the current in Eq. 16.5. The inductor feeding the DC/AC inverter can then be considered a constant-current source. Assume then that dI of Eq. 16.6 is ±20% of the current I_{cav} in Eq. 16.5. Then from the above value of 434 mA for I_{cav},

$di = 0.4 \times 434 = 174$ mA, and from Eq. 16.6, $L_{CF} = \int_A^C V_1\,dt/0.174$ H. In Figure 16.14*b*1, $\int_A^C V_1\,dt$ is the area (in volt-seconds) of the region lying between V_{ct} and the V_{dc} line.

By "eyeball" integration, that area is about 800×10^{-6} Vs. Then from Eq. 16.6, $L_{CF} = 800 \times 10^{-6}/0.174 = 4.6$, or about 4.0 mH. It will be wound on either a powered iron or gapped ferrite core, so it does not saturate at the maximum current it will draw. Although the normal current has been calculated above as 434 mA, it should be designed for about twice that to allow for turn-on transients.

Examination of waveforms in Figure 16.14*b*1, *b*2, and *b*3 offers insight into the characteristic sinusoidal wiggle at the top of the current waveforms in Figure 16.14*b*4 and *b*5. Consider that the voltage across L_{CF} is $V_l = L_{CF}\,di/dt$. At points *A* and *C* in Figure 16.14*b*1, this voltage is zero, so di/dt is zero at those points, and they correspond to points *I* and *K* in Figure 16.14*b*4, where di/dt is also zero. At point *B* in Figure 16.14*b*1, V_1 is a maximum and hence di/dt is also a maximum. Point *B* in Figure 16.14*b*1 thus corresponds to point *J* in Figure 16.14*b*4, where di/dt is also a maximum.

16.6.4 Specific Core Selection for Current Feed Inductor

The 4.0-mH inductor L_{CF} can be designed with MPP cores (Section 4.6.3), KoolMu cores[9] that are less expensive and more modern versions of MPPs (Figure 16.15), gapped ferrite cores (Section 4.9.6), or inexpensive but lossy powdered iron (Micrometals)[10] cores (Figure 16.17). The choice will be made on the basis of cost, core losses, and size. Table 16.1 serves as a guide to the choice.

The high price of MPP cores rules them out for this application, but the cost/loss comparison at 1000 G in Table 16.1 is deceptive for 4.0 mH, because the peak flux density for the large number of turns is so low that core losses even for the inexpensive Micrometals iron powder material are not significant. The following calculations will demonstrate the final core selection.

The required number of turns (N_r) will be calculated from A_l (millihenries per 1000 turns) as

$$N_r = 1000\frac{\sqrt{L}}{A_1} = 1000\sqrt{\frac{4}{A_1}} \qquad (16.7)$$

The peak flux density B_m can be calculated from Faraday's law

$$V_1 = NA_e\,dB/\,dt \times 10^{-8}$$

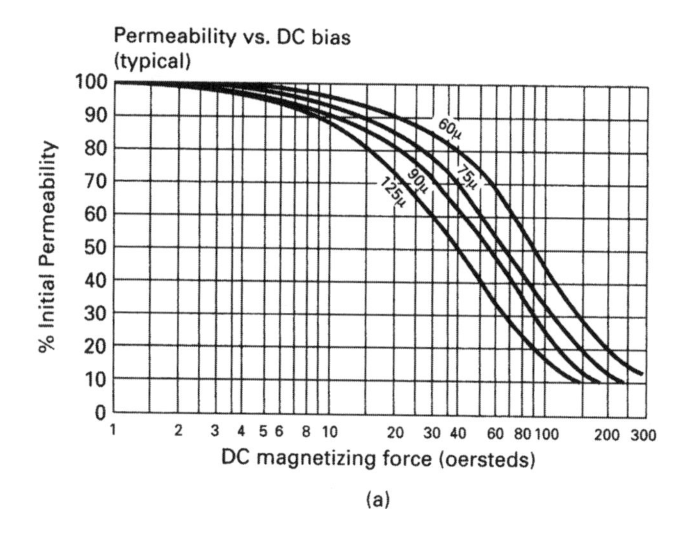

(a)

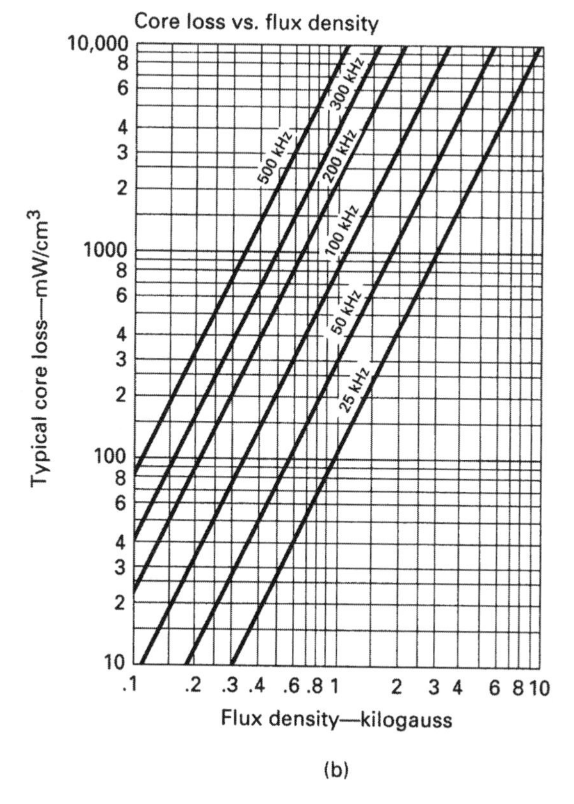

(b)

FIGURE 16.15 Characteristics of KoolMu, a powdered magnetic core that is less lossy than powdered iron. (*Courtesy Magnetics, Inc.*)

Core type	Cost (500 quantity), $	Core loss, mW/cm^3, at 1000 G, 50 kHz
MPP	14.00	180
KoolMu	4.20	300
Gapped ferrite (3C85)	2.20 (two halves, 3019 pot)	30
Micrometals	0.34 (26 material)	2000

TABLE 16.1 Cost/Loss Comparison of Various Core Materials

or

$$B_m = \int_A^C V_1\, dt / NA_e \times 10^{+8}$$

where $\int_A^C V_1\, dt$ is the area in volt-seconds between points A and C in Figure 16.14*b*1. That area is 800×10^{-6} Vs as estimated above. Then

$$B_m = \frac{800 \times 10^{+2}}{NA_e} \tag{16.8}$$

Core losses will be calculated from the manufacturer's curves of losses versus B_m and frequency. Frequency will be taken at 50 kHz, as Figure 16.14*b*1 shows that for 25-kHz oscillation frequency, inductor frequency is 50 kHz.

Tentative cores will be selected for the three types—KoolMu, Micrometals, and gapped ferrite. The initial selection will use Eq. 16.8, seeking an A_l that yields minimum turns without saturating the core, as determined from the manufacturer's curves of percentage inductance falloff versus magnetizing force H (Figure 16.16*a*). Cores with higher values of A_l (higher permeability) that minimize the number of turns, saturate at lower H. Maximum H will be calculated from

$$H_m = \frac{0.4\,\pi\, NI}{l_m} \tag{16.9}$$

where l_m is the magnetic path length in centimeters and I is twice the 434 mA calculated in Section 16.6.2, because of the possibility of turn "on" transients of twice the normal maximum current (Section 16.6.3).

One of the tentatively selected cores will be chosen that yields minimum core losses as calculated from N and B_m of Eq. (16.8) and the manufacturer's curves of core loss versus B_m and frequency (Figure 16.15*b*). We will first deal with KoolMu cores. See Table 16.2.

The inductor can be built with any of the above four cores for which total core losses are insignificant. The percentage falloff in inductance

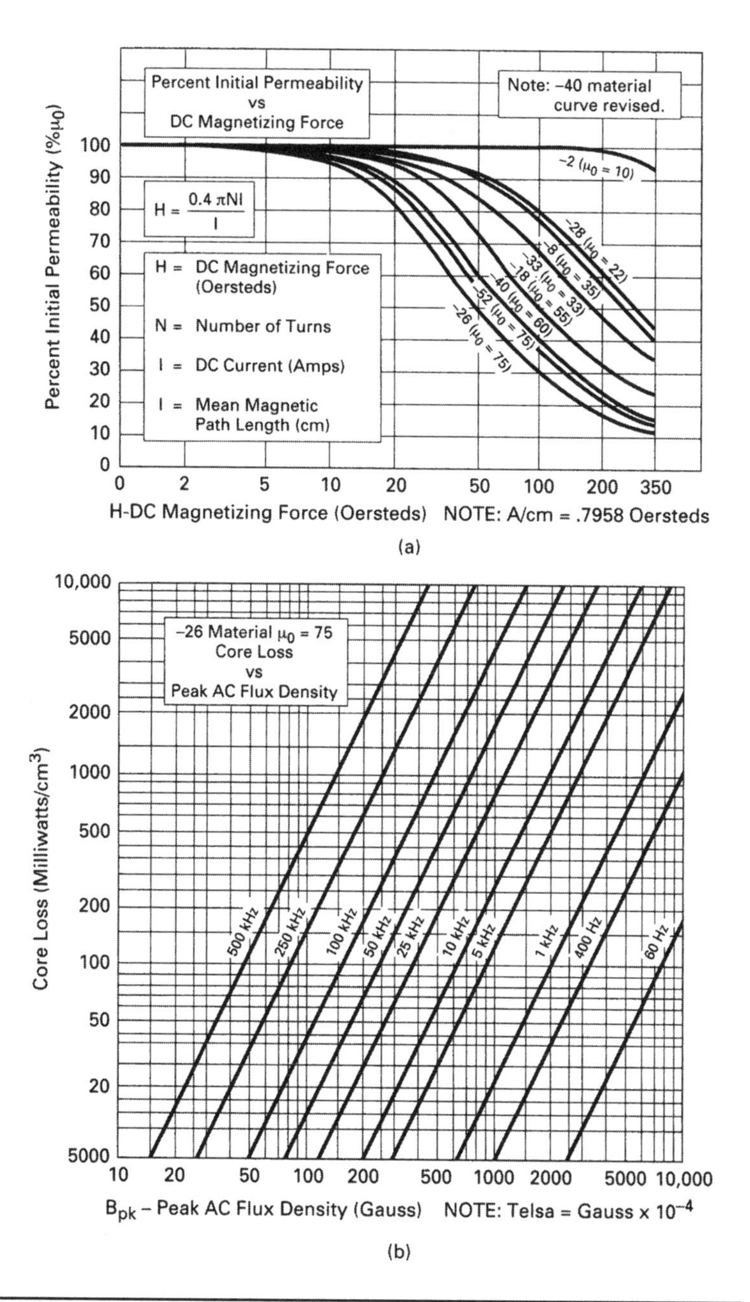

FIGURE 16.16 Characteristics of Micrometals powdered iron; this material is lossy but less expensive than KoolMu or MPP materials.

KoolMu type	A_l, mH/kT	N turns	A_e, cm^2	B_m, G	l_m, cm	H_m, Oe	%Fall	Loss, mW/cm^3	Volume, cm^3	Loss, total mW
77110	75	231	1.44	121	14.3	17.6	8	4	20.6	84
77214	94	206	1.44	136	14.3	15.7	11	5	20.6	103
77094	107	193	1.34	159	11.6	18.5	13	8	15.6	125
77439	135	172	1.99	117	10.74	17.4	9	4	21.3	85

TABLE 16.2 Possible KoolMu Cores for Current Feed Inductor

is negligible, and can be recovered by increasing the number of turns by the square root of the falloff percentage. Core 77439 would be the best choice as it has the fewest turns.

Repeating this procedure for Micrometals toroidal cores, for which loss versus B_m and frequency is taken from Figure 16.16*b* and the percentage fall in inductance is taken from Figure 16.16*a*, gives the results shown in Table 16.3.

Table 16.3 shows that total core losses with the inexpensive Micrometals cores are five to seven times higher than those with KoolMu cores, but not prohibitively high. The required number of turns does not differ greatly. The Micrometals 157-26 and 175-26, having smaller magnetic path length and more turns than the first two Micrometals types, have higher peak H, and consequently a higher "swing" or falloff in inductance.

The best choice among the Micrometals cores is probably the 250-26 despite the higher core dissipation, because of its lower inductance falloff.

The final choice, then, is based on a cost versus engineering performance comparison. The KoolMu 77439 (Figure 16.17*a*) is smaller, and has lower dissipation, but costs more. Its outside diameter (OD)

Micrometals type	A_l, mH/kT	N turns	A_e, cm^2	B_m, G	l_m, cm	H_m, Oe	%Fall	Loss, mW/cm^3	Volume, cm^3	Loss, total mW
225-26B	160	158	2.59	98	14	12.3	10	17	38	646
250-26	242	129	3.84	81	15	9.38	6	15	57	855
157-26	100	200	1.06	190	10	21.8	20	60	11	660
175-26	105	195	1.34	154	11	19.3	18.0	40	15	600

TABLE 16.3 Possible Micrometals Cores for Current Feed Inductor

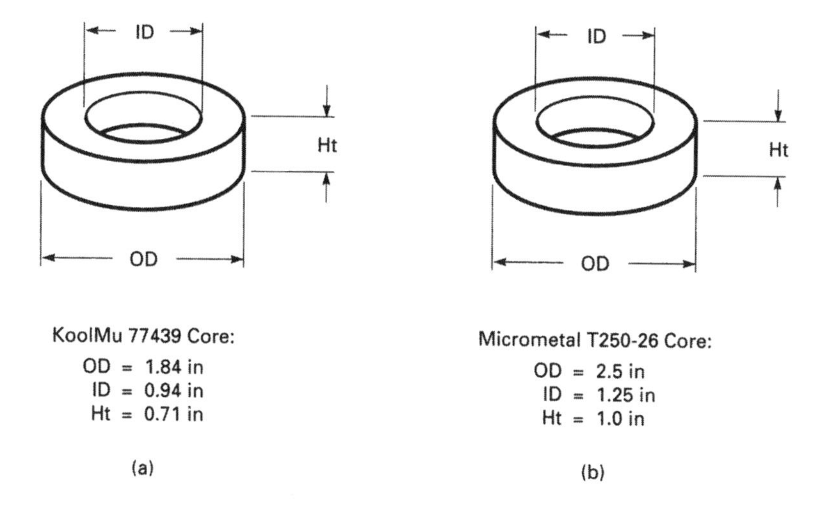

FIGURE 16.17 Candidate geometries for current feed inductor core L_{CF} of Figure 16.14*a*.

is 1.84 in, it dissipates 85 mW, and it costs $4.20 in 500 quantity. The Micrometals 250-26 (Figure 16.17*b*) has an OD of 2.5 in, and dissipates 855 mW, but costs 34 cents in 500 quantity.

A gapped-ferrite core might be the best compromise. A core can be selected by making an initial educated guess and then repeating calculations two to three times. Curves showing A_l versus ampere-turns for various air gaps, and the *cliff point* in ampere-turns at which the DC bias commences, are required. Example curves are shown in Section 4.6.2 in Figure 4.3.

The procedure is as follows. Guess at a core and thus its A_l, and then calculate the required number of turns for the desired inductance (4 mH) from Eq. 16.7. Calculate the maximum number of ampere-turns at the anticipated maximum current (1 A). If the ampere-turns exceed the saturation cliff point, the core is too small or its gap is not sufficiently large. Try again for the same core with larger gap (smaller A_l). Proceed until a core or gap or A_l is found for which $(NI)_{max}$ is less than the saturation cliff point.

For this selected core and A_e, calculate the peak flux density B_m from Eq. 16.8. Then from the manufacturer's curves (Figure 16.18), read the core loss in mW/cm^3 at the calculated B_m in Gauss and the current ripple frequency (50 kHz for 25-kHz switching frequency).

Thus, guessing first at the 2616 pot core and proceeding to the 3019 pot core yields Table 16.4.

Table 16.4 shows that the 2616 cannot be used because even with a 32-mil gap that requires 200 turns, the maximum ampere-turns at

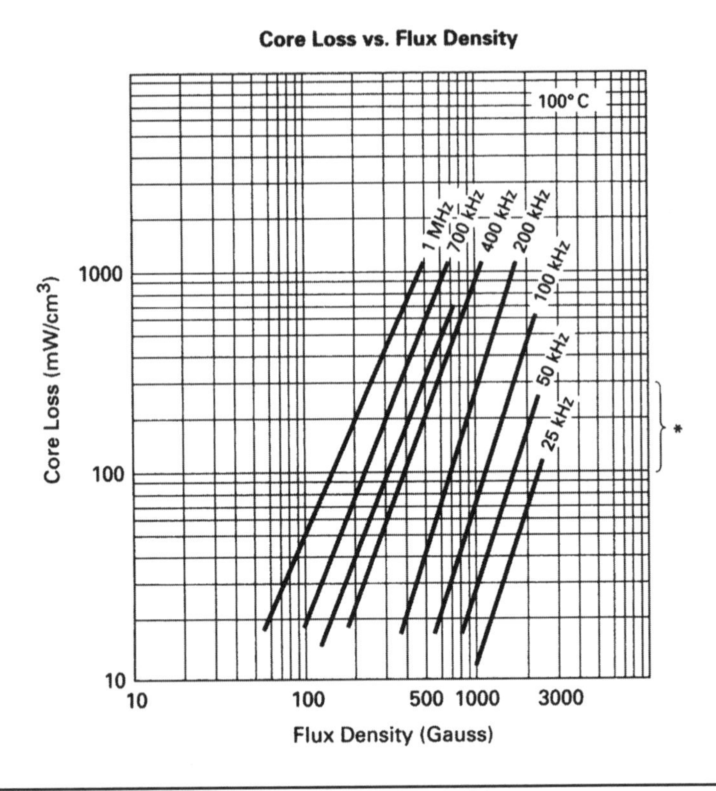

FIGURE 16.18 Core loss vs. peak flux density and frequency, Ferroxcube/Philips Ferrite 3C85 material. (*Courtesy Ferroxcube/Philips, Inc.*)

1 A of current just barely rest at the saturation cliff. Increasing the gap would probably put NI_{max} inside the cliff, but would require significantly more than 153 turns. Most likely, the core bobbin could not then accommodate the required turns of the appropriate wire size.

Core Philips	A_l, mH/kT	Gap, mils	N turns	NI_{max}, A × turns	Cliff point, A × turns
2616	170	32	153	153	129
2616	100	64	200	200	200
3019	500	11	89	89	60
3019	210	35	138	138	170

TABLE 16.4 Possible Gapped-Ferrite Cores for Current Feed Inductor

Core type	Cost (500 qty.), $	Total core loss, mW	OD, in	Height, in
KoolMu 77439	4.20	85	1.84	0.71
Micrometals 250-26	0.34	850	2.5	1.00
Ferrite pot 3019	2.20	31	1.18	0.74

TABLE 16.5 Comparison of Contending Current Feed Inductor Cores

It is then seen that the next-larger pot core—the 3019—would work with an appropriate gap. For an 11-mil gap, it doesn't make it, as it is subjected to 89 A × turns at 1 A, but the saturation cliff is at 60 A × turns. But with a 35-mil gap, it requires 138 turns, and NI_{max} at 1 A is 138 A × turns, which is safely inside the saturation cliff point of 170 A × turns.

The 3019 has an iron area of 1.38 cm^2, and from Eq. 16.8 has a peak flux density of 423 G. At that flux density, Figure 16.18 shows its loss is about 5 mW/cm^3. For its 6.19-cm^3 volume, its total loss is 31 mW.

Finally, the cost/performance comparison of the three possible cores is seen in Table 16.5. If minimum cost is the major criterion, the Micrometals 250-26 is the best choice despite its 850-mW core loss. If that core loss is not acceptable, the ferrite 3019 pot core is the best choice despite its higher cost.

16.6.5 Coil Design for Current Feed Inductor

The RMS current is the constant current in the coil, and was calculated above at 434 mA. At 500 circular mils/A, the required wire area is $500 \times 0.423 = 217$ cmils. Number 26 wire of 253-cmil area is adequate. If the 3019 ferrite core is selected, its bobbin width is 0.459 in, and height is 0.198 in. For a # 26 wire diameter of 0.0182 in, the number of turns per width is $0.459/0.0182 = 25$. The number of layers per bobbin height is $0.198/0.182 = 10$. Thus the 138 turns could be accommodated within six layers.

If any of the above toroidal cores were selected, the 138 turns could easily be accommodated in three layers.

The skin effect is no problem as the AC amplitude is small, and Table 7.6 shows that the AC-to-DC resistance ratio is unity at 50 kHz.

16.6.6 Ferrite Core Transformer for Current-Fed Topology

For two 40-W lamps as shown in Figure 16.14*a*, transformer primary input power at an assumed efficiency of 85% is 94 W.

The transformer core most likely will be ferrite, and its size will be selected from Table 7.2*a*. An attempted toroidal KoolMu core design will be shown that proved to be less flexible and lossier than others. Table 7.2*a* shows the maximum power available in a forward converter from a core at various frequencies and a peak flux density of 1600 G. For a push-pull topology, the available power is twice that.

Table 7.2*a* shows that at 24 kHz, the smallest core that can deliver more than 94 W in a push-pull topology is the *E*21, which is an international standard size and is available from a number of manufacturers. Its maximum available output power in a push-pull configuration is $2 \times 69.4 = 138$ W at a maximum flux density B_m of 1600 G. At $B_m = 200$ G it should be able to deliver $(200/16{,}000)138 = 172$ W. It has an iron area A_e of 1.49 cm^2.

Because of the unique nature of this resonant converter, it is only marginally usable. Let us see why this is so. Refer to Figures 16.14*a* and 16.14*b*1.

The number of primary turns N_p is calculated from Faraday's law—$E = N_p A_e dB/dt \times 10^{-8}$—or

$$N_p = \int_0^{\pi} \frac{V\,dt(10^{+8})}{A_e\,dB} = \int_0^{\pi} \frac{V\,dt(10^{+8})}{2A_e\,B_m} \qquad (16.10)$$

Here $\int_0^{\pi} V\,dt$ is the area of a half period in volt-seconds, and dB is the total flux change in that time, 4000 G for a peak flux density B_m of 2000 G. The area of a half sinusoid of peak V_p is $(2/\pi)(V_p)(T/2)$V × s. Then from Eq. 16.11, for $V_p = 322$ V as shown in Figure 16.14*b*1, *b*2, and for a 20-μs half period:

$$N_p = \frac{(2/\pi)(322)(20 \times 10^2)}{1.49 \times 4000}$$
$$= 69 \text{ turns}$$

The wire size will be calculated on the basis of 500 cmil/A. Figure 16.14*b*1 shows that the peak center tap voltage at the center of the "on" time is 322 V. The transformer is delivering its power at an RMS voltage of $0.707 \times 322 = 228$ V. For the above-calculated 94 W of input power, the RMS input current is then $94/228 = 0.412$ A. But each half primary carries this RMS current only for a half period out of every period. Hence the RMS current for each half primary should be $0.412 \times 0.5 = 0.291$ A. And at 500 cmil/A, wire of 146-cmil area is required.

For 146 cmil, # 27 gage wire of 202 cmils and 0.0164-in diameter would be used. The E21 core bobbin has width of 0.734 in and a height of 0.256 in. Its width can accommodate $0.734/0.164 = 44$ turns, and its height can accommodate $0.256/0.0164 = 15$ layers of # 27 wire. Thus each half primary of 69 turns will consist of two layers of 35 and 34 turns, and the full primary will occupy only four layers. Assuming

$N_s = 2N_p$ and a # 27 wire secondary, the primary and secondary together occupy only half the bobbin height. This would easily leave sufficient room for the filaments or a primary for a separate filament transformer, as shown in Figure 16.14*a*.

Transformer core material could be either Magnetics Inc. type P or Ferroxcube/Phillips 3F3, both of which have 75 mW/cm^3 loss at 2000 G and 25 kHz (Figure 16.18). The E21 size core (Figure 16.19) at a volume of 11.5 cm^3 will dissipate only $0.075 \times 11.5 = 863$ mW.

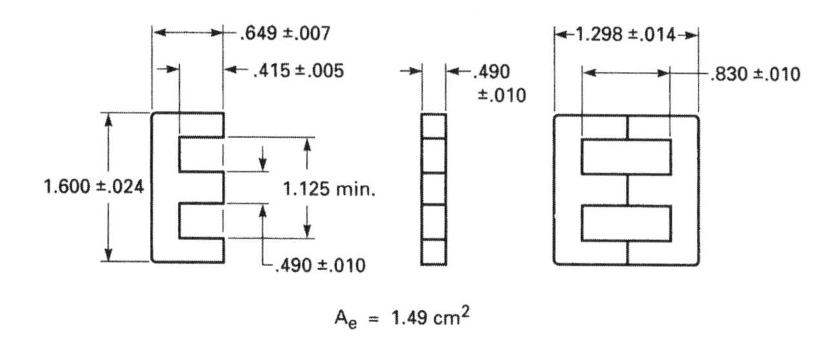

A_e = 1.49 cm^2

Standard Bobbin

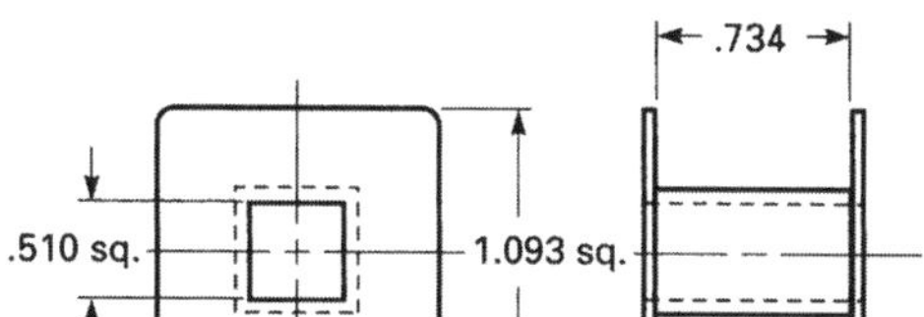

PART NO. E21F1N
Material: Nylon
Max. Operating Temp.: 105 °C
Winding Area: .188 in^2.
Mean Length of Turn: 3.08 in.
Flammability: UL94-V2

FIGURE 16.19 Dimensions of the E21 core and bobbin. This core is a candidate for the current-fed parallel resonant push-pull topology of Figure 16.14*a*. Its magnetizing inductance is the resonant inductor of the resonant circuit. Since there are constraints on how small the resonant capacitor is, this limits the magnetizing inductance to a relatively small value for resonance at 25 kHz. With a small magnetizing inductance, the circulating tank current is large, requiring such large wire as to make it impossible to fit the coil inside the bobbin. Thus, larger cores than might be expected are required for the given power level.

This appears to be a reasonable design, and the E21 core appears usable. The fallacy is the assumption that the primary carries current needed only to supply the output load power. The above calculation showed that to supply the load power of 94 W, the current in each half primary was a half sinusoid of 0.412-A RMS.

It will be seen below that the current actually carried by the primary is considerably greater than this. The two series half primaries are in shunt with C1 and the reflected ballast capacitors C4A, C4B (Figure 16.14*a*) to form the resonant tank circuit. The primary current is then fixed by the amplitude of the circulating resonant tank current, which is considerably more than the 0.412 A RMS shown above.

The actual primary current, which is the circulating tank current, is calculated as follows:

The voltage across this tank and the two half primaries is sinusoidal with a peak amplitude of 644 V and an RMS voltage of $0.707 \times 644 = 455$ V. Thus each half primary carries current during the full period rather than only one-half period. The current is $I_{\mathrm{RMS}} = V_{\mathrm{RMS}}/X_l$, where $X_l = 2\pi f L_t$, and L_t is the inductance of both primary halves in series. Thus the primary current depends on L_t, which cannot be made arbitrarily large to keep its current low and permit small wire. This is because L_t, together with its total shunt capacity C_t, sets the resonant frequency at $1/(2\pi\sqrt{L_t C_t}) = 25$ kHz.

Total capacitance C_t is the sum of C1 and the reflected ballast capacitors C4A, 4B, and C_t cannot be made small to permit a large value for L_t and consequent low current in L_t. At initial turn-on, when the lamp impedances are high, C4A and C4B are essentially out of the circuit, and C1 alone and L_t set the resonant frequency at its highest value. When the lamps light, the total resonating capacitance increases to C1 + C4A + C4B (assuming $N_s = 2N_p$) and the frequency drops to its desired value of 25 kHz. If one lamp is defective and out of the circuit, only one ballast capacitor reflects into the primary, and the operating frequency is somewhere in between. Thus to keep the frequency from changing too much under these three operating conditions, C1 should be large—or at least not small compared to C4A + C4B.

Equations 16.1 and 16.3 set restrictions on C4A, 4B. In Eq. 16.1, V_{ns} may be any value greater than the highest striking voltage for the given lamps. The higher it is, the greater the impedance X_b of the ballast capacitor C_4 must be to limit lamp current to its specified value. If V_{ns} is set high to permit a high value for X_b and hence low value for C_b, this makes it easier to achieve the desired goal of having C1 large compared to the reflected ballast capacitors. For high V_{ns}, however, $N_s/(2N_p)$ is high and any ballast capacitors reflect across into the primary as larger capacitors. There is thus no advantage in going to any turns ratio higher than $N_s/(2N_p) = 1$.

Thus for a specified lamp operating voltage V_{op} of 101 V RMS, an operating current of 430 mA RMS, and $V_s = 455$ V RMS as above, Eq. 16.1 shows

$$X_b = \frac{455 - 101}{0.430} = 823\ \Omega$$

And Eq. 16.3 shows

$$C_b = \frac{1}{2\pi F_r X_b} = \frac{1}{(2\pi)(25 \times 10^{+3})(823)} = 0.0077\ \mu\text{F}$$

Thus the two ballast capacitors reflect across into the primary as $2 \times 0.0077 = 0.015\ \mu$F. To achieve the goal of having not too great a frequency change under the above operating conditions, $C1$ is chosen equal to the sum of the two reflected ballast capacitors.

For $C_t = C1 + C4A + C4B = 0.03\ \mu$F and a resonant frequency of 25 kHz

$$L_t = \frac{1}{4\pi^2 F_r^2 C_t} = \frac{1}{4\pi^2(25000)^2(0.03 \times 10^{-6})} = 1.35\ \text{mH} \tag{16.11}$$

We round this up to 1.5 mH.

The core would be gapped so that the above-calculated 69 turns per half primary (Eq. 16.10) or 138 turns for both halves would have an inductance of 1.5 mH. This corresponds to an A_l of $(1000/138)^2 \times 1.5 =$ 79 mH/1000 T. Figure 16.20 shows this could be obtained with an air gap of 80 mils and with 455 RMS V across L_t; it draws a current of

$$I_{\text{RMS}} = \frac{455}{2\pi(2.5 \times 10^4)(0.0015)} = 1.93\ \text{A RMS}$$

At 500 cmil/A, this would require a wire of 966 cmil area.

Possible wire choices are shown in Table 16.6.

It is seen in Table 16.6 that the T_1 primary could only barely fit inside the E21 bobbin using # 20 wire, which has somewhat more than the required circular-mil area. Since it requires 136 turns of # 20 wire, the bobbin can hold only $20 \times 7 = 140$ turns. With # 21 wire, each half primary could be handled in three layers of 23 turns. This would leave $0.256 - (6 \times 0.0314) = 0.068$ in for the secondary plus any filament windings. Even though the secondary carries only the operating current for two lamps ($2 \times 0.43 = 0.86$ A) and can use wire smaller than # 21, there is far from enough space for the secondary, filaments, and insulation between primary and secondary. Thus a larger core must be chosen.

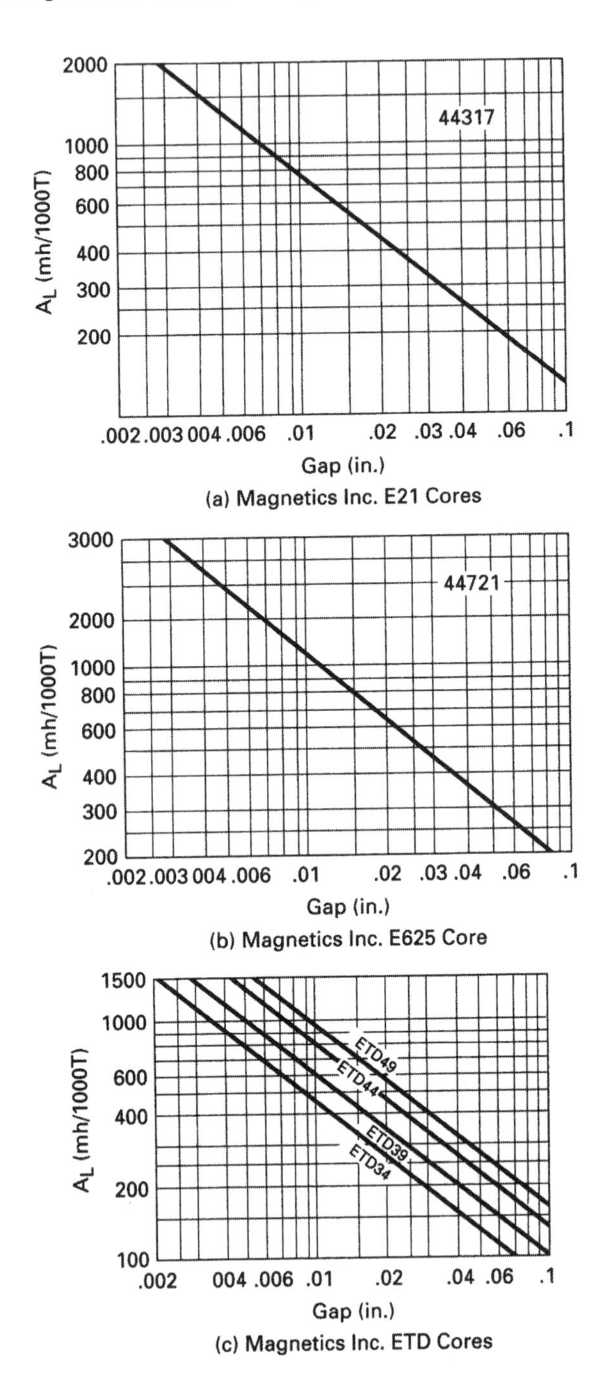

FIGURE 16.20 A_l (Inductance in mH per 1000 turns). (*Courtesy Magnetics, Inc.*)

Wire #	Area, cmil	Diameter, in	Bobbin width, in	Bobbin height, in	Turns/ width	Layers/ height
20	1020	0.0351	0.734	0.256	20	7
21	812	0.0314	0.734	0.256	23	8

TABLE 16.6 Possible Coil Design for an E21 Transformer Core

The core selection procedure is as follows:

Step 1 A tentative core selection will be made with an A_e large enough to yield a reasonably small number of half primary turns (N_p) as calculated from Eq. 16.10 for a peak flux density of 2000 G [dB in Eq. 16.10 = 4000 G].

Step 2 Values of resonating inductor L_t and capacitor C_t will be 1.5 mH and 0.03 μF as calculated above for the E21 core. From L_t and N_p, the A_l will be calculated. From A_l and the manufacturer's curves of A_l versus air gap (Figure 16.22), the air gap will be read.

Step 3 The current in L_t will be the same 1.93 A RMS as calculated above for the E21 core since the voltage across L_t is the same 455 V RMS. Wire size then will be either # 20 or 21, based on a current density of 500 cmil/A.

Step 4 From the manufacturer's data sheets, bobbin width and height will be read, and the total number of turns in L_t ($2N_p$) will be calculated to see if it can be accommodated in the bobbin with enough space remaining for the secondary and filament windings.

Step 5 If the windings fit within the bobbin, the choice of core will be made on the basis of its core loss (Figure 16.18) and cost.

This procedure was followed for three tentatively usable cores with the results shown in Table 16.7. All three cores could be used, but the E625 is questionable because there is marginally sufficient vertical height left in the bobbin for the secondary and filament windings after laying down the two half primaries. The secondary feeds two lamps at 0.43 A RMS each. For a total of 0.86 A RMS, at 500 cmil/A, the required area for the secondary wire is 430 cmil. Number 24 wire of 404-cmil area and 0.0227-in diameter would be adequate. A summary follows in the table.

	E625	783E608	ETD44
Secondary turns ($=2N_p$)	86	112	116
Max. turns/layer (# 24 wire)	37	45	51
Secondary layers	3	3	3
Total height primary + secondary, in (4 layers # 20 + 3 layers (# 24)	0.208	0.208	0.208
Remaining height in bobbin, in	0.043	0.062	0.075

The remaining height in the bobbin must accommodate primary-to-secondary insulation plus the filament wires, about one turn of # 22 wire at 0.0281-in diameter. The E625 core is marginal; of the other two, the ETD 44 appears preferable.

	E625	783E608	ETD44
A_e, cm^2	2.34	1.81	1.74
B_m, G	2000	2000	2000
N_p, turns/half primary	43	56	58
L_t, mH	1.5	1.5	1.5
C_t, μF	0.03	0.03	0.03
A_l, mH/1000T	203	120	111
Gap, mils	72	110	120
Bobbin width, in	0.85	1.024	1.165
Bobbin height, in	0.251	0.270	0.283
Turns/W, # 20 wire	24	29	33
Full primary layers	4	4	4
Layers/height for no. 20 wire	7	7	8
Core loss, mW/cm^3 at 2000 G, 25 kHz	80	80	80
Core volume, cm^3	20.8	17.8	18.0
Total core loss, W	1.66	1.42	1.44

TABLE 16.7 Characteristics of Possible Cores for $T1$ (Figure 16.14*a*)

16.6.7 Toroidal Core Transformer for Current-Fed Topology

The circumstance of the inside diameter (ID) of a toroid $C = \pi \times ID$ is much greater than the bobbin width of an EE core of roughly equal A_e. Hence a toroid permits more turns per layer than the EE and will have fewer winding layers. In many cases, only two layers will suffice.

This almost completely eliminates *proximity-effect* losses (Section 7.5.6). It also results in a more reliable design, since turns having a high voltage between them are spread farther apart and the possibility of arcing is far less. Further, the entire bobbin width in an EE core must not be utilized if VDE European safety specifications must be observed. This means more winding layers and a larger core. In some cases, VDE specifications permit using the entire bobbin width if triple-insulated wire is used, but this may still require a larger core.

A toroidal core (KoolMu or MPP) is more expensive and more lossy, however, and has less design flexibility. This can be seen from the following.

The number of turns per half primary N_p is calculated from Eq. 16.10 once a tentative core selection has been made, A_e has been established, and a B_m has been chosen that yields acceptable core loss (Figure 16.15*a*). The number of turns on the full primary $2N_p$ must yield the desired inductance L_t calculated from Eq. 16.7. Thus with L_t and $2N_p$ fixed, A_l is fixed, but KoolMu MPP and Micrometals toroidal cores come in discrete values for A_l, which are proportional to core permeability, available in only five or six discrete values.

This differs from the situation with EE cores, where the number of turns per half primary N_p can be set to yield a desired peak flux density B_m from Eq. 16.10. With that fixed value of $2N_p$ any desired value of A_l can be set with the air gap (Figure 16.22).

This problem—that A_l with the above toroidal cores is available in only five to six discrete values for any given A_e—can be solved by using gapped toroids. KoolMu and Micrometals cores are available with customized gaps to yield any desired A_l, but if a toroid is gapped, it has no significant advantage over a gapped ferrite EE core that is less expensive. The sole reason for considering a gapped toroid is that the coil winding length is longer than with an EE core. This offers fewer coil layers and makes it easier to meet VDE safety specifications.

16.7 Voltage-Fed Push-Pull Topology[6,7,8,11]

This topology is shown in Figure 16.21. The $T1$ center tap is fed directly from the rectified input line voltage, or the output of the power factor corrector building block following it, with no intervening inductor as in the current-fed topology.

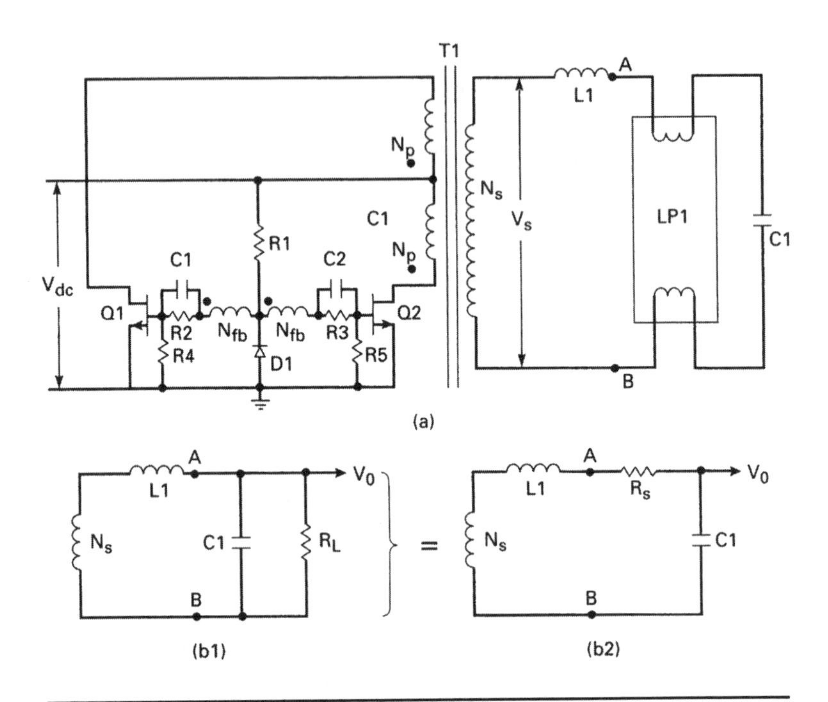

FIGURE 16.21 Voltage-fed push-pull topology. With no inductor between the DC input and the $T1$ center tap, voltage stress on the "off" transistor is 2 V_{dc}, rather than πV_{dc} as in the case for the current-fed push-pull topology.

This circuit is also a resonant oscillator rather than a driven inverter. Windings N_{FA} and N_{FB} on transformer $T1$ provide the positive feedback to keep the circuit oscillating. Resistor $R1$ draws a small current from V_{dc} to start the oscillation. After the start, the feedback windings supply current through the base emitter of the transistors and D1 to keep them "on."

The $T1$ secondary feeds the series resonant combination of $L1$ and $C1$ in series with the lamp filaments. The lamp resistance is in shunt with $C1$. Voltages at the collectors are square waves moving between V_{cesat} (about 1 V) and $2V_{dc}$. With square waves at the primary, voltage at the secondary is also a square wave. Secondary current is sinusoidal with the resonant LC circuit, and hence so is the primary current. When the secondary current reverses, so does the current in the feedback windings and the "on" transistor turns "off." This circuit thus also achieves turn "off" and turn "on" at the zero-current points of collector current and consequently has negligible switching losses.

With no inductor at the center tap, the voltage across the "on" transistor half primary is V_{dc}, as is the voltage across the other half primary. Thus the "off" transistor is subjected to $2V_{dc}$ rather than πV_{dc} as for the current-fed circuit of Figure 16.14*a*. For a V_{dc} of 205 V as assumed for the current-fed circuit, this means a maximum voltage rating for the transistor of somewhat more than 410 V rather than 644 V for the current-fed circuit. The reduced cost of the lower-voltage transistor and saving the input inductor cost might seem a decisive advantage, but high transient currents at turn "on" are a significant drawback to the voltage-fed topology. The higher required transistor current ratings outweigh the lower voltage ratings. This can be seen as follows:

The equivalent circuit beyond points *AB* is as shown in Figure 16.21*b*1, because the impedance of the lamp $L1$ filaments can be ignored. The effective lamp impedance R_L is very high before the lamp has lighted, and falls to a low value afterward. The circuit behavior can be understood by converting the parallel combination of $R_L C1$ in Figure 16.21*b*1 to its equivalent series $R_s C_s$ circuit of Figure 16.21*b*2. The impedance across points *AB* is

$$Z_{AB} = \frac{R_L X_{C1}}{R_L + X_{C1}} = \frac{R_L / j\omega C1}{R_L + 1/(j\omega C1)}$$

$$= \frac{R_L}{1 + j\omega R_L C1}$$

And multiplying the numerator and denominator by $1 - j\omega R_L C1$ yields

$$Z_{AB} = \frac{R_L}{(1 + \omega^2 R_L^2 C1^2)} - \frac{j\omega R_L^2 C1}{1 + \omega^2 R_L^2 C_1^2}$$

Now set $\omega R_L C1 = Q$. Then

$$Z_{AB} = \frac{R_L}{1 + Q^2} - \frac{j\omega R_L^2 C1}{1 + Q^2}$$

For $Q \gg 1$

$$Z_{AB} = \frac{R_L}{Q^2} + \frac{1}{j\omega C1} \tag{16.12}$$

Equation 16.12 and Figure 16.21*b*2 show that the transformer secondary N_s drives a series *LC* circuit that resonates at a frequency of $1/(2\pi\sqrt{L1C1})$ and has an equivalent series resistance R_L/Q^2. For $Q \gg 1$, that series resistance is very small. Now in a series resonant circuit at resonance, with input voltage V_{in}, current is $V_{in}/(R_L/Q^2)$.

At turn "on" before the lamp lights, resistance R_L is high, Q is high, and the equivalent series resistance R_L/Q^2 is very low. This results in high turn "on" currents that may be 5 to 10 times higher than operating currents.[7] With such high turn "on" currents, the normal base drive currents may be insufficient to keep the "on" collector in saturation and transistor failure may occur. Further, there is excessive voltage across the lamp before it lights, and lifetime may be reduced with frequent turn "ons."

Despite its drawbacks, some ballast manufacturers have used the voltage-fed circuit because of its lower required voltage rating on the transistors. With the current availability of inexpensive high-voltage transistors, however, the current-fed scheme of Figure 16.14*a* is preferable.

16.8 Current-Fed Parallel Resonant Half Bridge Topology[7]

This topology is shown in Figure 16.22. It is used when the AC input is 230 V and a power factor correction building block (Chapter 15) boosts rectified output to a voltage higher than the highest peak of the rectified 230 V AC.

For a ±15% tolerance on the AC input, that peak rectified voltage is $1.15 \times 1.41 \times 230 = 373$ V DC. The usual power factor correction circuit boosts that to 400 V, and the resonant half bridge is needed to cope with such a high voltage, which is more than a current-fed push-pull can handle.

For an AC input of 120 V and ±15% input tolerance, the peak rectified input is $1.15 \times 1.41 \times 120 = 195$ V, and the usual power factor correction building block boosts that to about 205 V (Section 16.6.2). It was seen in that section that the "off" transistor in current-fed resonant push-pull topology is subjected to $\pi(V_{dc}) = \pi(205) = 644$ V. For the 230 V AC, a current-fed push-pull circuit would subject the "off" transistor to $\pi(400) = 1257$ V DC, which would require too expensive a transistor.

It will be seen below that the current-fed parallel resonant half-bridge topology subjects the "off" transistor to only $(\pi/2)V_{dc} = (\pi/2)400 = 628$ V. There are numerous inexpensive candidates for such a transistor.

This circuit also is self-oscillating with windings $T2A$, $T2B$ on $T2$ providing the positive feedback. Here $T1$ is the main power transformer, and its magnetizing inductance, in shunt with C_r and the reflected ballast capacitor C_b, forms the parallel resonant circuit. Inductors $L1$ and $L2$ are the constant-current drive elements for the tank. The one lamp shown is a rapid-start type driven by the secondary for isolation.

Filament current for the lamp (or paralleled lamps, as the topology permits driving lamps in parallel) is taken in series from the transformer secondary and is limited by C_f and C_b. The circuit starts oscillating when the voltage across C_s has risen above the breakdown

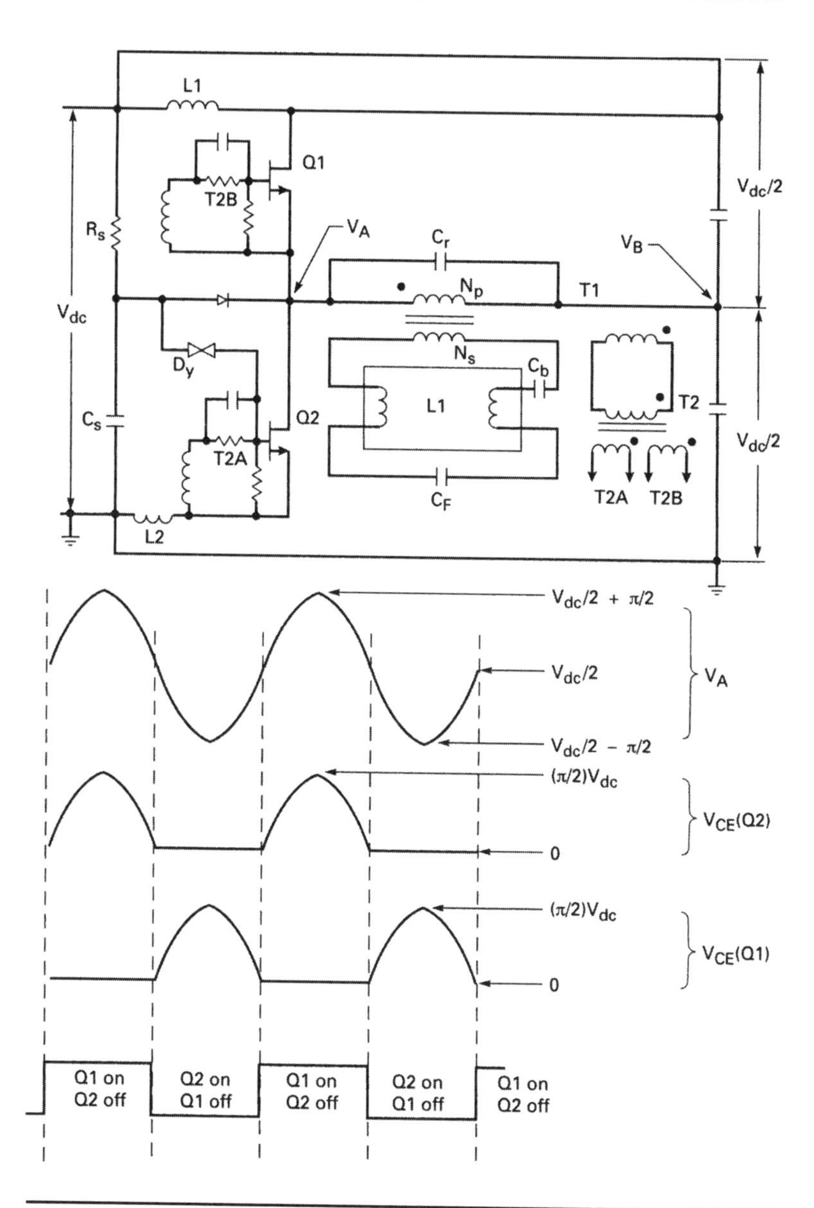

FIGURE 16.22 Current-fed half bridge topology.

voltage of diac D_y. When the diac fires, R_s supplies current into the base of Q_2, turning it "on." Thereafter, whenever $Q2$ turns "on," it discharges C_s, keeping it from interfering with the normal turn "on" voltage of the half sinusoid across $T2A$.

Waveforms at significant points are shown in Figure 16.14*b* to *e*. When $Q1$ turns "on," it produces a half sinusoid of positive voltage at V_A. When it turns "off," it produces a half sinusoid of negative voltage at V_A. The peak-to-peak voltage across the tank from V_A to V_B is $\pi V_{dc} = \pi(400) = 1257$ V. The RMS voltage across the tank is $V_{RMS} = 0.707 \times 1257/2 = 444$ V. Current in the primary is $I_{RMS} = V_{RMS}/X_{Lt}$, where L_t is the inductance of the transformer primary.

Transformer primary inductance L_t and C_t ($= C_t +$ reflected ballast capacitor C_b) are calculated as in Section 16.6.6 for the current-fed push-pull circuit. The number of primary turns N_p is calculated from Faraday's law (Eq. 16.10) for a tentatively selected core, and as high a peak flux density B_m as possible with still reasonably low core losses. Once N_p is selected, the core gap is chosen from curves (as in Figure 16.22) so that the selected N_p yields the chosen L_t. The ballast capacitor C_b is chosen from Eqs. 16.1, 16.2, and 16.3.

Figures 16.22*c*, *d*, and *e* show the maximum voltage stress across an "off" transistor is $(\pi/2)V_{dc} = 400(\pi/2) = 628$ V. There are many inexpensive 700-V transistors to meet this requirement.

16.9 Voltage-Fed Series Resonant Half Bridge Topology[5,6,7,8]

This topology is shown in Figure 16.23. It is used for an AC input line voltage of 230 V. Its advantage is that by eliminating any inductors in series with the rectified AC input or the power factor–corrected DC voltage, voltage stress on the "off" transistor is only V_{dc} instead of $(\pi/2)V_{dc}$, as in the current-fed half-bridge circuit. For a power factor–corrected voltage of 400 V, this means a maximum stress on the "off" transistor of 400 rather than 628 V and a much lower transistor cost.

Figure 16.23 shows a transformer driving the lamp to provide DC isolation. The circuit has been widely discussed in the literature in its non-isolated version, but transformerless circuits currently are not widely accepted.

It is apparent from Figure 16.23 that the "off" transistor is subjected to a maximum voltage stress of V_{dc}. The price paid for this advantage is that this series resonant circuit has the same problem of large-amplitude current spikes at turn "on," as discussed for the voltage-fed series resonant push-pull topology.

It is seen in Figure 16.23 that the series resonant circuit is in the $T1$ secondary and comprises L_r, C_r, the primary of current transformer, and $C1$ shunted by the lamp resistance. The high-current turn "on" spikes occur because at turn "on," the lamp resistance R_L is high and the equivalent series resistance $R_s = R_L/(1 + Q^2)$ of Figure 16.21*b*2 is

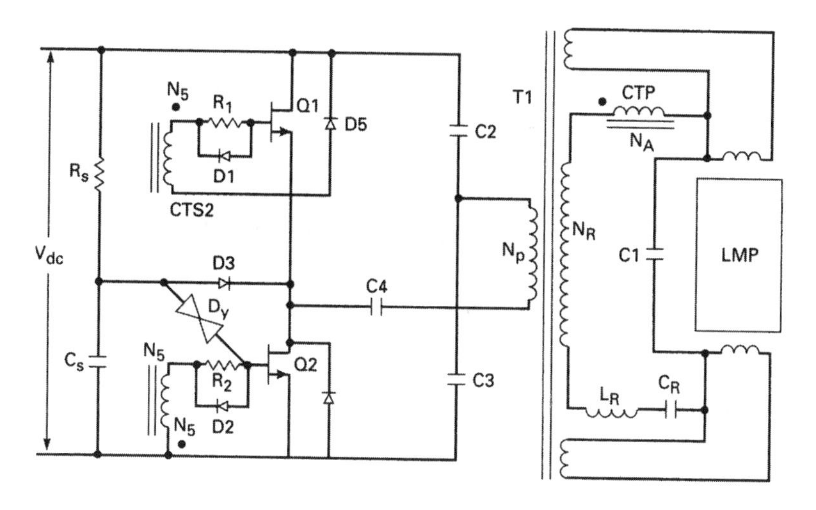

FIGURE 16.23 Voltage-fed series resonant half bridge topology. The series resonant circuit comprises L_r, and $C_r + C1$ in series at turn "on." When the lamp is lit, its low impedance shorts $C1$ out and the resonant frequency decreases. Current transformer primary CTP acts as the current limiting ballast impedance. It also acts as a proportional base drive transformer, as its turns ratio N_A/N_S is set equal to the minimum beta of the transistor.

very low. Resistance R_s is low because $Q = R_L C1$ is high (Eq. 16.12). Thus at turn "on," the impedance of the series resonant circuit is the low R_s alone, neglecting for the moment the impedance of the current transformer primary. It is this low R_s at turn-on that is the cause of the high-current spikes. After the lamp is lit, its impedance R_L falls, Q goes down, and R_s goes up, resulting in the normal current pulses for desired output power.

CTP in series with the resonant elements adds impedance at turn "on" and helps reduce the amplitude of the turn "on" current spikes. The current transformer is a proportional base drive transformer (Section 8.3.5). Its turns ratio N_A/N_S is set to the minimum transistor beta. This ensures adequate base drive at all current levels, as the collector/base current ratio will always equal the primary/secondary turns ratio. This guarantees adequate base drive at high output current and a reduced base drive at low current levels, which minimizes storage time.

At turn "on," the circuit oscillates at a frequency of $1/(2\pi\sqrt{L_r C_e})$, where C_e is the capacitance of C_r and $C1$ in series. When the lamp has lit, its low impedance shorts out $C1$ and the oscillation frequency drops to $1/(2\pi\sqrt{L_r C_r})$. The start circuit comprising R_s, C_s, diac D_y, and $D3$ works just as for the parallel resonant half bridge of Section 16.8, and C_4 is simply a DC blocking capacitor.

Instant Start

High Performanc Electronic Ballast

High performance features and maximum savings combined with the convenience of parallel wiring. Designed and manufactured to Sigma Six quality standards.

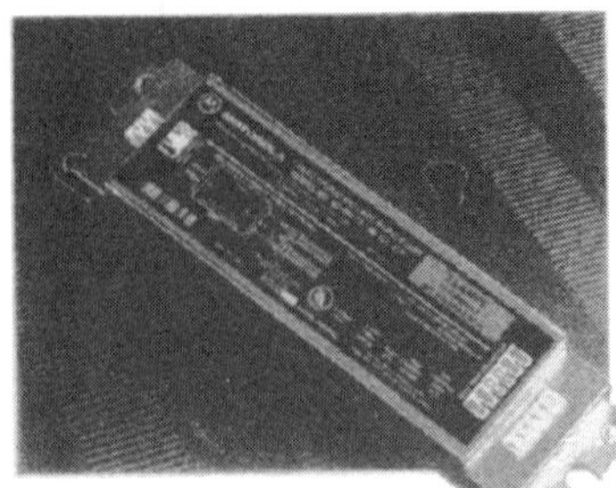

T5 Fluorescent

High Performanc Electronic Ballast

Electronic ballast benefits for the versatile T5 compact fluorescent lamp. These proven performers offer cooler operation than magnetic and up to 28 percent energy savings.

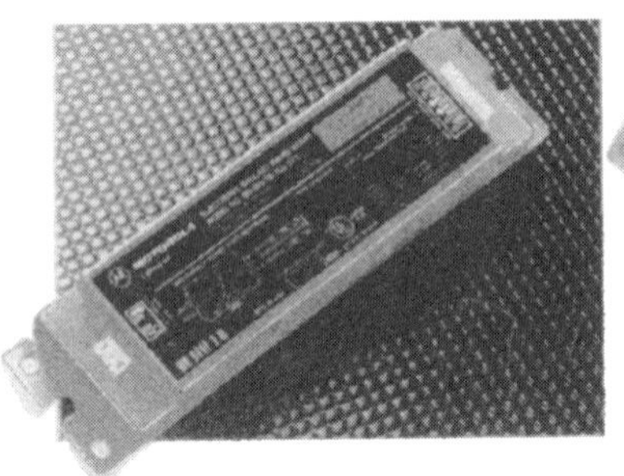

Rapid Start

High Performanc Electronic Ballast

Motorola Lighting High Performance ballasts and T8 lamps are designed to be the optimum combination for energy efficiency, power line quality and lighting system performance.

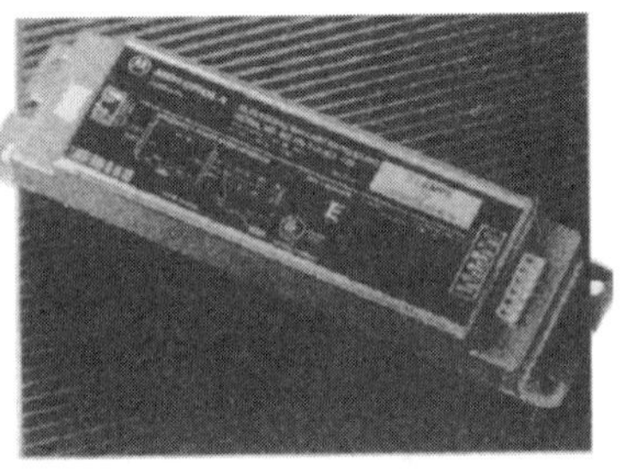

Rapid Start T12, T10

High Performanc Electronic Ballast

Highest performance choice for standard T12 systems. Also operates four foot T10 lamps.

FIGURE 16.24 Packaged ballast from one major manufacturer. (*Courtesy Motorola Lighting Inc.*)

In general, the circuit is not as easy to analyze and design as the current-fed circuit. It is not as easy to ensure that a worst-case design has been obtained and all units coming off a production line will be alike. This is so because of the odd transient conditions at turn "on" and because exact equivalent circuits for various phases of the operating cycle are not easily definable, and the exact lamp current is not easily calculable.

16.10 Electronic Ballast Packaging

The entire electronic ballast—the input rectifier, power factor–correction building block, and DC/AC converter—must be packaged to fit within the standard housing of a conventional magnetic ballast. Figure 16.24 shows that packaged ballast of one major manufacturer.

After Pressman *Modern electronic ballasts normally incorporate boost-type power factor correction. Many control chips now incorporate the control circuits for the power factor correction. The manufacturer's application notes provide typical examples and very few extra external components are required. ~K.B.*

References

1. "Fluorescent Lamps," General Electric Bulletin, General Electric Lighting, Cleveland, OH.
2. Motorola Lighting Inc., Technical Publication, Buffalo Grove, IL.
3. G. Meyers and J. C. Heffernan, "The Role of Crest Factor in Fluorescent Lamp Starting," Sylvania Lighting Products, Danvers, MA.
4. GTE/Sylvania Publication, Sylvania Lighting Center, Danvers, MA.
5. *ANSI Fluorescent Lamp Specifications,* American National Standards Institute, New York.
6. "Efficient Fluorescent Lighting Using Electronic Ballasts," Philips Semiconductor Inc., Saugerties, NY.
7. R. J. Haver, "Electronic Ballasts—Power Conversion and Intelligent Motion," April 1987.
8. R. J. Haver, "Solid State Ballasts Are Here," *Electronic Design News,* November 1976.
9. "KoolMu Powder Cores," Magnetics Inc., Butler, PA.
10. "Iron Powder Cores," Micrometals Inc., Anaheim, CA.
11. M. Bairanzade, J. Nappe, and J. Spangler, "Electronic Control of Fluorescent Lamps," Motorola Inc., Application Note AN 1049.
12. "Electronic Ballast Fundamentals," Motorola Lighting Inc., Buffalo Grove, IL.

CHAPTER 17

Low-Input-Voltage Regulators for Laptop Computers and Portable Electronics

17.1 Introduction

The explosion in the use of laptop computers and portable electronics in recent years has led to the formation of a new sector of the power conversion industry. This sector consists of low-input-voltage, battery-fed, boost, buck, and polarity-inverting configurations (Sections 1.3 to 1.5). They are almost entirely contained in one integrated-circuit (IC) package, and externally most require only a single inductor, capacitor, and diode, plus about three to five small resistors.

Since these designs operate at frequencies from 60 to 500 kHz, their external capacitors and inductors are small. They differ from the commonly used PWM control circuits in that they have the main power switch transistors inside the package. Since they are battery-operated, the output ground need not be isolated from input ground. This eliminates components such as optocouplers, pulse transformers, and housekeeping supplies on output ground (previously associated with sensing a voltage on output ground and controlling a pulse width on the input ground).

The output powers range from 0.5 up to 100 W, depending on the topologies and input and output voltages in the manufacturer's specific type numbers. Efficiencies range from 80 to 95%, thus minimizing heat sink size and, for a large range of output powers, obviating their need entirely. For the manufacturers' various specific types, input voltages for boost regulators range from 3 V (two battery cell minimum) up to 60 V, and for buck regulators from 4 to 60 V.

In addition to their use as stand-alone, battery-operated, DC/DC converters, these devices make a quick turnaround design possible at relatively low cost without too much effort. Typically, the application requires multi-output off-line power supplies in a distributed power supply system. Thus, a conventional off-line power supply can be designed to generate the usual +5-V, high-current isolated secondary. Other slave voltages can be generated at the point of use in either of two ways, depending on their output powers. Relatively low-power slave outputs can be generated by busing the +5-V output to the point of use and boosting it to the required output voltage with one of the above IC regulators. The boost regulators can also generate negative slave voltages from the +5-V input. For higher slave powers, it is more efficient though slightly more expensive to add another slave of about +24 V and bus it around to the points of use. There, IC buck regulators can convert it to the desired slave outputs.

Compared to the conventional scheme of generating slave outputs from added slave secondary windings on the main power transformer, this may appear at first glance more expensive and less efficient, but the advantages may outweigh the drawbacks. This will be discussed in detail below.

17.2 Low-Input-Voltage IC Regulator Suppliers

IC building blocks are available from several major U.S. manufacturers, in particular Linear Technology Corporation (LTC) in Milpitas, California, and Maxim Integrated Products in Sunnyvale, California. Their products will be discussed in detail here. Texas Instruments and Motorola also offer some products for this market, but they are not covered here.

The discussion here will cover only the applications of these devices, with some description of their internal design as necessary. The basic material comes from the suppliers' catalogs, which have an enormous number of products—boost and buck regulators and polarity inverters. Some devices have variable input and adjustable outputs, fixed input and fixed output (+5-V input, +3.3-V output), variable input and fixed outputs (bucking anything from +8 to +40-V input down to voltages of +5, +12, or +15 V). A large group are tailored for boosting low battery voltages from one to four series cells of 1.5 V. The devices are most often pulse-width-modulated, fixed-frequency types, operating at 40 to 500 kHz. Some operate with a fixed "on" time and vary frequency to achieve regulation.

17.3 Linear Technology Corporation Boost and Buck Regulators[1]

Examples of typical LTC boost and buck regulators are shown in Figure 17.1*a* and *b*. Boost regulators will be considered first.

LTC offers two families of boost regulators—a high-current one with switch output currents ranging from 1.25 to 10.0 A, and a micro-power family with output switch currents ranging from 95 to 350 mA.

The boost regulator in the discontinuous and continuous conduction modes is described in Sections 1.4.1 to 1.4.4 and 15.3.2 to 15.3.3. Over most of the current range of the LTC chips, the circuits operate in the continuous mode whose V_o/V_{in} relation is $V_o = V_{in}/(1 - T_{on}/T)$ (Eq. 15.1). Here T_{on} is the time the internal power switch is "on" out of a total period T. In the continuous mode, the "on" time remains constant (Figure 15.5). If the output load current decreases sufficiently, the circuit moves from the continuous to the discontinuous mode. That is no problem, for if the feedback loop has been stabilized for the continuous mode (by selecting $R3$ and $C1$ in Figure 17.1*a*), it will remain stable after entering the discontinuous mode.

To build a boost regulator with an LTC chip, the only external components required are shown in Figure 17.1*a*—the inductor $L1$, capacitor $C2$, diode $D1$, sampling resistors $R1$ and $R2$, and feedback loop stabilizing components $R3$ and $C1$. The design of the complete circuit requires only selection of these components.

Recalling how a boost regulator works (Sections 1.4.2 and 1.4.3), we see that a sink for current into ground is required at the bottom end of $L1$ (Figure 15.5*a*). This is provided in the LTC chip as an *NPN* power switch. Thus *boost* chips have an *NPN* collector at the output terminal marked V_{sw}, with the emitter at the GND terminal, as shown in Figure 17.1*a*.

Recalling how a buck regulator works (Section 1.3.1), we see in Figure 1.4 that a transistor switch is required at the input to $L1$ to interrupt the current coming from the source voltage, which is to be bucked down. As seen in Figure 17.1*b*, this is provided in the LTC buck chips by an *NPN* power switch transistor whose emitter is internally connected to the V_{sw} pin, and whose collector is connected internally to the V_{in} pin.

Although boost chips can be connected as buck regulators, this requires additional components and it is best to buck with buck chips. Buck chips can also be used to boost a negative voltage to a more negative voltage. These alternate configurations will be shown below.

The devices within each family have nearly identical internal block diagrams and differ only in operating frequency, maximum switch and input voltages, and maximum switch current ratings. The later devices in each family operate at higher frequencies (100 to 500 kHz)

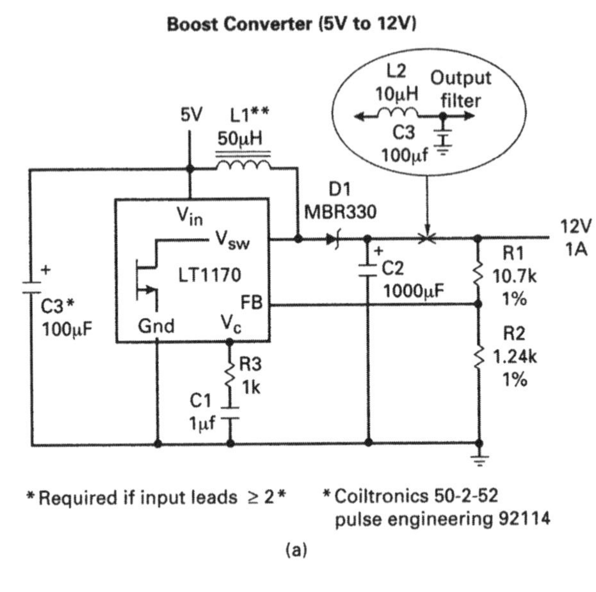

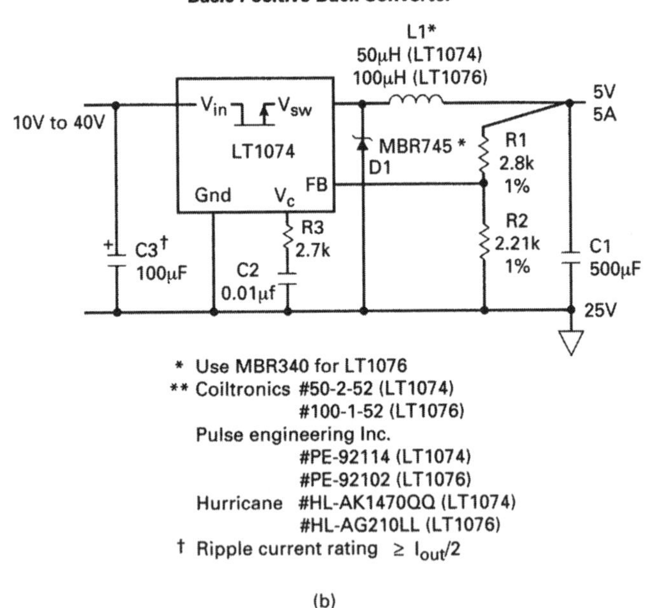

FIGURE 17.1 Basic boost and buck regulators from Linear Technology Corporation, Inc. These regulators have all the usual PWM control circuitry, plus the power output switch transistor built in. For most applications, the only external components required are $L1$, $D1$, $C1$, and $C3$.

to minimize the inductor size, as that inductor is usually the largest component of the regulator. Since the devices are typically fed from rechargeable batteries, a major objective is to maximize efficiency and maximize time between recharging. Most of the dissipation in these devices comes from the volt/ampere drop across the internal *NPN* power transistor and the external diode (Figure 17.1*a*). Hence in the newer devices, the switch is often a low R_{ds} power MOSFET.

The discussion to follow will cover in detail only the most often used members of the buck and boost families. Other members of each family will be listed only in tabular form with their significant specifications shown to permit quick device selection.

17.3.1 Linear Technology LT1170 Boost Regulator[3]

A basic boost regulator is shown in Figure 17.1*a*, and its internal circuitry in block diagram form is shown in Figure 17.2. The device uses current-mode topology, whose advantages are discussed in Section 5.3. Essentially, the power switch "on" time commences when the oscillator pulse sets a flip-flop in the logic section, and ends when it is reset by the comparator output. This reset instant is determined by the peak current in the power transistor, hence the description as *current mode*.

The comparator compares the DC output of the voltage error amplifier to the ramp-on-a-step voltage output of the current amplifier. At the instant the peak of the current waveform, which is converted to a voltage by R_S, exceeds the voltage error amplifier output, the comparator output goes positive and resets the flip-flop in the logic block, and the power switch turns "off."

There are thus two feedback loops. The voltage error amplifier senses output voltage to keep it constant by setting the threshold level which the ramp-on-a-step voltage waveform from the current amplifier must cross to reset the flip-flop. The second feedback loop monitors peak power switch current on a per cycle basis and keeps it constant.

The power switch current has a ramp shape because when it turns "on," there is a fixed voltage across inductor $L1$ (Figure 17.1*a*) and current rises at a rate of $dI/dT = (V_{in} - V_{cesat})/L1$. The current amplifier has a gain of 6. Its purpose is to increase the slope of the ramp without having to increase R_S, as this would increase dissipation. A larger signal at the comparator input is desirable to increase the signal to noise ratio, as small-amplitude noise spikes on a shallow slope can prematurely reset the flip-flop and shorten the power transistor "on" time, thereby producing instability in the output voltage.

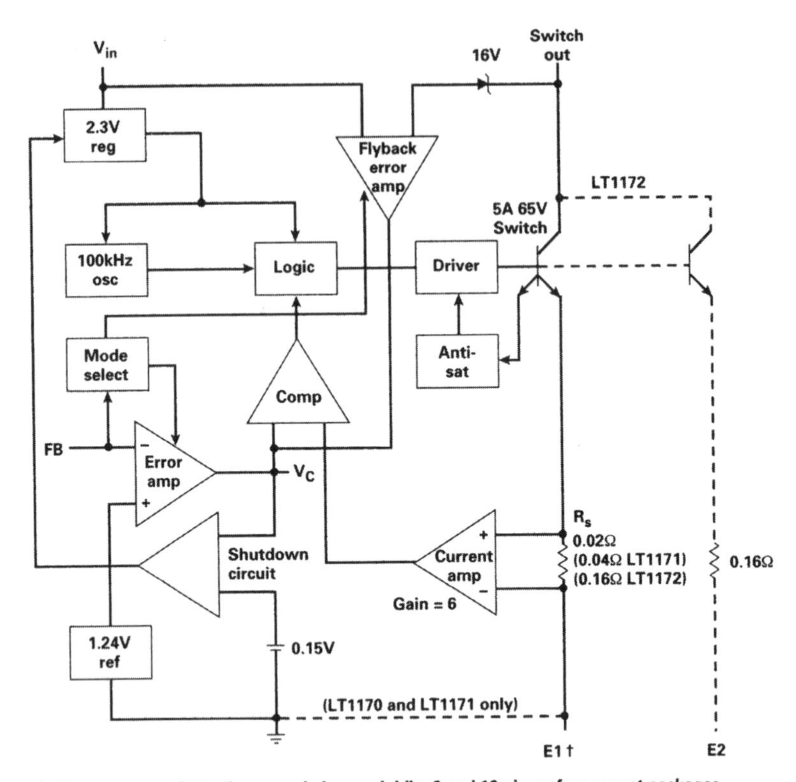

(a)

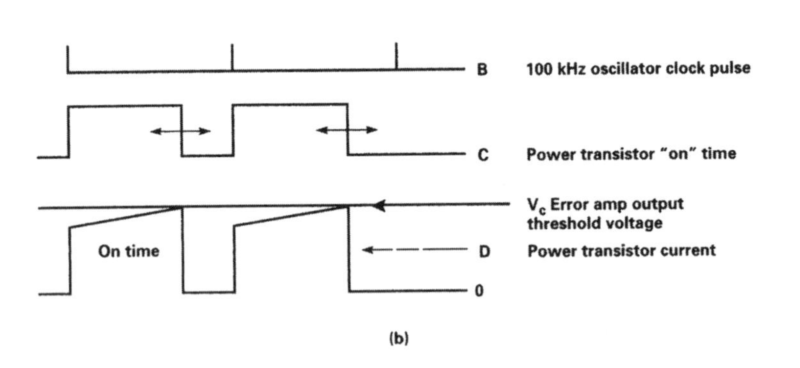

(b)

FIGURE 17.2 (*a*) LT1170 100-kHz, 5-A boost regulator; (*b*) LT1170 boost regulator waveforms. This is a classical current-mode boost regulator. Output transistor "on" time is initiated by the clock pulse, and is terminated when the ramp-on-a-step voltage waveform on sensing resistor R_S crosses the threshold V_c, set by the output voltage error amplifier.

The error amplifier output voltage, which controls the instant that the flip-flop resets and the power transistor turns "off," is brought out at the V_C pin. The voltage V_C ranges from 0.9 to 2.0 V, and is internally clamped at 2 V to limit the power switch peak current. This peak current limit point can be reduced externally by clamping V_C to a regulated voltage less than 2 V through a Schottky diode. This peak current should be selected for the maximum duty cycle, (which occurs at minimum input voltage), when average transistor current and hence transistor dissipation are maximum. This current is determined by thermal considerations which are described below.

The reference voltage at the error amplifier input is 1.24 V. Note that in all LT1170 applications, the output voltage sampling resistor string has a resistance of 1.24 kΩ from FB pin to ground ($R2$, Figure 17.1a). This is done to facilitate the calculation of the value of the resistor from the FB pin to the output voltage node ($R1$, Figure 17.1a). Since the DC voltage at the FB pin must equal the 1.24-V reference, current drawn by $R2$ is $1.24/1240 = 1$ mA, and since the same 1 mA must flow through $R1$, the voltage across it is $0.001R1$ and the output voltage is $V_o = 1.24 + 0.001\,R1$.

The LT1170 operates at a switching frequency of 100 kHz, and has a maximum power switch current rating of 5A. Minimum and maximum voltage ratings at the V_{in} pin are 3 and 40 V, respectively. Maximum switch output voltage is 65 V.

17.3.2 Significant Waveform Photos in the LT1170 Boost Regulator

It is instructive to examine some voltage and current waveforms in an LT1170 test circuit, and note how clean and glitch-free they are at the 100-kHz switching rate. Figure 17.3 shows the boost regulator with +12-V output from which the waveforms were taken. Waveforms show line regulation at maximum and minimum loads, for input variation from 4 to 8 V (Figure 17.4) as well as load regulation at +5-V input (Figure 17.5) and a 10:1 load variation of 0.082 to 0.82 A.

Figures 17.4a and b show power switch voltage and current at +12 V output and a minimum current of 82 mA for input voltages of +4 to +8 V. Note the current waveforms have ramps that are characteristic of discontinuous-mode operation. Output voltage increased by only 0.02 V for an input voltage increase from +4 to +8 V. Efficiency was 84%, which is reasonably good at an output power level of 1.2 W.

Figures 17.4c and d show transistor power switch voltage and current at +12-V output and maximum output current of 823 mA for input voltages of +4 to +8 V. The output voltage changed by only 0.06 V under these conditions. Waveforms were still glitch-free, and

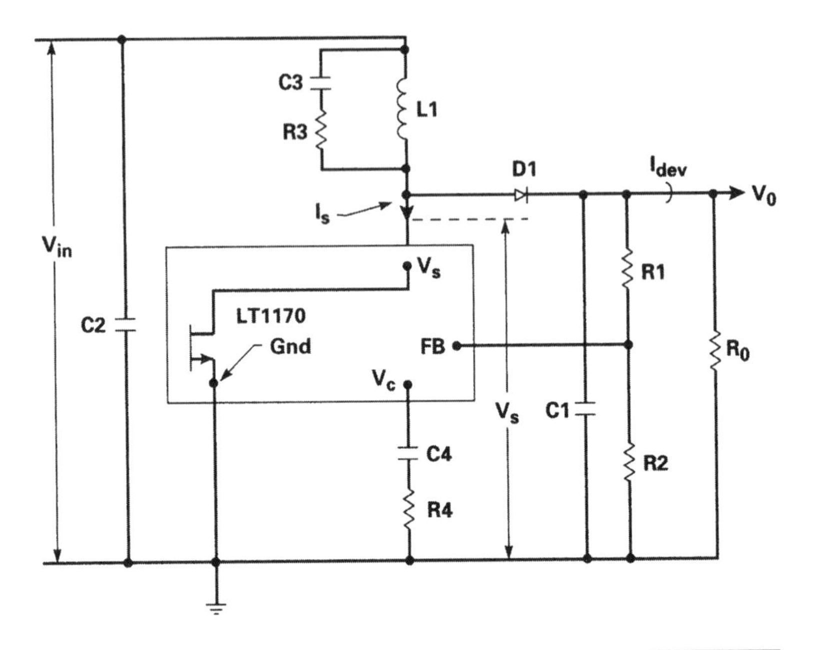

FIGURE 17.3 Actual LT1170 test circuit from which waveforms of Figures 17.4 and 17.5 were taken:
L1–50 μH, 18 turns #20 on MPP 55930 core
R1–13.35 K
R2–1.5 K
R3–2.2 K
R4–440 V
C1–1000 μF, 25 V
C2–1000 μF, 16 V
C3–1000 pF
C4–1.0 μF
D1–MBR340P

the worst-case efficiency was 81%. The switch current waveform has a ramp-on-a-step that is characteristic of continuous-mode operation, and the transistor "on" times are defined exactly by the relation $V_o = V_{in}/(1 - T_{on}/T)$.

Figure 17.4*e* shows transistor switch current during the "on" time, and output diode current during the "off" time. These are classical waveforms characteristic of continuous-mode operation. Output diode current at the start of the "off" time is exactly equal to the transistor switch current at the end of the "on" time. Also, output diode current at the end of the "off" time is exactly equal to the transistor current at the start of the "on" time. The ramp slopes are determined by the inductor, and are discussed below.

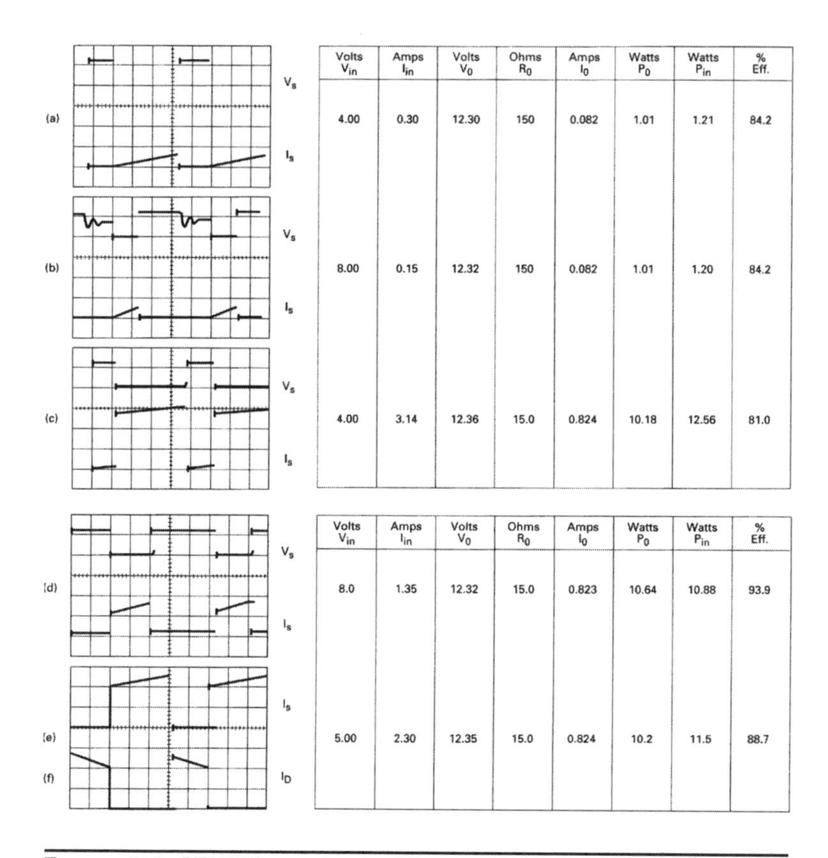

	Volts V_{in}	Amps I_{in}	Volts V_O	Ohms R_O	Amps I_O	Watts P_O	Watts P_{in}	% Eff.
(a)	4.00	0.30	12.30	150	0.082	1.01	1.21	84.2
(b)	8.00	0.15	12.32	150	0.082	1.01	1.20	84.2
(c)	4.00	3.14	12.36	15.0	0.824	10.18	12.56	81.0

	Volts V_{in}	Amps I_{in}	Volts V_O	Ohms R_O	Amps I_O	Watts P_O	Watts P_{in}	% Eff.
(d)	8.0	1.35	12.32	15.0	0.823	10.64	10.88	93.9
(e), (f)	5.00	2.30	12.35	15.0	0.824	10.2	11.5	88.7

FIGURE 17.4 LT1170 boost circuit of Figure 17.3: (*a*, *b*) line regulation at minimum load; (*c*, *d*) line regulation at maximum load; (*e*) transistor switch current; (*f*) output diode *D*1 current. All V_s waveforms at 10 V/cm, 2 μs/cm. All I_s waveforms at 1 A/cm, 2 μs/cm.

Figures 17.5*a* through *e* show transistor currents and voltages over a 10:1 load change from 823 to 82 mA for a constant input voltage of +5 V, boosted to an output of +12 V. Load regulation is excellent—output voltage varies only 0.03 V over the 10:1 load change. In Figures 17.5*a* through *d*, the "on" time remains constant, with the step part of the ramp-on-a-step amplitude decreasing as DC load decreases. In Figure 17.5*e*, the step has been lost, and the transistor "on" time has decreased, because operation entered the discontinuous mode.

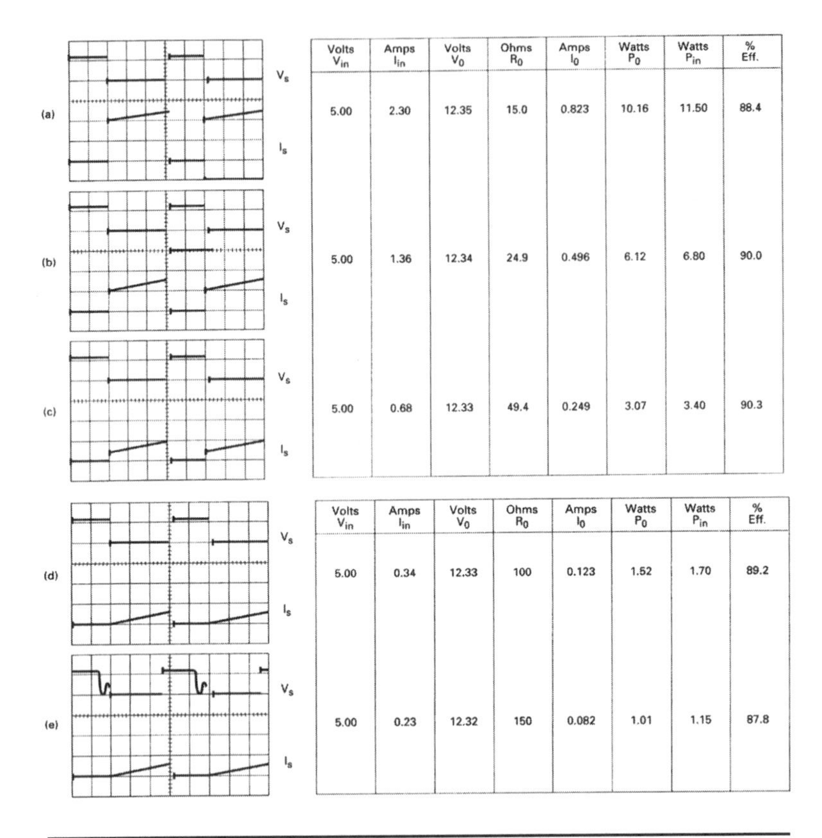

	Volts V_{in}	Amps I_{in}	Volts V_O	Ohms R_O	Amps I_O	Watts P_O	Watts P_{in}	% Eff.
(a)	5.00	2.30	12.35	15.0	0.823	10.16	11.50	88.4
(b)	5.00	1.36	12.34	24.9	0.496	6.12	6.80	90.0
(c)	5.00	0.68	12.33	49.4	0.249	3.07	3.40	90.3
(d)	5.00	0.34	12.33	100	0.123	1.52	1.70	89.2
(e)	5.00	0.23	12.32	150	0.082	1.01	1.15	87.8

FIGURE 17.5 Load regulation and efficiency for boost test circuit of Figure 17.3 at $V_{in} = 5.0$ V. All V_s waveforms at 10 V/cm, 2 μs/cm. All I_s waveforms at 1A/cm, 2 μs/cm. Note: "on" time remains constant in all continuous mode waveforms *a* to *d*.

17.3.3 Thermal Considerations in IC Regulators[3]

The integrated-circuit regulators considered here differ from the widely used PWM control chips (UC3525 family and similar voltage and current mode controllers) only in that they contain the power switch transistor inside the chip. This transistor is the major source of power dissipation for devices carrying more than about 1 A. Operating the regulator at its maximum specified transistor switch current can result in the need for a heat sink that is too large for the allocated space. Generally, however, the LTC regulators with transistor current ratings under 1 A operate quite safely with little or no heat sinking.

Thermal calculations to determine total regulator power dissipation are quite simple, and should be done early in the design. This consists of the following two parts:

1. Switch transistor dissipation = $I_{SW} V_{cesat} \times$ duty cycle
2. Internal control circuit dissipation = $V_{in} \times$ average current drawn by the V_{in} pin

$$I_{av} = 0.006 + I_{SW}(0.0015) + \frac{I_{SW}}{40}(\text{duty cycle})$$

Here 6 mA is the steady-state current drawn by the internal control circuitry, and the 0.0015 I_{SW} term is an increase in this steady-state current which is proportional to I_{SW}. The $(I_{SW}/40) \times$ duty cycle term is the average of the power switch base drive current, assuming an average switch transistor beta of 40.

Total chip dissipation (PD_{tot}) is the sum of parts 1 and 2. Most of these LTC regulators are rated at an absolute maximum operating temperature of 100°C, but this should be derated to about 90°C for greater margin.

Assume a maximum operating ambient temperature of 50°C and calculate, for example, the thermal resistance of the heat sink required for the peak specified transistor switch current of 5 A for an LT1170 in a TO220 package.

Assume the boost regulator output varies from +5 to +15 V. Then from Eq. 15.1, duty cycle T_{on}/T is 0.67 for a boost factor of 3. Part 1 above gives the power transistor dissipation as

$$PD_{SW} = I_{SW} V_{cesat} \times \text{ duty cycle}$$

LTC data sheets give V_{cesat} for all their regulators. This is shown in Figure 17.6 for the LT1170, and for 5-A peak current at 100°C, it is 0.8 V. Then for a duty cycle of 0.67, the power transistor dissipation is

$$PD_{SW} = 5 \times 0.8 \times 0.67 = 2.7 \text{ W}$$

For the LT1170 in a TO220 plastic package, the thermal resistance from junction to case θ_{jc} is 2°C/W.

For a transistor dissipation of 2.7 W, the transistor junction is $2 \times 2.7 = 5.4$°C above the case temperature. For 90°C maximum junction temperature, the transistor case must be at $90 - 5.4 = 85$°C.

Now from item 2 above, the average current drawn from V_{in} is

$$0.006 + 5 \times 0.0015 + {}^{5}\!/_{40} \times 0.66 = 0.096 \text{ A}$$

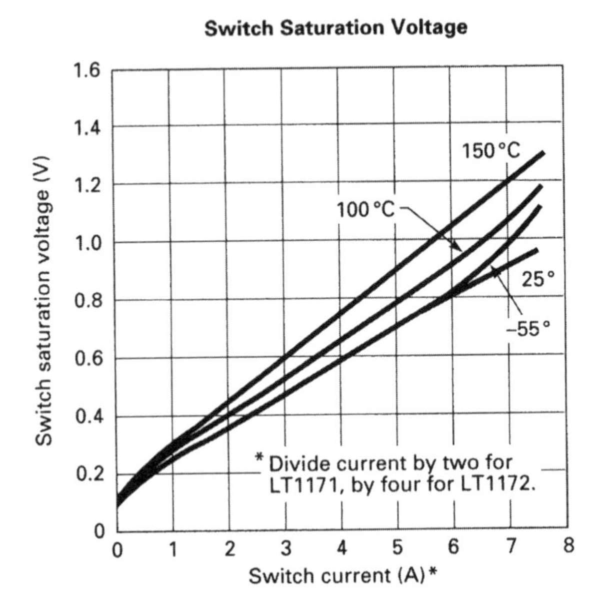

FIGURE 17.6 Power switch "on" voltage, LT1170 boost regulator. The "on" dissipation ($V_{on} I_{on} T_{on}/T$), rather than the peak current rating, determines how much peak current may be drawn. Attempting to operate at the device peak current could require a heat sink much larger than the device package itself.

For $V_{in} = 5$ V, dissipation due to this average current is $5 \times 0.096 = 0.70$ W. Total dissipation in the chip is then the sum of parts 1 and 2, or $2.7 + 0.7 = 3.4$ W.

Finally, for a TO220 case at the same temperature as that of the heat sink on which it is mounted, and for an ambient temperature of 50°C, the heat sink to ambient thermal resistance must be

$$\theta_{\text{hs amb}} = \frac{85 - 50}{3.4} = 10.3^\circ\text{C/W}$$

Referring to an Aham heat sink catalog, we see this could be achieved with a heat sink like the 342 – 1PP, which has a footprint area of 1.69 in by 0.75 in with four vertical fins of 0.75-in width and 0.60-in height. This is significantly larger than the device package itself.

This shows that the heat sink size, rather than the transistor peak current rating, determines the permissible peak operating current level.

17.3.4 Alternative Uses for the LT1170 Boost Regulator[5]

17.3.4.1 LT1170 Buck Regulator

Although intended as a boost converter, the LT1170 can perform other types of voltage conversion. Some are shown in Figure 17.7 (courtesy of Linear Technology Corporation).

Figure 17.7*a* shows a positive buck regulator, but it requires more external components than a device designed specifically as a buck converter (Figure 17.1*b*). The switch in a positive buck converter must source current to the inductor. Thus in Figure 17.7*a*, the emitter of the internal power transistor must supply current to $L1$, but that emitter is connected internally to the GND pin, which is the negative end of the internal +1.24-V reference voltage. Thus the +1.24-V reference voltage at the input to the internal error amplifier switches between converter common and V_{in}. Hence to use the internal error amplifier, a sample of the output voltage that moves up and down with the GND pin must be provided.

This is achieved in Figure 17.7*a* with $R1$, $R2$, $R4$, $C2$, and $D2$. When $Q1$ inside the chip turns "off," GND falls to one diode ($D1$) drop below common, and $R4$ and $D2$ in series charge $C2$ to one diode drop below V_o. Thus if the drops in $D1$ and $D2$ are equal, $C2$ is charged to a voltage equal to V_o. The voltage across $C2$, which is referenced to the negative end of the internal reference voltage (GND), is the voltage that is regulated. This means that the output voltage differs from the regulated voltage on $C2$ by the difference in the drops on $D1$ and $D2$, which is load sensitive.

Regulators like those in Figure 17.1*b* are designed as buck regulators with the internal power transistor emitter connected to the V_{sw} pin so that its voltage can switch up and down to drive the external inductor. The internal reference is connected to the GND pin. Hence the output voltage sample, which is also referenced to ground, can be tied directly to the FB pin. The devices thus do not require the baggage of Figure 17.7*a* to function as a buck regulator. Devices like the LT1075 are discussed below.

17.3.4.2 LT1170 Driving High-Voltage MOSFETS or NPN Transistors

In Figure 17.7*b*, a MOSFET gate is connected to V_{in}, which is fixed with respect to ground and may have any value between +10 V and the maximum gate-to-source voltage of the MOSFET of about 15 V.

When the internal transistor in the regulator turns "on," it pulls the MOSFET source to ground. The gate is held at $+V_{in}$, and the MOSFET turns "on" with a gate-to-source voltage of V_{in}. Turn "on" is fast because a large current is available from V_{in} to charge the gate-to-source

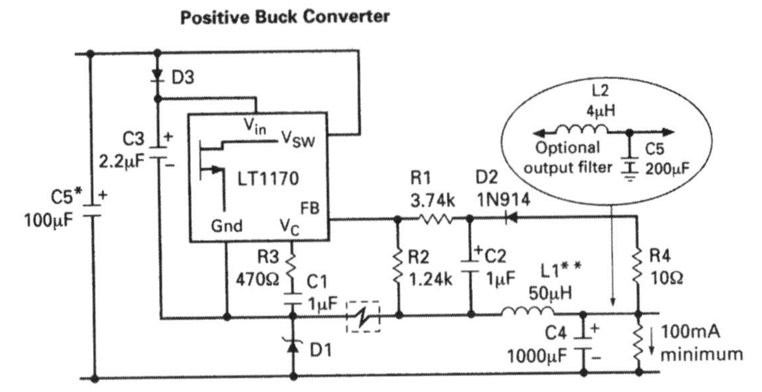

* Required if input leads ≥ 2'

** Pulse engineering 92114 coiltronics 50-2-52

(a)

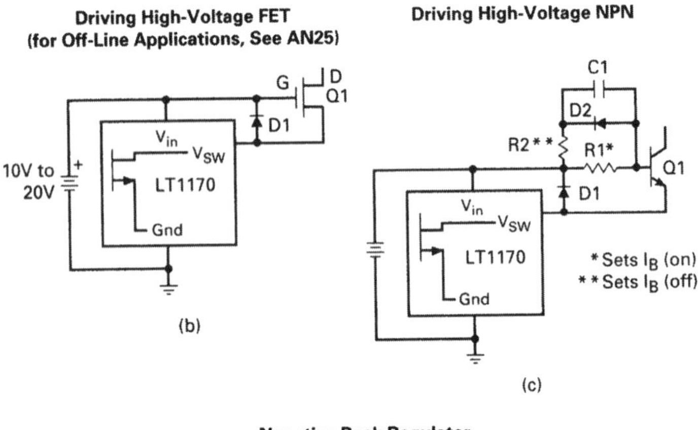

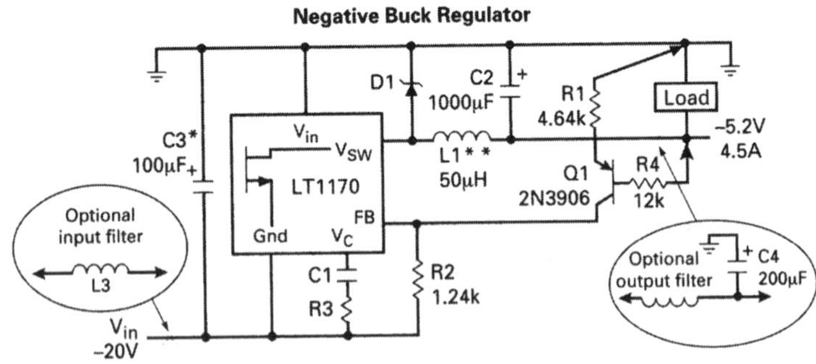

* Required if input leads ≥ 2'

** Pulse engineering 92114 coiltronics 50-2-52

(d)

FIGURE 17.7 Alternative uses for the LT1170 boost regulator.

Negative-to-Positive Polarity Converter

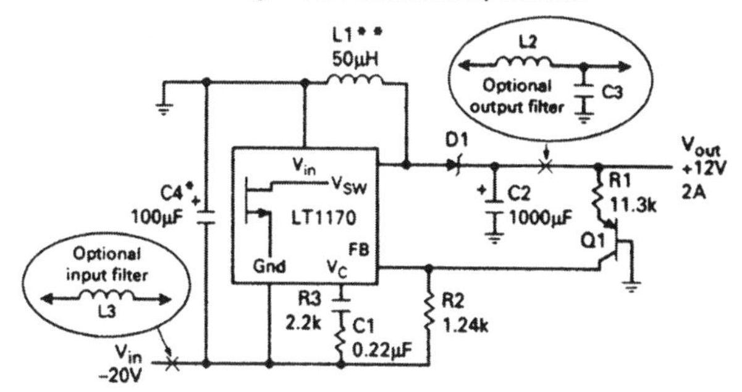

* Required if input leads ≥ 2'

** Pulse engineering 92114 coiltronics 50-2-52

This circuit is often used to convert –48V to 5V to guarantee full short-circuit protection the current limit circuit shown in AN19. Figure 39 should be added with C1 reduced to 200pF.

(e)

Positive-to-Negative Polarity Converter

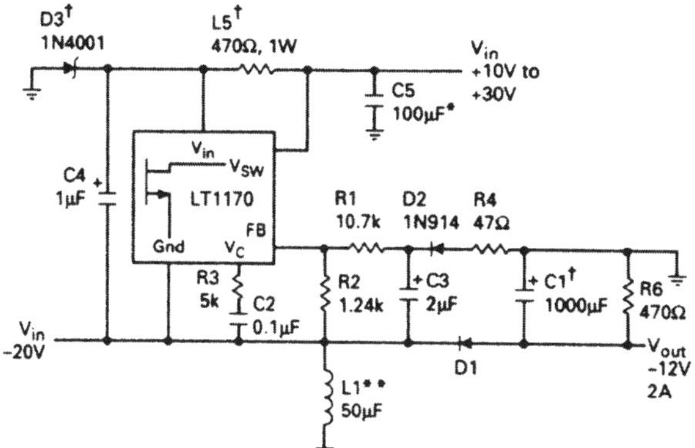

* Required if input leads ≥ 2'

** Pulse engineering 92114 coiltronics 50-2-52

† To avoid startup problems for input voltages below 10V. Connect anode of D3 to V_{in}, and remove R5. C1 may be reduced for lower output currents. C1 = (500μF)(I_{out}). For 5V outputs, reduce R3 to 1.5k, increase C2 to 0.3μF, and reduce R6 to 100Ω.

(f)

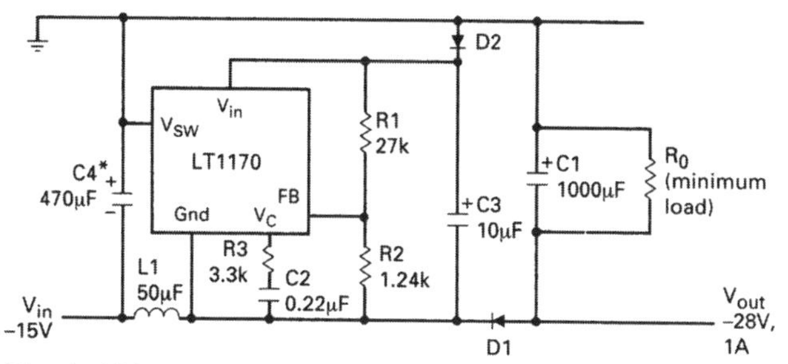

* Required if input leads ≥ 2'

(g)

FIGURE 17.7 *Continued.*

capacitance of the MOSFET. Turn "off" is slower— the gate-to-source capacitance holds the MOSFET "on" as the internal transistor turns "off."

The chip can drive an NPN transistor as in Figure 17.7*c*. The base of the transistor is driven by V_{in} through $R1$, which limits the base current. When the internal transistor is turned "on," $C1$ is charged via $R2$ to V_{in} with its right-hand end negative. When the internal transistor turns "off," the negative charge on $C1$ reduces the external transistor turn "off" time.

17.3.4.3 LT1170 Negative Buck Regulator

When the internal power transistor in Figure 17.7*d* turns "on," V_{SW} is shorted to GND at a voltage of $-V_{in}$. Its collector load is $L1$, which is bridged between $-V_o$ and $-V_{in}$, and the left-hand end of $L1$ is negative with respect to its right-hand end. When the internal power transistor turns "off," the voltage across $L1$ reverses polarity and the left-hand end of $L1$ is clamped to ground by $D1$. Thus the input end of $L1$ switches between $-V_{in}$ and ground. For an "on" time of T_{on} and period T, the average voltage at the input to $L1$ is $-V_{in}(T_{on}/T)$. The $L1$, $C2$ filter functions as in positive buck regulators, and is selected as in Sections 1.3.6 and 1.3.7. Constant current source $Q1$ and $R1$ force current in $R2$ that is proportional to V_o. The voltage across $R2$ is referenced to the chip's ground pin, and is the voltage that is regulated. Any change in the V_{be} of $Q1$ due to changes in operating point or temperature cannot be removed by the feedback loop, and is thus reflected into the output voltage as a small error.

17.3.4.4 LT1170 Negative-to-Positive Polarity Inverter

The converter in Figure 17.7*e* behaves like a positive boost regulator, but the 1.24-V reference voltage at the error amplifier input is referred to the GND terminal, which is at a voltage of $-V_{in}$ with respect to common. A sample of the output voltage, which is positive with respect to common, must be level-shifted to appear between pins FB and GND.

This is done with current source *PNP* transistor $Q1$, $R1$, and assuming the beta of $Q1$ is high, we get $I_{R1} = I_{R2}$. Neglecting the base-emitter drop of $Q1$, $V_{R2} = R2 I_{R1} = R2 V_o / R1$, this voltage is referred to the GND pin. Note that there are output voltage errors due to temperature variation of $Q1$'s beta and base-to-emitter voltage.

The output/input voltage relation is derived as in Section 15.3.2 for the case of a positive voltage boosted to a higher positive voltage. Since the volt-second product of $L1$ when the power transistor is "on" ($V_{in} T_{on}$) must be equal to that when the transistor is "off" ($V_o T_{off}$)

$$V_o = \frac{V_{in} T_{on}}{T_{off}} = \frac{V_{in} T_{on}}{T - T_{on}} = \frac{-V_{in}}{T/T_{on} - 1} \tag{17.1}$$

17.3.4.5 Positive-to-Negative Polarity Inverter

When the internal power transistor is "on" in Figure 17.7*f*, GND is pulled up to V_{in} via the saturated transistor, and current flows from V_{in} through $L1$ into ground. When the transistor turns "off," the voltage across $L1$ reverses polarity and pulls the bottom end of $C1$ negative via $D1$. Again the output/input relation is fixed by equating the volt-second product of $L1$ when the transistor is "on" to that when the transistor is "off." This yields the same expression as Eq. 17.1.

Here, as in Figure 17.7*a*, since the GND pin of the chip moves up and down between common and V_o, a sample of the output voltage must be shifted to between the FB and GND pins, because the internal reference voltage is referred to the GND pin. This is achieved by $R4$, $D2$, and $C3$. When the power transistor turns "off," the top end of $L1$ goes negative, pulls the bottom end of $C3$ negative, and clamps it with $D1$ to one diode drop below $-V_o$. The top end of $C3$ is clamped one diode drop below ground by $D2$, so assuming equal drops in those diodes, the voltage across $C3$ equals $-V_o$ and moves up and down with the GND pin.

As in Figure 17.7*a*, with $R2 = 1.2\ \text{k}\Omega$ across the FB and GND pins, the sampling circuit $R1$, $R2$ is a 1-mA circuit, and $V_o = 1.2 + 0.001R1$ volts.

17.3.4.6 LT1170 Negative Boost Regulator

When the internal power transistor turns "on" in Figure 17.7*g*, its emitter at GND pulls the top end of $L1$ up to ground, storing current in it. When the transistor turns "off," the voltage across $L1$ reverses polarity and pulls the bottom end of $C1$ negative via $D1$. Thus the GND pin swings between common $-V_{cesat}$ and one diode drop below $-V_o$. Again, equating the positive and negative volt-second products of $L1$ yields the output/input voltage relation:

$$V_o = \frac{-V_{in}}{1 - T_{on}/T} \tag{17.2}$$

The same circuit as in Figures 17.7*a* and *f* is used to transfer a sample of V_o to pins FB and GND, as the GND pin switches between roughly ground and $-V_o$.

17.3.5 Additional LTC High-Power Boost Regulators[5]

As mentioned in Section 17.3, Linear Technology offers a large number of boost regulators that differ only in frequency, voltage, and current ratings. A number of them are presented here in tabular form to permit quick selection. See Table 17.1.

LTC boost regulator	Input voltage, V		Switch voltage, V, max.	Frequency kHz	Switch	
	Min.	Max.			Current, A, max.	Resistance, Ω
LT1170	3.0	60	75	100	5	0.15
LT1172	3.0	60	65	100	1.25	0.60
LT1171HV	3.0	60	75	100	2.5	0.30
LT1270	3.5	30	60	60	8.0	0.12
LT1270A	3.5	30	60	60	10.0	0.12
LT1268	3.5	30	60	150	7.5	0.12
LT1373	2.4	30	35	250	1.5	0.50
LT1372	2.4	30	35	500	1.5	0.50
LT1371	2.4	30	35	500	3.0	0.25
LT1377	2.4	30	35	1000	1.5	0.50

TABLE 17.1

The specific selection is made on the basis of voltage and current ratings. The lower-current devices have a higher "on" resistance. They are generally less expensive but require a larger, perhaps more expensive heat sink. The higher-current devices, though more expensive, may be able to operate at the desired current with no heat sink at all. The next selection criterion is the operating frequency. Higher-frequency devices use smaller inductors (*L*1, Figure 17.1*a*), which are the largest and most expensive components, next to the regulators.

17.3.6 Component Selection for Boost Regulators[3]

Once the regulator chip has been selected, the major components to be chosen are (Figure 17.1*a*), *L*1, *D*l, and *C*2.

17.3.6.1 Output Inductor *L1* Selection

Somewhat poorer load regulation and greater input line current ripple in the discontinuous mode make it desirable to keep the circuit in the continuous mode down to minimum load. As seen in Figures 17.5*a* through *d*, as load current decreases, the transistor "on" time remains constant while the step part of the ramp-on-a-step decreases. Below this current, when the step has disappeared (Figure 17.5*e*) and discontinuous mode has started, the "on" time begins to decrease.

To decrease the "on" time, the sampling voltage at the internal error amplifier input must change somewhat, and hence the output voltage must change. In most cases, the output voltage changes are acceptable—they may amount to only 10 to 30 mV, as the error amplifier DC gain is very large.

The inductor $L1$ is selected as follows to maintain continuous mode operation down to minimum load. In Figure 17.4*e*, the input current is the sum of the transistor current when it is "on" and the $D1$ current when the transistor is "off." The total input current is at the center of the ramps in Figure 17.4*e*. Thus in Figure 17.5*a* or 17.5*d*, transistor current is a triangle starting from zero and the DC input is the average current at the center of the triangle at the low current limit of the continuous mode. This current is the input current ($I_{\text{dc min}}$) at minimum specified input voltage ($V_{\text{in min}}$).

Thus the change in the transistor input current, dI in Figure 17.5*d*, is $2I_{\text{dc min}}$ and

$$L = V_{\text{in min}}\frac{dt}{dI} = \frac{V_{\text{in min}}T_{\text{on}}}{2I_{\text{dc min}}}$$

From Eq. 5.1, $T_{\text{on}} = T(V_o - V_{\text{in}})/V_o$, so

$$L = \frac{V_{\text{in min}}(V_o - V_{\text{in min}})T}{2V_o I_{\text{dc min}}} \tag{17.3}$$

The current $I_{\text{dc min}}$ at the low end of continuous mode is usually set to 10% of the current at maximum input power.

Thus for the circuit of Figure 17.3, in which +5 V was boosted to +12 V, $T_{\text{on}} = T\,(12.3 - 5)/12.3 = 0.59T$, and from Eq. 17.3

$$L = \frac{5 \times 0.59T}{2 \times 0.1 \times 2.3} = 64\ \mu\text{H}$$

the closest value, a 50-μH inductor, was used—the exact value is not critical; it only sets the minimum current for continuous mode.

17.3.6.2 Output Capacitor C1 Selection[3]

The output filter capacitor $C1$ in a boost regulator is selected for minimum equivalent series resistance (ESR) to minimize peak-to-peak ripple at the switching frequency. This ripple is relatively large in a boost, as can be seen as follows.

When the internal transistor switch turns "on," diode $D1$ is reverse-biased and all the load current $I_{\text{dc }o}$ flows in $C1$. This causes a dip in output voltage of ESR $\times I_{\text{dc }o}$ volts, whose duration is T_{on}. When the transistor turns "off," there is a step in amplitude of ESR $(I_{\text{dc }o}\, V_o/V_{\text{in}})$. For $V_o/V_{\text{in}} = 3$, the peak-to-peak ripple is $4I_{\text{dc }o}$ ESR.

The simplest way to minimize this ripple voltage is to select $C1$ with minimized ESR, but capacitor vendors generally do not list that parameter for their devices.

It is of interest to calculate the peak-to-peak ripple voltage in a specific case. LTC Application Note AN 19-60 offers an empirical approximation to ESR as follows:

Mallory type VPR aluminum electrolytics:

$$\mathrm{ESR} = \frac{200 \times 10^{-6}}{CV^{0.6}}\,\Omega$$

Sprague 673D, 674D aluminum electrolytics:

$$\mathrm{ESR} = \frac{400 \times 10^{-6}}{CV^{0.6}}\,\Omega$$

If we assume a regulator boosting +5 to +15 V with 25-W (1.66-A) output, and a Mallory VPR 200 μF capacitor rated at 25 V, then from the above relation

$$\mathrm{ESR} = \frac{200 \times 10^{-6}}{200 \times 10^{-6} \times 25^{0.6}} = 0.145\ \Omega$$

The peak-to-peak ripple is thus $V_{o\,\mathrm{rp}} = 4I_{\mathrm{dc}\,o}\,(\mathrm{ESR}) = 4 \times 1.66 \times 0.145 = 0.963$ V.

Modern tantalum capacitors may have lower ESR and yield lower ripple. Ripple may be reduced by increasing capacitance, using capacitors with higher voltage ratings, or paralleling capacitors. All these techniques increase the required space. Compared to any of the above approaches, a smaller volume may result if a smaller capacitor with larger ESR is used and the resulting larger peak-to-peak ripple is eliminated with a small *LC* filter. Since the ripple frequency is twice the switch frequency, such a filter would be very small.

In a buck regulator, the switching frequency ripple is not as serious. There, as seen in Figure 17.1*b*, the output capacitor never supplies all the output load current by itself, whether the transistor is "on" or "off." The majority of that current is always being supplied by $L1$. Ripple current in the inductor flows in a loop through $C1$, diode $D1$, and back into the input end of $L1$, and that ripple current is minimized by selection of a relatively large inductor, as described in Section 1.3.6.

Because the output capacitor in a boost regulator supplies all the load current every time the power transistor turns "on," it is important to verify that the capacitor does not exceed its ripple current rating. Manufacturers often do not specify maximum ripple current limits for their devices.

17.3.6.3 Output Diode Dissipation

The output diode ($D1$, Figure 17.3) is most often a Schottky type, as its dissipation is second only to the power transistor inside the chip. Its power dissipation, assuming a 0.5-V drop when it conducts, is

$$\mathrm{PD_{D1}} = \frac{0.5 I_{\mathrm{in\ max}} T_{\mathrm{off}}}{T} = \frac{0.5 P_{\mathrm{in}}}{V_{\mathrm{in\ min}}} \frac{T_{\mathrm{off}}}{T} \quad \mathrm{W}$$

17.3.7 Linear Technology Buck Regulator Family

Buck regulators (Figure 1.4) were the earliest type of switching regulators, and are discussed in detail in Section 1.3. The theory of operation, significant waveforms, and component selection are described. The concept of continuous- and discontinuous-mode operation is introduced, and waveform photos of the transition between the two modes as the output load current is decreased are shown (Figures 1.6 and 1.7).

The basic operation of LTC integrated-circuit regulators is much like that of the circuit of Figure 1.4. They produce V_o from a higher V_{in} voltage by introducing a low-impedance saturating transistor switch between the source and the output filter. By modulating its T_{on}/T ratio, this yields a DC output voltage $V_o = V_{\mathrm{in}}(T_{\mathrm{on}}/T)$ after LC filtering. These integrated-circuit buck regulators have all the circuitry of Figure 1.4, including the power switching transistor, inside the package. In the simplest case, the only external components are L_o, C_o, $D1$, R_1, and R_2. They operate at fixed frequencies from 100 to 1 MHz and regulate by modulating T_{on}, as shown in Figure 1.4, or by operating with a constant "off" time and varying frequency. Most use current-mode topology (Section 5.2) rather than the voltage-mode scheme of Figure 1.4.

LTC offers an enormous number of new designs, and there is a driving force for them to improve efficiency, as their major use is for rechargeable, battery-operated equipment—laptop computers and portable consumer-type electronics. For such equipment, any improvement in efficiency offers a increase in the time between battery rechargings. Since the volt-ampere drop across the saturated switch is one of the major dissipators, it is replaced by a low R_{ds} power MOSFET in new designs.

17.3.7.1 LT1074 Buck Regulator

A typical example of high-power LTC positive buck regulator, the LT1074 is shown in Figure 17.1*b*, and its internal block diagram is depicted in Figure 17.8*a*. It is seen in Figure 17.7*a* that a "boost" regulator

Block Diagram

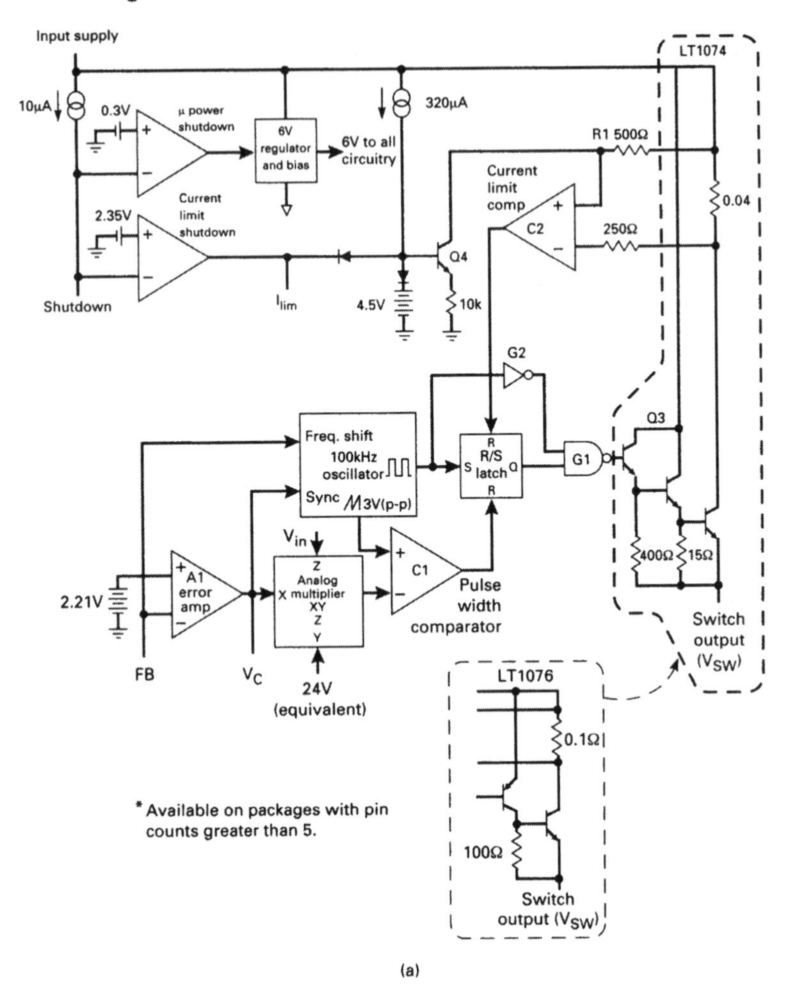

FIGURE 17.8 Linear Technology LT1074 100-KHz, 5-A buck regulator.

chip can also be used as a positive buck regulator but at the cost of added components (*D*2, *C*2, *R*1, and *R*2). The power transistor emitter in a chip designed primarily as a buck, such as the LT1074, is not fixed to ground and is available at the V_{SW} pin. There it can switch up and down, and source current into *L*1 from the internal collector connection to V_{in}. It thus does not need the extra circuitry of Figure 17.7*a* to function as a buck. The essentials of the LT1074 are given below (see Figure 17.8).

Typical Application

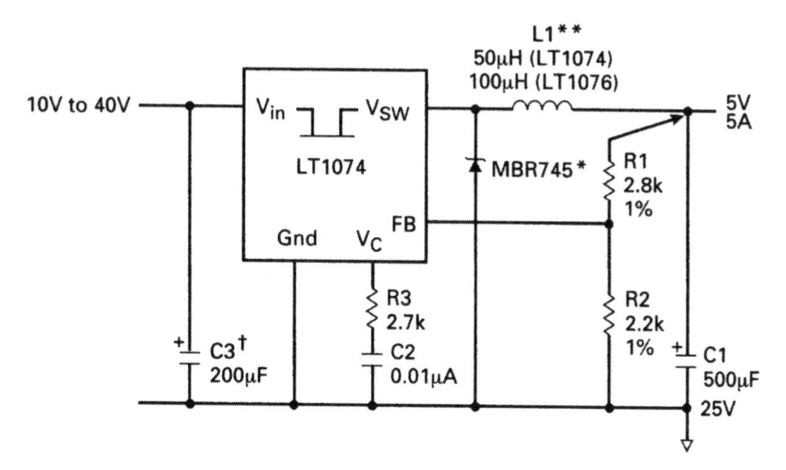

* Use MBR340 for LT1076
** Coiltronics #50-2-52 (LT1074)
#100-1-52 (LT1076)
Pulse Engineering, Inc.
#PE-92114 (LT1074)
#PE-92102 (LT1076)
Hurricane #HL-AK147QQ (LT1074)
#HL-AG210LL (LT1076)
† Ripple current rating $\geq I_{out}/2$

(b)

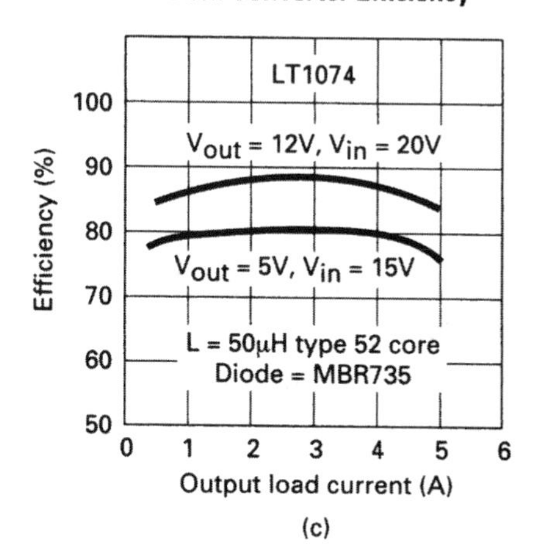

(c)

FIGURE 17.8 *Continued.*

The Darlington power transistor supplies the high specified output current of 5 A. The Darlington base needs a source of current, which is supplied by *NPN* transistor *Q*3. *Q*3 is driven by positive NAND gate *G*1, whose output goes low when both its inputs go high. That occurs after the R/S latch has been set and its *Q* output has gone high. The output of *G*1 is inhibited by a "low" at the *G*2 output for the duration of the positive trigger pulse from the oscillator, which sets the latch. This limits the maximum "on" time of the power transistor to a full period less the trigger pulse width, as a number of harmful things can occur if the power transistor never turns "off." This limits how close V_{in} may come to V_o without losing regulation.

The "on" time is terminated when the output of comparator *C*1 goes positive to reset the latch. The comparator compares the multiplier output to an internally generated 3-V triangle. The instant the triangle exceeds the multiplier output, the comparator output goes positive and resets the latch, turning "off" the power transistor. This is a *voltage-mode* circuit—the power switch "on" time is a function of only output voltage, and not of peak switch current as well (Section 5.2).

The multiplier output voltage is proportional to the voltage error amplifier output and inversely proportional to the input voltage. Thus an increase in V_o and the voltage at the FB pin, or in V_{in}, shortens the "on" time and keeps V_o constant. Feed-forward, introduced by making the multiplier directly responsive to a V_{in} change, instead of a change in V_o, results in faster correction of V_{in} changes.

Power transistor current is limited with *C*2. The threshold at which current limiting occurs is set by the small negative bias across *R*1, and that is controlled by the current source *Q*4. That current is kept relatively constant by the relatively constant voltage at the *Q*4 base. When the voltage drop across R_s exceeds the bias across *R*1, the output of comparator *C*2 goes positive, resets the latch, and turns "off" the power transistor.

Significant waveforms (idealized but quite accurate) in a typical LT1074 are shown in Figure 17.9.[6]

17.3.8 Alternative Uses for the LT1074 Buck Regulator

17.3.8.1 LT1074 Positive-to-Negative Polarity Inverter

The output/input relation for the inverter shown in Figure 17.10*a* can be derived by equating the volt-second product of *L*1 when the power transistor is "on" to that when the transistor is "off." When the transistor is "on," it brings the V_{sw} pin up to V_{in} and the volt-second product of *L*1 is $V_{in}T_{on}$. When the transistor turns "off," the polarity of the voltage across *L*1 reverses is clamped by *D*1 to V_o at the bottom

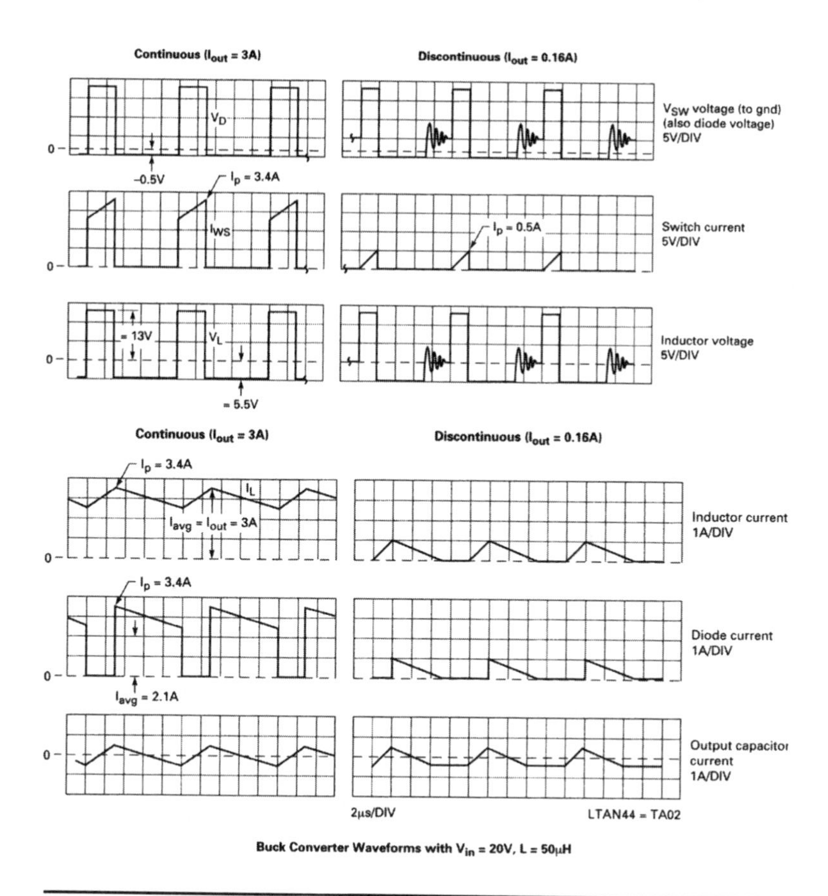

FIGURE 17.9 Significant waveforms for LT1074 buck regulator shown in Figure 17.8.

end of $C2$ for a time T_{off}. Then

$$V_{\text{in}}T_{\text{on}} = V_o T_{\text{off}} = V_o(T - T_{\text{on}})$$

17.3.8.2 LT1074 Negative Boost Regulator

A negative boost regulator is shown in Figure 17.10*b*. Again, we equate $L1$ volt-second products in the "on" and "off" states. When the power transistor is "on," V_{sw} is brought up to ground and the $L1$ volt-second product is $V_{\text{in}}T_{\text{on}}$. When the transistor is "off," the polarity of the voltage across $L1$ reverses and is clamped to the bottom end of $C1$ by

Positive-to-Negative Converter with 5V Output

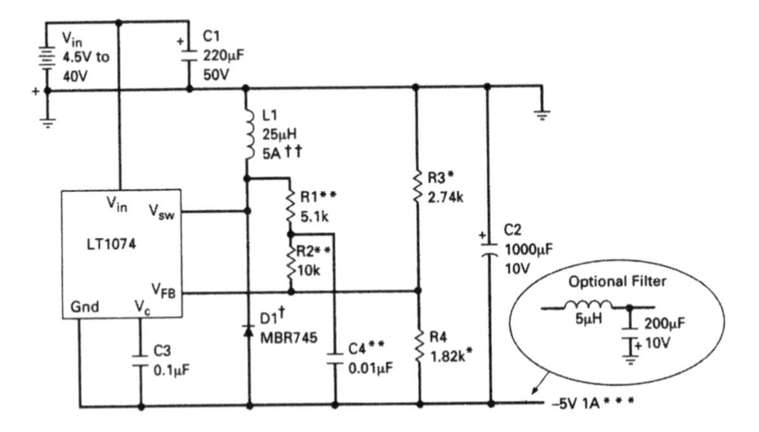

† Lower reverse voltage rating may be used for lower input voltages. Lower current rating is allowed for lower output current. See AN44.

†† Lower current rating may be used for lower output current. See AN44.

** R1, R2, and C4 are used for loop frequency compensation with low input voltage, but R1 and R2 must be included in the calculation for output voltage divider values. For higher output voltages, increase R1, R2, and R3 proportionately. For input voltage > 10V, R1 R2, and C4 can be eliminated, and compensation is done totally on the V_C pin.
$R3 = V_{out} - 2.37$ (KΩ)
R1 = (R3) (1.86)
R2 = (R3) (3.65)

*** Maximum output current of 1A is determined by minimum input voltage of 4.5V higher minimum input voltage will allow much higher output currents. See AN44.

* = 1% film resistors
D1 = Motorola-MBR745
C1 = Nichicon-UPL1C221MRH6
C2 = Nichicon-UPL1A102MRH6
L1 = Coiltronics-CTX25-5-52

(a)

FIGURE 17.10 Alternative uses for the LT1074 buck regulator. Note the triple Darlington output configuration of Figure 17.8 yields a 2.2-V "on" drop at 5 A, and may thus require a large heat sink.

*D*1 such that

$$V_{\text{in}} T_{\text{on}} = (V_o - V_{\text{in}}) T_{\text{off}} = (V_o - V_{\text{in}})(T - T_{\text{on}})$$
$$= V_o(T - T_{\text{on}}) - V_{\text{in}} T + V_{\text{in}} T_{\text{on}}$$

or

$$V_o = \frac{V_{\text{in}} T}{T - T_{\text{on}}} = \frac{V_{\text{in}}}{1 - T/T_{\text{on}}} \tag{17.4}$$

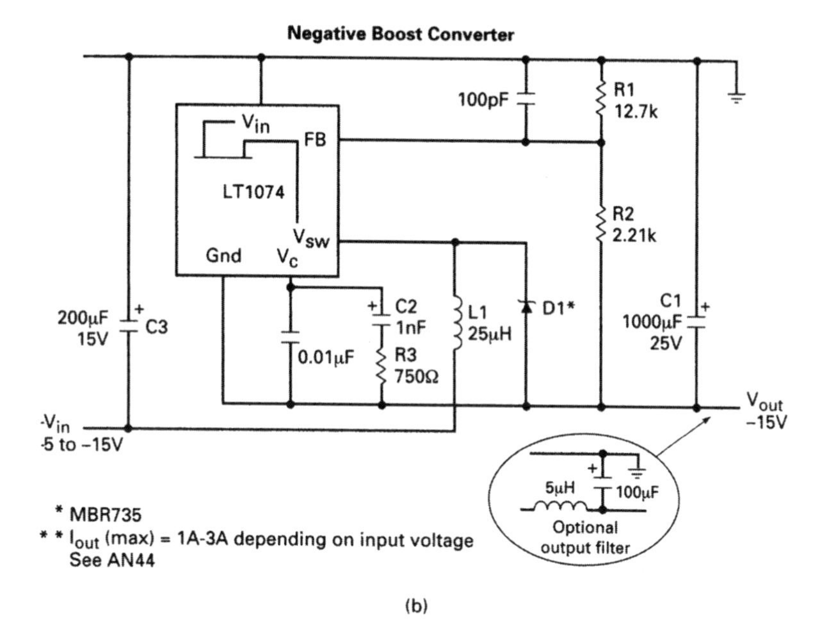

(b)

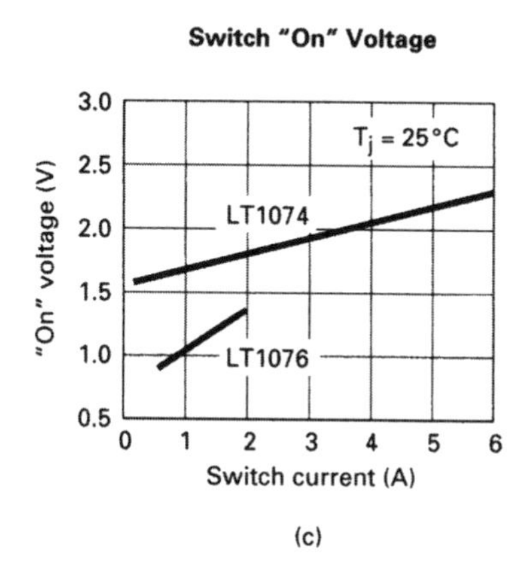

(c)

FIGURE 17.10 *Continued.*

17.3.8.3 Thermal Considerations for LT1074[5]

Let us consider a buck regulator for +24-V input, and +15-V, 5-A output. The prospect of doing this 75-W regulator with an LT1074 in a TO220 package with only L,C, and D components, as in Figure 17.1*a*, might seem very attractive at first glance. Thermal calculations will show that internal power dissipation, rather than the power transistor peak current rating, is often the limiting factor to a practical design.

Although internal dissipation varies with input and output voltages, attempting to operate at the transistor's peak current often results in a heat sink many times larger than the TO220 regulator package itself. This can be seen as follows:

For $I_o = 5$ A, Figure 17.10*c* shows the transistor's "on" voltage is 2.2 V. For $V_{\text{in nominal}} = +24$ V, $V_{\text{in min}} = +22$ V, duty cycle (DC) $= T_{\text{on}}/T = V_o/V_{\text{in min}} = 15/22 = 0.68$. Transistor switch dissipation is

$$PD_{SW} = V_{ce} I_o \times DC = 2.2 \times 5 \times 0.68 = 7.5 \text{ W}$$

The average dissipation of the control circuitry in addition to this is

$$PD_{cc} = V_{\text{in min}} I_{\text{in cc}}$$

where

$$I_{\text{in cc}} = 0.007 + 0.005 \times \text{ DC } + 2 I_o T_s F$$

Here the first term is the steady-state current drawn from V_{in}, and the second term is the increase in that proportional to output current. The last term is the average of the spikes of current lasting for the switching time T_s which are drawn from V_{in} at the instants of turn "on" and turn "off." For $F = 100$ kHz and $T_s = 0.06$ μs,

$$PD_{cc} = 22(0.007 + 0.005 \times 0.68 + 2 \times 5 \times 0.06 \times 10^{-6} \times 1 \times 10^{-5})$$

$$= 1.7 \text{ W}$$

The total internal dissipation is then $7.5 + 1.7 = 9.2$ W.

Assuming 50°C ambient temperature, calculate the size of the heat sink required for a desired maximum power transistor junction temperature of 90°C. With a TO220 package whose thermal resistance is 25°C/W, transistor case temperature is $90 - (7.5 \times 2.5) = 71$°C. If there is no temperature difference between the transistor case and heat sink, the permissible heat sink temperature rise above ambient is $71 - 50 = 21$°C.

Referring to an AHAM heat sink catalog for a 21°C temperature rise above ambient, a typical heat sink is type S1100 5.5. This heat sink has eight fins of 0.461-in height and a footprint area of 5.5 in by 4.5 in, so there is no significant advantage to putting the power transistor inside the package.

LTC has addressed this problem by offering other regulators (LT1142, 1143, 1148, 1149, 1430) which have less internal dissipation by use of external MOSFET transistors for the transistor switch and freewheeling diode. These have very low R_{ds}, and consequent lower "on" voltage and dissipation. They are available in surface-mounted packages and so still permit a small overall regulator size. These and other high-efficiency regulators are described below.

17.3.9 LTC High-Efficiency, High-Power Buck Regulators

17.3.9.1 LT1376 High-Frequency, Low Switch Drop Buck Regulator

A typical application in Figure 17.11*a* and block diagram in Figure 17.11*d* is a current-mode circuit (Sections 5.1 to 5.5), in which both the DC output voltage and transistor switch peak current are controlled, and determine the transistor "on" time.

For output currents from 1 to 1.5 A, it has greater efficiency and requires a smaller heat sink than the LT1074. For output currents under 1 A it may require no heat sink at all. Because it operates at 500 kHz, rather than 100 kHz for the LT1074, its output inductor and capacitor ($L1$ and $C1$ in Figure 17.11*a*) are much smaller.

It achieves greater efficiency primarily because of the low voltage drop across the output transistor of 0.5 V at 1.5 A in Figure 17.11*c*. This compares to 1.7 V at 1.5 A for the LT1074, or 1.25 V for the LT1076 (Figure 17.10c).

There are two reasons for this low drop. First, the output transistor is a single transistor rather that a triple Darlington (Figure 17.8). Second, the output transistor is driven harder into saturation by a voltage above V_{in}. This higher voltage is produced inexpensively by $D2$ and $C2$ (Figure 17.11*a*). When the internal power transistor turns "off," the V_{SW} node falls to one diode drop ($D2$) below ground, and $C2$ is charged to one diode drop below $+V_o$. When the internal power transistor turns back "on," $C2$ provides the positive boost voltage for the internal output transistor driver.

17.3.9.2 LTC1148 High-Efficiency Buck with External MOSFET Switches

A typical application is shown in Figures 17.12*a* and *c*. It achieves high efficiency because of the low-R_{ds}, low "on" drop P-channel MOSFET ($Q1$) which switches the input end of L up to V_{in} during the "on" time. During the "off" time, when $Q1$ is "off," the low-R_{ds}, low "on" drop N-channel MOSFET $Q2$ is "on" and acts as the freewheeling diode in shunt with $D1$ pulling the input end of L closer to ground than $D1$ alone can.

Efficiency is close to 95% because of these low "on" time drops, as can be seen in Figure 17.12*b*.

The circuit is unusual in that unlike most other LTC products, it operates with a fixed "off" time, and regulates by varying the switching frequency to control the duty cycle.

Historically, a constant-frequency, "on" time-modulated regulation scheme has been preferred. Often the switching power supply in a large system was fed a synchronizing pulse, and it was required that

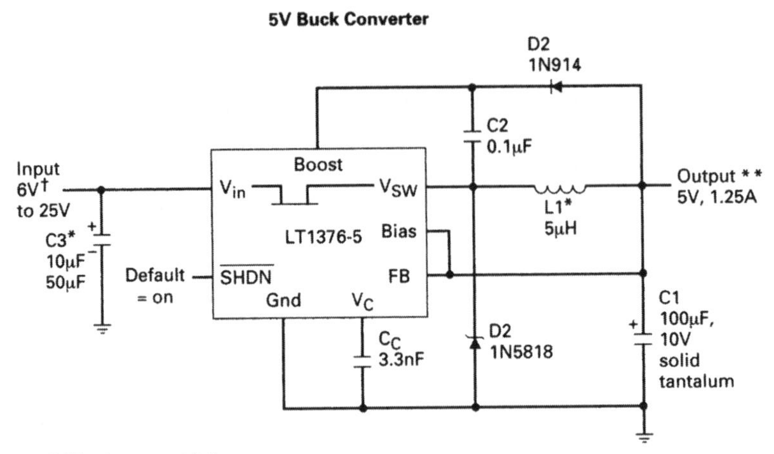

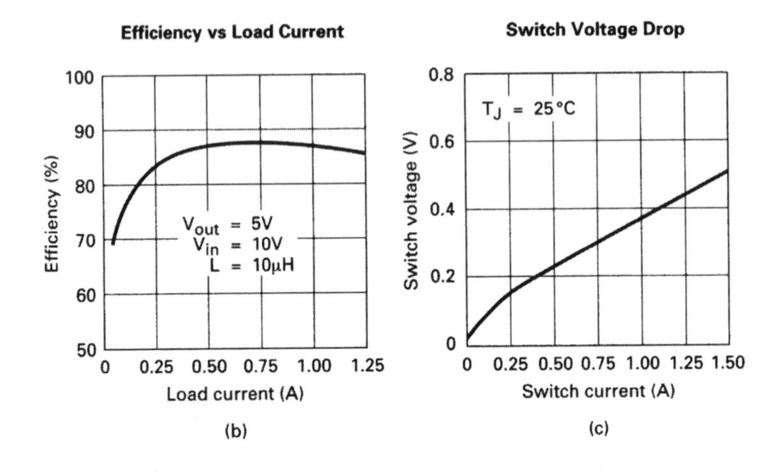

FIGURE 17.11 Linear Technology LT1375 500-kHz high-efficiency buck regulator. The single output transistor and the voltage boost provided by $D2$, $C2$ result in low-output transistor "on" drop and efficiency close to 90% over a large current range.

the power supply switching frequency be locked to a submultiple of this pulse frequency, which was synchronized and locked in phase either to a central computer clock or to the horizontal line rate in a CRT display.

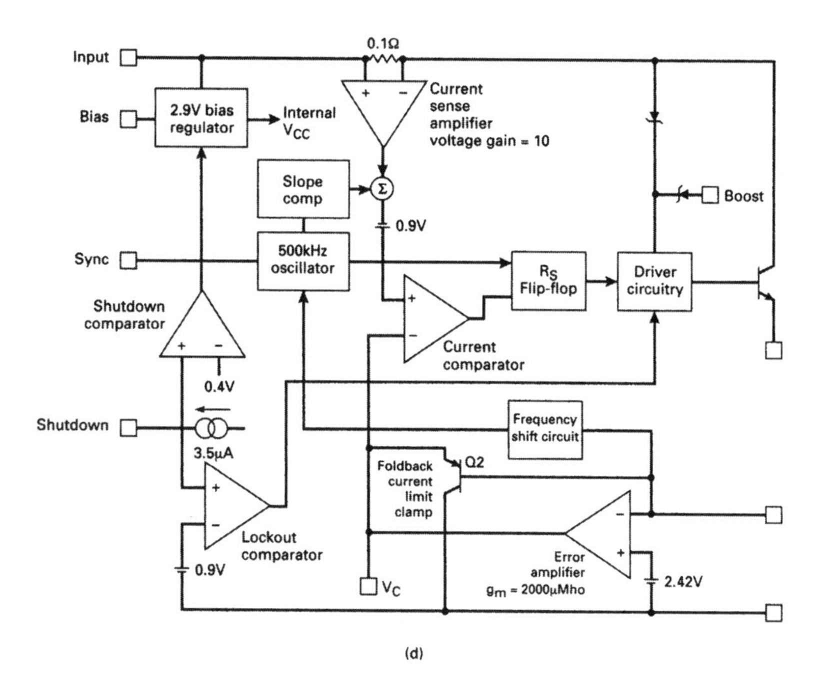

FIGURE 17.11 *Continued.*

There was fear that power supply conducted or radiated RFI could be picked up and interfere with nearby electronics such as CRT displays or computers. Any generated noise picked up by a CRT display in a synchronized system would be stationary on the screen, and not as disconcerting to the operator as if it wandered across the screen. Also, unsynchronized noise picked up by a computer would have a greater probability of falsely turning 1s into 0s. This is because the Fourier spectrum of variable-frequency noise is much wider than fixed-frequency.

For low-power supplies for laptop computers and portable electronics, where most often there is no other electronic equipment nearby, there is no valid reason for rejecting a variable-frequency voltage-regulating scheme.

17.3.9.3 LTC1148 Block Diagram

Despite regulating the output voltage by varying the frequency, rather than by varying the pulse width at a constant frequency, the output/input voltage relation is exactly the same as that for a PWM scheme.

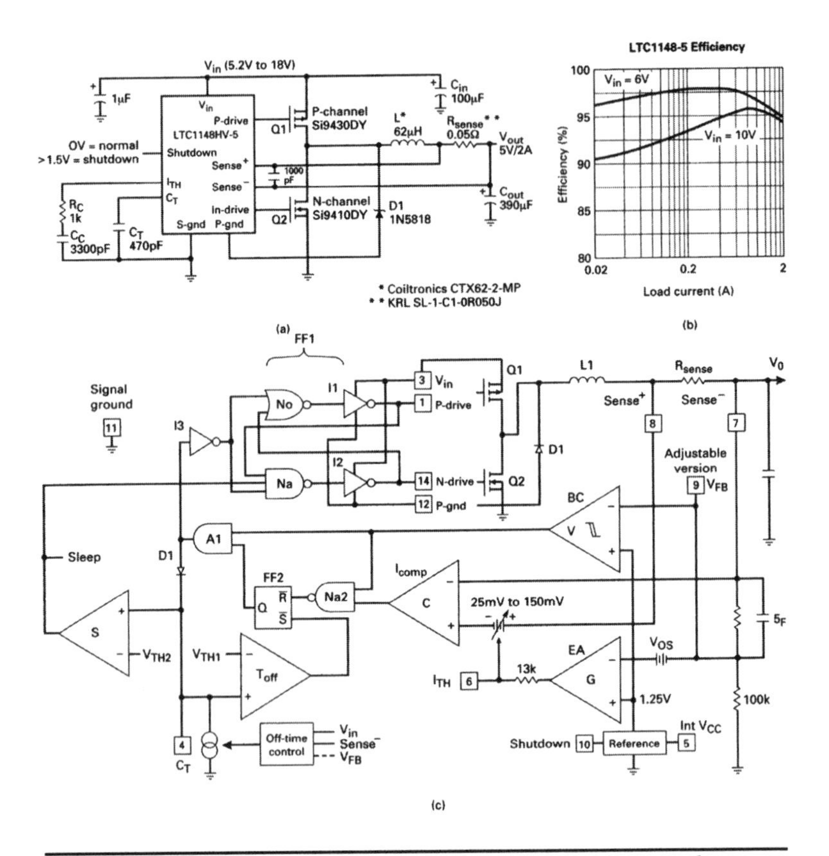

FIGURE 17.12 LIC1148 high-efficiency buck regulator with external MOSFETs. This regulator operates with fixed "off" time set by C_t. It regulates by varying switching frequency. The external P and N channel MOSFETs, with their low "on" drops, offer efficiencies exceeding 90% over a large current range.

This can be seen (Figure 17.12*a*) by equating the L volt-second products. When $Q1$ is "on," $Q2$ is "off" and the input end of L is essentially at V_{in}. When $Q2$ is "on" and $Q1$ is "off," the input end is at ground. Then

$$(V_{\text{in}} - V_o)t_{\text{on}} = V_o t_{\text{off}} = V_{\text{in}} t_{\text{on}} - V_o t_{\text{on}}$$

or

$$V_o = \frac{V_{\text{in}} t_{\text{on}}}{t_{\text{on}} + t_{\text{off}}} = V_{\text{in}} \frac{t_{\text{on}}}{T} \tag{17.5}$$

From this, the frequency versus output/input relation is

$$V_o = \frac{V_{\text{in}} t_{\text{on}}}{T} = \frac{V_{\text{in}}(T - t_{\text{off}})}{T} = V_{\text{in}}\left(1 - \frac{t_{\text{off}}}{T}\right)$$
$$= V_{\text{in}}(1 - f t_{\text{off}}) \quad (17.6)$$

which can be written

$$f = \frac{1 - V_o}{V_{\text{in}}/t_{\text{off}}} \quad (17.7)$$

It can be seen from Eq. 17.7 or 17.8 that with a constant t_{off}, to keep V_o constant as V_{in} goes up, frequency f goes up. Regulation can be seen as follows (Figure 17.12).

When $Q1$ is "on," P drive pin 1 is low, keeping the P-channel MOSFET "on." N drive pin 14 is also low, keeping the N-channel MOSFET "off." The four elements N_o, $I1$, N_a, $I2$ form a set-reset flip-flop FF1, and hence the P and N drive outputs remain locked in the low state, keeping $Q1$ "on" and $Q2$ "off" until the flip-flop is reset.

Current in L ramps up with the usual ramp-on-a-step waveform characteristic of an LC output filter. This inductor current is monitored by sensing the voltage drop across R_{sense}. That voltage drop is added to the negative bias voltage at the non-inverting input to comparator C, which is the output of voltage error amplifier G that compares a fraction of V_{out} to the internal 1.25-V reference.

When the voltage across R_{sense} exceeds the bias voltage at pin 6, output of comparator C goes high, and the output of NAND gate N_a goes negative, since its other input is high at this time. This resets flip-flop FF2, causing its Q output to go low. The output of AND gate A1 follows suit and commences the "off" time, which requires both the P and N drive outputs to go high. That turns the P MOSFET "off" and the N MOSFET "on," bringing the input end of L down to ground.

When the output of AND gate $A1$ went low at the start of the "off" time, it had two significant effects. First, the output of inverter $I3$ went high. Element N_o is a positive logic NOR gate, and element N_a is a positive logic NAND gate. Thus when the output of $I3$ went high, the N_o output went low and $I1$ output went high, causing P drive to go high to turn "off" the P MOSFET. Also, when the output of $I3$ went high, the output of N_a went low as all its other inputs were already high. That caused the N_a output to go low, the $I2$ output to go high, and N drive to go high. That finally turned "on" the N MOSFET. The flip-flop thus remained locked in the set state with $Q1$ "off" and $Q2$ "on" because of the cross-coupling. This "off" state remains until flip-flop FF2 is reset.

The design requires a constant "off" time, which comes about as follows. When the $A1$ output went low, resetting FF1, the second significant result was that the anode of diode $D1$ went low, disconnecting it from timing capacitor C_t at pin 4. Before the anode of $D1$ went low, that diode clamped the voltage across $C4$ to a reference of 3 V and also forced that same voltage to the non-inverting input to the "off" time comparator $T_{\text{off}\,c}$. Meanwhile, the inverting input of $T_{\text{off}\,c}$ was fixed at a threshold voltage V_{th1} of about 0.5 V. Thus the output of $T_{\text{off}\,c}$ was high.

Then when the $D1$ anode went low, turning $Q1$ "off" and $Q2$ "on," and putting a reverse bias on $D1$, this started the "off" time. The capacitor C_t started discharging with a nominal current of about 0.25 mA. When it discharged to below V_{th1}, the output of $T_{\text{off}\,c}$ went low and set FF2. This drove the output of $A1$ high again and via FF1 drove P drive and N drive low again, turning the P MOSFET "on" and the N MOSFET "off," ending the "off" time.

Thus the "off" time is the time required to discharge C_t about 3 V with the internal 0.25 mA, which is modulated internally to keep the frequency from going to low at low input voltage (Eq. 17.8). C_t is selected as follows: Choose the desired switching frequency f at nominal input voltage V_{in}. Then choose t_{off} from Eq. 17.8 for nominal input and output voltages V_{in}, V_o, and find C_t from

$$C_t = i\frac{dt}{dV} = \frac{0.0025t_{\text{off}}}{3}$$

17.3.9.4 LTC1148 Line and Load Regulation

From the foregoing, it can be seen that line and load regulation occurs by changing the switching frequency and the "on" time with a fixed "off" time. Consider again in detail how the "on" time changes.

Suppose V_{in} increases, and V_o does the same. The inverting input to voltage error amplifier G rises, and its negative bias from ramp voltage across R_{sense} decreases. Then the positive ramp at the non-inverting terminal of C crosses the threshold at its inverting terminal sooner. Thus comparator C output goes positive sooner, the "on" time decreases, frequency increases as t_{off} is constant, and output voltage goes back up.

The same "on" time modulation occurs for load changes, except that those changes are temporary and revert to the "on" time called for by Eq. 17.6. The temporary changes in the "on" time permit the step part of the ramp-on-a-step waveform (Figure 1.6*a*) to build up or down over a number of switching cycles, as the center of the ramps in Figure 1.6*a* is the output current.

17.3.9.5 LTC1148 Peak Current and Output Inductor Selection

The threshold voltage variation at the output of error amplifier G (Figure 17.12*c*, pin 6) ranges between –0.025 and –0.15 V. Recall that the peak output inductor current is reached when the voltage across

R_{sense} equals the threshold voltage. Then to consider tolerances, assume a maximum threshold of 0.100 V. For a specified maximum output current of I_{max},

$$R_{sense} = \frac{0.100}{I_{max}} \tag{17.8}$$

The output inductor is sized so that it is at the threshold of the continuous mode (Section 1.3.6) with the minimum bias of 0.025 V. Then the inductor "on" current is a triangle rising from 0 to a peak of $0.025/R_{sense}$ amperes in a time t_{on}, and then falls to zero in the t_{off} calculated from Eq. 17.8. Then

$$L = V\frac{dt}{dI} = \frac{V_o t_{off}}{0.025/R_{sense}} \tag{17.9}$$

17.3.9.6 LTC1148 Burst-Mode Operation for Low Output Current

At low output currents, rather than permitting the inductor to go into the discontinuous mode, the circuit is designed to stop switching completely. Load current is then supplied entirely from the output capacitor. This discharges the output capacitor after a time, and when it falls back to the desired output voltage, switching commences again.

This is achieved with comparator *BC*. In normal operation its output is high, as its inverting input is below its non-inverting one. This enables *A*1, and its output is then controlled by the state of FF2 *Q* output. When load current falls and V_o starts rising above its regulated value, the inverting terminal of *BC* rises above the reference at the non-inverting terminal. The *BC* output goes low, the *A*1 output goes low, the *I*3 output goes high and forces the *P* drive high via N_o and *I*1 to turn *Q*1 "off." *N* drive is kept high to keep *Q*2 "on" via N_a and *I*2. To lower the current drain in this non-switching mode as C_t falls below V_{TH2}, it turns the *N* drive "off" also, and the circuit is in the *sleeping* mode.

Thus no switching occurs, and this "off" time persists even though C_t discharges below V_{TH1}, because the low at the *BC* output keeps the *A*1 output low.

When the output capacitor discharges back down to the regulated value, the *BC* inverting input goes low, its output goes high, it releases the inhibit on *A*1, and the circuit returns to its switching mode. In the sleep mode, the *P* drive is high and *N* drive is low, reducing internal dissipation, which permits high efficiency down to a negligibly small load current.

17.3.10 Summary of High-Power Linear Technology Buck Regulators

The three buck regulators discussed above are typical and probably the most useful in the LTC buck family. The numerous others are specialized versions of the ones discussed, and include higher-voltage, fixed-output-voltage, and various lower-peak-current types. They use the same block diagram and component selection as discussed in previous chapters (Sections 1.3.6 and 1.3.7 for buck regulators, Sections 15.4.6 and 15.4.7 for boost regulators).

A tabular summary of the available LTC types is shown in Figure 17.13.

LOAD CURRENT	DEVICE	SYNCHRONOUS	SHUTDOWN CURRENT	EFFICIENCY
200mA to 400mA	LTC1174-3.3	NO	1μA	90%
200mA to 425mA	LTC1574-3.3	NO	2μA	90%
450mA	LTC1433	NO	15μA	93%
0.5mA to 2A	LTC1147-3.3	NO	10μA	92%
1A	LTC1265-3.3	NO	5μA	90%
1A to 5A	LTC1148-3.3	YES	10μA	94%
5A to 10A	LTC1266	YES	30μA	94%
5A to 20A	LTC1158	YES	2.2μA	91%

Step-Down Switching Regulator Selection Guide

DEVICE	INPUT VOLTAGE MIN	INPUT VOLTAGE MAX	MAXIMUM SWITCH VOLTAGE	MAXIMUM RATED SWITCH CURRENT (A)	PACKAGES AVAILABLE	COMMENTS
LT1074	8	40	65	5.5	K, T, Y	Integrated 5A Switch
LT1074HV	8	60	75	5.5	K, T, Y	64V Max Input Voltage
LT1076	8	40	65	2.0	K, Q, R, T, Y	Integrated 2A Switch
LT1076HV	8	60	75	2.0	K, R, T, Y	64V Max Input Voltage
LTC1142	5	16	*	*	28-SSOP	Dual Output Synchronous Switching Controllers
LTC1142HV	5	20	*	*	28-SSOP	Dual Output Synchronous Switching Controllers
LTC1143	5	16	*	*	N16, S16	Dual 3.3V/5VLTC1147
LTC1147	4	16	*	*	N8, S8	Ext. P-Channel MOSFET, 90% Efficiency
LTC1148	4	16	*	*	N14, S14	Synchronous Switching, 93% Efficiency
LTC1148HV	4	20	*	*	N14, S14	Synchronous Switching, 93% Efficiency
LTC1149	5	60	*	*	N16, S16	Synchronous Switching, 48V Input
LTC1158**	5	30	*	*	N16, S16	Up to 15A at 90% Efficiency
LTC1159	4	40	*	*	N16, S16, G20	Synchronous Switching, 94% Efficiency
LT1160**	8.8	20	*	*	N14, S14	Up to 10A at 90% Efficiency
LTC1174	4	13.5	13.5	1	N8, S8	Integrated Switch, 90% Efficiency
LTC1174HV	4	18.5	18.5	1	N8, S8	Integrated Switch

Step-Down Switching Regulator Selection Guide

DEVICE	INPUT VOLTAGE MIN	INPUT VOLTAGE MAX	MAXIMUM SWITCH VOLTAGE	MAXIMUM RATED SWITCH CURRENT (A)	PACKAGES AVAILABLE	COMMENTS
LT1176	8	38	50	1.2	N8, S20	Integrated 1.25A Switch, SO Package
LTC1265	3.5	13	V_{IN}-13	1.6	S14	Integrated Switch, 90% Efficiency
LTC1266	4	20	*	*	N16, S16	Synchronous Switching Uses N-Channel MOSFETs
LTC1267	4	40	*	*	28-SSOP	Dual Output Synchronous Switching Controller
LT1375/LT1376	6	25	30	1.5	N8, S8	500kHz Buck Switcher
LTC1430	4	9	*	*	S8, S16, N16	Synchronous Switching w/High Efficiency Up to 50A
LTC1435	3.5	36	*	*	S16	Constant Frequency w/Wide Input Voltage Range
LTC1436	3.5	36	*	*	24-SSOP	Constant Frequency, AUX Linear Regulator Controller
LTC1436-PLL	3.5	36	*	*	24-SSOP	Synchronizable Constant Frequency
LTC1437	3.5	36	*	*	28-SSOP	Synchronizable Constant Frequency, AUX Linear Regulator Controller
LTC1438	3.5	36	*	*	28-SSOP	Constant Frequency, Dual Output Synchronous Switching Regulator
LTC1439	3.5	36	*	*	36-SSOP	Constant Frequency, Dual Output Synchronous Switching Controller
LT1507	4	16	25	1.5	N8, S8	500kHz Buck Switcher, 5V to 3.3V
LTC1538	3.5	36	*	*	28-SSOP	LTC1438 with 5V Standby Active in Shutdown
LTC1539	3.5	36	*	*	36-SSOP	LTC1439 with 5V Standby Active in Shutdown
LTC1574	4.5	18.5	18.5	*	S16	1A Schottky Diode in Package

* Uses external FET switch

** Half-bridge driver requires external control circuit

FIGURE 17.13 Linear Technology buck regulators.

17.3.11 Linear Technology Micropower Regulators

Regulators with all semiconductors in one package for battery-powered laptop computers represent a large segment of the market for low-input voltage regulators. Linear Technology has a wide variety of products, but their topologies and circuitry are no different from what has been discussed above. Topologies are boost, buck, and polarity inverters, and they differ mainly in that output currents are all under 1 A and come in surface-mounted packages.

Since no new circuitry is involved, a tabular listing of the devices available is presented in Figures 17.14*a* and *b*.

17.3.12 Feedback Loop Stabilization[3]

Feedback loop stabilization was discussed mathematically in Chapter 12. There the significance of poles and zeros and their locations on the frequency axis to stabilize the feedback loop was described.

LTC prefers an empirical approach because the mathematical analysis depends on assumptions about various quantities that are not precisely specified by parts manufacturers. One such quantity is the filter capacitor and its ESR, which may vary with use. LTC prefers to observe the power supply output voltage response to a step change in load current on an oscilloscope. By optimizing the waveshape for various potential *RC* combinations (zeros) at the error amplifier output, the loop is stabilized. The stabilization process is thus free of assumptions.

All LTC boost, buck, and inverting configurations shown in their data sheets are stabilized this way. The scheme is described by C. Nelson and J. Williams in LTC Application Note 19.

The stabilization scheme is shown in Figure 17.15*a*. An additional step load (Figure 17.15*b*) of about 10% (*R*1) is AC-coupled and added to the nominal output current through a large capacitor *C*1. The usual step square wave generator frequency of 50 Hz is not critical. The regulator transient response to the steps is observed with the oscilloscope through the filter shown to keep switching frequencies out of the display.

The series *RC* network used to stabilize the supply is connected from the output of the internal voltage error amplifier (V_c pin) to ground. As a starting point, set *C*2 to 2 μF and *R*3 to 1 kΩ. This almost always yields a stable DC loop, but with *C*2 so large, the supply responds to the step load with large transient overshoots and a slow decay back to the nominal output voltage, as seen in Figure 17.15*c*. Now *C*2 is decreased in steps, yielding the response of Figure 17.15*d*. The overshoots are smaller in amplitude and fall back more quickly to

Battery-Powered DC/DC Conversion Solutions Selection Guide

The following tables are a short form component selection guide for a collection of commonly used battery-powered DC/DC conversion applications. No design is required since inductor, capacitor and resistor values are completely specified.

Choose the appropriate LTC DC/DC converter for your application from the following tables.

The LT1073, LT1107, LT1108, LT1110, LT1111, LT1173, LTC1174 and LT1303, LT1307 all have low-battery detection capability.

Step-Up From One Cell (1V)

V_{out} (V)	I_{out} (mA)	DEVICE	I_Q (μA)	L (μH)	C (μF)	R (Ω)	Fig	COMMENTS
3.3V	75	LT1307	50	10	10	–	4	
5	40	LT1073-5	95	82	100	0	1	Lowest I_Q
	40	LT1110-5	350	27	33	0	1	Best For Surface Mount
12	40	LT1307	50	10	10	–	4	
	15	LT1073-12	95	82	100	0	1	Lowest I_Q
	15	LT1110-12	350	27	33	0	1	Best For Surface Mount

Adjustable versions also available for V_{out} up to 50V

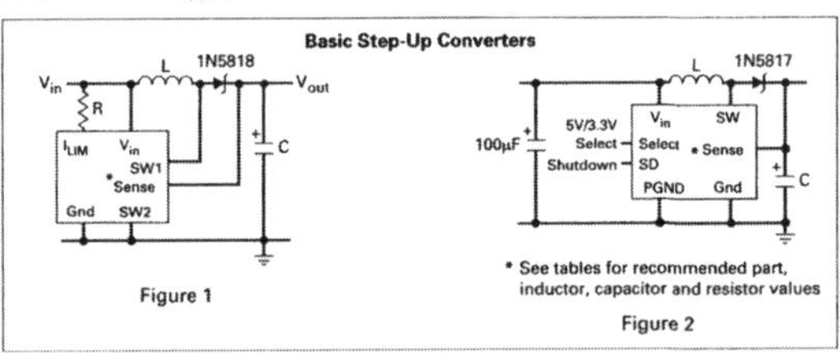

Step-Up From Two Cell (2V)

V_{out} (V)	I_{out} (mA)	DEVICE	IQ (μA)	L (μH)	C (μF)	R (Ω)	Fig	COMMENTS
3.3	400	LT1300**	120	10	100	–	2	Selectable 3.3V/5V Out
	250	LT1500	200	22	220	–	2	Low Spectral Noise
5	90	LT1173-5	110	47	100	47	1	Lowest I_Q
		LT1111-5	300	18	33	47	1	Surface Mount
	150	LT1107-5	300	33	33	47	1	Surface Mount
		LT1108-5	110	100	100	47	1	Lowest I_Q
	200	LT1304	120	22	100	–	2	Low-Battery Detector
		LT1501	200	22	220	–	2	Low Spectral Noise
	220	LT1300**	120	10	100	–	2	Selectable 3.3V/5V Out
		LT1301**	120	10	100	–	2	Selectable 5V/12V Out
	400	LT1305	120	10	220	–	2	High Output Power
	600	LT1302	200	10	100	–	*	Highest Power Output
12	20	LT1173-12	110	47	47	47	1	Lowest I_Q
		LT1111-12	300	18	22	47	1	Surface Mount
	40	LT1107-12	300	27	33	47	1	Surface Mount
		LT1108-12	110	82	100	47	1	Lowest I_Q
	50	LT1301**	120	10	100	–	2	Selectable 5V/12V Out
	120	LT1302	200	3.3	66	–	*	Highest Output Power

* See LT1302 data sheet ** For low-battery detection use LT1303 or LT1304

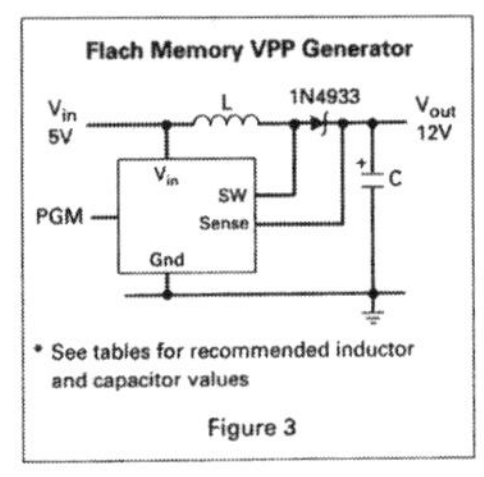

Battery-Powered DC/DC Conversion Solutions Selection Guide (Continued)

Step-Up From 5V to 12V

V_{in} (V)	I_{out} (mA)	DEVICE	I_Q (μA)	L (μH)	C (μF)	R (Ω)	Fig	COMMENTS
12	90	LT1173-12	110	120	100	0	1	Lowest I_Q
		LT1111-12	300	47	33	0	1	Surface Mount
	175	LT1107-12	300	60	32	0	1	Surface Mount
		LT1108-12	110	180	100	0	1	Lowest I_Q
	200	LT1301**	120	33	47	–	2	True Shutdown
		LT1500	200	22	220	–	2	Low Spectral Noise
	350	LT1373	1000	22	100	–	4	250kHz

** For low-battery detection use LT1303

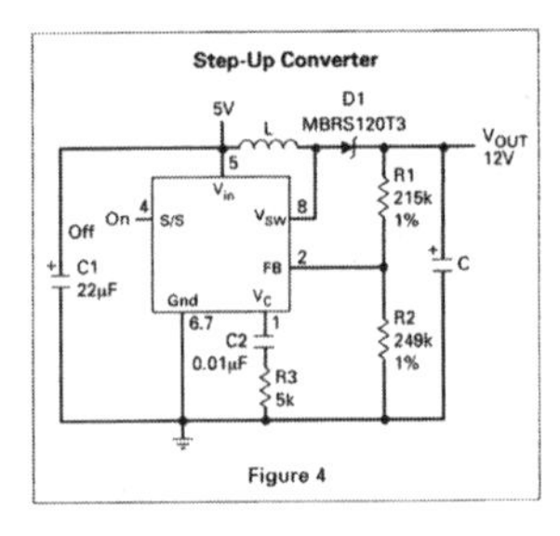

Step-Up From 5V to 12V

V_{in} (V)	I_{out} (mA)	I_{out} (mA)	DEVICE	I_Q (μA)	L (μH)	C (μF)	Fig	COMMENTS
3 or 5	12	60	LT1106	750	10	1	***	TSSOP
			LT1309	500	10	1	3	Small, Shutdown
5	12	60	LT1109-12	320	33	22	3	Small, SMT
		120	LT1109A-12	320	27	47	3	Small, SMT
		200	LT1301**	120	27	47	2	True Shutdown
		60	LT1109A-12	320	10	22	1	All Surface Mount
2 Cells	12	80	LT1301**	120	10	47	2	True Shutdown

** For low-battery detection use LT1303 *** See LT1106 data sheet

FIGURE 17.14 (*a*) Linear Technology micropower (I_o under 1 A) boost regulators.

Step-Down Conversion to 3.3V

V_{in} (V)	I_{out} (mA)	DEVICE	I_Q (μA)	L (μH)	C (μF)	IPGM	Fig	COMMENTS
4.5 to	200	LTC1174-3.3	450	50	2 x 33	To Gnd	6	Low Dropout
12.5	425	LTC1174-3.3	450	50	2 x 33	To V_{in}	6	Surface Mount
4.5 to	200	LTC1574-3.3	600	50	2 x 33	To V_{in}	6	Low Component Count
16	425	LTC1574-3.3	600	50	2 x 33	To V_{in}	6	Low Component Count
5 to	2A	LTC1148-3.3	160	–	–		–	See Step Down
16						–		Switchers
5 to	450	LTC1433	470	100	100		6	Constant Frequency
12						–		
12 to	2A	LTC1149-3.3	600	–	–		–	See Step Down
60						–		Switchers

Step-Down Conversion to 5V

V_{in} (V)	I_{out} (mA)	DEVICE	I_Q (μA)	L (μH)	C (μF)	IPGM	Fig	COMMENTS
5.5 to	200	LTC1174-5	450	100	2 X 33	To Gnd	6	Low Dropout,
12	400	LTC1174-5	450	100	2 X 33	To V_{in}	6	Surface Mount
5.5 to	200	LTC1574-5	600	100	2 X 33	To V_{in}	6	Low Component Count
16	425	LTC1574-5	600	100	2 X 33	To V_{in}	6	Low Component Count
5.8 to	450	LTC1433	470	100	100	–	6	Constant Frequency
12								
12 to	300	LT1107-5	300	60	100	100	5	Surface Mount
20	300	LT1108-5	110	180	330	100	5	Lowest IQ
20 to	300	LT1173-5	110	470	470	100	5	Lowest IQ
30	300	LT1111-5	300	180	220	100	5	Surface Mount
6 to	2A +	LTC1147/8-5	160	–	–	–	–	See Step Down
16								Switchers
12 to	2A +	LTC1149-5	600		–		–	See Step Down
60								Switchers

Adjustable output voltages up to 6.2V can be obtained with the adjustable versions If LT1173. LT1111, LT1107. LT1108, or LT1110.

Step-Down Converters

* See tables for recommended part inductor capacitor and resistor values

Figure 5

* Diode not required for LTC1574

Figure 6

Positive-to-Negative Voltage Conversion

V_{in} (V)	V_{out} (V)	I_{out} (mA)	DEVICE	I_Q (μA)	L (μH)	C (μF)	R (Ω)	Fig	COMMENTS
5	–5	75	LT1108-5	110	100	100	100	7	Lowest I_Q
			LT1107-5	300	33	33	100	7	Surface Mount
		150	LTC1174-5	450	50	2 x 33	–	8	Surface Mount
		290	LTC1433	470	68	100	–	8	Constant Frequency
12	–5	250	LT1173-5	110	470	220	100	7	Lowest I_Q
		250	LT1111-5	300	180	82	100	7	Surface Mount
1	–5	110	LTC1574-5	450	50	100	–	8	SMT, No Ext.
8		170							Schottky Diode
12.5		235							Required

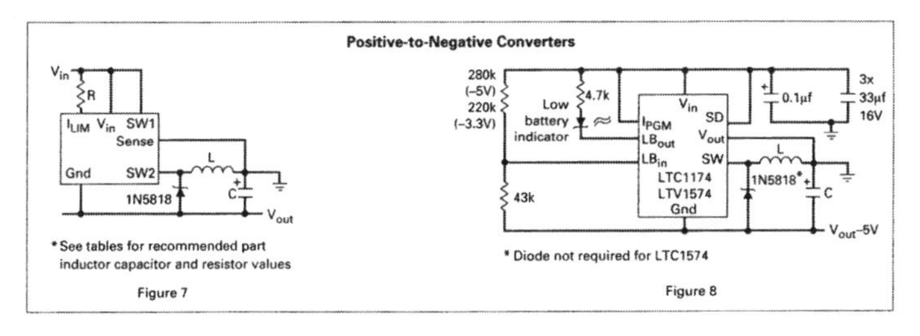

FIGURE 17.14 (*b*) Linear Technology micropower buck regulators.

the quiescent level. After the overshoot, there is a reverse-polarity *ring*. Now if *R*3 is increased, the waveshape of Figure 17.15*e* with the reverse ring eliminated results. Now decreasing *C* may decrease the amplitude of the overshoot. More details are provided in LTC Application Note 19.

In terms of the feedback analysis of Chapter 12, the combination of *C*2 and *R*3 provides only a single zero, but no pole in the amplifier transfer function (Sections 12.3 and 12.19). Recall in Section 12.3

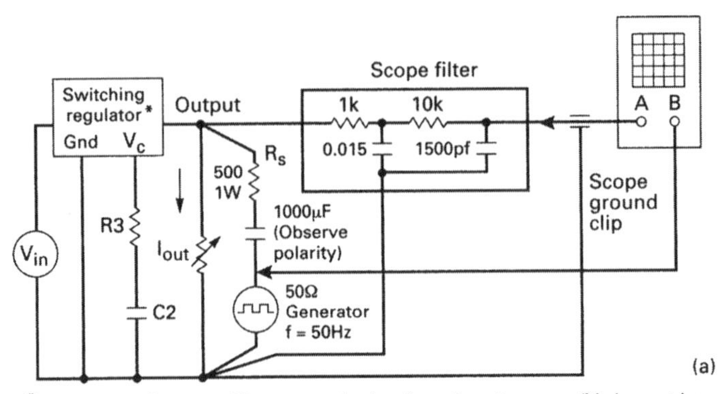

*All input and output filters must be in place. Input source (V_{in}) must be actual source used in final design to account for finite source impedance.

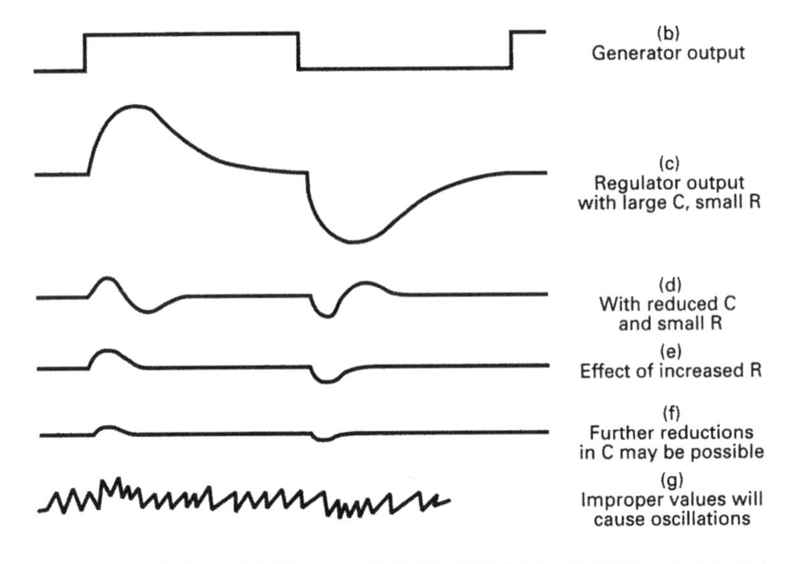

FIGURE 17.15 Stabilizing the feedback loop by varying R3C2 product, to optimize the transient response waveshape when subjected to a step load current via $R_s C_b$.

that a pole, comprising a shunt capacitor across the series *RC* combination, was added to reduce gain at high frequencies, so that any high-frequency noise spikes picked up would not get through to the output.

The LTC analysis and data sheets show no poles or shunt *C* across the series *RC* combination. Maybe there is already some shunt *C* from V_c to ground internally, or there are no such noise spikes in these relatively low-power devices with no large dV/dt or di/dt sources. If spikes are observed at the output, they can very likely be eliminated by empirically selected small capacitors from V_C to ground.

17.4 Maxim IC Regulators

Maxim Inc. is another major manufacturer of devices discussed here. Its product line also consists of boost, buck, and polarity inverters, and matches many of the LTC devices in maximum voltage and current ratings. Generally, they tend to specialize in devices with lower current ratings than those of LTC.

Since no new circuit techniques are involved, a tabular listing of the devices, showing all their significant voltage and current ratings, is shown in Figure 17.16.

The fact that most of the discussion here has concerned LTC devices, with only a tabular presentation of Maxim types, should not be taken as the author's assessment of the relative merits of the two manufacturers' devices. This discrepancy is due only to the author's earlier familiarity with LTC products.

17.5 Distributed Power Systems with IC Building Blocks[7]

Figure 17.17*a* shows a conventional off-line, multi-output power supply. The alternating current is rectified with or without power factor correction, and some topology—half, full bridge, forward converter, or flyback—is used to generate a precisely controlled *master* output voltage on output ground. This master is generally the highest current output—usually +5 V—and is very well regulated against line and load current changes. It is very well regulated because the feedback loop is closed on this output, and controls the "on" time of the power switching devices on input ground.

Additional outputs referred to as *slaves* are obtained by adding more secondaries whose turns are selected to yield the desired secondary voltages. Since the slave secondary switching times are the same as that of the master secondary, the DC voltages of the slaves after their *LC* filters are also well regulated against line input voltage changes.

Slaves are not well regulated against load current changes in either the master or the slave outputs, however (Sections 2.2.1 to 2.2.3), and are generally only ±5 to 8%. This can change as much as 50% if either the master or slave inductors are permitted to go into discontinuous mode (Section 2.2.4). Further, the absolute slave voltages cannot be set

Part Number	Input Voltage Range (V)	Output Voltage (V)	Supply Current (mA), max (typ)	Output (mA typ)	Control Scheme	Package Options *	EV Kit	Temp. Ranges **	Features	Price † 1000-up ($)
STEP-UP/STEP-DOWN SWITCHING REGULATORS										
MAX877/878/879	1 to 6.2	5/(3.3 or 3)/Adj.	0.310(0.220)	240	PFM	DIP.SO	Yes	C.E.M.	Gives regulated outout when input above and below the output; no transformer	2.95
STEP-UP SWITCHING REGULATORS										
MAX4193	2.4 to 16.5	Adj.	0.200(0.090)	300mW	PFM	DIP.SO		C.E.M.	Improved RC 4193 2nd source	1.74
MAX630	2 to 16.5	Adj.	0.125(0.070)	300mW	PFM	DIP.SO		C.E.M.	Improved RC 4193 2nd source	2.88
MAX631	1.5 to 5.6	5, Adj.	0.4(0.135)	40	PFM	DIP.SO		C.E.M.	Only 2 external components	2.56
MAX632	1.5 to 12.6	12, Adj.	2.0(0.5)	25	PFM	DIP.SO		C.E.M.	Only 2 external components	2.56
MAX633	1.5 to 15.6	15, Adj.	2.5(0.75)	20	PFM	DIP.SO		C.E.M.	Only 2 external components	2.56
MAX641	1.5 to 5.6	5, Adj.	0.4(0.135)	300	PFM	DIP.SO		C.E.M.	PFM controller	2.87
MAX642	1.5 to 12.6	12, Adj.	2.0(0.5)	550	PFM	DIP.SO		C.E.M.	PFM controller	2.87
MAX643	1.5 to 15.6	15, Adj.	2.5(0.75)	325	PFM	DIP.SO		C.E.M.	PFM controller	2.87
MAX731	1.8 to 5.25	5	4(2)	200	PWM	DIP.SO	Yes	C.E.M.		2.60
MAX732	4 to 9.3	12	3(1.7)	200	PWM	DIP.SO	Yes	C.E.M.	Flash memory programmer, ±4% output voltage tolerance	2.76
MAX733	4 to 11	15	3(1.7)	125	PWM	DIP.SO	Yes	C.E.M.		2.60
MAX734	1.9 to 12	12	2.5(1.2)	120	PWM	DIP.SO	Yes	C.E.M.	Flash memory programmer	2.23
MAX741U	1.8 to 15.5	5, 12, 15, Adj.	3.5(1.6)	5W	PWM	DIP.SSOP	Yes	C.E.M.	PWM step-up controller, $3V_{in}$ to $5V_{out}$ at 1A, 85% efficient	3.64
MAX751	1.2 to 5.25	5	3.5(2)	175	PWM	DIP.SO	Yes	C.E.M.		2.35
MAX752	1.8 to 16	Adj.	3(1.7)	2.4W	PWM	DIP.SO	Yes	C.E.M.		2.94
MAX756/757	1.1to 5.5	(3.3 or 5)/Adj.	0.060(0.045)	250	PFM	DIP.SO	Yes	C.E	Best combination of low I_Q & high 86% efficiency	1.95
MAX761/762	2 to 16.5	12/15 or Adj. to 16.5	0.1(0.080)	120	PFM	DIP.SO	Yes	C.E.M.	12V flash programmer, high efficiency over wide I_{out} range	2.23
MAX770/771/772	2 to 16.5	5/12/15 or Adj.	0.1(0.085)	1A	PFM	DIP.SO	Yes	C.E.M.	Controllers, high efficiency over wide I_{out} range	1.80
MAX1771	2 to 16.5	12 or Adj.	0.1(0.085)	1A	PFM	DIP.SO	Yes	C.E.M.	Same as MAX771, but only 100mV current-sense limit	1.80
MAX773	2 to 16.5	Adj. to 48	0.1(0.085)	1A	PFM	DIP.SO	Yes	C.E.M.	Controller, high-voltage output, high efficiency over wide I_{out} range	1.80
MAX777/778/779	1 to 6	5/(3 or 3.3)/Adj.	0.31(0.220)	300	PFM	DIP.SO	Yes	C.E.M.	On-chip active diode, true turn off in shutdown	2.65
MAX856/857	0.8 to 6	5/(3.3 or 5)/Adj.	0.060(0.025)	100	PFM	DIP.SO,μMAX	Yes	C.E	Best combination of low I_Q & high 85% efficiency	1.72
MAX858/859	0.8 to 6	5/(3.3 or 5)/Adj.	0.060(0.025)	25	PFM	DIP.SO,μMAX	Yes	C.E	Small, best combination of low I_Q & high efficiency	1.72

Part Number	Input Voltage Range (V)	Output Voltage (V)	Quiescent Supply Current (mA), max (typ)	Output (mA typ)	Control Scheme	Package Options *	EV Kit	Temp. Ranges **	Features	Price † 1000-up ($)
STEP-DOWN SWITCHING REGULATORS										
MAX638	2.6 to 16.5	5, Adj.	0.6(0.135)	75	PFM	DIP.SO		C.E.M.	Only 3 external components	2.56
MAX639/640/653	4 to 11.5	5/3.3/3 or Adj.	0.02(0.01)	225	PFM	DIP.SO	Yes	C.E.M.	>90% efficiencies over wide I_{out} range (1mA to 225mA)	2.96
MAX649/651/652	4 to 16.5	5/3.3/3 or Adj.	0.1(0.080)	2A	PFM	DIP.SO	Yes	C.E.M.	>90% efficiencies over wide I_{out} range, drives external P-channel FET	1.60
MAX649/1651	4 to 16.5	5/3.3/3 or Adj.	0.1(0.080)	2A	PFM	DIP.SO	Yes	C.E.M.	Same as MAX649/651 but 96.5% duty cycle and only 100mV current sense limit	1.60
MAX724/724H	3.5 to 40/60	Adj. (2.5 to 40)	12(8.5)	5A	PWM	TO/220, TO-3		C.E.M.	High power, few external components	4.52/6.83
MAX726	3.5 to 40/60	Adj. (2.5 to 40)	12(8.5)	2A	PWM	TO/220, TO-3		C.E.M.	High power, few external components	3.00
MAX727	3.5 to 40/60	5	12(8.5)	2A	PWM	TO/220, TO-3		C.E.M.	High power, few external components	3.00
MAX728	3.5 to 40/60	3.3	12(8.5)	2A	PWM	TO/220, TO-3		C.E.M.	High power, few external components	3.00
MAX729	3.5 to 40/60	3	12(8.5)	2A	PWM	TO/220, TO-3		C.E.M.	High power, few external components	3.00
MAX730A	5.2 to 11	5	3(1.7)	300	PWM	DIP.SO	Yes	C.E.M.	90% efficiencies, no subharmonic switching noise	2.15
MAX738A	6 to 16	5	3(1.7)	750	PWM	DIP.SO	Yes	C.E.M.	>85% efficiencies, no subharmonic switching noise	2.60
MAX741D	2.7 to 15.5	5, Adj.	4.25(2.8)	3A	PWM	DIP.SSOP	Yes	C.E.M.	PWM step-down controller, $6.5V_{in}$ to $5V_{out}$ at 3A, 90% efficient	3.74
MAX744A	4.75 to 16	5	2.5(1.2)	750	PWM	DIP.SO	Yes	C.E.M.	Optimized cellular communications, no subharmonic switching noise	2.90
MAX746	4 to 15	5/Adj.	1	2.5A	PWM	DIP.SO	Yes	C.E.M.	5V to 3.3V green PC apps., drives external N-channel FET	2.25
MAX747	4 to 15	5/Adj.	1.3(0.8)	2.5A	PWM	DIP.SO	Yes	C.E.M.	5V to 3.3V green PC apps., drives external N-channel FET	2.25
MAX748A	3.3 to 16	3.3	3(1.7)	750	PWM	DIP.SO	Yes	C.E.M.	>85% efficiencies, no subharmonic switching noise	2.60
MAX750A	4 to 11	Adj.	3(1.7)	1.5W	PWM	DIP.SO	Yes	C.E.M.	90% efficiencies, no subharmonic switching noise	2.15
MAX758A	4 to 16	Adj.	3(1.7)	3.75W	PWM	DIP.SO	Yes	C.E.M.	>85% efficiencies, no subharmonic switching noise	2.60
MAX763A	3.3 to 11	3.3	2.5(1.4)	500	PWM	DIP.SO	Yes	C.E.M.	>85% efficiencies, no subharmonic switching noise	2.15
MAX767	4.5 to 5.5	3.3, 3.45(R, or 3.6(S)	0.75	7A	PWM	SSOP	Yes	C.E	Green PC apps., high efficiency, small-size controller	3.40
MAX787/787H	3.5 to 40/60	5	12(8.5)	5A	PWM	TO-220, TO-3		C.E.M.	High power, few external components	††
MAX788/788H	3.5 to 40/60	3.3	12(8.5)	5A	PWM	TO-220, TO-3		C.E.M.	High power, few external components	††
MAX789/789H	3.5 to 40/60	3	12(8.5)	5A	PWM	TO-220, TO-3		C.E.M.	High power, few external components	††
MAX796-799	4.5 to 30	5.05/3.3/2.9 Adj.	1(0.7)	50W	PWM	DIP.SO	Yes	C.E.M.	Synchronous rectifier, secondary output regulation, Idle-mode PWM, high efficiency	††
MAX830-833	3.5 to 40	Adj./5/3/3.3	11(8)	1A	PWM	SO	Yes (MAX831)	C	High power, SOIC	††

Part Number	Input Voltage Range (V)	Output Voltage (V)	Quiescent Supply Current (mA), max (typ)	Output (mA typ)	Control Scheme	Package Options *	EV Kit	Temp. Ranges **	Features	Price † 1000-up ($)
INVERTING SWITCHING REGULATORS										
MAX4391	4 to 16.5	up to -20	0.25(0.09)	400mW	PFM	DIP.SO		C.E.M.	Improved RC4391 2nd source	2.09
MAX634	2.3 to 16.5	up to -20	0.15(0.07)	400mW	PFM	DIP.SO		C.E.M.	Improved RC4391 2nd source	2.61
MAX635	2.3 to 16.5	-5, Adj.	0.15(0.08)	50	PFM	DIP.SO		C.E.M.	Only 3 external components	2.56
MAX636	2.3 to 16.5	-12 Adj.	0.15(0.08)	40	PFM	DIP.SO		C.E.M.	Only 3 external components	2.56
MAX637	2.3 to 16.5	-15 Adj.	0.15(0.07)	25	PFM	DIP.SO		C.E.M.	Only 3 external components	2.56
MAX650	-64 to -42	5	10(0.5)	250	PFM	DIP.SO		C.E.M.	Telecom applications	3.50
MAX735	4 to 6.2	-5	3(1.6)	275	PWM	DIP.SO		C.E.M.	>80% efficiencies	2.15
MAX736	4 to 8.6	-12	3(1.6)	125	PWM	DIP.SO	Yes	C.E.M.	>80% efficiencies	2.75
MAX737	4 to 5.5	-15	4.5(2.5)	100	PWM	DIP.SO	Yes	C.E.M.	>80% efficiencies	2.75
MAX739	4 to 15	-5	3(1.6)	500	PWM	DIP.SO	Yes	C.E.M.	>80% efficiencies	2.75
MAX741N	2.7 to 15.5	-5, -12, -15, Adj.	4.0(2.2)	5W	PWM	DIP.SSOP		C.E.M.	PWM inverting controller, high efficiency	3.64
MAX749	2 to 6	Adj.	0.06	5W	PFM	DIP.SO	Yes	C.E.M.	Digital adjust for negative LCD	2.49
MAX755	2.7 to 9	Adj.	3.5(1.8)	1.4W	PWM	DIP.SO		C.E.M.	80% efficiencies	2.15
MAX759	4 to 15	Adj.	4(2.1)	1.5W	PWM	DIP.SO	Yes	C.E.M.	LCD driver, >80% efficiencies	2.75
MAX764/765/766	3 to 16.5	-5,/-12/-15 or Adj. to 21ΔV	0.1	200	PFM	DIP.SO	Yes	C.E.M.	High efficiency over wide I_{out} range	2.38
MAX774/775/776	3 to 16.5	-5/-12/-15 or Adj.	0.1	1A	PFM	DIP.SO	Yes	C.E.M.	Controllers, high efficiency over wide I_{out} range	2.20

* Package Options: DIP = Dual-In-Line Package, SO = Small Outline, SSOP = Shrink Small-Outline Package, TO___ = Can

** Temperature Ranges: C = 0°C to +70°C, E = -40°C to +85°C, M = -55°C to +125°C

† Prices provided are for design guidance and are FOB USA. International prices will differ due to local duties, taxes, and exchange rates.

†† Future product—contact factory for pricing and availability. Specifications are preliminary.

FIGURE 17.16 Maxim IC regulators. Devices similar to Linear Technology regulators are available from other major suppliers requiring only a minimum number of components external to the package.

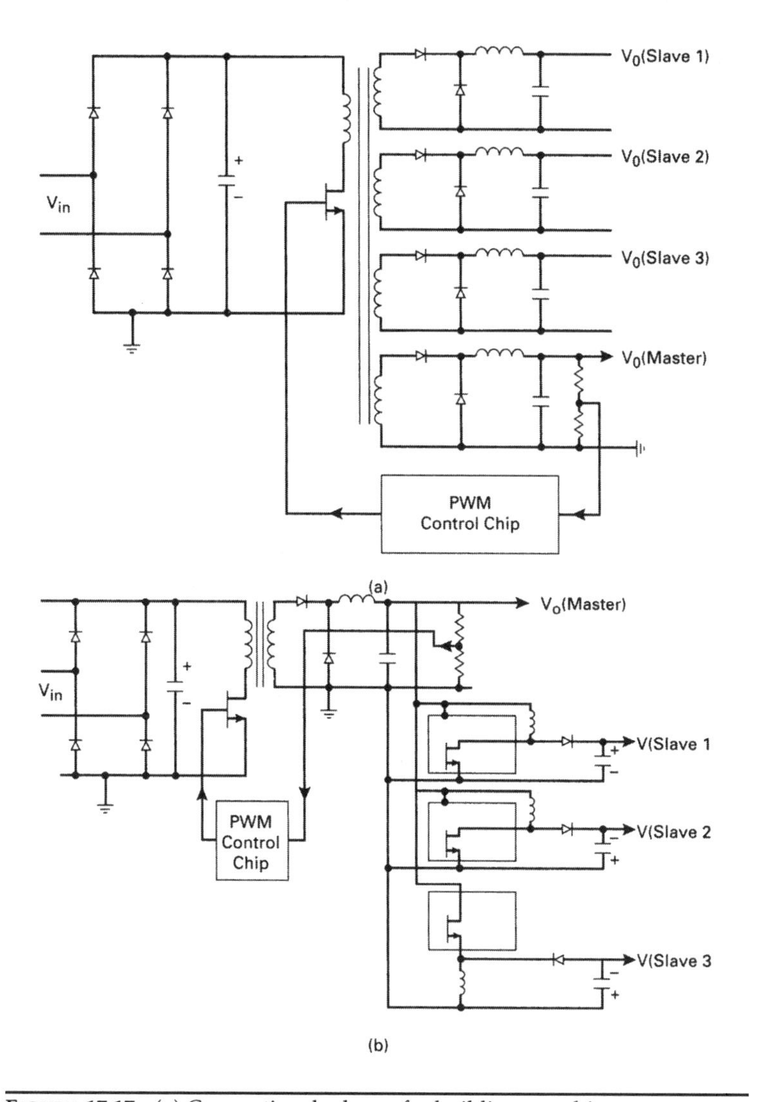

FIGURE 17.17 (*a*) Conventional scheme for building a multi-output power supply. A feedback loop around a master output regulates it against line and load changes. Secondaries on the power transformer yield slaves which are well regulated against line changes but only ±8% for load changes.
(*b*) Distributed power scheme. Here the transformer has only one secondary which is well regulated against line and load changes. That output is bussed around and is used to generate slave voltages at the point of use with standardized boost, buck, or polarity inverter DC/DC converters. The master may not even require regulation by PWM control of the primary side power transistor. That transistor may operate at a fixed "on" time, and the master regulation may be derived from its own DC/DC converter for sufficiently low output currents.

precisely, as the smallest amount by which they can be changed corresponds to adding or removing a single secondary turn. Since volts per turn is proportional to the switching frequency from Faraday's law $[E/N = A_e\,(dB/t_{on})(10^{-8}) = A_e\,(dB \times 0.4f)(10^{-8})]$ at high frequencies, E/N can amount to 2 to 3 V per turn depending on the flux change and core area.

These slave voltages are generated physically close to the main power transformer and are piped around to their points of use.

Generally the poor regulation and inability to set the voltages are not too much of a problem. Slaves are used to power operational amplifiers or motors for various computer peripherals, and these can tolerate large supply voltage changes.

In cases where slaves must be accurate and well regulated, they must be controlled by a dedicated feedback loop. This often is done by following the poorly regulated slave with a linear regulator for low current, a buck regulator for higher current, or a magnetic amplifier (Section 10.3).

Distributed power is an alternative which corrects this problem, and it has additional significant advantages, as shown in one of its versions in Figure 17.17*b*.

The essence of distributed power is that it generates a common DC voltage, which is not necessarily well regulated, at a central point and buses it around to the points of use. There, standardized well-regulated DC/DC converters—bucks, boost, or polarity inverters—convert it to the desired voltages. The availability of the above-described LTC and Maxim standard regulators makes them immediate candidates for such a distributed power scheme.

Consider the following advantages of one version of distributed power shown in Figure 17.17*b*. In that figure, the highest current output, usually +5 V, is generated directly by the main power transformer, which has only the one secondary shown. It is controlled by a feedback loop from a sample on the output ground around to control the "on" time of the power transistor on the input ground. All other outputs are derived from boost or polarity inverters, each having their own feedback loop. In the figure, all the slaves are generated by LTC1174 boost regulators. The advantages are as follows:

1. Simpler and less expensive main power transformer. That transformer is generally the largest and most expensive element in a switching power supply. With fewer secondary windings, it is easier to meet VDE safety specifications.
2. Ease of changing electrical parameters of the transformer. In the conventional scheme of Figure 17.17*a*, the initial transformer design often requires repeated versions. Some windings may require adding or removing turns. Leakage or magnetizing

inductances may be improper in an initial prototype. Winding sequence may have to be changed to improve coupling between windings or reduce proximity-effect losses. In the scheme of Figure 17.17*b*, each output has its own feedback loop, and the exact input voltage is unimportant as the output voltages are constant over a 3:1 input voltage range.

3. Ease of changing output voltages or currents without changing the main transformer design.
4. Ease of adding one or a number of new output voltages. Often in a large system design, it is found late in the design that some new voltages must be added.
5. Possibility of totally eliminating feedback on the master. This avoids all the problems of sensing an output voltage on output ground and controlling a pulse width on input ground, and makes unnecessary optocouplers with their troublesome gain variation with temperature, low-power housekeeping supplies on output ground, and small power transformers to couple pulses on output ground around to the power transistor on input ground.

All this becomes possible by generating an unregulated voltage of about +20 to +24 V, and bussing this around to the points of use, and generating the usual +5-V high current output from a standard buck converter, such as a high-efficiency LT1270A or LTC1159. Then the lower-current slaves can be produced with LT1074-type buck regulators.

This scheme using only secondary side regulation does not need pulse width modulation of the power transistor. It just requires setting the controller pulse width to about 85% of a half period, and using peak rectification with only a capacitor filter in the single secondary. Variation of the peak rectified voltage due to line and load changes and ripple at the filter capacitor is taken care of by the secondary regulators.

In any such distributed power scheme, it is best to choose a relatively high bus voltage such as +20 to +25 V and buck it down to the desired values, rather than a low bus of +5 V and boost it up to the desired voltages. The only time a low bus of +5 V makes sense is when the +5-V current is over 10 to 100 A, as such high currents are not usually generated using DC/DC converters.

Although a distributed power scheme may be more expensive and may dissipate somewhat more power, since the power is handled twice, the above advantages and the quick turnaround time in developing a new multi-output power supply may outweigh its disadvantages.

References

1. Linear Technology LT 1070 and LT1074 data sheets.
2. LT1170 data sheet.
3. Carl Nelson and Jim Wilson, LTC Application Note 19, Linear Technology.
4. LT 1170 data sheet.
5. Linear Technology Power Solutions, 1997.
6. Carl Nelson, LTC Application Note 44, Linear Technology.
7. R. Mammano, "Distributed Power Systems," Unitrode Corp. publication.

Appendix

Symbols, Units, and Conversion Factors

Symbols Frequently Used in This Book	
A_b	Winding area of a core bobbin (usually given in square inches)
A_e	Effective core area (usually given in square centimeters)
A_L	Inductance of core (usually given in millihenries per 1000 turns)
B_r	Remanence flux density of a core material (flux density at zero oersteds)
B_S	Saturation flux density of a core material
D_{CMA}	Current density in wire (usually expressed in circular mils per rms ampere)
l_m	Effective path length of a magnetic core (usually quoted in centimeters)
Φ	Flux in a magnetic core (in CGS/EMU units, it is usually given in maxwells and is equal to flux density in gauss × core area in square centimeters)

Quantity	Quantity symbol	Unit name	Unit symbol	Conversion
Electric				
Capacitance	C	farads	F	As/V
Charge	Q	coulombs	C	As
Current	I	amperes	A	
Energy	U	joules	J	Ws
Impedance	Z	ohms	Ω	V/A
Inductance, self-	L	henries	H	Wb/A
Potential difference	V	volts	V	
Power, real (active)	P	watts	W	VA
Power, apparent	S	volt amperes	VA	
Reactance	X	ohms	Ω	
Resistance	R	ohms	Ω	V/A
Resistivity, volume	ρ		Ωcm	
Magnetic				
Field strength	H		A/m	
Field strength (CGS)	H	oersteds	Oe	$1000/4\pi$ A/m
Flux	Φ	webers	Wb	Vs
Flux density	B	teslas	T	Wb/m^2
Permeability	μ		H/m	Vs/Am
Other				
Angular velocity	ω		rad/s	$2\pi f$
Area	A		m^2	
Frequency	f	hertz	Hz	s^{-1}
Length	l	meters	m	
Temperature	T	degrees Celsius	°C	
Temperature, absolute	T	kelvins	K	
Time	t	seconds	s	

Quantity	CGS/EMU units	MKS units	To convert from	
			CGS/EMU to MKS multiply by	MKS to CGS/EMU multiply by
Flux	maxwell	weber	10^{-8}	10^{8}
Flux density	gauss	tesla	10^{-4}	10^{4}
Flux density	gauss	millitesla	10^{-1}	10^{1}
Flux density	gauss	weber/meter^2	10^{-4}	10^{4}
Magnetic field intensity	oersted	ampere turns/meter	79.5	1.26×10^{-2}

TABLE A.1 Conversion Table for Frequently Used Magnetic Units

Quantity	A	B	To convert from	
			A to B multiply by	B to A multiply by
Area	circular mils	square inches	7.85×10^{-7}	1.27×10^{6}
Area	circular mils	square centimeters	5.07×10^{-6}	1.98×10^{5}

TABLE A.2 Other Conversion Factors Used in This Book

SI Base Units			
Quantity	Quantity symbol	Unit name	Unit symbol
Mass	m	kilogram	kg
Length	l	meter	m
Time	t	second	s
Electric current	I	ampere	A
Temperature	T	kelvin	K

Bibliography

References on Switchmode Power Supplies and Related Subjects

1. B. Bedford and R. Haft, *Principles of Inverter Circuits,* Wiley, New York, 1964.
2. K. Billings, *Switchmode Power Supply Handbook,* McGraw-Hill, New York, 1989.
3. G. Chryssis, *High Frequency Switching Power Supplies Theory and Design,* 2nd ed., McGraw-Hill, New York, 1989.
4. C. McLyman, *Transformer and Inductor Design Handbook,* Marcel Dekker, New York, 1978.
5. D. Middlebrook and S. Cuk, *Advances in Switchmode Power Conversion,* 3 vols., Tesla, Pasadena, California, 1981.
6. A. Pressman, *Switching and Linear Power Supply, Power Converter Design,* Switchtronix Press, Waban, Mass., 1977.
7. R. Severns and G. Bloom, *Modern DC-to-DC Switchmode Power Converter Circuits,* Van Nostrand Reinhold, New York, 1984.
8. K. Sum, *Switch Mode Power Conversion Basic Theory and Design,* Marcel Dekker, New York, 1988.
9. P. Wood, *Switching Power Converters,* Van Nostrand, New York, 1981.
10. *Unitrode Power Supply Design Seminar Handbook,* Unitrode Corp., Watertown, Mass., 1988.
11. J. R. Nowicki, *Power Supplies for Electronic Equipment,* Leonard Hill, London, 1971.
12. A. Pressman, *Switching and Linear Power Supply Power Converter Design,* Hayden, New York, 1977.
13. E. C. Snelling, *Soft Ferrites-Properties and Applications,* Iliffe, London, 1969.
14. Col. W. T. McLyman, *Transformer and Inductor Design Handbook,* Marcel Dekker, New York, 1978.
15. Col. W. T. McLyman, *Cut Core Inductor Design Manual,* Arnold Catalog No. SC-142A, 1978.
16. Col. W. T. McLyman, *Magnetic Core Selection for Transformers and Inductors,* Marcel Dekker, New York, 1982.
17. H. W. Bode, "Relations Between Attenuation and Phase in Feedback Amplifier Design," *Bell System Technical Journal,* 1940.
18. H. W. Bode, *Network Analysis and Feedback Amplifier Design,* Van Nostrand Reinhold, New York, 1945.
19. B. C. Kuo, *Automatic Control Systems,* Prentice-Hall, Englewood Cliffs, N.J., 1962.
20. J. J. D'Azzo and C. H. Houpis, *Linear Control System Analysis and Design: Conventional and Modern,* McGraw-Hill, New York, 1975.
21. K. Murdock, *Handbook of Electronic Design and Analysis Procedures Using Programmable Calculators,* Van Nostrand Reinhold, New York, 1979.
22. W. J. Cunningham, *Nonlinear Analysis,* McGraw-Hill, New York, 1958.
23. S. Austen Stigant, *The Elements of Determinants, Matrices and Tensors for Engineers,* MacDonald and Co. (Publishers) Ltd., 1959.

24. R. Bellman, *Perturbation Techniques in Mathematics, Physics and Engineering,* Holt, Rinehart and Winston, New York.
25. C. Desoer and E. Kuh, *Basic Circuit Theory,* McGraw-Hill, New York, 1969.
26. J. Abadie and J. Carpenter, "Generalization of the Waite Reduced Gradient Method to the Case of Nonlinear Constraints," in *Optimization,* R. Fletcher, ed., Academic Press, New York, 1969.
27. D. A. Pierre, *Optimization Theory with Applications,* Wiley, 1969, pp. 36–43.

Research Material

References, Articles, Application Notes, Extracts, and Patents

1. G. C. Uchrin and W. O. Taylor, "A New Self-Excited Square-Wave Transistor Power Oscillator," *Proc. IRE 43,* January 1955.
2. J. L. Jensen, "An Improved Square-Wave Oscillator Circuit," *IRE Trans. on Circuit Theory,* CT-4, September 1957.
3. J. Meyerhoff and R. M. Tillmaw, "A High Speed Two-Winding Transistor-Magnetic-Core Oscillator," *IRE Trans. on Circuit Theory,* CT-4, 1957.
4. R. E. Morgan, "A New Control Amplifier Using a Saturable Current Transformer and a Switching Transistor," Paper 58-858, *AIEE Summer General Meeting and Air Transportation Conference,* Buffalo, June 1958.
5. T. D. Towers, "Practical Design Problems in Transistor DC/DC Converters and DC/AC Inverters," *Proc. IEE (London),* 106B, Suppl. 18, May 1959.
6. R. E. Morgan, "A New Magnetic-Controlled Rectifier Power Amplifier with a Saturable Reactor Controlling On Time," Paper T-121, *AIEE Special Technical Conference on Nonlinear Magnetics and Magnetic Amplifiers,* Philadelphia, October 26–28, 1960.
7. R. W. Sterling et al., "Multiple Cores Used to Simulate a Variable Volt-Second Saturable Transformer for Application in Self-Oscillating Inverters," *IEEE International Conference on Nonlinear Magnetics,* Washington, D.C., April 17–19, 1963.
8. T. Roddam, *Transistor Inverters and Converters,* Iliffe, London, 1963.
9. B. D. Bedford and R. G. Hoft, *Principles of Inverter Circuits,* Wiley, New York, 1964.
10. R. E. Morgan, "High-Frequency Time Ratio Control with Insulated and Isolated Inputs," *IEEE Trans on Magnetics,* March 1965.
11. S. Lindena, "The Current-Fed Inverter: A New Approach and a Comparison with the Voltage-Fed Inverter," *20th Power Sources Conference Proc.,* May 1966.
12. J. P. Vergez, Jr., and V. Glover, *Low Power Solid-State Inverters for Space Applications,* WESCON 1966 Record.
13. A. Kossov, "Comparative Analysis of Chopper Voltage Regulators with LC Filters," *IEEE Trans. on Magnetics,* September 1968.
14. A. Kossov, "Comparative Analysis of Chopper Voltage Regulators with LC Filters," *IEEE Trans. on Magnetics,* MAG-4, December 1968.
15. R. E. Morgan, "Conversion and Control with High Voltage Transistors with Isolated Inputs," Paper 7.8, *1968 INTERMAG Conference,* Washington, D.C., April 3–5, 1968.
16. H. P. Hart and R. J. Kakalec, "The Derivation and Application of Design Equations for Ferroresonant Voltage Regulators and Regulated Rectifiers," *IEEE Trans. on Magnetics,* March 1971.
17. T. G. Wilson et al., "Regulated DC to DC Converter for Voltage Step-Up or Step-Down with Input-Output Isolation," *IEEE Fall Electronics Conference,* October 1971.
18. F. F. Judd and C. T. Chen, "Analysis and Optimal Design of Self-Oscillating DC to DC Converters," *IEEE Trans. Circuit Theory,* CT-18, November 1971.

19. J. J. Pollack, "Advanced Pulse Width Modulated Inverter Techniques," *IEEE Trans. Ind. Appl.*, IA-8, No. 2, March/April 1972.
20. F. C. Lee et al., "Analysis of Limit Cycles in a Two-Transistor Saturable-Core Parallel Inverter," *IEEE Trans. on Aerospace and Elec. Syst.*, AES-9, No. 4, July 1973.
21. D. H. Wolaver, "Requirements on Switching Devices in DC to DC Converters," *Electronics Specialists Conference Record*, 1973.
22. F. C. Lee and T. G. Wilson, "Analysis of Starting Circuits for a Class of Hard Oscillators: Two-Transistor Saturable-Core Parallel Inverters," *IEEE Trans. on Aerospace and Elec. Syst.*, AES-10, No. 1, January 1974.
23. F. E. Lukens, "Linearization of the Pulse-Width Modulated Converter," *Power Electronics Specialists Conference Record*, June 1974.
24. F. C. Lee, "Analysis of Transient Characteristics and Starting of a Family of Power Conditioning Circuits: Two-Transistor Saturable-Core Parallel Inverters," Ph.D. Dissertation, Duke University, Durham N.O., 1974.
25. F. C. Lee and T. G. Wilson, "Voltage-Spike Analysis for a Free-Running Parallel Inverter," *1974 Digest of the INTERMAG Conference*, May 1974.
26. F. C. Lee and T. G. Wilson, "Nonlinear Analysis of a Family of LC Tuned Inverters," *IEEE Trans. on Aerospace and Elec. Syst.*, AES-11, No. 2, March 1975.
27. T. G. Wilson et al., "Relationships Among Classes of Self- Oscillating Transistor Parallel Inverters," *IEEE Trans. on Aerospace and Elec. Syst.*, AES-11, No. 2, March 1975.
28. D. E. Nelson and N. O. Sokal, "Improving Load and Line Transient Response of Switching Regulators by Feed-Forward Techniques," *POWERCON 2, the Second National Solid-State Power Conversion Conference Record*, 1975.
29. Y. Yu et al., "Formulation of a Methodology for Power Circuit Design Optimization," *Power Electronics Specialists Conference Record*, June 1976.
30. E. T. Calkin and B. H. Hamilton, "A Conceptually New Approach for Regulated DC to DC Converters Employing Transistor Switches and Pulse Width Control," *IEEE Trans. Ind. Appl*, July/August 1976.
31. N. O. Sokal, "Feed-Forward Control for Switching Mode Power Converters—A Design Example." *POWERCON 3, the Third National Solid-State Power Conversion Conference Record*, 1976.
32. S. M. Cuk and R. D. Middlebrook, "A New Optimum Topology Switching DC to DC Converter," *Power Electronics Specialists Conference Record*, 1977.
33. S. M. Cuk and R. D. Middlebrook, "Coupled Inductor and Other Extensions of a New Optimum Topology Switching DC to DC Converter," *IEEE Ind. Appl. Society Annual Meeting*, 1977 Record, October 2–6, 1977.
34. S. M. Cuk and R. Erickson, "A Conceptually New High-Frequency Switched-Mode Amplifier Technique Eliminates Current Ripple," *POWERCON 5, the Fifth National Solid State Power Conversion Conference Record*, May 1978.
35. S. M. Cuk, "Discontinuous Inductor Current Mode in the Optimum Topology Switching Converter," *Power Electronics Specialists Conference Record*, June 1978.
36. S. M. Cuk, "Switching DC to DC Converter with Zero Input or Output Current Ripple," *IEEE Ind. Appl. Society Annual Meeting*, 1978 Record, October 1–5, 1978.
37. Y. Yu et al., "Development of a Standardized Control Module for DC-DC Converters," *NASA Contract Report NAS3-18918*, TRW Defense and Space Systems, 1977.
38. R. D. Middlebrook and S. M. Cuk, "Isolation and Multiple Output Extensions of a New Optimum Topology Switching DC to DC Converter," *Power Electronics Specialists Conference, 1978 Record*, June 1978.
39. F. Mahmoud and F. C. Lee, "Analysis and Design of an Adaptive Multi-Loop Controlled Two-Winding Buck-Boost Regulator," *Inter. Telecom. Energy Conference*, 1979.
40. Y. Yu et al., "Power Converter Design Optimization," *IEEE Trans. on Aerospace and Elec. Syst.*, AES-15 No. 3, May 1979.

41. L. Rensink et al., "Design of a Kilowatt Off-Line Switcher Using a Cuk Converter," *POWERCON 6, the Sixth National Solid-State Power Conversion Conference Record*, May 1979.
42. S. M. Cuk, "General Topological Properties of Switching Structures," *Power Electronics Specialists Conference Record*, 1979.
43. S. P. Hsu, "Problems in Analysis and Design of Switching Regulators," Ph.D. dissertation, California Institute of Technology, September 1979.
44. L. Rensink, "Switching Regulator Configurations and Circuit Realizations," Ph.D. dissertation, California Institute of Technology, December 1979.
45. S. M. Cuk and R. D. Middlebrook, "Advances in Switched-Mode Power Conversion, Part 1," *Robotics Age*, 1, No. 2, Winter 1979.
46. S. M. Cuk and R. D. Middlebrook, "Advances in Switched-Mode Power Conversion, Part 2," *Robotics Age*, 2, No. 2, Summer 1980.
47. F. C. Lee et al., "A Unified Design Procedure for a Standardized Control Module for DC-DC Switching Regulators," *Power Electronics Specialists Conference*, 1980 Record.
48. C. J. Wu et al., "Design Optimization for a Half-Bridge DC-DC Converter," *Power Electronics Specialists Conference Record*, 1980.
49. Y. Yu and F. C. Lee, "Application Handbook for Standardized Control Module for DC-DC Converters," *NASA Contract Report NAS3-20102*, prepared jointly by TRW Defense and Space Systems and Virginia Polytechnic Institute and State University, 1980.
50. F. C. Lee and Y. Yu, "An Adaptive Control Switching Buck Regulator Implementation, Analysis and Design," *IEEE Trans. on Aerospace and Elec. Syst.*, January 1980.
51. R. Erickson et al., "Characterization and Implementation of Power MOSFETs in Switching Converters," *POWERCON 7, the Seventh National Solid-State Power Conversion Conference Record*, March 24–27, 1980.
52. R. Redl and N. O. Sokal, "Push-Pull Current-Fed Multiple Output DC-DC Power Converter with Only One Inductor and with 0 to 100% Switch Duty Ratio," *Power Electronics Specialists Conference*, 1980 Record.
53. M. S. Makled and M. M. Fahmy, "An Analytical Investigation of a Ferroresonant Circuit," *IEEE Trans. on Magnetics*, March 1980.
54. R. Redl and N. O. Sokal, "Push-Pull Current-Fed Regulated Wide-Input-Range DC-DC Power Converter with Only One Inductor and with 0 to 100% Switch Duty Ratio: Operation at Duty Ratio Below 50%," *Power Electronics Specialists Conference Record*, 1981.
55. V. J. Thottuvelil et al., "Analysis and Design of a Push-Pull Current-Fed Converter," *Power Electronics Specialists Conference Record*, 1981.
56. L. H. Dixon and C. J. Baranowski, "Designing Optimal Multi-Output Converters with a Coupled-Inductor Current-Driven Topology," *POWERCON 8, the Eighth National Solid-State Power Electronic Conference Record*, April 1981.
57. R. D. Middlebrook, "Predicting Modulator Phase Lag in PWM Converter Feedback Loops," *POWERCON 8, the Eighth National Solid-State Power Electronics Conference Record*, April 1981.
58. J. N. Park and T. R. Zalom, "A Dual Mode Forward/Flyback Converter," *Power Electronics Specialists Conference Record*, 1982.
59. G. W. Wester and R. D. Middlebrook, "Low-Frequency Characterization of Switched DC-to-DC Converters," *Power Electronics Specialists Conference Record*, 1972.
60. R. D. Middlebrook, "Describing Function Properties of a Magnetic Pulse-Width Modulator," *Power Electronics Specialists Conference*, 1972 Record.
61. T. G. Wilson and F. C. Lee, "Analysis and Modeling of a Family of Two-Transistor Parallel Inverters," *IEEE Trans. on Magnetics*, MAG-9, No. 3, September 1973.
62. R. D. Middlebrook, "A Continuous Model for the Tapped Inductor Boost Converter," *Power Electronics Specialists Conference Record*, 1975.

63. D. J. Packard, "Discrete Modeling and Analysis of Switching Regulators," Ph.D. Dissertation, California Institute of Technology, May 1976.
64. R. D. Middlebrook and S. M. Cuk, "A General Unified Approach to Modeling Switching Converter Power Stages," *Power Electronics Specialists Conference Record*, 1976.
65. R. D. Middlebrook and S. M. Cuk, "Modeling and Analysis Methods for DC-to-DC Switching Converters," *IEEE International Semiconductor Power Converter Conference Record*, 1977.
66. S. M. Cuk and R. D. Middlebrook, "A General Unified Approach to Modeling Switching DC-to-DC Converters in Discontinuous Conduction Mode," *Power Electronics Specialists Conference Record*, 1977.
67. Y. Yu et al., "Modeling and Analysis of Power Processing Systems," *NAECON Record*, 1977.
68. R. P. Iwen et al., "Modeling and Analysis of DC-DC Converters with Continuous and Discontinuous Inductor Current," *Second IFAC Symposium on Control in Power Electronics and Electrical Devices*, Dusseldorf, October 1977.
69. Y. Yu et al., "Modeling and Analysis of Power Processing Systems," *NASA Contract Report NAS3-19690*, TRW Defense and Space Systems, November 1977.
70. R. P. Iwens et al., "Generalized Discrete Time Domain Modeling and Analysis of DC-DC Converters," *IEEE Trans. Ind. Electron. Contr. Instrum.*, IECI-26, No. 2, May 1979.
71. S. P. Hsu et al., "Modeling and Analysis of Switching DC-to-DC Converters in Constant Frequency Current Programmed Mode," *Power Electronics Specialists Conference Record*, 1979.
72. Y. Yu, F. C. Lee, and J. Kolecki, "Modeling and Analysis of Power Processing Systems (MAPPS)," *Power Electronics Specialists Conference Record*, 1979.
73. F. C. Lee, "Discrete Time Domain Modeling and Linearization of a Switching Buck Converter," *International Symposium on Circuits and Systems*, Tokyo, July 1979.
74. Y. Yu et al., "Modeling of Switching Regulator Power Stages with and without Zero-Inductor-Current Dwell Time," *IEEE Trans. Ind. Electron. Contr. Instrum.*, IECI-26, No. 3, August 1979.
75. R. D. Middlebrook, "A Continuous Model for the Tapped Inductor Boost Converter," *Power Electronics Specialists Conference Record*, 1975.
76. D. J. Packard, "Discrete Modeling and Analysis of Switching Regulators," Ph.D. dissertation, California Institute of Technology, May 1976.
77. R. D. Middlebrook and S. M. Cuk, "A General Unified Approach to Modeling Switching Converter Power Stages," *Power Electronics Specialists Conference Record*, 1976.
78. R. D. Middlebrook and S. M. Cuk, "Modeling and Analysis Methods for DC-to-DC Switching Converters," *IEEE International Semiconductor Power Converter Conference Record*, 1977.
79. S. M. Cuk and R. D. Middlebrook, "A General Unified Approach to Modeling Switching DC-to-DC Converters in Discontinuous Conduction Mode," *Power Electronics Specialists Conference Record*, 1977.
80. Y. Yu et al., "Modeling and Analysis of Power Processing Systems," *NAECON Record*, 1977.
81. R. P. Iwen et al., "Modeling and Analysis of DC-DC Converters with Continuous and Discontinuous Inductor Current," *Second IFAC Symposium on Control in Power Electronics and Electrical Devices*, Dusseldorf, October 1977.
82. Y. Yu et al., "Modeling and Analysis of Power Processing Systems," *NASA Contract Report NAS3-19690*, by TRW Defense and Space Systems, November 1977.
83. R. P. Iwens et al., "Generalized Discrete Time Domain Modeling and Analysis of DC-DC Converters," *IEEE Trans. Ind. Electron. Contr. Instrum.*, IECI-26, No. 2, May 1979.

84. S. P. Hsu et al., "Modeling and Analysis of Switching DC-to-DC Converters in Constant Frequency Current Programmed Mode," *Power Electronics Specialists Conference Record*, 1979.
85. Y. Yu, F. C. Lee, and J. Kolecki, "Modeling and Analysis of Power Processing Systems (MAPPS)," *Power Electronics Specialists Conference Record*, 1979.
86. F. C. Lee, "Discrete Time Domain Modeling and Linearization of a Switching Buck Converter," *International Symposium on Circuits and Systems*, Tokyo, July 1979.
87. Y. Yu at al., "Modeling of Switching Regulator Power Stages with and without Zero-Inductor-Current Dwell Time," *IEEE Trans, Ind. Electron. Contr. Instrum.*, IECI-26, No. 3, August 1979.
88. S. Pro, "Toroid Design Analysis," *Electro-Technology*, August 1966.
89. P. L. Dowell, "Effects of Eddy Currents in Transformer Windings," *Proc. IEE*, 113, No. 8, August 1966.
90. B. Castle, "Optimum Shapes for Inductors," *IEEE Trans. Parts, Materials and Packaging*, PMP-5, No. 1, March 1969.
91. F. C. Schwarz, "An Unorthodox Transformer for Free-Running Inverters," *IEEE Trans. on Magnetics*, MAG-5, 1969.
92. S. Y. M. Feng and W. A. Sander, III, "Optimum Toroidal Inductor Design Analysis," *20th Electronic Component Conference Proceedings*, 1970.
93. R. Lee and D. S. Stephens, "Gap Loss in Current Limiting Transformers," *Electromechanical Design*, April 1973.
94. R. Lee and D. S. Stephens, "Influence of Core Gap in Design of Current Limiting Transformers," *IEEE Trans. on Magnetics*, September 1973.
95. J. R. Woodbury, "Design of Imperfectly Coupled Power Transformers for DC-to-DC Conversion," *IEEE Trans. Ind. Elec. and Contr. Instrum.*, IECI-2I, No. 3, August 1974.
96. W. Dull, A. Kusko, and T. Knutrud, "Designers' Guide to Current and Power Transformers," *EDN Magazine*, March 5, 1975.
97. V. B. Guizburg, "The Calculation of Magnetization Curves and Magnetic Hysteresis Loops for a Simplified Model of a Ferromagnetic Body," *IEEE Trans. on Magnetics*, March 1976.
98. K. Ohri et al., "Design of Air-Gapped Magnetic Core Inductors for Superimposed Direct and Alternating Currents," *IEEE Trans. on Magnetics*, September 1976.
99. W. V. Manka, "Design Power Inductors Step by Step," *Electronic Design*, December 20, 1977.
100. N. R. Grossner, "The Geometry of Regulating Transformers," *IEEE Trans. on Magnetics*, March 1978.
101. W. A. Martin, "Simplify Air Gap Calculating with a Hanna Curve," *Electronic Design*, April 12, 1978.
102. T. Gross, "Multistrand Litz Wire Adds 'Skin' to Cut AC Losses in Switching Power Supplies," *Electronic Design*, February 1, 1979.
103. T. Konopinski and S. Szuba, "Limit the Heat in Ferrite Pot Cores for Reliable Switching Power Supplies," *Electronic Design*, June 7, 1979.
104. T. Gross, " Little Understanding Improves Switching Inductor Designs," *EDN Magazine*, June 20, 1979.
105. J. Mas, "Design and Performance of Power Transformers with Metallic Glass Cores," *Power Conversion International*, July/August 1980.
106. P. E. Thibodeau, "The Switcher Transformer: Designing it in One Try for Switching Power Supplies," *Electronic Design*, September 1, 1980.
107. C. J. Wu et al., "Minimum Weight El Core and Pot Core Inductor and Transformer Designs," *IEEE Trans. on Magnetics*, September 1980.
108. S. A. Chin et al., "Design Graphics for Optimizing the Energy Storage Inductor for DC-to-DC Power Converters," *Power Electronics Specialists Conference Record*, 1982.

109. J. R. Leehey et al., "DC Current Transformer," *Power Electronics Specialist Conference Record*, 1982.
110. W. E. Rippel and Col. W. T. McLyman, "Design Techniques for Minimizing the Parasitic Capacitance and Leakage Inductance of Switched Mode Power Transformers," *POWERCON 9, the Ninth National Solid-State Power Electronic Conference Record*, 1982.
111. H. S. Black, "Stabilized Feedback Amplifiers," *Bell System Technical Journal*, January 1934.
112. L. R. Poulo and S. Greenblatt, "Research Investigations on Feedback Techniques and Methods for Automatic Control," *Contract ECOM-0520-F*, Bose Corp., Natick, Mass., April 1969.
113. D. E. Combs, "Stability Analysis of a Pulse-Width Controlled DC to DC Regulated Converter Using Linear Feedback Control System Technique," *Nat. Elec. Conf. Record*, 26, 1970.
114. R. P. Iwens et al., "Time Domain Modeling and Stability Analysis of an Integral Pulse Frequency Modulated DC to DC Power Converter," *Power Electronics Specialists Conference Record*, 1975.
115. C. Griffin, "Optimizing the PWM Converter as a Closed Loop System," *POWERCON 4, the Fourth National Solid-State Power Electronics Conference Record*, 1977.
116. H. D. Venable and S. R. Foster, "Practical Techniques for Analyzing, Measuring and Stabilizing Feedback Control Loops in Switching Regulators and Converters," *POWERCON 7, the Seventh National Solid-State Power Electronics Conference Record*, 1980.
117. J. J. Beiss and Y. Yu, "A Two-Stage Input Filter with Nondissipatively-Controlled Damping," *INTERMAG Conference Record*, April 1971.
118. Y. Yu and J. J. Beiss, "Some Design Aspects Concerning Input Filters for DC-DC Converters," *Power Conditioning Specialists Conference Record*, 1971.
119. N. O. Sokal, "System Oscillations from Negative Input Resistance at Power Input Port of Switching-Mode Regulator, Amplifier, DC/DC Converter or DC/AC Inverter," *Power Electronics Specialists Conference Record*, 1973.
120. R. D. Middlebrook, "Input Filter Considerations in Design and Application of Switching Regulators," *IEEE Ind. Appl. Soc. Annual Meeting Record*, 1976.
121. R. D. Middlebrook and S. M. Cuk, "Design Techniques for Preventing Input Filter Oscillations in Switched-Mode Regulators," *POWERCON 5, the Fifth National Solid-State Power Electronics Conference Record*, 1978.
122. F. C. Lee and Y. Yu, "Input Filter Design for Switching Regulators," *IEEE Trans. on Aerospace and Elec. Syst.*, AES-15, No. 5, September 1979.
123. T. K. Phelps and W. S. Tage, "Optimizing Passive Input Filter Design," *POWERCON 6, the Sixth National Solid-State Power Conversion Conference Record*, 1979.
124. S. S. Kelkar and F. C. Lee, "A Novel Input Filter Compensation Scheme for Switching Regulators," *Power Electronics Specialists Conference Record*, 1982.
125. S. S. Kelkar and F. C. Lee, "Adaptive Feedforward Input Filter Compensation for Switching Regulators," *POWERCON 9, the Ninth National Solid-State Power Conversion Conference Record*, 1982.
126. C. McIntyre, "SR-52 Solves Network Equations by Finding Complex Determinant," *Electronics*, May 12, 1977.
127. F. M. Lilienstein, "Analyze Switcher Stability, Bandwidth and Gain with a Programmable Calculator," *Electronic Design*, June 7, 1979.
128. B. Przedpelski, "Eliminate Bandwidth Calculation Drudgery with a Universal Calculator Program," *Electronic Design*, October 11, 1979.
129. W. A. Geckle, "Compute S-Function/Time-Domain Response Quickly with a Programmable Calculator," *Electronic Design*, December 6, 1979.
130. F. W. Hauer, "Speed Ferromagnetic Inductor Designs with a Programmable Calculator," *Electronic Design*, December 20, 1979.

131. C. Gyles, "Analyze Complex Linear Networks with a Building Block Calculator Program," *Electronic Design*, April 26, 1980.
132. C. J. McCluskey, "TI-59 Calculator Analyzes Complex Ladder Networks," *Electronic Design*, May 10, 1980.
133. F. Cornelissen, "TI-59 Solves Network Equations Using Complex Matrices," *Electronics*, July 31, 1980.
134. B. K. Erickson, "Ladder Network Calculations," *IEEE Trans. Cons, Electronics*, CE-26, November 1980.
135. G. West, "Use a Programmable Calculator to Ease Transformer Design," *EDN Magazine*, November 24, 1982.
136. *Contract ECOM-0520-F*, Bose Corp., Natick, Mass., April 1969.
137. V. E. Legg, "Magnetic Measurements at Low Flux Densities Using the Alternating Current Bridge," *Bell System Technical Journal*, January 1936.
138. R. A. Homan, "DC Power System Dynamic Impedance Measurements," *National Electronics Conference Record*, October 1964.
139. R. D. Middlebrook, "Measurement of Loop Gain in Feedback Systems," *International Journal of Electronics*, 38, No. 4, 1975.
140. R. D. Middlebrook, "Improved Accuracy Phase Angle Measurement," *International Journal of Electronics*, 40, No. 1, 1976.
141. P. C. Todd, "Automating the Measurement of Converter Dynamic Properties," *POWERCON 7, the Seventh National Solid-State Power Conversion Conf., Record*, 1980.
142. F. Barzegar et al., "Using Small Computers to Model and Measure Magnitude and Phase of Regulator Transfer Functions and Loop Gain," *POWERCON 8, the Eighth National Solid-State Power Conversion Conference Record*, 1981.
143. B. A. Wells et al., "Analog Computer Simulation of a DC-to-DC Flyback Converter," *Suppl. to IEEE Trans. on Aerospace and Elec. Syst.*, AES-3, November 1967, pp. 399–409.
144. S. Y. M. Feng et al., "A Computer Aided Design Procedure for Flyback Step-Up DC-to-DC Converters," *IEEE Trans. on Magnetics, MAG-8*, No. 3, September 1972.
145. D. Y. Chen et al., "Computer Aided Design and Graphics Applied to the Study of Inductor Energy Storage DC-to-DC Electronic Power Converters," *IEEE Trans. Aerospace and Elec. Syst.*, AES-9, No. 4, July 1973.
146. W. A. Schnider, "Verify Network Frequency Response with This Simple BASIC Program," *EDN Magazine*, October 5, 1977.
147. Y. Yu, "Computer Aided Analysis and Simulation of Switched DC-DC Converters," *IEEE Southeastern Proceedings*, April 1978.
148. R. Keller, "Closed-Loop Testing and Computer Analysis Aid Design of Control Systems," *Electronic Design*, November 22, 1978.
149. N. P. Episcopo and R. P. Massey, "Computer Predicted Steady State Stability of Pulse-Width-Controlled DC/DC Converters," *POWERCON 6, the Sixth National Solid- State Power Conversion Conference Record*, 1979.
150. F. C. Lee and Y. Yu, "Computer Aided Analysis and Simulation of Switched DC-DC Converters," *IEEE Trans. Ind. Appl.*, IA-15, No. 5, September/October 1979.
151. V. G. Bello, "Computer Modeling of Pulse-Width Modulators Simplifies Analysis of Switching Regulators," *Electronic Design*, January 18, 1980.
152. G. H. Warren, "Computer Aided Design Program Supplies Low-Pass Filter Data," *EDN Magazine*, August 20, 1980.
153. J. E. Crowe, "Mains Hold-up Performance in Switched Mode PSUs," *Electronic Engineering*, November 1980.
154. E. Niemeyer, "Network Analysis Program Runs on Small Computer System," *EDN Magazine*, February 4, 1981.
155. V. G. Bello, "Computer Program Adds SPICE to Switching Regulator Analysis," *Electronic Design*, March 5, 1981.
156. V. G. Bello, "Using the SPICE 2 CAD Package for Easy Simulation of Switching Regulators in Both Continuous and Discontinuous Conduction Modes,"

POWERCON 8, the Eighth National Solid-State Power Conversion Conference Record, 1981.
157. H. T. Meyer, "Matrix Statements Define Complex Variables, Perform Complex Math in BASIC," *Electronic Design*, July 23, 1981.
158. S. Hageman, "Program Analyzes Six-Element Active RC Networks," *Electronic Design*, January 7, 1982.
159. W. N. Waggener, "Analyze Complex Circuits with a Matrix Inversion Program," *EDN Magazine*, March 17, 1982.
160. H. W. Bode, Amplifier, U.S. Pat. No. 2,123,178, July 12, 1938.
161. F. C. Schwarz, Analog Signal to Discrete Time Interval Converter (ASDTIC), U.S. Pat. No. 3,659,184, April 25, 1972.
162. E. T. Calkin, B. H. Hamilton, and F. C. Laporte, Regulated DC-to-DC Converter with Regulated Current Source Driving a Nonregulated Inverter, U.S. Pat. No. 3,737,755, June 5, 1973.
163. H. D. Venable, Regulated DC-to-DC Converter, U.S. Pat. No. 3,925,715, December 9, 1975.
164. P. W. Clarke, Converter Regulation by Controlled Overlap, U.S. Pat. No. 3,938,024, February 10, 1976.
165. P. Kotlarewsky, Master-Slave Voltage Regulator Employing Pulse Width Modulation, U.S. Pat. No. 9,174,539, November 13, 1979.
166. S. M. Cuk and R. D. Middlebrook, DC-to-DC Switching Converter, U.S. Pat. No. 4,184,197, January 15, 1980.
167. S. M. Cuk, Push-Pull Switching Power Amplifier, U.S. Pat. No. 4,186,437, January 29, 1980.
168. S. M. Cuk, DC-to-DC Switching Converter with Zero Input and Output Current Ripple and Integrated Magnetics Circuits, U.S. Pat. No. 4,257,087, March 17, 1981.
169. G. E. Bloom and A. Eris, DC-to-DC Converter, U.S. Pat. No. 4,262,328, April 14, 1981.
170. S. M. Cuk and R. D. Middlebrook, DC-to-DC Converter Having Reduced Ripple Without Need for Adjustments, U.S. Pat. No. 9,279,133, June 16, 1981.
171. E. T. Calkin and B. H. Hamilton, "Circuit Techniques for Improving the Switching Loci of Transistor Switches in Switching Regulators," *IEEE Ind. Appl. Society Conference Record*, 1972.
172. F. C. Lee and T. G. Wilson, "Voltage Spike Analysis for a Free-Running Parallel Inverter," *IEEE Trans. on Magnetics, MAC-10*, No. 3, September 1974.
173. J. M. Peter, ed., "The Power Transistor in Its Environment," Thomson-CSF Semiconductor Division Publication, 1978.
174. F. C. Lee and T. G. Wilson, "Nonlinear Analysis of Voltage Spike Suppression Networks for a Free-Running Parallel Inverter," *IEEE Ind. Appl. Annual Meeting Record*, 1979.
175. W. McMurray, "Selection of Snubbers and Clamps to Optimize the Design of Transistor Switching Converters," *Power Electronics Specialists Conference Record*, 1979.
176. W. J. Shaughnessy, "LC Snubber Networks Cut Switcher Power Losses," *EDN Magazine*, November 20, 1980.
177. H. F. Baker, "On the Integration of Linear Differential Equations," *Proc. London Math. Soc.*
178. B. Van der Poi, "Forced Oscillations in a Circuit with Non-linear Resistance," 34, 1902, pp. 347–360; 35, 1903, pp. 333–374; second series 2, 1904, pp. 293–296, Phil. Meg., 7–3, 1927, pp. 65–80.
179. V. Fleece and G. P. McCormick, "Computational Algorithm for the Sequential Unconstrained Minimization Technique for Nonlinear Programming," *Management Science*, 10, July 1964, pp. 601–617.
180. W. C. Mylender, R. L. Holmes, and G. P. McCormick, "A Guide to SUMT-Version 4," *Paper RAC-P-63*, Research Analysis Corp., October 1971.
181. Motorola Application Note AN460.

182. S. M. Cuk and R. D. Middlebrook, "Modeling Analysis and Design of Switching Converters," *Report No. NASA CR-135174, Contract No. NAS3-19690 and NAS3-20102.*
183. "Loop Gain Measurements with HP Wave Analyzers," Hewlett- Packard Application Note 59.
184. "Low Frequency Gain Phase Measurements," Hewlett-Packard Application Note 157.
185. "Parallel Inverter with Resistive Load," *Electrical Engineering*, November 1935.

Index

B

D

I

J

K

L

M

N

O

P

R

S

T

Z